Goodman's Medical Cell Biology

FOURTH EDITION

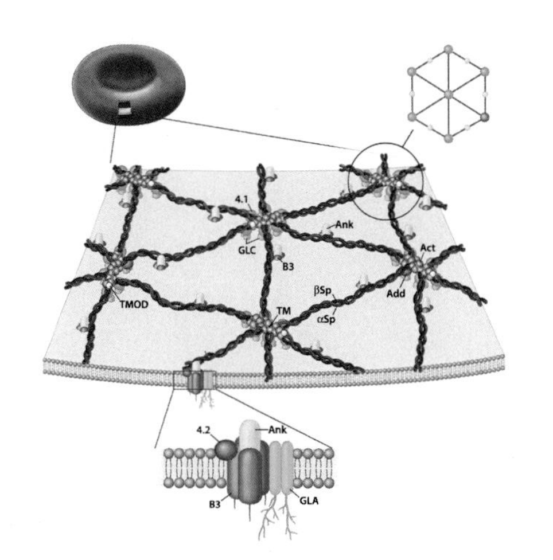

Goodman's Medical Cell Biology

FOURTH EDITION

Edited by Steven R. Goodman, PhD

University of Tennessee Health Science Center, Memphis, TN, United States

Academic Press is an imprint of Elsevier
125 London Wall, London EC2Y 5AS, United Kingdom
525 B Street, Suite 1650, San Diego, CA 92101, United States
50 Hampshire Street, 5th Floor, Cambridge, MA 02139, United States
The Boulevard, Langford Lane, Kidlington, Oxford OX5 1GB, United Kingdom

Notices
Knowledge and best practice in this field are constantly changing. As new research and experience broaden our understanding, changes in research methods, professional practices, or medical treatment may become necessary.

Practitioners and researchers must always rely on their own experience and knowledge in evaluating and using any information, methods, compounds, or experiments described herein. In using such information or methods they should be mindful of their own safety and the safety of others, including parties for whom they have a professional responsibility.

To the fullest extent of the law, neither the Publisher nor the authors, contributors, or editors, assume any liability for any injury and/or damage to persons or property as a matter of products liability, negligence or otherwise, or from any use or operation of any methods, products, instructions, or ideas contained in the material herein.

Library of Congress Cataloging-in-Publication Data
A catalog record for this book is available from the Library of Congress

British Library Cataloguing-in-Publication Data
A catalogue record for this book is available from the British Library

ISBN: 978-0-12-817927-7

For information on all Academic Press publications
visit our website at https://www.elsevier.com/books-and-journals

Publisher: Andre Gerhard Wolff
Acquisitions Editor: Ana Claudia Garcia
Editorial Project Manager: Pat Gonzalez
Production Project Manager: Punithavathy Govindaradjane
Cover Designer: Matthew Limbert

Typeset by SPi Global, India

I dedicate the fourth edition of *Goodman's Medical Cell Biology* to those who have assisted on my lifelong odyssey to understand and communicate biomedical science. This includes the following:

My family: wife Cindy, sister Sue, and children Laela, Gena, Jessie, David, Christie, and Laurie.

My friends who made more pleasurable the journey: Aaron Ciechanover, Peter Stambrook, Rob Williams, Steve and Lisa Youngentob, Ian Zagon, Warren, and Danna Zimmer, and far too many others to name.

Those who have supported my creative professional aspirations: Steve Schwab (chancellor, UTHSC); the UTHSC Office of Research Associate Vice-Chancellors (SLY, Gabor, Tiffany, Phil, Steve, and Sarah) and Staff; the SEBM Leadership, Council, and Publication Committee; the EBM Associate Editors and Editorial Board; and the CTN2 Board and Leadership Group (Ari, Phil, Bob, Karen, Richard, and Bill).

And the visiting professors, postdoctoral fellows, and graduate and undergraduates students that graced my laboratory and taught me so much while making my research career of value and fun.

Contents

Contributors

John G. Burr, PhD (Ch. 1)
Department of Biological Sciences
The University of Texas at Dallas
Richardson, TX, United States

Frans A. Kuypers, PhD (Ch. 2)
Children's Hospital Oakland Research Institute
Oakland, CA, United States

Warren E. Zimmer, PhD (Ch. 3, Ch. 6)
Texas A&M College of Medicine
Texas A&M Health Science Center
Texas A&M University
College Station, TX, United States

Michael A. Whitt, PhD (Ch. 4)
Department of Microbiology, Immunology and Biochemistry
GME Research
The University of Tennessee Health Science Center
Memphis, TN, United States

John V. Cox, PhD (Ch. 4)
Department of Microbiology, Immunology, and Biochemistry
College of Graduate Health Sciences
University of Tennessee Health Science Center
Memphis, TN, United States

Taosheng Huang, MD, PhD (Ch. 5)
Division of Human Genetics
Cincinnati Children's Hospital Medical Center
Cincinnati, OH, United States

Jesse Slone, PhD (Ch. 5)
Division of Human Genetics
Cincinnati Children's Hospital Medical Center
Cincinnati, OH, United States

R.K. Rao, PhD, AGAF (Ch. 7)
Department of Physiology, College of Medicine
University of Tennessee Health Science Center
Memphis, TN, United States

Zhaohui Wu, MD, PhD (Ch. 7)
Department of Radiation Oncology, College of Medicine
University of Tennessee Health Science Center
Memphis, TN, United States

Danna Zimmer, PhD (Ch. 8)
Experimental Biology and Medicine
Rxploration, LLC
College Station, TX, United States

Leszek Kotula, MD, PhD (Ch. 9)
Basic and Translational Research Program, Upstate Cancer Center
Cancer Biology and Prostate Cancer Research Program
Department of Urology
Department of Biochemistry and Molecular Biology
SUNY Upstate Medical University
Syracuse, NY, United States

Angelina Regua, PhD (Ch. 9)
Department of Cancer Biology
Wake Forest University School of Medicine
Winston-Salem, NC, United States

Peter J. Stambrook, PhD (Ch. 10)
University of Cincinnati College of Medicine
Department of Molecular Genetics
Biochemistry and Microbiology
Cincinnati, OH, United States

Santosh R. D'Mello, PhD (Ch. 11)
Department of Biological Sciences
Southern Methodist University
Dallas, TX, United States

Mohammad A.A. Ibrahim, MBBCh, MSc, PhD, FRCPath, FRCP (Ch. 12)
School of Immunology and Microbial Sciences, King's College London
King's College Hospital NHS Foundation Trust, King's Health Partners
Immunological Medicine, Denmark Hill, London, United Kingdom

Y. James Kang, PhD (Ch. 13)
Regenerative Medicine Research Center, Sichuan University, West China Hospital, China
Memphis Institute of Regenerative Medicine, University of Tennessee Health Science Center
Memphis, TN, United States

Wenjing Zhang, MD, PhD (Ch. 13)
Memphis Institute of Regenerative Medicine
University of Tennessee Health Science Center
Memphis, TN, United States

Robert W. Williams, PhD (Ch. 14)
Department of Genetics, Genomics and Informatics
University of Tennessee Health Science Center
Memphis, TN, United States

Valeria R. Mas, PhD (Ch. 14)
Transplant Research Institute
James D. Eason Transplant Institute
Department of Surgery
School of Medicine
The University of Tennessee Health Science Center
Memphis, TN, United States

John F. Cryan, PhD (Ch. 15)
Department of Anatomy and Neuroscience
APC Microbiome Institute
University College Cork
College Road, Cork, Ireland

Kenneth O'Riordan, PhD (Ch. 15)
APC Microbiome Institute
University College Cork
College Road, Cork, Ireland

Thomaz Bastiaanssen, MSc (Ch. 15)
Department of Anatomy and Neuroscience
APC Microbiome Institute
University College Cork
College Road, Cork, Ireland

Ari VanderWalde, MD, MPH, FACP (Clinical Cases)
West Cancer Center
UTHSC
Germantown, TN, United States

Karen C. Johnson, MD, MPH (Clinical Cases)
Department of Preventive Medicine
University of Tennessee Health Science Center
Memphis, TN, United States

Preface

I am pleased to provide the fourth edition of *Goodman's Medical Cell Biology*. The popular focus and approach of the prior three editions has been maintained. We provide a concise focus on eukaryotic cell biology as it relates to human and animal disease, all within a manageable approximately 400-page format. This is accomplished by explaining general cell biology principles in the context of organ systems and disease. Our target audience for this textbook are students (medical, osteopathic, dental, veterinary, nursing, and related disciplines) preparing for a clinical career, graduate students, and advanced undergraduates who are our future health professionals and researchers.

The Editor has once again recruited a team of outstanding experts in the various fields covered in *Goodman's Medical Cell Biology* 4th Edition. While the vision remains the same, the fourth edition is very different from its predecessors:

- We have 15 chapters in the fourth edition.
- We have new chapter topics Mitochondria and Diseases (Chapter 5); Cell Biology of the Immune System (Chapter 12); Stem Cells and Regenerative Medicine (Chapter 13); Omics, Informatics, and Precision Medicine (Chapter 14); and The Microbiome (Chapter 15).
- We have new authors: Mike Whitt and John Cox (Chapter 4, Organelle Structure and Function), Taosheng Huang and Jesse Slone (Chapter 5, Mitochondria and Diseases), R.K. Rao (Chapter 7, Cell Adhesion and the ECM), Leszek Kotula and Angelina Regua (Chapter 9, Cell Signaling Events), Mohammad A. A. Ibrahim (Chapter 12, Cell Biology of the Immune System), James Kang and Wenjing Zhang (Chapter 13, Stem Cells and Regenerative Medicine), Rob Williams and Valeria Mas (Chapter 14, Omics, Informatics, and Precision Medicine), and John Cryan, Kenneth O'Riordan and Thomaz Bastiaanssen (Chapter 15, The Microbiome).
- The Clinical Vignettes for Chapters 2–15 have been written by Ari VanderWalde and Karen Johnson, two leading clinical researchers. Each chapter contains two Clinical cases that provide clinical relevance to the cell biology being discussed.

We are proud to present the fourth edition of *Medical Cell Biology*. The first three editions were very well received by the global educational community, and we feel that the fourth edition will be of great value to educators and students worldwide. We hope that lecturers will find the textbook to be an outstanding educational tool and that students will enjoy the readability of our book while learning this fascinating material. As always, we welcome and appreciate your comments that help us make each edition better for future students. Ancillary materials can be found on Companion Website: https://www.elsevier.com/books-and-journals/book-companion/9780128179277.

I thank all of the authors of *Goodman's Medical Cell Biology* 4th edition, Pat Gonzalez (Senior Editorial Project Manager), Tari Broderick (Sr. Acquisitions Editor), and the staff for Elsevier, who have put great effort and skill into creating a unique and beautifully crafted textbook.

Steven R. Goodman, Editor

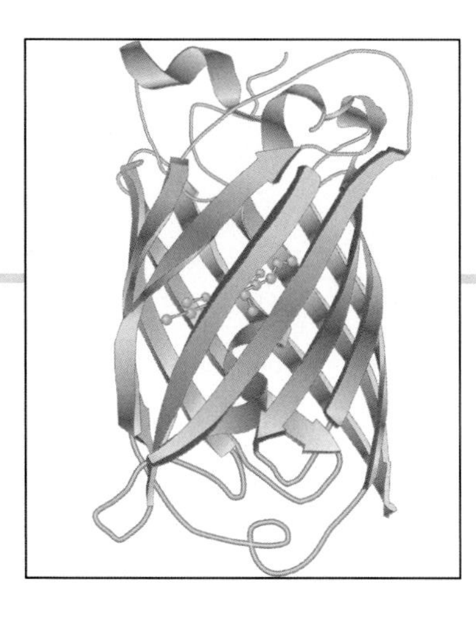

Chapter 1

Tools of the Cell Biologist

The Human Genome Project has revolutionized the study of cell biology, and it will continue to have a large impact on the practice of medicine in the decades to come. Approximately, 20,000 protein-coding genes have been identified in the human genome.

Based on amino acid sequence homologies with proteins of known structure and function, some predictions can be made about the cellular roles of approximately 60% of these genes. But researchers are completely ignorant about the function of the proteins encoded by the remaining 40% of human genes because they have no identifiable sequence homologies to other proteins in the database. A major task for the future, therefore, will be to work out the functions of these thousands of novel proteins.

Because the cell is the fundamental unit of function in the organism, this project translates into searching for the functions of these newly discovered proteins in the life of a cell. The project, therefore, will be largely the task of cell biologists, using the powerful tools of modern molecular and cell biology. This chapter provides a brief review of some of these tools.

One of the first questions a cell biologist might ask in his or her search for a protein's function would be, "Where is it located in the cell?" Is it in the nucleus or the cytoplasm? Is it a surface membrane protein or resident in one of the cytoplasmic organelles? Knowing the subcellular localization of a protein provides significant direction for further experiments designed to learn its function. The structure and function of cells is explained throughout this textbook.

The cell is the basic unit of life. Broadly speaking, there are two types of cells: prokaryotic and eukaryotic. Prokaryotes (eubacteria and archaea) do not have a nucleus; that is, their DNA is not enclosed in a special, subcellular compartment with a double membrane. Eukaryotic cells do have a nucleus; they are also much larger than prokaryotic cells and have numerous organelles and certain substructural elements not found in prokaryotes. This textbook will for the most part be focused on eukaryotic cells, except in the chapter dealing with the microbiome (Chapter 15). The structural features of a generalized eukaryotic cell are shown in Figure 1–1.

MICROSCOPY: ONE OF THE EARLIEST TOOLS OF THE CELL BIOLOGIST

Microscopy, in its various forms, has historically been the primary way in which investigators have examined the appearance and substructure of cells and increasingly in recent decades the location and movement of biological molecules within cells (Box 1–1). We may speak broadly of two kinds of microscopy, **light microscopy** and **electron microscopy** (EM), although the field of microscopy recently has been broadened by the advent of **atomic force microscopy**.

Goodman's Medical Cell Biology. https://doi.org/10.1016/B978-0-12-817927-7.00001-6

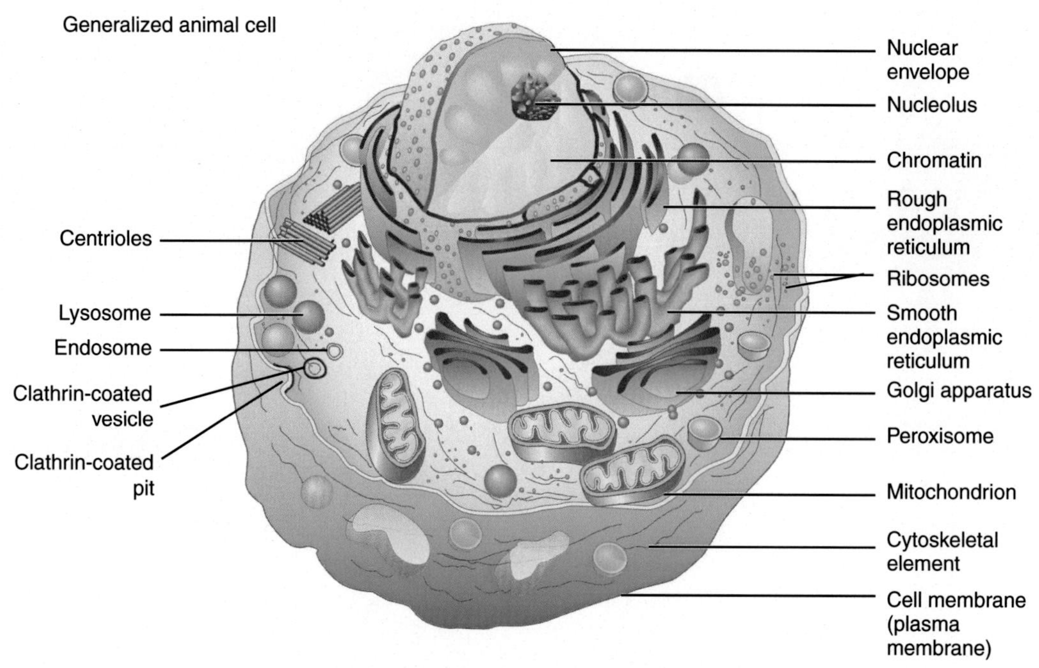

Figure 1–1. **Structural features of animal cells.** Summary of the functions of cellular organelles. Mitochondria: (1) site of the Krebs (citric acid) cycle, produce ATP by oxidative phosphorylation; (2) can release apoptosis-initiating proteins, such as cytochrome *c*. Cytoskeleton: made up of microfilaments, intermediate filaments, and microtubules; governs cell movement and shape. Centrioles: components of the microtubule-organizing center. Plasma membrane: consists of a lipid bilayer and associated proteins. Nucleus: contains chromatin (DNA and associated proteins), gene regulatory proteins, and enzymes for RNA synthesis and processing. Nucleolus: the site of ribosome RNA synthesis and ribosome assembly. Ribosomes: sites of protein synthesis. Rough ER, Golgi apparatus, and transport vesicles: synthesize and process membrane proteins and export proteins. Smooth ER: synthesizes lipids and, in liver cells, detoxifies cells. Lumen: Ca^{2+} reservoir. Clathrin-coated pits, clathrin-coated vesicles, and early and late endosomes: sites for uptake of extracellular proteins and associated cargo for delivery to lysosomes. Lysosomes: contain digestive enzymes. Peroxisomes: cause β-oxidation of certain lipids (e.g., very long chains of fatty acids). *(Modified from Freeman S.* Biological Science, *1st ed. Upper Saddle River, NJ: Prentice Hall, 2002.)*

BOX 1–1. Resolution and Magnification in Microscopy

The two properties that define the usefulness of a microscope are **magnification** and **resolution**. Light microscopes use a series of glass lenses to magnify the image; electron microscopes use a series of magnets to produce the magnified image (Fig. 1–2).

However, because of the wave nature of light (and of electrons), light waves arriving at the focal point produce a magnified image in which different wave trains are either in or out of phase, amplifying or canceling each other to produce interference patterns. This phenomenon, known as **diffraction**, results in the image of straight edge appearing as a fuzzy set of parallel lines and that of a point as a set of concentric rings (Fig. 1–3).

This fundamental limit on the clarity of an optical image, known as the limit of **resolution**, is defined as the minimum distance (*d*) between two points such that they can be resolved as two separate points. In 1873 Ernst Abbé showed that the limit of resolution for a particular light microscope is directly proportional to the **wavelength of light** used to illuminate the sample. The smaller the wavelength of light, the smaller is the value of *d*; that is, the better is the resolution of the magnified image. Abbé also showed that resolution is affected by two other features of the system: (1) the **light-gathering properties** of the microscope's objective lens and (2) the **refractive index of the medium** (e.g., air or oil) between the objective lens and the sample. The light-gathering properties of the objective lens depend on its focal length, which can be characterized by a number called the **angular aperture**, α, where α is the half angle of the cone of light entering the objective lens from a focal point in the sample (Fig. 1–4). These three parameters were quantified by Abbé in the following equation:

$$d = 0.61\lambda/n \sin \alpha,$$

where d is the resolution, λ is the wavelength of illuminating light, n is the refractive index of the medium between the objective lens and the sample, and α is the angular aperture. The denominator term in this equation ($n \sin \alpha$) is a property of the objective lens termed its **numerical aperture (NA)**. Because $\sin \alpha$ has a limit approaching 1.0 and the refractive index of air is (by definition) 1.0, the best nonoil objective lenses will have numerical apertures approaching 1.0 (e.g., 0.95); because the refractive index of mineral oil is 1.52, the numerical apertures of the best oil-immersion objective lenses will approach 1.5 (typically 1.4).

With a light microscope under optimal conditions, using blue light (λ approximately 400 nm [0.4 μm]), and an oil-immersion lens with a numerical aperture of 1.4, the limit of resolution will therefore be approximately 0.2 μm. This is approximately the diameter of a lysosome, and a resolution of 0.2 μm is approximately 1000-fold better than the resolution that can be attained by the unaided human eye.

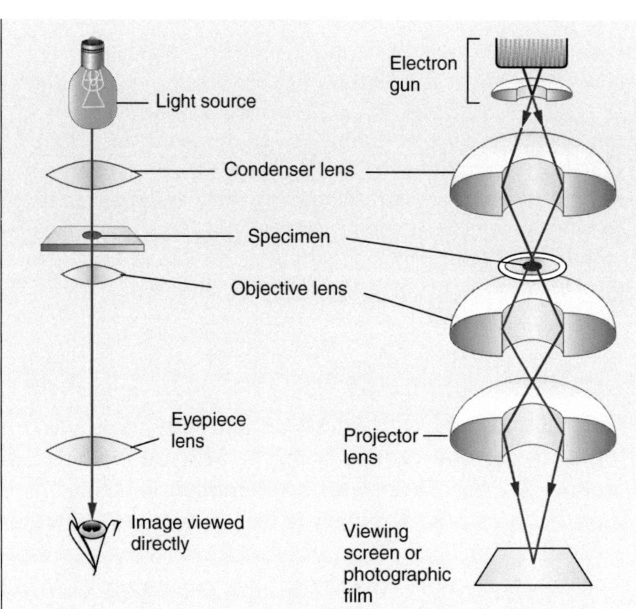

Figure 1–2. **Comparison of the lens systems in a light microscope and a transmission electron microscope.** In a light microscope (left), light is focused on the sample by the condenser lens. The sample image is then magnified up to 1000 times by the objective and ocular lenses. In a transmission electron microscope (right), magnets serve the functions of the condenser, objective, and ocular (projection) lenses, focusing the electrons and magnifying the sample image up to 250,000 times. *(Modified from Alberts B, et al.* Molecular Biology of the Cell, *4th ed. New York, NY: Garland Science, 2002.)*

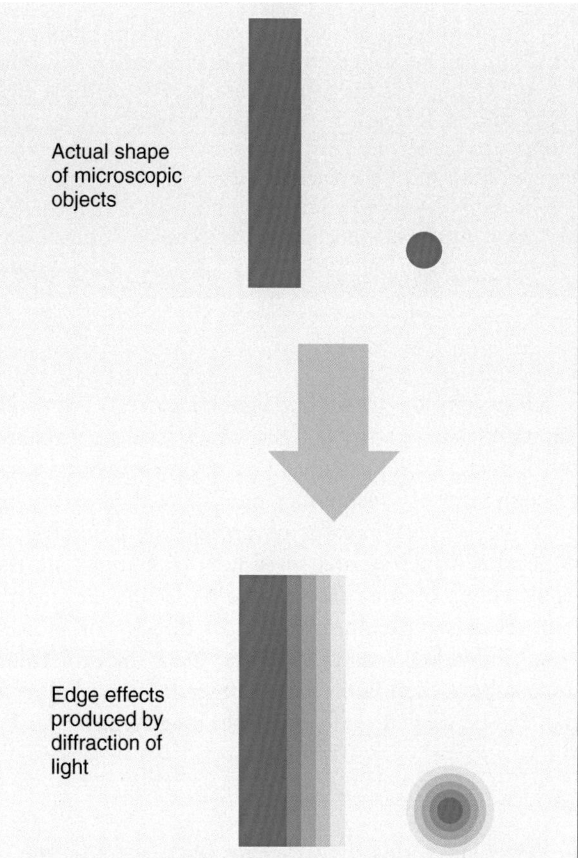

Figure 1–3. **Light passing through a sample is diffracted, producing edge effects.** When light waves pass near the edge of a barrier, they bend and spread at oblique angles. This phenomenon is known as diffraction. Diffraction produces edge effects because of constructive and destructive interference of the diffracted light waves. These edge effects limit the resolution of the image produced by microscopic magnification. *(Modified from Alberts B, et al.* Molecular Biology of the Cell, *4th ed. New York, NY: Garland Science, 2002.)*

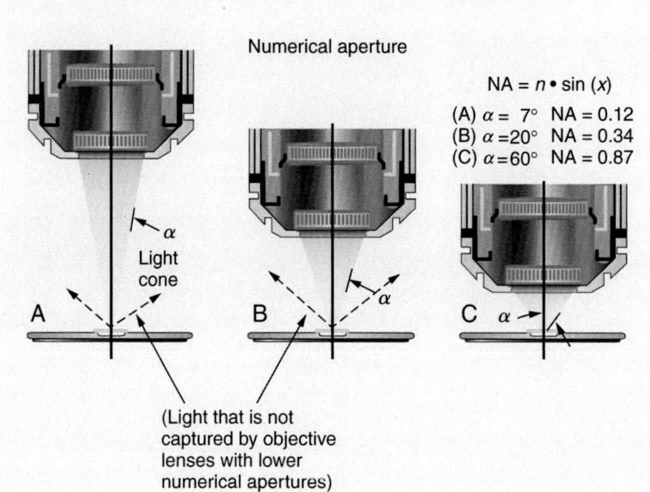

Figure 1–4. **Numerical aperture (NA).** The NA of a microscopic objective is defined as $n \sin \alpha$, where n is the refractive index of the medium between the sample and the objective lens (air), and α, the angular aperture, is the half angle of the cone of light entering the objective lens from a focal point in the sample. Objective lenses with increasingly high NA values (A, B, and C) collect increasingly more light from the sample. *(Modified from* http://www.microscopyu.com/articles/formulas/formulasna.html.*)*

Continued

The best light microscopic images, with a resolution of 0.2 μm, can be magnified to any desired degree (i.e., photographically), but no further information will be gained. No further increase in resolution beyond 0.2 μm can be obtained with a standard light microscope, and any further magnification of the sample image would be **empty magnification**, devoid of additional information content.

In a *transmission electron microscope* with an accelerating voltage of 100,000 V, electrons are produced with wavelengths of approximately 0.0004 nm. The effective numerical aperture of an electron microscope is small (on the order of 0.02), but from Abbé's equation, one would predict a resolution on the order of 0.1 nm. Practical limitations related to sample preparation, sample thickness, contrast, and so forth result in actual resolutions on the order of 2 nm (20 Å). This is approximately *100-fold better than the resolution of light microscopy.*

The resolution of standard light microscopy is limited by the wavelength of visible light, which is comparable with the diameter of some subcellular organelles, but a variety of contemporary techniques now exist that permit light microscopic visualization of proteins and nucleic acid molecules. Chief among these new techniques are those using either organic fluorescent molecules or quantum nanocrystals ("quantum dots") to directly or indirectly "tag" individual macromolecules. Once the molecules of interest have been fluorescently tagged, their cellular location can be viewed via **fluorescence microscopy** (Fig. 1–5).

Fluorescence Microscopy

In many situations, fluorescence microscopy is the first approach one might take to identify the subcellular location of particular proteins. One widely used technique to fluorescently tag a protein is based on the great precision and high affinity with which an antibody molecule can bind its cognate protein antigen. This antibody-based approach has been termed **immunolabeling** (see Box 1–2 for a brief summary of the structure and function of antibodies). Because antibodies are relatively large molecules that do not cross the surface membrane of living cells, one must fix and permeabilize cells before an antibody can be used to view the location of a target protein.

In recent years, it has become possible to view the location and movement of fluorescently tagged proteins inside **living cells**, using an approach that has been broadly termed **genetic tagging**. With this approach, one uses genetic engineering to create a plasmid expressing the protein of interest, which has been fused at its amino or carboxy terminus with either a directly fluorescent tag, such as **green fluorescent protein (GFP)**, or an indirect fluorescent tag, such as tetracysteine. Tetracysteine-tagged proteins when expressed in cells can bind

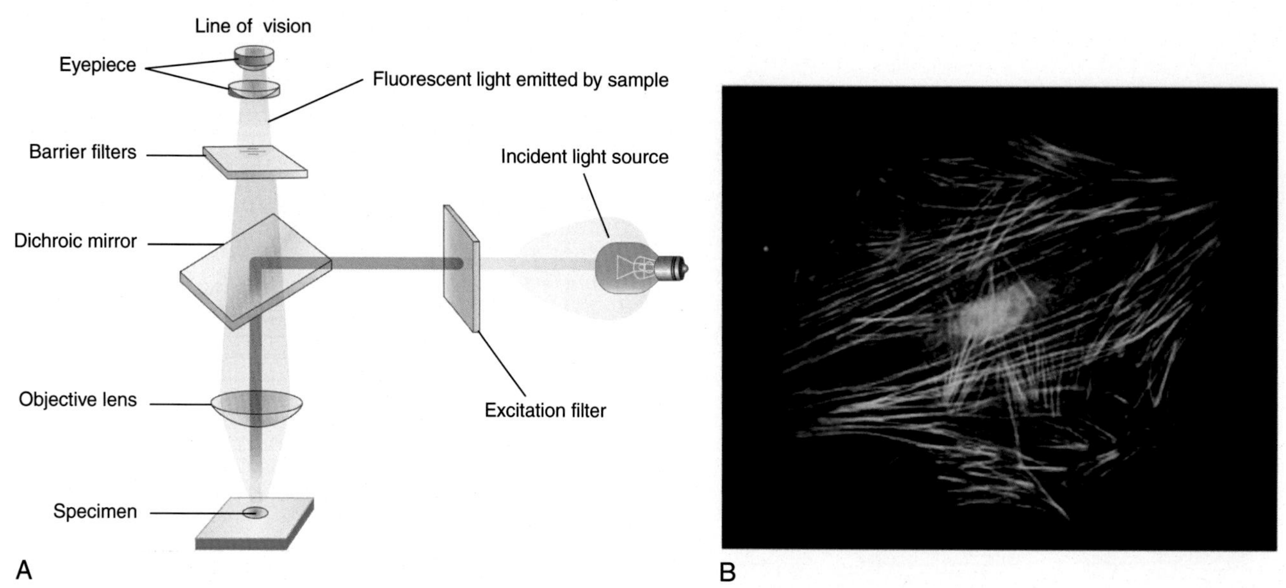

Figure 1–5. **Fluorescence microscopy.** (A) Optical layout of a fluorescence microscope. Incident light tuned to excite the fluorescent molecule is reflected by a dichroic mirror, and then focused on the sample; fluorescent light (longer wavelength than excitation light) emitted by the sample passes through the dichroic mirror for viewing. (B) Immunofluorescent micrograph of a human skin fibroblast, stained with fluorescent antiactin antibody. Cells were fixed, permeabilized, and then incubated with fluorescein-coupled antibody. Unbound antibody was washed away before viewing. *([A] Modified from Lodish H, Berk A, Zipursky SL, Matsudaira P, Baltimore D, Darnell J.* Molecular Cell Biology, *4th ed. New York, NY: W.H. Freeman, 2000; [B] courtesy E. Lazerides.)*

BOX 1–2. Antibodies

Antibodies, also known as **soluble immunoglobulins**, are specialized proteins that play an important role in immunity because of their ability to bind tightly to the foreign molecules (**antigens**) expressed by pathogens that infect an individual. An antibody molecule is a Y-shaped protein, consisting of two identical **heavy chains**, plus two identical **light chains** (Fig. 1–6). The disulfide-bonded, carboxyl-terminal halves of the heavy chains (the "tail" of the antibody) are jointly called the **Fc domain**; the two arms, which bind antigens at their tips, are called the **Fab** domains.

Immunoglobulins are synthesized by a type of lymphocyte called a **B cell** and are initially expressed as transmembrane proteins on the surface of each B cell, where they are termed **surface immunoglobulin M** (surface IgM) (a small amount of a surface immunoglobulin called **IgD** is also expressed by B cells). Each of the millions of B cells produced by the bone marrow each day makes an immunoglobulin with a unique binding specificity. The unique binding specificity of an immunoglobulin is determined by the unique amino acid sequence (called the *variable sequence*) located at the amino-terminal end of both the heavy and the light chains of each immunoglobulin molecule.

Should a particular B cell encounter its cognate antigen, that B cell first proliferates and then differentiates into an antibody-secreting **plasma cell**. Some of the proliferating B cells differentiate early into plasma cells and secrete soluble **IgM**. Soluble IgM is a pentameric molecule and often has relatively weak binding affinity; sibling B cells differentiate later, after undergoing the processes of **somatic cell hypermutation** and **class switching**. During the process of somatic cell hypermutation, the DNA encoding the variable regions of the immunoglobulin chains is selectively mutated, and cells expressing mutated, higher affinity immunoglobulin are then selected. "Class switching" refers to the process whereby the gene segment encoding an IgM-type Fc domain (Fcμ), initially expressed in all B cells, is switched out for a different gene segment, encoding a different Fc domain.Any one of three different gene segments, each encoding a different Fc domain, can be chosen to replace the Fcμ segment in the B-cell immunoglobulin gene, such that any one of three different kinds (classes) of antibody is secreted by the plasma cell after this process of class switching. These three classes of antibody are called **IgG**, **IgA**, and **IgE**. The class of antibody expressed depends on the identity of the pathogen causing the infection. IgE, for example, is most effective against many parasites; IgA protects against mucosal infections; and IgG is effective against many types of pathogens and is the most abundant immunoglobulin in blood. Each of these four classes of antibody (IgM, IgG, IgA, and IgE) has a characteristic amino acid sequence in its Fc domain that distinguishes it from the other three classes, and each of the four Fc domains has unique effector functions that activate specific features of the immune system after binding of the antibody to its cognate antigen. The differentiation of B cells into plasma cells occurs in **secondary lymphoid tissue** such as the **lymph nodes** and the **spleen**.

The particular molecular structure on an antigen to which an antibody binds is called an **epitope**. When the antigen is a protein, the epitope typically consists of several adjacent amino acids. Injection of a foreign protein into an experimental animal typically elicits the differentiation of multiple B cells into corresponding clones of descendant plasma cells, each member of a plasma cell clonal population secreting a particular antibody that binds to just one of the multiple possible epitopes on the surface of the antigenic protein. Serum collected from an immunized animal will therefore contain a mixture of antibodies against the immunizing foreign protein, and such serum is called a **polyclonal antiserum**. This polyclonal mixture of antibodies can be purified from an antiserum and used for a variety of experimental purposes, such as western blotting and immunofluorescence microscopy.

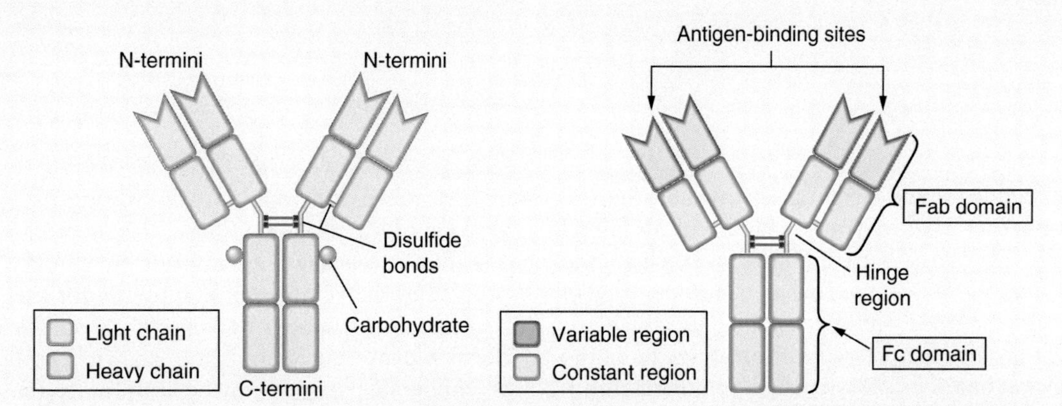

Figure 1–6. **Structure of an antibody molecule.** An antibody molecule consists of two identical heavy chains, plus two identical light chains. The disulfide-bonded, carboxyl-terminal halves of the heavy chains (the "tail" of the antibody) are jointly called the *Fc domain;* the two arms, which bind antigens at their tips, are called the *Fab domains.* Because all immunoglobulins are modified by the attachment of carbohydrate, they are examples of a type of protein termed a *glycoprotein.* The immunoglobulin shown here is an IgG molecule; class M, A, and E immunoglobulins are roughly similar, except IgM and IgE have larger Fc domains. The different immunoglobulin classes are also glycosylated at different sites. *(Modified from Parham P.* The Immune System, *2nd ed. New York, NY: Garland Publishing, 2005.)*

Continued

But for many medical and diagnostic purposes, it is useful to have a preparation of pure antibodies directed against a single epitope. Such antibodies could be obtained if one had a single clone of plasma cells, able to grow indefinitely in culture and secreting a single antibody (a **monoclonal antibody**). Because plasma cells or their B-cell precursors, or both, have a limited proliferation potential, primary cultures of plasma cells have limited usefulness for the routine production of monoclonal antibodies. However, one can fuse such cells with a special line of cancerous lymphocytes called **myeloma cells.** Such myeloma cells are "**immortal,**" that is, able to grow indefinitely in culture. The hybrid cells obtained from such a fusion, termed **hybridoma** cells, produce monoclonal antibody, like the B-cell/plasma cell parent, and yet proliferate indefinitely in culture, like the myeloma parent. In the practical application of this technique, a mouse is immunized with a particular antigen, for example, protein X; after several boosts the animal is killed, and the mix of activated B cells and plasma cell precursors in its spleen is harvested. After fusion with myeloma cells and selection for hybridoma cells in a special selection medium (in which unfused parent cells either die or are killed), the particular hybridoma colony producing a monoclonal antibody of interest is then identified (Fig. 1–7).

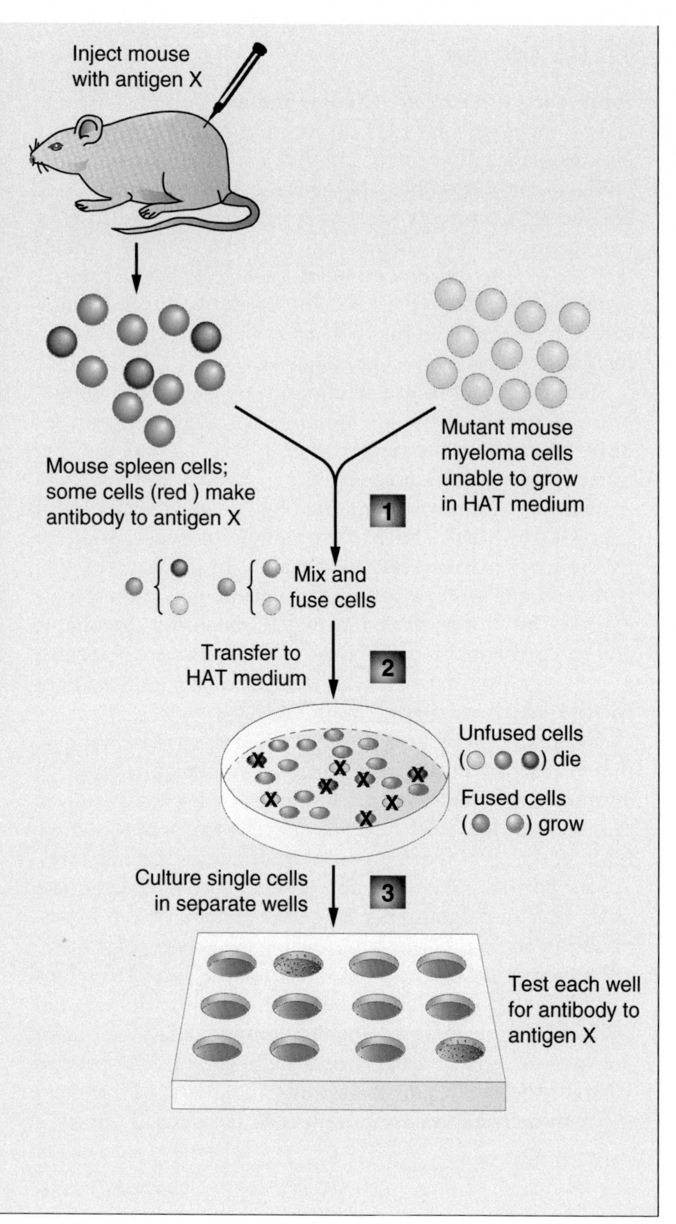

Figure 1–7. **Monoclonal antibodies.** (1) Myeloma cells are fused with antibody-producing cells from the spleen of an immunized mouse. (2) The mixture of fused hybridoma cells together with unfused parent cells is transferred to a special growth medium (HAT medium) that selectively kills the myeloma parent cells; unfused mouse spleen cells eventually die spontaneously because of their natural limited proliferation potential. Hybridoma cells are able to grow in HAT medium and have the unlimited proliferation potential of their myeloma parent. (3) After selection in HAT medium, cells are diluted and individual clones growing in particular wells are tested for production of the desired antibody. *(Modified from Lodish H, Berk A, Matsudaira P, Kaiser CA, Krieger M, Scott MP, Zipursky SL, Darnell J.* Molecular Cell Biology, *5th ed. New York, NY: W.H. Freeman, 2004.)*

subsequently added small, membrane-permeable fluorescent molecules such as the red or green biarsenicals FlAsH and ReAsH. The lines between immunolabeling and genetic tagging blur when one considers another type of genetic tagging, termed **epitope tagging,** in which the recombinant protein is expressed with an antigenic amino acid sequence at one of its ends, to which commercial antibodies are readily available, such as a "myc tag."

Let us first consider immunolabeling in more detail and then consider genetic labeling, using the example of GFP.

Immunolabeling

Specific antibodies directed against the protein of interest, used in combination with either light microscopy or EM, are useful tools for discovering the subcellular location of the protein.

A fluorescent tag (e.g., fluorescein) can be chemically coupled to the Fc domain of antibody for use in fluorescent light microscopy. For use in transmission EM, an electron-dense tag such as the iron-rich protein ferritin or nanogold particles can be coupled to the antibody. These two techniques are referred to as immunofluorescence microscopy and immunoelectron microscopy, respectively. Figure 1–5B shows an example of the use of immunofluorescence to visualize the actin "stress fibers" in a fibroblast; an example of immunoelectron microscopy is shown later in Figure 1–17.

So how would one go about obtaining antibodies to a particular protein?

Antipeptide Antibodies

One way to obtain antibodies here would be to chemically synthesize peptides corresponding to the predicted amino acid sequence of the protein product of the gene of interest. One would then chemically couple these peptides to a carrier protein, such as serum albumin or keyhole limpet hemocyanin (commonly used), and then immunize an animal such as a rabbit with the peptide-carrier complex.

This approach has one potential problem. If dealing with one of the newly discovered human genes whose protein product is completely uncharacterized, one would not have any information about the three-dimensional structure of this protein. Consequently, one would not know whether any particular amino acid sequence chosen for immunization purposes would be exposed on the surface of the native, folded protein as found in a cell. If the selected peptide corresponded to an amino acid sequence that is buried in the interior of the folded structure, antibodies directed against it would not be able to bind the native protein in the fixed cell preparations one would be using for microscopy. It turns out that amino- or carboxyl-terminal amino acid sequences are frequently exposed on the surface of many natively folded proteins; for this reason, peptides corresponding to these terminal sequences are frequently chosen for immunization of rabbits. Also, hydrophilic sequences are generally found on the surface of folded proteins, and if one or more such sequences can be identified in the predicted amino acid sequence of the protein of interest, they too would be good candidates for immunization (Box 1–3).

Because of the preceding considerations, antipeptide antibodies are not always successful in immunofluorescent localization experiments, where the target protein is in a native configuration. They are, however, often useful for the technique of **western blotting** (see Fig. 1–11).

A convenient feature of antipeptide antibodies is that excess free peptide competes for the protein in the binding of the antibody and provides a useful control for the specificity of any antibody-protein interaction observed.

Antibodies Against Full-Length Protein

The alternative to immunizing rabbits with synthetic peptides is to immunize them with either the entire protein or a stable subdomain (e.g., the extracellular

BOX 1–3. Standard Techniques for Protein Purification and Characterization

Proteins differ from each other in size and overall charge at a given pH (dependent on a property of a protein called its isoelectric point). Although other features of a protein can be used as a basis for purification (hydrophobicity, posttranslational modifications such as glycosylation or phosphorylation, ligand-binding properties, and so on), size and charge are the basis for several standard techniques for protein purification and characterization.

The most widely used technique for protein purification is **liquid chromatography**, in which an impure mixture of proteins containing a protein of interest (e.g., a cell extract in a buffered aqueous solvent with a defined salt concentration) is layered on top of a porous column filled with a packed suspension of fine beads with specific properties of porosity, charge, or both; the column itself is equilibrated in the same or a comparable solvent (Fig. 1–8). A "developing" solvent is then percolated through the column, carrying with it the mixture of proteins. Because of the properties of the beads, and/or the nature of the developing solvent, the various proteins pass through the column at differing rates, and the mixture is thereby resolved. Two commonly used types of beads employed resolve proteins either by size (**gel-filtration chromatography**) or by charge (**ion-exchange chromatography**). In a third approach (**affinity chromatography**), the beads can be derivatized with a molecule to which the protein of interest specifically binds; if that protein were an enzyme, for example, the beads could be coated with a substrate analog to which the enzyme tightly binds; more commonly, genetically engineered proteins have tags such as "6× histidine" or "glutathione *S*-transferase" (GST), which specifically bind to beads derivatized with

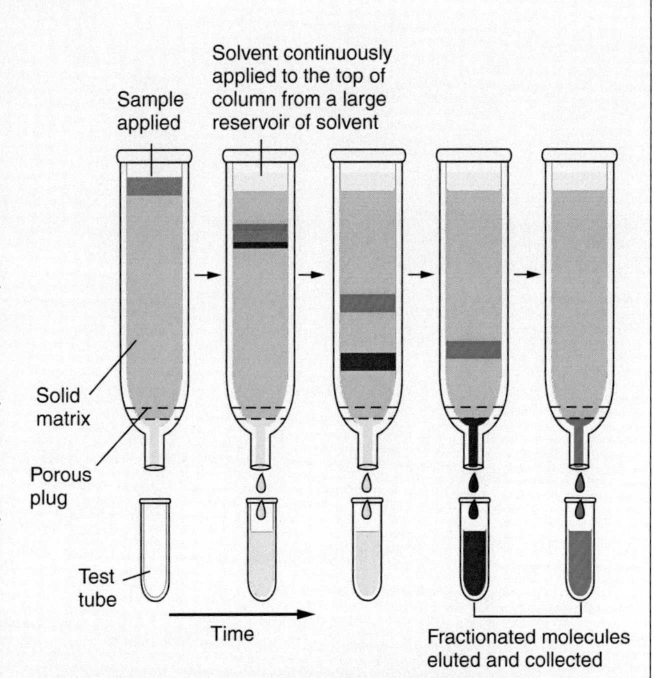

Figure 1–8. **Column chromatography.** A porous column of beads equilibrated in a particular solvent is prepared, and a sample containing a mixture of proteins is applied to the top of the column. The sample is then washed through the column, and the column eluate is collected in a succession of test tubes. Because of the properties of the beads in the column, proteins with different properties elute at varying rates off the column. *(Modified from Alberts B, et al.* Molecular Biology of the Cell. *New York, NY: Garland Science, 2002.)*

Continued

Ni^{2+}-nitriloacetic acid and glutathione, respectively. All three of these types of beads are illustrated in Figure 1–9. In other cases the beads might be covered with an antibody directed against the desired protein. This is called **immunoaffinity chromatography.**

A related analytic application of these principles of protein resolution (size and charge) is the several techniques of gel electrophoresis. Electrophoresis is the movement of molecules under the influence of an electric field. A widely used analytic gel electrophoresis technique is called **sodium dodecyl sulfate polyacrylamide gel electrophoresis (SDS-PAGE).** Polyacrylamide gels can be cast with any desired degree of porosity such that a small protein with a net charge migrates readily through the gel matrix toward an electrode, whereas a larger protein migrates more slowly through the matrix. The native charge on a protein can be the basis for its electrophoretic mobility in a gel, but it is more convenient to denature proteins with the negatively charged detergent **SDS.** SDS molecules have a hydrophobic hydrocarbon "tail" and a hydrophilic, anionic sulfate "head." The SDS molecules unfold proteins by binding via their tails at closely spaced intervals along the length of the polypeptide chain. The negative charge of the many bound SDS molecules overwhelms the intrinsic charge of a protein and thereby gives all

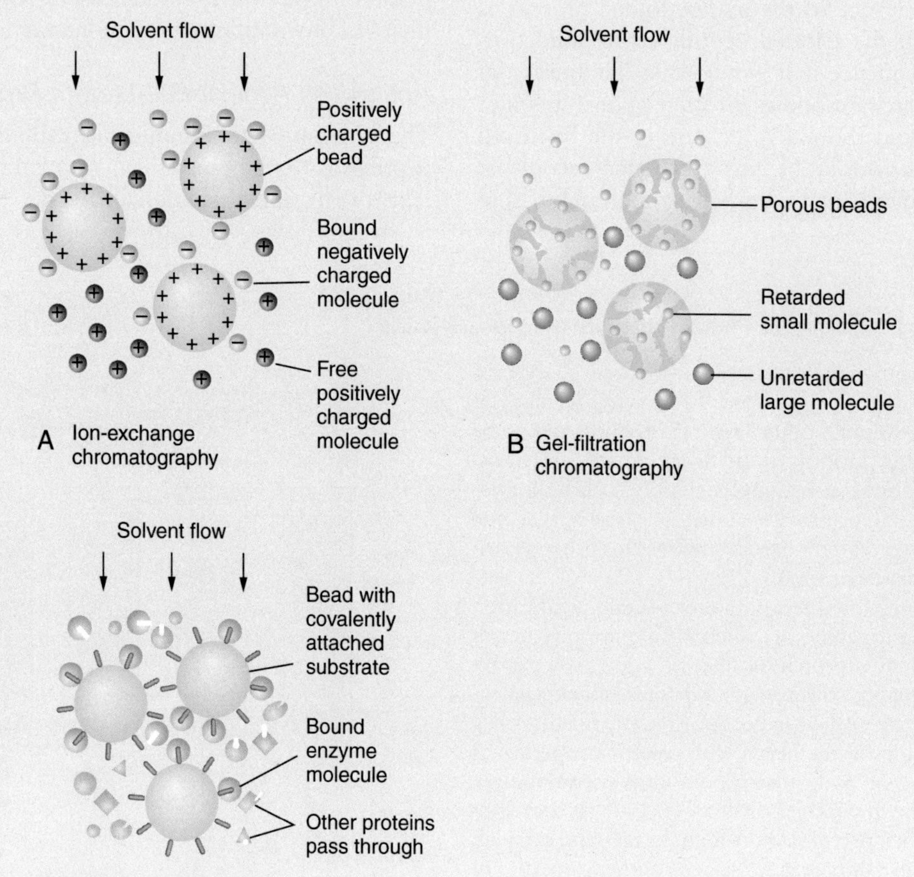

Figure 1–9. **Three types of beads used for column chromatography.** (A) The beads may have a positive or a negative charge. (The positively charged beads shown in the figure might, for example, be derivatized with diethylaminoethyl groups, which are positively charged at pH 7.) Proteins that are positively charged in a pH 7 buffer will flow through the column; negatively charged proteins will be bound to the beads and can be subsequently eluted with a gradient of salt. (B) The beads can have cavities or channels of a defined size; proteins larger than these channels will be excluded from the beads and elute in the "void volume" of the column; smaller proteins of various sizes will, to varying degrees, enter the beads and pass through them, thereby becoming delayed in their elution from the column. Such columns, therefore, resolve proteins by size. (C) The beads can be derivatized with a molecule that specifically binds the protein of interest. In the example shown, it is a substrate (or substrate analog) for a particular enzyme; the beads could also be derivatized with an antibody to the protein of interest, in which case this would be called *immunoaffinity chromatography. (Modified from Alberts B, et al.* Molecular Biology of the Cell. *New York, NY: Garland Science, 2002.)*

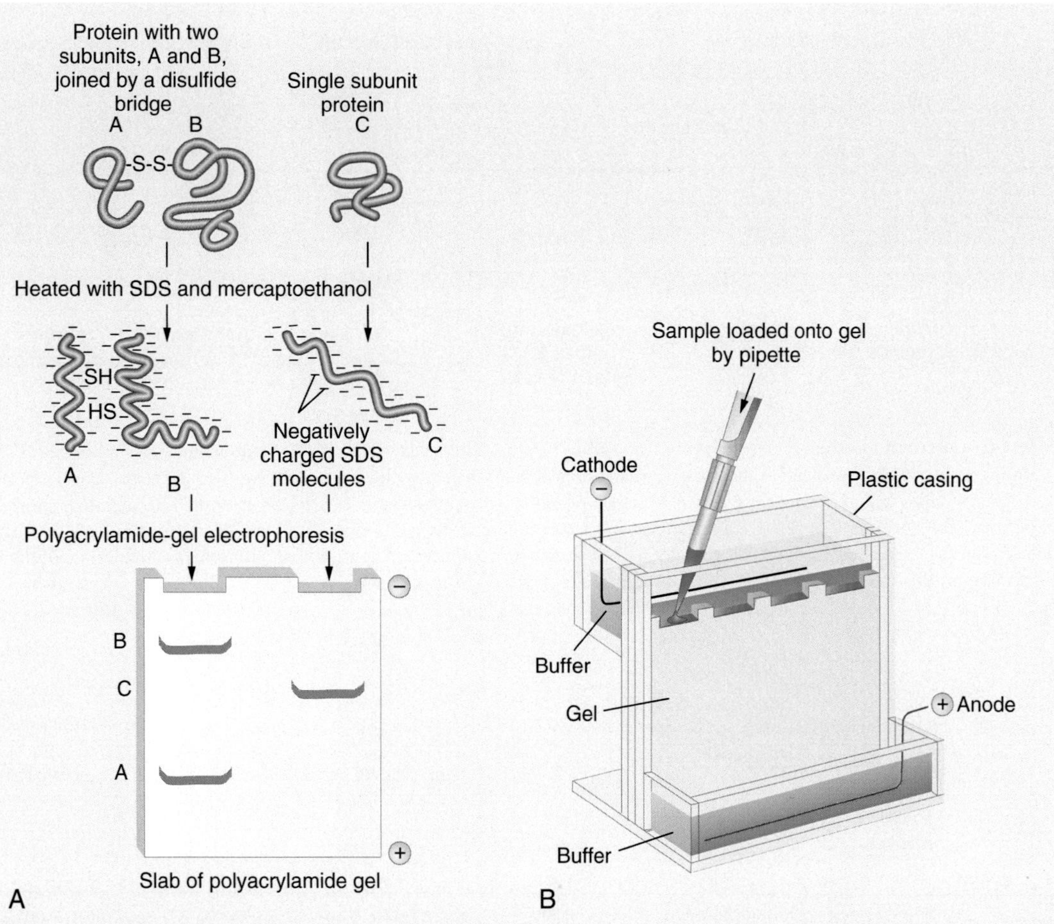

Figure 1–10. **Sodium dodecyl sulfate polyacrylamide gel electrophoresis (SDS-PAGE).** (A) Proteins in the sample are heated with the negatively charged detergent SDS, which unfolds them and coats them with a uniform negative charge density; disulfide bonds (S—S) are reduced with mercaptoethanol. (B) The sample is applied to the well of polyacrylamide gel slab, and a voltage is applied to the gel. The negatively charged detergent-protein complexes migrate to the bottom of the gel, toward the positively charged anode. Small proteins can move more readily through the pores of the gel, but larger proteins move less readily, so individual proteins are separated by size, smaller toward the bottom and larger toward the top. *(Modified from Alberts B, et al.* Molecular Biology of the Cell, *4th ed. New York, NY: Garland Science, 2002.)*

proteins a uniform negative charge density. SDS-denatured proteins therefore migrate as polyanions through a polyacrylamide gel *by their size* toward the positive electrode (the anode). At the end of the electrophoretic separation, smaller proteins will be found near the bottom of the gel and larger proteins near the top (Fig. 1–10).

A useful application of SDS-PAGE is the technique of **western blotting (immunoblotting)**. This application resolves a mixture of proteins by SDS-PAGE and then transfers the resolved set of proteins to a special paper, such as nitrocellulose paper. Proteins adsorb strongly and nonspecifically to nitrocellulose, so that the nitrocellulose paper with the adsorbed set of proteins can subsequently be bathed in a solution containing an antibody (the 1° antibody) specific to one of the proteins in the resolved set. The antibody will bind to the protein of interest; unbound 1° antibody is then washed away, and a second enzyme-linked antibody (the 2° antibody) that binds the 1° antibody is added. The enzyme linked to the 2° antibody (e.g., alkaline phosphatase) is able to catalyze the conversion of a substrate molecule to colored or fluorescent products, or to products that release light as a by-product of their formation. In this way the 1° immune complex on the nitrocellulose sheet can be detected (Fig. 1–11).

A complex mixture of proteins (e.g., a whole-cell extract) will have many proteins that by chance have similar or even identical molecular weights, such that they are indistinguishable by SDS-PAGE. In this case, one can use a technique with a higher degree of resolution, namely, **two-dimensional gel electrophoresis** (Fig. 1–12). For the *first dimension* of this process, one begins by resolving the mixture of proteins based on their individual **isoelectric points**, using the method of **isoelectric focusing (IEF)**. *(The isoelectric point of a protein can be defined as the pH at which the protein has no net*

Continued

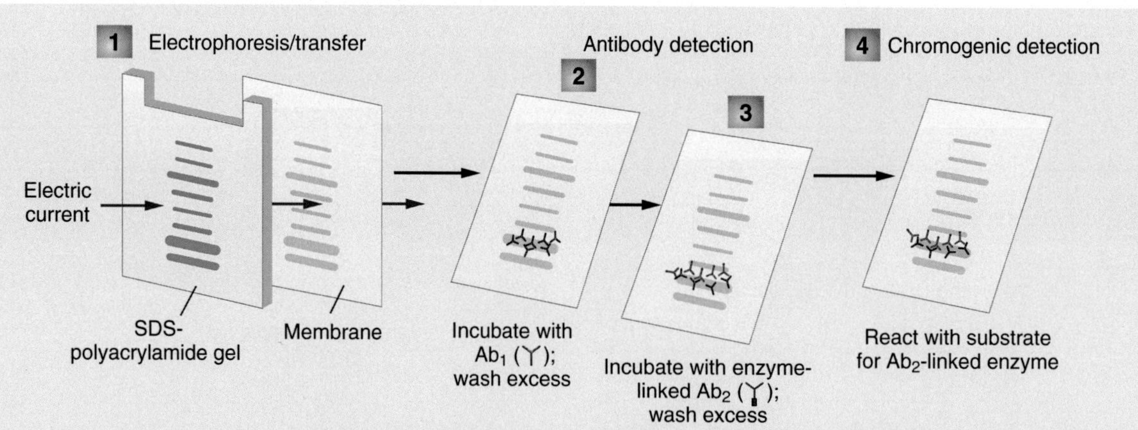

Figure 1–11. **Western blotting.** (1) Proteins are resolved by sodium dodecyl sulfate polyacrylamide gel electrophoresis (SDS-PAGE). The gel with the resolved set of proteins is then placed in an apparatus that permits electrophoretic transfer of the proteins from the gel to the surface of a special paper (e.g., nitrocellulose paper) to which proteins strongly adsorb. (2) After transfer the nitrocellulose sheet is incubated with an antibody (the "primary" antibody) directed against the protein of interest. (Before this incubation [not shown], the surface of nitrocellulose paper is "blocked" by incubating it with a nonreactive protein such as casein, to prevent nonspecific binding of the 1° antibody to the nitrocellulose; this casein block leaves the sample proteins still available for antibody binding.) (3) After washing away unbound 1° antibody, an enzyme-linked 2° antibody is added, which binds the 1° antibody and (4) can generate a colored product for detection. *(Modified from Lodish H, et al.* Molecular Cell Biology, *5th ed. New York, NY: W.H. Freeman, 2004.)*

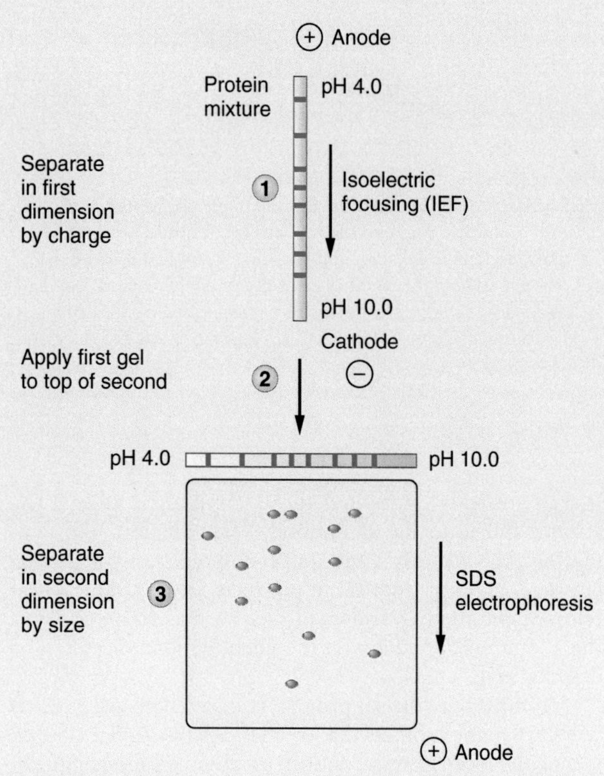

Figure 1–12. **Two-dimensional gel electrophoresis.** (1) Proteins in the sample are first separated by their isoelectric points in a narrow diameter tube gel with a fixed pH gradient, by a technique called isoelectric focusing (IEF). This is the "first-dimensional" separation. (2) The IEF gel is then soaked in SDS and laid on top of a slab SDS polyacrylamide gel for the "second-dimensional" separation of SDS-PAGE (3), which resolves proteins based on their size. *(Modified from Lodish H, et al.,* Molecular Cell Biology, *5th ed. New York, NY: W.H. Freeman, 2004.)*

charge. Many of the amino acids that comprise a protein have side chains that function as acids or bases; at a low pH, basic amino acids will be positively charged; at high pH values, acidic amino acids will be negatively charged. For every protein, there will be a pH at which the number of positively charged amino acids equals the number of negatively charged ones, such that the protein has no net charge. This is the isoelectric point of that protein.)

In the application of IEF used for two-dimensional gel electrophoresis, one first completely denatures the proteins with 8 M urea. One then applies the sample to a glass tube containing a high-porosity polyacrylamide gel that has the same 8 M concentration of urea, together with a mix of hundreds of small molecules (ampholytes) each with a unique isoelectric point. When a voltage is applied to the gel, the ampholytes migrate in the electric field, setting up a fixed pH gradient. The proteins in the sample migrate in the field until they reach the pH in the gradient corresponding to their isoelectric point, at which point they cease moving; that is, they become focused as a band in the gel. After all of the proteins have banded (focused) at their individual isoelectric points, the IEF gel is extruded from the tube and soaked in SDS buffer. It is then laid on top of an SDS polyacrylamide slab, and electrophoresed in the presence of SDS. This is the *second dimension* of resolution, where proteins are resolved by size. This sequential resolution of proteins, *first by charge, then by size*, produces good resolution of complex mixtures of proteins.

globular domain of a single-pass transmembrane protein). Immunization with the whole protein requires purification of relatively large amounts of the protein of interest (tens or hundreds of milligrams). Production of large amounts of protein (overexpression) from a cloned gene is greatly facilitated by the use of any of several plasmid- or virus-based **protein expression vectors.** Insertion of the coding sequence into an expression vector also allows creation of a "run-on" protein with a carboxyl-terminal "tag" sequence that permits subsequent rapid and efficient affinity purification. Commonly used tag sequences are "6× histidine" and glutathione *S*-transferase ("GST") tags. Such tags permit rapid and efficient affinity purification of the overexpressed protein.

Escherichia coli is often used for the expression of cloned genes, but because of different codon usage between prokaryotes and eukaryotes (and corresponding differences in the levels of the various cognate tRNA), human genes are sometimes not satisfactorily expressed in *E. coli.* Furthermore, overexpressed proteins in *E. coli* often form insoluble aggregates called **inclusion bodies**, and posttranslational modifications such as glycosylation cannot occur in bacteria. For these reasons a human gene might preferably be expressed in a eukaryotic expression system, using either a highly inducible expression vector in yeast or the insect baculovirus, *Autographa californica*, in insect Sf9 cells.

Once sufficient amounts of the protein have been purified, a **polyclonal antiserum** can be obtained by immunizing rabbits; alternatively, mice can be immunized for the production of **monoclonal antibodies** (Fig. 1–7).

Genetic Tagging

Green Fluorescent Protein

GFP was first identified and purified from the jellyfish *Aequorea victoria,* where it acts in conjunction with the luminescent protein aequorin to produce a green fluorescence color when the organism is excited. In brief, excitation of *Aequorea* results in the opening of membrane Ca^{2+} channels; cytosolic Ca^{2+} activates the aequorin protein, and aequorin, in turn, uses the energy of ATP hydrolysis to produce blue light. By quantum mechanical resonance, blue light energy from aequorin excites adjacent molecules of GFP; these excited GFP molecules then produce a bright green fluorescence. Thus the organism can "glow green in the dark" when excited. The resonant energy transfer between excited aequorin and GFP is an example of a naturally occurring **fluorescence resonance energy transfer** (**FRET**) process (see later).

The gene for GFP has been cloned and engineered in various ways to permit the optimal expression and fluorescence efficiency of GFP in a wide variety of organisms and cell types. Cloning has furthermore permitted the GFP coding sequence to be used in protein expression vectors such that a chimeric construct is expressed, consisting of GFP fused onto the amino- or carboxyl-terminal end of the protein of interest. Variant GFP proteins and related proteins from different organisms are now available that extend the range of fluorescence colors that are produced: blue (cyan) fluorescent protein (CFP), yellow fluorescent protein, and red fluorescent protein.

GFP is a β-barrel protein (its structure is shown in Fig. 1–13). Within an hour or so after synthesis and folding, a self-catalyzed maturation process occurs in the protein, whereby adjacent serine, glycine, and tyrosine side chains in the interior of the barrel react with each other and with oxygen to form a fluorophore covalently attached to a through-barrel α-helical segment, near the center of the β-barrel cavity. The GFP fluorophore thus produced is excited by the absorption of blue light from the fluorescence microscope and then decays with the release of green fluorescence.

Because the amino- and carboxyl-terminal ends of GFP are free and do not contribute to the β-barrel structure, the coding sequence for GFP can be incorporated into expression vector constructs, such that chimeric

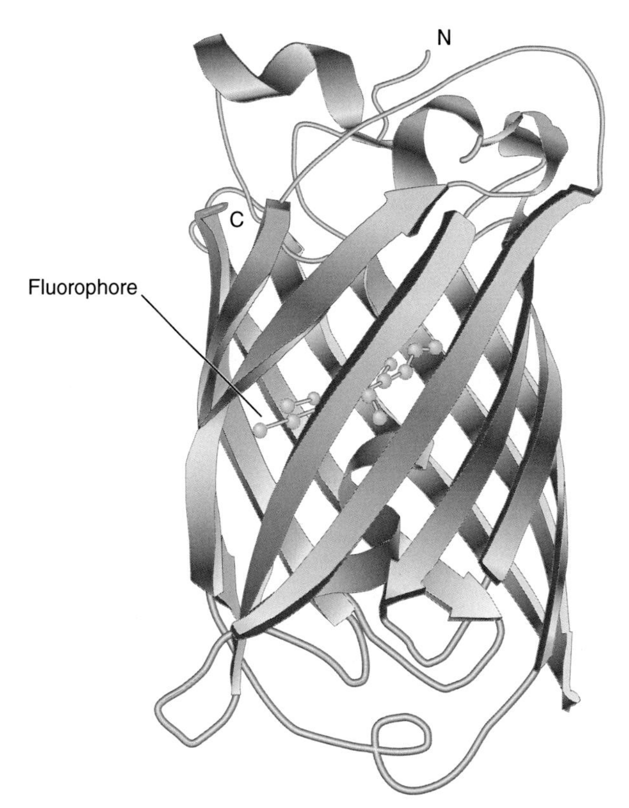

Figure 1–13. **The structure of green fluorescent protein (GFP).** GFP is an 11-strand β-barrel, with an α-helical segment threaded up through the interior of the barrel. The amino- and carboxyl-terminal ends of the protein are free and do not participate in forming the stable β-barrel structure. Within an hour or so after synthesis and folding, a self-catalyzed maturation process occurs in the protein, whereby side chains in the interior of the barrel react with each other and with oxygen to form a fluorophore covalently attached to the through-barrel α-helical segment, near the center of the β-barrel cavity. *(Modified from Ormö M, et al.* Science *1995;273:1392–1395.)*

fusion proteins can be expressed with a GFP domain located at either the amino- or carboxyl-terminal ends of the protein of interest. As mentioned earlier the great advantage of genetic tagging of proteins with fluorescent molecules such as GFP is that this technique permits one to visualize the subcellular location of the protein of interest in a living cell. Consequently, one can observe not only the location of a protein but also the path it takes to arrive at that location. For example, using a GFP-tagged human immunodeficiency virus (HIV) protein, it was discovered that after entry into cells the HIV reverse transcription complex travels via microtubules from the periphery of the cell to the nucleus.

The FRET technique can be used to monitor the interaction of one protein with another inside a living cell. As discussed earlier in this chapter, in *Aequorea,* blue light energy from aequorin is used to excite GFP by the quantum mechanical process of resonance energy transfer. Energy transfer like this can occur only when donor and acceptor molecules are close to each other (within 10 nm). Investigators are able to take advantage of this process to detect when or if two proteins in the cell bind each other under some circumstance. Both proteins of interest need merely be tagged with a pair of complementary (donor-acceptor) fluorescent proteins, such as CFP and GFP, and then coexpressed in the cell. CFP is excited by violet light and then emits blue fluorescence. If the two proteins do not bind each other in the cell, only blue fluorescence will be emitted on violet light excitation; if, however, the two proteins do bind each other, resonant energy transfer from the donor CFP will be captured by the GFP-tagged partner, and green fluorescence will be detected (Fig. 1–14).

Technologies That Enhance the Clarity and Resolution of Fluorescence Microscopy

If we wish to view the subcellular location of a particular protein, we might well begin by making the cellular population of that particular protein fluorescent, by techniques described previously in this chapter, and then, as again previously described, view it in the cell using fluorescence microscopy.

Now, speaking broadly, various subpopulations of our fluorescent protein of interest may well be found throughout the cell, perhaps for instance in association with subcellular structures, such that, as we focus our microscope from top to bottom throughout the cell, we will find it in differing amounts and in different locations in the different focal planes we successively view.

But a problem can arise here. Namely, that with a standard fluorescence microscope, and especially when viewing thicker specimens, light from the protein of interest in any particular focal plane will be mixed with the light coming from all the other fluorescent proteins located both above and below it. This produces a background glow that degrades the image.

Confocal Microscopy

One way to get around this dilemma, enabling high clarity images of the fluorescently labeled population of a target protein in a cell, is to use a special type of microscope, called a confocal microscope. Such microscopes use optical methods to obtain images from a specific focal plane within the cell and exclude light from other

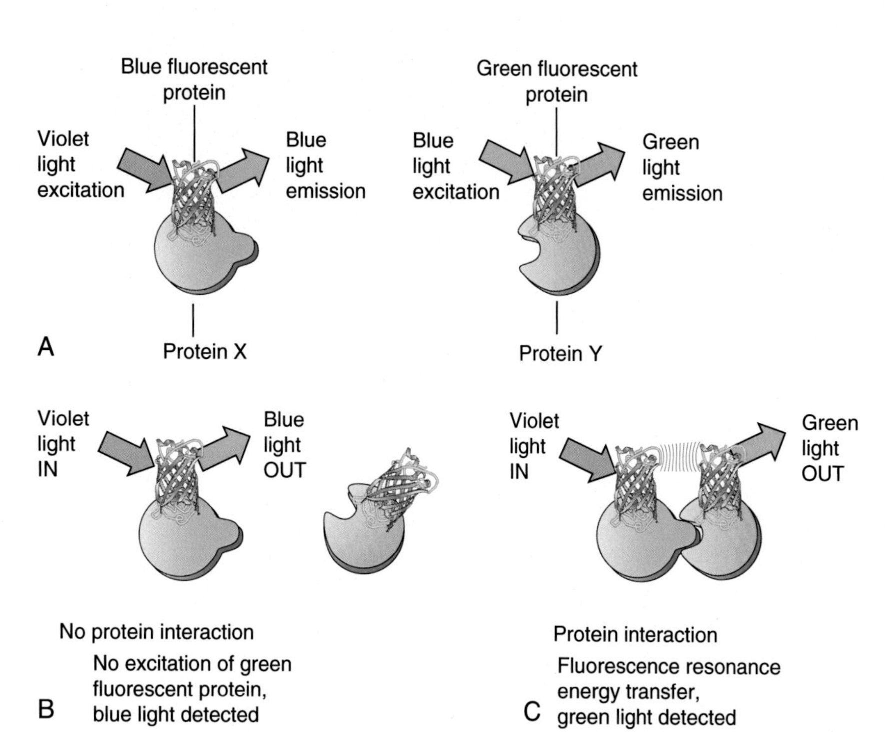

Figure 1–14. **Fluorescence resonance energy transfer (FRET).** (A) The two proteins of interest are expressed in cells as fusion proteins with either blue fluorescent protein (BFP) (protein X) or GFP (protein Y). Excitation of BFP with violet light results in the emission of blue fluorescent light by BFP; excitation of GFP with blue light yields green fluorescence. (B) If the two proteins do not bind each other inside the cell, excitation of the BFP molecule with violet light results simply in blue fluorescence. If, however, (C) the two proteins do bind each other, they will be close enough to permit resonant energy transfer between the excited BFP molecule and the GFP protein, resulting in green fluorescence after violet excitation. *(Modified from Alberts B, et al.* Molecular Biology of the Cell, *4th ed. New York, NY: Garland Science, 2002.)*

planes. There are two types of such confocal microscopes: (1) "point scanning" (also known as "laser scanning") and (2) "spinning disk" confocal microscopes. The fundamental principle in both types is the use of a pinhole aperture in the microscope (or many such pinholes in the case of "spinning disk" confocal microscopes). The pinhole aperture is located such that fluorescent light emitted from a particular location in a focal plane within the cell is brought to focus precisely at that pinhole aperture. Thus the pinhole aperture and the fluorescing proteins in the focal plane are confocal. Any fluorescent light coming from regions either above or below the focal plane is excluded from entering the pinhole aperture. So only the fluorescing proteins in the focal plane are viewed. One is then able to view the images of many successive focal planes from top to bottom of the cell and thereby construct a three-dimensional image (Fig. 1–15).

Superresolution Fluorescence Microscopy

As was discussed in Box 1–1, the limit of resolution in light microscopy is approximately 0.2 μm (200 nm). But in recent years, techniques have been developed that permit one to achieve resolutions with fluorescence light microscopy that achieve much better resolutions—As low as 20 nm! As a group, such techniques yield what is referred to as "superresolution fluorescence microscopy."

One example of such techniques involves single protein detection and localization and is called "photoactivated localization microscopy" (**PALM**). A closely related technique is termed stochastic optical reconstruction microscopy" (**STORM**). At the heart of these techniques was the discovery of a mutant green fluorescent protein (GFP). This variant GFP, called "photoactivatable GFP" (**PAGFP**), is not able to fluoresce, but it can be activated by illumination with 413-nm ultraviolet

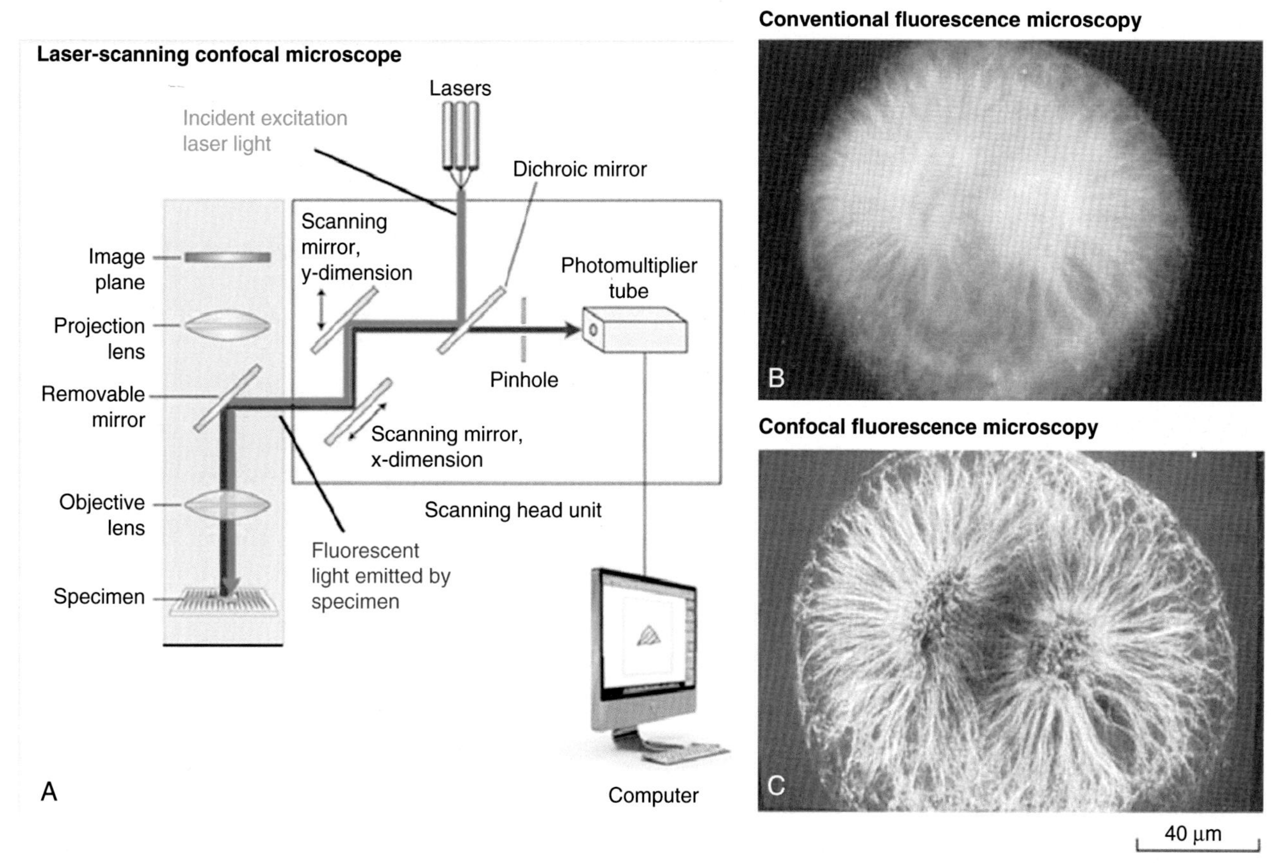

Figure 1–15. **Laser scanning confocal fluorescence microscopy.** (A) Optical layout of a laser scanning confocal microscope. Incident laser light tuned to excite the fluorescent molecule *(green)* is reflected off a dichroic mirror and is then guided by two scanning mirrors and the objective lens to the focal plane to illuminate a spot in a focal plane in the specimen. The scanning mirrors rock back and forth so that the light sweeps across the specimen in raster fashion. The fluorescence *(green)* emitted by the fluorescently tagged molecules in the specimen is then sent back to be captured by a photomultiplier tube; on the way back, it must pass through a pinhole that is confocal with the specimen focal plane. The confocal pinhole excludes light from out-of-focus focal planes in the specimen. The fluorescence signals from photomultiplier tube are then sent to a computer that reconstructs the confocal image. (B) An image of fluorescent tubulin in a fertilized sea urchin egg as viewed with conventional fluorescence microscopy. (C) The same fertilized egg viewed by confocal microscopy in a focal plane that passes through the equator of the egg. One sees clearly the mitotic spindle apparatus as the egg is dividing into two daughter cells. *([A] Modified from Lodish H, Berk A, Kaiser C, Krieger M, Bretscher A, Ploegh H, Amon A, Martin C.* Molecular Cell Biology, *8th ed. New York: W.H. Freeman, 2016; [B] and [C] White J, et al.* J Cell Biology *1987;105:41–48, by permission of the Rockefeller University Press.)*

light; following activation, it will then fluoresce in the usual way upon excitation with 488-nm light.

One can take advantage of the properties of PAGFP to do superresolution microscopy in the following way: After the population of the protein of interest in the cell has been tagged with PAGFP, one can now activate a portion (not all) of these proteins in the focal plane by exposing them to a brief pulse of activating ultraviolet light and then follow that with a pulse of light at the excitation wavelength. In this circumstance, one will view some individual points of fluorescent light, each corresponding to a single molecule of the PAGFP-tagged protein of interest. If occasionally these individual fluorescing molecules happen to be too close together (remembering the resolution limit of visible light), one would not of course be able to resolve their two exact locations. But in the protocol, most of the relatively rare fluorescing molecules in the population will be sufficiently far apart that they will appear as individual blurred glowing spots, each spot having a Gaussian distribution of light intensity, and of course, the location of the fluorescing protein will be in the center of that diffuse spot. And these exact fluorescing locations (centers of the diffuse spots) in the focal plane are recorded digitally.

The process is then performed over and over on the same focal plane for hundreds of imaging cycles. In these many cycles the fluorescent spot of a PAGFP protein less than 10–20 nm away from the previously recorded location of an earlier identified protein will be recorded, and in this way, computer-assisted image analysis will build up a very highly resolved image (tens of nanometer resolution, as we said at the outset) (Fig. 1–16).

ELECTRON MICROSCOPY

There are two broad categories of EM: **transmission EM** and **scanning EM**. First, we discuss the topic of transmission EM, including the special techniques of **cryoelectron microscopy**.

Transmission Electron Microscopy

Transmission electron microscopes use electrons in a way that is analogous to the way light microscopes use visible light. The various elements in a transmission electron microscope that produce, focus, and collect electrons after their passage through the specimen are all related in function to the corresponding elements in a light microscope (see Fig. 1–2). Rather than a light source, there is an electron source, and electrons are accelerated toward the anode by a voltage differential. In an electron microscope the electrons are focused not by optical lenses of glass, but instead by **magnets**. Because electrons would be scattered by air molecules, both the electron trajectory and the sample chambers must be maintained in a **vacuum**.

We are perhaps more accustomed to thinking of an electron as a particle-like object rather than as an electromagnetic wave, but of course, quantum mechanically, electrons can behave as either particles or waves. As is the case for all waves, the frequency (and hence wavelength) of an electron is a function of its energy, which in turn is a function of the accelerating voltage that drives an electron from its source in an electron microscope. Typical electron microscopes are capable of producing accelerating voltages of approximately 100,000 V, producing electrons with energies that correspond to **wavelengths of subatomic dimensions**. This would, in theory, permit subatomic resolutions! A number of factors such as lens aberrations and sample thickness, however, limit the practical resolution to much less than this. Under usual conditions with biological samples, *electron microscopic resolution is approximately **2 nm***, which is still more than 100-fold better than the resolution with standard light microscopy. This increased resolution, in turn, permits much larger useful magnifications, up to 250,000-fold with EM, compared with approximately 1000-fold in a light microscope with an oil-immersion lens.

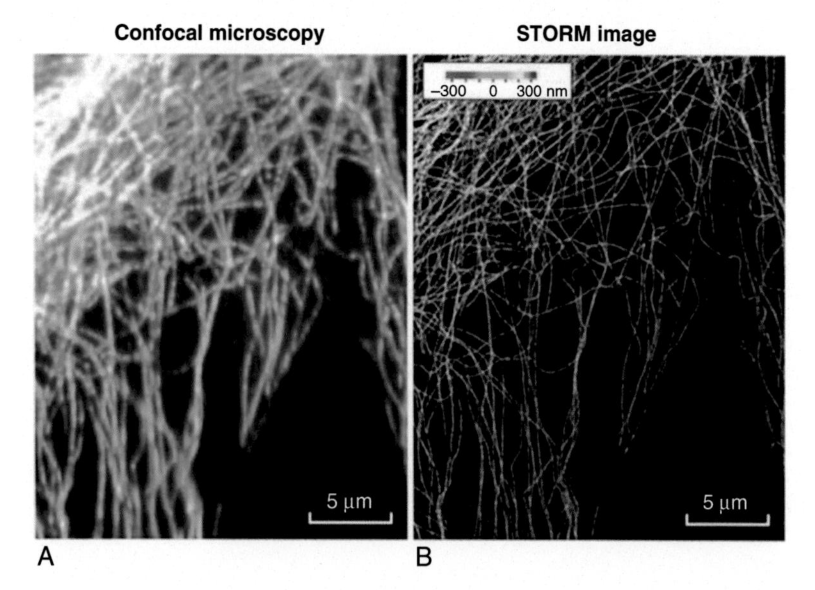

Figure 1–16. **Superresolution fluorescence microscopy.** (A) A confocal image of fluorescently labeled microtubules located near the periphery of a fibroblast (BSC-1 cells). (B) A corresponding superresolution image of the same microtubules. *(From Xiaowei Zhuang: Huang et al.* Science *2008;319:810–813, with permission from AAAS).*

Because of the high vacuum of the EM chamber, living cells cannot be viewed, and typical sample preparation involves fixation with covalent cross-linking agents such as glutaraldehyde and osmium tetroxide, followed by dehydration and embedding in plastic. Because electrons have poor penetrating power, an ultramicrotome is used to shave off extremely thin sections from the block of plastic in which the tissue is embedded. These ultrathin sections (50–100 nm in thickness) are laid on a small circular grid for viewing in the electron microscope.

Electrons would normally pass equally well through all parts of a cell, so membranes and various cellular macromolecules are given contrast by "staining" the tissue with heavy metal atoms. For example, the **osmium tetroxide** used as a fixing agent also binds to carbon-carbon double bonds in the unsaturated hydrocarbons of membrane phospholipids. Because osmium is a large, heavy atom, it deflects electrons, and osmium-stained membrane lipids appear dark in the electron microscope image. Similarly, **lead and uranium salts** differentially bind various intracellular macromolecules, thereby also staining the cell for EM.

The preceding discussion has been about how one would go about viewing the overall layout of the cell under an electron microscope, but most often, we are interested in the subcellular location of a particular molecule, usually a protein. Here again a specific antibody against the protein can be brought into play, this time tagged with something electron dense; most often, this electron-scattering tag will be commercially available nanoparticles of colloidal **gold**, coated with a small antibody-binding protein, called **Protein A** (Fig. 1–17). Gold-tagged antibodies can also be used to stain various genetically tagged proteins containing tags such as GFP, a myc tag, or any other epitope.

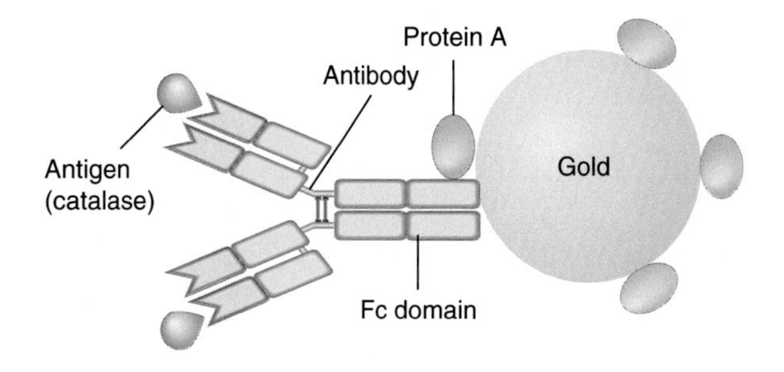

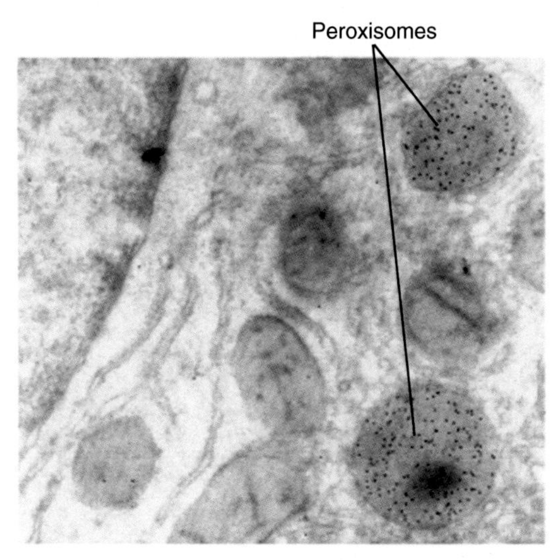

Figure 1–17. **Protein A–coated gold particles can be used to localize antigen-antibody complexes by transmission electron microscopy (EM).** (A) Protein A is a bacterial protein that specifically binds the Fc domain of antibody molecules, without affecting the ability of the antibody to bind antigen (the enzyme catalase, in the example shown here); it also strongly adsorbs to the surface of colloidal gold particles. (B) Anticatalase antibodies have been incubated with a slice of fixed liver tissue, where they bind catalase molecules. After washing away unbound antibodies, the sample was incubated with colloidal gold complexed with protein A. The electron-dense gold particles are thereby positioned wherever the antibody has bound catalase, and they are visible as black dots in the electron micrograph. It is apparent that catalase is located exclusively in peroxisomes. *([A] Modified from Lodish H, Berk A, Matsudaira P, Kaiser CA, Krieger M, Scott MP, Zipursky SL, Darnell J.* Molecular Cell Biology, *5th ed. New York, NY: W.H. Freeman, 2004; [B] from Geuze HF, et al.* J Cell Biol *1981;89:653, by permission of the Rockefeller University Press.)*

In some circumstances, one may wish to obtain a more three-dimensional sense of a surface feature of the cell, or of a particular object such as a macromolecular complex. Two different techniques can be used to do this with a transmission electron microscope: one is called **negative staining**, and the other is called **metal shadowing**. In the case of negative staining, the objects to be viewed (e.g., virus particles) are suspended in a solution of an electron-dense material (e.g., a 5% aqueous solution of uranyl acetate), and a drop of this suspension is placed on a thin sheet of plastic, which, in turn, is placed on the EM sample grid. Excess liquid is wicked off, and when the residual liquid dries, the electron-dense stain is left in the crevices of the sample, producing images such as that shown in Figure 1–18A.

The second technique, metal shadowing, is illustrated in Figure 1–19. The chemically fixed, frozen, or dried specimen, on a clean mica sheet, is placed in an evacuated chamber, and then, metal atoms, evaporated from a heated filament located at an overhead angle to the specimen, coat one side of the elevated features on the surface of the sample, creating a **metal replica**. When subsequently viewed in the electron microscope, electrons are unable to pass through metal-coated surfaces but are transmitted through areas in the sample that were in the shadow of the object and were therefore not metal coated. The resulting image, usually printed as the negative, is remarkably three-dimensional in appearance (see Fig. 1–18, B).

In situations that involve a frozen sample (see the following section), after metal shadowing, the entire surface of the sample can then be coated with a film of carbon. After removal of the original cellular material, the metal-carbon replica is viewed in the electron microscope. When used in conjunction with a method of

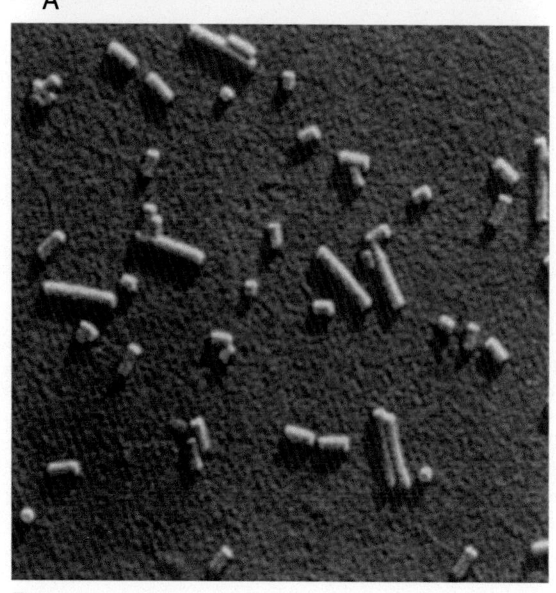

Figure 1–18. **Electron microscopic images of negatively stained versus metal-shadowed specimens.** A preparation of tobacco rattle virus was either (A) negatively stained with potassium phosphotungstate or (B) shadowed with chromium. *(Courtesy M. K. Corbett.)*

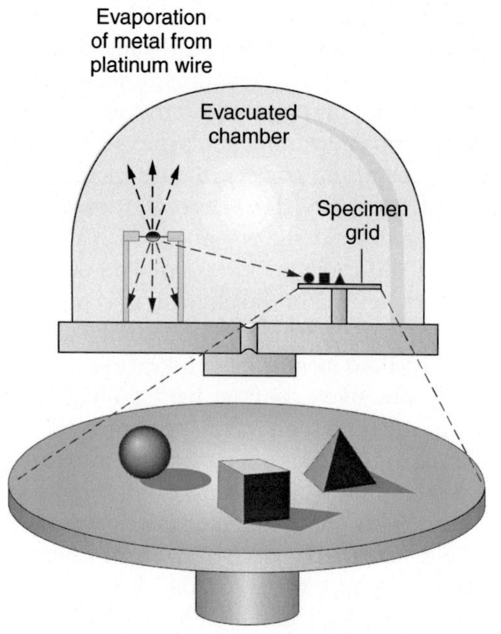

Figure 1–19. **Procedure for metal shadowing.** The specimen is placed in a special bell jar, which is evacuated. A metal electrode is heated, causing evaporation of metal atoms from the surface of the electrode. The evaporated metal atoms spray over the surface of the sample, thereby "shadowing" it. *(Modified from Karp G.* Cell and Molecular Biology, *3rd ed. New York: John Wiley & Sons, 2002.)*

sample preparation called *freeze fracture,* metal shadowing has been useful in visualizing the arrangement of proteins in cellular membranes.

Cryoelectron Microscopy

The dehydration of samples that accompanies standard fixation and embedding procedures denatures proteins and can result in distortions if one wishes to view molecular structures at high levels of magnification in the electron microscope. One solution to this difficulty is the technique of cryoelectron microscopy. Here the sample (often in suspension in a thin aqueous film on the sample stage) is rapidly frozen by plunging it into liquid propane (−42°C) or placing it against a metal block cooled by liquid helium (−273°C). Rapid freezing results in the formation of microcrystalline ice, preventing the formation of larger ice crystals that might otherwise destroy molecular structures. The frozen sample is then mounted on a special holder in the microscope, which is maintained at −160°C. In some cases, surface water is then lyophilized off ("freeze-etch") from the surface of the sample, which is then metal shadowed, producing images such as that in Figure 1–20.

In other cases, when there are many identical structures such as virus particles, computer-based averaging techniques, in combination with images from multiple planes of focus, can produce tomographic three-dimensional images with single nanometer resolution (Fig. 1–21). This technique showed, for example, a previously unsuspected "tripod" structure for the HIV virus envelope spike (Fig. 1–10).

Scanning Electron Microscopy

The surfaces of metal-coated specimens can also be viewed to good advantage with another type of electron microscope, the scanning electron microscope. Unlike the case with metal shadowing in transmission EM, in this case, the entire surface of the specimen is covered with metal. The source of electrons and focusing magnets in a scanning electron microscope are like those of a traditional transmission electron microscope, except that an additional magnet is inserted in the path of the electron beam. This latter magnet is designed to sweep (scan)

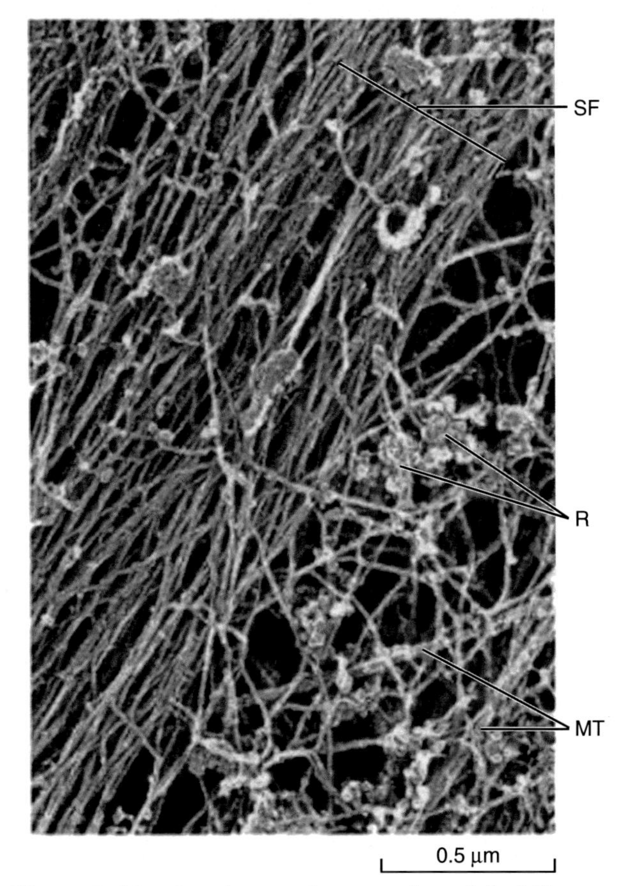

Figure 1–20. **Cryoelectron microscopy of cytoskeletal filaments, obtained by deep etching.** A fibroblast was gently extracted using the nonionic detergent Triton X-100 (Sigma, St. Louis), which dissolves the surface membrane and releases soluble cytoplasmic proteins, but has no effect on the structure of cytoskeletal filaments. The extracted cell was then rapidly frozen, deep etched, and shadowed with platinum, then viewed by conventional transmission electron microscopy. *MT*, microtubules; *R*, polyribosomes; *SF*, actin stress fibers. *(From Heuzer JE, Kirschner M.* J Cell Biol *1980;86:212, by permission of Rockefeller University Press.)*

the focused, narrow, pencil-like electron beam in parallel lines (a raster pattern) over the surface of the specimen. Back-scattered electrons, or secondary electrons ejected from the surface of the metal-coated specimen (usually coated with gold or gold-palladium), are collected and focused to generate the scanned image. The resolving power of a scanning electron microscope is a function of the diameter of the scanning beam of electrons. Newer machines can produce extremely narrow beams with a resolution on the order of 5 nm, permitting remarkably detailed micrographic images (Fig. 1–22).

ATOMIC FORCE MICROSCOPY

Atomic force microscopy (AFM) was developed in the 1980s, and it has become an increasingly useful tool for cell biology. The principle of AFM is illustrated in Figure 1–23. A nanoscale cantilever/tip structure moves over the surface of the sample, and the up and down deflections of the cantilever tip are detected by a laser beam focused on its upper surface. Deflections on the order of a nanometer can be detected, producing resolutions comparable with or exceeding those of the best scanning electron microscopes.

Samples to be scanned by AFM need not be metal plated and put in a vacuum as is required for scanning EM, and a particular advantage of AFM over scanning EM is that samples immersed in aqueous buffers, or even living cells in culture medium, can be scanned by an AFM device. In this way, for example, the real-time opening and closing of nuclear pores in response to the presence or removal of Ca^{2+} (in the presence of ATP) has been demonstrated by AFM of isolated nuclear envelopes (not shown), and a novel cell surface structure called

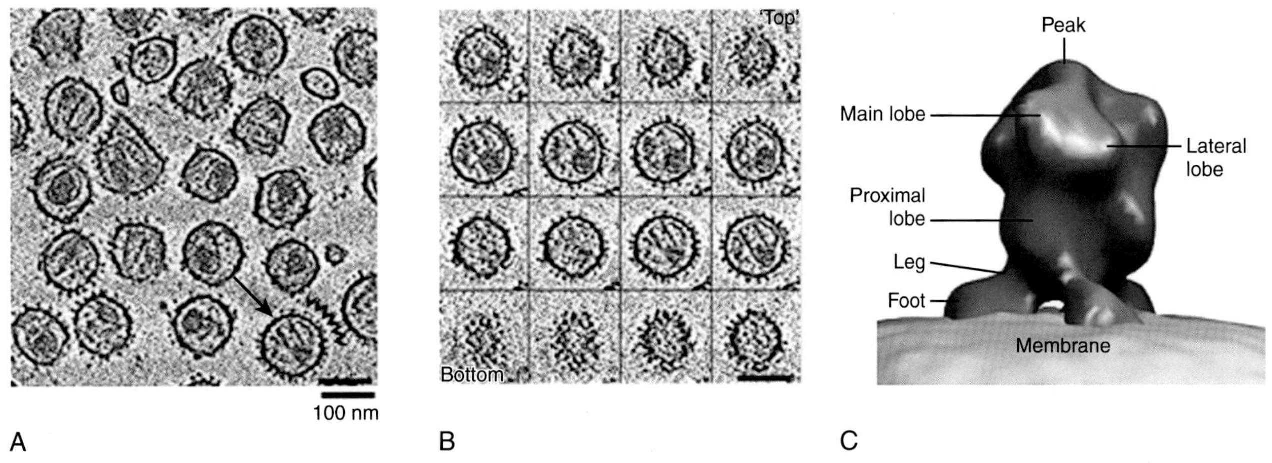

Figure 1–21. **Cryoelectron microscopy and tomography of human immunodeficiency virus (HIV).** Concentrated virus (HIV or SIV) in aqueous suspension was placed on a grid and rapidly frozen by plunging the grid into liquid ethane at −196°C. The frozen sample was then placed in a cryoelectron microscopy grid holder for viewing at a magnification of ×43,200. The sample holder was tilted at a succession of angles for consecutive images, from which tomograms were computed. (A) Sample virus field; the virus shown in this field is simian immunodeficiency virus (SIV), which has a higher density of surface spike proteins than HIV. The virus particle indicated by the *arrow* was chosen for tomographic analysis. (B) Computationally derived transverse sections through the selected virus particle (from top to bottom). (C) Tomographic structure of the virus envelope spike complex, which is a trimeric structure of viral gp120 (globular portion of the spike) and gp41 (transmembrane "foot") proteins, in the form of a twisted tripod. *(From Zhu P, et al.* Nature *2006;441:847, by permission.)*

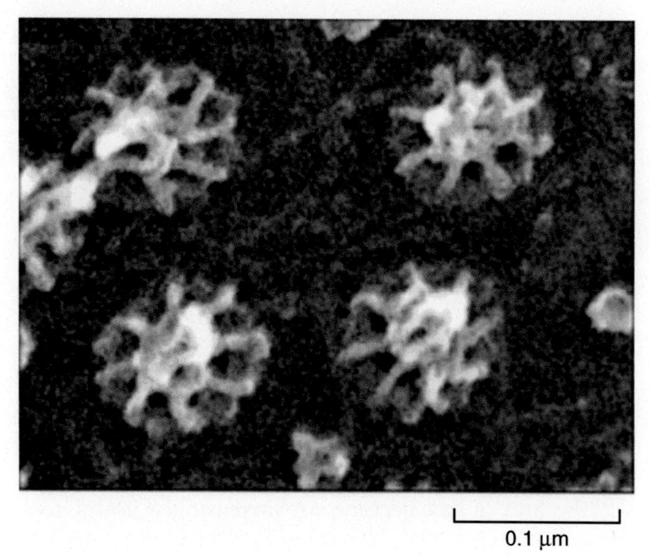

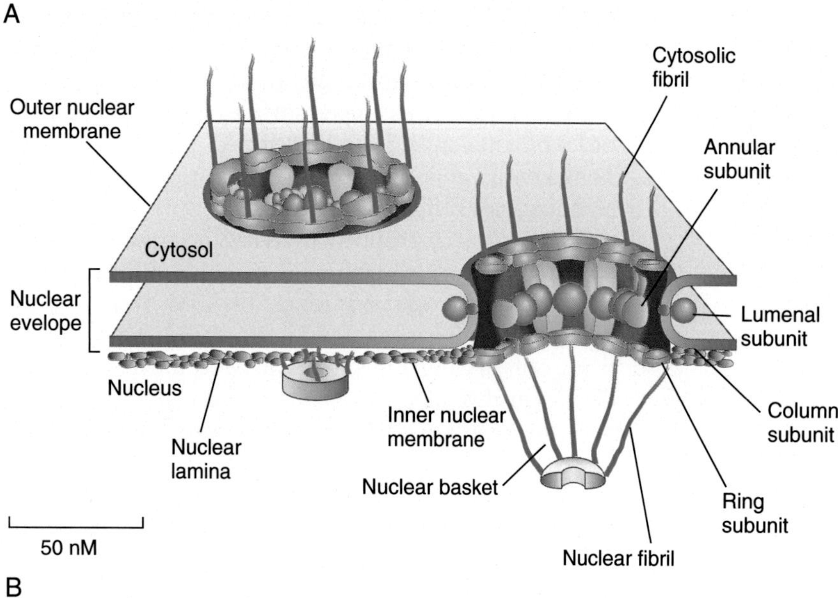

Figure 1–22. **High-resolution scanning electron micrograph of nuclear pore complexes.** (A) Purified nuclear envelopes were prepared for scanning electron microscopy. The electron micrograph shows the image of nuclear pores as viewed from the nuclear side of the pore. (B) Current model for the structure of a nuclear pore. *([A] From Goldberg MW, Allen TD.* J Cell Biol *1992;119:1429, by permission of Rockefeller University Press; [B] modified from Alberts B, et al.* Molecular Biology of the Cell, *4th ed. New York, NY: Garland Science, 2002.)*

the *fusion pore* was identified on the apical surface of living pancreatic acinar cells (Fig. 1–24).

Not all applications of AFM technology are topologic. For example, by increasing the downward force of the probe tip on the sample, nanodissections can be performed, such as taking a "biopsy" sample from a specific region of a single chromosome (Fig. 1–14)!

In addition, a set of nontopologic uses of AFM technology exist that might be regarded as biophysical but which have cell biological ramifications. In these cases the cantilever tip is used to measure interactive or deforming forces (Fig. 1–25). For example, ligands or reactive molecules can be attached to the tip of the cantilever. After the binding of the tip to the sample, one can measure the force required to either lift the tip or move the object to which the tip is bound. Experiments such as these yield insights into such processes as the force required to unfold modular protein domains, the strength of lectin-glycoprotein interactions, and so forth.

MORE TOOLS OF CELL BIOLOGY

In a search for the functions of novel genes demonstrated by the Genome Project, there are, of course, many other techniques in addition to microscopy that might be brought into play. The techniques of **animal cell culture, flow cytometry,** and **subcellular fractionation** are considered in the following sections.

Cell Culture

Many bacteria (auxotrophs) can be successfully grown in a medium containing merely a carbon source (e.g., sugar) and some salts. Animals (heterotrophs) have lost the

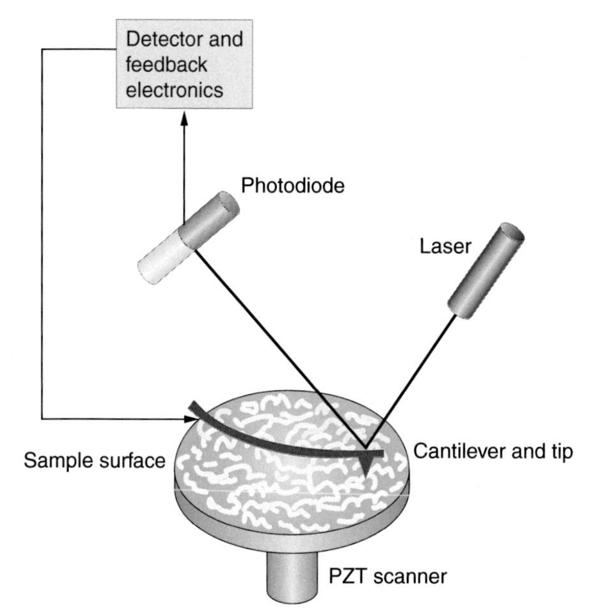

Figure 1–23. **Atomic force microscopy (AFM).** In AFM the sample is scanned by a microscale probe, consisting of a sharp tip attached to a flexible cantilever. The deflection of the probe as it moves over the sample is measured by the movement of a laser beam reflected from the top of the cantilever onto an array of photodiodes. *(Modified from the Wikipedia article "Atomic Force Microscope,"* http://en.wikipedia.org/wiki/Atomic_force_microscopy.*)*

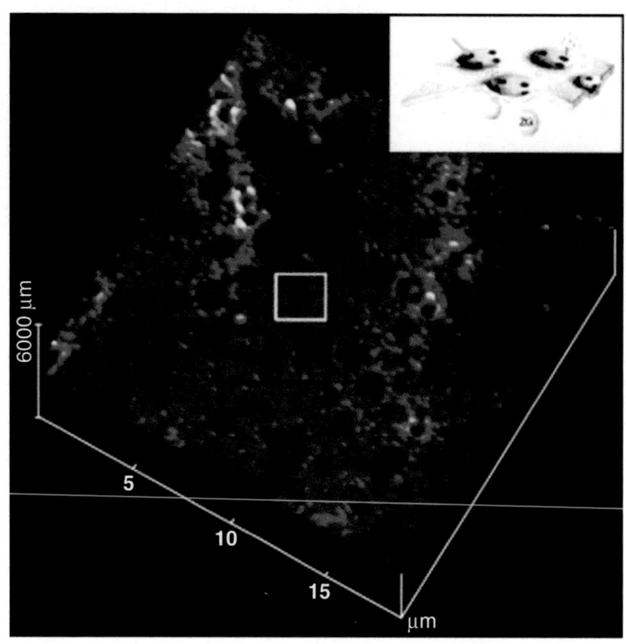

Figure 1–24. **Atomic force microscopy (AFM) image of fusion pores in the membrane of a living cell.** The apical plasma membrane of living pancreatic acinar cells was scanned by AFM, producing this image of multiple pore structures. Pores are located in permanent pit structures (one of which is framed by the *white box*) on the apical surface membrane. Inset: Schematic depiction of secretory vesicle docking and fusion at a fusion pore. Fusion pores *(blue arrows)*, 100–180 nm wide, are present in "pits" *(yellow arrows)*. *ZG*, zymogen granule. *(From Hörber J, Miles M.* Science *2003;302:1002, reprinted with permission from AAAS.)*

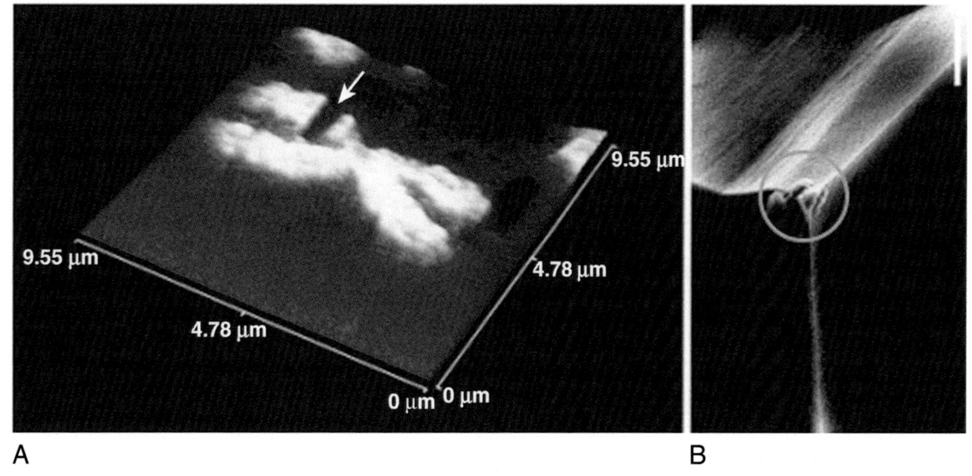

Figure 1–25. **Atomic force microscopy (AFM) "biopsy" of a human chromosome.** Metaphase chromosome spreads were prepared and fixed on glass microscope slides by standard techniques. Air-dried, dehydrated chromosomes were first scanned by AFM in noncontact mode; for dissection (A) the probe was dragged through a previously identified location on a selected chromosome with a constant applied downward force of 17 micronewtons. (B) Scanning electron microscopic image of the tip of the probe used for the dissection shown in A; the material removed from the chromosome on the tip of the probe is *circled.* DNA in the sample could subsequently be amplified by polymerase chain reaction. *(From Fotiadis D, et al.* Micron *2002;33:385 by permission of Elsevier Science, Ltd.)*

ability to synthesize all their amino acids, vitamins, and lipids from scratch and require many such nutrients to be provided preformed in their diet. Mammals, for example, require 10 amino acids in their diet. Mammalian cells grown in culture require the same 10 amino acids, plus three others (cysteine, glutamine, and tyrosine) that are normally synthesized from precursors by either gut flora or by the liver of the intact animal. By the 1960s all the micronutrient growth requirements for mammalian cells had been worked out (amino acids, vitamins, salts, and trace elements), and yet, it was found that it was still necessary to supplement the growth medium with serum (typically 5%–10%) to achieve cell survival and growth. Eventually, it was shown that *serum provides certain essential proteins and growth factors:* (1) **extracellular matrix proteins** such as cold-insoluble globulin (a soluble form of fibronectin), which coat the surface of the petri dish and provide a physiologic substrate for cell attachment; (2) **transferrin** (to provide iron in a physiologic form); and (3) three **polypeptide growth factors:** platelet-derived growth factor, epidermal growth factor, and insulin-like growth factor. It is now possible to provide all the required components of serum in purified form to produce a completely defined growth medium. This can be useful in certain circumstances, but for routine growth of cells, serum is still used.

Embryonic tissue is the best source of cells for growth in culture; such tissue contains a variety of cell types of both mesenchymal and epithelial origin, but one cell type quickly predominates: cells of mesenchymal origin, resembling connective tissue fibroblasts. These fibroblastic cells proliferate more rapidly than the more specialized organ epithelial cells and hence soon outgrow their neighbors. Special procedures must be used if one wishes to study other differentiated cell types from either embryonic or adult tissue, such as liver epithelial cells and breast epithelial duct cells, and it is often not easy to maintain the differentiated phenotype of these cells after prolonged growth in culture. Cultured fibroblasts, however, have proved useful for explorations of the fundamental details of mammalian molecular and cell biology.

To obtain cells for growth in culture, a tissue source is gently treated with a diluted solution of certain proteolytic enzymes, such as trypsin and collagenase, often in the presence of the chelating agent ethylenediaminetetraacetic acid (EDTA). This procedure loosens the adhesions between cells and breaks up the extracellular matrix, thereby producing a suspension of individual cells. The cells are suspended in growth medium and transferred to clean glass or (more commonly) specially treated plastic petri dishes. After transfer the cells settle to the bottom of the dish, where they attach, flatten out, and begin both moving around on the surface and proliferating. Eventually the cells (fibroblasts) cover the bottom surface of the dish, forming a monolayer; at this point the growth and movement of the cells greatly slow or cease. This is known as **contact inhibition of growth.** At this point, typically 3–5 days after seeding, the cells can again be treated with a trypsin solution to remove them from the dish; an appropriate aliquot of the cells is then resuspended in fresh growth medium and reseeded into a new set of petri dishes. This process of cell transfer is called *trypsinizing* the cell cultures.

Cell Strains Versus Established Cell Lines

Cells freshly taken from the animal initially grow well in culture, but eventually, their rate of proliferation slows and stops. Depending on the animal of origin and its age, this typically occurs after anywhere from 20 to 50 cell doublings. This phenomenon is termed **cellular senescence**, and the slowing of proliferation that precedes it is termed *crisis* (Fig. 1–26). In some cases, especially with rodent cells, rare variants arise in the culture that has escaped the senescent restriction on cellular proliferation and now grow indefinitely. Such cells are termed **established cell lines.** In the case of mouse embryo cells, for

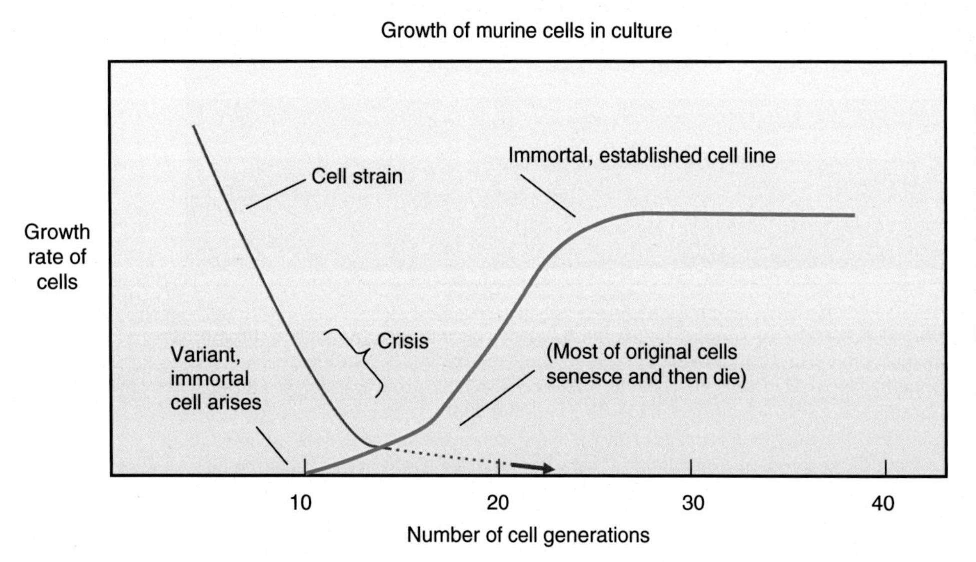

Figure 1–26. **Cell strain versus established cell lines.** Murine cells (e.g., mouse embryo cells) initially grow well in culture, and during this period of growth, such cells are termed a "cell strain." But the growth rate falls after several generations, and the cells enter "crisis," following which almost all cells senesce and die. Often, however, a rare variant cell will arise in the culture, capable of indefinite growth (i.e., "immortal"). The descendants of this variant cell become an "established cell line." These immortalized cells are typically aneuploid. *(Modified from Todaro GJ, Green H.* J Cell Biol *1963;17:299–313.)*

example, this frequently occurs, and a well-known cell line derived from mouse embryo cells in this way is called the *3T3* cell line. Cell lines are sometimes referred to as being "immortal" because of their ability to proliferate indefinitely in culture. Spontaneously arising cell lines such as mouse 3T3 cells usually have abnormalities in chromosome content and can have precancerous properties.

In the case of primary human cell cultures, this escape from crisis to form an established cell line never occurs. Human and mouse cells before crisis are referred to as **cell strains**, or sometimes, more colloquially, "primary cells." The latter term, however, is more appropriately used for cells freshly taken from the animal, before trypsinization, to produce a secondary culture.

At the heart of cellular senescence are repetitive, noncoding sequences called **telomeres**, which are found at both ends of the linear chromosomal DNA molecules of eukaryotic cells. Because of the biochemistry of DNA replication, terminal sequence information is lost each time a linear DNA molecule is replicated. The telomeric sequences of eukaryotic chromosomes protect coding DNA, because it is the "junk" telomeric DNA at the ends that shortens when chromosomal DNA is replicated. In very early embryo cells and in adult germline cells and certain stem cells, an enzyme called **telomerase** is expressed, which maintains the length of the telomeres during cell proliferation. In most somatic cells, however, telomerase is not expressed; as a result, each time the cell replicates its DNA and divides, the telomeric DNA sequences shorten. After a certain number of cell doublings, the shortened telomeric DNA reaches a critical size limit that is recognized by the cellular machinery responsible for activating the senescence program (i.e., the cessation of further cell proliferation).

One of the critical steps in the conversion of a normal cell into a cancer cell is reactivation of telomerase expression. Because cancer cells are therefore able to maintain telomere length, they escape senescence and are "immortal." Consequently, cancer cells, if adapted to growth in culture, grow as established cell lines. For many years, there were no established lines derived from normal human cells; the one established human cell line that was available was the **HeLa** cell line. These widely used cells were derived in the 1950s from the cervical cancer tissue of a woman named Henrietta Lacks. Normal animal cells must attach and spread out to grow (the **"anchorage requirement" for growth**), but HeLa cells, like some other established lines derived from cancer cells, have lost the anchorage requirement for growth and can be grown in suspension like bacteria or yeast cells.

Flow Cytometry

Flow cytometry is a method to count and sort individual cells based on cell size, granularity, and the intensity of one or another cell-associated fluorescent marker. The device that is most commonly used to perform the analysis is called a **fluorescence-activated cell sorter (FACS)**, and the layout of a typical FACS instrument is shown in Figure 1–27. In the device, cells pass single file into sheath liquid, which in turn passes through a special vibrating nozzle that creates roughly cell-sized droplets. Most droplets contain no cell, but some droplets contain a single cell (droplets that contain no cell or aggregates of two or more cells are detected and discarded). Just before the cells enter the nozzle, each cell is illuminated by a laser beam that causes any cell-associated dye to fluoresce. Forward- and side-scattered light is also measured. Based on these measurements, individual droplets are given either no charge (i.e., empty droplets or droplets with clumps of cells) or a positive or negative charge, and are then deflected (or not, if uncharged) by a strong electric field, which sends them to a particular sample collector.

The FACS device can be used simply to measure the characteristics of a population of cells (**cytometry**), or to sort and isolate subpopulations of cells (**cell sorting**). Earlier devices had a single laser source, and four light detectors, one each for forward scatter (a measure of cell size), side scatter (cellular granularity), and red or green fluorescence. Coupled with the use of red and green fluorescently tagged monoclonal antibodies directed against particular surface proteins, devices such as these played a large role in working out the role of various

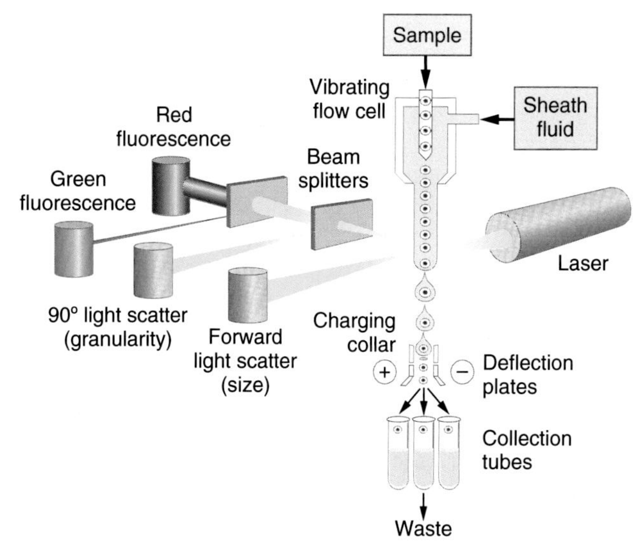

Figure 1–27. Fluorescence-activated cell sorting (FACS). Antibodies tagged with red or green fluorescent molecules and specific each for one of two different cell surface proteins (e.g., CD4 and CD8) are used to label a population of cells (e.g., a population containing the CD4 and/or the CD8 protein on their surface). The labeled cells pass into a vibrating flow cell, from which they emerge within individual fluid droplets. The droplets are excited by a laser beam. Forward-scattered laser light, side-scattered laser light, and red and green fluorescent light from the droplet are measured. Based on these measurements, individual droplets will be given a positive (+) or negative (−) charge and then diverted to collection tubes via charged deflection plates. *(Modified from Roitt IM, Brostoff J, Male D.* Immunology, *5th ed. St. Louis: Mosby Year-Book, 1998.)*

populations of precursor cells in the process of lymphocyte differentiation.

Second- and third-generation instruments now use as many as three lasers and can detect and sort cells based on as many as 12 different fluorescent colors. FACS analysis is quite useful in studies of, for example, cytokine production by individual T-cell populations, expression of activation markers, and apoptosis induction in cell population subsets. FACS not only uses monoclonal antibodies for staining of surface markers but also can be used to sort and clone hybridoma cells present at low frequency in a postfusion population. This can permit the rescue of rare hybridoma clones expressing useful monoclonal antibodies that might otherwise be lost to overgrowth by nonproducing hybrids. When coupled with reporter gene constructs, such as those expressing proteins tagged with GFP, rare cells expressing the reporter can be identified and captured for further growth and analysis. Additional applications of flow cytometry are discussed in Chapter 2.

Subcellular Fractionation

Subcellular fractionation is a set of techniques that involve cell lysis and centrifugation. These techniques were intimately involved in the discovery, over the course of the last half of the preceding century, of all the various compartments, membrane structures, and organelles that are now known to make up the internal structure of a cell. They are also part of the working repertoire of any contemporary cell biologist.

Cell Lysis

Cells can be lysed in any of a variety of ways; the optimum method depends on the cell or tissue type and the intent of the investigator. One common way to gently break open tissue culture cells, for example, is a device called a *Dounce homogenizer.* A **Dounce homogenizer** consists of a glass pestle with a precision-milled ball at the end; the dimensions of the ball are such that it slides tightly into a special tube in which the cell suspension is contained. Several up and down strokes of the pestle suffice to break open the majority of the cells while leaving nuclei and most organelles intact.

Centrifugation

After cell lysis, centrifugation can then be used to separate the various components of the cellular homogenate, based on their particular size, mass, and/or density. In one common approach, used for the rough fractionation of a homogenate, the lysate is centrifuged in a stepwise fashion at progressively greater speeds and longer times, collecting pelleted material after each step. A low-speed spin will pellet unbroken cells and nuclei; centrifugation of the supernatant from the low-speed spin at a higher, intermediate speed and for a longer duration will bring down organelles such as the mitochondria; centrifugation of the supernatant from the intermediate speed pellet at yet greater speeds and even longer durations will pellet microsomes (ER) and other small vesicles (Fig. 1–28). This type of procedure is termed **differential centrifugation.** Differential centrifugation can be usefully applied to separate subcellular components that differ greatly in size or mass. But the pelleted materials thus obtained are usually contaminated with many different components of the cell; in the case of Dounce homogenates, for example, the low-speed nuclear fraction contains not only unbroken cells but also large sheets of plasma membrane wrapped around the nuclei; mitochondrial pellets contain lysosomes and peroxisomes.

Further purification or more detailed analyses can be obtained by two other techniques of centrifugation: **rate-zonal centrifugation** (also known as **velocity sedimentation**) and **equilibrium density-gradient centrifugation** (sometimes called **isopycnic density-gradient centrifugation**). In both of these techniques, an aliquot of cellular material (whole-cell lysate or a resuspended pellet from differential centrifugation) is added as a thin layer on top of a gradient of some dense solute such as sucrose.

In the case of **rate-zonal centrifugation** (Fig. 1–29, A), the sample is layered on a relatively shallow sucrose gradient (e.g., 5%–20% sucrose) and then spun at an appropriate speed (based on the size and mass of the material in the sample); in this case, cellular material is not pelleted; instead the centrifugal field is used to separate materials based on their size, shape, and density; the shallow sucrose gradient serves simply to stabilize the sedimenting material against convective mixing. After the sample components have been resolved based on their sedimentation velocity (but typically before any of the material has actually formed a pellet on the bottom of the centrifuge tube), the centrifuge is stopped, the bottom of the tube is pierced, and sequential fractions of the resolved material are collected for assay. In this way, for example, ribosomes and polyribosomes were first isolated and characterized. The velocity at which a particle moves during centrifugation can be characterized by a number called its "**sedimentation coefficient,**" often expressed in **Svedbergs (S)**. The value of S is a function of the mass, buoyant density, and shape of an object. Large and small mammalian ribosomal subunits, for example, have sedimentation coefficients of 60S and 40S, respectively, whereas the whole ribosome has a sedimentation coefficient of 80S.

In the previously described applications of centrifugation, objects are separated based largely on their relative mass and size. Alternatively, cellular materials can be resolved based on their **buoyant density**. Various

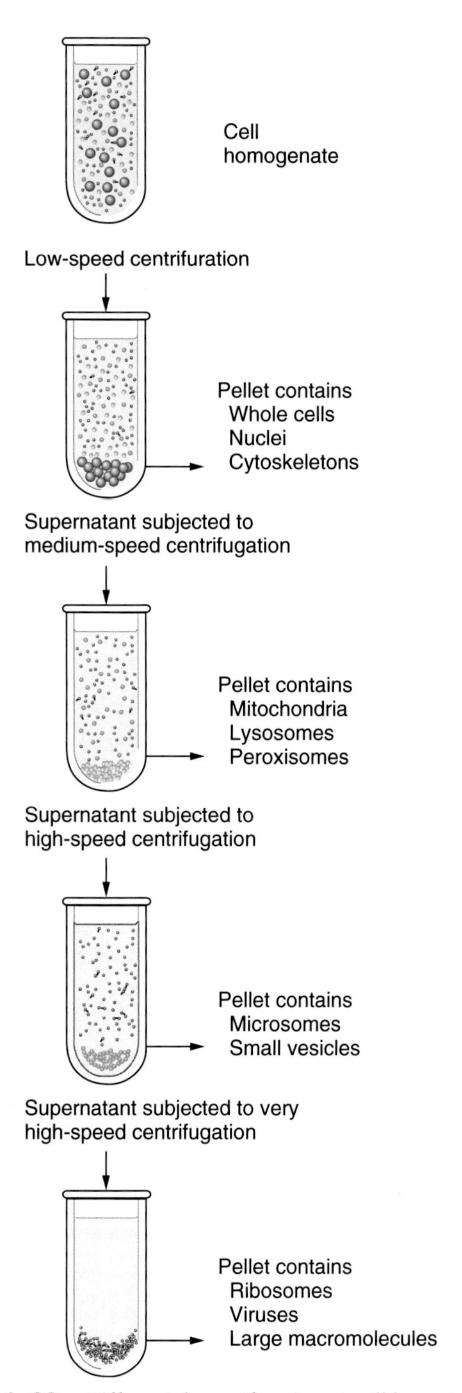

Figure 1–28. **Differential centrifugation.** A cell lysate is placed in a centrifuge tube, which, in turn, is mounted in the rotor of a preparative ultracentrifuge. Centrifugation at relatively low speed for a short time (800 *g*/10 min) will suffice to pellet unbroken cells and nuclei. The supernatant of the low-speed spin is transferred to a new tube, and centrifuged at a greater speed and longer time (12,000 *g*/20 min) will pellet organelles (mitochondria, lysosomes, and peroxisomes); centrifugation of that supernatant at high speed (50,000 *g*/2 h) will pellet microsomes (small fragments of endoplasmic reticulum and Golgi membranes); centrifugation at very high speeds (300,000 *g*/3 h) will pellet free ribosomes or viruses or other large macromolecular complexes. *(Modified from Alberts B, et al.* Molecular Biology of the Cell, *4th ed. New York, NY: Garland Science, 2002.)*

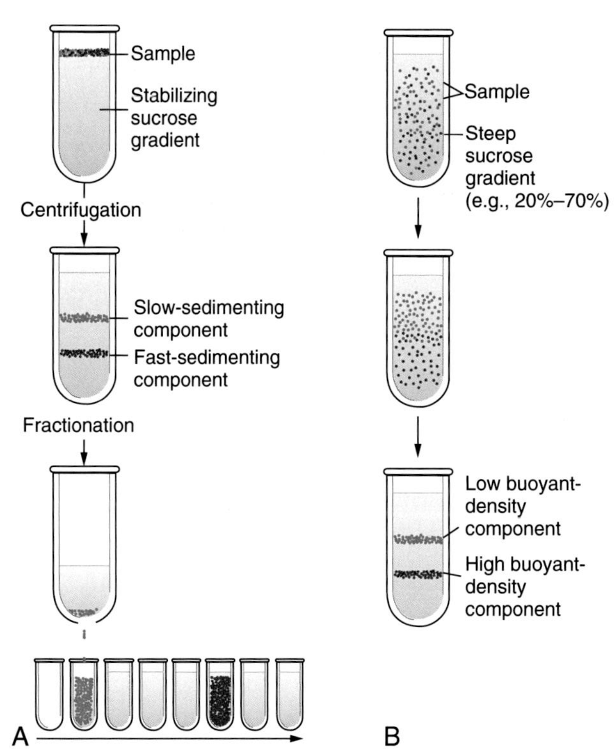

Figure 1–29. **Rate-zonal centrifugation versus equilibrium density gradient centrifugation.** (A) In rate-zonal centrifugation the sample is layered on top of a shallow sucrose gradient. During centrifugation the various components in the sample then move toward the bottom of the tube based on their sedimentation coefficients. After resolution of the components, the bottom of the plastic tube is pierced and fractions are collected. (B) Equilibrium density centrifugation resolves components in the sample based on their molecular density. The sample is either layered onto or incorporated into a steep sucrose gradient; during centrifugation, individual components move in the centrifugal field until they reach a density in the gradient that is identical to the buoyant density of the sample component. At this point, each component stops moving and forms a band in the gradient. *(Modified from Alberts B, et al.* Molecular Biology of the Cell, *4th ed. New York, NY: Garland Science, 2002.)*

proteins, for example, can differ widely in molecular mass, but all proteins have approximately the same buoyant density (approximately 1.3 g/cm^3); carbohydrates have densities of approximately 1.6 g/cm^3; RNA has a density of about 2.0 g/cm^3; membrane phospholipids have densities on the order of 1.05 g/cm^3; and cellular membranes, composed of both lipid and protein, have densities of approximately 1.2 g/cm^3. These differences in intrinsic molecular densities permit the resolution of a variety of cellular substituents by the technique of **equilibrium density-gradient centrifugation** (see Fig. 1–29, B). Here again the sample would be layered on top of a gradient of dense solute. For resolving cellular membranes and organelles, the solute would be sucrose, and a 20%–70% sucrose gradient typically would be used, generating densities ranging from 1.1 to 1.35 g/cm^3. For resolving proteins and nucleic acids, higher density gradients made with cesium chloride would be used. During centrifugation over the course of several hours, cellular components migrate in the tube until they reach a point in the density gradient equal to their own buoyant density, at which point they cease moving and form a disk or "band" at their equilibrium position in the gradient. Rough ER membranes, smooth ER membranes, lysosomes, mitochondria, and peroxisomes all have unique buoyant densities, for example, and are readily separated from each other by this method.

THE TECHNIQUES OF PROTEOMICS AND GENOMICS ARE DISCUSSED IN LATER CHAPTERS

Identification of the functions of the many novel proteins revealed by the Genome Project is, of course, not the only project of contemporary cell biology. Other important research goals to which cell biologists are making contributions include a deeper understanding of the molecular basis of cancer and embryologic development; stem cell properties and function; and, perhaps most formidable of all, the "neural correlates of consciousness," to use the phrase of Francis Crick. In these open-ended kinds of investigations, a number of other techniques in addition to the ones described in this chapter are required. Two of the most important such techniques are **mass spectrometry** and **microarrays**, or "gene chips" as the latter is sometimes called. Both of these techniques are discussed in Chapters 6 and 14.

An important component of the solution to these problems is a subdiscipline called **systems biology**, whereby one seeks to understand all the interrelations between individual signaling pathways and genetic regulatory mechanisms, and how they function as an integrated whole to produce the overall behavior of the cell. A systems biology attitude is integral to the approach we have taken to all the topics discussed in this text.

SUMMARY

Cell biologists have many powerful and sophisticated tools to deploy in their investigations of the function of uncharacterized cellular proteins. Microscopy techniques, in the forms of fluorescence microscopy, EM, and AFM, are among the most useful of these tools, as are the allied techniques of immunology. Tissue culture techniques provide a source of defined, uniform cell types for protein expression and analysis, and flow cytometry technology permits rapid and extremely sensitive analysis of cell populations. Epitope tagging of the proteins encoded by cloned complementary DNA molecules permits their efficient affinity purification, especially in conjunction with the standard techniques of subcellular fractionation and liquid chromatography. Two-dimensional gel electrophoresis and Western blotting are powerful analytic methods for resolving and characterizing complex mixtures of proteins.

Suggested Readings

Alberts B, Johnson A, Lewis J, Morgan D, Raff M, Roberts K, Walter P. In: Ways of Working With Cells. *Molecular Biology of the Cell*, New York, NY: Garland Science, 2015.

Fotiadis D, Scheuring S, Muller SA, Engel A, Muller DJ. Imaging and manipulation of biological structures with the AFM. *Micron* 2002;33:385–397.

Giepmans BNG, Adams SR, Ellisman MH, Tsien RY. The fluorescent toolbox for assessing protein location and function. *Science* 2006;312:217–224.

Harlow E, Lane D. *Using Antibodies. A Laboratory Manual.* Cold Spring Harbor, NY: Cold Spring Harbor Laboratory Press, 1999.

Hayat MA. *Principles and Techniques of Electron Microscopy*, 4th ed. Cambridge: Cambridge University Press, 2000.

Herzenberg LA, Parks D, Sahaf B, Perez O, Roederer M, Herzenberg LA. The history and future of the fluorescence activated cell sorter and flow cytometry. *Clin Chem* 2002;48:1819–1827.

Hörber JKH, Miles MJ. Scanning probe evolution in biology. *Science* 2003;302:1002–1005.

Jonkman J & Brown CM. Any Way You Slice It—A Comparison of Confocal Microscopy Techniques. *J Biomol Tech* 2015;26(2): 54–65.

Lodish H, Berk A, Kaiser C, Krieger M, Bretscher A, Ploegh H, Amon A, Martin K. In: Culturing and Visualizing Cells. *Molecular Cell Biology*, W.H. Freeman, 2016.

Schermelleh L, Heintzmann R, Leonhardt H. A guide to super-resolution fluorescence microscopy. *J Cell Biol* 2019;190:165–175.

Tsien RY. The green fluorescent protein. *Annu Rev Biochem* 1998;67:509–544.

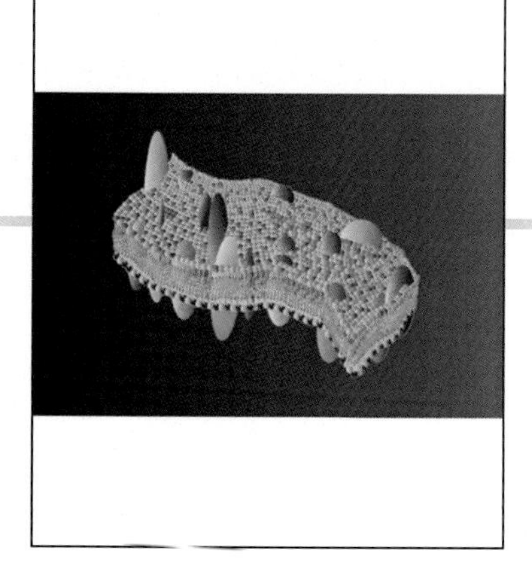

Chapter 2

Cell Membranes

Biological membranes are complex mixtures of lipids and proteins. They form barriers between different cellular compartments and define inside from outside. Without this separation, cells would not be able to function, and as such, proper membrane structure and function are essential for life as we know it. Lipids are the primary constituents of membranes as they separate the water compartments on both sides. Phospholipids spontaneously arrange themselves such that their polar head groups face the water and the hydrocarbon fatty acid tails face the hydrophobic core (Fig. 2–1).

The forces that determine this arrangement are collectively called the *hydrophobic effect*. It can be argued that the hydrophobic effect is perhaps the most important single factor to organize molecules of living matter into complex structural entities such as cellular plasma membranes or organelles. It is equally important in the formation of detergent micelles and a number of other phenomena that occur in aqueous solutions. If you have witnessed the beading of oil and its separation from the water in the kitchen, you have witnessed the hydrophobic effect. Water is a polar molecule in which the hydrogen is partially electropositive and oxygen partially electronegative. Because of their polar nature, water molecules tend to form clusters by hydrogen bonding to each other. Hydrocarbons or lipids dropped into an aqueous solution disrupt the hydrogen bonding of the water molecules. The coalescing of the hydrocarbons is driven by reestablishment of the hydrogen-bonding pattern of the water molecules. Phospholipid molecules are amphipathic, meaning they have hydrophilic (water-loving) polar head groups and hydrophobic (water-fearing) fatty acyl chains. When placed in water, these amphipathic phospholipids try to bury their hydrophobic fatty acyl chains away from water by forming spherical micelles in which the fatty acyl chains face the center of the sphere, and the polar head groups are at its surface. Alternatively, bilayers are formed in which sheets of two phospholipid monolayers or leaflets separate water compartments (see Fig. 2–1). The bilayers eventually must form a spherical vesicle so that there is no hydrophobic edge facing the water. When phospholipids are put on top of water they can form monolayers, where the polar head groups face the water and fatty acids stick up in the air. This characteristic was used by the Dutch pharmacists Gorter and Grendel in 1925 in their description that lipids can form bilayers. They extracted the lipids from human red cell membranes and placed them on a water surface. When the phospholipids were compressed with a movable barrier, the surface area covered by the phospholipids on the water surface was twice the surface area of the erythrocyte membranes from which they were extracted. Based on this experiment, they concluded that the lipids in the red cell membrane must be organized in a double layer, a bilayer. In 1965 Alec Bangham and colleagues at the Agricultural Research Council Institute of Animal Physiology in Cambridge, the United Kingdom, showed that phospholipids can self-assemble into small bags. These microscopic phospholipid "vesicles" or "liposomes" share the properties

Goodman's Medical Cell Biology. https://doi.org/10.1016/B978-0-12-817927-7.00002-8

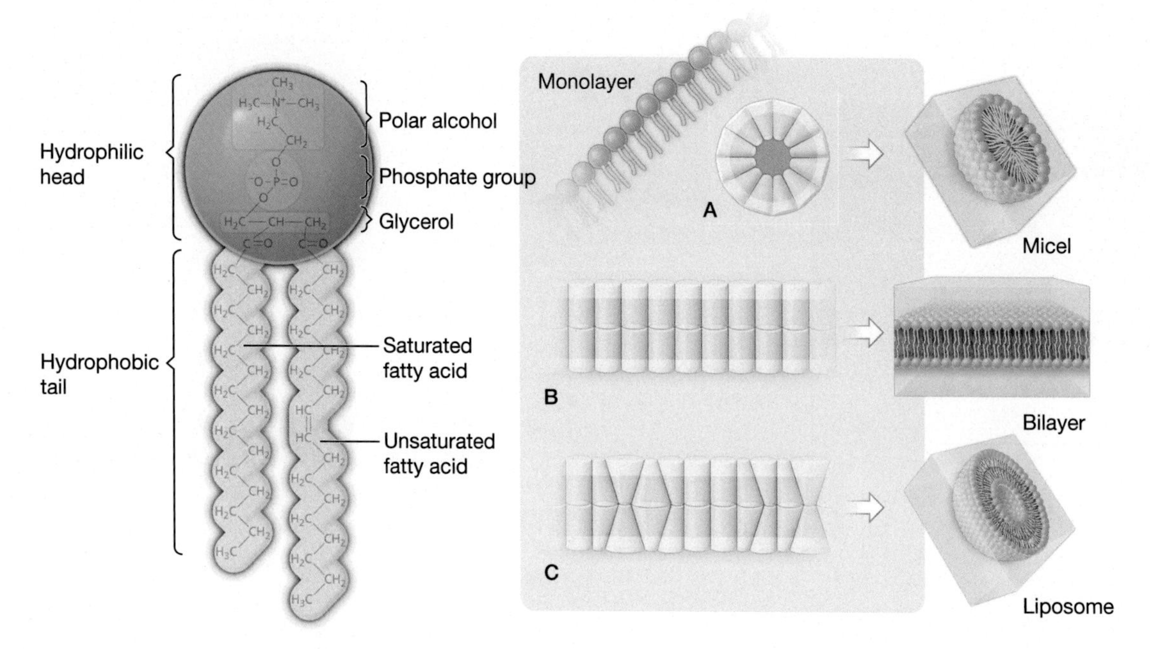

Figure 2–1. **The hydrophobic effect drives rearrangement of lipids, including the formation of bilayers.** The driving force of the hydrophobic effect is the tendency of water molecules to maximize their hydrogen bonding between the oxygen and hydrogen atoms. Phospholipids placed in water would potentially disrupt the hydrogen bonding of water clusters. This causes the phospholipids to bury their nonpolar tails by forming micelles, bilayers, or monolayers. Which of the lipid structures is preferred depends on the lipids and the environment. The shape of the molecules (size of the head group and characteristics of the side chains) can determine lipid structure. (A) Molecules that have an overall inverted conical shape, such as detergent molecules, form structures with a positive curvature, such as micelles. (B) Cylindrical-shaped lipid molecules such as some phospholipids preferentially form bilayer structures. (C) Biological membranes combine a large variety of lipid molecular species. The combination of these structures determines the overall shape of the bilayer, and a change in composition or distribution will lead to a change in shape of the bilayer. Similarly a change in shape needs to be accommodated by a change in composition and organization of the lipid core.

of cell membranes, as water compartments are separated by a bilayer of lipids. In 1974 Efraim Racker and Walther Stoeckenius of Cornell University incorporated bacteriorhodopsin and an ATPase into a lipid membrane and were able to generate ATP from light, showing that simple (artificial) membranes that mimic biological membranes can be generated in the laboratory.

In 1972 Singer and Nicolson proposed the fluid mosaic model of membranes (Fig. 2–2). The basic principles of this model suggest that membrane proteins are embedded within the bilayer, with hydrophobic portions of the proteins buried within the hydrophobic core of the lipid bilayer and hydrophilic portions of the protein exposed to the polar head groups and aqueous environment. In this model, one can imagine membrane proteins to be mobile "icebergs within a lipid sea." Most of the basic principle of this bilayer model remains intact today. However, the structure is much more complex and highly dynamic, characteristics that are difficult to represent by a simple cartoon. Figure 2–18 shows a more detailed model of one membrane protein embedded in a lipid bilayer with water on both sides.

As shown in Figure 2–1, phospholipid molecules can vary in shape based on their chemical structure. When combined in complex mixtures, the combined shape will determine the final shape of the bilayer. Altering the lipid composition on either side of a bilayer will result in bending of the entire structure. Similarly, bending the bilayer by an outside force will lead to a rearrangement of the molecules in this structure. In addition, membranes are continuously remodeled; lipid-lipid, protein-protein, and lipid-protein interactions in the membrane, as well as with constituents on either side of the membrane, determine membrane structure and function. The movements of membrane components in the plane of the bilayer and across the bilayer are highly dynamic, but these movements are not random. Some membrane

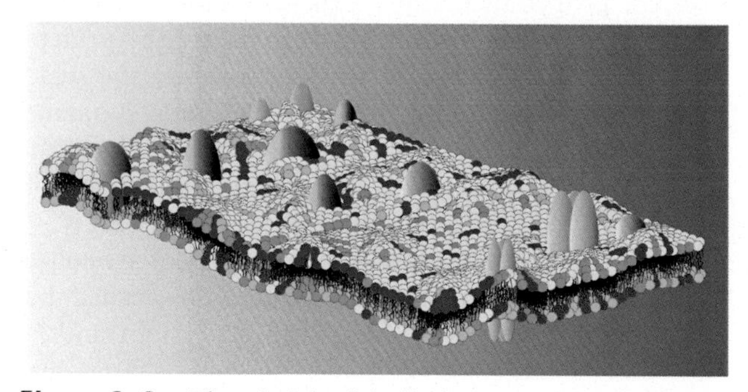

Figure 2–2. **The principle of the fluid mosaic model of biological membranes as proposed by Singer and Nicolson.** In this model, globular integral membrane proteins are freely mobile within a sea of phospholipids and cholesterol.

proteins are freely mobile, but others and lipids aggregate in specific areas, leading to a heterogeneous lipid and protein distribution in the plane of the bilayer. Both lipids and proteins are also highly asymmetrically distributed across the bilayer. Differences in protein structures on either side of the membrane allow directional functionality. The different molecular species of the lipid bilayer on either side accommodate protein function and can also change rapidly, modulated by the cellular function needed. Maintenance of protein and lipid composition and organization by an intricate system of enzyme and transport activities play a crucial role in the functionality of the bilayer. Together, biological membranes serve as a selective permeability barrier separating cellular compartments and inside from outside. Consequently the composition of the extracellular fluid can differ from that of the cytoplasm, and the polar molecules within the cytoplasm can differ from those within the lumen of organelles like the endoplasmic reticulum, the Golgi complex, or mitochondria. This in turn indicates that dynamics of lipid composition and organization are extremely important parameters for membrane and cellular function. It also points at the difficulty of lipid or membrane research as it pertains to biological systems. The molecular structures of the individual lipids that form biological membranes were solved many decades ago. Individual proteins can be modified by molecular biology techniques. The function of lipids and membranes needs to be assessed by studying large and complex mixtures of molecules.

This chapter focuses primarily on the plasma membrane, the lipid core of membranes, and basic transport mechanisms that allow compounds to transfer across lipid bilayers. Other chapters in this book describe membranes from organelles such as the mitochondria and the nucleus. In this chapter the red blood cell membrane is often offered as a model to describe a typical plasma membrane, and normal and abnormal red cell membranes illustrate issues that are common for all membranes. The reasons are simple. Red cells are cells easy to collect and purify, and the adult red cell has only one membrane. Hence the plasma membrane does not have to be purified and separated from other membranes. Separation of the plasma membrane in other cells changes the original structure. In addition, no de novo lipid synthesis or transport between different membranes alters the lipid composition of the red cell bilayer and obscures the results. These factors facilitate structure analysis, enzymology, and proteomic analysis of a plasma membrane composed of hundreds of different molecules. The reticulocyte, the very young red cell just born from the bone marrow, still contains RNA, allowing molecular biology techniques to identify proteins involved in membrane maintenance. Studies of red cell precursors, from the bone marrow stem cell to the reticulocyte, allow assessment of factors that link genetics to the final product, the adult red blood cell.

MEMBRANE LIPIDS

The Lipid Composition of Human and Animal Biological Membranes Includes Phospholipids, Cholesterol, and Glycolipids

The cellular plasma and organelle membranes within the body contain 40%–80% lipid. Among these lipids the phospholipids are the most prevalent. Four of the major phospholipids found in human and animal membranes are the glycerophospholipids phosphatidylcholine (PC), phosphatidylserine (PS), and phosphatidylethanolamine (PE) and the sphingosine-based phospholipid sphingomyelin (SM). Many minor species exist in the membrane, such as phosphatidylinositol (PI) or phosphatidic acid (PA). These latter species may be minor in their abundance, but they have important physiological functions. The polar head group for the phospholipids can be choline, serine, ethanolamine, or inositol, linked by a phosphate ester bond to carbon 3 of the glycerol or sphingosine backbone (Fig. 2–3).

The hydrophobic portion of the glycerophospholipids contains two hydrocarbon fatty acyl chains linked to carbons 1 and 2 of the glycerol backbone, or in sphingomyelin, a fatty acid bound to the sphingosine backbone. The charge of the polar head group differs based on its nature and the pH of the surrounding medium. At neutral pH, PS contains a net negative charge. PC, in contrast, has both a negative (on the phosphate) and positive (on the

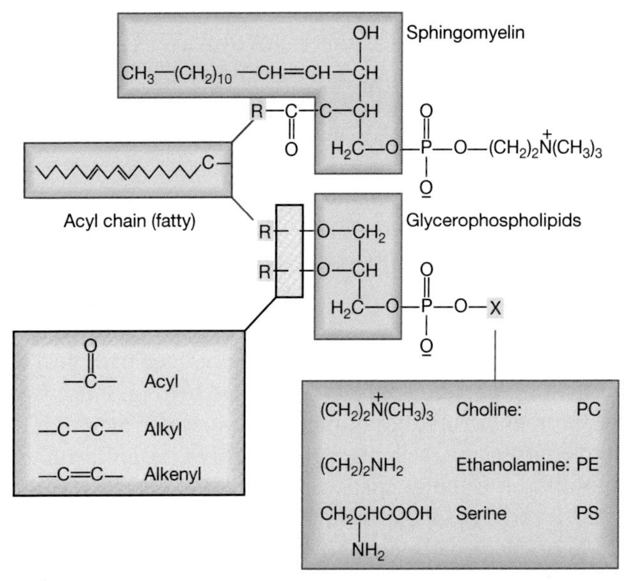

Figure 2–3. **Structure of phospholipids.** All phospholipids have a polar hydrophilic head group and nonpolar hydrophobic hydrocarbon tails. Glycerophospholipids are characterized by their glycerol backbone. Long carbon chains connected to the first and second carbon of glycerol provide the hydrophobic part of the molecule. The phosphate and additional head group structure provide the hydrophilic portion of the molecule. In sphingomyelin the backbone is sphingosine. A long-chain fatty acid provides the second hydrophobic tail. Note that both phosphatidylcholine and sphingomyelin have a choline-containing polar head group.

choline) charge and therefore behaves as a zwitterion at neutral pH. These different phospholipid species are not only distinguished by their head groups or backbone (glycerol or sphingosine). They also differ by the way the fatty acyl chains are bound to the glycerol backbone, which can be via an ether, vinyl ether, or ester bond, and the fatty acyl chains can be 14–26 carbon atoms in length with 0–6 double bonds.

The fatty acids of phospholipids in mammalian plasma membranes typically contain an even number of carbon atoms (16, 18, or 20). Often one of the fatty acids (in particular on the carbon 2 position in glycerophospholipids) is unsaturated (contains at least one double bond). The saturated fatty acid is straight and flexible, whereas the unsaturated fatty acid (which most typically contains one *cis* double bond) has a kink at the site of the double bond. These fatty acyl tails interact with each other by van der Waals forces. This interaction depends significantly on the level of unsaturation of the acyl chains. As an example, when phospholipids have two palmitic acids (16 carbon atoms, 0 double bonds) as side chains, they will interact much more strongly with their palmitic acid neighbor as compared with species that have 1 palmitic acid and 1 oleic acid (18 carbon atoms, 1 double bond).

As can be expected, all these molecules, or "molecular species," have different chemical and physical-chemical characteristics. Their relative abundance in a membrane will determine membrane characteristics. Although all possible combinations of head group, kind of bond, and fatty acyl chain could lead to the presence of thousands of different molecules randomly distributed in a given biological bilayer, this is not the case. The molecular species composition is well defined and specific for each cell type and membrane. As an example, approximately 25% of the phospholipids in the human red cell membrane are PC.

Of PCs, approximately 10% is diacylated (an ester bond on both the sn-1 and sn-2 glycerol position) with two palmitic acids. This molecule is shown as an example in Figure 2–4, containing two palmitic acid moieties (16 carbon atoms in length, 0 double bonds). The relative level of this particular molecule is maintained over the life of the red cell, and changes in its relative abundance will result in abnormalities in the function of the membrane. In the human red cell, more than 90% of the glycerophospholipid molecular species of both PC and PS is of the diacyl variety. In contrast, PE, which makes up about 25% of the glycerophospholipids of the red cell, contains approximately 40% of molecular species with a vinyl ether on the sn-1 position. An example of one of these plasmalogen phospholipids (1-palmitenyl, 2-linoleoyl PE) is shown in Figure 2–4. These two examples not only make the argument that a specific molecular composition is essential, for a specific membrane, but also point at the complexity in how to define their role in a highly dynamic structure. In addition to phospholipids, cholesterol is also a major component of biological membranes (see Fig. 2–4). The structure of cholesterol indicates that it is amphipathic, with a polar hydroxyl group and a hydrophobic planar steroid ring. Cholesterol intercalates between the phospholipids, with its hydroxyl group near the polar head groups and its steroid ring parallel to the fatty acid chains of the phospholipids and perpendicular to the membrane surfaces. Interactions between the different acyl groups of neighboring phospholipids are modulated by cholesterol. This is linked to the diverse characteristics of membrane phospholipids and their different behavior in pure lipid bilayers. As an example, dipalmitoylphosphatidylcholine (DPPC) exhibits a solid or gel-like behavior at room temperature because of the strong van der Waals interactions of the palmitic acid side chains with its neighbors in a bilayer. This is not the case for palmitoyl-oleoyl-phosphatidylcholine (POPC), in which this interaction is much less between the oleic acid-palmitic acid pair with its neighbors, and a bilayer of this lipid has a more fluid behavior at room temperature. When DPPC and POPC are mixed together, their different interactions will lead to segregation of the two species in the bilayer, like ice shelves in the Arctic Ocean. When equimolar amounts of cholesterol are added, it will alter the van der Waals interactions between the acyl groups, and the segregation is abolished. Hence, one could argue that cholesterol has a major function as a "species mediator" in the lipid bilayer.

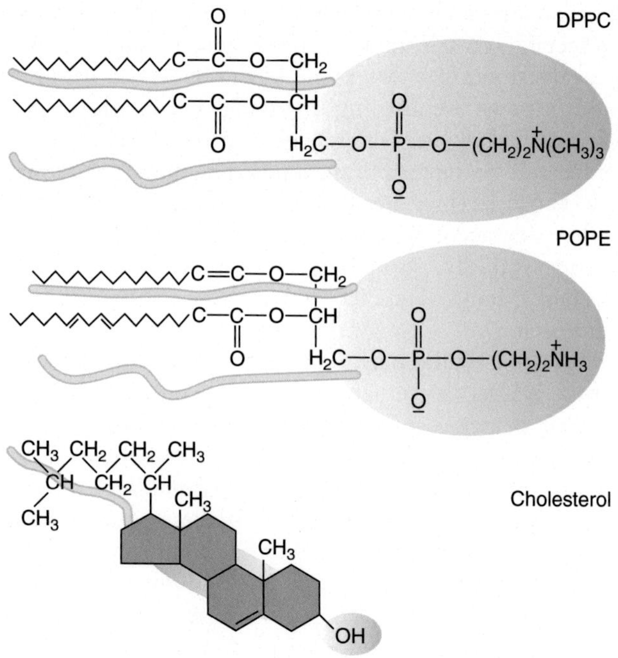

Figure 2–4. **Structure of the glycerophospholipids.** *DPPC*, dipalmitoylphosphatidylcholine; *POPE*, palmitoyl-oleoyl phosphatidylethanolamine; and cholesterol.

In addition to the major components of the lipid bilayer that determine its overall structure, many minor lipid components are present and essential for the normal function to the cell membrane. As an example, plasma membranes contain many different molecular species of glycolipids, lipids with attached sugar residues. The sugar residues of plasma membrane glycolipids almost always face the outside of the cell; that is, they have an asymmetric distribution, being found only in the outer leaflet of the bilayer. In human and animal cells, glycolipids are produced primarily from ceramide and are referred to as glycosphingolipids. These lipids can be neutral glycolipids with 1–15 uncharged sugar residues or gangliosides with 1 or more negatively charged sialic acid sugars. The glycolipids are important in cell-cell interactions, contribute to the negative charge of the cell surface, and play a role in immune reactions.

Together, it will be clear that the lipid bilayer is a complex mixture of many lipid molecular species. These molecules and thereby the membranes that they build are generated during development of the cell, by de novo synthesis of the different components and by remodeling of the phospholipids after they are assembled in the membrane.

Membrane Lipids Undergo Continuous Turnover

The composition of membrane lipids seems relatively constant, but the components are continuously turned over to accommodate functional aspects. Endocytosis or exocytosis processes that involve membrane fusion allow exchange of components between compartments separated by membranes. These membrane fusions also facilitate lipid flow between membrane fractions such as the ER, the Golgi complex, and the plasma membrane (Fig. 2–5). These events have the potential to randomize lipid compositions, and to counteract this, systems are in place to remodel and restore lipid composition and organization. The precise nature of these systems that "sense" membrane changes and act accordingly to restore the structure is poorly understood.

Another example of lipid turnover is the oxidative alteration and repair that has to take place in all membranes, as illustrated in Figure 2–6. Our ability to use oxygen for life comes at a high price. Oxygen will attract electrons whenever it can. Unpaired electrons in reactive oxygen species (ROS) are highly reactive, and polyunsaturated fatty acyl groups are vulnerable to attack. ROS will react with the double bonds, inserting a polar oxygen moiety. This will alter the normal packing in the bilayer and compromise its barrier function. The now more polar acyl chain will bend toward the water phase; phospholipases will recognize this defect in membrane packing and cleave the ester bond of the fatty acid (see Fig. 2–6). The resulting "lysophospholipid" with a free alcohol needs to be restored to a phospholipid by a reacylation process. Fatty acid is activated by acyl-CoA synthetase (ACSL), which binds coenzyme A (CoA) to the carboxyl group, at the expense of ATP. This acyl-CoA molecule, with energy stored in the thioester bond, is then used by another enzyme, acyl-CoA acyltransferase (LAT), to generate a new ester bond between the fatty acid and the lysophospholipid.

This releases CoA for a next cycle and generates a new phospholipid. Given that ROS-induced damage is

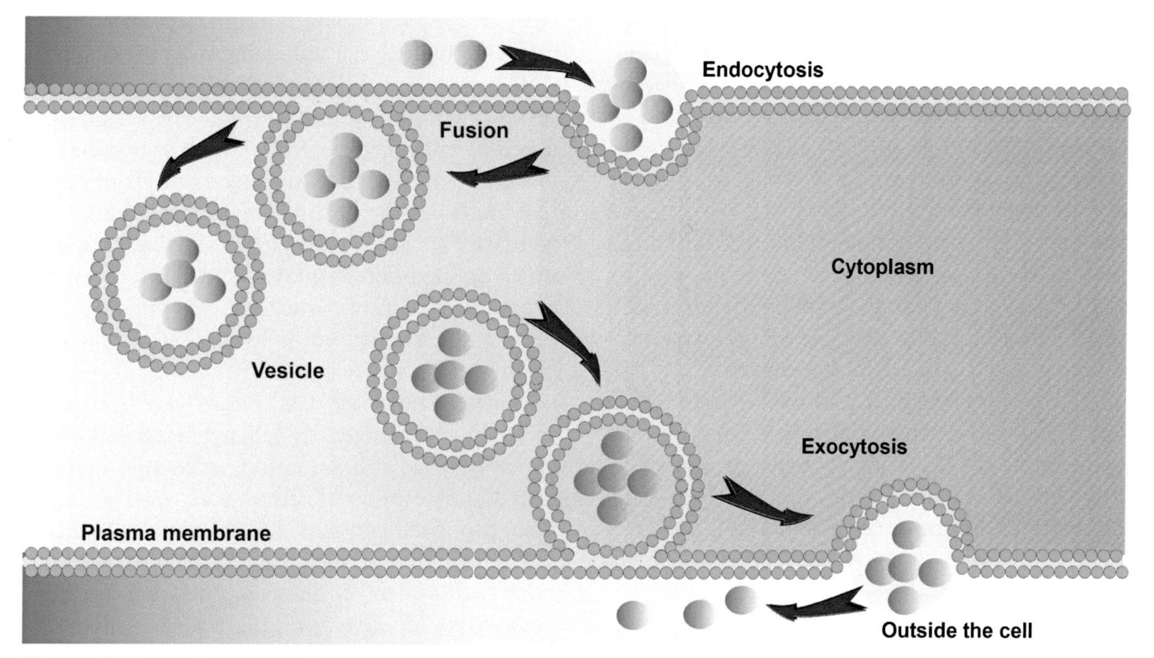

Figure 2–5. **Endocytosis and exocytosis.** Particles and other entities can be taken up by the cell by an active process called *endocytosis*. The plasma membrane rearranges its lipids and encloses the particle to be taken up. As a last step the membrane fuses and closes. Lipids in the membrane have to be remodeled to restore the lipid bilayer to its original composition. Examples are resorption processes in the gut, or phagocytosis. Exocytosis is a similar process in the reverse direction. Examples are secretion of enzymes and hormones and release of neurotransmitters.

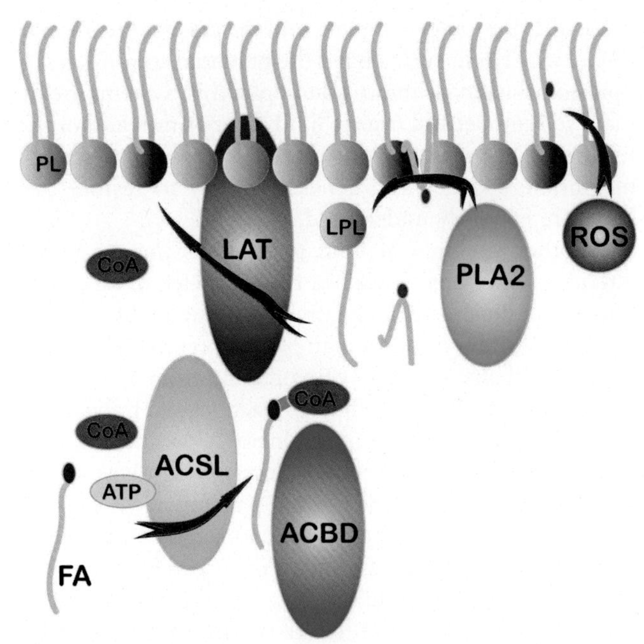

Figure 2–6. **Repair of an oxidatively damaged phospholipid.** Reactive oxygen species (ROS) oxidize unsaturated fatty acid in phospholipids (PL). This changes the polarity of the fatty acyl chain and the phospholipid tilts toward the water phase. Phospholipase A_2 recognizes this breach in the structure and hydrolyzes the phospholipid to lysophospholipid (LPL). FAs are activated to acyl coenzyme A (FA-CoA) by acyl-CoA synthetase (ACSL) using ATP. FA-CoA and LPL are used by LPL acyl-CoA acyltransferase (LAT) to form phospholipids, releasing CoA for the next cycle. Lipid-binding entities like acyl-CoA binding domain proteins (ACBD) modulate this process.

random in nature, a diverse pool of fatty acids is available, and the requirement to restore the molecular species composition provides an important biological dilemma, the repair cycle needs to be highly selective. This deacylation/reacylation of phospholipids originally described by William Lands 60 years ago is essential to maintain or remodel phospholipid molecular species. The so-called "Lands pathway" can be part of a repair process, as described, or active to alter molecular species in different membranes depending on the functional needs. In addition to the involvement of a complex family of enzymes and isoforms involved in these reactions, protein-protein and protein-lipid interactions modulate this repair processes. Molecules like acyl-CoA or lysophosphatidylcholine (LPC) can act as detergents, and their presence in membranes needs to be highly controlled. Lipid-binding proteins like acyl-CoA binding proteins as shown in Figure 2–6 play an essential role and by interacting with the substrates in these reactions modulate these processes.

In addition to structural rearrangements and repair processes, membrane lipids are also metabolized as normal steps in physiological processes. As an example, this section examines the action of phospholipases and their action on ester bonds. From a chemical point of view, the bond between a fatty acid and an alcohol (like glycerol) is simply an exchange of water (Fig. 2–7).

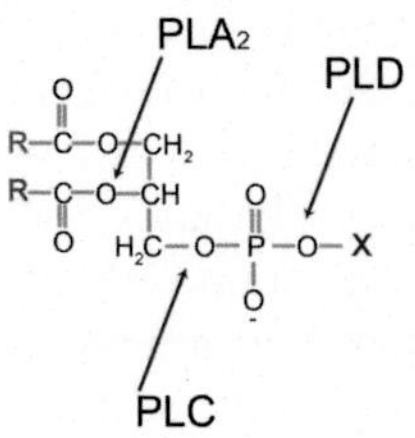

PLA2
PLD
PLC

Figure 2–7. **Phospholipases hydrolyze phospholipids.** The ester bond hydrolyzed by phospholipases determines the nomenclature of these enzymes. Phospholipase A_2 (PLA_2), phospholipase D (PLD), and phospholipase C (PLC) are shown.

Phospholipases are a large family of proteins with different locations, sizes, and characteristics that catalyze this exchange. They can be grouped based on the ester bonds that they are able to cleave. Figure 2–7 shows a number of examples. Phospholipase D cleaves the ester bond between the phosphate moiety and the head group. Hence, if it hydrolyzes PC, the choline group will be released, and PA will result from this action. Phospholipase C cleaves at the third carbon in the glycerol backbone to generate diacylglycerol (DAG). When a similar enzyme cleaves sphingomyelin, ceramide is formed. Hydrolysis at the second carbon by phospholipase A_2 will result in a lysophospholipid (see Fig. 2–6). The products of these reactions can have important physiological consequences, depending on the products released. The release of arachidonic acid by phospholipase A_2 action is an important step in the synthesis of leukotrienes and thromboxanes, molecules important for cellular signal functions in processes such as inflammation. Hydrolysis of PC to PA by phospholipase D, followed by phospholipase A_2 hydrolysis, will generate lysophosphatidic acid (LPA), a powerful lipid mediator involved in many processes, including wound healing. The action of phospholipase C will generate DAG, which in turn has a function in signal transduction. Another example of a set of bioactive lipids are the ceramides, the basic building block of SM (see Fig. 2–3). The sphingosine backbone connected by an amide bond to a fatty acid chain of varying length provides structural integrity. Sphingomyelin can be disassembled to ceramide by sphingomyelinase (a phospholipase C), which can further be metabolized to sphingosine by ceramidase. Ceramides have been shown to act as second messengers in cell signaling processes, including apoptosis, inflammation, cell cycle arrest, and the heat shock response. Furthermore, various recent studies have proposed ceramides as potential novel biomarkers and reported ceramide modulation in pathological conditions such as cancer, diabetes, Alzheimer's disease, coronary artery disease, multiple sclerosis, and depression. Sphingosine-1-phosphate has structural similarities to LPA and is similarly involved in signal transduction pathways. When the target of phospholipase C is PIP_2, the phosphorylated form of PI, the inositol triphosphate (IP_3) acts as an important mediator in cellular signal

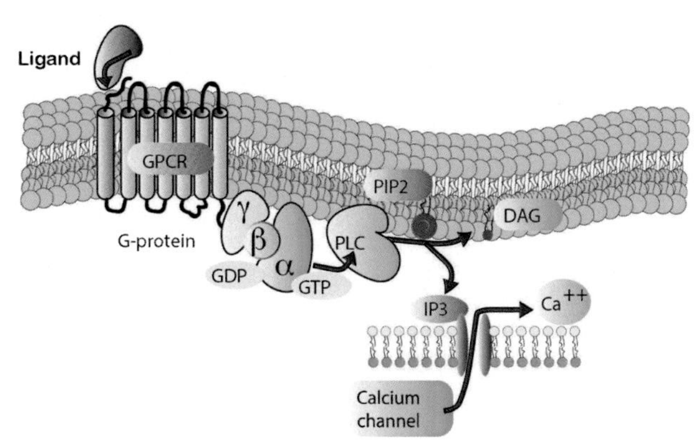

Figure 2–8. **G protein-mediated signal transduction.** A ligand binds to a G protein-coupled receptor (GPCR) in the membrane. This, in turn, activates a phospholipase C, which hydrolyzes phosphatidylinositol biphosphate (PIP_2) to form diacylglycerol (DAG) and inositol triphosphate (IP_3). IP_3 acts to increase cytosolic calcium as part of a signal transduction cascade.

transduction. One example of lipid-mediated signal transduction is illustrated in Figure 2–8.

G protein-coupled receptor (GPCR) acts on binding of ligands to its extracellular portion. GPCRs are known to play a crucial role in the development and progression of major diseases such as cardiovascular, respiratory, gastrointestinal, neurological, psychiatric, metabolic, and endocrinological disorders. GPCRs can be divided into groups distinguished by the G-protein to which they bind. They will trigger different pathways, including the activation of phospholipase C, which in turn generates bioactive molecules including DAG and IP_3.

Together, phospholipid turnover plays an important role in the maintenance of membrane phospholipid composition and many important physiological pathways.

Membrane Lipids Are Constantly in Motion

In addition to the complex composition of lipid bilayers, it is essential to realize that membranes are highly dynamic systems, often defined by terms such as *fluidity*, *mobility*, and *packing*. Membrane phospholipids are capable of several types of motion within the biological membrane (Fig. 2–9). The phospholipids can rotate rapidly around a central long axis. The fatty acid chains of phospholipids are flexible and will "wiggle" in all directions, with greatest flexion toward the center of the hydrophobic bilayer core. Phospholipids can move laterally across a biological membrane at a rate of approximately 1×10^8 cm^2/s at 37°C, which means that they exchange places with their nearest neighbor about 10^7 times per second. In addition, lipids can move (flip and flop) across the bilayer. In pure phospholipid bilayers, in the absence of proteins, this flip-flop is a slow process as a polar entity, the head group, has to transfer through an apolar layer (the fatty acyl chains). However, in the

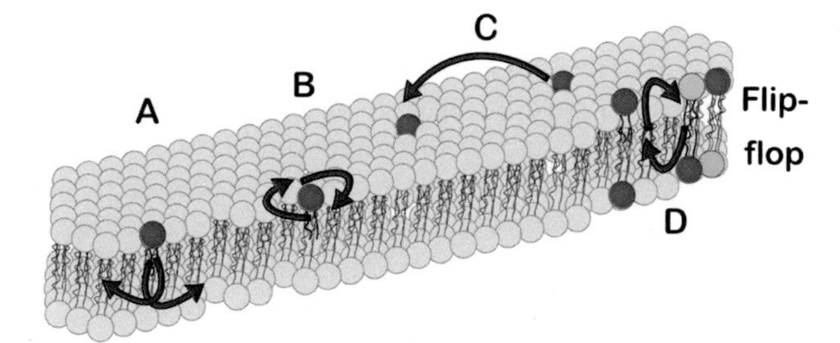

Figure 2–9. **The motion of phospholipids within the lipid bilayer.** (A) The fatty acyl tails undergo constant flexion as they interact with their neighbors. (B) Phospholipids can rotate rapidly around a central axis. (C) They are able to move in the plane of the bilayer at very fast rates. (D) Lipids are capable of transbilayer movement (flip-flop).

presence of proteins, this transfer can be facilitated in both directions.

The overall movement of membrane components is often described as *fluidity*, a rather vague term that describes how a particular molecule or sets of molecules can move around, wobble, and rotate. This mobility is governed by the molecular interactions between the neighboring molecules, be it lipids or proteins. Because the composition is complex, the interactions that govern the "packing" of the molecules in the membrane structure can differ significantly. Therefore the description of the average movement can be deceiving because it will differ significantly from molecule to molecule. One example that illustrates how packing can change mobility is the finding that cholesterol, which alters the packing of the phospholipids (see earlier), also tends to slow their lateral mobility. Despite the finding that phospholipids move rapidly in the plane of the bilayer, this is not a random process. Both lipid and protein organization will lead to domains in the membrane where certain lipids and proteins are enriched. Lipid molecular species need to accommodate the protein structures in the bilayer. In addition, there are "neighborhoods" of lipids and proteins involved in specific functions. As an example, these microdomains, or "rafts," are involved in specific physiological processes such as signal transduction (see Fig. 2–8). It makes intuitive sense to aggregate different proteins and lipids closely together when they have to act in concert. Alternatively, this means that if this domain organization is disrupted, these processes will not proceed properly. The rafts in plasma membranes are, in general, enriched in molecules such as sphingomyelin, saturated glycerol phospholipids, and cholesterol. Lowering of the cholesterol content of the membrane tends to "dissolve" these rafts, leading to an altered function of the membrane. It will be obvious that all lipids are in a tightly organized equilibrium not only in the bilayer but also with the environment.

Although the hydrophobic effect forces these molecules in the direction of bilayer structures, this still is an equilibrium and not an absolute "all-or-nothing" distribution. Depending on the nature of the lipid, they will be able to exchange between bilayers. When the acyl chains are shorter than 16 carbon atoms, the hydrophobicity decreases such that they will be able to relatively easy leave the bilayer and exchange to other membranes through a water phase. Hence, mammalian plasma membranes contain few molecular species with shorter carbon chains because the stability of the membranes would be compromised. Free fatty acids, lysophospholipids, and cholesterol can much more easily leave the membrane and exchange with other bilayers. Such exchanges are facilitated by lipid-binding proteins that "shield" the fatty acyl groups from water and "carry" these molecules in the transport process. Examples are fatty acid-binding proteins (including albumin) and acyl-CoA binding proteins (see Fig. 2–6). In addition, the transfer of long-chain phospholipids between membranes can be facilitated by lipid exchange proteins, the cargo vessels for lipid transport. They bind lipids such that the hydrophobic portions are kept away from the water phase. Some of these proteins can be quite specific with respect to the lipids they bind, whereas others are less specific. Serum albumin that reaches concentrations of about 2% in blood will bind a large variety of lipid molecules and as such can function as a transporter. As indicated in the description of the Lands pathway, it is also important to consider that molecules such as lysophospholipids or acyl-CoA are powerful detergent-like entities. If not properly bound and kept at low free concentrations, they would "dissolve" the cell membrane leading to the demise of the cell. Lipoproteins are structures (as the name implies) that are made up of lipids and specific proteins, with a function to transport these molecules in blood from the liver to other spots in the organism where they are needed. Changes in any of these pathways can affect the lipid composition of cells, including the plasma membranes.

As an example, blood cells exchange lipids with lipoproteins in plasma. Therefore changes in lipoproteins will affect the membranes of these cells. Individuals with significant liver problems are often characterized by changes in their blood cell membranes as the result of changes in lipoproteins. A direct interaction of plasma lipoproteins with blood cells is illustrated by the transport function of high-density lipoprotein (HDL) important in reverse cholesterol transport, an essential process in maintaining a proper lipid profile of "bad" and "good" cholesterol, linked to atherosclerosis.

Figure 2–10 shows that cholesterol is acylated to cholesterol ester by lecithin-cholesterol acyltransferase (LCAT). This enzyme transfers a fatty acid from HDL-PC to cholesterol, and as a result, LPC is formed, and the more hydrophobic cholesterol ester moves to the inner part of the HDL particle. The PC substrate for LCAT needs to be replaced, and LPC needs to be removed as its detergent characteristics would potentially destroy the HDL particle. The red blood cell membrane, simply as the result of its high abundance in the circulation, provides a place for LPC to relocate. The red cell

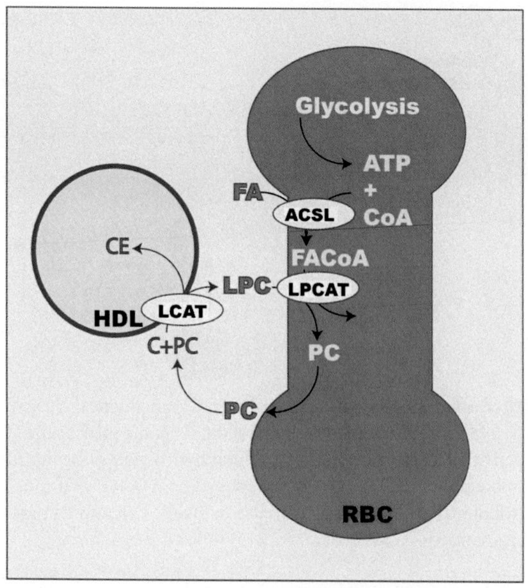

Figure 2–10. **Red cell lipid turnover and lipoproteins.** LCAT acts on phosphatidylcholine (PC) and cholesterol (C) in HDL to generate cholesterol ester (CE) and lysophosphatidylcholine (LPC). LPC in the red cell is reacylated to PC by an ATP consuming process fueled by RBC glycolysis (see Fig. 2–6). PC is transported back to HDL and is used for the next cycle to make cholesterol ester.

generates ATP by glycolysis as long as glucose is present and will reacylate PC from LPC using the Lands pathway as described earlier (Fig. 2–6). The PC pool in HDL is restored, and LCAT can resume its action in reverse cholesterol transport. Hence, one could argue that the red blood cell plays an essential role in the pathways linked to atherosclerosis. Such relationships can be shown by following the fate of a radioactive fatty acid added to plasma and RBC, which leads to the generation of radioactive HDL cholesterol ester. Another example of lipid transport related to disease is the fact that cells with too high concentrations of cholesterol will be negatively affected in their function. One could argue that plasma membranes have significant amounts of cholesterol to start with and a bit extra would not be so bad. Small variations are indeed tolerated, but they are bound to well-defined limits; outside these limits the altered lipid-protein interactions that occur can have dire consequences for the cell.

In addition to the heterogeneous distribution in the plane of the bilayer, phospholipids also transfer from the outer to the inner monolayer and back. Again, this flip-flop is not random; it is orchestrated by proteins in the bilayer. Figure 2–11 shows the normal and scrambled distribution of phospholipids in the human red cell membrane. The choline-containing phospholipids (PC) and sphingomyelin (SM) are normally mainly found in the outer monolayer, whereas the amino phospholipids are predominantly (PE) or exclusively (PS) found in the inner monolayer (Fig. 2–11, A). This highly asymmetrical plasma membrane distribution is typical for most mammalian cells. PS (and PE) is actively transferred from

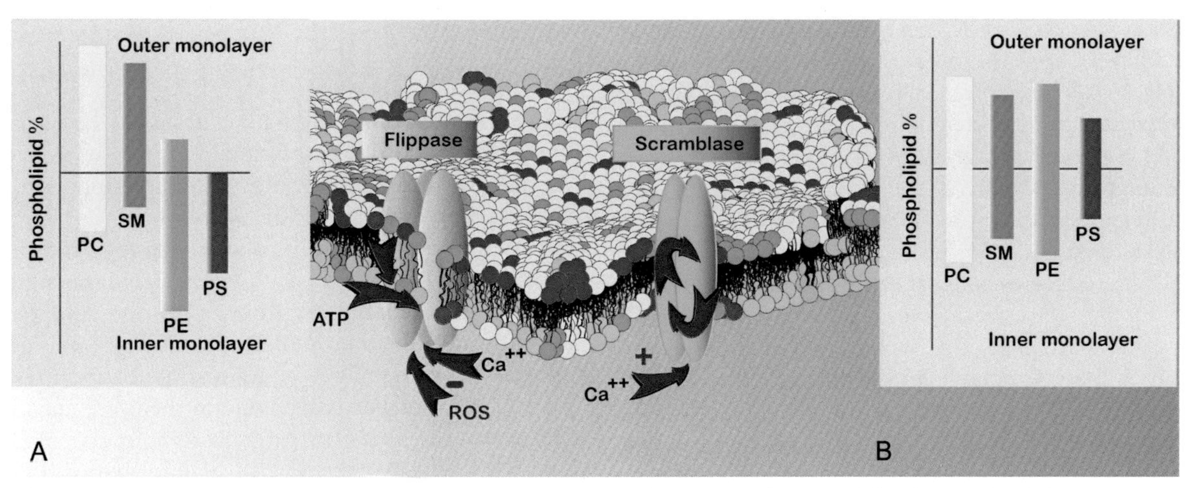

Figure 2–11. **The distribution of phospholipids in the human red cell membrane.** (A) *Normal distribution:* The choline-containing phospholipids, phosphatidylcholine (PC) and sphingomyelin (SM), are mainly found in the outer monolayer, whereas the amino phospholipids are predominantly [phosphatidylethanolamine (PE)] or exclusively [phosphatidylserine (PS)] found in the inner monolayer. (B) *Scrambled distribution:* Deactivation of the flippase and activation of a scrambling process will lead to the exposure of PS on the surface of the cell.

the outer to inner monolayer by an aminophospholipid translocase, or flippase, that consumes one Mg-ATP for each PS molecule transported. The flippase, an Mg-ATPase, can be inhibited by calcium and is sensitive to sulfhydryl modification. The scramblase is activated by the presence of calcium.

When the cell is not able to generate ATP or pump calcium out or is damaged by oxidative modification of its sulfhydryl groups, phospholipid asymmetry is lost (Fig. 2–11, B). In the laboratory, this can be rapidly accomplished by using a calcium ionophore A23187 and a sulfhydryl agent like NEM. These molecules added to the red cell make it permeably for calcium and kill the flippase. The calcium pump is overwhelmed, the cell runs out of ATP, the scramblase randomizes the phospholipid organization, and PS is exposed. In addition to active movement from the outer to inner monolayer, several proteins have been identified that can transport lipids from the inner to outer monolayer by either a directional active (ATP-consuming) process or a bidirectional scrambling process. ATP-binding cassette (ABC)-containing transporters have a special role in this regard. Important members of the ABC transporter superfamily are the multidrug resistance (MDR) ATPase and the ABCA1, which are involved in lipoprotein metabolism.

The MDR ATPase was initially discovered when oncologists realized that tumors were often resistant to a broad range of distinct anticancer drugs. The reason for this resistance was that tumor cells often overexpress the MDR ATPase, which uses energy of ATP hydrolysis to pump hydrophobic molecules out of the cell. Because a broad range of anticancer drugs are hydrophobic, the effective concentration was difficult to establish. Normally the MDR ATPase is expressed in liver, kidneys, and intestines, where it is thought to pump toxic substances into the bile, urine, and intestinal lumen, respectively. Therefore liver cancer (hepatoma) is, in part, difficult to treat because the affected cells are resistant to a wide range of chemotherapeutic drugs.

Both the inward movement of PS and the scrambling of the bilayer such that PS is exposed are highly controlled in the cell. Even when the scrambling is activated, as long as the flippase is active, PS exposure will be minimal. However, when both the flippase is deactivated and scrambling of the lipids across the bilayer is proceeding, PS will be exposed on the surface of the cell, with important physiological consequences. Exposure of PS is also an important trigger for recognition and removal of cells. Early in apoptosis or programmed cell death, PS is exposed on the surface of the cell. Macrophages recognize this abnormal surface and will engulf the cell as it undergoes its apoptotic program, ensuring that the cell is processed before it expels into the environment its potentially noxious breakdown products. Therefore PS exposure has become a powerful indicator of the onset of apoptosis. The binding of annexin, a molecule that binds to a cell with PS on its surface, is a routine measurement of an apoptotic cell (Fig. 2–15). Removal of cells is a normal, if well regulated, event in tissue development. It also removes cells that are not able to function normally as is evident in ineffective erythropoiesis in thalassemia where a large portion of the developing red cells undergo apoptosis before they make it out of the bone marrow. Together with a decreased lifespan of the thalassemic red cell in the circulation, these "eryptotic" events will lead to anemia. Dysregulation in these apoptotic programs may also lead to major problems. Cells that should be removed but are not recognized as such (cancerous cells) hide as their membranes appear normal.

In these processes, movement of calcium across membrane bilayers appears to be a common theme. As an example, in the human red cell, the calcium ATPase, an active calcium transporter, pumps out calcium against a strong concentration gradient (see later for a description of ion transport). Inside the red cell the calcium concentration is kept at nanomolar levels, whereas in plasma it is approximately 2 mM. When the cell becomes "leaky" for calcium, either by natural causes or induced by calcium ionophore to the membrane, calcium levels will increase. In particular, this will be the case when the calcium ATPase is unable to pump calcium out efficiently. This will induce lipid scrambling, downregulate the flippase, and PS will be exposed. Entry of calcium is often linked to other membrane damage, and oxidative stress plays an important role. The flippase is sensitive to (oxidative) sulfhydryl modification, and obviously, decline in ATP will affect this ATPase; PS is exposed and not pumped back efficiently. Hence a cell that is (oxidatively) damaged and will not be able to maintain ATP and calcium homeostasis is sentenced to be removed. During platelet activation, exposure of PS is essential because this will form the docking site for hemostatic factors such as the prothrombinase complex. The proteins in this complex assemble on the PS surface, and prothrombin is cleaved to thrombin, an essential step in blood coagulation. In contrast, random, unwanted exposure of PS on plasma membranes leads to a prothrombotic state and imbalance of the normal hemostatic processes (see Fig. 2–15).

In addition to an asymmetrical distribution of phospholipids based on their head groups, species will move and distribute asymmetrically based on their fatty acyl chains. The inner monolayer is often enriched in fatty acyl chains with higher levels of unsaturation. In addition, cholesterol, a major component in the bilayer, exhibits different properties with respect to transbilayer movement as compared with phospholipids. Because the polar head group of cholesterol is a small hydroxyl group, cholesterol (unlike phospholipids) can readily flip-flop from inner to outer monolayer and back, regardless of the presence of proteins. Therefore cholesterol distributed on both sides of the bilayer can move across the bilayer in response to shape changes within the plasma membrane.

Together the dynamic movement of lipids in the plane and across the bilayer, combined with enrichment in certain domains in the plane of the bilayer or across the bilayer, plays an important role in protein-lipid interaction and therefore physiological function. Many details on the significance of the heterogeneous or asymmetric distribution of proteins and phospholipids across the membrane need to be explored. Nevertheless, the clues of this dynamic organization point at a simple significance: it allows the two sides of the membrane to be functionally distinct and gives opportunities for proteins and enzymes in distinct pathways to act more efficiently.

Membrane Protein-Lipid Interactions Are Important Mediators of Function

From the preceding discussion, it should be clear that lipid composition, organization, and lipid-protein interactions are essential for the structure, function, biogenesis, and trafficking of membranes. Lipid-protein interactions are at the heart of many biological processes such as lipolysis, blood coagulation, signal transduction, and the maintenance of proper cytosolic composition. As indicated before, molecular biology and genetic approaches only indirectly address lipid structures. The protein mechanisms that maintain them are encoded in DNA, not the lipids or their macromolecular organization. Nevertheless, the wealth of information that has become available through the major advances in understanding the genome has only further increased interest in lipid-protein interactions.

The discovery through genome analysis of the abundance of membrane proteins, the identification of the many specific functions that lipids have, and the detection of lipids at specific sites in atomic structures of membrane proteins are just three issues that will likely lead to a better understanding at the molecular level of how the biomembrane functions. Bridging the specifics of the world of membrane lipids and the special features and difficult handling of proteins that are present in membranes has been and will continue to be extremely challenging. Nevertheless, a better understanding of how these interactions drive the functionality of membranes will be essential to understanding the function of cells.

Membrane Protein Structure and Function

The two basic types of membrane proteins, integral and peripheral, are distinguished by their operational definition, based on the ability to extract them from the lipid bilayer. Integral membrane proteins are embedded in the lipid bilayer and can be removed only by disrupting the bilayer. To do so, detergents are used. These amphipathic molecules disrupt the bilayer by forming mixed phospholipid-detergent micelles. In these micelles the integral membrane proteins are coated by the hydrophobic domains. Peripheral membrane proteins can be removed from membrane without dissolving the bilayer. Most frequently, these peripheral proteins are removed by shifting the ionic strength or pH of the aqueous solution, thereby dissociating the ionic interactions of the peripheral protein with either phospholipid polar head groups or other membrane proteins.

Various subtypes of integral and peripheral membrane proteins exist (Fig. 2–12). Some integral membrane proteins are transmembrane and make a single pass through the membrane. This type of protein usually has a hydrophilic section containing charged and polar amino acids in the aqueous environment outside the cell, a

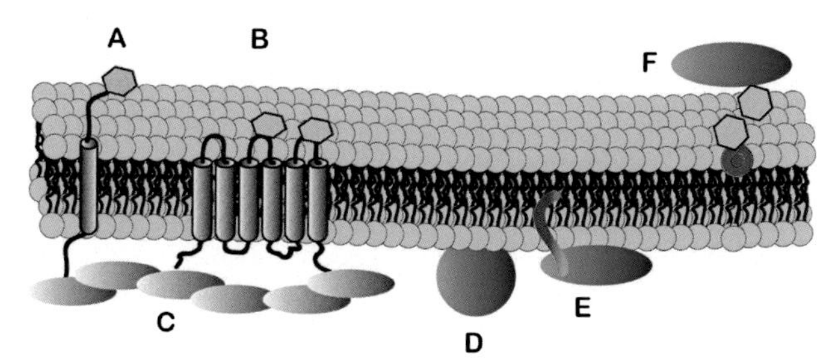

Figure 2–12. **Integral and peripheral membrane proteins.** Integral and peripheral membrane proteins can interact with the lipid bilayer in many different ways. The following situations are presented: (A) a single-pass glycosylated integral membrane protein (note that a single α-helical segment of the protein crosses the bilayer); (B) a multipass glycosylated integral membrane protein (this structure is found in transporters and membrane channels); (C) membrane proteins can interact with membrane skeleton protein structures to stabilize the membrane; (D) a peripheral membrane protein associated with the polar head groups of phospholipids by an ionic interaction; (E) a membrane protein for which the protein itself does not enter the bilayer but instead is covalently linked to a fatty acid tail; and (F) a membrane protein for which the protein itself does not enter the bilayer but instead is covalently linked by sugars to phosphatidylinositol.

hydrophobic stretch of 20–25 nonpolar amino acids forming an α-helix within the hydrophobic core of the bilayer, and a hydrophilic portion within the aqueous interior of the cell.

An example of a single-pass transmembrane protein is glycophorin A, the major sialoglycoprotein in the erythrocyte plasma membrane. Other proteins make multiple α-helical passes through the membrane. Transporters and ion channel are examples of multipass transmembrane proteins. Unlike single-pass transmembrane proteins, multipass transmembrane proteins can contain polar and even charged amino acids within the bilayer core. These polar amino acids, when facing one side of the α-helix, contribute to the formation of aqueous pores. An example of an important multipass transmembrane protein is the erythrocyte anion-exchange protein called *band 3*. Band 3 is responsible for the one-for-one exchange of HCO_3^- for Cl^- across the erythrocyte membrane that allows the release of CO_2 in the lungs. The protein makes several α-helical passes through the membrane, which contains a small hydrophilic COOH terminus and a longer hydrophilic NH_2-terminal domain extending into the cytoplasm. The NH_2-terminal domain has binding sites for glycolytic enzymes, hemoglobin, and regions that link band 3 to the membrane skeleton. The carbohydrate moieties associated with band 3 are on the outside surface of the red blood cell membrane.

Notably, all transmembrane proteins are integral membrane proteins, but not every integral membrane protein is a transmembrane protein. Some proteins are linked to the bilayer by covalently bound fatty acids or phospholipids. Examples are the so-called glycosylphosphatidylinositol (GPI)-anchored proteins. These proteins are linked to sugar moieties that, in turn, are linked to PI, which is (as all phospholipids are) embedded in the hydrophobic core with its acyl chains. An example of this class of proteins is acetylcholinesterase, an enzyme essential for the breakdown of acetylcholine (see Fig. 2–26). This structure, by which the protein itself is outside the bilayer but firmly connected to the bilayer by its phospholipids, gives these proteins a relatively high mobility compared with other proteins. With the large head that sticks out in the water phase, the mobility may be less than that of lipids, but it is much faster than proteins with amino acid chains embedded in the bilayer. Therefore it is logical to expect such structures for proteins that must be able to move rapidly, such as certain receptors or enzymes. The peripheral proteins can attach to the membrane surface by ionic interactions with an integral membrane protein (or another peripheral membrane protein) or by interaction with the polar head groups of the phospholipids. Many proteins also use fatty acids to increase their interaction with the hydrophobic portion of the bilayer. These fatty acids are mainly myristic acid (C14:0) or palmitic acid (C16:0) and can be covalently bound to amino acid moieties providing an "anchor" for the binding to membranes. These posttranslational protein modifications are essential for the proper function of these acylated proteins and point at important metabolic pathways and the close interaction of several proteins. As an example, the N-myristoyltransferase (NMT) facilitates acylation of a 14-carbon chain to a glycine residue. The process starts similar as described for the Lands pathway (Fig. 2–6) as an acyl-CoA is the substrate for NMT. Similarly as LAT transfers the fatty acyl group from acyl-CoA to a lysophospholipid (Fig. 2–6), NMT transfers the C14 from myristoyl-CoA to a protein using the energy in the thiol bond of myristoyl-CoA. Interestingly, while the 14-carbon chain should be the preferred substrate, NMT lacks specificity. In its purified state, NMT binds but very poorly transfers acyl-CoAs other than myristic acid. As a result the unwarranted occupation of the active site on NMT by acyl-CoA species that are much more abundant than C_{14}-CoA blocks the protein acylation cycle. In this case a member of the acyl-CoA binding domain protein family (ACBD) plays an essential role to provide specificity to NMT and enhances the myristoylation process. ACBD6 interacts with NMT and modulates the use of C_{14}CoA, such that this minor species becomes the preferred substrate to modify (myristoylate) proteins.

The interaction of proteins with membranes is highly regulated. Similar to lipids, membrane proteins are capable of lateral movement within the plane of the membrane. A quantitative measurement of protein lateral mobility can be obtained by fluorescence recovery after photobleaching (FRAP). In FRAP studies a surface protein is fluorescently labeled. A focused laser beam bleaches a small selected region of the plasma membrane, decreasing fluorescence in that spot. The fluorescence of the bleached area returns with time because unbleached labeled surface molecules diffuse into it. The percentage recovery in the bleached spot is proportional to the fraction of that specific integral membrane protein that is mobile. The rate of recovery of fluorescence allows a diffusion coefficient to be calculated. From this approach, we know that integral proteins within artificial lipid vesicles diffuse at a rate of 10^{-9}–10^{-10} cm^2/s, whereas in a biological membrane they diffuse at a rate of 10^{-10}–10^{-12} cm^2/s. The slower movement in the biological membrane is caused by protein-protein and lipid-protein interactions. One example is band 3, the anion transporter of the red cell membrane. A major fraction of these molecules cannot move laterally within this membrane, but these molecules are rapidly mobile when purified and placed within an artificial lipid vesicle. The reason is that the band 3 molecules are restricted by their interaction with the membrane skeleton on the cytoplasmic membrane surface. Therefore, although band 3 is capable of movement, its lateral mobility is restricted by its interaction with the membrane skeleton. As indicated earlier, GPI-anchored proteins move rapidly, but as they "bump" into other proteins, their movement is slowed. Despite their movements and similar to membrane lipids, membrane proteins are not randomly distributed in the plane of the bilayer. They aggregate in areas depending on their function. Similar to lipids, integral membrane proteins can also rotate along their axis within the membrane, but in contrast with lipids, they do not flip-flop from one leaflet to the other, and their asymmetry is more absolute.

Spectrin, a peripheral membrane protein, is always associated with the inner cytoplasmic leaflet of the erythrocyte membrane. Glycophorin A always has its NH_2-terminal domain outside the red cell and its COOH-terminal domain within the cytoplasm. Similar to the glycolipids, the carbohydrate portion of glycoproteins has an asymmetric distribution across biological membranes. The sugar residues are almost always found on the outside of the membrane. The carbohydrate moieties of glycolipids and glycoproteins, as well as glycosaminoglycans, which are oligosaccharides bound together by small protein cores, make up a fuzzy coat observed on electron microscopy of the outer surface of the plasma membrane, often referred to as a **glycocalyx**, meaning "sugar chalice." This fuzzy coat, on the surface of plasma membranes, discovered by Martinez and Palamo in 1970, contributes to cell-cell communication and intercellular adhesion.

Important Changes in the Plasma Membrane Occur in Sickle Cell Disease

A disease that affects humans was well known to the African tribes for the past 5000–10,000 years. In 1910 the Chicago-based physician James B. Herrick described, "Peculiar elongated and sickle-shaped red corpuscules in a case of severe anemia." This gave sickle cell disease (SCD) its name in western medicine, but it was another 40 years before it became apparent what the underlying molecular mechanism of the presence of these "sickled cells" was. In the late 1940s Linus Pauling perfected his technique to separate proteins and was able to show that hemoglobin collected from a sickle cell patient had a different protein mobility in his separation compared with normal hemoglobin. Pauling's article "Sickle Cell Anemia: A Molecular Disease," published in the journal *Science* in 1949, set the stage for molecular medicine that we are familiar with today. Pauling made the (first) direct link between an abnormal protein and a disease. A few years later, after the discovery of the DNA structure by Watson and Crick, it became apparent that a simple point mutation from a T to an A in one codon of the beta-globin locus on chromosome 16 led to a switch from glutamic acid to valine on the sixth position of beta-globin. This simple mutation results in abnormal behavior of hemoglobin, in particular, in its homozygous state (SS) when the mutation is inherited from both sickle gene-carrying parents (AS). Figure 2–13 summarizes the effect of this simple point mutation on the viability of the sickle red blood cell.

Hemoglobin S polymerizes under low oxygen, resulting in the "sickle shapes" that Herrick described, which puts a high mechanical stress on the membrane. This leads to loss of membrane material (microvesicles, a.k.a. membrane dust) and a mechanical stress that leads to alterations in membrane structure. To allow its function as an oxygen carrier, the normal red cell is well equipped to counter ROS attack with compounds such as superoxide dismutase, catalase, glutathione, and vitamin E. Despite these antioxidant systems that maintain a proper redox status of the cell, damage will occur, and repair mechanisms are in place to counter this. One example is described in Figure 2–6. These antioxidant systems are important since, following the entry in the circulation, the adult red blood cell lacks the ability to make new proteins or lipids. However, red cell metabolism (glycolysis) provides the energy to counter free radical attack and fuel repair processes if damage incurred. Nevertheless, after around 120 days, the cell is not able to function properly anymore and is removed from the circulation and replaced by fresh red blood cells from the bone marrow. This constant cycle of birth and removal of red cells is essential for proper blood function. In the sickle red blood cell, the abnormal hemoglobin increases the formation of ROS adding to the stress on the membrane.

Both the mechanical and oxidant stress on the membrane lead to loss of normal membrane structure and inability to repair and initiate a process akin to apoptosis (a.k.a. eryptosis) that much earlier than in normal red cells, leads to early recognition and removal. The imbalance of rapid removal and inability of the marrow to generate sufficient amounts of new RBC results in anemia of

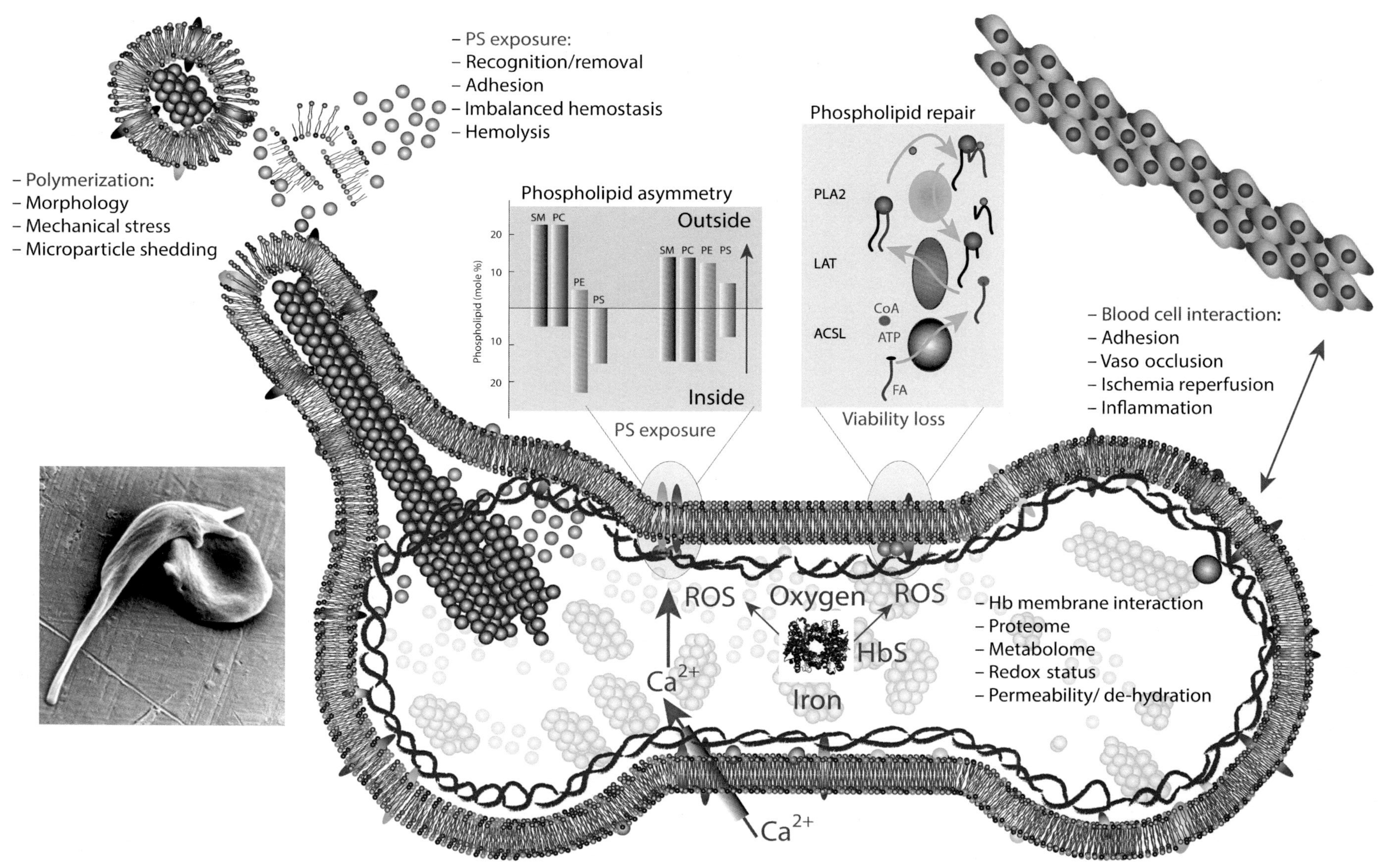

Figure 2–13. **Hemoglobin S and membrane changes.** Polymerization of sickle hemoglobin under low oxygen tension changes red cell morphology, separates the lipid bilayer from the membrane skeleton, puts mechanical stress on the membrane, and results in the shedding of PS-exposing microparticles. The unstable character of sickle hemoglobin increases oxidant stress, alters the metabolome and proteome, changes the redox status of the cytosol, and damages membrane lipids and proteins. Oxidized lipids are repaired as shown in Figure 2–11. Damage to the proteins involved in this repair system results in impaired repair, membrane viability is lost and the sickle cell membrane takes a central role in adhesion, vaso occlusion, ischemia reperfusion, and inflammatory processes. The increase in cytosolic calcium and oxidant stress leads to apoptotic plasma membrane processes including the loss of phospholipid asymmetry. PS exposure results in recognition and removal of the sickle red blood cell, increased adhesion, imbalanced hemostasis, and hemolysis. All these processes contribute to the vasculopathy that characterizes sickle cell disease.

the patient. While anemia is an obvious concern, the altered membrane of red cells in SCD patients determines many of the clinical problems, including increased interaction with other blood cells and endothelium, the prothrombotic state, release of hemoglobin and heme in the circulation, the blockage of the microvasculature leading to vasoocclusive crisis, stroke-like events, kidney damage, and the resulting ischemia/reperfusion and high inflammatory state. The inability of the highly orchestrated removal process to remove these cells with a damaged membrane cells quickly enough plays an important role in the disease.

As shown above, PS exposing membranes imbalance hemostasis. In addition, PS exposure makes these cells vulnerable to membrane hydrolysis by the inflammatory mediator secretory phospholipase A_2 ($sPLA_2$). This in turn results in intravascular hemolysis. The release of hemoglobin and heme in the circulation exacerbates the inflammatory state. Sickling and red cell endothelial interaction leads to vasoocclusion, pain crisis, and ischemia reperfusion injury. When this occurs in the microvasculature of the bone marrow, apoptotic and necrotic cells may enter the circulation, the system reacts as it would in the case of sepsis, inflammation increases to extreme levels, the apoptotic and necrotic material is hydrolyzed by $sPLA_2$, and the products interact with the vascular bed in the lung. This in turn may lead to the devastating "acute chest" syndrome in which the lungs fill with fluid, an obvious dire medical condition (Clinical Case 2–1).

Clinical Case 2–1

Wendell Washington is a 17-year-old black boy who was admitted to the Sacramento City Hospital on June 2 for "excruciating total body pain." He had just graduated as the valedictorian of his high school class and was on a weekend trip with his class to Lake Tahoe to celebrate graduation and the Memorial Day holiday. The weather was unseasonably warm. In the middle of the picnic, he felt feverish and extraordinarily thirsty, despite drinking large amounts of water. By late afternoon, all his joints had started to hurt, and he was asking the school nurse/chaperone for pain medication. A painful cough had also developed. By 10 p.m., he was exquisitely uncomfortable and anxious, and he asked to be taken to the hospital. Recognizing that he was seriously ill, the nurse immediately called for ambulance transport with oxygen and arranged for Wendell to be taken to the Sacramento hospital rather than to Reno, which was appreciably closer. On admission to the ED, Wendell presented as an obviously dehydrated thin black teenager in acute distress. He was crying out in pain and complaining that his legs, ribs, and abdomen hurt more than he could bear. He could not be examined until he had received 5-mg morphine sulfate intravenously.

His blood pressure was 110/60 mmHg, his pulse was 144 beats per minute and regular, his respirations were 24 breaths per minute, and his temperature was 41.2°C. Pertinent findings on physical examination were dry, pale mucous membranes, an extensive area of dullness and rales covering most of his right upper lung field, and markedly tender legs bilaterally. He also was guarding his abdomen, but there were no masses or focal tenderness. Neither the liver nor the spleen could be palpated, and there was no splenic dullness by percussion. A stat X-ray confirmed extensive right upper and midlobar pneumonia, and a sputum sample obtained with difficulty showed gram-positive diplococci. Immediate blood analysis showed a white blood cell count of $21,000 \times 10^9$/L (normal range = $4000–11,000 \times 10^9$/L) with 92% neutrophils, a hemoglobin level of 8.6 g/dL (normal = 14–18 g/dL), and a platelet count of $210,000 \times 10^9$/L. Blood smears showed high percentage of band neutrophils and Döhle bodies in the neutrophils, with poikilocytosis and anisocytosis in the red cells. Reticulocytes made up 8% of red cells, and 10% of red cells were sickle shaped in the air-dried smears. A scanty urine specimen showed a specific gravity of 1.010 and extensive microscopic hematuria. Wendell was given high-flow oxygen by mask, whereas major efforts to hydrate him with intravenous fluids begun. He was immediately started on intravenous antibiotics for his presumed bacterial pneumonia with empiric coverage for pneumococcus, *Staphylococcus aureus*, and mycoplasma. A sample was sent to the blood bank for typing and matching of four units of blood, and he was promptly transfused. His pain was treated with ketorolac (a nonopiate analgesic) to avoid the respiratory depressive effects of opiate analgesics. A history subsequently obtained from Wendell's mother disclosed that he was the second of her six children with sickle cell disease. Previously, he had had only mild problems, with no more than one yearly admission for a moderately painful crisis as he grew up. He did not require opiates between pain crises and had not been on any preventive medications such as hydroxyurea. Although he had limited his athletic activities in school, he had done well academically and was highly regarded by his classmates.

Cell Biology, Diagnosis, and Treatment of Sickle Cell Disease

Ever since the path-breaking biochemistry of Pauling, sickle cell disease has been recognized as the quintessential molecular disease. The homozygous inheritance of mutant beta-globin genes (SS) leads to the dramatic decrease in quality of life of the sickle cell patient. A change in protein characteristics defined by Pauling and the formation of hemoglobin polymers when red cells are exposed to a medium where oxygen is chemically bound have led to early diagnostic tests for sickle cell. In the late 1960s the Black Panthers supported testing for sickle cell disease on the street corners of Oakland, and in 1972 the sickle cell act was signed into law. The recognition of this disease, which in its homozygous form affects roughly 100,000 Americans and many more worldwide, has made a big difference for those individuals affected. Newborn screening identifies every child born in the United States with the sickle gene, and similar efforts are underway in other countries. This allows early intervention, and proper medical care has made a big difference for the patients.

Wendell had the poor fortune to inherit the aberrant sickle globin gene from each of his heterozygous parents, as proven by PCR testing revealing homozygous SS disease.

As such, he was vulnerable to the oxygen-induced changes in his red cells as indicated earlier and suffered from many of the secondary problems that result from it. These include the entrapment of red cells in the microcirculation in regions of low oxygen tension, the exquisite painful crises when this entrapment induces anoxia in bones, the gradual infarction of the spleen and the immunologic defects consequent to that, and the infarction of the renal papillae, which removes the ability to concentrate urine. This, in turn, makes patients hypersensitive to dehydration, which in turn can increase hemoglobin polymerization. In some severe cases, perhaps from secondary membrane stickiness, there is a tendency toward cerebral infarction and cumulative brain damage. Ironically, both low oxygen tension and dehydration increase the likelihood of in vivo sickling. The elevated altitude of Lake Tahoe, together with the dehydration caused by the weather and Wendell's fever, induced his sickling crisis. A combination of pulmonary vasoocclusion by sickle cells and fat emboli released by infarcted bone resulted in acute chest syndrome, manifesting in life-threatening hypoxia. The poorly perfused lungs, combined with the underlying immunodeficiency, allowed for secondary infection of his lungs by bacterial pathogens. The infection added to his hypoxia and threatened to establish a lethal hypoxia-sickling-hypoxia cycle. Fortunately the good judgment of the nurse in promptly sending Wendell to a responsive ED at a lower altitude saved his life.

The tools available to the physician to specifically treat SCD patients are limited. In addition to efforts to combat infection, inflammation, imbalanced hemostasis, and pain in acute episodes, treatment includes blood transfusion and efforts to lower sickling. Transfusion is limited as a chronic treatment since available matching blood units decline as the result of patient-induced antibody response. In addition, iron overload in regularly transfused patients needs to be carefully monitored and modulated with iron chelation. Transfusion remains potentially useful in acute crises. Hydroxyurea is widely used in the last few decades as it reverses the hemoglobin switch to upregulate gamma globin relative to beta-globin, and the resulting increase in fetal hemoglobin lowers sickling. While relatively effective, not all patients respond as well, and it also may have side effects. Other antisickling agents are currently under development and show promise. A potential cure for a genetic disorder is gene therapy (possible through bone marrow transplant), and several approaches are in routine use or in clinical trials. While replacing bone marrow stem cells can potentially cure the genetic disorder, it still is a major clinical procedure and may be of limited availability to the majority of patients outside of the care of modern western medicine.

To understand why the sickle gene is present in the human population, it is important to note that sickle globin in the heterozygous state (AS) provides a better survival in areas where malaria is endemic. Plasma membranes are the entry doors for intracellular parasites including bacteria, viruses, or protozoa. Inside the cell, they use their host to proliferate. Examples are *Chlamydia trachomatis* (a bacterium), dengue (a virus), and malaria (protozoa). These three infectious agents are the cause of huge health problems for the human population of our planet. As can be expected from parasites, they use their host system to proliferate. This assistance of the host includes help with the formation of their lipid bilayers. While their lipids may be unique, proteins in the host cell support the formation of lipid bilayers as well as generation of lipid modification of parasite proteins using similar pathways as discussed earlier.

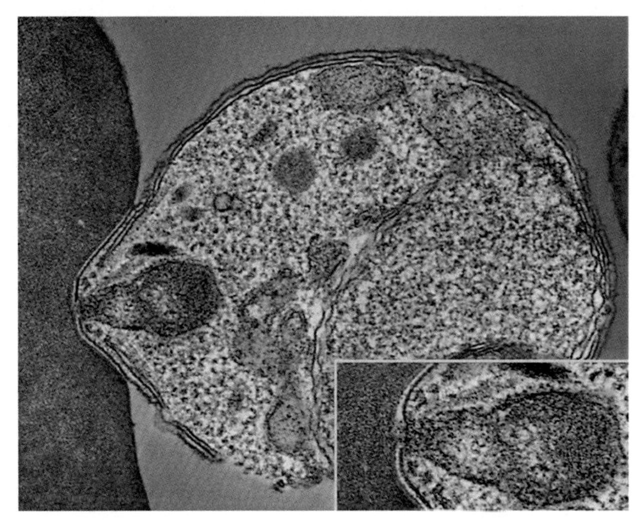

Figure 2–14. **A colorized electron microscopic picture of a malaria parasite (right, *blue*) attaching to a human red blood cell.** The inset shows a detail of the attachment point at higher magnification, used by the parasite to penetrate the red cell membrane and find home inside the cell to hide from the immune system.

The link between pressure on the human genome and resulting mutations is well illustrated by malaria, endemic throughout most of the tropics, and still one of the most important infectious diseases in the world. Roughly 300 million cases of malaria are recorded each year and 1 million result in death, mainly children in Africa. In addition to a major health issue, the economic impact of the disease is also highly significant, as it severely impacts the function of individuals infected. When an infectious anopheline mosquito bites in sub-Saharan Africa, it injects sporozoites of *Plasmodium falciparum*. They move to the liver and invade hepatocytes, and after 1 or 2 weeks, merozoites are released, and red blood cells become the target for invasion (Fig. 2–14). Inside the red blood cell, where it hides from the immune system, the parasite proliferates in several stages. It forms ring forms, mature trophozoites, and multinucleated schizonts, and after a few days the red cell ruptures and releases more merozoites that in turn seek to invade red blood cells.

The red cell stage of infection is responsible for all clinical aspects of malaria. The repeated cycles of red cell invasion and rupture lead to chills, fever, headache, fatigue, other nonspecific symptoms, and with severe malaria organ dysfunction. Some parasites develop into gametocytes, which may be taken up by mosquitoes during a blood meal. They develop inside the mosquito to a new set of infectious sporozoites. While this seems a complex life cycle, between the parasite, the vector (mosquito), and the victim, it is highly effective, and malaria has been the scourge of human kind.

Since part of the life of the malaria parasite happens to be inside the red blood cell, many mutations that decrease the ability of the malaria parasite to survive and thereby benefit the human have occurred and are maintained in areas of high malaria. From the perspective of the malaria parasite, it is very important to be able to hide inside the red blood cell from the immune system. Alterations in the red cell membrane make it harder for the parasite to invade and hide from the white cells. Inside the cells the parasite needs conditions for proliferation and survival, and alterations in cytosolic conditions may affect its well-being. In addition, the host (the red cell) should not hemolyze before the planned parasite maturation cycle is complete. Hence, anything that interferes with this part of its life cycle is bad for the parasite but beneficial for the human.

Anthony Allison, during a university expedition in 1949, noted high frequencies of sickle cell heterozygotes (AS) in the human populations living near the coast of Kenya and Lake Victoria in contrast to the highlands where malaria was much less prevalent. This led to his hypothesis that red cell mutations were favorable for humans in areas where malaria is endemic. This has been proven a correct assumption and genetic alterations, including hemoglobin mutations (HbS, HbC, HbE, HbD, and thalassemias), membrane disorders (spherocytosis and ovalocytosis), specific blood group antigens (Duffy), or metabolic alterations (G6PD deficiency and pyruvate kinase deficiency); all have their roots in the fact that they make life for the parasite more difficult, thereby giving an advantage to the infected individual. The numbers of individuals affected by these malaria-linked red cell aberrations are staggering. The sickle cell mutation in its heterozygote form (AS) is thought to affect more than 300 million individuals worldwide. These numbers pale when the other red cell aberrations such as the thalassemias are added, and the impact on global health is huge. It is important to note that not only the sickle mutation is found in Africa but also SCD is apparent in the Middle East and India and the sickle gene is found in Europeans. In addition, combinations of these genes can have unexpected results. As an example, HbE and beta-thalassemia, both prevalent in Southeast Asia, may not result in major clinical implications by themselves. However, the combination HbE-beta-thal when inherited from either parent is a red cell disorder that may lead to transfusion dependency. Currently, assessment of DNA plays an important role in the diagnosis of inherited diseases, and recent developments of DNA analysis have allowed DNA assessment to study the history of these human mutations. Analysis of the DNA from 90 mummies who were buried from 1380 BCE to 425 CE provided reliable results to define the ancestry of peoples who inhabited the Nile Delta and diseases that they carried. Tutankhamen became pharaoh of Egypt in 1332 BCE, over 3000 years ago. He reigned for a brief 10 years and died at age 19. Presence of malaria in his body was established. Forensic evidence indicated a strong probability that the young king suffered from SCD and may have died as a result of its complications. When confirmed, King Tut would be the earliest molecularly identified sickle cell patient.

The malaria parasite comes in different forms, and red cell membrane alterations determine the global distribution of these different malaria species. As an example, *P. vivax* is much more prevalent in Southeast Asia as compared with Africa where we find mainly *P. falciparum*. The Duffy blood group is the receptor for *P. vivax* and its absence in persons of African descent provides resistance to *P. vivax*. This has become a well-known example of innate resistance to an infectious agent because of the absence of a receptor for the agent on target cells.

While the term "beneficial" for the human with malaria protective mutations is loosely used, it is important to note that mutations that provide a level of protection against malaria may also come at a cost. The combination of HbE and beta-thal as described earlier is one example. Also, carrying the sickle gene (AS) as is the case for more than 300 million individuals worldwide may provide an advantage against malaria, but it does not make the individual immune or completely protected. While sickling events in AS individuals are rare, they can occur at higher altitudes where the reduced oxygen tension may promote the tendency of red blood cells to sickle even with one copy of the sickle gene. As should be clear from the description earlier, in its homozygote form, inheritance of the sickle gene (SS) is devastating. SCD patients have an extreme poor survival rate in many countries where modern medical care is lacking (Clinical Case 2–2).

Clinical Case 2–2

Henry, an American-born 26-year-old male presents with a history of progressive dyspnea for over a week and fever as well as a 2-day history of jaundice. He was diagnosed 3 days before his presentation led as having a viral syndrome. The patient returned 3 weeks earlier from a 2-month stay in Benin, West Africa. Before his travel, he received immunizations against yellow fever, was provided pills to prevent malaria until his return to the United States, and was advised to use mosquito repellant and a mosquito net. The physical examination shows mild jaundice, moderate respiratory distress, and diffuse pulmonary crackles. His vital signs included a respiratory rate of 33 breaths per minute, oxygen saturation of 87% while breathing ambient air, and a temperature of 39.8°C. Abnormal results of laboratory tests include a hematocrit of 32.2%, platelet count of 78 per cubic millimeter, total bilirubin level of

4.2 mg per deciliter, and a creatinine level of 2.2 mg per deciliter. The physician ordered a Giemsa-stained blood that showed numerous ring forms of *P. falciparum*, with a parasitemia estimated at 2%. He is immediately hospitalized, and an infectious disease consultant recommends that the Centers for Disease Control and Prevention (CDC) be contacted to obtain intravenous artesunate for his treatment. Of note, Henry traveled with a friend, Adam, an African American 25-year-old male, who accompanied Henry to his doctor visit. Adam remembers receiving multiple mosquito bites in Benin, never slept under a mosquito net, and never took antimalarial pills. However, Adam did not develop any of the symptoms affecting Henry. On questioning, Adam reveals that he has a sister with sickle cell disease (SS). Adam himself does carry the sickle gene (AS), but unlike his sister, his heterozygous state did not lead to any typical sickle cell disease symptoms.

Cell Biology, Diagnosis, and Treatment of Malaria

As with all mosquito-borne diseases, control of the vector is effective to avoid infection, and treatment can be highly effective if diagnosed properly. Before Henry visited Benin, he was immunized for yellow fever, took malaria prophylactics, and did bring a mosquito net. However, visiting rural areas, he did not strictly keep to sleeping under the net every night and did "forget" to take the prophylactics as he should have, in part because they resulted in "weird dreams." Although it is important to consider malaria in all febrile patients with a history of travel to areas where malaria is endemic, American physicians frequently do not do so. Several thousand episodes of malaria are diagnosed each year in Americans after return from travel, and approximately 5%–10% of these cases are estimated to meet criteria for severe malaria. Henry's friend Adam may have been infected by mosquito bites, but if that was the case, the parasites did obviously not take hold. Thus Henry was fortunate that his physician noted this as an option and used a simple microscopic assessment to find proof of malaria in his blood. His artemisinin and quinine treatment was effective. These compounds act both short and long term, respectively, on a specific pathway essential for the parasite inside the red cell. When the parasite proliferates, it uses hemoglobin as its main source of energy and concentrates the toxic hemoglobin breakdown products in "hemozoin" that can be seen under the microscope as small black pits in the food vacuole of the parasite. Compounds that interfere with this pathway (quinine) or lead to the generation of oxidant stress through reactions with the iron in hemozoin (artemisinin) will lead to the demise of the parasite. Qinghaosu first described in the 1960s is the Chinese name for a compound purified from *Artemisia annua*, a.k.a. sweet wormwood, or sweet Annie. The plant has been used to make a fever-reducing (antimalarial) tea, as a mainstay in Chinese herbal medicine. Artemisinin derivatives (artesunate, artemether, and artemotil) have become an important class of compounds recommended by the WHO as antimalarials. A full understanding of their effectivity is lacking, but the compound contains a large ring structure with a peroxide group (double oxygen) spanning it, and its activity may be related to free radical production in the parasite food vacuole and inhibition of a parasite calcium ATPase leading to the demise of the parasite.

Optical Technologies Such as Microscopy and Flow Cytometry Have Revolutionized the Study of Membranes

A myriad of techniques has been developed over the years to study membranes. Proteins, lipid, or sugar components can be labeled to identify certain cells and to study the function of different membrane building blocks. The optical technologies are currently in use to visualize these extremely small entities started about 300 years ago with the development of the science of microscopy by a pioneer in cell biology. Anthoni van Leeuwenhoek (1632–1723) was a tradesman of Delft, the Netherlands. He had no higher education or university degrees and knew no languages other than his native Dutch. However, in a letter of June 12, 1716, he described:

> My work, which I've done for a long time, was not pursued in order to gain the praise I now enjoy, but chiefly from a craving after knowledge, which I notice resides in me more than in most other men. And therewithal, whenever I found out anything remarkable, I have thought it my duty to put down my discovery on paper, so that all ingenious people might be informed thereof.

He did not "invent" the multilens microscope, as is often thought, but his simple yet very powerful "magnifying glasses" allowed him to significantly increase magnification. Importantly, he described and communicated his observations in detail. He discovered and described bacteria, free-living and parasitic microscopic protists, blood cells, sperm cells, microscopic nematodes and rotifers, and much more. His studies, which were widely circulated, exposed an entire world of microscopic life to the awareness of scientists.

Today, light microscopy is joined by electron microscopy, fluorescent and confocal microscopy, and many other technologies that pry information out of cells. In addition, as described in Chapter 1, research projects in cell biology, particularly studies of biomembranes, use flow cytometry, as it allows quick assessment of membrane characteristics of thousands of cells. Since the 1980s flow cytometers have developed from complex machines that could easily fill a room to desktop workhorses for cell biology. Similar to microscopy, this technology enables the analysis of single cells. In contrast with microscopy, many millions of cells are analyzed in a short time. This renders valuable information on cell populations or rare events in that population. The intense data flow and analysis in real time, as well as evaluation of the data sets after collection, require significant computing power, which currently can be performed easily with commercially available computers.

The rapid development of computer technology has also led to "image flow cytometry" that generates a regular picture of each flow cytometric "event" or cell.

The flow cytometric "events" recorded by detectors by measurement of the amount of light detected at different angles are now shared by a camera that makes a picture, and tens of thousands of pictures can be analyzed in addition to the presence of fluorescently labeled probes (such as antibodies or dyes) as described in Chapter 1. Figure 2–15 shows the morphological assessment of a blood sample of a sickle cell patient exposed to low oxygen. The dot plot in Figure 2–15A shows the events after blood was exposed to low oxygen leading to morphological changes and the formation of bizarre shapes in the population. Each event (picture) identified as a dot in the plot can by analyzed by algorithms of pixel analysis and thereby defined as normal or abnormal.

Figure 2–15B shows the changes in blood over time when exposed to 1% oxygen. This is an approximate of the oxygen tension encountered in the periphery as compared with the 20% oxygen in the lung. In blood collected from a patient under room air (time zero), a percentage of cells show an abnormal morphology. These are the so-called irreversible sickled cells, cells that have not been able to revert back to their normal biconcave shape after the endured hypoxic stress in the circulation and have not been removed yet by the systems in the body that recognize these abnormal red cells.

In time, under hypoxia, the percentage of abnormal cells increases to a maximum value. In the example shown, 80% of the population of cells are "sickled" after 20 min. It is important to note that it does not mean that the 20% of normal shaped cells do not have any polymer inside; it simply shows that the polymer has not changed their shape. When exposed to air the morphology instantaneously reverts to the starting value. The results of this sickling kinetics experiment vary from patient to patient and can be used to link to clinical symptoms or the effect of treatment. Figures 2–15 and 2–16 show that flow cytometry can also be used to measure plasma membrane lipid structure and movement. Figure 2–16 shows cells labeled with fluorescent diannexin, a probe that binds to cells with PS on its surface. When PS is exposed on the surface of red cells in the circulation, they act similarly as activated platelets and provide a surface for the prothrombin complex to form (Fig. 2–16). Prothrombin becomes thrombin under release of fragment 1.2 (F1.2). The level of F1.2 in plasma of sickle cell patients is correlated to the number of PS-exposing RBC (Fig. 2–16, C), suggesting that altered RBC membranes are related to the prothrombotic state in these patients that in turn is related to the inability to move PS from the outside of the cell to the inner leaflet.

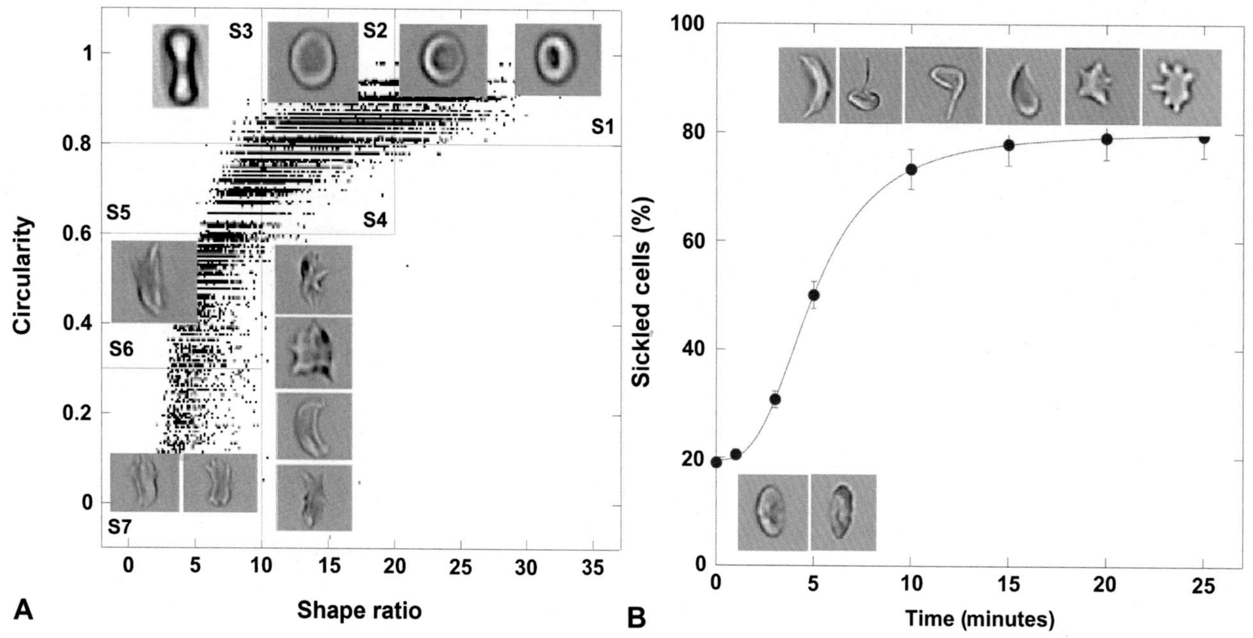

Figure 2–15. **Image flow cytometry and sickling kinetics.** (A) Red cells enter the flow cell in a single file. A high-resolution microscope produces detailed bright-field, dark-field, and fluorescence imagery and intensity of each "event" in the flow cell. A typical flow cytometry dot plot is shown. Pixel analysis algorithms such as "circularity" and "shape ratio" can be used to characterize the morphology of the red cell. Each "*dot*" represents an image, and examples are shown. Events in sectors S1 and S2 show images of normal red cell shape. Sectors S5 and S6 show highly distorted (sickled) cells. (B) Sickle blood exposed to 1% oxygen. At each time point, 10,000 events are analyzed by image flow cytometry, and the percentage of abnormal shaped (sickled) cells is calculated. The curve fit provides a sigmoid relation of morphology changes in time, which defines the sickling kinetics of sickle blood exposed to low oxygen.

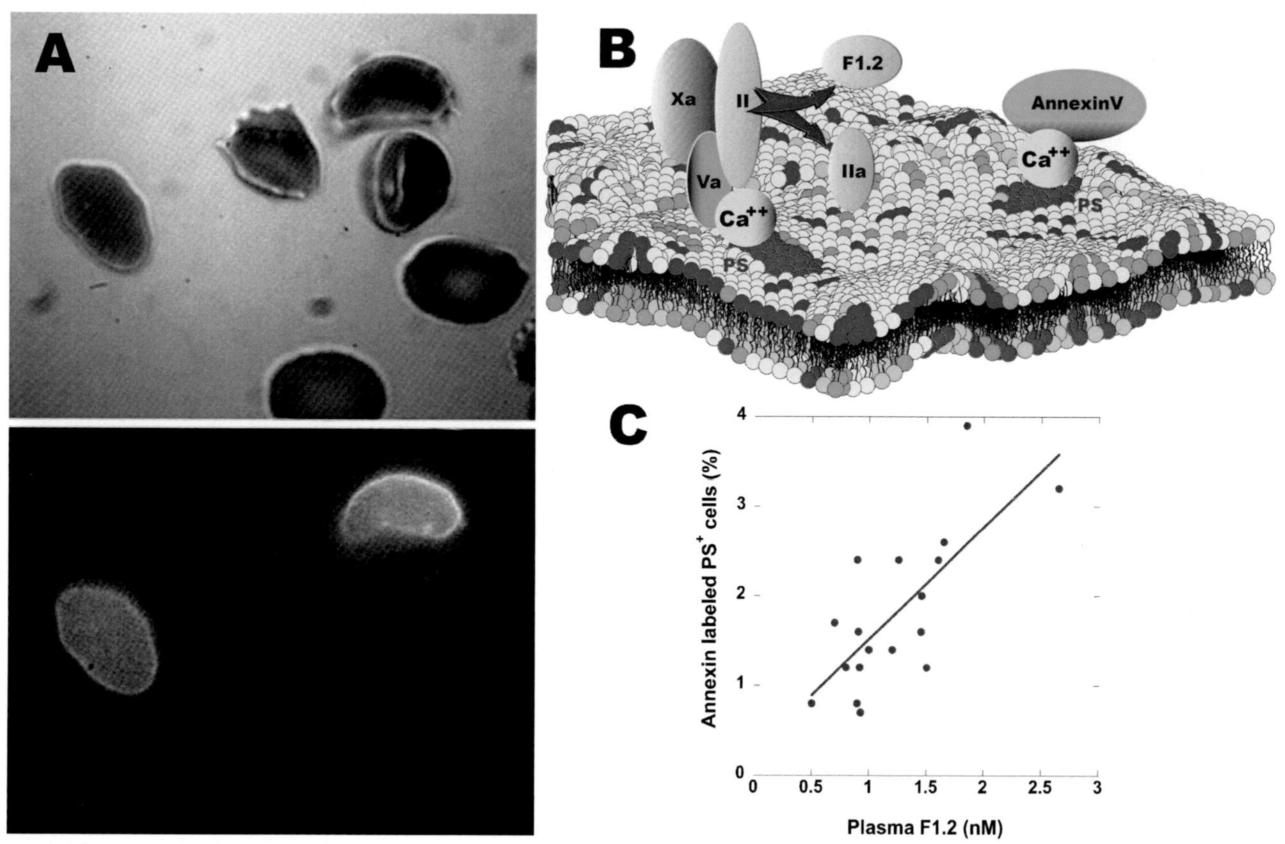

Figure 2–16. **(A) An example of sickle cells incubated with fluorescent annexin V.** The top micrograph is in normal light; the bottom is in fluorescence. Some cells bind fluorescently labeled annexin V, a protein that binds to phosphatidylserine (PS) on the surface of cells in the presence of calcium. (B) Both the prothrombinase complexes of annexin V bind to PS on the surface of the cell in the presence of calcium. (C) The percentage of red cells that expose PS, as measured by flow cytometry in relation to the level of fragment 1.2 (F1.2) in the plasma of the patient. PS exposure on platelets is essential for the normal action of the hemostatic system as it provides a surface where coagulation factors attach and prothrombin is cleaved to release thrombin and F1.2. PS-exposing RBC that are not quickly enough removed lead to an imbalance in the normal pathways, and indeed, PS-exposing red cells are related to the prothrombotic state in these patients.

The Cell Membrane Is a Selective Permeability Barrier That Maintains Distinct Internal and External Cellular Environments

As we have discussed aspects of membrane structure, let us examine some of the functions of the (plasma) membrane. As it separates inside from outside, transport of molecules or ions across this barrier is important. This section reviews different modes of transport—diffusion, osmosis, and active transport. Other modes of transport are also possible, such as endocytosis and exocytosis (see Fig. 2–5).

Small uncharged molecules can pass through a lipid bilayer by simple diffusion (Fig. 2–17, A). For example, small gaseous molecules, such as O_2, CO_2, and N_2, and small uncharged polar molecules, such as ethanol, glycerol, and urea, can simply diffuse down their concentration gradient across the lipid bilayer as particles of a substance dissolved in liquid or gas solvent are in continuous random (Brownian) movement. They tend to spread from areas of high concentration to areas of low concentration until the concentration is uniform throughout the solution, and if separated by a permeable membrane, particles will move in both directions, but because there are many more particles on the high-concentration side, the net flux will be from high to low concentration until equilibrium has been reached as concentrations on both sides of the membrane are the same. Particles will continue to move across the bilayer, but no further *net* diffusion will occur.

When small molecules are present in two solutions at concentrations outside (C_o) and inside (C_i) the cell ($C_o > C_i$), separated by a membrane permeable to the molecule, but not to solvent, Fick's first law of diffusion states that the net rate of diffusion is $J = -DA\, dc/dx$. The rate of diffusion will be proportional to the surface area of the membrane *(A)*, the concentration gradient across the membrane *dc/dx*, and a factor called the diffusion coefficient *(D)*, the rate at which the molecule can permeate the membrane. Importantly, this describes the movement of a particle between *two* compartments (inside and outside), and the minus sign in this equation is because diffusion is positive in the direction of higher to lower concentration. When considering the diffusion of a molecule *through* a biological membrane, one must also

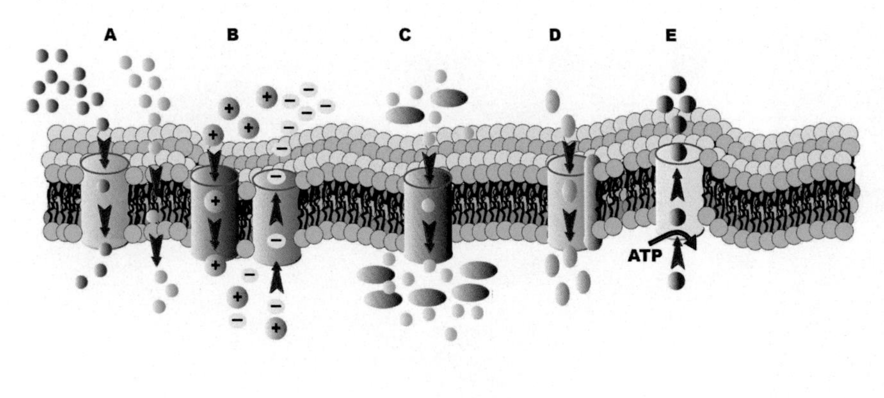

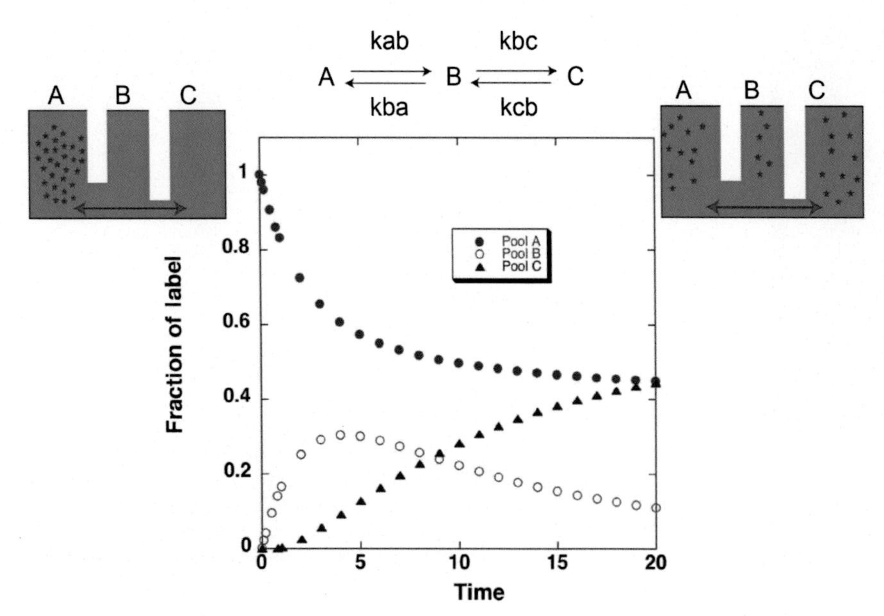

Figure 2–17. **Examples of transport across a biological membrane.** (A) Diffusion of small hydrophilic or hydrophobic particles driven by a concentration gradient. (B) Diffusion of hydrophilic or charged particles driven by a voltage gradient. (C) Osmosis, diffusion of solute driven by a concentration gradient of a nonpermeable compound. (D) Facilitated diffusion. (E) Active transport against a concentration gradient. (B) Diffusion among the three compartments *A, B,* and C. At time zero, label is added to follow the diffusion between the different compartments. In time the fraction of label will decrease in compartment *A,* as compared with start. Similarly the concentration in the other two compartments will alter in time, defined by the rate constants (kab, kba, kbc, and kcb) that modulate to movement between the compartments. Since kab > kbc, the label in pool B appears to increase and then decreases as label moves into pool C. Label moves in both directions, and at equilibrium, no net flow occurs. The size of the compartments can be deducted from the concentration of label in each at equilibrium, and the rates can be deducted from the time curves.

consider the bilayer as a compartment or pool of a "size" that depends on the particle that moves between the outside and inside aqueous compartments. This becomes obvious if we think about the lipid solubility of the diffusing molecule. The permeability of the plasma membrane to a particular molecule increases with its partition coefficient (the movement *into* and *out* of the lipid bilayer). Hence, transport takes place between three pools or compartments. For a substance diffusing through a biological membrane, its concentration at the outer face of the bilayer will be bC_o, in which b represents the partition coefficient. Its concentration at the inner face will equal bC_i. The concentration gradient within the membrane will be $dc/dx = b(C_o - C_i)/dx$. The more lipid soluble the substance, the larger the partition coefficient, b; therefore the larger the effective concentration gradient within the membrane is. If we substitute the concentration gradient across a biological membrane into Fick's equation, the result is $J = Db/dx * A(C_o - C_i)$, with Db/dx referred to as the permeability coefficient, or P. The permeability coefficient for a diffusing molecule is proportional to the partition coefficient, b, and to the diffusion coefficient within the membrane, D, but is inversely proportional to the distance across the membrane, dx. Therefore Fick's equation applied to biological membranes simplifies to $J = -PA(C_o - C_i)$ and applies to small, uncharged molecules. The diffusion of charged molecules across biological membranes is determined not only by the concentration gradient but also by the membrane potential; that is, the movement of small charged molecules across the membrane is related to the electrochemical gradient, not simply to the chemical gradient.

Another way to explain this movement through a membrane is that the membrane is a third compartment or pool in the diffusion process. The "size" is dependent on the pool size in the bilayer and the affinity of the compound to be in that compartment. Compounds equilibrate between the "three" compartments, as visualized in Figure 2–17, lower section. A compound is added to compartment A. The constant "*k*" describes the rate of movement between the compartments A, B, and C. Although particles keep moving in both directions, the concentrations will equilibrate. By adding, for example, a radiolabeled molecule in one compartment, following its fate will provide information on both the diffusion rates and the relative pool sizes for that particular compound.

Water Movement Across Membranes Is Based on Osmosis

The movement of water across a lipid bilayer has long been a difficult issue to understand. The whole reason of a membrane is to separate two aqueous compartments, and its formation is driven by the hydrophobic effect (see Fig. 2–1). A major development in understanding how water can move across a lipid bilayer came with the discovery of aquaporin (Fig. 2–18), for which Peter Agre received the Nobel Prize in Chemistry in 2003, which he shared with Roderick MacKinnon for his studies of ion channels.

Aquaporins are membrane water channels that play critical roles in controlling the water contents of cells. They are widely distributed in all kingdoms of life, including bacteria, plants, and mammals. Several diseases, such as nephrogenic diabetes insipidus and congenital cataracts, are connected to the impaired function of these channels. Aquaporins form tetramers in the cell membrane and facilitate the transport of water and, in some cases, other small solutes across the membrane. Interestingly and quite paradoxical, these water pores are completely impermeable to charged species, such as protons. This seems paradoxical because protons are usually transferred readily through the water phase. This remarkable property is critical for the conservation of membrane's electrochemical potential.

Osmosis is the flow of water across a membrane from a compartment where solute concentration is lower to one in which the solute concentration is higher. When a semipermeable membrane (permeable to water) separates a solution with a low concentration of salt from a solution with a high-concentration salt, water flows to equilibrate the concentrations. The amount of pressure that would have to be applied on one compartment to keep water from entering is called the **osmotic pressure**. The osmotic pressure of a solution depends on the number of particles in solution; therefore it is referred to as a **colligative property**. Because it depends on the actual number of particles, the degree of ionization must be considered. In calculating osmotic pressure, one molecule of NaCl yields two particles, whereas

A

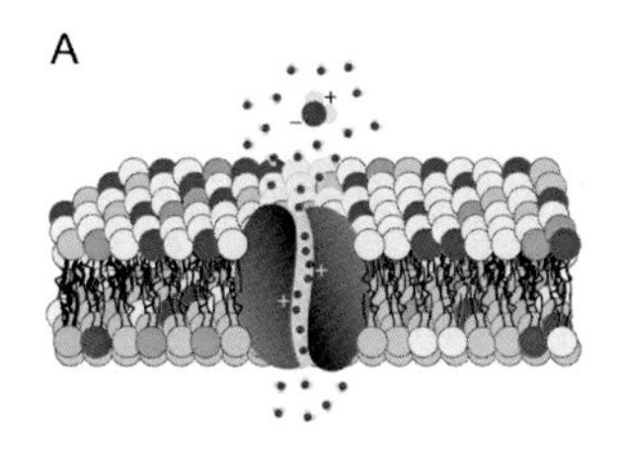

B

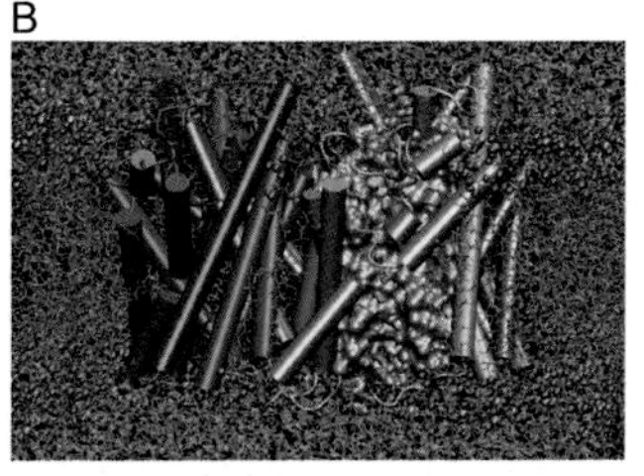

Figure 2–18. **The movement of water through a membrane, facilitated by the water channel aquaporin.** (A) the water channel allows water to move rapidly through the apolar region of the bilayer. (B) A more detailed depiction of aquaporin (www.ks.uiuc.edu/Research/aquaporins).

Na_2SO_4 yields three particles. By van't Hoff's law, osmotic pressure $= iRTc$, where $I =$ number of ions formed by dissociation of a solute molecule, $R =$ ideal gas constant, $T =$ absolute temperature, and $c =$ molal or molar concentration. At physiologic concentrations of solutes, such as NaCl, the values obtained for osmotic pressure differ from theoretical values based on van't Hoff's law. Therefore a correction factor called the **osmotic coefficient** (θ) is inserted into the van't Hoff formula rendering Osmotic pressure $(\pi) = \phi \times iRTc$. The factor ϕ approaches a value of 1 as the solution becomes increasingly dilute. The term in the equation $\phi \times ic$ is referred to as the **osmolar concentration** with units of osmoles per liter (Osm/L).

To calculate the osmotic pressure (at 37°C) of saline (154-mM NaCl) ($\phi = 0.93$) solution, we find that $\pi = (0.93)\times(2)\times(8.2\times10^{-2})\times(310\,\text{K})\times(0.154) = 7.28\,\text{atm}$. To calculate the osmolarity of this solution: $\phi \times ic = (0.93)\times(2)\times(0.154) = 0.286\,\text{Osm/L}$ or $286\,\text{mOsm}$, so roughly twice the molar concentration. When solutions are *compared,* they can be defined as hypoosmotic, hyperosmotic, or isosmotic, or hypo-, iso-, and hypertonic. If the osmotic pressures of two solutions are equal, they are called *iso*smotic. A solution that has a lower osmotic pressure in comparison is *hypo*osmotic and is *hyper*osmotic when the osmotic pressure is higher. Water moves rapidly through the aquaporin water channels in cellular membranes, and the driving force of osmosis is to maintain *iso*smotic conditions across the membrane. Hence, if the NaCl concentration drops on one side of the membrane, water moves rapidly to the other side to regain that equilibrium across the membrane.

Red blood cells are useful when discussing the osmotic properties of cells because they behave as almost perfect osmometers. In plasma, human erythrocytes have their normal shape and volume (Fig. 2–19). The intracellular substances within the erythrocyte that produce its cytosolic osmotic pressure include hemoglobin, K^+, organic phosphate, and Cl^-. In total the tonicity of the red cell cytosol is 286 mOsm. Hence the medium must have the same osmolar concentration of 286 mOsm and an osmotic pressure of 7.28 atm at 37°C to maintain the normal volume of the cell. When plasma is replaced with 286-mOsm NaCl, the **isotonic** condition is maintained. When the NaCl concentration increases (**hypertonic**), water will leave the cytosol, and the red cell shrinks. When NaCl decreases (**hypotonic**), water will enter the cell, and the cell swells. Obviously, this swelling can only go that far. The cells change their shape from their discoid form to spherical, and when the red cell reaches 1.4 times its original volume, the membrane cannot handle the increasing pressure, bursts, and releases hemoglobin and other cytosolic components.

Not all cells in the population behave identically. Some cells reach their critical volume earlier (at higher osmolarities) and other cells later (at lower osmolarities) before they burst (hemolyze). The relation between the

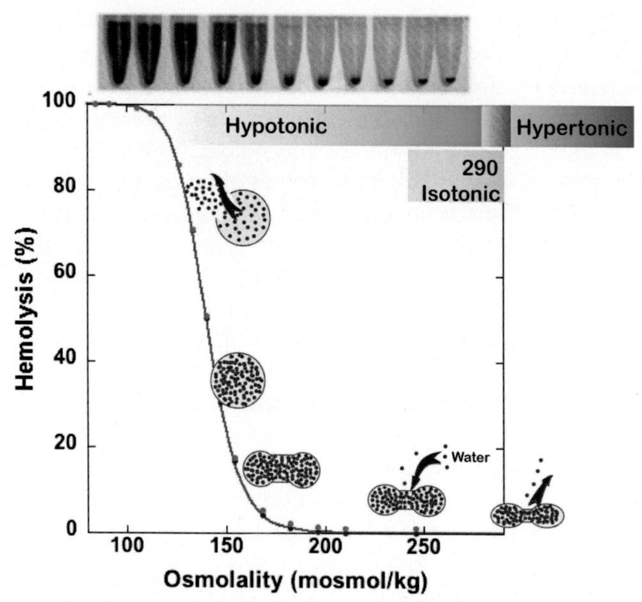

Figure 2–19. **The red cell as an osmometer.** When red cells are placed in a buffer of isotonicity (290 mOsm), they have their normal volume. Water will leave the cell when the tonicity of their surrounding medium increases to hypertonicity, and the cell shrinks. A decrease in tonicity toward hypotonicity will lead to the influx of water, and the cells swell. Their shape changes from the typical biconcave shape to a sphere. The surface area of the membrane is limited, and the pressure inside the cell increases as more water enters the cytosol to compensate for the lower tonicity. Cells in the population start to burst (hemolyze), and hemoglobin leaves the cell. The picture on top shows samples after centrifugation. The supernatant contains more and more hemoglobin as osmolality decreases. At 150 mOsm, approximately 50% of the cells have hemolyzed, a process that rapidly progresses to 100% at lower tonicity of the surrounding medium.

osmolarity and hemolysis, also called the *osmotic fragility curve*, is shown in Figure 2–19. Alterations in the pumps that maintain the ionic balance between inside and outside will affect swelling or shrinking of the cell.

DONNAN EFFECT AND ITS RELATION TO WATER FLOW

Cells contain many negatively charged ions in the cytoplasm, such as protein and RNA, that cannot diffuse through the membrane. These impermeable anions cause a redistribution of permeable anions in a manner that Donnan and Gibbs predicted (Fig. 2–20). The two compartments, A and B, are separated by a semipermeable membrane. Both compartments have NaCl. Compartment A contains a protein with a net negative charge that cannot diffuse through the membrane.

The total concentration of cations and anions are equal in compartments A and B. The Cl^- initially will be at a higher concentration in compartment B than in A. Because the membrane is permeable to Cl^-, it will move down its chemical gradient toward compartment A. Na^+ will follow Cl^- into compartment A to maintain electroneutrality on side A. After these events the concentration of particles will be higher in compartment

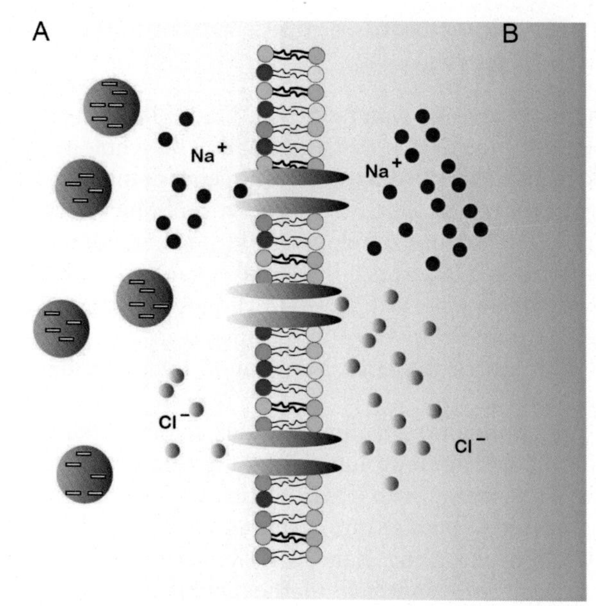

Figure 2–20. The Donnan effect. A semipermeable membrane will allow the diffusion of Na^+ and Cl^- ions, but not the negatively charged protein, in chamber A. Initially, there is an equal number of cations and anions in chambers A and B, which means that the Cl^- is at much greater concentration in B than in A to balance the anionic charge of the impermeable protein. With time, Cl^- will flow down its concentration gradient from B to A. To maintain electroneutrality, Na^+ would also move from B to A. However, this would cause the osmotic concentration to be greater in chamber A than in chamber B; therefore water moves from B to A.

A as compared with compartment B. For single cations and anions with identical valence, their distribution across the semipermeable membrane at equilibrium can be expressed by the Donnan equation: $[Na]_A[Cl]_A = [Na]_B[Cl]_B$. Because the total number of particles in compartment A will be greater at equilibrium than in compartment B, water will flow toward compartment A. For this same reason the Donnan effect tends to cause water to flow into cells but is balanced by the active outward pumping of ions by the Na/K-ATPase.

FACILITATED TRANSPORT

Both diffusion and osmosis are passive transport properties across biomembranes, governed by mechanisms as described earlier. However, many molecules may exhibit very low permeability coefficients *(P)*; therefore they diffuse slowly, slower than needed to maintain functionality of the cells. Hence, this process needs to be facilitated. Molecules apparently can cross biological membranes far more rapidly than would be predicted by Fick's equation because of a process called **facilitated diffusion.** Transmembrane proteins play an essential role in this process. As an example, let us consider the uptake of glucose into the RBC, essential for its metabolic processes (glycolysis). The blood plasma glucose concentration of approximately 5 mM is much higher than the concentration of glucose within the RBC. However, because of its

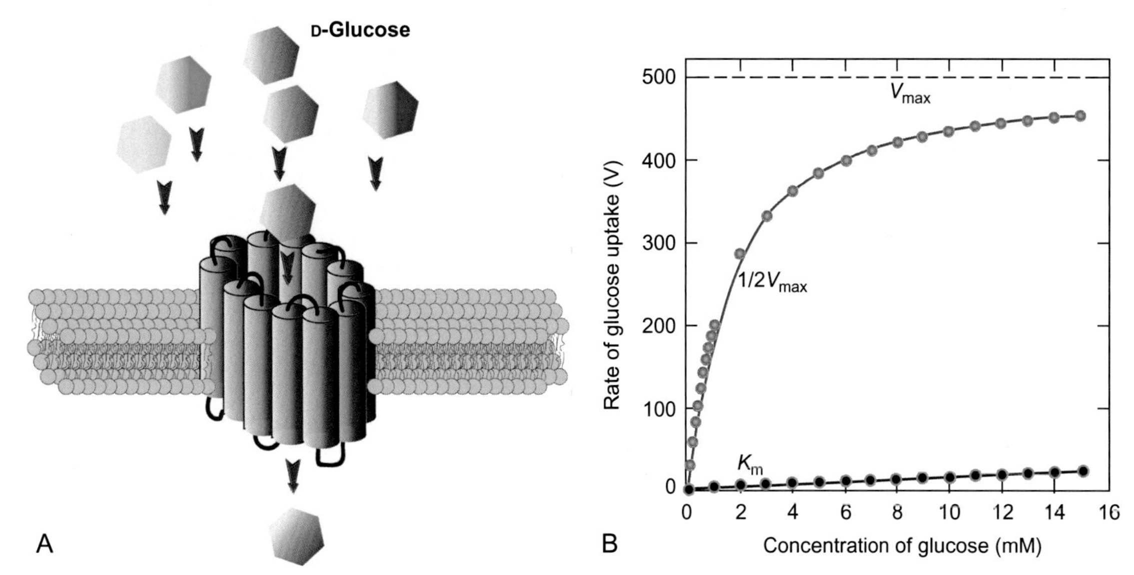

Figure 2–21. **Glucose transport into red blood cells: facilitated diffusion.** (A) A glucose transport protein (permease) for glucose. (B) Glucose moves down its concentration gradient into red blood cells at a rate much faster than would be predicted by simple diffusion through the lipid bilayer *(black line)*. The plot for rate of glucose uptake versus external glucose concentration is hyperbolic *(red line)*. The rate increases with external glucose concentration until it reaches a maximal velocity (V_{max}). The K_m is the concentration of external glucose at which half maximal velocity is reached.

large size, glucose cannot simply diffuse through the lipid bilayer (Fig. 2–21). It is moved rapidly across the erythrocyte membrane, down its concentration gradient, by associating with a glucose transporter or permease that changes conformation and allows passage of glucose. As indicated in Figure 2–21B, the rate of uptake of glucose is saturable. At high concentrations of glucose, all of the erythrocyte glucose transporters are occupied so that a maximal velocity is reached (V_{max}). The K_m represents the concentration of glucose at which the rate of uptake $V = {}^1/_2 V_{max}$. It is a measure of the affinity of glucose to bind to its transporter.

For example, whereas the K_m for D-glucose is 1.5 mM, the K_m for its stereoisomer L-glucose is greater than 3000 mM. This illustrates that facilitated diffusion, unlike simple diffusion, is highly (stereo) specific. The glucose transporter will bind D-glucose, but it binds L-glucose very poorly providing a specificity for the D form.

Other six-carbon sugar molecules that are similar in structure to D-glucose can be transported by the glucose transporter, but they have a greater K_m value. For example, the K_m values for D-mannose and D-galactose are 20 and 30 mM, respectively. These sugars can competitively inhibit the uptake of D-glucose into erythrocytes. The curve in Figure 2–21B fits the Michaelis-Menten equation for enzymatic activity: $V = V_{max}/(1 + K_m/C)$, where C is the concentration of substrate. Although the transporter is not an enzyme, it carries out a function similar to that of an enzyme. Instead of chemically converting a substrate, as an enzyme does, transporters move a substrate across a biological membrane in a saturable manner. The glucose transporter is a multipass transmembrane protein (see Fig. 2–12) with 12 α-helical transmembrane segments. Multipass transmembrane proteins have a higher proportion of polar amino acids within the bilayer core than do the single-pass ones. The binding of D-glucose to an extracellular domain of the glucose transporter is thought to cause a conformational change in the protein, which allows the polar amino acids within the bilayer core to hydrogen bond with the hydroxyl groups of glucose, thereby facilitating its movement down its concentration gradient. Facilitated diffusion is sometimes called **passive transport** because it requires no external source of energy. D-Glucose simply moves down its concentration gradient.

ACTIVE TRANSPORT

Active transport and selective permeability of the plasma membrane for ions create large differences in the ionic composition of the cytosol and extracellular fluid surrounding mammalian cells. As an example, the Na^+ concentration is maintained at approximately 10- to 20-fold greater concentration outside than inside cells, whereas K^+ is at a 20- to 40-fold greater concentration inside the cells. Whereas the calcium concentration in plasma is about 2 mM, inside the red cell, it is maintained at nanomolar concentrations by a pump that uses energy (ATP) to expel calcium from the cells. Similarly the enzyme Na/K-ATPase uses the energy of ATP hydrolysis to pump three Na^+ ions out of the cell against their electrochemical gradient and two K^+ ions into the cytosol against their electrochemical gradient (Fig. 2–22).

The Na/K-ATPase is an important example of **primary active transport**, for which energy derived from ATP

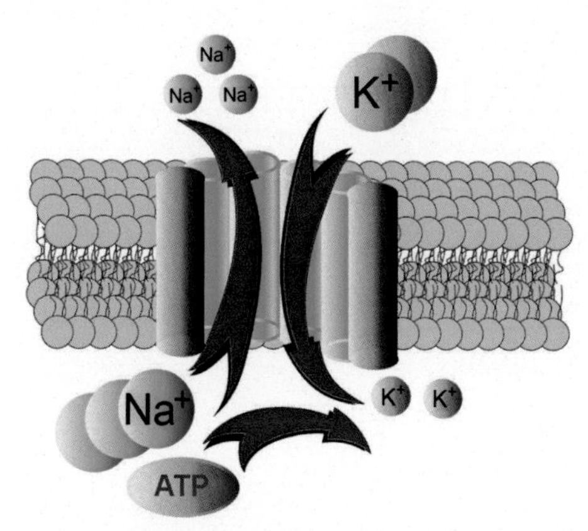

Figure 2–22. **The Na/K-ATPase is an electrogenic pump.** It moves three Na^+ ions out of the cell and two K^+ ions into the cytoplasm at the expense of ATP hydrolysis to ADP and inorganic phosphate.

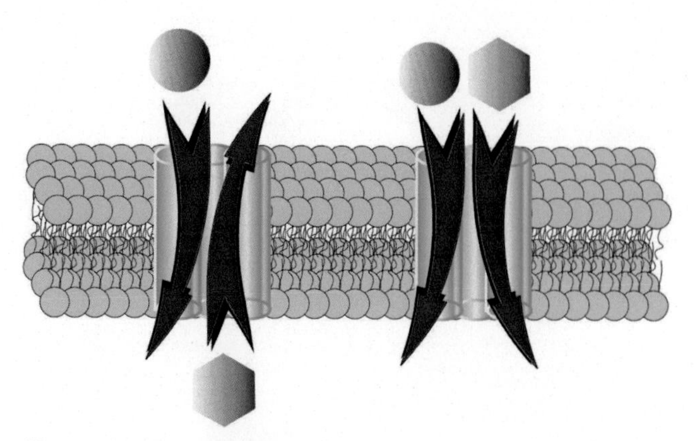

Figure 2–23. **Coupled transport.** The transport of one molecule across a biological membrane can be coupled by the transport protein to the movement of another molecule. If the movement of both molecules is in the same direction, the cotransport is referred to as *symport.* If the molecules are being moved in opposite directions, the cotransport is referred to as *antiport.*

hydrolysis directly moves molecules across membranes against their electrochemical gradient. The extracellular domains of the α subunit of mammalian Na/K-ATPases contain binding sites for two K^+ ions and the inhibitor ouabain, a compound related to the cardiac glycoside digitalis. The intracellular domains contain binding sites for three Na^+ ions, ATP, and a phosphorylation site. The α-subunit is autophosphorylated by ATP on a single aspartate residue, conformationally converting the ATPase from a form that transports K^+ to a form that transports Na^+. At the outer surface of the membrane, binding of K^+ promotes hydrolysis of the phosphate group from the α-subunit. Cleavage of the phosphate converts the carrier back to a form that preferentially transports K^+. That three Na^+ ions are pumped out and two K^+ ions enter the cell with hydrolysis of one ATP molecule are important in several ways. First, Na/K-ATPase is defined to be electrogenic because each cycle leads to one net positive charge to the outside surface of the membrane, thereby contributing in a small way to the development of the membrane potential. Second, three osmotic particles are pumped out, whereas only two osmotic particles are pumped in for each cycle of ATP hydrolysis. This counteracts the swelling of the cells induced by the large impermeable anions within cells. Similar ATPases pump Ca^{2+} out of cells or back into the sarcoplasmic reticulum of muscle or pump protons into the lumen of the stomach. These are called the *P class of ATPases* because they contain nearly identical sequences surrounding the phosphorylated aspartate. Similarly the flippase moves lipid components against its concentration gradient hydrolyzing one ATP per PS molecule transported (see Fig. 2–11).

SECONDARY ACTIVE TRANSPORT

The movement of Na^+ down its electrochemical gradient can be coupled to the movement of another molecule against its gradient. This is referred to as **secondary active transport**, because the Na^+ electrochemical gradient is maintained by Na/K-ATPase. The transported molecule and cotransported molecule can move in the same direction, which is called **symport**, or they can move in opposite directions across the membrane, which is referred to as **antiport** (Fig. 2–23).

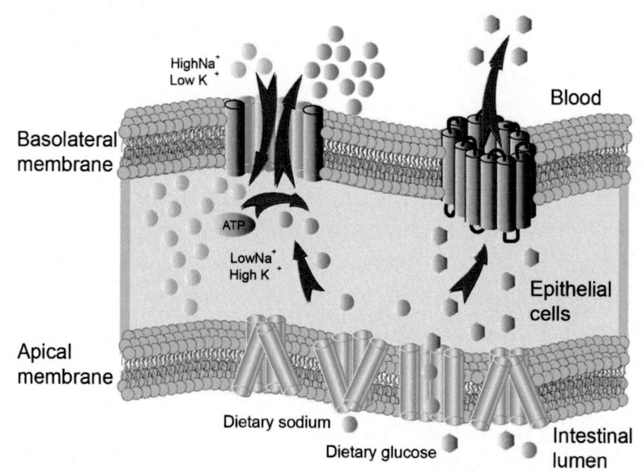

Figure 2–24. **Transport of glucose into and out of intestinal epithelial cells.** Glucose enters from the intestinal lumen through the apical membrane of the epithelial cells by a glucose Na^+ symport transporter. The binding of one Na^+ molecule and one glucose molecule to separate sites on the extracellular domain of a glucose Na^+ symport protein causes the transporter to change conformation. The change of conformation creates a channel through which Na^+ and glucose can be transported into the cytosol. The symport protein then returns to its original conformation. The Na^+ ions are pumped back out of the cell by Na/K-ATPase, located on the basolateral membrane. Glucose then exits the cell by facilitated diffusion, by permeases located in the basolateral membrane.

An example to illustrate such cotransport is the uptake of amino acids and glucose from the diet. Glucose is transported against its concentration gradient from the intestinal lumen to the intestinal epithelial cell across the apical membrane by cotransport with sodium (Fig. 2–24).

Another example is the antiport of sodium and calcium ions, which plays an important role in cardiac muscle contraction. The uptake of Ca^{2+} into cardiac muscle triggers contraction. The accumulated Ca^{2+} is then moved out of the cardiac muscle cell by an antiport protein, which is powered by the simultaneous movement of Na^+ down its electrochemical gradient into the heart muscle cells. Again the Na^+ electrochemical gradient is maintained by Na/K-ATPase; therefore this is another example of secondary active transport. Drugs such as digoxin (Crystodigin and Lanoxin) and ouabain increase the force of heart muscle contraction. These drugs inhibit Na/K-ATPase, causing an increase in intracellular Na^+. This dissipates the Na^+ electrochemical gradient and inhibits the Na^+-Ca^{2+} antiport. This results in increased intracellular Ca^{2+} and, consequently, stronger heart contraction.

ION CHANNELS AND MEMBRANE POTENTIALS

The hydrophobic environment of the phospholipid membrane bilayer is virtually impermeable to hydrated ions in aqueous solution. Ions are immiscible with the nonpolar hydrocarbon fatty acid chains of membrane phospholipids and, therefore, are unable to cross the lipid bilayer by simple diffusion. Permease proteins allow transporting of ions and other hydrophilic substances across membranes, but another major class of transport proteins works differently and highly efficiently. These transporters form aqueous ion channels that traverse the lipid bilayer. Ion channels are distinguished physiologically from carrier proteins by the high velocity of ion flux they allow across the membrane, without expending energy. Whereas permease-mediated transport operates at a maximum of 10^5 ions/s, ions flow at a rate 100–1000 times faster through channels. This mediates not only rapid electrical signaling of nerve impulses and muscle contraction but also many biological responses common to nonexcitable cells.

An important principle is that ion channels determine the rate but not the direction of ion flow across cell membranes. Movement of a nonionic solute through a channel is strictly passive, determined only by its concentration gradient across the membrane. For an ion the direction of flow depends on both the chemical concentration gradient *and* the electrical potential across the membrane.

The concept of the ion channel is 150 years old. However, a better understanding with respect to their basic properties was not gained until the 1960s. A major breakthrough came with the discovery that ion channels could be artificially introduced into cell membranes by treating them with small hydrophobic proteins, termed **ionophores.** Made primarily by different fungi, some ionophores have been used as antibiotics, but many more have become important tools for the cell biologist. The loss of phospholipid asymmetry can be induced by activating phospholipid scrambling by increasing cytosolic calcium. In the laboratory, this can be achieved by treating cells with the calcium ionophore A23187, as described in relation to Figure 2–11. The addition of this compound to membranes makes them highly permeable to calcium. The calcium pump, which normally keeps intracellular calcium low by actively pumping calcium out at the expense of ATP, is completely overwhelmed as the ionophore equilibrates calcium across the bilayer at near to diffusion rates. Under these conditions, ATP runs out, and the Mg-ATPase stops working, and the calcium-induced scrambling of the membrane exposes PS. Using this approach in the lab mimics a situation where the cell is not able to maintain proper Ca^{2+} homeostasis, or ATP levels, and the apoptotic program is initiated. This process can be further exacerbated by using a sulfhydryl reagent like NEM that mimics oxidative damage of the flippase.

Gramicidin A, as another example, is a linear polypeptide with 15 amino acids that, like all ionophores, is readily miscible with the phospholipid bilayer. This ionophore, when inserted into the membrane, forms an unusual α-helix. Alternating D-and L-amino acid residues in the molecule orient the polar (hydrophilic) carbonyl oxygen atoms and amide nitrogens of the peptide bonds toward the hollow center of the helix, where they form the wall of the channel pore. Hydrophobic amino acid side chains radiate outwardly from the helix to anchor the ionophore in surrounding lipid (Fig. 2–25).

Two helical molecules of gramicidin A are thought to align end to end, to completely span the lipid bilayer. The different flow rates of various ion species across membranes punctuated with these simple channels indicate that gramicidin pores prefer small cations to anions and that partial dehydration is necessary for ions to traverse the channel in single file, as they do through many animal membrane channels.

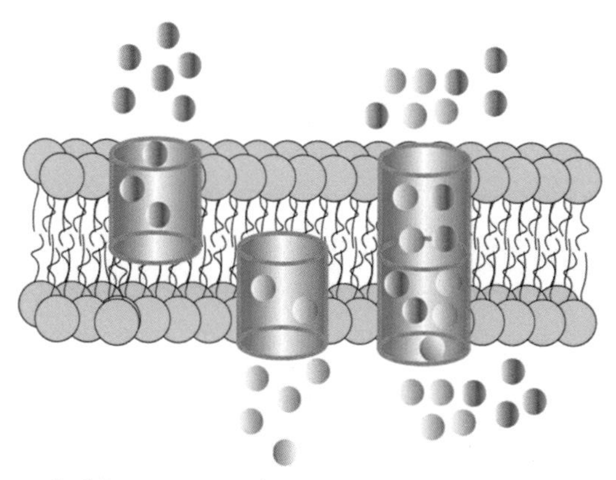

Figure 2–25. **Structure of the helix and transmembrane channel formed by gramicidin A.** Two gramicidin A peptides dimerize head to head to span the lipid bilayer.

Whereas channel-forming proteins expressed by mammalian cells also form helices in membranes, they differ fundamentally from ionophores such as gramicidin A. Because membrane-spanning proteins in eukaryotes consist solely of L-amino acids, their α-helix lacks a central pore. Hydrogen bonding between the carbonyl oxygens and amide nitrogens of the polypeptide backbone prevents them from interacting with water. As a result, these α-helices are naturally hydrophobic and highly suited for association with the lipid bilayer. By organizing several membrane-spanning domains such that hydrophilic amino acid residues point toward the aqueous channel and hydrophobic residues face the lipid bilayer, channels can be formed as illustrated for the acetylcholine receptor in Figure 2–26. These channels are found in special junctions, termed **synapses**, between nerve terminals and apposed stretches of the muscle cell membrane. Acetylcholine (ACh) released from a stimulated nerve terminal diffuses across a cleft in the synapse and functions at the muscle cell as a neurotransmitter molecule that binds to external sites on the acetylcholine receptor channel protein. The binding of two acetylcholine molecules to the α-subunits evokes a conformational change that briefly opens the channel for about 1 millisecond before the channel recloses, still remaining bound to acetylcholine. Once the channel is closed, acetylcholine dissociates from the channel, which returns to an unbound conformation. At that point, **acetylcholinesterase** (AChE) also present on the postsynaptic membrane as a GPI-anchored protein catalyzes the breakdown of acetylcholine terminates the signal transmission by hydrolyzing ACh, and the liberated choline is taken up again by the presynaptic neuron, and ACh is synthesized by combining with acetyl-CoA through the action of choline acetyltransferase, ready for the next action. Rapid breakdown of ACh is essential to avoid continuous ACh action, and AChE has a very high catalytic activity as each molecule of AChE degrades about 25,000 molecules of ACh per second, approaching the limit allowed by diffusion of the substrate. The ability of AChE to move quickly in the plane of the bilayer due to its GPI linkage further facilitates its rapid deployment.

Many different ion channels have been reported to date. Channels are often glycoproteins that contain several α-helical membrane-spanning regions, flanked by hydrophilic portions protruding into the extracellular space and cytoplasm. Given the high flux of ions through these channels, regulation of their permeability status appears to be important to maintain proper ion balance across the membrane. Indeed, ion channels, in general, exist in two or more conformations, including a brief but stable open state and a stable closed state.

Different types of stimuli can control the opening and closing, or **gating**, of ion channels. Three major varieties of rapidly gated ion channels and a fourth slowly gated type, the gap junction, can be distinguished (Fig. 2–27). **Voltage-gated** channels are responsible for the propagation of electrical impulses over long distances in nerve and muscle, and they open specifically in response to a change in the electric field that exists across the plasma membrane of cells at rest. **Ligand-gated** ion channels are insensitive to voltage change, but they are opened by the noncovalent reversible binding of a chemical ligand. These substances include neurotransmitters or drugs that bind to the extracellular portion of the receptor.

Alternatively an intracellular second messenger or enzyme interacting with the cytoplasmic face of the channel can also influence its conformational state. Ligand-gated ion channels enable rapid communication between different neurons and between neurons and muscle or glandular cells across synapses. A few cell types have **mechanically gated** ion channels for which the opening is controlled by cellular deformation. A fourth class of ion channel, the **gap junction**, enables ions to flow between adjacent cells without traversing

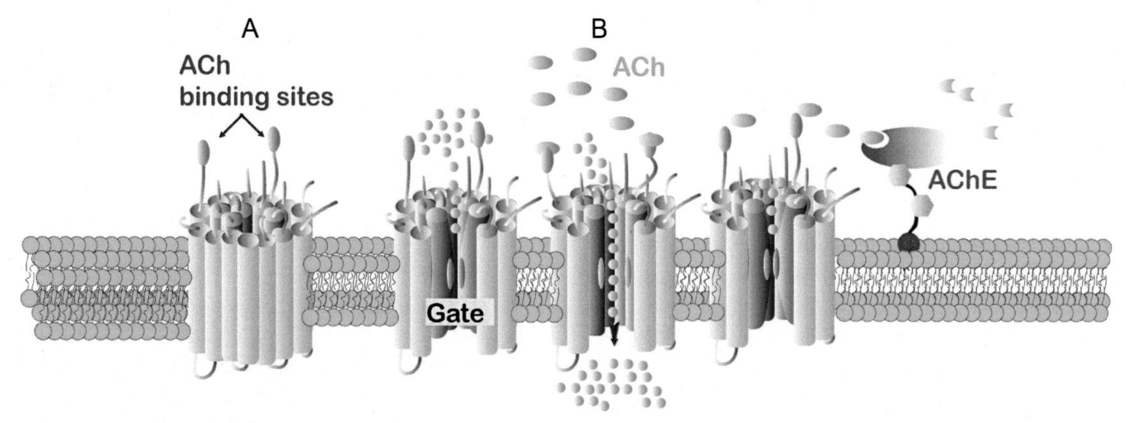

Figure 2–26. **A model of the acetylcholine receptor.** (A) The pentameric configuration of the receptor is typical of many other members of the ligand-gated ion channel family. Two of the subunits are identical, and three are different; each of the two α-subunits contains an extracellular binding site for acetylcholine (ACh). (B) Three conformational states of the acetylcholine-gated ion channel. In the absence of ACh, the gate is closed. ACh is released, and the binding of two acetylcholine molecules alters the protein conformation to open the channel pore. However, the effect is only transient; the pore soon closes with acetylcholine still bound to the receptor sites. Once the ligand dissociates from the receptor, the channel can return to a closed but receptive conformation, and ACh is rapidly degraded by acetylcholinesterase (AChE), a GPI-anchored protein that can move very rapidly to scavenge its substrate.

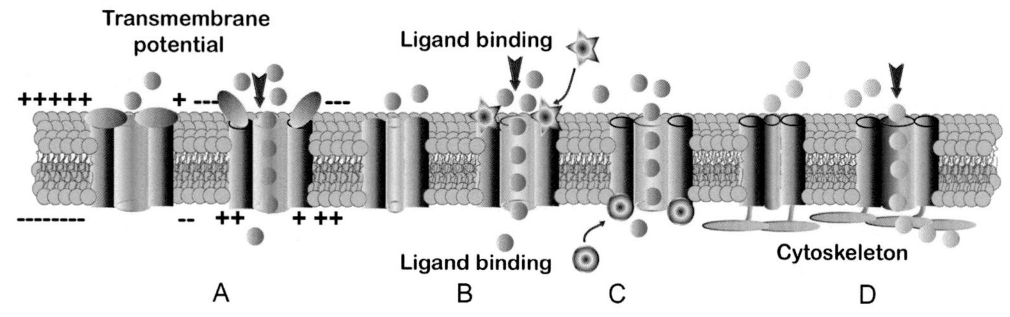

Figure 2–27. **Classes of ion channels stimulated by different gating mechanisms.** Ion channels are distinguished according to the signal that opens them. (A) Voltage-gated channels require a deviation of the transmembrane potential. (B,C) Ligand-gated receptors respond to the binding of a specific ligand, either an external neurotransmitter molecule or an internal mediator such as a nucleotide or ion. (D) Mechanically gated channels can sense movement of the cell membrane linked by cytoskeletal filaments to the channel protein. Each effector causes an allosteric change that opens the channel, thereby causing an ion flux across the membrane.

the extracellular space. Gap junctions are not rapidly gated but rather open and close in response to changes in the intracellular concentration of Ca^{2+} and protons. An example of the ligand-gated class is the channel gated by acetylcholine on skeletal muscle cells as shown in Figure 2–26. Many different neurotransmitters other than acetylcholine operate elsewhere, primarily by binding to transmitter-gated ion channels. As a rule, each of these ligands has its own specific ion channel receptor that, when opened, is selectively permeable to a certain ion.

Ion channels in mammalian cell membranes are selective for the type of ions that flow through them. The acetylcholine receptor at the neuromuscular junction is permeable to small cations (K^+, Na^+, and Ca^{2+}), not anions. Other channels are permeable to a single cation, either K^+, Na^+, or Ca^{2+}. Similarly, channels permeable only to Cl^- have been identified. The voltage-gated K^+ channel, for example, is too small to pass Ca^{2+}. Interestingly, even though Na^+ is the smaller ion, this channel shows a 100-fold greater selectivity for K^+ than for Na^+. Hydration of cations in solution appears to play an important role. Because Na^+ is smaller than K^+ in ionic diameter, its charge density and electric field are stronger. Hence, Na^+ interacts more strongly with surrounding water molecules compared with K^+. This reduces its mobility, and the larger shell of electrostatically bound water molecules results in a larger "virtual" diameter such that hydrated Na^+ cannot permeate the smaller-diameter K^+ channels. Although ion channels are most often recognized in the role of electrical signaling in nerve and muscle cells, they exist, to some degree, in all cells to mediate other functions. The selectivity of a cell membrane for permeant ions depends on the relative proportions of various types of ion channels. A single neuron can have as many as four different ion-selective channels. The most common one among neural and nonneural cell types is termed the **K^+ leak channel** because its opening does not require a specific gating stimulus. These channels enable all cells in the body to maintain a voltage difference across their plasma membrane, the **membrane potential.**

The Membrane Potential Is Caused by a Difference in Electric Charge on the Two Sides of the Plasma Membrane

Cells maintain slightly more negative than positive ions in the cytosol and more positive than negative ions in the extracellular fluid (Fig. 2–28, A). Membranes with their hydrophobic lipid bilayer are poor conductors of ionic current, and the transmembrane voltage across the plasma membrane resembles an electrical capacitor. As a result an accumulation of negative charges along the cytosolic side of the membrane attracts positively charged

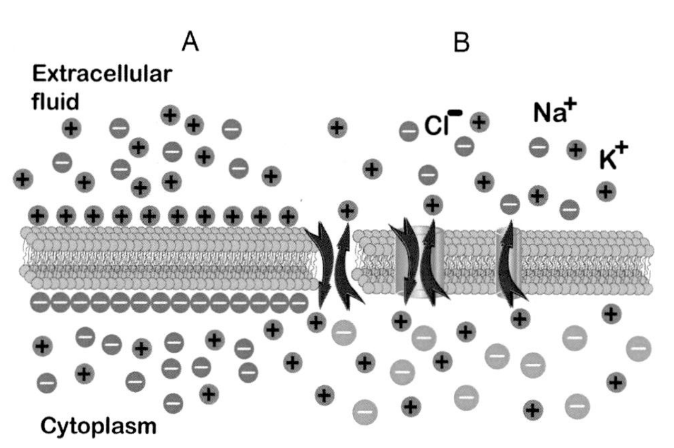

Figure 2–28. **The separation of charge across the cell membrane forms a membrane potential.** (A) A net excess negative charge inside the membrane and a matching net excess positive charge outside the membrane form a transmembrane potential difference that is maintained across an impermeable lipid bilayer. Charge on either side of the membrane is concentrated in a thin layer (1 nm thick) and formed by an extremely small percentage of the total ions in the cell. (B) Opposing forces regulate K^+ flux across the plasma membrane. Membrane potential is primarily based on four ion species: K^+, Na^+, Cl^-, and organic anions (−), such as amino acids and other metabolites. The resting membrane potential of a cell permeable only to K^+ depends on the passive diffusion of K^+ out of the cell down its concentration. If left unchecked, K^+ efflux would eventually create an excess negative charge in the cell (an overbalance of organic anions) and a buildup of $[K^+]_o$, were it not for an electrical driving force moving K^+ in the opposite direction. An equilibrium results when these two opposing forces counterbalance each other.

ions on the extracellular side of the bilayer. The voltage gradient that arises across the membrane (5.0 nm) is nearly 200,000 V/cm. Therefore membrane potential and transmembrane ionic gradients provide a driving electrical force for many biological processes.

A membrane potential V_m is defined as $V_m = V_i - V_o$, where V_i is the voltage inside the cell and V_o is the voltage outside the cell. Because V_o is arbitrarily set to zero, V_m in the undisturbed, or **resting**, cell becomes negative as the result of the slightly negative net ionic charge of the cytoplasm. Membrane potential is primarily based on four ion species: K^+, Na^+, Cl^-, and organic anions (A^-), such as amino acids and other metabolites. Of these, Na^+ and Cl^- are concentrated in the extracellular fluid, whereas K^+ and A^- are preponderant inside the cell. The active extrusion of Na^+ from the cell by Na/K-ATPase maintains the osmotic balance of the cytosol by preventing the influx of water that would otherwise occur. In exchange for pumping three Na^+ ions out of the cell, two K^+ ions enter the cytosol to counterbalance organic anions that do not permeate the plasma membrane.

Neurons or muscle cells have many Na^+ channels in their membranes. They remain closed, however, when the cell is at rest. Therefore Na^+ extruded by Na/K-ATPase cannot readily reenter the cell down its steep concentration gradient. Only nongated K^+ leak channels remain open. As a result, K^+ will tend to passively leak out of the cell through the K^+ leak channels down its steep concentration gradient until the force of outward diffusion is counterbalanced by a second and opposing inward electrical force created by the attraction that organic anions in the cytosol have for K^+ (see Fig. 2–28, B).

Together the concentration gradient and the voltage gradient for a particular ion across a membrane determine the net electrochemical gradient that drives the flow of that particular ion species through a membrane channel. When these forces balance, the electrochemical gradient is zero, and no net flow of this ion occurs across the membrane. Given the concentration of an ion inside and outside the cell, the voltage necessary to achieve this equilibrium, termed the **equilibrium potential**, can be calculated from the Nernst equation. For K^+, this would be

$$E_k = RT/ZF^* \ln[K^+]_o/[K^+]_I$$

where E_k is the value of the equilibrium potential of K^+; R is the gas constant ($2\,\text{cal}\,\text{mol}^{-1}\,\text{K}^{-1}$); T is the absolute temperature in °K; F is the Faraday constant ($2.3 \times 10^4\,\text{cal}\,\text{V}^{-1}\,\text{mol}^{-1}$); and $[K^+]_o$ and $[K^+]_i$ are the concentrations of K^+ outside and inside of the cell, respectively, such as K^+. The charge $(Z) = +1$, for a monovalent cation such as K^+, and at 37°C, RT/ZF is 27 mV. By using typical values for $[K^+]_o$ and $[K^+]_i$ in mammalian tissue, the equation renders $E_k = -96$ mV.

The equilibrium potential for any of the other ion species across the membrane can be similarly calculated. Ultimately the resting membrane potential (V_R) for a cell is determined by permeability of the membrane to specific ions and their concentrations inside and outside the cell. The Goldman equation is a summation of contributions of fluxes of different ions, determined by their concentrations and permeabilities (P):

$$V_R = RT/F \;* \ln\left(P_k[K^+]_o + P_{Na}[Na^+]_o + P_{Cl}[Cl^-]_i\right)/\left(P_k[K^+]_i + P_{Na}[Na^+]_i + P_{Cl}[Cl^-]_o\right)$$

Based on the individual contributions, this equation can be simplified drastically. Because Ca^{2+} does not contribute significantly, it is not included. Chloride equilibrates across the membrane through nongated Cl^- channels, but most remains in the extracellular fluid to counterbalance nonpermeable intracellular anions. That is, if the permeability to one ion far supersedes that of others ($P_k \gg P_{Cl}$, P_{Na}), the Goldman equation is reduced to the Nernst equation for that ion. When the cell is at rest, the ratio of open K^+ leak channels to Na^+ channels is high, making the cell much more permeable to K^+ than to Na^+. The contribution of Na^+ influx to V_R is minimal, and a balance of −80 to −90 mV for V_R is attained by mammalian cells, a value *near* E_k and far from E_{Na}.

Together, for a nerve or muscle cell with multiple ion-selective channels in its membrane, V_R is affected by the permeability of the membrane for each diffusible ion species (K^+, Na^+, and Cl^+) and the concentration of each inside and outside the cell. Consider, for instance, the glial cell, having only nongated K^+ leak channels in its membrane. V_R is essentially equal to E_k. What would happen when we open a few Na^+ channels in the glial cell membrane, as illustrated in Figure 2–29?

Because Na^+ is more concentrated outside than inside the cell, it will flow passively into the cell through the opened Na^+ channels. The electrical force of attraction

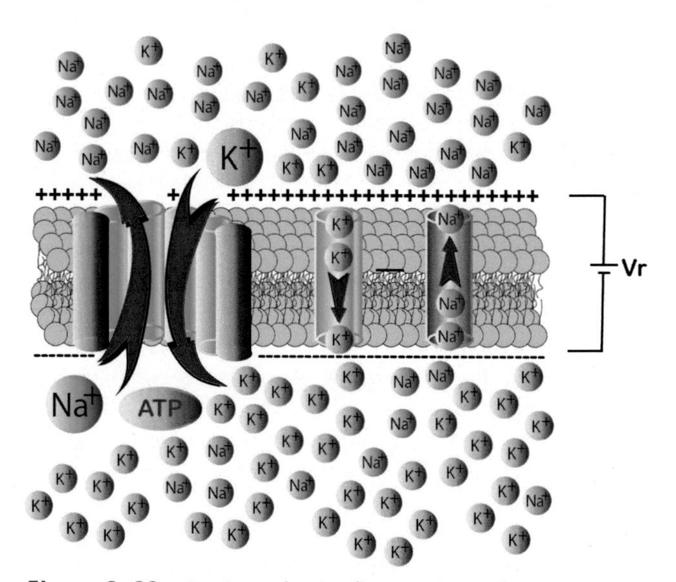

Figure 2–29. **Passive and active fluxes maintain the resting membrane potential.** The cell at rest maintains a steady state whereby Na^+ influx and K^+ efflux defined by passive diffusion are balanced by active transport of these ions in the opposite direction by Na/K-ATPase.

of a slightly electronegative membrane potential generated by the Na/K-ATPase and E_k (−96 mV) will also attract Na^+ into the cell. The magnitude of the latter can be calculated from the Nernst equation: $E_{Na} = +67\,mV$. This tells us that E_{Na} is 163 mV away from the V_R established by E_k alone (−96 mV). As a result, both the electrical and chemical forces work in the same inward direction to form a strong electrochemical gradient driving Na^+ into the cell. The Na^+ influx should then depolarize the cell; that is, it should reduce the charge separation across the membrane by making the interior less electronegative relative to outside of the cell. Indeed, if K^+ efflux and the smaller Na^+ influx were allowed to continue unchecked across the plasma membrane of a neuron, transmembrane gradients for both ions would eventually dissipate, because $[K^+]_i$ would plummet and $[Na^+]_i$ would gradually increase, thereby reducing the V_R. Opposing this effect is Na/K-ATPase, which continues to pump three Na^+ ions out of the cell for every two K^+ ions pumped in, and thereby maintains the V_R constant. Because there is a net transfer of positive charge out of the cell, the pump is said to be **electrogenic**, creating the slight excess of negative charge inside the cell membrane. Thus passive channel-mediated ion fluxes occurring by simple diffusion are balanced by active fluxes that require energy. A steady state between the two processes is reached where the *net* ion flux across the membrane is zero to define the resting membrane potential.

Any disturbance to a cell that increases the membrane permeability for an ion will drive the membrane potential away from V_R in the direction of the equilibrium potential for that ion. These transient deviations from V_R and the opening of gated ion channels that cause them are the basis of electrical signals that convey information not only along and between nerve cells but also from nerve to muscle.

Neurons convey signals along their length by generating passively spreading local potentials and self-propagated action potentials. A local potential arises when the cell membrane is stimulated to become more permeable to certain ions than it is at rest. If Na^+ influx results, the voltage drop across the membrane is reduced, and the signal is said to be **depolarizing**. Alternatively, enhanced Cl^- influx will increase the voltage drop in some neurons and **hyperpolarize** the cell relative to its resting state. When current enters the neuron through open membrane channels, it diffuses in all directions for a distance that depends on the intrinsic properties of the neuron. The voltage signal decreases exponentially as it travels away from the site of entry as ions will leave the cell by nongated K^+ leak channels. This rapid decrease in the spreading wave of depolarization is sufficient to convey electrical signals toward the cell body along relatively short input fibers called **dendrites**. However, such signals would dissipate well before leaving the cell along its **axon**, which is typically a much longer structure (Fig. 2–30).

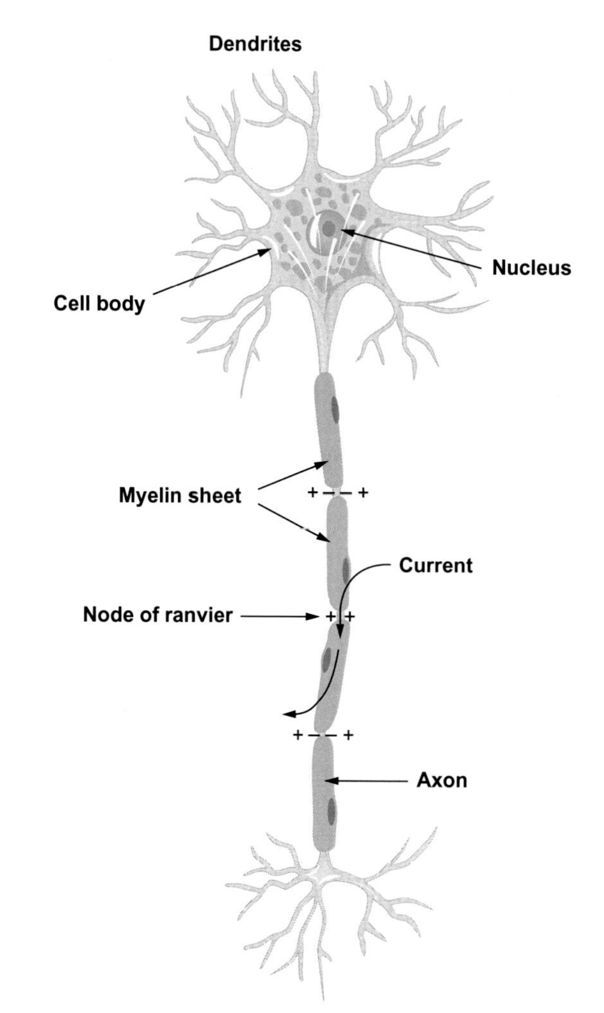

Figure 2–30. **Prototypical neuron.** Incoming local potentials from dendrites converge at the cell body and reach the axon at its origin. The action potential, typically triggered at a region known as the *axonal hillock*, is propagated down the length of the axon, where it will initiate a chain of events leading to neurosecretion from the terminal branches.

Action Potentials Are Propagated at the Axon Hillock

"Nerve impulses" are transmitted along most neurons as local potentials that travel toward the cell body through dendritic processes and propagate away from the cell body along the axon as **action potentials**. The action potential is typically triggered at a membrane segment where the axon emanates from the cell body, a region called the **axon hillock**.

In this region, electrical signals encounter voltage-gated Na^+ channels at sufficient density to trigger the action potential. At appropriate current load a few local voltage-gated Na^+ channels in the axon will open, allowing a small amount of Na^+ to enter the axon down its steep electrochemical gradient. Increased intracellular Na^+ further depolarizes the cell and recruits more voltage-sensitive channels to open, resulting in more Na^+ entry. This progressive self-amplifying mechanism for depolarization reaches a limit when the membrane potential (V_R) has shifted from about −90 mV for mammalian myocytes and neurons to about −50 mV, when the equilibrium potential for Na^+ (E_{Na}) is nearly reached.

Current will diffuse longitudinally along the axon. In this way, what started as a local depolarization reaches the threshold to initiate a new action potential at a neighboring membrane segment.

Two mechanisms next engage to return the same membrane patch to its original resting potential. First, opened voltage-gated Na^+ channels rapidly close in the depolarized membrane because they are conformationally unstable in the open state. Once closed again the channels remain in an inactivatable state until after the membrane is repolarized. Similarly, as described for the acetylcholine receptor (see Fig. 2–26), voltage-gated Na^+ channels can exist in three different conformations: closed, but activatable; open; and closed, but inactive. In this case, not a ligand but voltage drives the transition of recruited channels through each of these states (Fig. 2–31).

The second mechanism contributing to the decay of the action potential is the opening of voltage-sensitive K^+ channels. These channels respond more slowly to depolarization than do voltage-gated Na^+ channels and are not opened until the action potential is near its peak. With a membrane potential of −50 mV, a strong electrochemical gradient for K^+ efflux builds inside the cell and is released when the voltage-sensitive K^+ channels transiently open. As a result the sudden loss of positive charge repolarizes the cell more quickly. The voltage-gated Na^+ channels are closed now, and the voltage-sensitive K^+ channels aid strongly the nongated K^+ leak. The return of the membrane potential toward the K^+ equilibrium potential closes voltage-gated K^+ channels and allows the inactivated Na^+ channels to regain their activatable state. Thus, by accelerating the decay of the action potential and membrane repolarization, voltage-gated K^+ channels reduce the length of a refractory period (<1 ms) before another action potential can be triggered at the same site. Thus the action potential self-propagates as a transient depolarizing wave form racing toward the axonal terminal at speeds in vertebrates ranging from 1 to 100 m/s, depending on the type of axon. The all-or-none nature of the action potential overcomes the problem of keeping a proper signal over relatively long distances. However, just as with your cable television signal, internet 4 or 5G connection, or the electrical power grid, additional properties of the axon determine the efficiency at which these nerve impulses travel.

As the action potential propagates from one patch of axolemma to the next, local current diffuses within the axon to depolarize neighboring segments to the threshold value. The effective distance over which this occurs is determined by the same principles that govern any flow of electricity. In this case, important parameters are the internal resistance (r_i) of the axonal cytosol and the membrane resistance (r_m) to current (K^+) escaping through open channels. Both a decrease in r_i and an increase in r_m will allow larger currents to flow longer distances

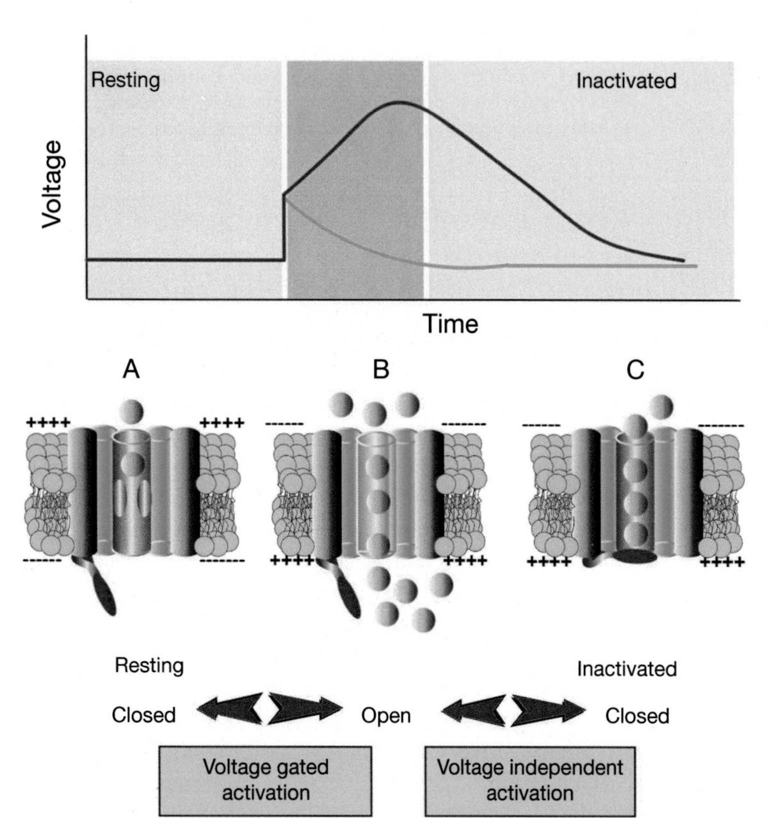

Figure 2–31. **Three conformation states of the voltage-gated Na^+ channel.** (A) In response to a brief pulse of current, depolarization of a mammalian neuron to a threshold of about 50 mV *(red curve)* triggers the opening of the voltage-gated Na^+ channel by opening an activation gate formed by the channel protein lining. (B) Ions flow, and the membrane potential increases *(red curve)*. The open state is metastable and is inactivated rapidly by a separate inactivation-gating mechanism provided by the cytosolic portion of the channel protein. (C) After reaching a maximum of 50 mV, the membrane potential declines to its resting state because Na^+ channels inactivate, and the efflux of K^+ ions through nongated leak channels continues unabated. Another action potential is not possible until the Na^+ channels have returned to the closed but activatable state. In the absence of voltage-gated Na^+ channels, the modest depolarization evoked by the current stimulus would have immediately begun to decay *(green curve)*.

inside the axon, thereby charging the capacitance of neighboring membranes at a faster rate. Consequently a mere doubling of the axonal diameter (reducing r_i) can significantly speed the conduction rate of nerve impulses. Although r_m would actually decrease as a result of doubling the axonal circumference and thereby the number of open ion channels per segment, this is compensated by a fourfold increase in cross-sectional area of the axon, which reduces r_i by a factor of 4. The net effect of enlarging the axonal diameter is to increase the propagation velocity of the action potential. Although such rapid signaling along giant axons evolved by the squid works well, it would not be sufficient in humans, because it would require a spinal cord with the size of a tree trunk. In vertebrates, considerable savings of energy and space are achieved by insulating many axons with a **myelin sheath**.

The formation and maintenance of myelin is the task of two glial cell types: oligodendrocytes in the central nervous system and their counterparts in peripheral nerves, the Schwann cells. These cells generate enormous quantities of flattened plasma membrane and wrap it around axons in a concentric fashion to form a biochemically specialized sheath up to 200 layers in thickness. Every myelinating Schwann cell invests a single axon to form a segment of sheath, termed an **internode**, that occupies approximately 1 mm of axon length. By contrast a single oligodendrocyte extends branches from its cell body that flatten and expand to cover as many as 40 different axons.

The extremely high lipid-to-protein ratio and the stacking of the individual myelin membranes result in a high transmembrane resistance (r_m), greatly improving the electrical characteristics of the axon. Between one internode of myelin and the next are regions of axon, varying in length from 0.5 to 20 mm, termed **nodes of Ranvier** (see Fig. 2–29). Almost all membrane channels, including voltage-gated Na^+ channels, are confined to nodes of Ranvier along a myelinated axon. Current entering the axon through Na^+ channels at nodes of Ranvier spreads with greater efficiency than would otherwise occur in a nonmyelinated axon, because the increased r_m minimizes current leakage.

Moreover, in a nonmyelinated axon, the buildup of opposite charges on either side of the membrane occurs along its entire length. For a myelinated axon, this is confined to the nodes of Ranvier. Having the lower capacitance, myelinated axons require the influx of fewer positive charges to reduce the transmembrane potential to the threshold for an action potential. Hence, in addition to speeding the nerve impulse, myelination drastically reduces the amount of energy needed for axonal conduction. In the unmyelinated axon, Na/K-ATPase is needed to restore the membrane potential along the entirety of the fiber. In the myelinated axon, this occurs only at the depolarized nodes of Ranvier.

Because myelin speeds the conduction of nerve impulses, diseases or toxins that injure the myelin sheath or myelin-producing cells can cause significant neurologic problems. Axons stripped of their myelin sheath conduct impulses slowly or not at all. Conduction deficits resulting from damage to the brain and spinal cord are particularly severe because oligodendrocytes, unlike Schwann cells, fail to regenerate sufficiently in most circumstances. Moreover, damage to a single oligodendrocyte is of greater physiological consequence because it produces internodes of myelin that surround segments of several different axons. Multiple sclerosis is the most prevalent and is the prototype of demyelinating diseases that affect the central nervous system. The symptoms of this disease that are usually manifested clinically are the result of dysfunctional sensory and motor neuron systems that are most dependent on rapid neurotransmission.

SUMMARY

Membranes form the boundaries of cells and cell organelles. They separate inside from outside and allow compartmentalization of DNA, RNA, proteins, and other molecules or ions. Without membranes, life as we know it would not be possible. It has been argued that lipids and the formation of entities that separated compartments provided the earliest structures that made cellular life develop. The prokaryotic plasma membrane often surrounded by a protective cell wall encloses a single cytoplasmic compartment. The eukaryotic cell is surrounded by a plasma membrane that defines its outer boundaries. Inside this cell, we find cell organelles, including the nucleus, mitochondria, lysosomes, peroxisomes, Golgi apparatus, and ER, all of which by themselves are defined by membranes that allow them to perform their specific functions. In a multicellular organism, cells differentiate to develop particular features. Mammalian cells are different depending on the tissue that they form (e.g., the heart, kidney, and liver). Moreover, within a certain tissue, we find cells that are radically different depending on their function.

All cells in a human have descended from the same fertilized egg and are generated with the same genetic information embedded in DNA. The variation in gene expression that determines the characteristics of each cell in each tissue also determines the characteristics of each membrane. Although plasma membranes or membranes of cell organelles differ with their different functions, their basic structure is similar. They all form a barrier between water-containing compartments and are specific with respect to the compounds that they let permeate. Therefore, whether a cell is (or will become) a neuron or kidney cell, the basic structure of its membranes is similar, whereas the actual composition of the lipids and proteins that form the biological membranes will differ drastically depending on the specific function of the membrane. This variation in composition and organization gives the different cell types their typical structure and shape and allows them to communicate with other cells and transport specific ions, proteins, and other compounds across the membrane.

This chapter describes a number of characteristics of cell membranes that form the basis for the function of membranes in all mammalian cells with their dizzying variety of characteristics and function.

Suggested Readings

de Kruijff B. Biomembranes. *Biochim Biophys Acta* 2004;1666 (1–2):1–290.

Embury S, Hebbel R, Mohandas N, Steinberg M, eds. *Sickle Cell Disease, Basic Principles and Clinical Practice*. New York: Raven Press; 1994.

Hoffman R, Benz JE, Shattil S, Furie B, Cohen H, Silberstein L, McGlave P. *Hematology, Basic Principles and Practice*. New York: Elsevier; 2005.

Shah S. *The Fever, How Malaria Has Ruled Humankind for 500,000 Years*. New York: Sarah Crichton Books, Farrar, Straus and Giroux; 2018.

Shapiro H. *Practical Flow Cytometry*. New York: Wiley-Liss; 2005.

Singer SJ, Nicolson GL. The fluid mosaic model of the structure of cell membranes. *Science* 1972;175:720–731.

Tanford C. *The Hydrophobic Effect*. New York: John Wiley & Sons; 1980.

Yeagle P, ed. *The Structure of Biological Membranes*. London: CRC Press; 1992.

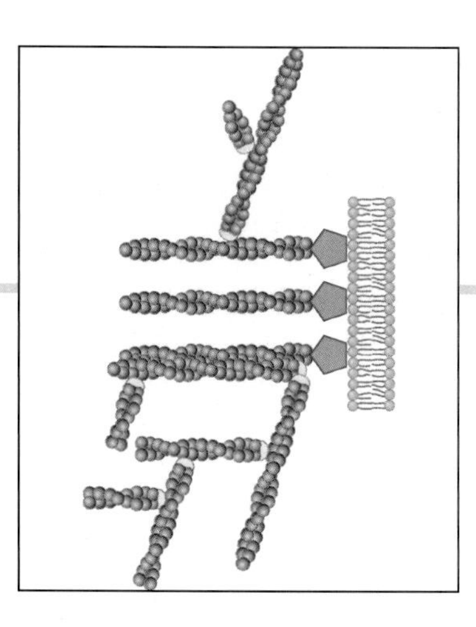

Chapter 3

Cytoskeleton

An intriguing feature of eukaryotic cells is the ability of extracts that contain cytosol, devoid of organelles, to roughly maintain the shape of the cell and even to move or contract, depending on how the extracts are prepared. This maintenance of structure by the cytosol arises from a complex network of protein filaments that traverse the cell cytoplasm, called the **cytoskeleton**. The cytoskeleton is not simply a passive feature of the cell that provides structural integrity; it is a dynamic structure that is responsible for whole-cell movement, changes in cell shape, and contraction of muscle cells—it provides the machinery to move organelles from one place to another in the cytoplasm. In addition, recent studies have provided evidence that the cytoskeleton is the master organizer of the cell's cytoplasm, furnishing binding sites for the specific localization of ribonucleic acids (RNAs) and proteins that were once thought to diffuse freely through the cytoplasm.

Amazingly the many activities of the cytoskeleton depend on just three principal types of protein assemblies: **actin filaments, microtubules**, and **intermediate filaments** (IFs). Each type of filament or microtubule is formed by a specific association of protein monomers. The dynamic aspects of the cytoskeletal structures arise from accessory proteins that control the length of the assemblies; their position within the cell; and the specific-binding sites along the filaments and microtubules for association with protein complexes, organelles, and the cell membrane. Thus, although the protein filaments and microtubules define the cytoskeleton, the participation of accessory or regulatory proteins conveys its diverse activities. This chapter discusses the structures built from the interaction of proteins with the individual cytoskeletal assemblies, beginning with an examination of actin filaments. The initial focus is on their well-defined role in muscle cell contraction; then, their participation in the membrane skeletal complex and structures formed in nonmuscle cells is described. Through a discussion of cell motility, we consider how the different components of the cytoskeleton work together as an integrated network that leads to an essential cellular function. Finally the IFs and microtubule components of the cytoskeleton are discussed.

MICROFILAMENTS

Actin-Based Cytoskeletal Structures Were First Described in Muscle Tissue

Actin, first isolated from skeletal muscle, was originally thought to be a protein found exclusively in muscle tissue. However, actin is a component of all cells, representing 5%–30% of the total protein in nonmuscle cells. Although present in all eukaryotic cells, actin isolated from nonmuscle cells is different from that found in skeletal muscle. Six different isoforms of actin have been described in human and animal cells: α-skeletal, in skeletal muscle; α-cardiac, in heart muscle; α-vascular, in smooth muscle of the vasculature; γ-enteric, in smooth muscle

Goodman's Medical Cell Biology. https://doi.org/10.1016/B978-0-12-817927-7.00003-X

of the viscera; and β-cytoplasmic and γ-cytoplasmic, preponderantly in nonmuscle cells. Actin is an extremely conserved protein, with greater than 80% identity of amino acid sequence between the different isoforms. The major difference in amino acid sequence occurs at the NH_2-terminal end of the actin isoform and appears to have little effect on the rate of actin monomer polymerization into filaments, but it is essential for the association of specific actin-binding and regulatory proteins (see detailed discussion later in this chapter).

Many of the other protein components that are common to actin-based cytoskeletal structures in all cells were also first isolated from muscle tissue. In muscle, these proteins demonstrate a rigorous organization, forming the specialized contractile machinery of the muscle cell. Therefore we examine the role of actin filaments and associated proteins in muscle cells first to lay the groundwork for our understanding of actin-based structures in nonmuscle cells.

Skeletal Muscle Is Formed From Bundles of Muscle Fibers

The organization of the skeletal muscle from the gross level to the molecular level is depicted in Figure 3–1. Skeletal muscle is composed of long, cylindrical, multinucleated cells that can be several centimeters in length. The individual muscle fibers are surrounded by a delicate, loose connective tissue, called the **endomysium**, that carries the capillary network of blood supply for the muscle. Bundles of the individual muscle fibers are grouped together, forming the muscle fasciculi, which are bounded by a layer of connective tissue, the **perimysium**. The fasciculi are grouped to form the definitive muscle tissue that is covered by a thick, tough connective tissue layer, the **epimysium**. The three connective tissue layers of muscle tissue contain fibers of collagen and elastin and differ from each other primarily by their thickness. Skeletal muscle causes specific movements of the body by their attachment to tendons that are usually attached to the skeleton or bone.

The Functional Unit of Skeletal Muscle Is the Sarcomere

Each skeletal muscle cell, or myofiber, contains many bundles of regularly arranged filaments, called *myofibrils*. It is the highly structured arrangement of filaments within the **myofibrils** that give skeletal muscle its characteristic striped or striated appearance. Skeletal muscle is the best biological example of the relation of structure, as viewed through the microscope, with function. Longitudinal sections of skeletal muscle, viewed under the light and electron microscopes, demonstrate an ordered banding pattern (Fig. 3–2). These are called the **A band, I band,** *and* the **Z disk** *or* **Z line.**

The A band is the dark-staining region of the myofilaments, and it contains the thick filaments, composed of the protein myosin II, and overlapping thin filaments. The light-staining I band contains the thin filaments, of which the main protein component is actin. The Z disk appears as a dark line that bisects the I band. In electron micrographs of skeletal muscle, the dark-staining A band is observed to have distinct regions, termed the *H band* and *M line*. The H band is a zone of lighter staining within the central region of the A band, which is bisected by a dark-staining M line. This region of the A band is where the assembly of the myosin thick filaments occurs.

The segment of the myofibril between two Z disks, containing a complete A band and two halves of adjoining I band regions, is called the **sarcomere.** The sarcomere is the functional contractile unit of the myofibril. The myosin thick filaments mark the A band, which is equidistant from the two Z disks of the sarcomere. The thin filaments of the sarcomere are joined to the Z disk and extend through the light-staining I band region and partially into the A band, where they interdigitate with the myosin thick filaments. The Z disk functions to anchor the thin filaments of the sarcomere. Cross sections through different portions of the sarcomere provide additional information about the organization of the thick and thin filaments (see Fig. 3–1). A cross section through the I band shows only thin filaments, arranged in a hexagonal pattern. A section through the H zone of the A band demonstrates only thick filaments, whereas a section through the M band zone of the A band shows a network of coiled filaments, representing the assembly of the bipolar myosin thick filaments. The segment of the A band at which the thin filaments interdigitate with the myosin thick filaments shows that each thick filament is surrounded by six thin filaments. This arrangement of thick and thin filaments is an essential structural feature of the sarcomere and is required for the sliding of filaments during contraction.

Thin Filaments Are Built From the Proteins Actin, Tropomyosin, Troponin, and Tropomodulin

All eukaryotic cells appear to contain filaments 7–8 nm in diameter, called **microfilaments**, that are polymers of the protein actin. These filaments, referred to as **filamentous** or **F-actin**, are built from polymerization of a globular actin monomer, called **G-actin**, which has a relative molecular mass (M_r) of 43,000 (43 kDa). Each F-actin microfilament appears as two helically intertwined chains of G-actin monomers for which a complete turn of the helix occurs over a distance of 37 nm, or 14 G-actin monomers (Fig. 3–3).

Each G-actin monomer must have an adenosine triphosphate (ATP) molecule bound to polymerize onto an actin filament. ATP hydrolysis is slow for actin monomers but greatly accelerates once the monomer

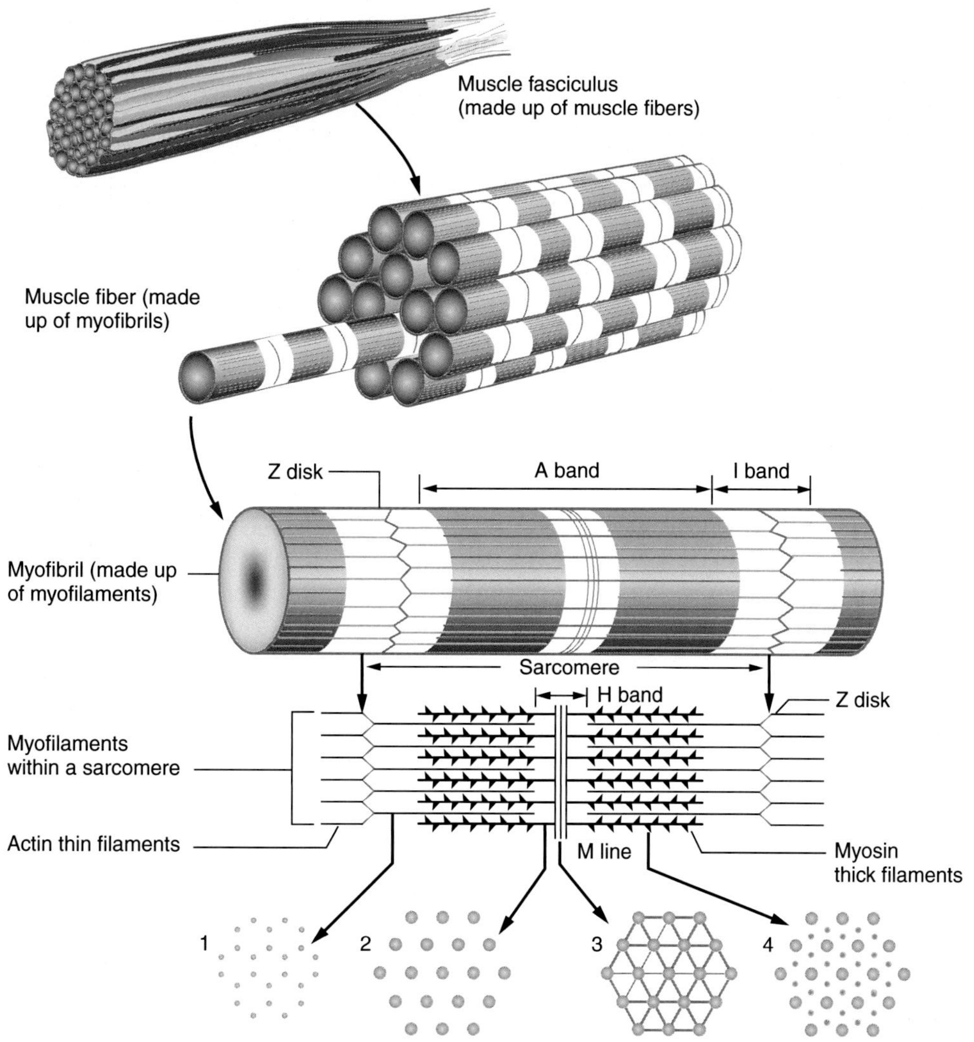

Figure 3–1. Organization of a skeletal muscle. A skeletal muscle consists of bundles of fibers called fasciculi. Each fasciculus consists of a bundle of long, multinucleated muscle fibers that are the cells of the muscle tissue. Within the muscle cells are the myofibrils, which are composed of highly organized arrangements of myosin II (thick) and actin (thin) filaments. The extreme structural organization of the myofilaments is the basis for the striated appearance of skeletal muscle. The myofilaments are organized into the functional units of skeletal muscle, the sarcomere, which extends from one Z disk to the next. Actin thin filaments extend from the Z disk *(light-staining I band)* toward the center of the sarcomere, where they interdigitate with the myosin thick filaments *(dark-staining A band).* Cross sections through the sarcomere near the Z disk (1) show the ~8-nm actin thin filaments, whereas sections in the regions of the A band (4) demonstrate that each –15-nm-thick filament is surrounded by a hexagonal array of six actin thin filaments. Sections through the sarcomere near the center in the segment of the A band referred to as the H band show the organization of the myosin thick filaments (2), whereas cross sections through the center of the H band demonstrate a network of filaments that participate in the assembly of the thick filaments to form the M line (3). *(Modified from Bloom W, Fawcett DW.* A Textbook of Histology, *10th ed. Philadelphia: WB Saunders, 1975.)*

is incorporated into the actin filament. If ATP and Mg^{2+} (or physiologic salt concentrations) were added to G-actin at a high enough concentration, it would spontaneously polymerize to F-actin. The polymerization would have several stages (Fig. 3–4). First would be a lag phase when three G-actin monomers form an actin trimer, which can then serve as a seed or nucleation site for the polymerization of G-actin monomers onto the actin filament during the polymerization phase. And the last would be a steady-state phase is reached, during which the rate of addition of G-actin monomers onto the filament equals the rate at which these monomers leave the filament. Actin microfilaments have a polarity, with a fast-growing plus (+) end and a slow-growing minus (–) end. At each end of the actin filament, there is a critical concentration of G-actin, at which the rate of addition to that end matches the rate of monomer removal from the same end. For the plus end, this concentration of G-actin is approximately 1 μM; it is 8 μM at the minus end. Therefore, at concentrations of G-actin between 1 and 8 μM in the presence of ATP and Mg^{2+}, a treadmill is formed by which actin is being added to the plus end

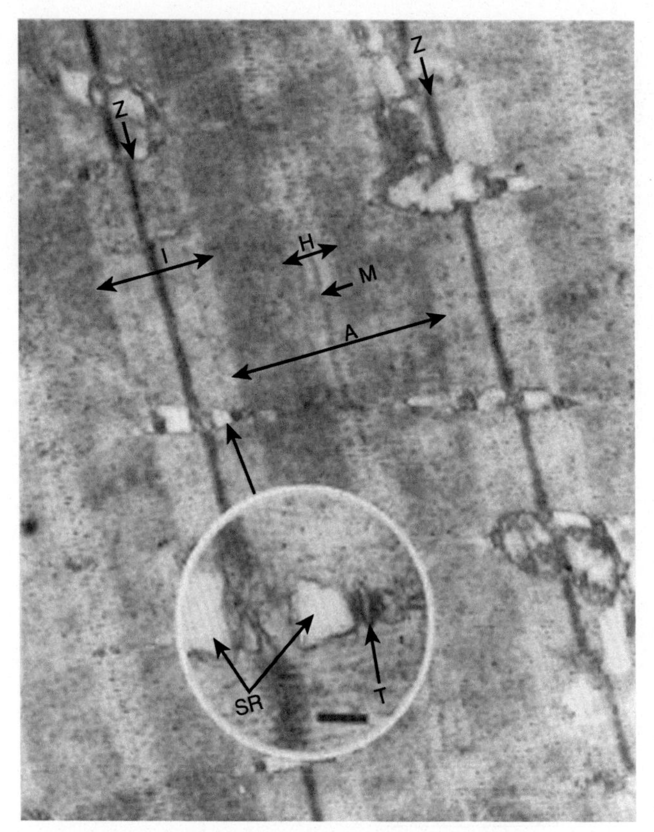

Figure 3–2. Electron microscopy of skeletal muscle. A longitudinal section through a skeletal muscle cell demonstrates the regular pattern of cross-striations derived from the myofibrils. As shown in this low-magnification electron micrograph, the skeletal muscle cell has many myofibrils aligned in parallel. In this repeating structure, one can easily discern the Z disk. Scale bar = 0.3 µm for A, I, and H bands and M line of the sarcomere. (inset) Terminal cisternae of the sarcoplasmic reticulum (SR) and associated transverse tubule (T). *(Courtesy Dr. Phillip Fields.)*

and subtracted from the minus end. If no energy was supplied to this system, this would be a perpetual motion machine, which is thermodynamically impossible. However, shortly after the addition of each G-actin monomer to the actin filament, ATP is cleaved to adenosine diphosphate (ADP), with release of inorganic phosphate (Pi). This raises several interesting questions. First the concentration of actin within the cytoplasm of muscle and nonmuscle cells is greater than 100 µm, suggesting that almost all of the actin within the cells of your body would be filamentous actin (because this is far above the critical concentration for the plus or minus end). However, most cells have a mechanism for maintaining a pool of G-actin monomers. Why is this the case? Second, does treadmilling occur within the living cell? The answer is yes, as we will discuss later in this chapter.

Although F-actin is the preponderant protein of the skeletal muscle thin filament, these filaments also contain other proteins, tropomyosin, troponin, and tropomodulin (Fig. 3–5). Tropomyosin is a long, rod-shaped molecule (~41 nm in length), so called because of its similarities with myosin, specifically the rodlike tail domain of the myosin molecule. Tropomyosin is formed from a dimer of two identical subunits. The individual subunit polypeptides are α-helical, and the two α-helical chains wind around each other in a coiled coil to form the rigid, rod-shaped molecule. Tropomyosin binds along the length of the actin filament, lining the grooves of the helical F-actin molecule, thereby stabilizing and stiffening the filament.

Another major accessory protein of the skeletal muscle thin filament is troponin. Troponin is a complex of three polypeptides: troponins T (TnT), I (TnI), and C (TnC). These polypeptides are named for their apparent

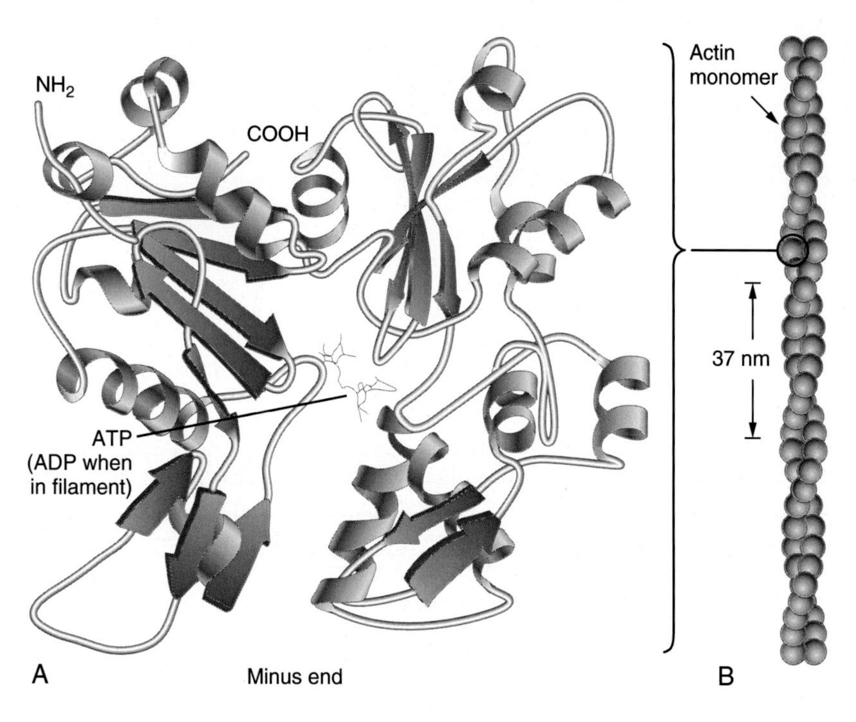

Figure 3–3. Structure of globular (G) and filamentous (F) actin. (A) G-actin is a 43-kDa monomer with four structural domains. ATP and ADP bind to G-actin within the groove separating domains 1 and 3. (B) F-actin is a helical filament composed of polymerized G-actin monomers (spheres). The filament undergoes a complete turn of the helix every 14 G-actin monomers, or 37 nm.

functions within the troponin complex: TnT for its tropomyosin binding, TnI for its inhibitory role in calcium regulation of contraction (see later discussion), and TnC for its calcium-binding activity. The troponin complex is elongated, with the I and C subunits forming a globular head region and the T polypeptide forming a long tail domain. The tail domain, formed from the T subunit, binds with tropomyosin, which is thought to position the complex on the actin thin filament. Because there is only one troponin complex for every seven actin

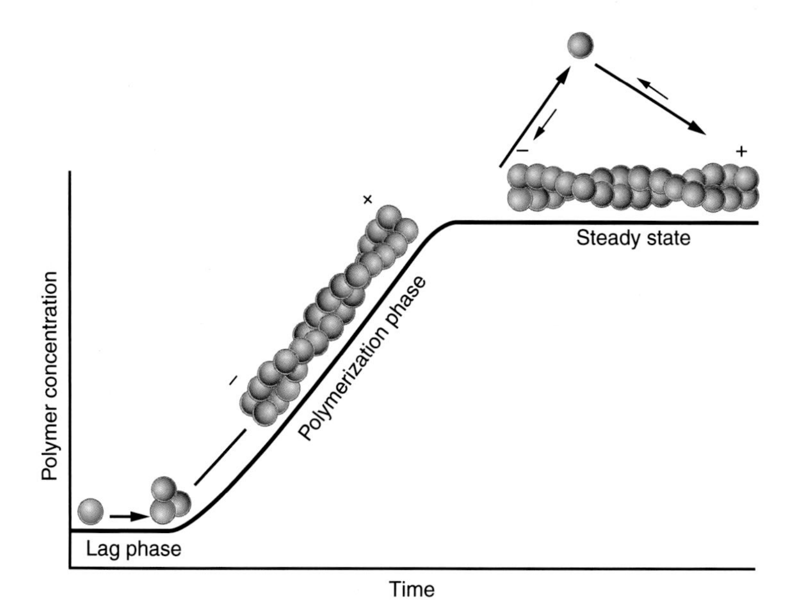

Figure 3–4. Polymerization of actin. The polymerization of actin occurs in three stages: (1) a lag phase in which an actin trimer nucleation site is formed; (2) a polymerization phase, during which G-actin monomers are added preferentially at the plus end of the actin filament; and (3) a steady state, at which actin monomers are being added at the plus end at the same rate they are being removed at the minus end.

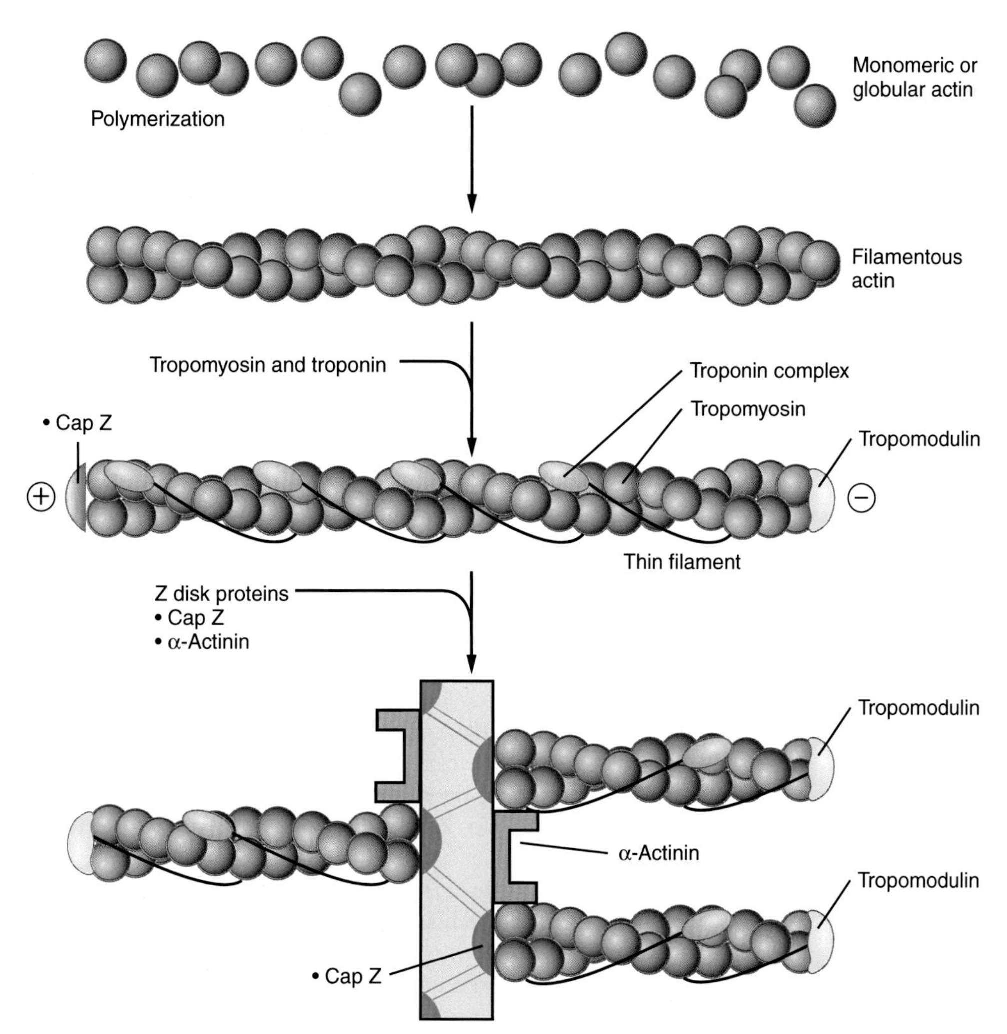

Figure 3–5. Formation of actin thin filaments and their arrangement in the sarcomere. Globular actin monomers polymerize through head-to-tail association to form the helical filamentous (F-actin) form of actin. Thin filaments are built from the specific association of the F-actin filaments with the rodlike tropomyosin molecule, which lines the grooves of the actin filament, and the troponin polypeptide complex. In the sarcomere the thin filaments are anchored at the Z disk through their interactions with binding proteins, principally cap Z and α-actinin. The exact structure of the Z disk is unknown; the protein interactions shown in this diagram are based on the in vitro capabilities of the isolated cap Z and α-actinin proteins. As illustrated the specific protein interactions of the Z-disk proteins immobilize the thin filaments at their plus (+) ends, thereby maintaining the polarity of the actin thin filaments in the sarcomere. The minus (−) ends are capped by tropomodulin.

monomers in an actin filament, the positioning of the complex by the specific interactions of the T subunit with the tropomyosin molecule is critical for its ability to regulate contraction.

Tropomodulin (a 43-kDa globular protein) binds tropomyosin and caps the minus end of the actin filament, thereby regulating the length of the actin thin filaments within the sarcomere.

Thick Filaments Are Composed of the Protein Myosin

Myosin was also first described in muscle cells but is now known to be a ubiquitous component of nonmuscle cells. The major form of myosin found in most cells, including skeletal muscle, is referred to as **myosin II** because it contains two globular heads or motor domains. Myosin II has an M_r of approximately 460 kDa, with two identical heavy chains of M_r of 200 kDa, which form a coiled-coil helical tail and two globular heads (Fig. 3–6).

For a coiled-coil helix to form, the myosin heavy chains must have a heptad amino acid repeat sequence—a, b, c, d, e, f, g, a, b, c, d, e, f, g—with hydrophobic amino acids in positions a and d. Because an α-helix makes a complete turn every 3.5 amino acids, such a repeat would create a hydrophobic stripe that slowly rotates around the helix. To bury this hydrophobic stripe away from the aqueous environment, two such α-helices would wind around each other into a coiled coil. The myosin molecule also contains two pairs of light chains with M_r of 20 and 18 kDa. These light chains are found associated with the myosin heads. If purified myosin is proteolytically cleaved with the enzyme papain, the globular heads (called SF1 *fragments*) can be separated from the myosin tails (see Fig. 3–6). The myosin tails brought to physiologic ionic strength and pH will spontaneously form thick filaments, similar to those found in skeletal muscle. The SF1 heads contain all of the myosin ATPase activity required for muscle contraction. If the purified

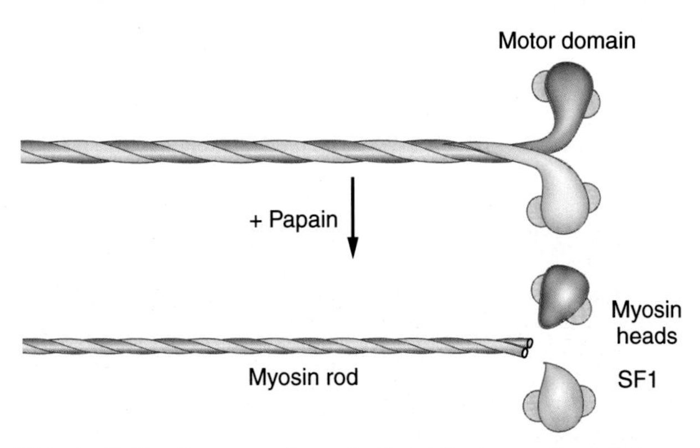

Figure 3–6. Structure of myosin II and its cleavage by papain. Myosin II is a 150-nm-long fibrous protein, with two globular heads. Treatment of myosin II with the proteolytic enzyme papain releases the two myosin heads, or SF1 fragments, from the myosin rod.

heads are added to preformed F-actin and viewed by electron microscopy, the SF1 fragments look like arrowheads that all face in one direction. The pointed end of the arrowheads faces the minus, or slow-growing, end of the filament, and the barbed end faces the plus, or fast-growing, end (Fig. 3–7).

The polarity of actin filament interaction with myosin SF1 fragments (see Fig. 3–7) is important for muscle contraction. Thick filament formation arises from association of the tail or rod segment of the myosin molecule, as demonstrated by the aggregation of isolated tail domains produced by proteolytic cleavage of myosin II. The association of the myosin II heavy chain dimers is due to hydrophobic interactions of the rodlike tail segments, and the formation of filaments depends on interactions between the coiled myosin II tail domains. In muscle the rodlike fibrous tails of 300–400 myosin II dimers pack together to form the bipolar 15-nm-diameter-thick filaments. This association of the myosin II tail segments results in the formation of a filament that has a bare central zone composed of an antiparallel array of myosin II tails (Fig. 3–8).

The globular myosin II head segments protrude from the filament at its terminal regions in a helical array, with a periodicity of 14 nm. Thick filaments then display a high degree of structure. They are symmetric about the bare central zone with the polarity of the filament determined by the arrangements of the globular head segments, which are reversed on either side of this central zone.

Accessory Proteins Are Responsible for Maintenance of Myofibril Architecture

In vertebrate skeletal muscle the structural orientation of the thick and thin filaments is crucial for contraction. Thus the maintenance of this structure is important for muscle function. Several proteins (but probably not all that are necessary) that interact with the thick and thin filaments and play a role in the maintenance of myofibril structure have now been identified.

The thin filaments terminate and are anchored to the Z-disk structures of the sarcomere. This immobilizes the thin filaments with their plus ends at the Z disk and their minus ends extending to the center region of the sarcomere. Therefore a **sarcomere unit** (defined as the distance between two adjacent Z-disk structures) contains actin filaments that extend from each Z disk and exhibit polarity that is opposite on either side of the central region of the sarcomere. Cap Z, a two-subunit protein (M_r 32,000 and 36,000) that binds selectively to the plus ends of actin filaments, is one of the proteins that helps with the anchoring of thin filaments to the Z disk. Because it binds to the fast-growing or plus end of the actin filament, cap Z is thought to prevent growth and depolymerization of F-actin, causing the filaments of the myofibril to be very

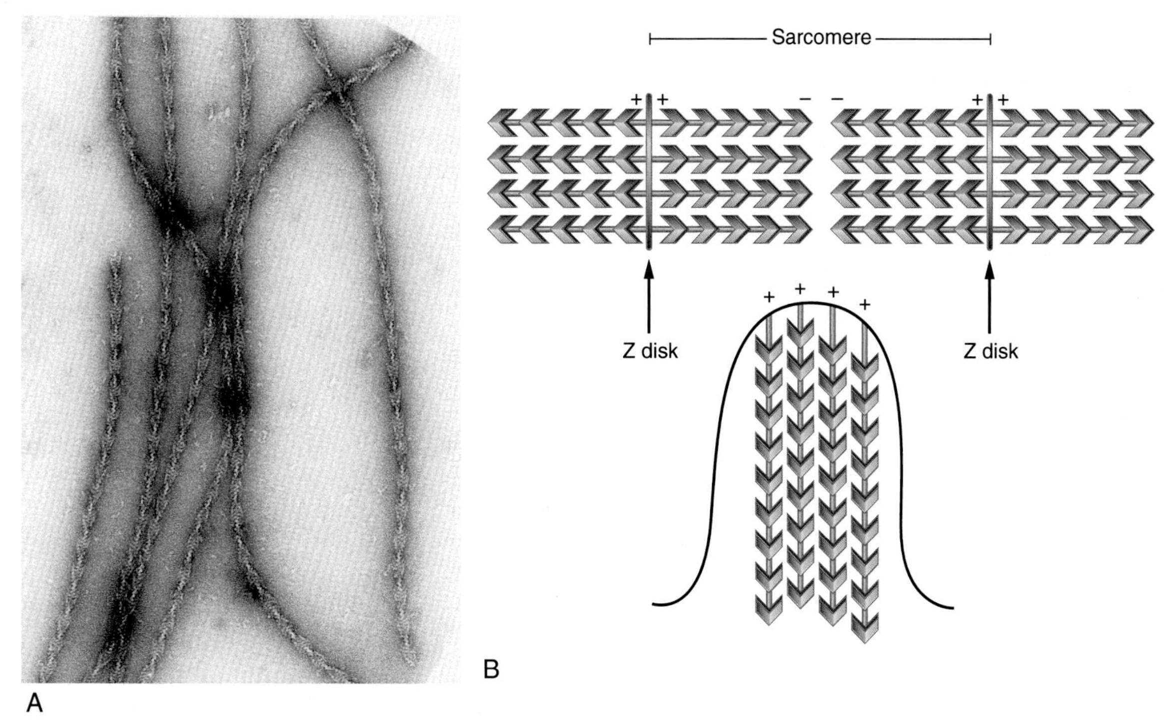

Figure 3–7. Actin filaments have a polarity. The polarity of actin filaments can be visualized by labeling with myosin SF1 fragments. (A) This is an electron micrograph of in vitro formed actin filaments that have bound myosin SF1 fragments. The myosin fragments bind to the actin filaments, demonstrating their polarity. The myosin heads look like arrowheads that all point to the minus (−) ends of the actin filament, the barbed ends facing the plus (+) ends of the filaments. (B) In a sarcomere the barbed or plus (+) ends are attached to the Z disk. When actin filaments are bound to the cytoplasmic surface of the plasma membrane, it is the plus end that is associated with the membrane. The example shown here is the attachment of actin filaments to the tip of the microvillus. *([A] Courtesy Dr. Roger Craig, University of Massachusetts.)*

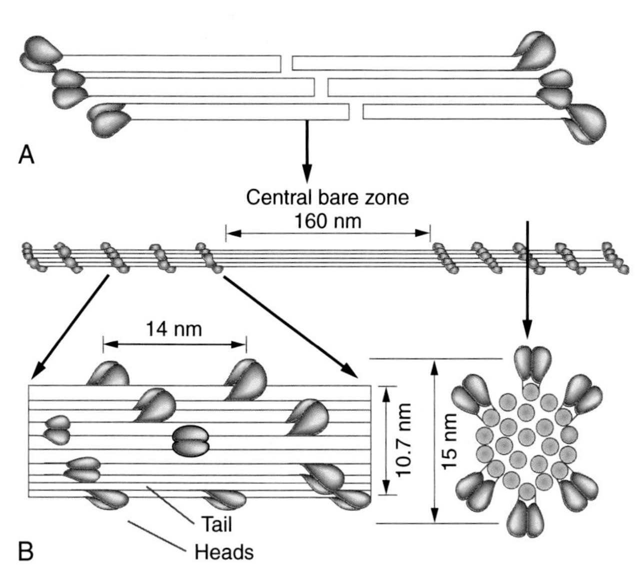

Figure 3–8. Formation of myosin thick filaments. (A) Thick filament formation is initiated by the end-to-end association of the rodlike tail domains of myosin II molecules. (B) This results in the formation of the bipolar thick filament, with globular heads at either end separated by a 160-nm central bare zone consisting of myosin II tail domains. At the filament ends the myosin globular head domains protrude from a 10.7-nm-diameter central core at intervals of 14 nm. The successive myosin heads rotate around the fiber, which forms a filament containing six rows of myosin head domains to contact the adjacent thin filaments of the sarcomere.

stable structures. Its localization at the Z disk suggests that cap Z may assist in the immobilization of the thin filaments (see Fig. 3–5), perhaps by interactions with other proteins of the Z disk. The major component of the Z disk is the protein α-actinin, a fibrous protein composed of two identical subunits (M_r 190,000). The NH_2-terminal domain of α-actinin bears a strong resemblance to NH_2-terminal domains of other cytoskeletal proteins (principally members of the spectrin supergene family) that function to bind and cross-link actin filaments. It is the NH_2-terminal domain of α-actinin that provides the ability of this protein to bind tightly to the sides of actin filaments, allowing the bundling together of adjacent thin filaments at the Z disk. Although the exact structure of the Z disk is unknown, evidence now indicates that it contains two sets of overlapping actin filaments of opposite polarity that originate in the two sarcomeres adjacent to the Z disk, and the thin filaments are anchored to the disk structure by interactions with proteins such as cap Z and α-actinin. As mentioned earlier, tropomodulin, a 43-kDa protein, binds to and caps the minus slow-growing ends of actin filaments, regulating their length.

In skeletal muscle, there are mechanisms that maintain the relative position of the myofilaments and regulate the length of the polymerized filaments. Two proteins, titin and nebulin, appear to be important for these functions (Fig. 3–9). Titin, a large fibrous protein, appears to

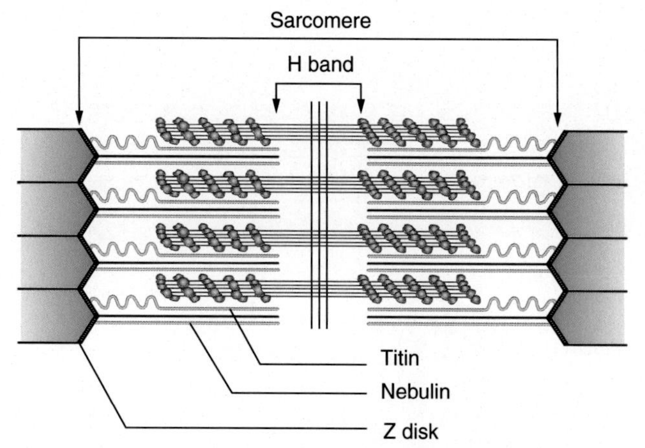

Figure 3–9. Titin and nebulin: accessory proteins of the skeletal muscle sarcomere. The location of the proteins titin and nebulin within the sarcomere is shown. Titin, a large protein that has elastic properties and links the myosin thick filaments to the Z disks, helps maintain their location in the sarcomere. Nebulin, a large filamentous protein anchored at the Z disk, is in close apposition to the actin thin filaments. Their close association with the thin filaments suggests that the nebulin fibers serve to organize the actin filaments of the sarcomere.

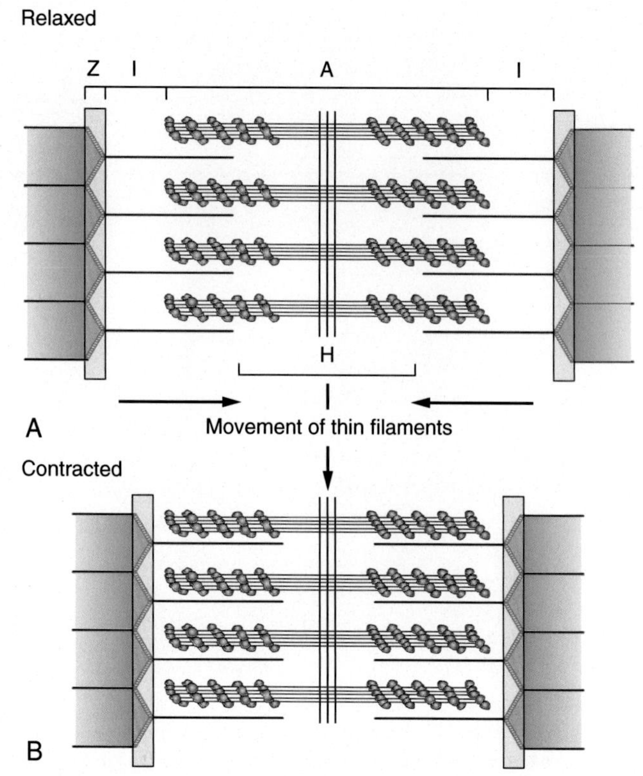

Figure 3–10. Sliding filament model of muscle contraction. Muscle contraction occurs by the sliding of the myofilaments relative to each other in the sarcomere. (A) In relaxed muscle the thin filaments do not completely overlap the myosin thick filaments, and a prominent I band exists. (B) With contraction, movement of the thin filaments toward the center of the sarcomere occurs, and because the thin filaments are anchored to the Z disks, their movement causes shortening of the sarcomere. The sliding of thin filaments is facilitated by contacts with the globular head domains of the bipolar myosin thick filaments.

connect the thick filaments to the Z disk. Titin is the largest protein described to date (~3.5 million daltons), and it contains a long series of immunoglobulin-like domains. It functions to keep the myosin thick filaments centered within the sarcomere structure. It may act as an elastic band to keep filaments in an appropriate orientation, also inhibiting the structural deterioration of the sarcomere during muscle contraction. Another large fibrous protein, nebulin, forms a long, inextensible filament that extends from the Z disk to the minus end of the thin filaments. Nebulin contains a 35-amino acid repeating actin-binding motif. Because of their exacting length and their repeating association with the actin filaments, nebulin fibers may regulate the number of actin monomers that polymerize into thin filaments and aid in the formation of the regular geometric pattern of thin filaments during muscle formation.

Muscle Contraction Involves the Sliding of the Thick and Thin Filaments Relative to Each Other in the Sarcomere

Measurements of sarcomere and A and I band lengths from electron micrographs of contracted and resting muscle firmly established the mechanism of muscle contraction: the sliding of actin thin and myosin thick filaments passed each other within the sarcomere unit. These measurements demonstrated that the lengths of the individual filaments do not change as a muscle contracts; yet the distance between two adjacent Z disks becomes shortened in contracted muscle relative to relaxed muscle. When the length of a sarcomere decreases in contracted muscle, the I band region shortens, whereas the length of the A band remains unchanged (Fig. 3–10).

Because the lengths of the thick and thin filaments do not change, the change in length of the I band could occur only if the thin filaments were to slide past the thick filaments. Therefore the reversed polarity of thick and thin filaments relative to the center line of the sarcomere (defined by the M line) would cause a shortening of the sarcomere during contraction by the sliding of thin actin filaments, which are attached to the Z disk, past the thick myosin filaments toward the center of the sarcomere. This model of muscle contraction, called the **sliding filament model**, was first proposed in 1954 and led to the dissection of molecular mechanisms of contraction.

Adenosine Triphosphate Hydrolysis Is Necessary for Cross-Bridge Interactions With Thin Filaments

Skeletal muscle contraction requires the interactions of myosin II head groups with the thin filaments. These interactions are governed by binding and hydrolysis of the high-energy molecule ATP by the ATPase activity resident in the globular myosin II head domain.

The ATP-driven interactions between myosin II and actin are illustrated in Figure 3–11.

When a myosin II head binds a molecule of ATP, it causes a weakening of the myosin-actin interaction. A dissociation of the myosin II head group binding to the thin filament occurs (step 1). The cleavage of ATP to ADP and Pi creates an "activated" myosin II head that has undergone a change in structure, facilitated by the flexible hinge regions of the molecule, such that the myosin II head is perpendicular with an adjacent actin thin filament (step 2). The conversion between these two stages is reversible, because the ADP and Pi remain bound to the myosin II head, and the energy released from ATP hydrolysis is stored in the strained bonds resulting from the rotation of the myosin II head group. The activated myosin II molecule then encounters a neighboring actin subunit, and this binding triggers the release of Pi, which, in turn, strengthens the myosin-actin interaction (step 3). This strong binding causes a conformational change in the myosin II head, generating a "power stroke," which pulls the actin filament relative to the fixed myosin II filament, resulting in contraction (step 4). The product of this step is the so-called rigor complex, in which the actin-myosin linkage is inflexible, and the thick and thin filaments cannot move past each other. If no ATP is available to the muscle (e.g., after death), the muscle will remain rigid, owing to the tight myosin-actin interactions. This condition is referred to as **rigor mortis**. Under normal circumstances a molecule of ATP will displace the bound ADP, causing release of the actin filament from the myosin head group, effectively relaxing the muscle and returning to step 1 of the cycle. The hydrolysis of the newly bound ATP then prepares the muscle for further rounds of myosin-actin interactions.

Each cycle of myosin-actin interaction would result in movement of an actin thin filament by about 10 nm. A coordination of multiple myosin II head group interactions to provide a concerted movement of filaments and a mechanism by which these interactions are regulated must exist to achieve the rapid rates of contraction for intact muscle fibers. Each thick filament is formed from the aggregation of multiple myosin II rod domains, which results in a bipolar filament, with each side of the filament containing approximately 300–400 head groups protruding in a spiral fashion. This arrangement provides multiple contacts of a thick filament (called **cross-bridge interactions**) with a thin filament. Along the length of the thin filament, there will be myosin II cross bridges at various points in the myosin-actin cycle (see Fig. 3–11), such that the collective actin cross-bridged contacts ensure the smooth and rapid movement of the thin filament relative to the thick filament. Each myosin head group cycles about five times per second, sliding the myosin thick filaments and actin thin filaments past each other at a rate of $\sim$15 μm/s. Sarcomeres can shorten by about 10% in length in about 20 ms.

To effectively coordinate the sliding of filaments from entire groups of myofibrils, leading to muscle contraction capable of producing mechanical work, a transient increase in calcium regulates the interactions of myosin

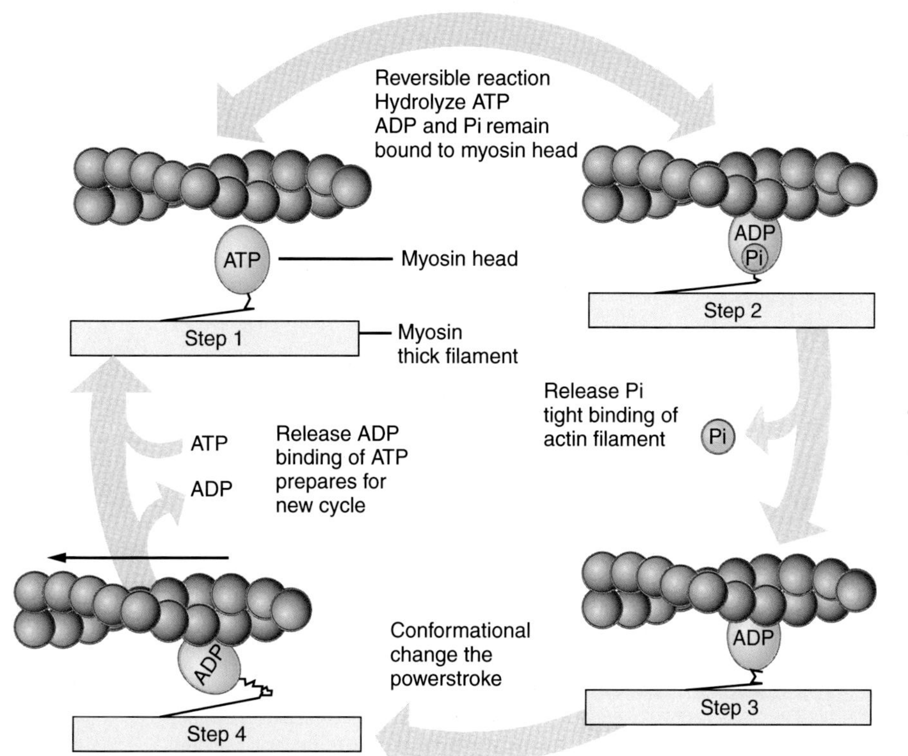

Figure 3–11. Illustration of the ATP-driven myosin-actin interactions during contraction. The binding of ATP to a myosin head group causes release from the actin filament (step 1). The hydrolysis of ATP to ADP+Pi readies the myosin head to contact an actin filament (step 2). The initial contact of the myosin with an actin filament causes the release of Pi and a tight binding of the actin filament (step 3). This tight binding induces a change in conformation of the myosin head, such that it pulls against the actin filament, the power stroke (step 4). This change in conformation is accompanied with the release of ADP. The binding of an additional ATP causes a release of the actin filament and a return of the myosin head to a position ready for another cycle.

II head group cross bridges at the cellular level. The calcium-based regulation of muscle contraction occurs by overcoming a block of myosin-actin interactions by the troponin-tropomyosin complexes on the thin filament. The specific interactions involved in this regulation are discussed in the following section.

Calcium Regulation of Skeletal Muscle Contraction Is Mediated by Troponin and Tropomyosin

When myosin is mixed with filaments made from purified actin, the myosin ATPase activity is stimulated to its maximal activity, independent of calcium addition to the reaction. If thin filaments, which contain actin, tropomyosin, and troponin, are added to purified myosin, the stimulation of the myosin ATPase activity is wholly dependent on the presence of calcium. The basis of calcium-dependent hydrolysis of ATP in this reaction is a reversal of the inhibition of the actin-myosin interaction caused by the position of tropomyosin and troponin on the thin filaments (Fig. 3–12).

Each rodlike tropomyosin molecule contacts seven actin monomers and lines the grooves of the F-actin helix. Bound to a specific site of each tropomyosin molecule is the troponin complex that comprises three polypeptides: TnT, TnI, and TnC. The elongated TnT molecule (M_r 37,000) binds the COOH-terminal region of tropomyosin and links both TnI and TnC to the tropomyosin. TnI (M_r 22,000) binds TnT, as well as actin, and in concert with tropomyosin causes a change in the conformation of F-actin such that it interacts only weakly with myosin head groups. This weak interaction cannot activate the myosin ATPase activity. Together with TnI the TnC ($M_r = 20{,}000$) subunit forms a globular domain of the troponin complex. TnC, the calcium-binding subunit, has a structure and function like that of the intracellular calcium receptor protein calmodulin. The binding of calcium ions at all four of the calcium-binding domains of TnC releases the TnI tropomyosin inhibition of actin activation of the myosin ATPase, thereby allowing contraction of the myofibril. The binding of calcium by TnC results in a shift or movement of the tropomyosin toward the center of the actin helix, which exposes a region of the actin monomer, allowing the binding of myosin head groups in such a way that activation of the myosin ATPase activity occurs. The hydrolysis of ATP permits cycling of cross-bridge interactions and the sliding of filaments. The myosin-activating sites of F-actin are blocked by the troponin-tropomyosin complex in the resting, but not in the active, state of the myofibril. Thus the contraction of skeletal muscle is regulated by the concentration of intracellular calcium ions.

Intracellular Calcium in Skeletal Muscle Is Regulated by a Specialized Membrane Compartment, the Sarcoplasmic Reticulum

In the resting or relaxed state, the concentration of calcium ions in skeletal muscle cells is low. Thus, to have contraction and relaxation cycles of muscle, a mechanism must exist by which the internal calcium ion concentration is regulated. Moreover the concerted contraction of a muscle to produce work depends on the

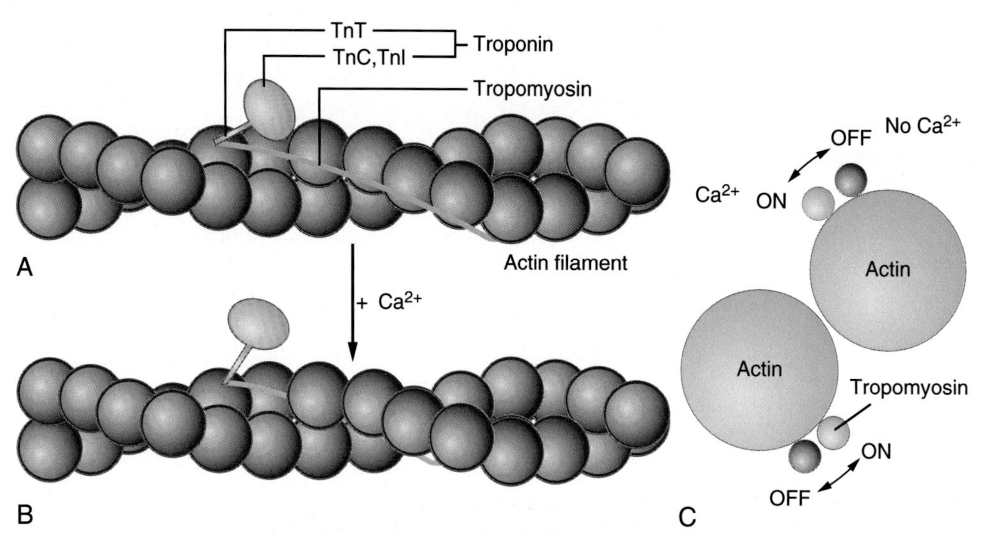

Figure 3–12. Diagram of Ca^{2+}-mediated movements of troponin and tropomyosin filaments during muscle contraction. (A) In the relaxed muscle the tropomyosin filament is bound to the outer domains of seven actin monomers along the actin filament. The troponin complex is bound to the tropomyosin by the rod-shaped troponin T (TnT) polypeptide. (B) In the presence of Ca^{2+}, troponin C (TnC) binds the calcium, causing the globular domain of troponin [TnC and troponin I (TnI)] to move away from the tropomyosin filament. (C) This movement permits the tropomyosin to shift to a position that is farther inside the groove of the helical actin filament, allowing the myosin heads to make contact with the released sites of the actin monomers.

simultaneous contraction of all of its constituent myofibers and their myofibrils. Therefore the rapid changes in calcium ion concentration that are needed along the entire length of the myofibril for contraction must be maintained by mechanisms other than simple diffusion, which would be too slow for simultaneous contraction of myofibrils in skeletal muscle cells. To deliver calcium in a uniform fashion throughout the muscle cell, there is a special membrane-bound tubule system, derived from the endoplasmic reticulum (ER) in these cells.

Electron microscopy of skeletal muscle shows a network of smooth membranes, called the *sarcoplasmic reticulum* (SR), surrounding the myofibrils. The SR forms a network of membrane-limited tubules and cisternae that surround the outer regions of the A band of each myofibril (Fig. 3–13). In addition the SR forms a more regular structure, called the **terminal sac** or **terminal cisternae,** which is a membrane-limited channel that surrounds the A-I junction of each individual myofibril. The terminal cisternae are near a specialized channel formed from delicate invaginations of the sarcolemma (plasma membrane of the muscle cell), called the **transverse tubules** (t-tubules). The t-tubules, in association with terminal cisternae from adjacent myofibrils, form a triad structure (see Fig. 3–13). These structures are

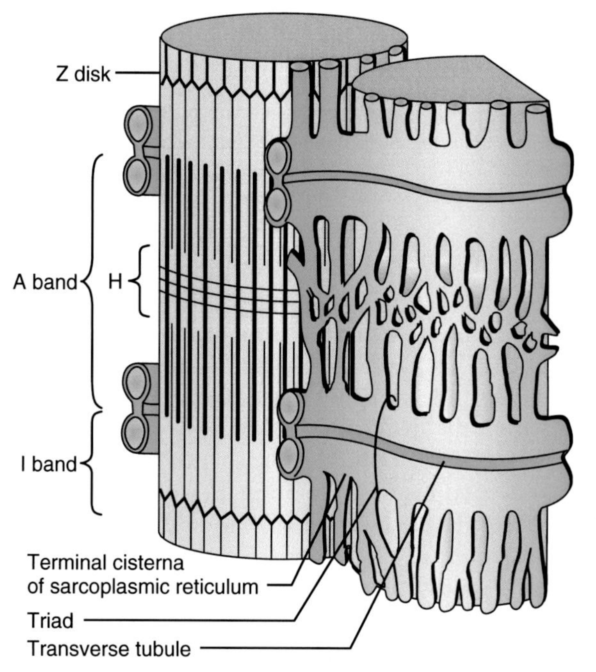

Figure 3–13. Diagram of part of a skeletal muscle fiber, illustrating the organization of the sarcoplasmic reticulum (SR) and transverse tubule (t-tubule) networks. The SR is a specialized smooth endoplasmic reticulum (ER) that in muscle serves as a store for Ca^{2+} ions. The SR forms a membranous tubule network that surrounds the myofibrils. At the A-I band junctions of the sarcomere, the SR forms a more regular channel, referred to as the terminal cisternae. Two terminal cisternae are separated by a second tubule system, the t-tubules, which are special invaginations of the sarcolemma. These three membrane-bound tubules form a structure known as the triad: a t-tubule flanked on either side by a terminal cisternae of the SR, at the region of the A-I junction of the sarcomere. *(Modified from Cormack DH.* Ham's Histology, *9th ed. Philadelphia: JB Lippincott, 1987.)*

important for the coupling of external stimuli (e.g., signals from motor neurons) to muscle contraction.

The SR forms a membranous compartment that occupies 1%–5% of the total muscle volume and serves as a reservoir of calcium ions sequestered away from the myoplasm and myofibrils. For its role in maintenance of calcium ion concentration, the SR membrane contains numerous proteins for the transport of calcium, including a Ca^{2+}-ATPase protein that pumps calcium from the cytosol into the lumen of the SR. For each 1 mol of ATP hydrolyzed by the ATPase activity of the calcium pump, 2 mol of calcium are sequestered into the lumen of the SR. This active transport mechanism is responsible for the maintenance of the low calcium ion concentration in resting muscle. The stored calcium is released from the SR into the sarcoplasm as the action potential spreads along the sarcolemma. The action potential, stimulating Ca^{2+} release, travels through the t-tubule system. A voltage-sensitive protein sensor located in the t-tubule membrane, termed the **dihydropyridine-sensitive receptor** (DHSR), feels the action potential and translates its presence to the SR through direct interaction with an SR calcium channel, the **ryanodine receptor.** These proteins, the DHSR and ryanodine receptor, are analogous to proteins found in other cells whose function is to release calcium from internal stores, the so-called IP_3 (inositol 1,4,5-triphosphate) receptor pathway (described in Chapter 9). The large complex formed by these proteins in muscle when viewed in the electron microscope is often referred to as "feet" on the SR (Fig. 3–14). The net effect is the release of a pulse of calcium into the sarcoplasm by the transit of an action potential. The released calcium stimulates contraction through binding the troponin complex on the thin filaments. After contraction the calcium is actively transported into the lumen of the SR by Ca^{2+}-ATPase, returning the muscle to the relaxed state (see Fig. 3–14).

Within the lumen of the SR are proteins that function to bind and store the internalized calcium ions (see Fig. 3–14). The best characterized example is calsequestrin. Although the binding affinity of Ca^{2+} by calsequestrin is low, each molecule of the protein binds 40–43 Ca^{2+} ions. Thus calsequestrin, together with other proteins that have similar properties, effectively reduced the SR luminal concentration of calcium from 20 to 30 mM (if all the Ca^{2+} ions were free in solution), to about 0.5 mM. The result of binding Ca^{2+} ions in the SR lumen is to greatly reduce the concentration gradient against which the membrane Ca^{2+} pump must act.

Three Types of Muscle Tissue Exist

In the preceding section, we focused on the contractile apparatus found in skeletal muscle. Two other major types of muscle are present in the vertebrates. Cardiac muscle forms the walls of the heart and is also found in walls of the major vessels adjacent to the heart. Smooth muscle is found in the hollow viscera of the body (e.g., the

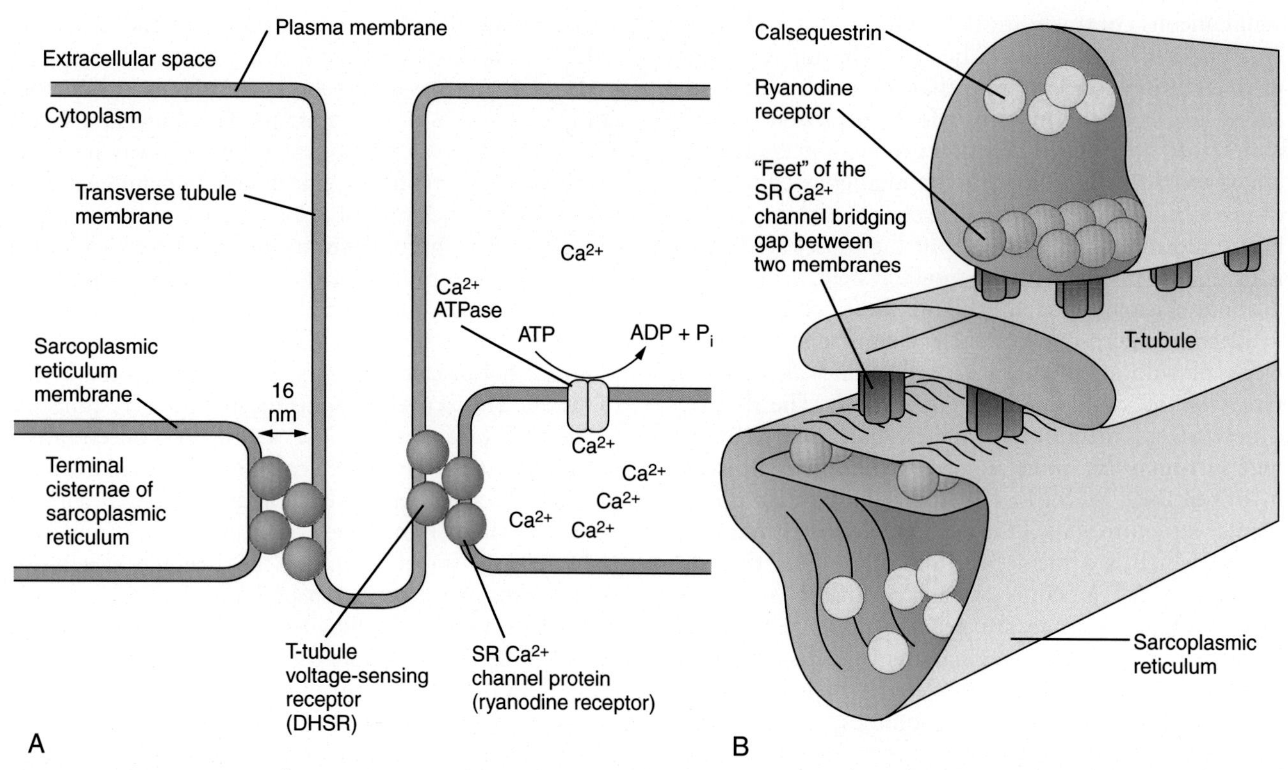

Figure 3–14. Model of Ca^{2+} ion regulation by the sarcoplasmic reticulum (SR) in muscle. (A) An illustration of the association of the transverse tubules (t-tubules) and the terminal cisternae of the SR. The SR Ca^{2+} channel is shown to make direct contact with the voltage-sensing Ca^{2+} channel of the t-tubule. When depolarized the t-tubule voltage-sensing protein (DHSR) undergoes a change in conformation and, because of its close association with the SR Ca^{2+} channel, causes the SR channel to open and release calcium to the cytoplasm. This Ca^{2+} release occurs with essentially no delay because of the direct interaction of the DHSR and the Ca^{2+} channel of the SR, the ryanodine receptor. The Ca^{2+} ions in the cytoplasm are returned to the lumen of the SR by the Ca^{2+}-ATPase pump in the SR membrane. (B) View of the t-tubule and SR terminal cisternae associations. The t-tubules and SR terminal cisternae are in close proximity; "feet" of the SR channel protein are shown bridging the gap between the t-tubule and SR membranes. Inside the lumen of the SR is the protein calsequestrin that weakly binds the internalized Ca^{2+} ions, reducing the effective internal concentration of free Ca^{2+} ions. *([A] Modified from Agnew WS.* Nature *1988;344:299–303. [B] Modified from Eisenberg BR, Eisenberg RS.* Gen Physiol *1982;79:1–17.)*

intestines) and in most blood vessels. All three types of muscle use actin-myosin structures for contraction by a sliding filament mechanism. However, some fundamental differences exist in the structural organization of the contractile apparatus and the regulation of contraction in the different muscle cells.

Myocardial Tissue: Striated Muscle Built From Individual Cells

Cardiac tissue consists of long fibers that, like skeletal muscle, exhibit cross-striations under the light microscope. The striated appearance of cardiac muscle derives from the highly organized arrangement of actin and myosin filaments of the contractile apparatus. Although cardiac muscle is similar in appearance to striated skeletal muscle, two main histologic criteria distinguish these two muscle types.

The first criterion is the positioning of the nuclei within the cells. In skeletal muscle, nuclei are located at the periphery of the cell, just under the sarcolemma, whereas in cardiac muscle the nuclei are found at the central regions of the cell. Thus cardiac cells have a bare or cleared zone surrounding the nucleus, the perinuclear space, which arises from the myofilaments arranging themselves such that they detour around the nuclear compartment.

The second major criterion that distinguishes cardiac from skeletal muscle is the appearance of dark-staining disk structures in cardiac muscle, the **intercalated disk.** These are specialized junctional complexes that separate one cardiac muscle cell from another. Thus cardiac muscle fibers are built from an arrangement of single cells, unlike skeletal muscle fibers that are built from the fusion of individual cells into a multinucleated fiber. Although at the light microscopic level the intercalated disk appears as straight lines, demarcating one cell from another, the view of these structures in the electron microscope reveals that they take an irregular steplike path, such that part of this cell-cell junction is horizontal and part is longitudinal. Therefore the individual cardiac muscle cells interdigitate with each other, forming the myocardial muscle fibers (Fig. 3–15). This arrangement of cell-cell

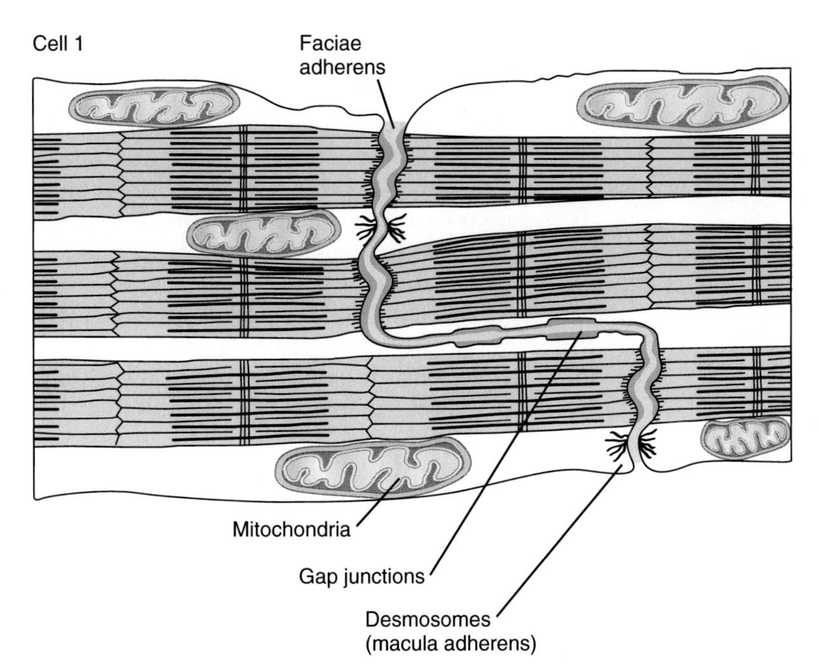

Figure 3–15. Diagram of an intercalated disk between two cardiac muscle cells. The intercalated disk is a steplike structure that allows the interdigitation of cardiac muscle cells. In the transverse sections of this structure are the desmosomes, which hold the cells together, and the junctional complexes, the fascia adherens, which function as Z-disk structures to anchor actin thin filaments from adjacent cells. In the longitudinal sections of the intercalated disk are the gap junctions. These junctional complexes allow communication between the cells such that adjacent cardiac cells are coupled electrically.

contact allows myocardial muscle to contain straight fibers and fibers that branch to effectively construct a hollow organ capable of pumping blood.

Different regions of the intercalated disk contain specific junctional complexes (see Fig. 3–15). In the transverse (or vertical) sections of the intercalated disk, two junctional complexes exist. The first is the **desmosomes**, which are sometimes referred to as the **macula adherens.** These junctions function as the "spot welds" that hold the adjacent cardiac cells together. In this region of the plasma membrane is a second type of junctional complex, termed the **fascia adherens,** which functions to connect the thin filaments of adjacent cells and to hold them in register with the myosin thick filaments (see later). In the longitudinal regions of the intercalated disk are junctional contacts called **gap junctions** (sometimes referred to as nexus). These contacts allow the cardiac cells to exchange small cytoplasmic solutes. The gap junction contacts permit electrical coupling of the cardiac muscle cells, such that synchronization of contraction exists among these cells.

Some of the muscle cells within the heart, the Purkinje fibers, are specialized to carry electrical impulses. These cells are grouped into bundles that form two branches, one to each ventricle. Histologically, these cells are larger and more irregular in shape than are the surrounding cardiac muscle cells. The Purkinje cells contain large glycogen deposits and have smaller bundles of myofibrils at their periphery. These special conducting fibers are responsible for the final distribution of electrical stimulus to the myocardium.

The Contractile Apparatus of Cardiac Muscle Is Similar to That of Skeletal Muscle

Cardiac muscle owes its striated appearance to the arrangement of thick and thin filaments that make up the contractile apparatus. Electron micrographs of cardiac muscle reveal a banding pattern of myofibrils similar to that observed for skeletal muscle. Like skeletal muscle, these bands are referred to as the A band, I band, and Z disk. The dark-staining A band is the region of the myofilament that contains the thick filaments composed of myosin and overlapping thin filaments. The I band contains the thin actin filaments and is bisected by the Z disk, to which the actin filaments are anchored. One notable difference in the structure of myofilaments in cardiac cells, compared with skeletal muscle cells, is the termination of some actin thin filaments at the region of the intercalated disk (Fig. 3–16).

The fascia adherens complex in the transverse segment of the intercalated disk functions to anchor actin thin filaments at the cell periphery. Although the molecular details of how these junctional complex binds and arranges actin thin filaments are unknown, these complexes function as a Z disk, in that they maintain the exacting arrangement of six actin filaments surrounding each myosin thick filament.

Cardiac muscle thick filaments are made from a cardiac isoform of myosin II, which has a subunit structure similar to that found in skeletal muscle. Cardiac myosin II has two heavy chains of approximately 200,000 M_r that assemble by association of a rodlike tail domain and fold into a globular head domain at their NH_2 terminus. There are four light chains, two pairs of M_r 18,000–20,000, with one polypeptide from each set bound with each head segment of the molecule. Associated with the globular head domain of cardiac myosin II is an actin-activated ATPase activity that functions in cross-bridge formation and contraction. However, the isozymes of myosin expressed in cardiac muscle have a lower ATPase activity than those in skeletal muscle. Familial hypertrophic cardiomyopathy is caused by defects in the cardiac β

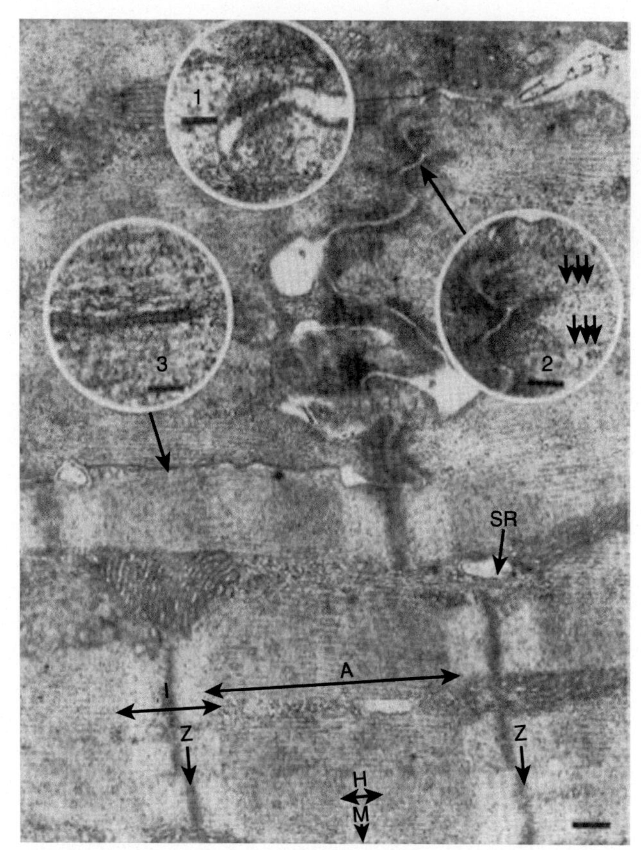

Figure 3–16. Electron micrograph of cardiac muscle. This electron micrograph of cardiac muscle cells shows the regular arrays of the myofibrils into the sarcomeres. In this arrangement, one can easily see the Z disk; the A, I, and H bands; and the M line of the sarcomere structure. The inserts show a higher magnification of the junctional compartments of the intercalated disk: (1) the macula adherens or desmosomes, (2) the fascia adherens, and (3) gap junctions. Notice the thin filaments that terminate in the fascia adherens complex. Scale bar = 0.2 μm; (insets) 0.05 μm. *(Courtesy Dr. Phillip Fields.)*

myosin (at or near the head or motor domain) or in myosin light chains, troponin, or tropomyosin. This disease affects approximately 2 of every 1000 people, with the clinical outcome being enlargement of the heart and cardiac arrhythmias.

The thin filaments of cardiac muscle are built from actin, tropomyosin, and troponin. Although these proteins form the same complex as that found in skeletal muscle, they are different from the polypeptides found in their skeletal muscle counterpart; that is, they are cardiac-specific isoforms. Cardiac muscle thin filaments exhibit the same stoichiometry and structure as those discussed for skeletal muscle. Thus, in cardiac muscle, there is an arrangement of six thin filaments surrounding each thick filament, and contraction or cross-bridge formation in cardiac muscle is regulated by Ca^{2+} by the thin filament-based troponin-tropomyosin complex. Defects in cardiac α-actin cause familial dilated cardiomyopathy. Patients with dilated cardiomyopathy demonstrate a defective left and/or right systolic pump function leading to cardiac enlargement and hypertrophy, which leads to early heart failure.

The Smooth-Muscle Cell Does Not Contain Sarcomeres

Smooth muscles are made of individual cells that can vary notably in size, from 20 μm in length in the walls of the small blood vessels to 200–300 μm in length in the intestine. The smooth-muscle cell is characterized by its fusiform shape. The cells are thickest at their midregion and taper at each end. Smooth muscles are built from sheets of cells that are linked together by various junctional contacts that serve as sites of cell-cell communication (e.g., gap junctions) and mechanical linkages. Cells of the smooth muscle are active in the synthesis and deposition of connective tissue matrix, which serves to embed the cells and acts in limiting the distension of the hollow viscera.

Smooth-muscle cells do not contain a highly ordered array of thick and thin filaments; thus they do not appear striated. In electron micrographs of smooth muscle, numerous dense-staining regions, known as **dense bodies**, are found throughout the cytoplasm of the cell. The major protein component of the dense body is the actin-binding protein, α-actinin, which indicates that they serve as the functional equivalent of a skeletal muscle Z disk. Indeed, actin thin filaments are found anchored to the dense bodies. Two proteins, desmin and vimentin, belonging to the IF class of proteins (discussed later), are expressed at high levels in smooth-muscle cells. The filaments formed from these proteins are prominent in these cells and appear to serve as links between the dense bodies and the cytoskeletal network of the cell. These links aid in contraction by maintenance of the dense body positioning (Fig. 3–17), allowing movement of the cell by an inward pulling of the plasma membrane.

The Contractile Apparatus of Smooth Muscle Contains Actin and Myosin

Actin and myosin can be isolated from smooth-muscle cells, and in vitro, these proteins demonstrate a sliding filament mechanism for contraction. However, the regulation of contraction in smooth muscle follows a path very different from that observed for striated muscle. The thin filaments of smooth and striated muscle have similar structures, except that the calcium regulatory protein troponin is not present in smooth muscle. The cellular content of actin and tropomyosin is greater in smooth muscle than in striated muscle (by about twofold). This, in combination with a reduced quantity of myosin in smooth muscle compared with striated muscle, produces a greater ratio of thin-to-thick filaments in smooth muscle (−12 thin per 1 thick) than that observed in the striated muscles (−6 thin per 1 thick).

Smooth muscle contains numerous thin filaments that approximately align along the long axis of the cell. These thin filaments are embedded into the cytoplasmic densities (dense bodies) and exhibit the same polarity relative to their attachment points. The thin filaments have their plus (+) ends at the dense body and their minus (−) ends

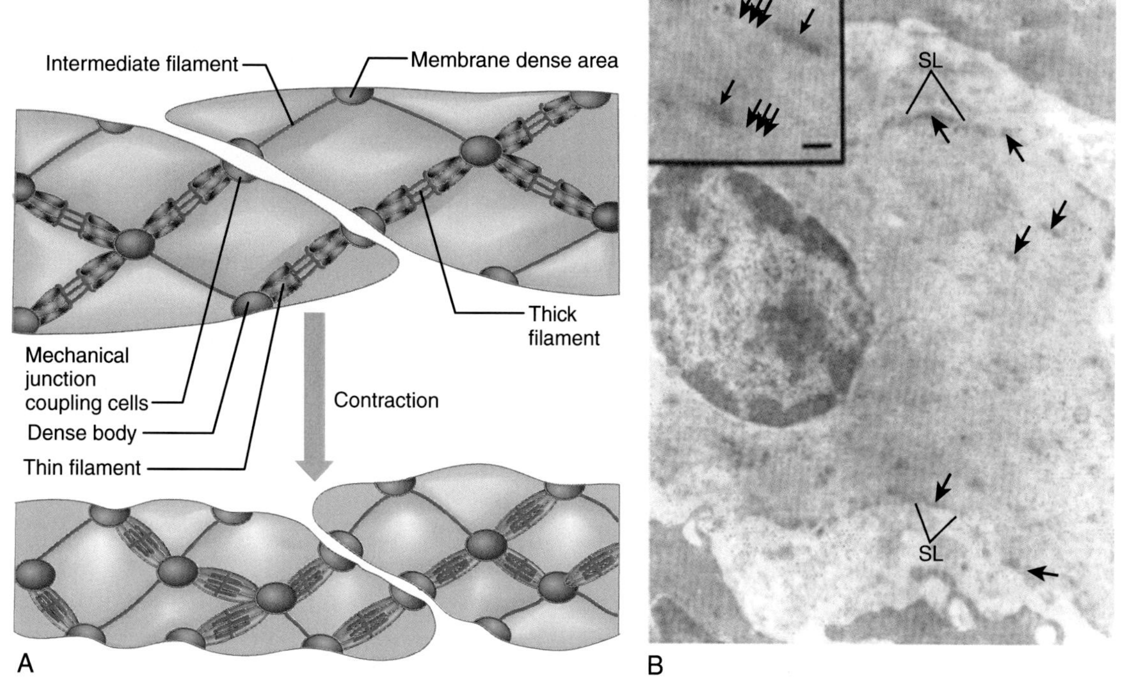

Figure 3–17. Organization of cytoskeletal and myofilament elements in smooth muscle. (A) Smooth-muscle cells contain small contractile elements that are not organized as in striated muscle. Numerous actin thin filaments are anchored into dense bodies within the smooth-muscle cytoplasm, which is the functional equivalent of the striated muscle Z disk. Intermediate filaments of desmin and vimentin form linkages between the dense bodies and the cytoskeleton of the cell. These links are important for contraction, which pulls the plasma membrane inward and changes the shape of the cell. (B) In this electron micrograph of a smooth-muscle cell, the dense bodies are seen throughout the cell and near the sarcolemma (SL). At higher magnification (insert), myofilaments *(small arrows)* are observed as they emanate from the dense bodies *(large arrows)*. Scale bar = 0.29 pm. *([B] Courtesy Dr. Phillip Fields.)*

extending into the cellular cytoplasm. Thus, although the filaments in smooth muscle are not as highly organized as those found in striated muscle, the polarity of actin thin filaments in smooth muscle is such that contraction by myosin cross-bridge cycling would cause a pulling of dense bodies toward one another. This is critical for smooth-muscle contraction, in that it would cause an inward pulling of the plasma membrane, creating force generation by essentially reshaping the cell. A change in shape of several coupled cells would generate the force of smooth-muscle contraction.

Myosin isolated from smooth muscle has properties different from those of striated muscle. Like skeletal muscle, smooth-muscle myosin consists of two heavy chains and four light chains. Two polypeptides of myosin light chains are associated with each globular head domain of the smooth-muscle myosin. However, smooth-muscle myosin will form filaments under only certain conditions. When myosin isolated from smooth-muscle cells is dephosphorylated, it remains fully soluble. Analysis of soluble myosin by sedimentation assays and electron microscopy shows that the dephosphorylated myosin folds up into a compact unit, with the tail domain reaching toward the globular head domain. In this configuration the isolated myosin resists the formation of thick filaments, and its actin-activated ATPase activity is essentially blocked. On phosphorylation of the 18-kDa light chain of smooth-muscle myosin by the enzyme myosin light chain kinase (MLCK), the tail segment is released from the head segment (Fig. 3–18). The resulting released myosin tails can form bipolar thick filaments. Moreover, the freeing of the tail domain to form bipolar filaments allows the activation of the head domain ATPase (permitting cross-bridge formation).

Smooth-Muscle Contraction Occurs via Myosin-Based Calcium Ion Regulatory Mechanisms

Smooth-muscle cells lack troponin, the Ca^{2+} regulatory protein found in the thin filaments of striated muscle, yet micromolar increases in intracellular Ca^{2+} concentrations are required for smooth-muscle contraction to occur. The Ca^{2+} regulation of smooth-muscle contraction occurs by changes in the phosphorylation state of the myosin molecule. The regulation of smooth-muscle contraction is said to be myosin based. When stimulated the Ca^{2+} concentration in the smooth-muscle cytoplasm increases, and the released Ca^{2+} first encounters the Ca^{2+}-binding protein, calmodulin. Calmodulin is present in all cells, and it is referred to as a **modulator protein.** The calmodulin molecule lacks enzymatic activity but exerts its effects by binding Ca^{2+}, and the Ca^{2+}-calmodulin complex is then able to bind with other

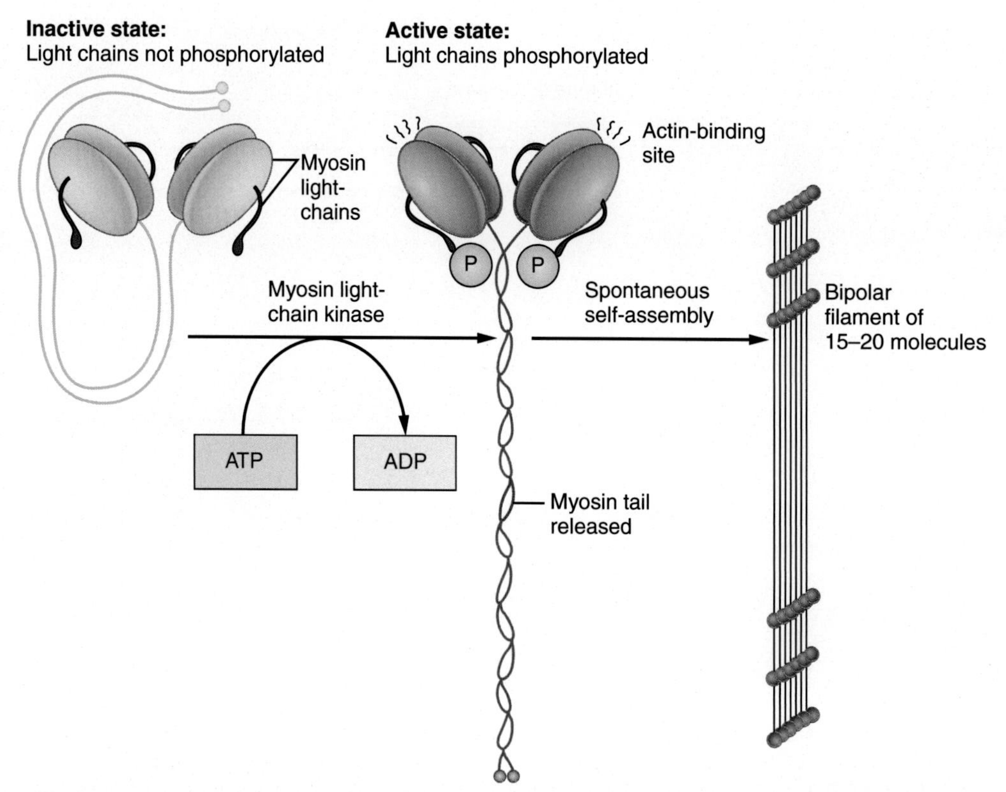

Figure 3–18. Model for assembly of smooth-muscle myosin thick filaments. Dephosphorylated myosin isolated from smooth-muscle cells is in an inactive state and does not readily form thick filaments because of the conformation of the tail domain that binds with the globular head domain. The phosphorylation of the 18-kDa light chain of myosin has two effects: it causes a change in the conformation of the myosin head, exposing its actin-binding site, and it releases the myosin tail from its inactive conformation, allowing the myosin molecules to assemble into bipolar thick filaments. *(Modified from Alberts B, Bray D, Lewis J, et al.* Molecular Biology of the Cell, *2nd ed. New York: Garland Publishing, 1989.)*

proteins and modulate their activity. One such calmodulin-regulated protein is smooth-muscle myosin light chain kinase (SmMLCK). Without Ca^{2+}-calmodulin the SmMLCK is in an inactive state. After the binding of Ca^{2+}-calmodulin, SmMLCK is active and phosphorylates the 18-kDa regulatory light chain of smooth-muscle myosin II (Fig. 3–19).

This phosphorylation permits myosin II to aggregate into thick filaments and allows cross-bridge formation of the thick filaments with the thin filaments of the smooth muscle. Thus the phosphorylation of myosin light chains is an obligatory event for cross-bridge formation and cycling in smooth muscle. During relaxation, Ca^{2+} ion concentration decreases, and a net dephosphorylation occurs. The reduction in intracellular Ca^{2+} concentration causes an inactivation of the SmMLCK (by a reversal of Ca^{2+}-calmodulin binding). The regulatory activity of the requisite phosphatase enzyme(s) that dephosphorylates the myosin light chain is not well defined.

Smooth-Muscle Contraction Is Influenced at Multiple Levels

Because it can be stimulated by a variety of sources—neuronal and hormonal inputs—smooth-muscle contraction can be regulated by several mechanisms. These include regulation by cyclic adenosine monophosphate (cAMP), diacylglycerol, and the protein caldesmon (see Fig. 3–19). Each of these pathways effects a negative regulation; they serve to maintain a relaxed state of the smooth muscle. For example, activation of β-adrenergic receptors on smooth-muscle cells causes an increase in intracellular cAMP levels, which, in turn, activates cAMP-dependent protein kinase. One of the targets for cAMP-dependent protein kinase in smooth muscle is SmMLCK, and phosphorylation of the MLCK results in a lower affinity of the kinase for the Ca^{2+}-calmodulin complex. As a result the SmMLCK does not phosphorylate myosin, and the myosin (and smooth muscle) remains in its relaxed state. Other hormones relax smooth muscle by activation of protein kinase C, which is mediated by Ca^{2+} and 1,2-diacylglycerol. The activation of protein kinase C allows it to phosphorylate SmMLCK, causing it to remain in an inactive state.

In addition to hormonal regulation of contraction, smooth-muscle cells contain Ca^{2+}-binding proteins that interact with the actin thin filaments, thereby affecting contraction. Caldesmon is an elongated calmodulin-binding protein. In the absence of Ca^{2+}, caldesmon will bind to the actin filaments of smooth muscle, restricting the ability of actin and myosin to interact. In the presence of increased Ca^{2+} concentrations, the Ca^{2+}-calmodulin

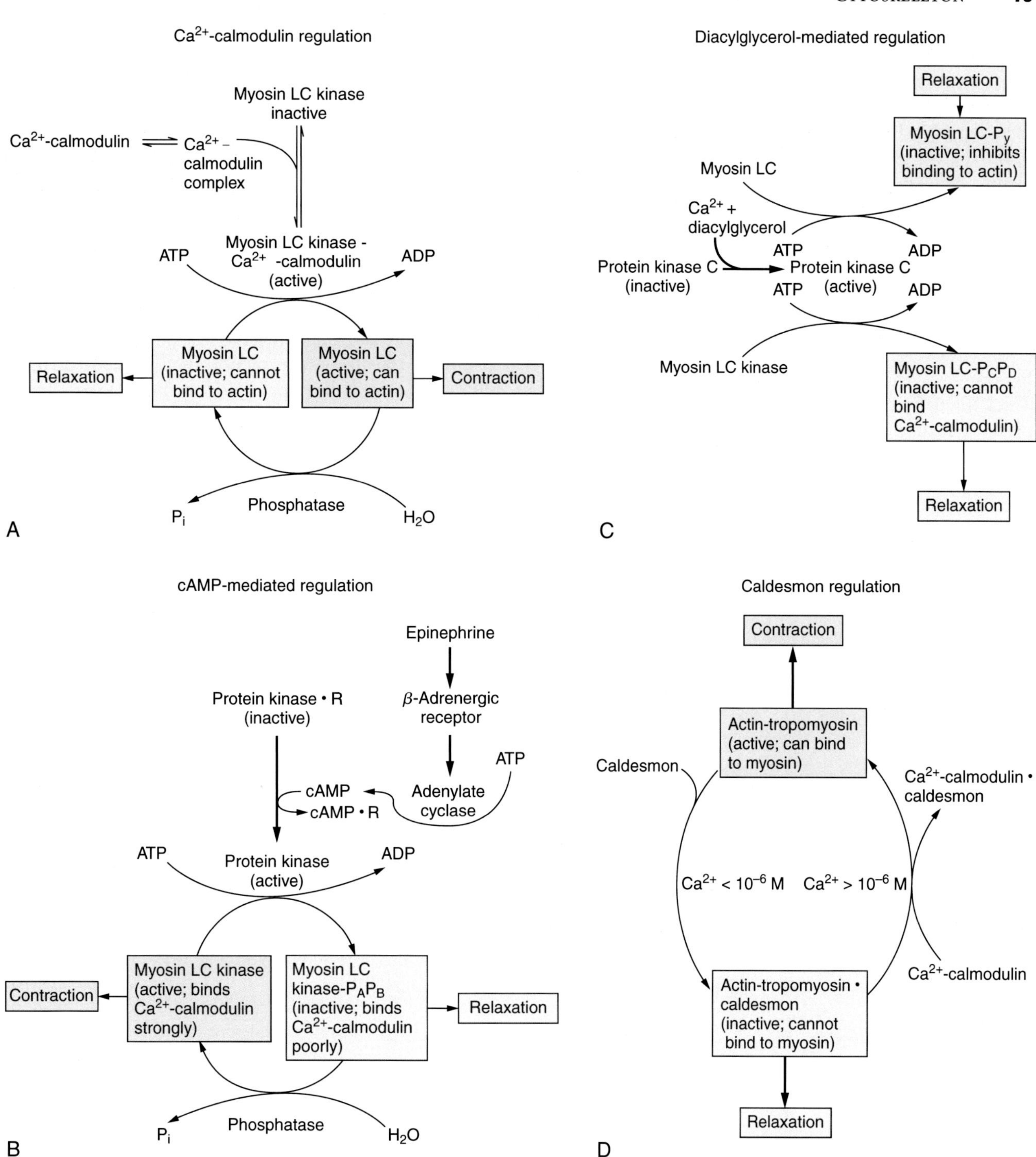

Figure 3–19. Mechanisms that regulate smooth-muscle contraction and relaxation. (A) Regulation by Ca^{2+}-calmodulin: as intracellular Ca^{2+} increases, excess Ca^{2+} is bound by calmodulin, and the Ca^{2+}-calmodulin complex binds to and activates myosin light chain (LC) kinase. The activated kinase phosphorylates the regulatory LC of myosin at site X, which leads to contraction. As the intracellular Ca^{2+} concentration declines to less than 0.1 μM, there is a dissociation of the Ca^{2+}-calmodulin complex from myosin LC kinase, rendering it inactive. Under these conditions the myosin LC phosphatase, which is not dependent on Ca^{2+} for activity, dephosphorylates myosin, causing relaxation. (B) Regulation by cyclic adenosine monophosphate (cAMP): stimulation of β-adrenergic receptors by catecholamines, such as epinephrine, causes the stimulation of adenylate cyclase and an increase in intracellular cAMP concentrations. This stimulates the cAMP-dependent protein kinase that phosphorylates myosin LC kinase at sites A and B near the calmodulin-binding domain of the molecule. This causes the myosin LC kinase to have a lower affinity for calmodulin, rendering it inactive, such that it does not phosphorylate the regulatory light chain of myosin, causing relaxation. Dephosphorylation of the myosin LC kinase restores its ability to bind Ca^{2+}-calmodulin for contraction. (C) Diacylglycerol-mediated regulation: diacylglycerol and Ca^{2+} stimulate the activity of protein kinase C, which phosphorylates myosin LC kinase at sites different from those of the cAMP-dependent kinase (sites C and D). In addition, protein kinase C phosphorylates the regulatory LC of myosin at a position different from the myosin LC kinase. Both of these events render the proteins inactive and cause relaxation. (D) Caldesmon regulation: at low concentrations of Ca^{2+} (>1 μM), caldesmon binds to tropomyosin and actin, inhibiting the binding of myosin, thereby keeping the muscle in a relaxed state. When the intracellular Ca^{2+} concentration increases, the Ca^{2+} is bound by calmodulin, and the Ca^{2+}-calmodulin complex binds with caldesmon, releasing it from the actin filament and allowing contraction. *(Modified from Adelstein RS, Eisenberg, E.* Annu Rev Biochem *1980;49:92–125; Rasmussen H, Takuwa Y, Park S.* FASEB J *1983;1:177–185.)*

complex binds to caldesmon, causing a release of the protein from the thin filaments. Thus the Ca^{2+}-calmodulin complex modulates contraction in smooth muscle by affecting myosin head group phosphorylation, in addition to releasing the caldesmon block on actin thin filaments. This dual control by Ca^{2+}-calmodulin allows the cell to regulate the duration and frequency of contractions.

Actin-Myosin Contractile Structures Are Found in Nonmuscle Cells

In nonmuscle cells the actin/myosin ratio is about 100:1. Thick filaments and microfilaments form within the cytoplasm, but they are in equilibrium with pools of nonpolymerized myosin and G-actin. Although the nonmuscle thick filaments are shorter than those of skeletal muscle and the myosin and actin filaments do not form the highly structured array found in skeletal muscle, they are still responsible for contraction in nonmuscle cells. Figure 3–20 gives two examples.

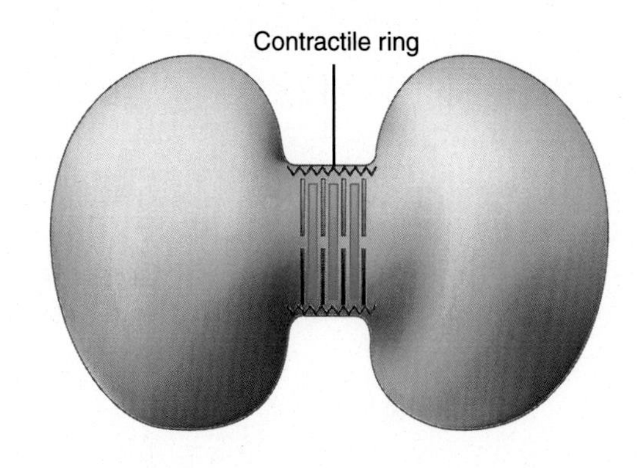

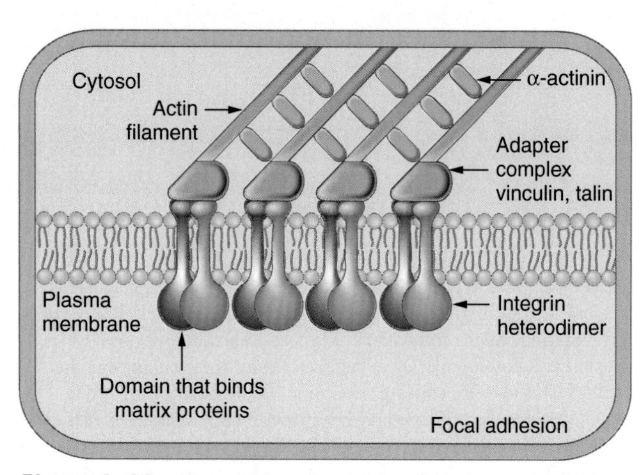

Figure 3–20. Nonmuscle actin and myosin have contractile functions. Two examples of a contractile function for nonmuscle actin and myosin are demonstrated. An assembly of actin and myosin creates a contractile ring (top) that draws in the center of a cell, leading to cell division. A simplified presentation of stress fibers (bottom) that interact with the plasma membrane at focal contacts and because of the contractile activity of actomyosin causes flattening of substrate-attached fibroblasts.

During telophase, the last stage in mitosis, a contractile ring forms on the cytoplasmic membrane surface at the cleavage furrow. This contractile ring contracts (like a belt pulled tightly around the waist), forming a cleft between two cells that are separating. Just before telophase, actin filaments begin to form at the site that will become the cleavage furrow. In addition, free myosin begins to polymerize at the same site and, together with actin and the actin-binding protein α-actinin, forms the contractile ring. Because the actin filaments that are attached to the plasma membrane have mixed polarity, the short myosin filaments can use the energy of ATP hydrolysis to cause a contraction that pulls the dividing cell into a dumbbell shape. Before cell division the actin and myosin filaments rapidly depolymerize.

A second example of nonmuscle contraction is pulling on the plasma membrane, created by stress fibers formed within fibroblasts. Fibroblasts are cells that synthesize and are in contact with extracellular matrix proteins throughout much of the connective tissue that surrounds the organs of your body. The plasma membrane of the fibroblast contacts extracellular matrix proteins, both within a tissue culture dish and within the body's connective tissue, at sites called **focal contacts** or **adhesion plaques.** At these contact sites an integral transmembrane protein of the integrin family binds an extracellular matrix protein, such as fibronectin, at the outer cell surface; this interaction pulls outward on the plasma membrane. The integrins are heterodimers that contain various isoforms of α and β subunits, and the distinct combination determines the specificity of binding various extracellular matrix proteins. The fibroblast is not pulled apart because the same integrin binds actin bundles called **stress fibers** at the cytoplasmic membrane surface. The stress fibers contain interdigitated actin bundles of mixed polarity that are linked together in a parallel array. The stress fibers bind the integrin through adapter proteins, called **talin** and **vinculin**, and a plus end-capping protein. Again, it is the plus end of the actin bundle that binds end-on to the plasma membrane at the focal contact and to plus end-capping proteins on that side of the fibroblast that is not attached to the substrate. The stress fibers also contain short myosin filaments, and they exert a contractile force on the actin bundles, which results in an inward pulling on the plasma membrane that counteracts the outward pull of the extracellular matrix and, in culture, leads to flattening of the fibroblast. The stress fibers rapidly assemble in response to fibroblast attachment to a substrate, and they rapidly depolymerize when the cells are detached. The depolymerization of actin bundles causes the cells to round up. A third example is the actin and myosin filaments that associate with the adhesion belt, characteristically located below the tight junction of an epithelial cell. Adjacent cells in an epithelial cell layer are held 15–20 nm apart by a Ca^{2+}-dependent transmembrane protein called *uvomorulin* or *E-cadherin.* Uvomorulin also binds, through the actin-binding proteins, α-actinin and vinculin, to the sides of actin filament bundles that form an adhesion belt around the cytoplasmic membrane surface. Myosin filaments and this circumferential F-actin contract thereby mediating an important process in human development: the folding of epithelial cells into tubes. In the neural plate, this contraction causes an apical narrowing, which leads to the plate rolling up to form the neural tube during human development.

Members of the Myosin Supergene Family Are Responsible for Movement of Vesicles and Other Cargo Along Actin Tracks in the Cytoplasm

We now know that a large family of myosins exists. The human genome contains 40 myosin genes. What all of these myosins have in common is a conserved motor domain, but they vary in their tail domains, which allows for the diversity of their functions. It will be instructive to look at four myosins (I, II, V, and VI) (Fig. 3–21).

After the discovery of muscle myosin (myosin II), which is involved in contraction, the next myosin described was myosin I. This myosin I was so named because it has only one motor head group and also has a short tail. Myosin I has the ability to walk along an actin filament, with the energy source being ATP hydrolysis, toward the plus end. In so doing, it can carry vesicular cargo through the cytoplasm, helping to establish intracellular organization. Many more myosins were then discovered and named for the order of their discovery (myosin III to XVIII to date). Myosin V has two heads and is involved in vesicular and organelle transport. Myosin VI is the only member of the myosin family that moves toward the minus end of actin filaments, because of a break in the conserved sequence of its motor domain. The diversity of directional movement and tail domains allows for different cargo to be moved throughout the cell cytoplasm on actin tracks.

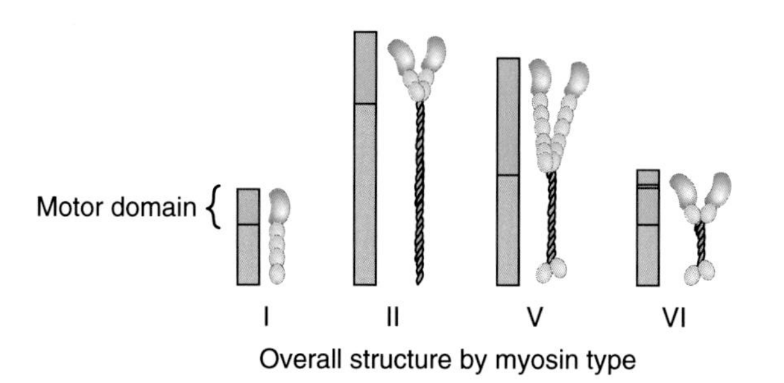

Figure 3–21. Four members of the myosin family. We demonstrate the domain structure of myosin I, II, V, and VI. All have a common N-terminal motor domain shown in *blue*, and a variable C-terminal domain shown in *red* that allow for their different functions. All of the myosins move toward the plus ends of actin filaments except for myosin VI, which moves toward the minus end. This is due to the small break (amino acid sequence change) in the structure of the motor domain (shown in *white*).

Bundles of F-Actin Form a Structural Support for the Microvilli of Epithelial Cells

Epithelial tissue, which lines the surfaces of the body, the internal organs, body cavities, tubes, and ducts, contains absorptive cells with numerous fingerlike projections, called *microvilli*, on their apical surface. These **microvilli** increase the surface area of the apical plasma membrane of the epithelial cell, thereby permitting a greater absorption of important nutrients. The microvilli, which are approximately 80 nm wide and 1 μm long, need a stable cytoskeletal scaffolding to maintain their shape and upright position. A stable and highly structured core of 20–30 bundled actin filaments, which run parallel to the microvilli and attach to the cytoplasmic surface of the plasma membrane, serves as this scaffolding (Fig. 3–22).

The actin filaments are bundled by two proteins, named *fimbrin* and *villin*. Actin-bundling proteins are characterized by having two binding sites for F-actin. As they bind to the sides of actin filaments in a helical

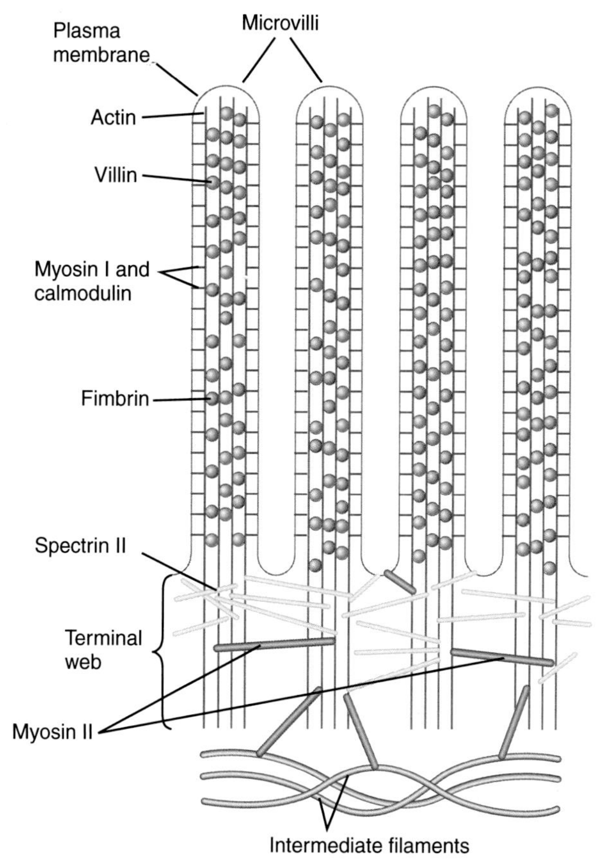

Figure 3–22. Bundled actin filaments have a structural function within microvilli. Actin bundles are attached at their plus ends to the tip of microvilli. The actin filaments are bundled by the proteins villin and fimbrin, and the bundles are attached to the side walls of the microvilli plasma membrane by association with myosin I and calmodulin. Within the terminal web, nonerythroid spectrin (spectrin II) and myosin II link adjacent actin bundles to each other and to intermediate filaments.

staircase, they group the filaments into parallel bundles. Villin has an interesting second function: at Ca^{2+} concentrations greater than 10^{-6} M, villin becomes an actin-severing protein. (This class of proteins is discussed later in this chapter.) The actin bundles are attached at their plus end to the tip of the microvilli plasma membrane by undefined proteins. The lateral attachments of the actin bundles to the side wall of the microvilli's plasma membrane are through a complex that contains calmodulin and myosin I (minimyosin). The core bundles of microvillar actin filaments end just below the surface of the apical plasma membrane in a region of the epithelial cell, called the *terminal web* because it contains a meshwork of actin filaments, actin-binding proteins, and IFs. The actin cross-linking protein, nonerythroid spectrin, and short myosin filaments run perpendicular to and attach adjacent actin core bundles. These attachments of the core bundles to nonerythroid spectrin and myosin are thought to hold the microvilli upright. Nonerythroid spectrin also cross-links the actin core bundles to IFs.

The Gel-Sol State of the Cortical Cytoplasm Is Controlled by the Dynamic Status of Actin

The cytoplasm of human and animal cells has regions that have the characteristics of a pseudoplastic gel and other regions that liquefy into the sol state. Gel-sol transformations of the cytoplasm are essential for altering the shape of cells and controlling their movement. The gel-sol conversion within the cytoplasm is regulated by the dynamic state of actin and its interaction with actin-binding proteins.

For example, in the cortical cytoplasm just below the plasma membrane, there is a thick, three-dimensional matrix of actin filaments that excludes organelles from this region of the cell cytoplasm. Long actin filaments tend to self-associate, causing a highly viscous solution. In the cortical cytoplasm, however, these actin filaments are cross-linked into a three-dimensional meshwork by long, fibrous, actin cross-linking proteins. The two most prevalent actin cross-linking proteins are nonerythroid spectrin and filamin, both of which are long fibrous proteins with two well-separated actin-binding sites at their ends. On occasion, it is essential that a region of the cortical cytoplasm becomes liquefied. For instance, when a macrophage contacts a bacterial cell, the cortical actin network must locally disassemble so that the cell surface can restructure to engulf the microorganism. This is conducted by a local increase in the cytoplasmic Ca^{2+} concentration, to 10^{-6} M, which stimulates a Ca^{2+}-sensitive, actin-severing protein, called **gelsolin**, to cut the actin filaments into short protofilaments. In the process the gelsolin molecule binds to the plus end of the severed actin filaments and caps that end. Gelsolin is removed from the plus end of actin filaments by association with phosphatidylinositol-4,5-bisphosphate (PIP_2). (This is discussed further in "Regulation of Actin Dynamics" section.)

Another protein that causes severing of actin filaments, by a different mechanism, is cofilin. Cofilin is a member of a family of proteins that are called *actin depolymerizing factors*. Cofilin binds to G-actin and the sides of actin filaments tightening the helix and increasing the torque. As a result, these more strained actin filaments are more easily severed, by mechanical stress, producing new plus ends for nucleation of actin growth.

Although the concentration of actin within nonmuscle cells is 50–200 μM, far greater than the critical concentration for the plus and minus ends of F-actin, only 50% of the actin is in polymerized form in most cells. The actin within nonmuscle cells is in a dynamic state, undergoing polymerization and depolymerization as required. The reason for the pool of G-actin within nonmuscle cells is a group of small actin-binding proteins of the thymosin family. Thymosin β_4 is a 5-kDa protein that binds to G-actin. G-actin bound to thymosin cannot hydrolyze or exchange its ATP, nor can it bind to the plus or minus end of F-actin. When rapid polymerization of actin is required, for example, at the leading edge of a motile cell, then activated profilin competes with thymosin for binding to G-actin. The profilin-actin complex then binds to the plus end of actin filaments, followed by the dissociation of profilin. Profilin binds to the second and fourth domains of G-actin on the opposite side from the ATP-binding cleft. The binding of profilin causes a conformational change in G-actin opening further the cleft region, leading to more rapid ATP-ADP exchange. Localized release of G-actin with ATP bound, in turn, promotes rapid polymerization. Profilin is activated by phosphorylation and binding to inositol phospholipids. Figure 3–23 provides a summary of thymosin and profilin function.

Cell Motility Requires Coordinated Changes in Actin Dynamics

The movement of cells, or **cell motility**, is essential for human development. Many cells are moving during embryogenesis. Growth cones, at the leading edge of motile axons, move toward their synaptic targets; macrophage and neutrophils move toward sites of infection; and fibroblasts migrate through connective tissue. Cell motility is also important in cancer biology. Cancer cells from primary tumors can crawl from the primary tumor sites to invade neighboring tissues and enter blood vessels. Therefore understanding how cells move is important. The steps in cell movement are protrusion of the plasma membrane on the leading edge, attachment to the substrate, tension on the actin filaments, and retraction of the tail of the cell (Fig. 3–24).

The protrusions of the plasma membrane include lamellipodia, which are sheetlike projections, and filopodia or

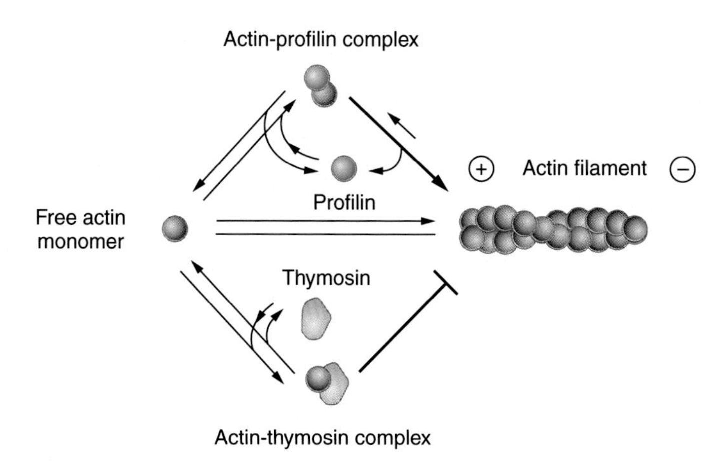

Figure 3–23. Role of thymosin and profilin in the dynamics of polymerization and depolymerization of actin. Summary of the interactions of thymosin and profilin with actin.

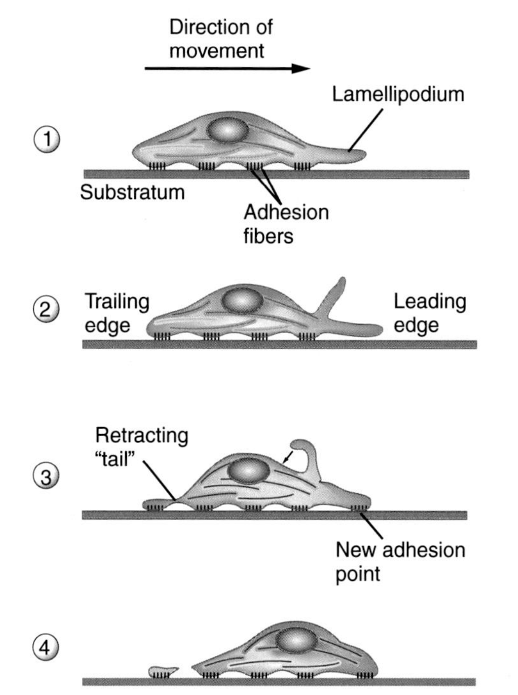

Figure 3–24. The steps in cell motility. The cell to the left (1) is moving toward the right. The cell first extends a flat, sheetlike projection of its plasma membrane [lamellipodia; (2)]. The projection then attaches to the substrate or travels back toward the cell soma (3). The protrusions then adhere to the substrate and form sites of attachment for actin filaments to the focal contact (4). The tension on the actin filaments cause the cell to be pulled forward, leaving behind remnants of the tail region.

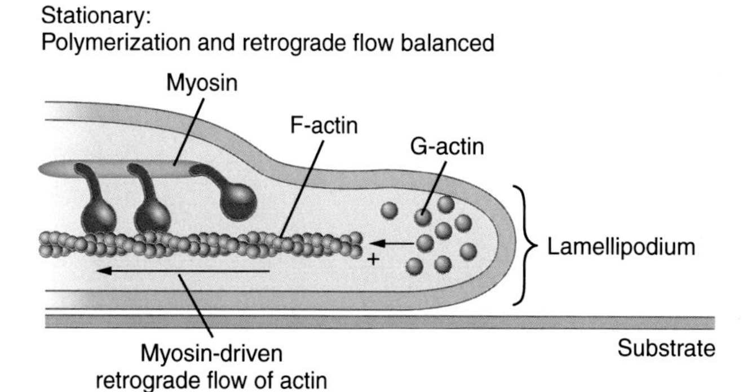

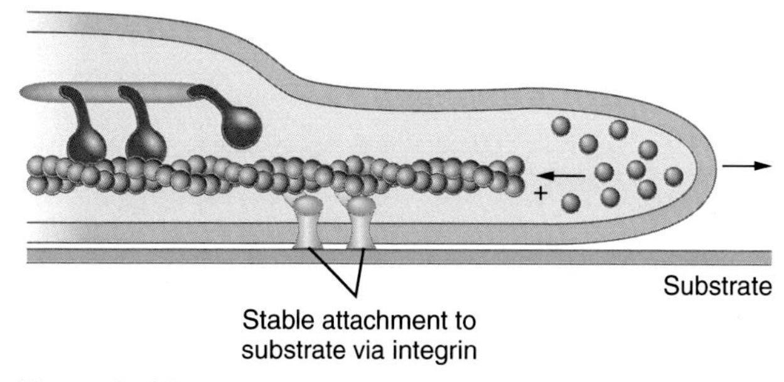

Figure 3–25. Protrusion in a motile cell is based on the balance of actin polymerization at the leading edge and retrograde flow of actin filaments toward the rear. *(Modified from Becker et al.* The World of the Cell, *6th ed. Pearson, Benjamin Cummings, 2006.)*

microspikes, which are pointed, narrow projections at the leading edge of the cell. In the lamellipodia and filopodia, two processes are occurring. They are the rapid polymerization of actin filament at the plus end, which is associated with the membrane and is pushing it forward, and the retrograde flow of the actin filaments toward the rear of the cell based on a myosin II-based contractile process. When polymerization at the plus end is more rapid than retrograde flow, then the cell moves forward. When the two are balanced, the cell is stationary (Fig. 3–25).

The lamellipodia contain branched actin filaments with their plus ends associated with the plasma membrane. The association to the cytoplasmic surface of the plasma membrane is through a protein family called *formins*. The formins have two actin-binding domains and can therefore remain attached to the actin filament while moving toward the plus end as new actin monomers are added. The branching is due to the Arp 2/3 nucleating complex. This complex contains two actin-related proteins (Arp 2 and 3), with 45% sequence identity when compared with actin, five other smaller proteins, and a nucleation-promoting factor. The Arp 2/3 nucleating complex can bind to the sides of existing actin filaments at its minus end and nucleate plus end growth of a new actin filament. Branches formed by the Arp 2/3

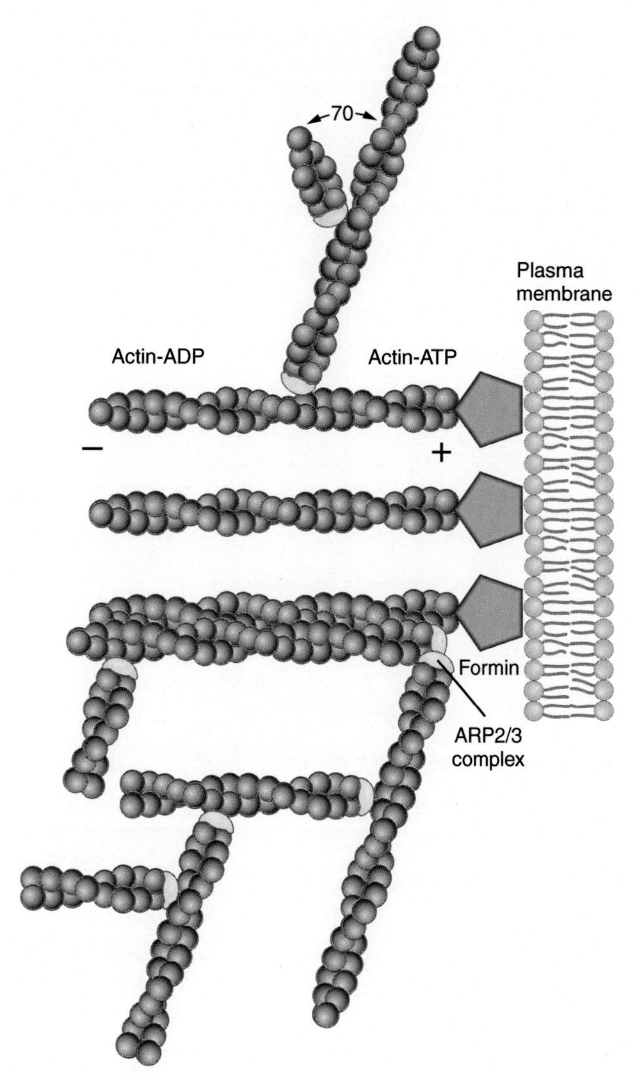

Figure 3–26. Role of formins and Arp 2/3 in the polymerization, branching, and membrane association of actin filaments within the lamellipodia.

complex are at a 70-degree angle relative to the original filament. Therefore the lamellipodia is being forced forward by a branched actin meshwork with its plus ends pushing out the plasma membrane (Fig. 3–26).

Filopodia, in contrast, are formed by bundled parallel actin filaments with their plus end polymerizing in association with the filopodia tip. The actin filaments are bundled within the filopodia by the actin-bundling protein called fascin. The result is a stiff microspike that serves as a feeler as the cell moves forward.

The lamellipodia must now make an attachment to the surface through interaction of the integrins with extracellular matrix proteins on the outside of the cell and linker or adapter proteins on the cytoplasmic side. These linker proteins (talin, vinculin, and α-actinin) attach actin filaments to the focal contact site as previously described.

An alternative form of cell motility is amoeboid movement. This form of motility is used by amoeba, slime molds, and leukocytes. In amoeboid movement the cell extends pseudopodia (derived from the Greek meaning "false feet"). The basis for movement is conversion of a thick, gelatinous, actin cross-linked ectoplasm at the periphery of the cytoplasm into a fluid bacterial infection. Receptors exist on the surface of neutrophils that allow them to detect low levels of N-formylated peptides. Because prokaryotes and not eukaryotes produce proteins with an N-terminal methionine, these peptides can be derived from only a bacterial source. The receptors on the neutrophil surface allow this white blood cell to move in the direction of the bacterial infection.

Inhibitors of Actin-Based Function

Cytochalasins, a group of chemicals excreted by various molds, block cell movement. The cytochalasins bind to the plus end of microfilaments; block further polymerization; and inhibit cell motility, phagocytosis, microfilament-based trafficking of organelles and vesicles, and the production of lamellipodia and microspikes. Latrunculin, extracted from a sea sponge, binds to and stabilizes actin monomers, the result being a net depolymerization of actin filaments. Latrunculin has similar effects on actin-based function as the cytochalasins have. Phalloidin, an alkaloid isolated from the toadstool *Amanita phalloides*, stabilizes microfilaments and does not allow endoplasm in the direction of the protrusion. Then the endoplasm congeals into ectoplasm, thereby turning the tip of the pseudopodia into a pseudoplastic gel. A series of this gel-sol-gel-sol interconversion force the cell forward. Meanwhile, at the rear of the cell, the gel-sol interconversion allows the rear of the cell to retract. Severing of actin filaments by gelsolin, due to increased local calcium concentrations, causes the gel-sol conversion. The reverse sol-gel conversion is caused by actin polymerization and cross-linking.

Cells in the body and in embryos demonstrate directional movement. This directional movement is based on diffusible molecules that are recognized by the cell surface. These diffusible molecules are called chemoattractants or chemorepellants based on whether the cell moves toward or away from this directional cue. The process of directed movement based on chemical gradients is called *chemotaxis*. An example of chemotaxis is the movement of a neutrophil toward a depolymerization. These chemicals also block cell movement, indicating that both actin filament assembly and disassembly are required for cell motility.

ACTIN-BINDING PROTEINS

Figure 3–27 and Table 3–1 summarize the functions of the various proteins that interact with actin. We have discussed intracellular actin-based assemblies (e.g., within the cytoplasm of the cell); now, we consider two mechanisms by which actin filaments bind to the cytoplasmic

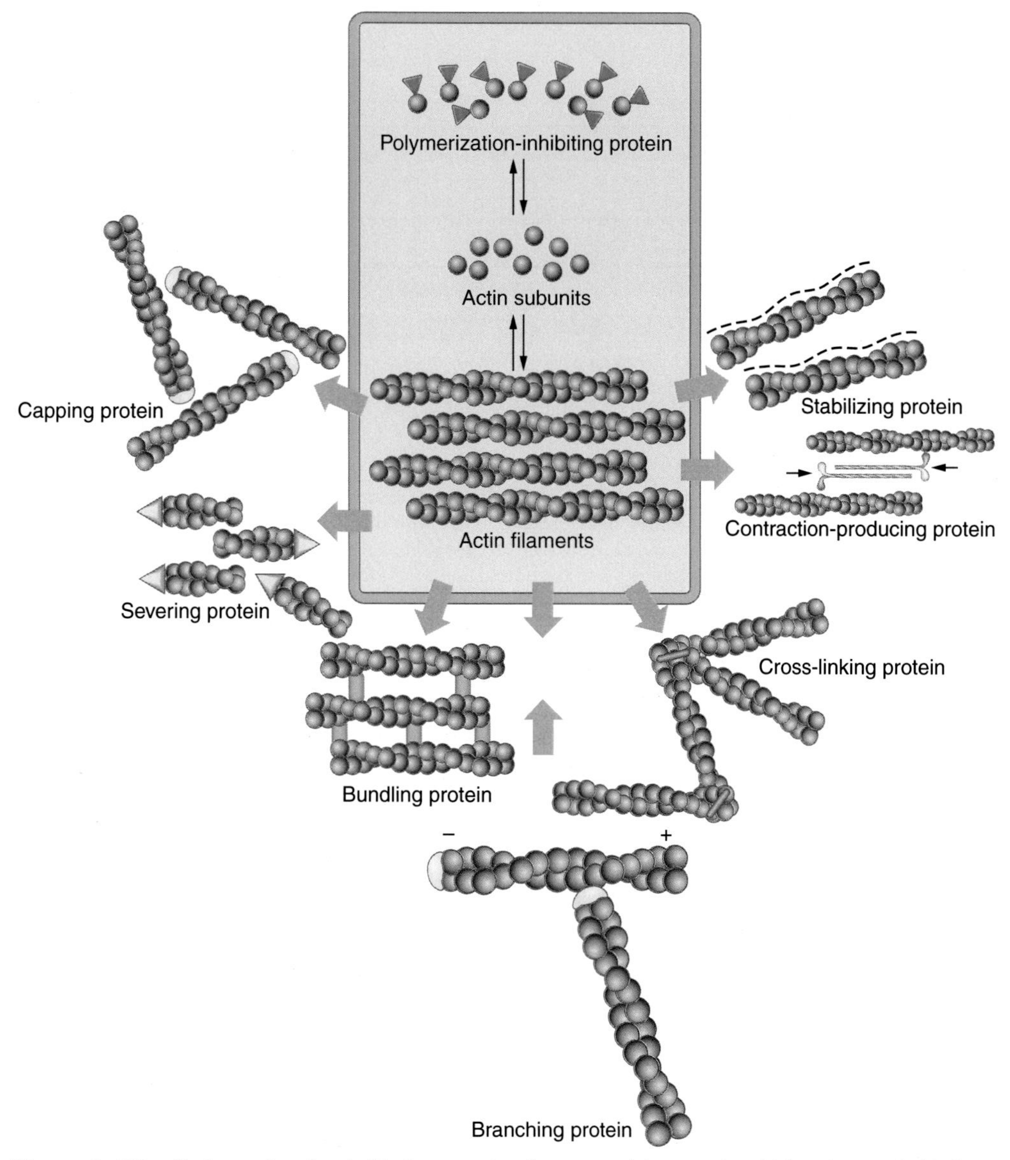

Figure 3–27. Various roles of actin-binding proteins. Summary of the ways in which various actin-binding proteins regulate the cellular organization of actin. *(Modified from Widnell CC, Pfenninger KH.* Essential Cell Biology. *Baltimore: Williams & Wilkins, 1990.)*

TABLE 3–1. Actin-Binding Proteins

Protein	Functions
Tropomyosin	Stabilizes filaments
Fimbrin, α-actinin, villin	Bundles filaments
Formin, Arp 2/3	Nucleates polymerization and forms branches
Filamin	Cross-links filaments
Spectrin I/II	Cross-links filaments in membrane skeleton
Gelsolin	Fragments filaments
Myosin II	Slides filaments in muscle
Myosin I	Moves vesicles on filaments
Cap Z	Caps plus ends of filaments
Profilin, thymosin	Binds actin monomers

surface of the cell membrane. First, we discuss the binding through the ezrin, radixin, and moesin (**ERM**) **family** and then associations found in the **spectrin membrane skeleton**.

The ERM Family Mediates End-on Association of Actin With the Cytoplasmic Surface of the Plasma Membrane

The ERM family, of protein 4.1-related proteins, attaches actin filaments to the plasma membrane in many cell types. The C-terminal domain of activated ERM proteins binds the side of preformed actin filaments, and the N-terminal domain binds the cytoplasmic domain of transmembrane proteins such as CD44 (the receptor for hyaluronan) (Fig. 3–28).

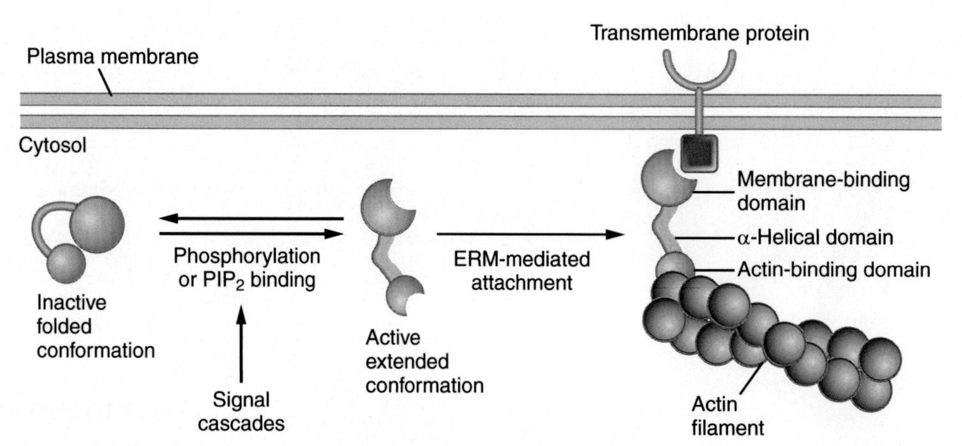

Figure 3–28. ERM proteins serve to link the plus end of actin filaments to the plasma membrane. When in their folded inactive form, the ERM proteins cannot associate with the plus end of actin filaments or the cytoplasmic domain to transmembrane proteins such as CD44. However, they become activated by phosphorylation or association with PIP_2, and the unfolding of the ERM proteins allows binding to transmembrane proteins and actin filaments.

Defects in the ERM protein, called merlin, lead to the human genetic disease called *neurofibromatosis*, where multiple benign tumors develop in auditory nerves and other parts of the nervous system. The ERM interactions are regulated by phosphorylation or binding to PIP_2.

SPECTRIN MEMBRANE SKELETON

The spectrin membrane skeleton, first described in erythrocytes, but now known to be a ubiquitous component of nonerythroid cells, is essential for maintaining cellular shape and membrane stability and for controlling the lateral mobility and position of transmembrane proteins within biological membranes.

The Structure and Function of the Erythrocyte Spectrin Membrane Skeleton Are Understood in Exquisite Detail

The spectrin membrane skeleton was first described and is best understood in the mammalian erythrocyte or red blood cell (RBC). The spectrin membrane skeleton of the human erythrocyte maintains the biconcave shape of the RBC, gives it its properties of elasticity and flexibility, stabilizes the plasma membrane, and controls the lateral mobility of integral membrane proteins. These are important properties for an 8-μm-diameter biconcave disk that must continuously deform as it passes through capillaries as small as 2 μm in diameter.

The major proteins of the erythrocyte skeleton are **spectrin, actin,** and **protein 4.1** [a nomenclature based on migration on sodium dodecyl sulfate-polyacrylamide gel electrophoresis (SDS-PAGE)]. Erythrocyte spectrin is composed of two large subunits of approximately 280 (α) and 246 kDa (β). The simplest form of spectrin is an antiparallel αβ heterodimer; however, on the cytoplasmic surface of the erythrocyte membrane, it is an $(\alpha\beta)_2$ tetramer, formed by head-to-head interaction of two heterodimers. Each end of the spectrin tetramer contains an actin-binding site, and RBC spectrin cross-links the actin filaments into a two-dimensional meshwork that covers the cytoplasmic surface of the plasma membrane. The actin filaments are short, approximately 14 actin monomers long (~33 nm); therefore they are called **actin protofilaments.** The actin protofilaments are stabilized by **tropomyosin** lining the grooves in the filament, the minus end is capped by **tropomodulin**, and each protofilament binds six spectrin tetramers, forming a hexagonal array. The spectrin-F-actin complex is stabilized by protein 4.1 and adducin, which also binds to the ends of the spectrin tetramer. The spectrin skeleton is attached to the bilayer by two types of linkages. A peripheral protein, called **ankyrin,** binds to the spectrin β subunit toward the junctional end of the heterodimers and links spectrin to the cytoplasmic NH_2-terminal domain of band 3. Protein 4.1, in addition to stabilizing the spectrin-actin interaction, binds to a member of the glycophorin family, thereby serving as a link to the bilayer. The structure of the membrane skeleton is shown in Figure 3–29.

In its most extended form, the spectrin tetramer is 200 nm in length. Because of the spectrin repeat units separated by flexible hinge regions, found through most of the length of both the α and β subunits, and described in more detail later, the spectrin tetramer can condense down to 50–60 nm. This is what provides the erythrocyte membrane skeleton and hence the RBC, with its properties of elasticity and flexibility that allow it to navigate the circulatory system for 120 days. The spectrin repeats are described later.

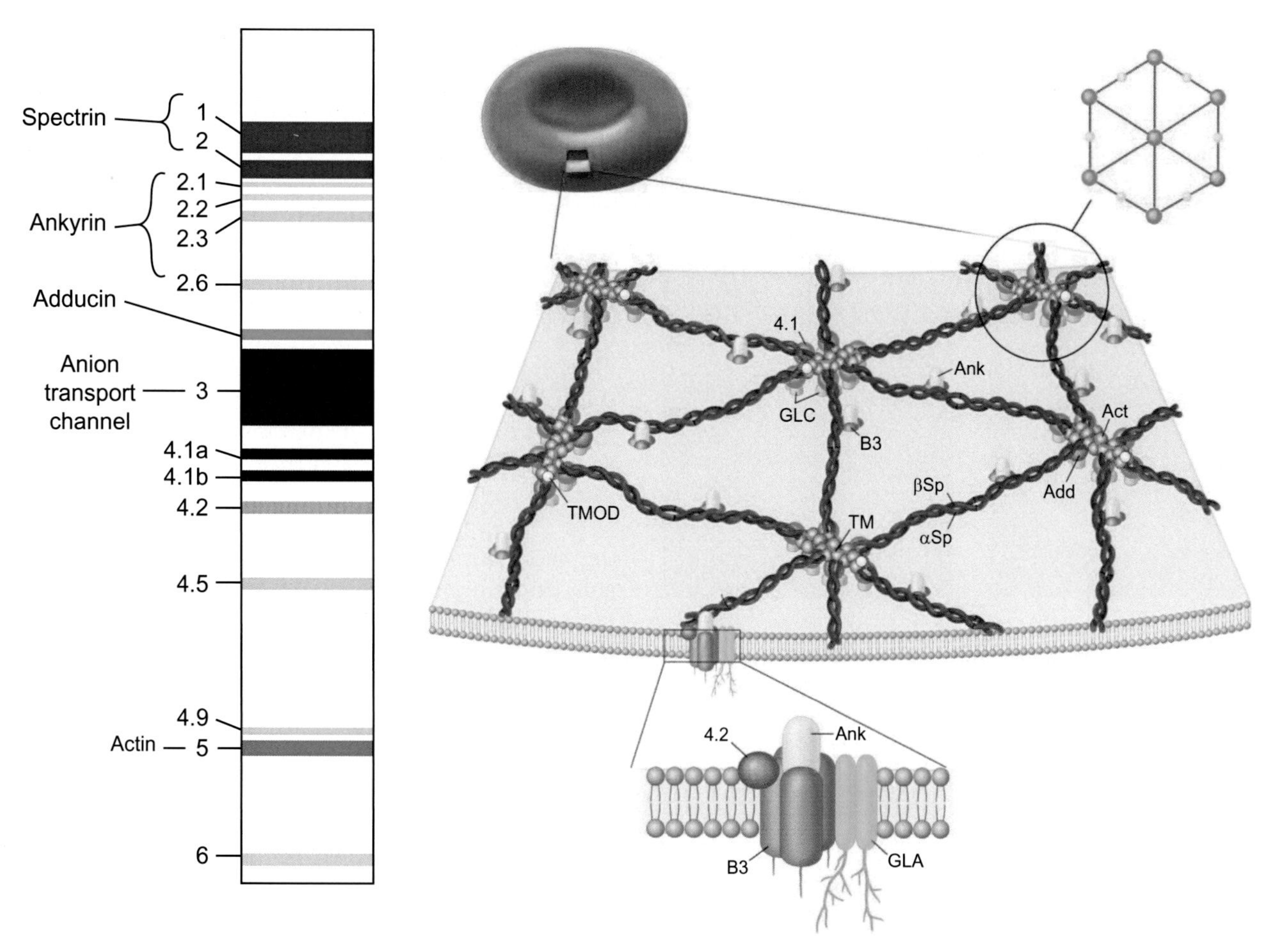

Figure 3–29. Protein interactions in the erythrocyte's spectrin membrane skeleton. This figure provides a view of the cytoplasmic surface of the red blood cell (RBC) plasma membrane with matching color-coded SDS-PAGE of RBC proteins to the left. The spectrin membrane skeleton shown in this diagram provides the RBC with its shape, elasticity, and deformability. It provides the bilayer with structural stability and controls the lateral mobility of transmembrane proteins. Spectrin tetramers crosslink actin assemblies containing short actin protofilaments, containing 13–14 G-actin (Act) monomers, tropomyosin (TM), and tropomodulin (TMOD), into what appears as a hexagonal array upon negative staining and EM of expanded membrane skeletons. The interaction of spectrin and actin filaments is strengthened by association of protein 4.1 (4.2) and adducin (Add) to the tails of beta spectrin (βSp). These are referred to as horizontal interactions. Spectrin associates with the membrane by association of βSp with ankyrin (Ank) which in turn associates with a band 3 (B3) complex which also includes protein 4.2 (4.1) and glycophorin A (GLA). The tails of spectrin are associated with the membrane via an interaction with 4.1 which in turn binds to glycophorin C (GLC). *(Modified from Goodman SR, Zagon IS.* Brain Res Bul *1984;13:813–832 and Liu SC, Derick LH, Palek J.* J Cell Biol *1987;104:527–536.)*

Phosphorylation of protein 4.1 and adducin by A kinase and C kinase, respectively, downregulates the formation of the spectrin-4.1-actin and spectrin-adducin-actin ternary complexes. Goodman and colleagues have demonstrated that α-spectrin has an E2/E3 ubiquitin conjugating and ligating activity that can ubiquitinate itself, near the tail regions of the tetramer, as well as various other target proteins. Ubiquitination of the C-terminal tail region of α-spectrin also downregulates the affinity of the spectrin-4.1-actin and spectrin-adducin-actin ternary complexes. As discussed later, this becomes important in RBCs from patients with sickle cell disease, in whom the level of ubiquitination of spectrin is substantially diminished.

Because the spectrin membrane skeleton is responsible for the normal biconcave shape of an erythrocyte, genetic defects in these proteins cause abnormal red cell shapes, stability, and life span in the circulation. Hereditary spherocytosis (HS) is a common hemolytic anemia in Caucasian populations, in whom the erythrocytes are spherical and fragile (Fig. 3–30). All patients with the common dominant form of HS show a small-to-moderate reduction in spectrin content, sometimes because of genetic defects in ankyrin, band 3, or both. A small number of HS subjects have a defective spectrin molecule that cannot bind protein 4.1 and, therefore, cannot form a stable spectrin-4.1-actin complex. In hereditary elliptocytosis, in which the red cells are elliptical and fragile, the most prevalent defect is a spectrin dimer that cannot form tetramers.

The primary genetic defect in sickle cell anemia is in the hemoglobin molecule. A subset of sickle cell RBCs locks into a dense irreversibly sickle cell, even when hemoglobin S is depolymerized, which is an important factor leading to vasoocclusion and the sickle cell crisis. The molecular bases of the irreversibly sickle cell are the following: (1) a posttranslational modification in β actin, in which a disulfide bridge is formed, leading to actin filaments that depolymerize slowly, and (2) diminished ubiquitination of spectrin, which leads to

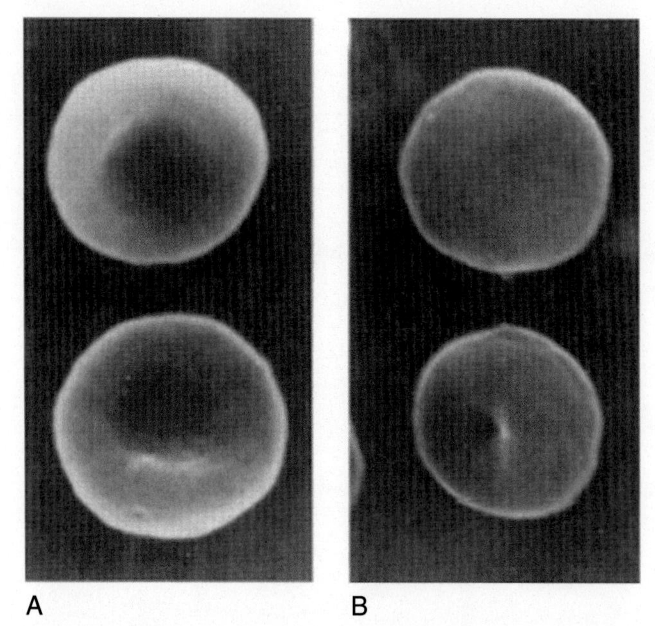

Figure 3–30. Scanning electron microscopy of normal and hereditary spherocytosis (HS) erythrocytes. (A) Biconcave erythrocytes from a normal subject. (B) A spherocyte and stomatospherocyte from an HS subject. *(From Goodman SR, Shiffer K.* Am J Physiol *1983;244: C134–141, by permission.)*

spectrin-4.1-actin and spectrin-adducin-actin ternary complexes that disassemble slowly. The result is a "locked" membrane skeleton leading to a cell that cannot change shape (Fig. 3–31; Clinical Case 3–1).

Spectrin Is a Ubiquitous Component of Nonerythroid Cells

Until 1981 spectrin and the membrane skeleton were thought to be components found only in erythrocytes. That year, Goodman and coworkers demonstrated that spectrin-related molecules were ubiquitous components of nonerythroid cells. The first demonstration that there were multiple spectrin isoforms, which in certain cases (such as neurons) can exist in the same cell, came in 1986. An erythroid isoform has been demonstrated in the neuronal soma, dendrites, and postsynaptic terminals, and a nonerythroid isoform was found axons and presynaptic terminals of the same cells in a wide variety of neurons. These isoforms were called brain spectrin 240/235 (nonerythroid) and brain spectrin 240/235E (for erythroid). Today, these isoforms would be called αSpIIΣ1/βSpIIΣ1 (nonerythroid spectrin isoform) and αSpIΣ1/βSpIΣ2 (the erythroid spectrin isoform).

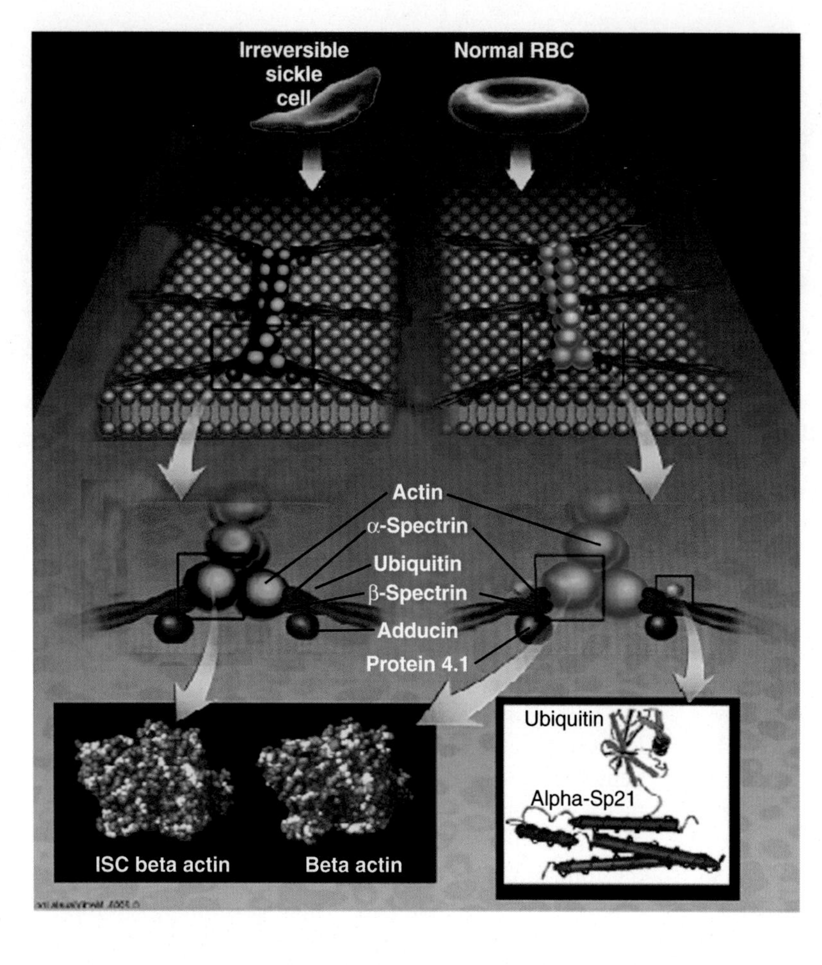

Figure 3–31. Molecular basis of the irreversible sickle cell (ISC). The inability of the ISC and its membrane skeleton to remodel is due to the inability of actin protofilaments to disassemble (indicated by the darkening of the *green* monomers) and the lack of α-spectrin ubiquitination leading to tightened spectrin-4.1-actin and spectrin-adducin-actin ternary complexes (indicated by the darkening of the *red* spectrin tails). *(Modified from Goodman SR.* Cell Mol Biol *2004;50:53–58, by permission.)*

Clinical Case 3–1

Anthony is a 10-year-old fifth-grade boy who has been visiting the school nurse on a regular basis complaining of vague abdominal pain. He didn't complain of nausea or vomiting, so the nurse generally provided reassurance and sent him back to class. This past Monday the nurse noticed that Anthony's eyes looked slightly yellow at the time of his complaint and called his parents. His mother brought him in to see his pediatrician.

At the time of the visit, Anthony complained of minor abdominal pain that was localized mostly on the left side of his belly. Nothing he did seemed to make it better or worse. He had no nausea or vomiting and did not complain of any recent illnesses. His mother reported that Anthony had jaundice as a newborn but it resolved after a few days and he was sent home with no additional needed follow-up. She also noted that she had a history of chronic anemia and took iron supplementation since she had been a child. Anthony's grandfather had died when his mother was a child due to complications after getting his gallbladder removed.

On exam, there was very slight scleral icterus on exam. Anthony's abdomen was mildly tender, and his spleen was easily palpable 2 fingerbreadths below the rib cage.

The physician ordered laboratory tests that showed a hemoglobin level of 10.2 (below normal), mean corpuscular hemoglobin concentration of 40 g/dL (high), slightly low mean corpuscular volume, a reticulocyte count of 16%, and normal white cell and platelet counts. The laboratory also reported multiple spherocytes on the blood smear. When the pediatrician reviewed the slides, he noticed a number of smaller-than-normal red blood cells that were missing the central pallor normally associated with red cells. Serum chemistry was normal except a total bilirubin of 5.2 (elevated), though the direct fraction was within normal limits. Liver function tests were within normal limits. Lactate dehydrogenase level was high, and haptoglobin level was absent. Coombs test (direct antiglobulin test) was negative.

Because of these results the doctor asked for an additional osmotic incubation test on Douglas's blood. This revealed an osmotic fragility curve that was markedly shifted to the right, with more than 50% of Anthony's red cells hemolyzing in 0.5 g% saline.

Anthony was then referred to a pediatric hematologist with a presumptive diagnosis of hereditary spherocytosis.

Cell Biology, Diagnosis, and Treatment of Hereditary Spherocytosis

Hereditary spherocytosis is an autosomally dominant hemolytic disorder (causing blood cell destruction). It is the most common hemolytic disorder in North American Caucasian populations. The abnormality is caused by mutations in genes relating to membrane proteins that allow for red blood cells to change shape. The normal genes code for a redundant lipid membrane on red cells, ultimately allowing red cells to deform to get through narrow capillaries. When this redundancy is lost due to mutations, red cells lose surface area and become more rounded and less deformable (called spherocytes). While normal RBCs appear as a biconcave disk, spherocytes are globular. Under a microscope, spherocytes appear smaller and do not have the normal central pallor associated with normal red cells (as the pallor is associated with a thinner area in the middle of normal red cells).

The rounded shape and lack of deformability ultimately results in inability to pass through capillaries, making cells more prone to rupture, and leads to increased clearance of these cells into the spleen, which functions to filter abnormal red cells from the microcirculation.

Specific genetic defects in any one of several proteins of the red cell membrane skeleton, which normally supports and stabilizes the lipid membrane, are responsible for the membrane loss. Depending on the gene involved (ANK1, SPTB, SPTA, SLC4A1, and EPB42), HS can be separated into five subtypes, all of which lead to membrane size loss, decreased surface area, and formation of spherocytes. In vitro the membrane loss can be nicely demonstrated by the increased osmotic fragility, because the cells act as perfect osmometers and their reduced membrane cannot accommodate the swelling that occurs in hypotonic saline.

The disorder can present at any age, though symptoms usually occur soon after birth, and include jaundice and anemia. Because of incomplete genetic penetrance, the severity of the disease can be widely varied, from symptom-free carriers to severe hemolysis resulting in death without therapy.

In older children like Anthony, symptoms often include fatigue and anemia with visible spherocytes on blood smear. Jaundice and splenomegaly are less common. Occasionally, because of the unconjugated bilirubin that is released by the destruction of red blood cells, pigmented gallstones (made of bilirubin) can occur. This was likely the cause of Anthony's grandfather's cholecystitis that led to his demise.

There is no cure for the genetic defect that causes HS. Treatment is therefore directed at minimizing symptoms and preventing complications. Because of the increased folate requirements needed to keep up with increased RBC production, folic acid supplementation is often appropriate. Blood transfusions are usually not required unless there is a concurrent infection that disrupts bone marrow function (such as parvovirus) and are rarely needed on a chronic basis. In patients with symptomatic pigment gallstones, cholecystectomy may be performed. In patients with severe hemolysis, splenectomy may be indicated, which often improves the anemia. Importantly, there are no specific activity restrictions in patients with HS, though there may be a slight increased risk of splenic rupture during contact sports in patients with very large spleens.

Because of Anthony's abdominal symptoms and his new jaundice, his parents elected to undergo splenectomy. He received vaccinations for encapsulated organisms prior to surgery. After recovery, Anthony no longer complained of abdominal pain, and his hemoglobin stabilized at 12.5.

At the electron microscope level of resolution, spectrin isoforms were observed not only on the cytoplasmic surface of the plasma membrane but also on the cytoplasmic surface of all organelle membranes (nuclear envelope, endoplasmic reticulum, mitochondria, etc.). The spectrin isoforms were also in the cytoplasm cross-linking cytoskeletal elements (actin filaments, intermediate filaments, and microtubules) to each other and to membrane surfaces. Brain spectrin 240/235 (αSpIIΣ1/βSpIIΣ1) was in the presynaptic terminal associated with the plasma membrane, cytoskeletal structures, and small spherical synaptic vesicles, while spectrin 240/235E (αSpIΣ1/βSpIΣ2) was found in the postsynaptic terminal associated with the plasma membrane, mitochondria, cytoskeletal elements, and postsynaptic densities.

Today, we know that there are two alpha spectrin genes in human, SPAT1 and SPTAN1, that encode the erythroid (αSpI) and nonerythroid (αSpII) spectrin isoforms, respectively. There are also five beta spectrin genes SPTB, SPTBN1, SPTBN2, SPTBN4, and SPTBN5 that encode the erythroid (βSpI) and nonerythroid (βSpII, βSpIII, βSpIV, and βSpV) spectrin isoforms. Table 3–2 gives the protein nomenclature, names in the literature, molecular weight, human gene symbol, and human chromosome locus for the seven spectrin genes and their products. Alternate splice forms of the spectrin genes give rise to isoforms indicated with the symbol Σ. Note that a common alternate name for the nonerythroid spectrin is fodrin.

Spectrin Family Functions

Spectrin family members are found on the cytoplasmic facing surface of the plasma membrane and all organelle membranes, in the cytosol interacting with cytoskeletal structures and cross-linking them to each other and membrane surfaces, and in the nucleus playing important functions in DNA repair and linking the nucleus to cytoplasmic signaling pathways. Since spectrin(s) are located on numerous cellular structures and seem to play important roles throughout the cell, Goodman and colleagues surveyed numerous databases to extract all known protein interactions and locations of the spectrin isoforms. This resulted in construction of the "spectrinome" (Fig. 3–32), a compilation of spectrin's interactions and locations that will give rise to a better understanding of the functions of this isoform family of proteins.

The functions of the spectrin family of isoforms, in addition to membrane skeleton functions, are as diverse as ER to Golgi vesicle-mediated transport, involvement in signaling pathways (insulin receptor, RAF/MAP kinase cascade, Ras guanyl-nucleotide exchange, SOS-mediated, SMAD pathway, etc.), axon guidance, and synaptic vesicle association with the active zone of the presynaptic plasma membrane, cell-cell adherence via NCAM and LCAM, DNA repair, maintaining telomere integrity, and being an important component of the nucleoskeleton.

Spectrins, α-Actinin, and Dystrophin Form the Spectrin Supergene Family

The complete sequences for the seven spectrin gene products have been determined. Both α and β spectrins contain triple-helical repeat units of approximately 106 amino acids, separated by flexible nonhelical regions, throughout most of their sequence. These repeats share approximately 20%–40% sequence identity. Interestingly the NH_2- and COOH-terminal ends of α- and β-spectrins do not contain the typical repeat structure. Furthermore a 140-amino acid stretch at the NH_2 terminus of the β subunit has been demonstrated to represent the actin-binding domain of spectrin (Fig. 3–33).

Whereas erythroid and nonerythroid spectrins share only approximately 60% sequence identity throughout the α and β sequence, they are more than 90% identical in the actin-binding domain. α-Actinin (an actin-bundling protein) and dystrophin (the protein missing in subjects with Duchenne muscular dystrophy [DMD]) have sequences that are highly related to spectrins. α-actinin, a 190-kDa dimer, is composed of two identical antiparallel subunits. Dystrophin is an 800-kDa homodimer, with

TABLE 3–2. Specrtrin Proteins

Spectrin Protein Nomenclature	Names in Literature	Molecular Weight	Human Gene Symbol	Human Chromosome
αSpIΣ1	α Erythrocyte spectrin	280 kDa	SPTA1	1q23.1
αSpIIΣ1,2,3,4	Nonerythroid α-spectrin, α spectrin II, α fodrin	280 kDa for Σ1	SPTAN1	9q34.11
βSpIΣ1	β Erythrocyte spectrin	246 kDa	SPTB	14q23.3
βSpIΣ2	β Erythroid spectrin	268 kDa	–	–
βSpIIΣ1,2	Nonerythroid β-spectrin, β-spectrin II, β fodrin	275 kDa for Σ1	SPTBN1	2p16.2
βSpIIIΣ1	–	271 kDa	SPTBN2	11q13.2
βSpIVΣ1,2,3,4,5	–	288 kDa for Σ1	SPTBN4	19q13.13
βSpVΣ1	–	417 kDa	SPTBN5	15q21

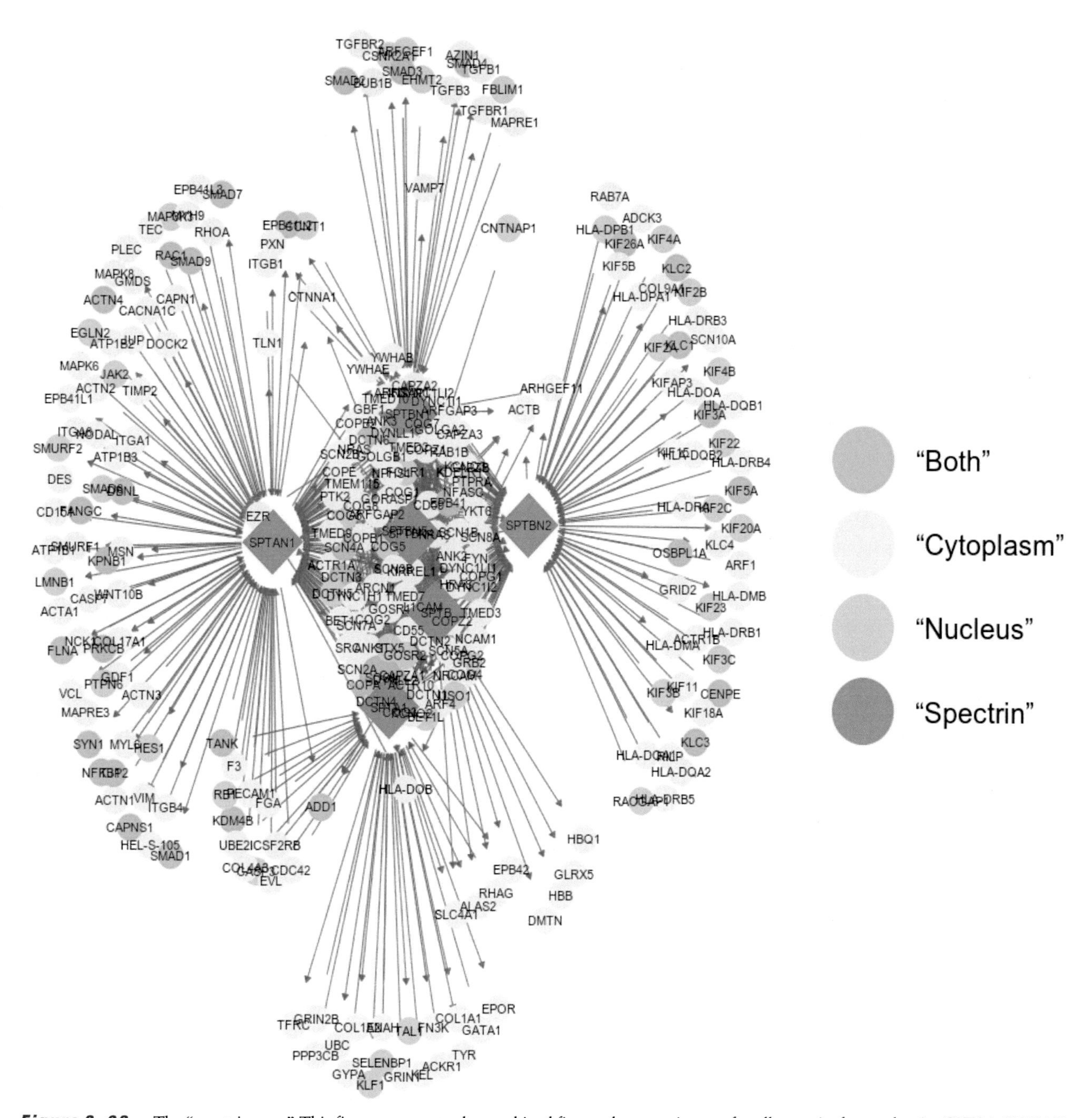

Figure 3–32. The "spectrinome." This figure represents the combined first-order spectrinomes for all seven isoform subunits (SPTA1, SPTAN1, SPTB, SPTBN1, SPTBN2, SPTBN4, and SPTBN5). The inner ring proteins have, for the most part, common interactions with the spectrin isoforms, while the outer rings contain the unique interactions. *(From Goodman SR, Johnson D, Youngentob SL, Kakhniashvili D.* Exp Biol Med *2019, by permission.)*

two antiparallel 400-kDa subunits. Both α-actinin and dystrophin contain the spectrin triple-helical repeat units, with about 10%–20% identity with the spectrin repeats. Both proteins contain a nonhelical region at their NH_2 terminus, with 60%–80% identity with the actin-binding domain of β-spectrin. This finding was important in determining the function of dystrophin and the cause of DMD. Because of its sequence identity with the actin-binding domain of spectrin, dystrophin was proposed and since demonstrated, to function in anchoring actin filaments to the plasma membrane in skeletal muscle. The common structure found for spectrins, α-actinin, and dystrophin has led to the concept that they are descendants of a common ancestral gene and thus to their being called the **spectrin supergene family** (Clinical Case 3–2).

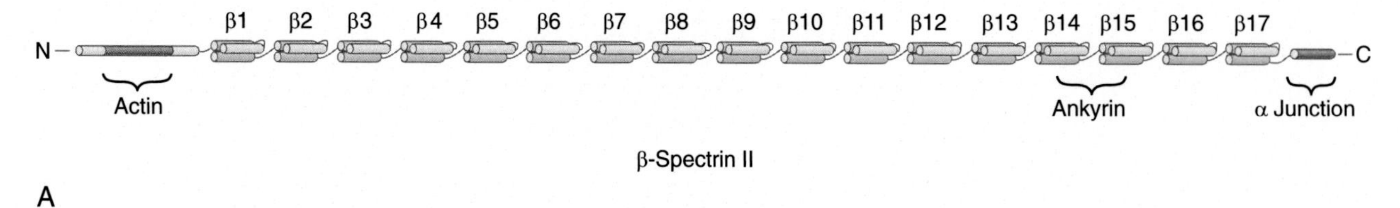

Figure 3–33. Structure of β-spectrin II. (A) β-Spectrin II is presented as an example of the structure of the members of the spectrin supergene family. β-Spectrin II has a nonhelical actin-binding domain at its NH2 terminus. There are 17 triple-helical spectrin repeats separated by flexible hinge regions. The COOH-terminus contains a nonhelical region involved in association of the α-spectrin subunit. (B) Detailed structure of one triple-helical repeat is presented. *([A] From Ma Y, Zimmer WE, Riederer BM, et al.* Mol Brain Res *1993;18:87–99, by permission.)*

Clinical Case 3–2

Michael is a 12-year-old boy who was in his usual state of health until recently, when he started complaining of difficulty climbing stairs. He tells the orthopedist that it is hard to stand up when he is sitting on the floor and that he has to use his hands to "walk himself up," and he says his calf muscles often feel "tight." He also tells the doctor that he does well in school without any problems but that he is not particularly athletic. His mother reports that Michael's father's side of the family is healthy and that she and her sisters have no medical problems, but that her mother's brother was wheelchair bound and died of a heart attack in his early 40s.

On physical examination the physician notes significantly enlarged calf muscles bilaterally. Michael has four out of five strength in his quadriceps muscles, but no noticeable weakness in his upper extremities. His pectoral muscles and biceps appear somewhat underdeveloped. Laboratory tests are notable for a mildly elevated aminotransferase (ALT and AST) but normal gamma-glutamyl transferase (GGT). The serum creatine kinase level is more than five times the upper limit of normal.

The orthopedist performs a quadriceps muscle biopsy. On histopathology, there is mild muscle fibrosis and occasional tissue necrosis. Immunohistochemistry for dystrophin expression shows only patchy staining and incomplete surrounding of muscle fibers.

A blood sample is sent out and mutational analysis performed that reveals an exon 3 deletion on the dystrophin gene on the X-chromosome.

Cell Biology, Diagnosis, and Treatment of Muscular Dystrophy

Michael has Becker muscular dystrophy (BMD), which is related to the more severe Duchenne muscular dystrophy (DMD). Both diseases result in muscle weakness, but while DMD is diagnosed early and is rapidly progressive, BMD can be diagnosed in adolescence or adulthood and often has a milder clinical course. Weakness selectively affects proximal muscles at first, and therefore patients have difficulty running, jumping, and climbing steps. Often, rising from the floor is only accomplished using hand support (known as Gower's sign). Calf enlargement can be observed as well, due to muscle hypertrophy to compensate for the proximal muscle weakness. Creatine kinase is almost uniformly elevated early in the disease course, representing muscle destruction. Aminotransferases may also be elevated, representing muscle destruction rather than liver abnormalities.

Both DMD and BMD are caused by a defective gene on the X-chromosome that is responsible for the production of dystrophin. Dystrophin is a huge protein that resides on the cytoplasmic face of the plasma membrane in muscle fibers, functioning as part of a large glycoprotein complex. Normal dystrophin provides support and reinforcement to the tubular sheath that envelops muscle fibers and helps prevent degradation of the glycoprotein complex. When dystrophin is absent (as in DMD) or decreased (as in BMD), the glycoprotein complex is digested by proteases, and this allows for muscle fibers to degenerate, causing muscle weakness. Because of the X-linked nature of inheritance, girls are virtually never affected, and boys inherit the disease from their mothers, who exhibit normal phenotypes.

While in DMD patients are almost universally wheelchair bound by the age of 12, BMD patients usually remain ambulatory until their late teens and commonly even into adulthood. While DMD is associated with intellectual disability and muscle contractures, these are not as common in BMD. Both diseases are commonly associated with cardiac muscle involvement leading to fibrosis and ultimately heart failure or heart conduction abnormalities.

Corticosteroids are a mainstay of treatment for DMD, and they improve motor function, strength, and pulmonary function and may delay the onset of cardiomyopathy. However, long-term use of steroids leads to immunosuppression, which combined with the progressive scoliosis and pulmonary dysfunction may contribute to pneumonia risk. Patients with BMD are often able to avoid steroids until late in the disease course. Novel therapies, including gene therapies with viral vectors that contain partial dystrophin genes, or therapies that inactivate myostatin, are being developed and are in various stages of clinical trials. All patients with DMD or BMD should undergo physical therapy. Regular cardiac surveillance and pharmacological preventive treatment is often recommended. In DMD, pulmonary complications are often fatal, and respiratory management is a critical component of care. Both types of dystrophy are relentlessly progressive. Patients with DMD usually die in their 20s or 30s of pulmonary complications. The most common cause of death in BMD is cardiac and often live into their mid to late 40s.

REGULATION OF ACTIN DYNAMICS

We mentioned earlier that contacts between spectrin and actin are regulated by the phosphorylation of protein 4.1 and adducin and the ubiquitination of spectrin within the RBC membrane skeleton. The regulation of actin dynamics with cells that contain a cytoskeleton is far more complex.

As discussed in Chapter 8, phosphatidylinositol and its derivatives are involved in cell signaling cascades. Several activated kinases convert phosphatidylinositol into PIP_2. PIP_2 then is a key regulator of actin microfilament polymerization, depolymerization, attachment of actin-binding proteins, and cleavage by gelsolin. PIP_2 associates with profilin at the cytoplasmic surface of the plasma membrane, preventing its association with actin. Hydrolysis of PIP_2 by a signal-activated phospholipase C releases profilin from the membrane, allowing it to bind G-actin. This promotes ADP-ATP exchange with subsequent polymerization of G-actin to F-actin (Fig. 3–34).

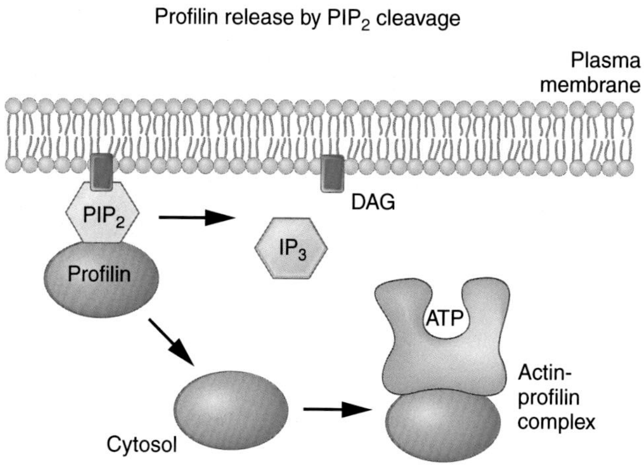

Profilin release by PIP_2 cleavage

Plasma membrane

PIP_2

Profilin

DAG

IP_3

ATP

Actin-profilin complex

Cytosol

Figure 3–34. Regulation of the profilin-actin interaction by PIP_2.

Gelsolin and cofilin, which sever actin filaments, are both inhibited by PIP_2. Once PIP_2 is released from these proteins, or hydrolyzed, filaments are severed, increasing the number of plus ends available, thereby stimulating actin polymerization.

The Wiskott-Aldridge syndrome protein (WASP, which is an inherited immune system disorder characterized by low levels of platelets and white blood cells) and Scar (or WAVE) bind to Arp 2/3, activating it and promoting actin polymerization. WASP and Scar also bind to PIP_2 and Rho family proteins.

The Rho protein family includes Cdc 42, Rac, and Rho, which are monomeric G-proteins. These GTPases cycle between an active guanosine triphosphate (GTP)-bound form and an inactive guanosine diphosphate (GDP)-bound form. Activation of Cdc 42 causes activation of Scar, leading to its binding the ARP 2/3 complex and rapid actin polymerization and bundling. This results in the formation of filopodia or microspikes. Activation of Rac causes activation of WASP and PI(4)P 5-kinase, leading to the formation of lamellipodia and membrane ruffles. The kinase generates a form of PIP_2 that uncaps plus ends of F-actin, leading to polymerization. Activation of Rho promotes bundling of actin filaments, with myosin II thick filaments, forming stress fibers interacting with focal contacts. Rho activates Rho kinase (ROCK), which phosphorylates myosin II light chains. This phosphorylation stimulates the association of the myosin heads with actin filaments, resulting in contraction. Therefore the Rho family of GTPases regulates the state of actin and is critical to actin-based cell motility. The effect of Rac, Rho, and Cdc 42 on a fibroblast is shown in Figure 3–35

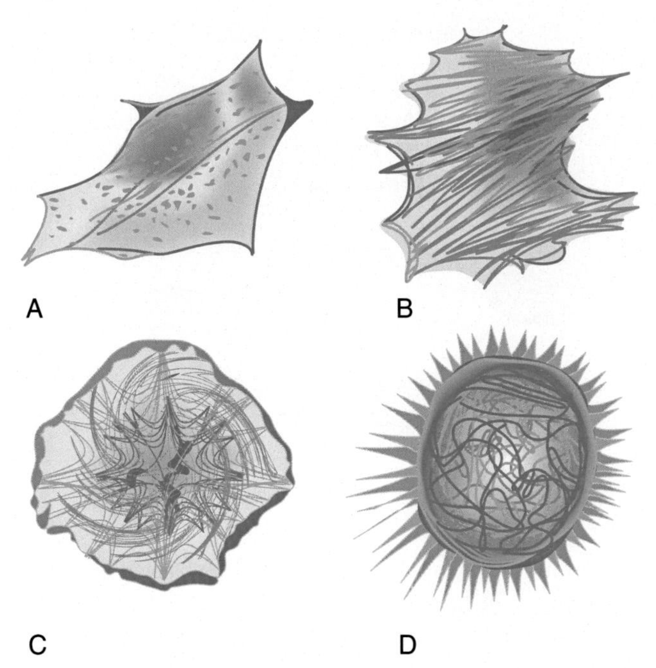

Figure 3–35. Effect of Rho family activation on actin organization in fibroblasts. (A) Quiescent cell, (B) Rho activation, (C) Rac activation, and (D) Cdc 42 activation.

CYTOSKELETON PARTICIPATION IN CELL SIGNALING

As mentioned previously the cytoskeleton plays a dynamic role in cellular metabolism. We have explored the activation of the actin skeletal components via the actions of the Rho family of signaling molecules. Recent studies have shown that actin and spectrin cytoskeleton networks can be direct, active participants in intracellular signaling that influence the activity of cells, including their shape as described earlier, and can lead to dramatic changes in cellular metabolism that may lead to disease. Two of the best examined examples of cytoskeletal components directly participating in cell signaling include actin filaments in regulation of gene expression profiles through G-/F-actin-binding proteins and the α-subunit of erythroid spectrin that contains a ubiquitin ligase activity that can regulate turnover of proteins in cells.

Myocardin is a protein that is expressed specifically in cardiac muscle cells and is absolutely required for appropriate development of the heart. It works through binding with the known transcription activator protein (detailed in Chapter 6) called serum response factor (SRF). There are proteins very similar to myocardin that are found to be expressed in mesenchymal cells other than cardiac muscle called myocardin-related transcription factors (MRTF) A and B. It has been demonstrated that activation of cell actin-based cytoskeleton via Rho kinase (Fig. 3–35) is accompanied by an increased activity of serum response factor (SRF)-regulated genes. It has also been observed that MRTF A/B in Rho-stimulated cells moves from the cytoplasm to the nucleus to participate in the SRF gene activation. MRTF A/B have been found to bind cytoplasmic G-actin, and when the G-actin is stimulated to form filaments from Rho kinase signaling, the MRTF A/B is released from the G-actin monomers and is then able to move into the nucleus and interact with SRF to stimulate the transcription of the specific genes. Thus the Rho signaling cascade affects not only the formation of actin filaments but also a change in the gene expression of the stimulated cells to maintain, and because the MRTFs are anchored in the cytoplasm via G-actin bound, the cytoskeleton can be viewed to directly participate in cellular signaling.

As discussed previously, spectrin is a dimer made up of nonidentical α- and β-subunits. It is the anchor for the spectrin membrane skeleton and participates in cellular shape and internal functions of the cell. It is formed by repeating domains of ~106 amino acids separated by flexible regions that allow spectrin to fold or compress depending upon cellular demands. The actin-binding segment of β-spectrin has been demonstrated to be an important determinant in the locking of the membrane in sickle cell disease (Fig. 3–31). Recently, Goodman and colleagues demonstrated that the α-subunit of erythroid spectrin functions in the ubiquitination pathway. The carboxyl-terminal segment of α-spectrin is found to contain enzymatic activities to transfer the short ubiquitin peptide (described in Chapter 4) to itself and to other proteins. This ubiquitination enzymatic activity is an important signal for protein degradation and for regulating protein-protein interactions. Therefore spectrin is a direct participant in cellular signaling leading to protein metabolism and regulating protein interactions.

These are the best characterized examples of how cytoskeletal proteins not only are static members of the cell architecture forming and maintaining the cell shape but also are very active components of the cellular signaling system leading to changes in cellular homeostasis. There are likely to be discovered more connections of cellular structure with function in the future.

INTERMEDIATE FILAMENTS

IFs are 10 nm in diameter and, therefore, intermediate in thickness between microfilaments and myosin thick filaments, or microfilaments and microtubules. Although much work is required to determine the functions of this ropelike filament, their role appears to be primarily structural; that is, the major function of IFs is to provide resistance to mechanical stress placed on the cell. IFs within muscle cells link together the Z disks of adjacent myofibrils. Neurofilaments within the axon serve as a structural support to resist breakage of these long, slender processes, and they increase in number as the caliber of an axon increases during development. IFs of epithelial cells interconnect spot desmosomes, thereby stabilizing epithelial sheets.

A Heterogeneous Group of Proteins Form IFs in Various Cells

The protein monomers that constitute IFs differ from the components of microfilaments and microtubules (see discussion later in this chapter) in several important ways. The IFs in various human and animal cells are composed of a heterogeneous group of proteins, but microfilaments are always composed of actin, and microtubules are always composed of tubulin. The IF subunits are fibrous proteins, although both G-actin and tubulin are globular. Almost all of the IF subunits are incorporated into stable IFs within various cells, whereas the same cells contain a substantial pool of unpolymerized G-actin and tubulin. No energy in the form of ATP or GTP hydrolysis is required for IF polymerization. IFs have no polarity, whereas microfilaments and microtubules have plus and minus ends. IFs are composed of a heterogeneous class of subunits (Table 3–3).

The keratin filaments found in epithelial cells always contain an equal number of subunits of acidic (type I) and neutral basic (type II) cytokeratins. In humans, there is a genetic disease, epidermolysis bullosa simplex, that arises from mutations in the keratin genes expressed in the basal cell layer of the epidermis. This disrupts the normal network of keratin filaments in these cells, and people afflicted with these keratin gene mutations are keenly sensitive to mechanical injury; even a gentle squeeze can cause disruption of this cell layer and blistering of the skin.

Vimentin, desmin, glial fibrillary acidic protein (GFAP), and neurofilament light chain (NF_L) are capable of forming homopolymeric IFs, but when together in a single cell (e.g., muscle cells and glial cells), they may copolymerize. However, in an epithelial cell where cytokeratins can be coexpressed with vimentin, they do not copolymerize, but instead form separate IFs. Within axons and dendrites, NF_L, neurofilament medium chains (NF_M), and neurofilament heavy chains (NF_H) copolymerize to form the neurofilaments. The cell-type specificity of IF proteins has been useful to pathologists, who use fluorescent IF-type-specific monoclonal antibodies to identify the tissue of origin of metastatic cancer cells. The nuclear lamina is composed of the IF-related proteins lamin A, lamin B, and lamin C. These proteins and the square lattice that they form on the inner nuclear envelope are discussed in Chapter 6 (see the section dealing with the nucleus).

TABLE 3–3. Intermediate Filaments of Human Cells

Intermediate Filament	Subunits (M_r)	Cell Type
Keratin filaments	Type I acidic keratins Type II neutral/basic keratins (40–65 kDa)	Epithelial cells
Neurofilaments	NF_L (70 kDa) NF_M (140 kDa) NF_H (210 kDa)	Neurons
Vimentin-containing filaments	Vimentin (55 kDa) Vimentin + glial fibrillary acidic protein (50 kDa) Vimentin + desmin (51 kDa)	Fibroblasts Glial cells Muscle cells
Nuclear lamina	Lamins A, B, and C (65–75 kDa)	All nucleated cells

NF_L, neurofilament light chain; *NF_M*, neurofilament medium chain; *NF_H*, neurofilament heavy chain.

How Can Such a Heterogeneous Group of Proteins All Form Intermediate IFs?

It is truly remarkable that proteins of the IF class, which range in M_r from 40 to 210 kDa, are all capable of forming IFs. The molecular basis for this common morphology is shown in Figure 3–36.

All IF proteins contain a subunit-specific NH_2 terminus of variable size, a homologous central α-helical region of approximately 310 amino acids (with three nonhelical gaps), and a subunit-specific COOH-terminus of variable size. Only the homologous 310-amino acid α-helical region is a portion of the 10-nm IF core. The variable regions extend from the core and are responsible for cross-linking IFs to other cytoskeletal structures. In the formation of the IFs, the first step is that the 310-amino acid α-helical region of two monomers wind around each other into a parallel coiled coil. The IF proteins contain the heptad repeat within the 310-amino acid α-helical region required for coiled-coil formation. Next, two dimers link side by side in an antiparallel conformation to form a tetramer. Because the tetramers have an antiparallel conformation, the IFs have no polarity. The IF tetramers attach laterally to each other in a staggered array until there are eight tetramers (32 monomers) making up the wall of the IF. The eight tetramers are wound to form the ropelike structure of the IF. Microfilaments have actin-binding proteins to allow their association with other cytoskeletal structures, and microtubule-associated proteins (MAPs) play a similar function for microtubules. There are also specific cross-linking proteins for IFs, such as filaggrin, which bundles keratin filaments; plectin, which bundles vimentin-containing IFs; and neurofilaments, which bundle to each other and to microtubules and microfilaments. The variable COOH-terminal regions of NF heavy and medium subunits bundle neurofilaments together, giving a structurally stable core to axons and dendrites.

The importance of these IF linkages is demonstrated in people who have a mutation in plectin. The result of such a mutation is a human disease that combines the phenotypes of epidermolysis bullosa simplex, muscular dystrophy, and neurodegeneration.

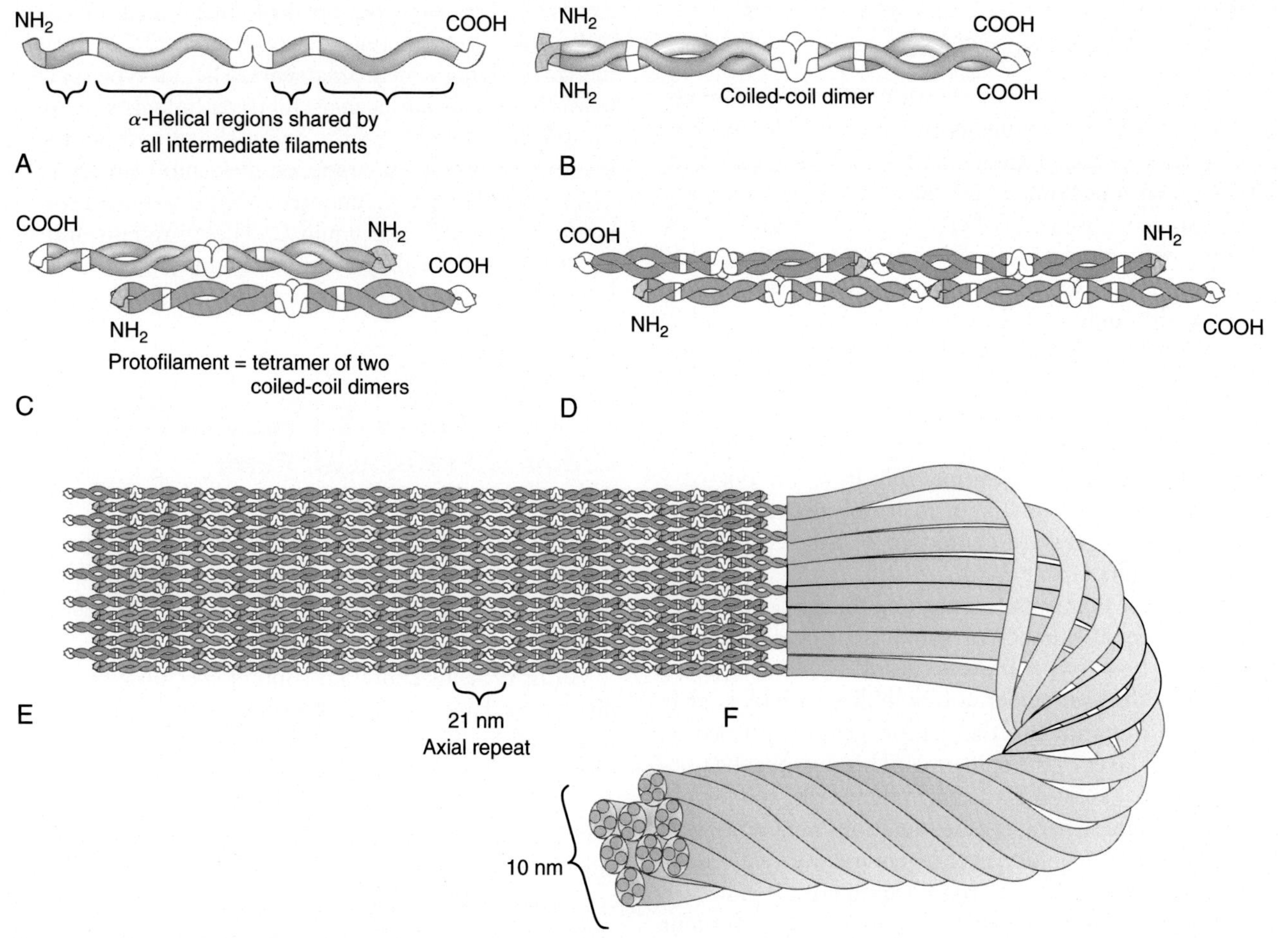

Figure 3–36. Assembly of intermediate filaments (IFs). (A) IF monomers. (B) Two monomers form a parallel coiled-coil dimer. (C) Two dimers form an antiparallel tetramer by side-to-side interaction. (D) The two dimers forming a tetramer are staggered, which allows the formation of higher-order structure. (E) The tetramers continue to associate in a helical array up to eight tetramers (protofilaments) wide. (F) The intermediate filaments became longer and wound into a ropelike structure. *(Modified from Alberts B, Bray D, Lewis J, et al.* Molecular Biology of the Cell, *3rd ed. New York: Garland Publishing, 1994.)*

In anaphase cells the IFs normally form a tight weave around the nucleus and then spread in wavelike fashion toward the plasma membrane. If the microtubules of an anaphase cell are depolymerized with colchicine (Col-Benemid) or demecolchicine (Colcemid), the IFs collapse around the nucleus; obviously, the IFs are highly integrated with microtubules. When antibodies against spectrin II were microinjected into fibroblasts, the IF network again collapsed around the nucleus, even though there was no obvious effect on microtubules or microfilament stress fibers. This suggests that spectrin II may also play an important role in linking IFs to other cytoskeletal structures. Indeed, immunoelectron microscopy has demonstrated such a role for spectrin II within the terminal web of epithelial cells and within the axons and dendrites of mammalian neurons. All of the studies described earlier suggest an association of IFs with the nuclear envelope. In addition, IFs attach to the plasma membrane by interactions with ankyrin and with spectrin.

MICROTUBULES

Microtubules Are Polymers Composed of Tubulin

Microtubules are the third type of cytoskeletal structure, and they have been implicated in a variety of cellular phenomena, including ciliary and flagellar motility, mitotic and meiotic chromosomal movements, intracellular vesicle transport, secretion, and several other cellular processes. Their principal component is the protein tubulin, a heterodimer composed of nonidentical α and β subunits, with each subunit having an M_r of nearly 50 kDa. In addition, several other proteins are associated with microtubules, and it is these accessory proteins that are responsible for many of the characteristics of microtubule-based motility. The MAPs are described later in this section. Through the electron microscope, microtubules are seen as hollow cylinders with an outer diameter of 24 nm. When viewed in cross section, the wall of each microtubule is seen to be composed of 13 tubulin dimers, which represent 13 protofilaments composed of tubulin subunits. As microtubules assemble, the tubulin molecules are added to the growing microtubule to form the 13 protofilaments. The individual protofilaments are organized such that α- and β-subunits alternate along the length of the protofilament, which provides a microtubule with an inherent polarity. The tubulin dimer α-subunits, for each protofilament, all face the minus slow-growing end. The β-subunits all face the plus end (Fig. 3–37).

Tubulin is one of the most highly conserved proteins known. The significance of this is currently unclear, but it is presumed that this results from the many essential functional subdomains within the tubulin molecule. Not only are there regions that are necessary for subunit interactions during microtubule assembly, but also tubulin contains regions for GTP binding, for interacting with MAPs, and sites for binding to several different drugs. Pharmacologic agents that are bound by tubulin, such as colchicine, vinblastine sulfate (Velban), nocodazole, and paclitaxel (Taxol), disrupt the normal dynamic behavior of microtubules. Because proper microtubule functioning is essential for spindle formation and cell division, microtubule inhibitors are commonly used for cancer chemotherapy.

Microtubules Undergo Rapid Assembly and Disassembly

Cytoplasmic microtubules are labile structures that have the capacity to undergo rapid assembly and disassembly. This characteristic is important for many microtubule functions. For example, cytoplasmic microtubules must be broken down rapidly as the cell enters mitosis. Likewise the disassembled microtubules must be reformed to assemble the mitotic spindle. The ability of the spindle microtubules to be broken down by the cell during mitosis also appears to be essential for chromosomal separation. If dividing cells are cultured in the presence of Taxol, a drug that blocks microtubule disassembly, chromosomal segregation is blocked. The ability of the microtubule cytoskeleton to reorganize rapidly may be important for many other cellular events, such as cell migration and the establishment of cellular polarity.

Similar to actin filaments, growing microtubules have an inherent structural polarity. This polarity occurs because of the orientation of the tubulin subunits in the microtubule polymer. When growing microtubules are analyzed in vitro, subunits add to one end of the elongating polymer faster (the plus end) than to the other end (the minus end) (Fig. 3–38).

Inside cells, however, the minus end of the microtubule is capped, owing to its association with the centrosome complex; therefore only the events that occur at the plus end will be considered. Current ideas concerning microtubule dynamics focus on the binding of GTP by tubulin subunits during microtubule assembly and the subsequent hydrolysis of the bound GTP to GDP (see Fig. 3–38).

For the tubulin dimer to add to an elongating microtubule polymer, the tubulin α and β subunits must each bind GTP. The GTP-tubulin can then add to the growing end of a microtubule, and sometime after adding to the microtubule, the bound GTP associated

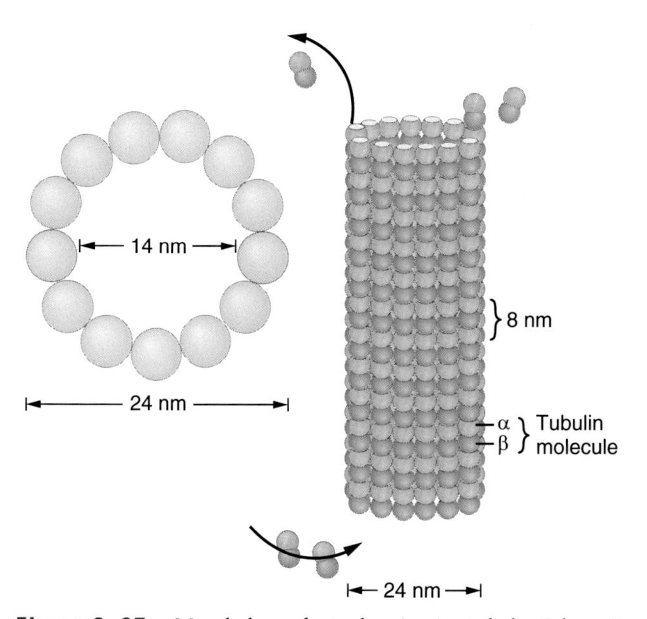

Figure 3–37. Morphology of cytoplasmic microtubules. Schematic representations of microtubules in cross and longitudinal sections.

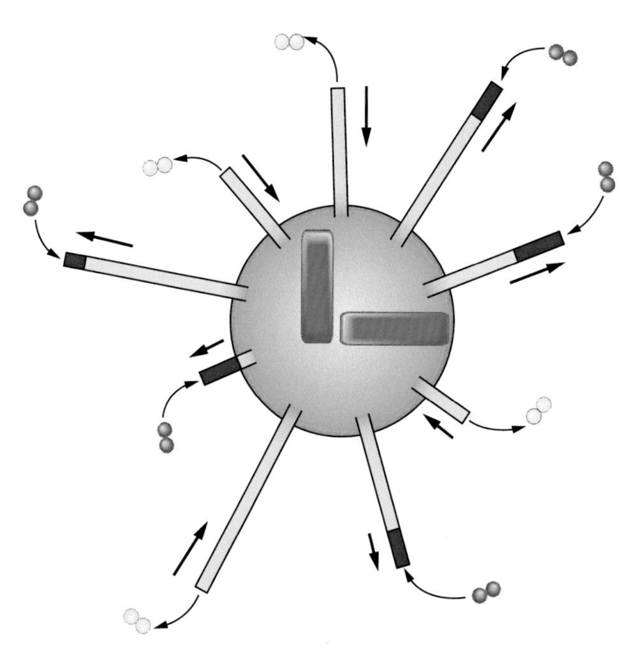

Figure 3–38. Dynamic instability of microtubules. Schematic demonstrating dynamic instability. The GDP-tubulin regions of microtubules are shown as *yellow* regions, and GTP-tubulin microtubule caps are shown as *purple* regions. Microtubules that are assembling from the centrosome contain GTP caps, whereas those that are catastrophically disassembling contain only GDP-tubulin.

with the β-subunit is hydrolyzed to GDP. In effect, this results in the presence of either a small GTP- or GDP-tubulin cap on the end of a microtubule, with the remainder of the microtubule polymer being composed of GDP-tubulin. As long as a microtubule continues to grow rapidly, tubulin subunits will be added to the tubule faster than the nucleotide can be hydrolyzed, and the GTP cap will remain intact. This is important because tubulin subunits add to GTP-capped microtubules much more efficiently than they bind to GDP tubules. If the rate of microtubule assembly slows, GTP hydrolysis can catch up, the GTP cap will be lost, and the entire length of the microtubule polymer will comprise GDP-tubulin. The GDP-capped microtubules are unstable and tend to lose GDP-tubulin subunits from the end of the microtubules, which results in microtubule shortening. Moreover, this rate of loss is rapid and is referred to as a catastrophic. If the hydrolysis of GTP-tubulin becomes slower than GTP-tubulin addition, then the same microtubule can resume growth; this is referred to as a "rescue." This formation and catastrophic breakdown of microtubules is called **dynamic instability**. Dynamic instability provides a partial explanation for how a cell is able to reorganize its microtubule cytoskeleton so rapidly.

By Capping the Minus Ends of Microtubules, the Centrosome Acts as a Microtubule-Organizing Center

Unlike other cytoskeletal filaments, which appear to be nucleated and oriented haphazardly throughout the cytoplasm, cytoplasmic microtubules are all nucleated by the centrosome complex. If cultured cells are fixed and processed for antitubulin immunofluorescence microscopy, a starlike array of microtubules is observed that originates near the nucleus and radiates throughout the cytoplasm (Fig. 3–39).

At the focal point of the astral microtubule array is the centrosome. Ultrastructurally the centrosome is composed of a centriole pair and an osmiophilic cloud of amorphous material, called *pericentriolar material*, which surrounds the centrioles (Fig. 3–40). The individual centrioles of the centriole pair are oriented at right angles to each other, and each centriole comprises nine triplets of short microtubules (0.4–0.5 μm in length).

Experimental analysis shows that the microtubule nucleating capacity of the centrosome complex is contained within the pericentriolar material and not in the centriole pair. The centrosome component that nucleates microtubules is a **γ-tubulin ring complex** found within the pericentriolar material.

The γ-tubulin ring complex is composed of γ-tubulin plus several accessory proteins. γ-Tubulin is a specialized member of the tubulin superfamily of proteins that is localized specifically to the pericentriolar material. It has been demonstrated that γ-tubulin forms a nucleation complex within the centrosome that allows microtubule formation. Moreover, by binding to the minus ends of microtubules, the γ-tubulin within the centrosome caps these microtubule ends. This means that all assembly and disassembly phenomena must occur at the plus ends of microtubules. In addition, the capping of the minus ends of the microtubules provides an explanation of how microtubule formation could occur inside a cell. Like actin microfilaments, free microtubules in vitro have an assembly, or plus end, and a disassembly, or minus end. Microtubules will assemble in vitro only if the concentration of tubulin is so high that tubulin subunits are adding to the assembly end more quickly than tubulin monomers are dissociating from the disassembly end. Apparently the concentration of tubulin inside a cell is below this critical concentration necessary for spontaneous assembly. Therefore the rate of tubulin loss from the minus end of a microtubule that is free in the cytoplasm would exceed the rate of addition of tubulin subunits to the plus end. As a result, microtubules cannot form freely in the cytoplasm and must be nucleated by the centrosome. By capping the minus end of the microtubule,

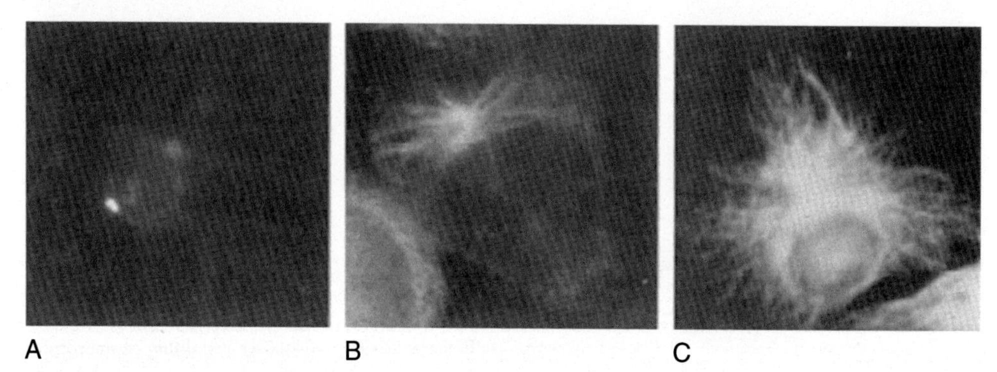

Figure 3–39. The centrosome nucleates cellular microtubules. Antitubulin immunofluorescent staining of cultured mammalian cells demonstrating that the centrosome is the microtubule-organizing center in mammalian cells. (A) Mammalian cells were experimentally treated so that all of the microtubules were disassembled. Only the microtubule-containing centrosomes could be identified. (B) When the experimental treatment was reversed, a starlike array of microtubules began to form off the centrosome. (C) With time the microtubules elongated until they eventually filled the cytoplasm. *(Courtesy R. Balczon.)*

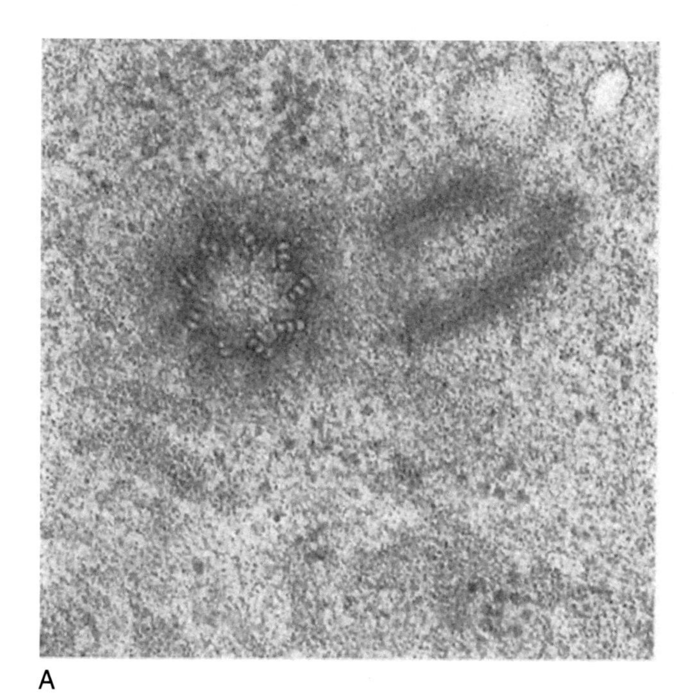

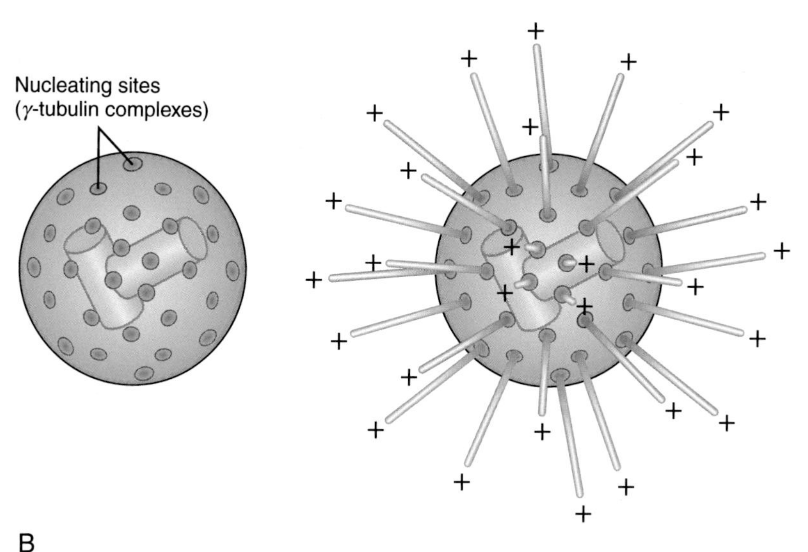

Figure 3–40. Morphology of the centrosome complex. (A) Electron micrograph of a centrosome complex. The centrioles of a centriole pair are oriented at right angles to each other. The centrioles are composed of nine short-triplet microtubules and are surrounded by weakly staining pericentriolar material. (B) γ-Tubulin ring complex nucleates microtubule polymerization from the centrosome. *([A] Courtesy R. Balczon.)*

the presence of a centrosome permits a cell to maintain its cytoplasmic tubulin concentration at levels that are too low to support spontaneous microtubule assembly. Because tubulin levels are so low, the only microtubules that can form in a cell under physiologic conditions are those capped at their minus ends by their associations with a centrosome.

The Behavior of Cytoplasmic Microtubules Can Be Regulated

The picture that develops when considering dynamic instability is one of a cytoplasm that is constantly changing because of the rapid turnover of microtubules. At any one instance, many microtubules would be rapidly growing, whereas others would be quickly and catastrophically disassembling. Although true to some degree the average life span of cytoplasmic microtubules is about 10 min, microtubules can exist for periods that are longer than might be expected because the cell has several mechanisms for stabilizing cytoplasmic microtubules.

One adaptation that cells can use for stabilizing microtubules is to cap the plus end. Because cytoplasmic microtubules are capped on their minus ends by the centrosome, if they were capped on their plus ends, they would not have a free end available for disassembly. This mechanism is, in fact, used by cells during mitotic spindle assembly. As a cell enters into mitosis, numerous microtubules begin to form off the centrosomes. Most of these microtubules disassemble rapidly because of dynamic instability. However, some of the growing microtubules contact the kinetochore regions of the

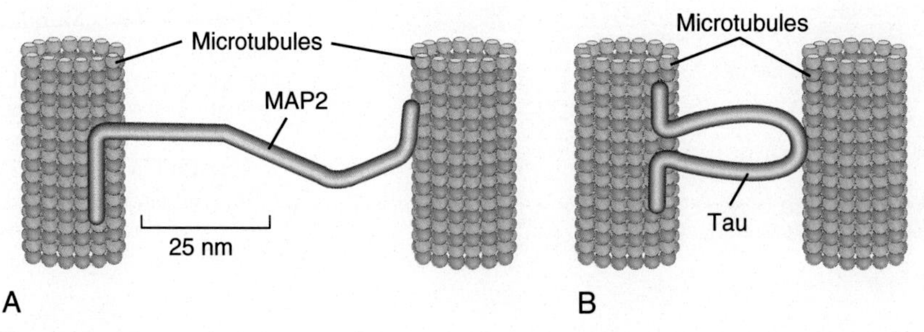

Figure 3–41. High-molecular weight microtubule-associated proteins MAPS (A) and tau (B) bundle microtubules. Because MAP2 has a longer cross-linking domain than tau, it creates a tighter bundle.

mitotic chromosomes and are capped on their plus ends by the proteins of the kinetochore. This microtubule capping selectively stabilizes these microtubules and is an important event in spindle morphogenesis. Other molecular and biochemical modifications can result in the increased stability of cytoplasmic microtubules. These changes in microtubule behavior can be caused either by posttranslational modifications of tubulin or by the interaction of microtubules with any one of several MAPs. The principal posttranslational modification of tubulin in cellular microtubules is the removal of the COOH-terminal tyrosine of the α-tubulin subunit by a detyrosinating enzyme that is present in cells. This detyrosination occurs after tubulin has been incorporated into a microtubule and results in the stabilization or maturation of cytoplasmic microtubules. However, detyrosination may have no direct effect on microtubule kinetics; microtubules that are formed in vitro, using either tyrosinated or detyrosinated tubulin, show no differences in their inherent stability. Therefore it is conceivable that detyrosination may act as a signal that induces the binding of a second protein to microtubules, which results in the increased stability that is observed in detyrosinated cytoplasmic microtubules. When a detyrosinated microtubule is disassembled, a cellular cytoplasmic enzyme is responsible for adding a tyrosine back onto the COOH-terminus of the α-tubulin polypeptide.

The interaction of tubulin with MAPs also results in considerable modifications in the behavior of microtubules. A protein is characterized as being a MAP if it binds to and copurifies with microtubules during their isolation from cellular homogenates. The individual types of MAPs appear to vary among cell types. However, considerable information on the functions of MAPs has been obtained by studying those that have been isolated from neural tissue. Two major classes of MAPs have been identified in neurons: the **high M_r MAPs**, a small family of proteins of 200–300 kDa, and the **tau proteins**, a group of polypeptides of 40–60 kDa. Experimental analysis has demonstrated that these proteins bind to tubulin monomers and assist with microtubule nucleation. In addition, they appear to be involved in the tight bundling of microtubules that is characteristic of the microtubular configurations seen in nerve axons and at other selected cellular sites (Fig. 3–41).

Other types of MAPs can also regulate microtubule length and polymerization rate. A pool of free tubulin dimers can be found within cells. One reason is the protein **stathmin**, which binds to tubulin dimers and prevents their addition to microtubules. The stathmin-tubulin interaction is dissociated on phosphorylation of stathmin. **Katanin** is a microtubule-severing protein and detaches them from the microtubule-organizing center. The severing activity of katanin is ATP dependent and results in rapid depolymerization of released microtubules.

Microtubules Are Involved in Intracellular Vesicle and Organelle Transport

One of the important functions of microtubules is the intracellular transport of organelles and vesicles. For the microtubule cytoskeleton to fulfill this role, microtubules must have a means of generating a force that allows such motile behavior to occur. Experimental analysis has allowed the identification and purification of the ATPases that appear to be involved in force generation.

Two families of microtubule-dependent ATPases have been identified that appear to be important in cytoplasmic transport. One of these families is kinesin and kinesin-related proteins (KRPs). The original kinesin, isolated from squid axons, is a large multisubunit protein that is involved in translocating vesicles along microtubules from the minus, or centrosome, end toward the distal plus ends. This allows the transport of vesicles from deep within the cytoplasm, where they are produced by budding from the Golgi apparatus, to the cell cortex where secretion can occur. In neurons, it allows for the transport of cargo from the soma to the presynaptic terminal of axons.

We now know that there are many families of kinesins and KRPs where the only common feature is a conserved motor domain. The original kinesin had two heavy chains, where the motor domain was close to the

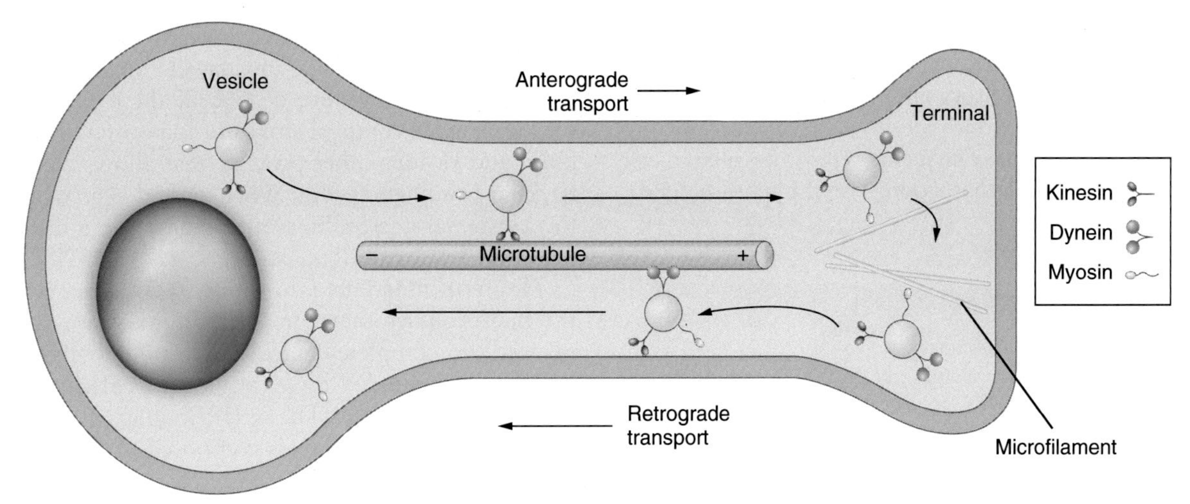

Figure 3–42. Vesicle transport along microtubule tracks. Schematic demonstrating how vesicles might be transported along cytoplasmic microtubules by microtubule-dependent ATPases. Microtubules are polar structures with defined plus and minus ends. Vesicles are thought to be transported in the anterograde direction (from the minus to the plus end of a microtubule) by certain members of the kinesin family. Vesicles and organelles are thought to be translocated in the retrograde direction (from the plus to the minus end of the microtubule) by cytoplasmic dynein and other kinesin-related proteins.

N-terminus, and two light chains. Its structure is remarkably similar to myosin II. Other members of the kinesin and KRP family that have the motor domain at or near the N-terminus move vesicles toward the plus ends of microtubules (an example is KIF 1B). Members of the kinesin and KRF families that have the motor domain close to the C terminus move cargo toward the minus ends of microtubules (an example is KIF C2). KIF2 is a member of an unusual family of kinesins that have their motor domain toward the center of the heavy chain and cannot translocate along microtubules. Instead, this family, named catastrophins, binds to the end of microtubules and stimulates dynamic instability. KIF 1B is a member of a family of monomeric kinesins that translocates cargo toward the plus end of microtubules (Fig. 3–42).

The other family of enzymes that are involved in cellular motile events is the cytoplasmic form of the ciliary enzyme dynein. Cytoplasmic dynein is a high M_r multisubunit protein complex that translocates structures along microtubules from their plus to their minus ends (Fig. 3–43). Microtubules appear to play a relatively passive role in most types of intracellular movement, with the active roles being performed by the microtubule-dependent ATPases. A good analogy for visualizing these events would be to consider a railroad: The microtubules would serve as the tracks, and the locomotory forces responsible for transporting the vesicular cargo would be generated by ATPases, such as kinesin and dynein.

How does the cargo associate with kinesin or dynein? Membrane-associated motor receptors (MAMRs) interact with the tails of kinesin family members. One example of a MAMR is the amyloid precursor protein (APP). The abnormal processing of APP has been linked to Alzheimer's disease. Cytoplasmic dynein interacts with membrane vesicles through the dynactin complex (Fig. 3–44).

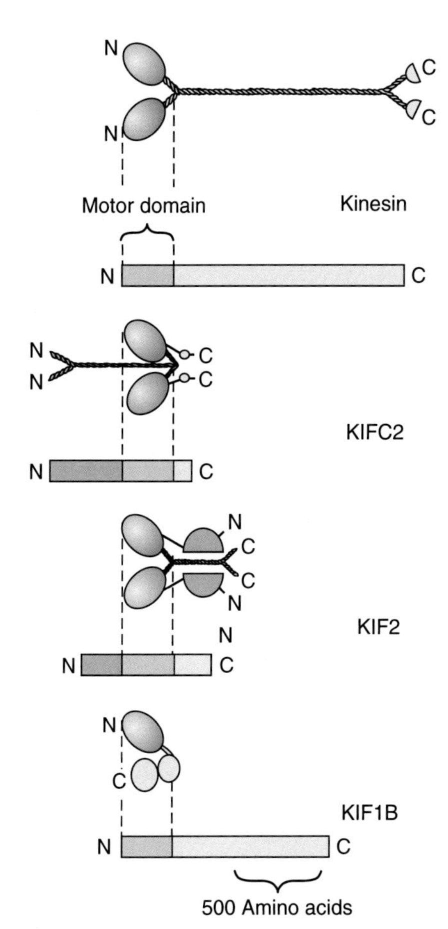

Figure 3–43. Kinesin and kinesin-related proteins.

Cilia and Flagella Are Specialized Organelles Composed of Microtubules

Cilia and flagella are specialized cellular appendages that extend from the surfaces of several different cell types. Cilia are prominent in the respiratory tract and on the apical surface of the epithelial cells that line the oviduct. In the respiratory tract, cilia are involved in clearing mucus from the respiratory and nasal passages, whereas those that line the oviduct are involved in transporting ova toward the uterus. The major type of flagellated cell in humans is the spermatozoon. For the mature sperm cell, the beating flagellum provides the force that allows the sperm to swim. Cilia and flagella are similar ultrastructurally. At the core of one of these organelles is the axoneme, a complex structure composed of microtubules and various other proteins that allows ciliary and flagellar bending to occur. When viewed in cross section, the axonemal microtubules are arranged in a distinctive nine-plus-two array (Fig. 3–45).

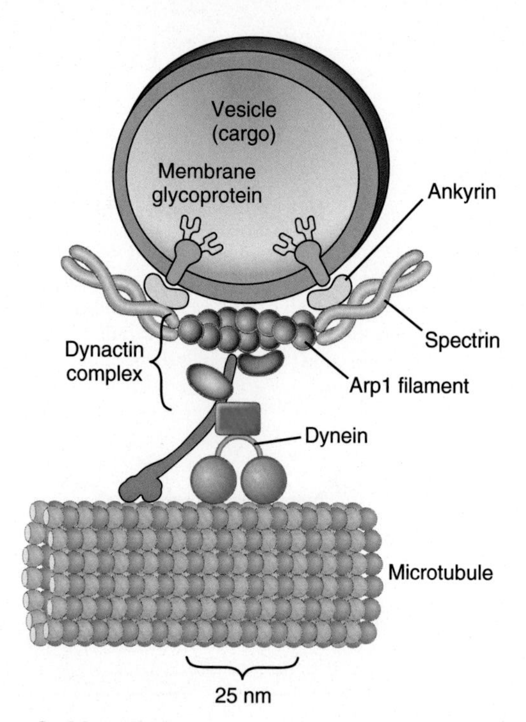

Figure 3–44. The dynactin complex links cargo to cytoplasmic dynein. The dynein tail region is linked by accessory proteins to an Arp 1 filament that is associated with actin protofilaments within a spectrin membrane skeleton linked to the cargo via ankyrin.

The term **nine-plus-two** refers to the orientation of the microtubules that comprise the axoneme. In axonemes, two complete central microtubules (the central pair) are surrounded by a circumferential ring of nine doublet microtubules. The outer doublets are arranged so that each doublet pair is composed of one complete microtubule (the A tubule), which consists of 13 protofilaments, and an incomplete microtubule (the B tubule), which is composed of only 11 protofilaments. The B tubule shares a portion of the A tubule wall. The nine-plus-two array of microtubules traverses the axoneme, extending from the specialized centrioles, called **basal bodies**, which are located at the base of the cilium or flagellum, all of the way out to near the tip of the cilium or flagellum.

In addition to the microtubules, several other important proteins can be found in axonemes. These accessory proteins are absolutely essential for normal ciliary function. Extending from the A tubule of each doublet toward the B tubule of the neighboring doublet are two proteinaceous arms (see Fig. 3–44). These arms are actually the enzyme dynein, the ATPase that is responsible for ciliary and flagellar motility. Also extending between the A and B tubules of the neighboring doublet is a protein called *nexin*. This attaches neighboring doublets to each

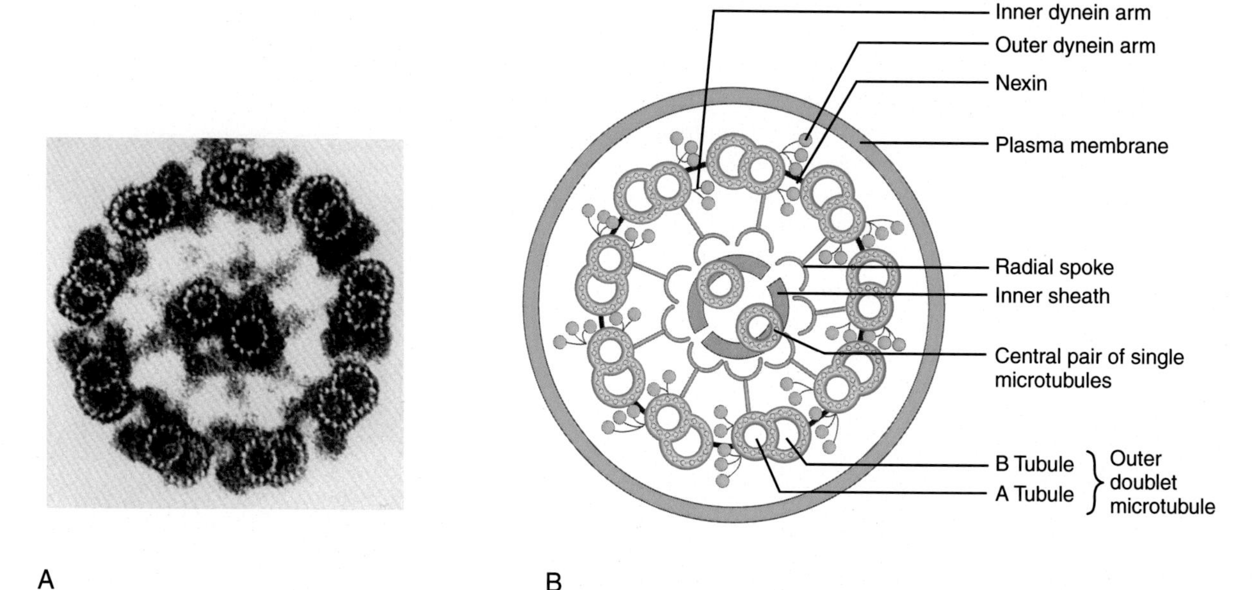

Figure 3–45. Axonemal nine-plus-two array of microtubules and other axoneme proteins. (A) An electron micrograph showing the nine-plus-two organization of microtubules in an axoneme. (B) Schematic showing the organization of the proteins in an axoneme. *([A] Courtesy Dr. W.L. Dentler; [B] modified from Alberts B, Bray D, Lewis J, et al.* Molecular Biology of the Cell, *2nd ed. New York: Garland Publishing, 1989.)*

other. Finally a radial spoke extends off the A tubule of each doublet and contacts an electron-dense sheath surrounding the central pair of microtubules, thereby connecting the doublet microtubules to the central pair. The dynein arms, nexin links, and spoke proteins exhibit a periodicity along the entire length of the axoneme. In addition to these prominent proteins, numerous other minor proteins are present in the axoneme.

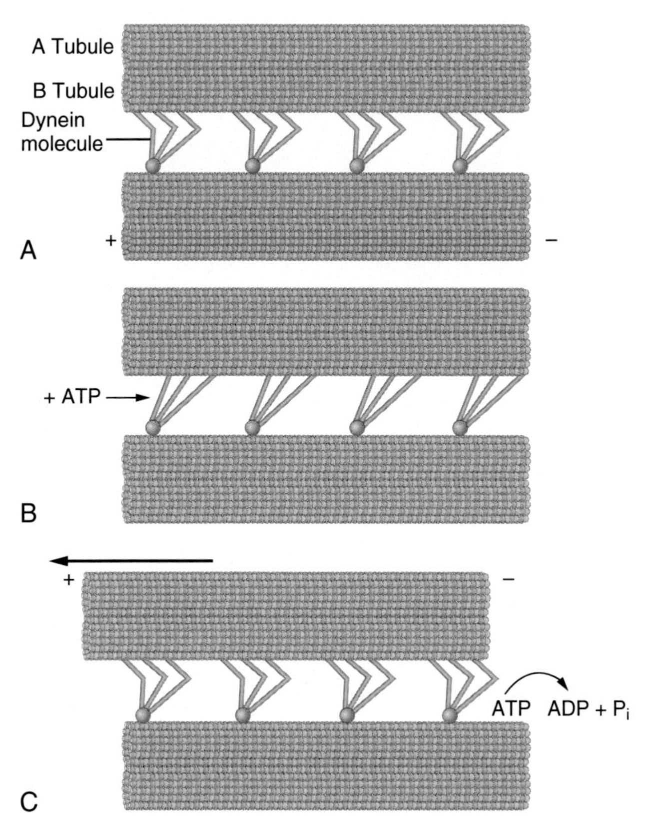

Figure 3–46. Dynein cross-bridge cycles lead to the bending of cilia and flagella. (A) Schematic showing how the binding of ATP causes a conformational change in the structure of the ATPase dynein. (B) The ATP binding causes dynein to release from the wall of the adjacent tubule of a microtubule pair and then reattach to that same microtubule pair farther down the length of that microtubule doublet. (C) Cleavage of ATP to ADP results in a force that causes the two microtubule pairs to slide past one another. In vivo the sliding of the microtubule doublets relative to one another is converted to bending by the nexin cross-links and spoke proteins.

Axonemal Microtubules Are Stable

Most cellular microtubules are labile structures that can be assembled and disassembled rapidly. Axonemal microtubules, however, are stable structures that resist breakdown. One of the modifications of axonemal microtubules that may contribute to this increased stability is the enzymatic acetylation of a lysine residue on α-tubulin subunits. Cytoplasmic microtubules, which are turning over rapidly, are nonacetylated. Like detyrosinated tubulin, acetylated tubulin shows in vitro kinetic behavior similar to unmodified tubulin, suggesting that the acetylation of axonemal α-tubulin subunits may serve as a signal for other proteins that bind to the microtubule wall to stabilize the axonemal tubules.

Microtubule Sliding Results in Axonemal Motility

Axonemes can be isolated from several sources quite easily. This is accomplished by shearing cilia or flagella from cells, selectively removing the residual plasma membrane that surrounds the axonemes, and extracting the axonemes in a buffer that contains a mild detergent. When isolated sperm flagellar axonemes are incubated in a buffer that contains ATP, the axonemes will continue to beat in a relatively normal fashion; therefore all of the information that is required for ciliary and flagellar motility is contained within the structure of the axoneme alone. Biochemical and genetic studies have been used to dissect the mechanism of axonemal motility.

When isolated axonemes are treated with a proteolytic enzyme, the nexin cross-links and radial spokes are selectively digested, whereas the microtubules and dynein arms remain intact. If these protease-treated axonemes are incubated with ATP, the axoneme elongates up to nine times its original length. Microscopic analysis has shown that this is because the nine outer doublet microtubule pairs actively slide past one another. That these treated axonemes lack cross-links and spokes suggests that the dynein arms are the ATPase that drives axonemal motility. Moreover, this result suggests that nexin and the spoke protein can convert the activity of dynein into the bending that results in ciliary and flagellar motility. In fact, this is true. When isolated axonemes are low-salt extracted, the dynein arms are released, whereas the microtubules, nexin links, and spoke proteins remain intact. Such extracted axonemes will not beat when ATP is added. However, if ATP is added to the salt extract that contains the dynein arms, the ATP is actively hydrolyzed. Therefore the following mechanism can be visualized for ciliary and flagellar activity (Fig. 3–46).

In the presence of ATP, dynein undergoes a conformational change. The net effect of this change is that the dynein arm releases from the B tubule of the adjacent microtubule doublet pair and then reattaches to that same doublet pair farther down the length of the doublet. This cycle is repeated when another ATP binds to the dynein arm. However, this "walking" of the dynein arms down the length of the microtubule wall is resisted by the nexin cross-links and radial spokes, and these two structures convert the sliding of the adjacent microtubule pairs into a bending motion. Such a complex pathway must be tightly regulated because at any one instance dynein arms on one region of an axoneme must be active, whereas in

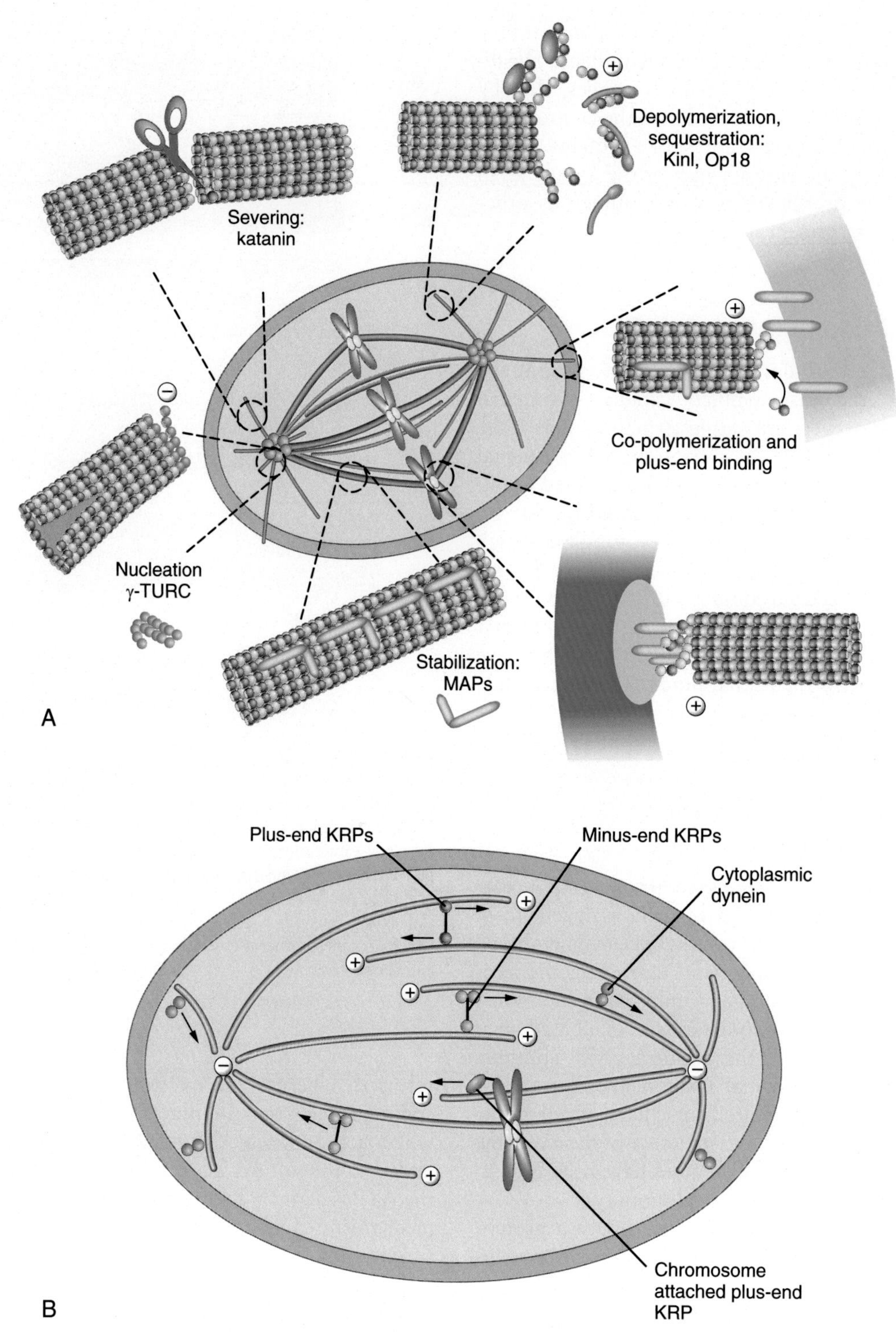

Figure 3–47. The role of microtubules, microtubule-associated proteins (MAPs), and motor proteins in mitosis. (A) The role of various classes of MAPs in the dynamics of the microtubules of the mitotic spindle. (B) The role of various classes of kinesin-related proteins and cytoplasmic dynein in mitosis. *(Modified from Gadde S, Heald, R.* Curr Biol *2004;14:R797–R805.)*

other areas the dynein arms must be relaxed for ciliary and flagellar beating to occur.

Several human diseases are the result of mutations in one of the genes that encodes axonemal proteins. As would be expected, male individuals with such conditions are sterile because their sperm are immotile. In addition, patients afflicted with one of these conditions show chronic respiratory tract problems because the respiratory cilia are unable to clear mucus from the bronchiole and nasal passages.

Microtubules and Motor Proteins Are Responsible for the Function of the Mitotic Spindle

Mitosis and the mitotic spindle are discussed in relation to the cell cycle in Chapter 10. Here, we discuss the role of microtubules, MAPs, and motor proteins in the proper functioning of the mitotic spindle. As mitosis begins, the centrosomes and chromosomes have duplicated, and during this prophase the nucleus breaks down, the chromosomes condense, and then the chromosomes become separated. The interphase microtubules disassemble based on an increase in catastrophic microtubule depolymerization and decreased rescue events. As a result, new microtubules are formed with their plus ends associating with the newly condensed chromosomes. At prometaphase, newly formed microtubules growing from the two centrosomes become attached to the kinetochore of the condensed chromatids, which move to the center of the microtubule array. The microtubules growing from the two centrosomes have opposite polarity or are antiparallel in polarity. At metaphase the chromosomes are aligned at the center of this antiparallel set of microtubules. A second class of microtubules is the interpolar microtubules that overlap at the center of the spindle, and the third class is the astral microtubules that polymerize from the centrosomes or spindle poles growing away from the kinetochore and interpolar microtubules. At anaphase A the sister chromosomes separate, moving toward the spindle poles, and at anaphase B the spindle poles separate farther with a formation of a central spindle. Finally, at telophase, cytokinesis occurs, and the nuclear envelope reforms.

As shown in Figure 3–47, the role of microtubules, MAPs, and motor proteins in the process described earlier is beginning to be understood. As discussed earlier in the chapter, microtubules grow from the γ-tubulin ring complex and undergo catastrophic shrinkage when the plus end contains a GDP cap while being rescued from shrinkage when a GTP cap exists. Catastrophins (like KinI), members of the KRP family, stimulate catastrophic shrinkage, whereas MAPs such as MAP-2 and tau protein stabilize microtubules. End-binding MAPs such as EB1 and CLIP 170 are thought to attach microtubules to kinetochores and cell membranes by linking to proteins called APC and CLASP. Katanin (see earlier discussion) severs microtubules, creating new GDP-capped ends and releasing the microtubules from the centrosome. Stathmin/Op18 bind to tubulin dimers, stimulating GTP hydrolysis and maintaining a pool of nonpolymerized tubulin. As shown in Figure 3–47B, motor proteins play several important roles in the mitotic process. Tetrameric plus end kinesin family members (BimC and Eg5 family) bind microtubules of opposite polarity, causing spindle pole separation. Minus end KRPs cross-link microtubules and focus astral microtubule minus ends at the spindle poles. Chromosome-attached plus end KRPs are involved in chromosomal attachment to the microtubules and movement of chromosomes toward the metaphase plate. The minus end cytoplasmic dynein serves as a "feeder" moving microtubules toward the spindle poles where they are fed into the "chipper" KinI family members that chew the plus ends of the microtubules in Pacman fashion. Cytoplasmic dynein is also thought to move kinetochore microtubules in a plus end direction toward the chipper. Although many players remain to be determined, we are now starting to see how microtubule dynamics, MAPs, and motor proteins work together to cause the detailed movements seen in mitosis.

SUMMARY

The cytoskeleton is responsible for contraction, cell motility, movement of organelles and vesicles through the cytoplasm, cytokinesis, establishment of the intracellular organization of the cytoplasm, establishment of cell polarity, and many other functions that are essential for cellular homeostasis and survival. It accomplishes these tasks through three basic structures: 7- to 8-nm-diameter microfilaments composed of actin, 10-nm IFs with cell-specific composition, and 24-nm-outer-diameter microtubules composed of tubulin dimers. The cytoskeleton is a dynamic structure where the three major filaments and tubules are under the influence of proteins that regulate their length, state of polymerization, and level of cross-linking. Members of the myosin family move vesicles along actin microfilaments with specific directionality, whereas members of the kinesin/KRP and dynein families move cargo along microtubule tracks and play essential roles in the formation and function of the mitotic spindle. Both forms of translocation require ATP hydrolysis. Contraction caused by the interaction of myosin heads with actin filaments also derives its energy from ATP hydrolysis.

The membrane skeleton was first described in erythrocytes, where it maintains the cellular biconcave shape and gives the cell its properties of elasticity and flexibility. Spectrins are now known to be present in all eukaryotic cells, and these membrane skeletons play roles as diverse as controlling membrane traffic, DNA repair, calcium release from internal stores, and synaptic transmission.

Suggested Readings

Bornens M. Centrosome composition and microtubule anchoring mechanisms. *Curr Opin Cell Biol* 2002;14:25–34.

Endow SA. Kinesin motors as molecular machines. *BioEssays* 2003;25:1212–1219.

Etienne-Manneville S, Hall A. Rho GTPases in cell biology. *Nature* 2002;420:629–635.

Gadde S, Heald R. Mechanisms and molecules of the mitotic spindle. *Curr Biol* 2004;14:R797.

Goodman SR, Johnson D, Youngentob SL, Kakhniashvili D. The spectrinome: the interaction of a scaffold protein creating nuclear and cytoplasmic connectivity and function. *Exp Biol Med (Maywood)* 2019;244:1273–1302.

Gundersen GG, Gomes ER, Wen Y. Cortical control of microtubule stability and polarization. *Curr Opin Cell Biol* 2004;16:1–7.

Herrmann H, Aebi U. Intermediate filaments and their associates: multi-talented structural elements specifying cytoarchitecture and cytodynamics. *Curr Opin Cell Biol* 2000;12:79.

Hsu YJ, Goodman SR. Spectrin and ubiquitination: a review. *Cell Mol Biol* 2005;(Suppl. 51):OL801–OL807.

Pollard TD, Borisy GG. Cellular motility driven by assembly and dis-assembly of actin filaments. *Cell* 2003;112:453–455.

Spudich JA. The myosin swinging cross-bridge model. *Nat Rev Mol Cell Biol* 2001;2:387–392.

Yin HL, Janmey PA. Phosphoinositide regulation of the actin cytoskeleton. *Annu Rev Physiol* 2003;65:761–789.

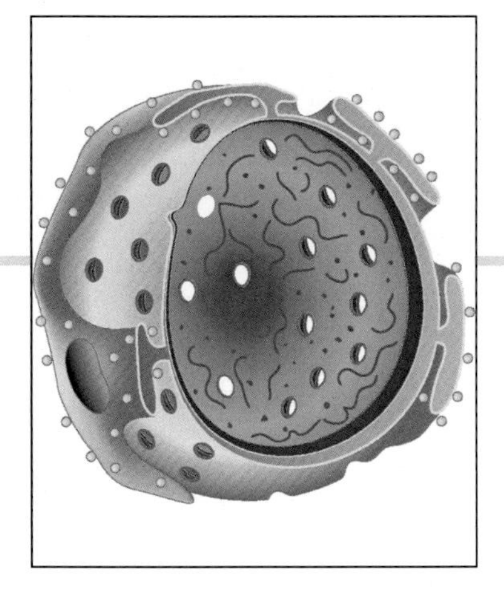

Chapter 4

Organelle Structure and Function

This textbook is focused primarily on eukaryotic mammalian cells that have dimensions in the 10- to 60-μm range and contain complex membrane-bound subcellular organelles essential for cell functions. Figure 4–1 presents an overview of the subcellular organelles of a typical mammalian cell, including the plasma membrane that surrounds the cell. Most cells in the human body are part of a solid organ and contain plasma membranes specialized to contact the plasma membranes of adjacent cells. This specialization results in two distinct types of plasma membrane that have different protein and lipid compositions that do not mix freely. Typically a portion of the plasma membrane faces a cavity, such as the lumen of the intestine, and is called the *apical membrane*. Projections called microvilli extend out from the apical surface into the lumen of the organ. The plasma membrane surface that faces adjacent cells on the side, and material on the opposite side of the apical face, is called the *basolateral plasma membrane*. Specialized junctions between the cells prevent proteins and lipids in apical and basolateral plasma membranes from diffusing past the junctions and mixing. Because the shapes of cells in a solid organ are not symmetric, they are referred to as polarized cells. In contrast, cell types such as red blood cells and lymphocytes that are not part of a solid organ have only one type of plasma membrane and are not polarized.

The nuclear membrane contains two distinct lipid bilayers, an inner and outer membrane (see Figs. 4–2 and 4–3). There are specialized structures called *nuclear pores* that form a channel connecting the inside of the nucleus (the nucleoplasm) with the cytoplasm. The outer nuclear membrane is continuous with the endoplasmic reticulum (ER), the membrane with the largest surface area in the cell. There are two types of ER: the rough ER (RER) and the smooth ER (SER), which have different functions (discussed later).

The Golgi complex is composed of a stack of multiple flattened cisternae. The stack has a forming face termed the *cis* Golgi network (CGN) that receives newly synthesized proteins and lipids from the ER and a *trans* face from which cargo depart for delivery to the plasma membrane or lysosomes, an organelle that contains hydrolytic enzymes that can degrade all classes of macromolecules. In addition to membrane traffic from the ER to distal compartments of the secretory pathway, extracellular cargo internalized via endocytic vesicles formed at the plasma membrane passage through early endosomes and late endosomes prior to undergoing delivery to the lysosome or other internal membrane compartments. The membrane-bound organelles described earlier are connected by pathways of membrane flow that involve membrane vesicles or membrane tubules that form at one membrane compartment and then fuse with the next

☆ Certain sections of the chapter were retained from the previous edition, which were written by Drs Gail A. Breen and Rockford K. Draper.

Goodman's Medical Cell Biology. https://doi.org/10.1016/B978-0-12-817927-7.00004-1

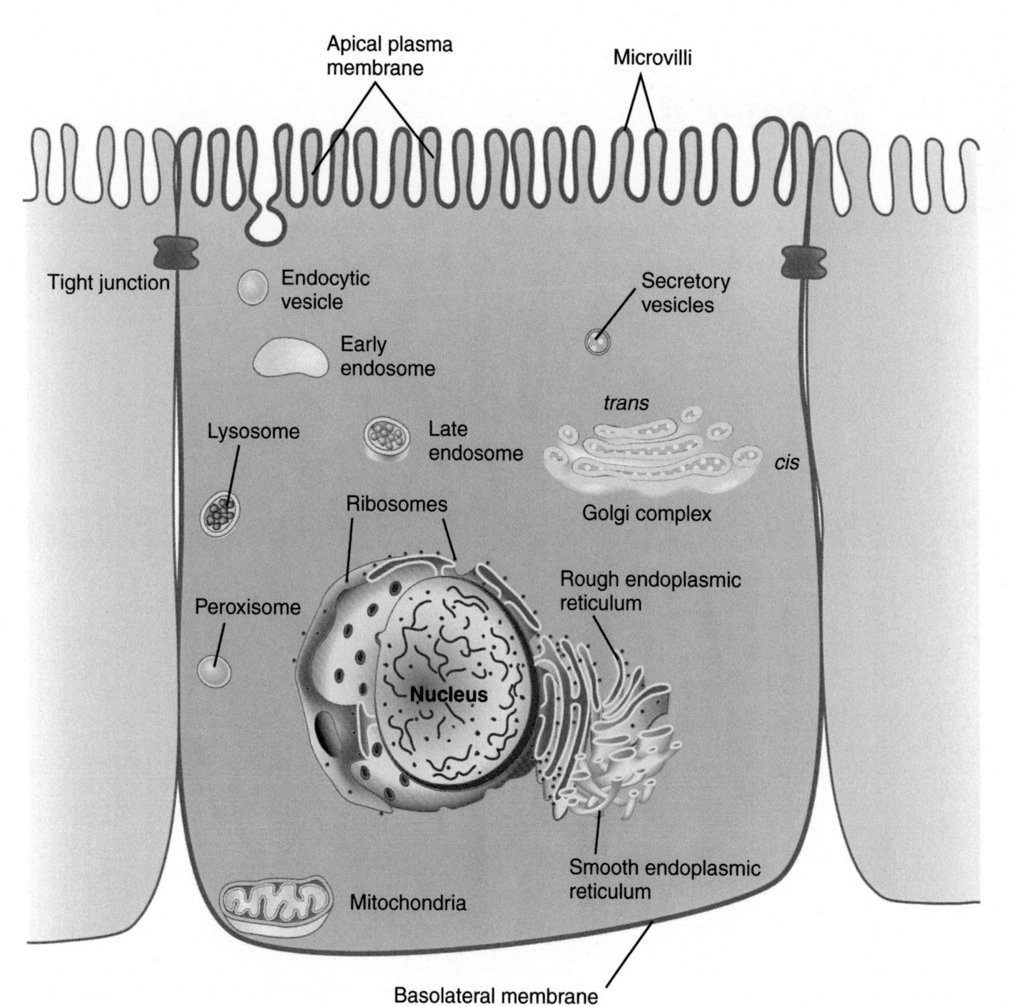

Figure 4–1. **Overview of plasma membrane and subcellular organelles of mammalian cell.** The subcellular organelles of a typical mammalian cell include the nucleus (surrounded by a double membrane); the rough endoplasmic reticulum (RER); the smooth endoplasmic reticulum (SER); the Golgi complex; secretory vesicles; various endosomes; lysosomes; peroxisomes; and mitochondria (contains an inner and an outer membrane). This example shows a simple polarized cell with apical and basolateral plasma membranes and the tight junctions that separate them. Other junctions between cells that are not shown here are discussed in Chapter 7.

compartment in the pathway. Most proteins and lipids are incorporated into a membrane in the ER, move to the Golgi complex, and then travel on to the plasma membrane, an endosome, or a lysosome. Membrane traffic from the ER out toward the plasma membrane is called *anterograde transport*. Membrane traffic in the reverse direction, called *retrograde transport*, also occurs to recover membrane components removed by anterograde transport, thereby maintaining homeostasis. Because the different organelles have different functions, they each have different protein and lipid compositions necessary to support their functions. How membranes maintain their unique compositions despite robust anterograde and retrograde membrane traffic is not well understood. Two organelles in mammalian cells, the mitochondria and peroxisomes, are for the most part not stations in these pathways of membrane flow, and specialized mechanisms exist to deliver proteins and lipids to them.

A typical mammalian cell makes about 20,000–30,000 different proteins, and approximately, two-thirds of these proteins remain in the cytosol following synthesis. Proteins that do not stay in the cytosol can pass through the ER membrane en route to secretion or become a membrane-spanning protein. Alternatively a subset of proteins undergo posttranslational delivery to other subcellular compartments, including the nucleus, mitochondria, and peroxisomes. Sorting signals (Table 4–1) encoded within the proteins serve as "zip codes" that are read by the cellular sorting machinery, which delivers the proteins to their correct cellular destination. Mutations that lead to the missorting of cellular proteins can lead to a variety of pathologic conditions that cannot be appreciated without an understanding of organelle cell biology. In the rest of this chapter, the structure and function of subcellular organelles is presented beginning with the nucleus and moving to the ER and outward, following the innate pathways of membrane flow that connect the organelles.

THE NUCLEUS

The largest organelle in most mammalian cells is the nucleus, bounded by the double nuclear membrane called the *nuclear envelope* (Fig. 4–2). Within the nuclear envelope is the nucleoplasm that contains the cell's DNA

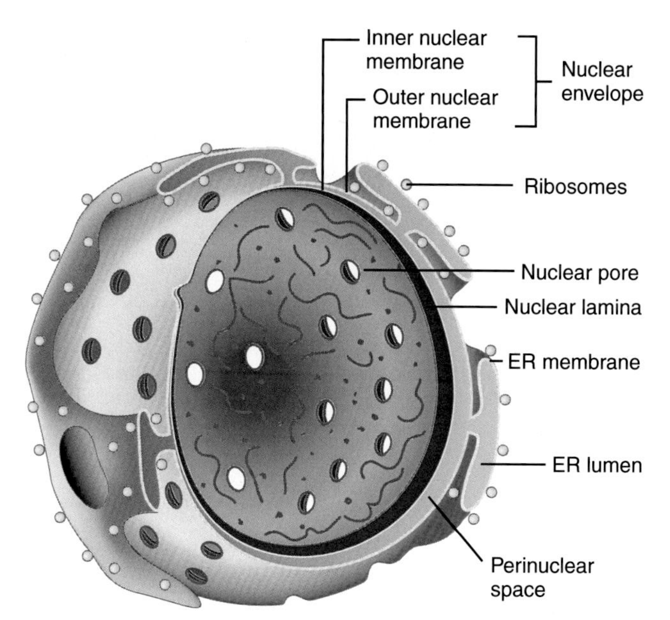

Figure 4–2. **Relationship of the nuclear envelope with cellular structures.** Diagram shows the double membrane that surrounds the nuclear compartment. The inner nuclear membrane is lined by the fibrous protein meshwork of the nuclear lamina. The outer nuclear membrane is contiguous with the membrane of the endoplasmic reticulum (ER). The outer nuclear membrane often has ribosomes associated with it that are actively synthesizing proteins that first enter the region between the inner and outer nuclear membranes, the perinuclear space, which is contiguous with the lumen of the ER. The double membrane of the nuclear envelope contains pores that have regulated channels for passage of material between the cytoplasm and the nucleoplasm.

(except for mitochondrial DNA). The nucleus also contains one or more nucleoli, which are readily seen by light microscopy and are sites where ribosomal RNA (rRNA) is synthesized and begins to assemble into ribosomes. Evidence also exists that the nucleoplasm contains a fibrous network called the *nuclear matrix* that may provide a scaffold for chromatin binding. The synthesis of DNA (replication) and RNA (transcription) occur in the nucleus (see Chapter 6 for further discussion).

The inner nuclear membrane faces the nucleoplasm and contacts an underlying nuclear lamina, a network made from fibrous proteins called *lamins*. The nuclear lamina helps shape the nuclear envelope and is a key element in the breakdown of the nuclear envelope at mitosis. During mitosis the nuclear membrane disassembles, allowing for segregation of replicated chromosomes to daughter cells. At the completion of mitosis, the nuclear membrane undergoes reassembly. The perinuclear space between the inner and outer nuclear membranes is about 20–40 nm across. The outer nuclear membrane is continuous with the ER membrane, so the perinuclear space is an extension of the ER lumen. The outer nuclear membrane and the surface of the RER contain bound ribosomes, which translate membrane bound and secretory protein into the ER and participate in translocating these proteins through the ER membrane. The nuclear envelope contains holes where the inner

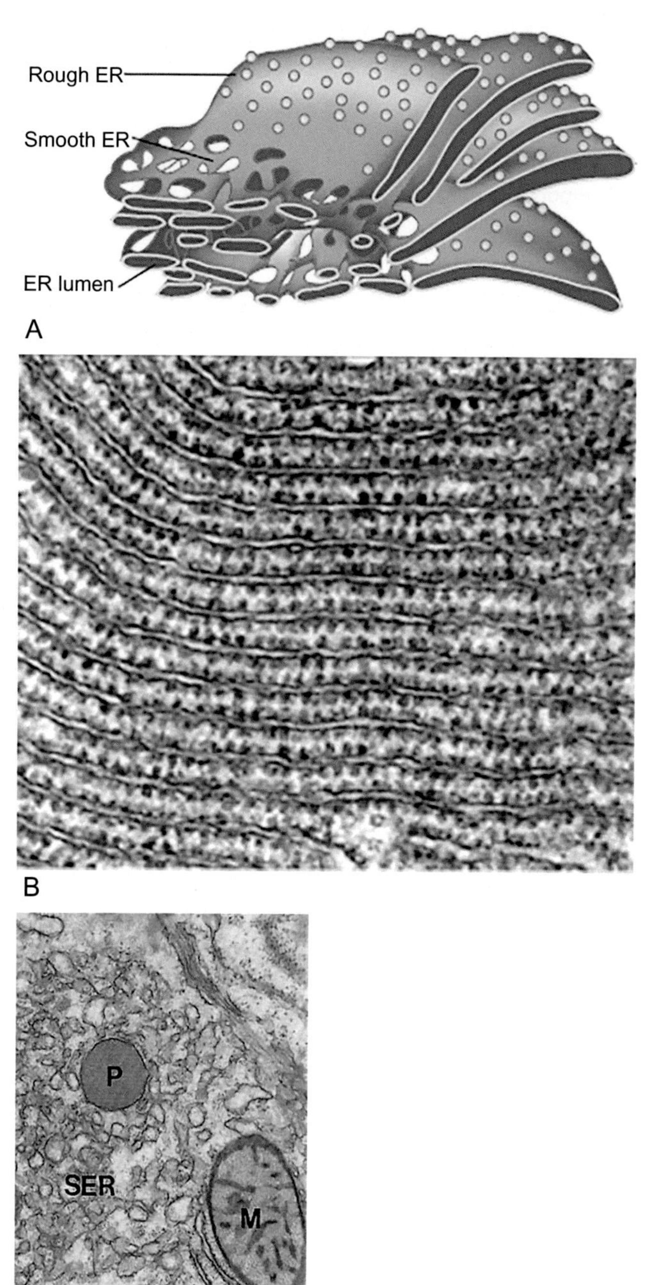

Figure 4–3. **Structure of the rough and smooth endoplasmic reticulum (ER).** (A) The rough endoplasmic reticulum (RER) consists of oriented stacks of flattened cisternae studded with ribosomes on their cytoplasmic surface. The luminal space is 20–30 nm. The smooth ER (SER) has no associated ribosomes and often appears as 30- to 60-nm-diameter membranous tubes that are connected with the RER, so the two ER compartments share the same luminal space. (B) Electron micrograph of the RER in secretory cell from the silk gland of the silkworm. (C) Electron micrograph of the smooth ER (SER) from a rat hepatocyte. P indicates a peroxisome and M indicates mitochondria. *([B] From Shibata Y, Voeltz GK, Rapoport TA. Rough sheets and smooth tubules.* Cell *2006;126(3):435–439. [C] Modified from Steegmaier M, Oorschot V, Klumperman J, Scheller RH. Syntaxin 17 is abundant in steroidogenic cells and implicated in smooth endoplasmic reticulum membrane dynamics.* Mol Biol Cell *2000;11(8):2719–2731.)*

TABLE 4–1. Some Typical Signal Sequence

Function of Signal Sequence	Example of Signal Sequence
Import into nucleus	-Pro-Pro-Lys-Lys-Lys-Arg-Lys-Val-
Export from nucleus	-Met-Glu-Glu-Leu-Ser-Gln-Ala-Leu-Ala-Ser-Ser-Phe-
Import into mitochondria	$^{+}H_3N$-Met-Leu-Ser-Leu-Arg-Gln-Ser-Ile-Arg-Phe-Phe-Lys-Pro-Ala-Thr-Arg-Thr-Leu-Cys-Ser-Ser-Arg-Tyr-Leu-Leu-
Import into plastid	$^{+}H_3N$-Met-Val-Ala-Met-Ala-Met-Ala-Ser-Leu-Gln-Ser-Ser-Met-Ser-Ser-Leu-Ser-Leu-Ser-Ser-Asn-Ser-Phe-Leu-Gly-Gln-Pro-Leu-Ser-Pro-Ile-Thr-Leu-Ser-Pro-Phe-Leu-Gln-Gly-
Import into peroxisomes	-Ser-Lys-Leu-COO^{-}
Import into ER	$^{+}H_3N$-Met-Met-Ser-Phe-Val-Ser-Leu-Leu-Leu-Val-Gly-Ile-Leu-Phe-Trp-Ala-Thr-Glu-Ala-Glu-Gln-Leu-Thr-Lys-Cys-Glu-Val-Phe-Gln-
Return to ER	-Lys-Asp-Glu-Leu-COO^{-}

Some characteristic features of the different classes of signal sequences are highlighted in color. Where they are known to be important for the function of the signal sequence, positively charged amino acids are shown in *red* and negatively charged amino acids are shown in *green*. Similarly, important hydrophobic amino acids are shown in *orange* and important hydroxylated amino acids are shown in *blue*. $^{+}H_3N$ indicates the N-terminus of a protein; COO^{-} indicates the C-terminus.

and outer membranes fuse together to create pores (Fig. 4–2). rRNA, transfer RNA (tRNA), and messenger RNA (mRNA) made in the nucleus must be exported to the cytoplasm, and proteins made in the cytoplasm must be imported into the nucleus. This bidirectional transport occurs through the nuclear pores (see Chapter 6 for a detailed description of bidirectional transport and nuclear function).

ENDOPLASMIC RETICULUM

The RER and SER together form the ER, the largest membrane system in mammalian cells (Fig. 4–3). No discontinuity exists in the luminal space enclosed by the RER and SER, but the two membranes are morphologically and functionally distinct. The RER is studded with ribosomes engaged in the synthesis of secretory and integral membrane proteins, and it is the bound ribosomes that give the RER its rough appearance in electron micrographs (Fig. 4–3, B). The SER most often appears by electron microscopy as tubules and cross-sectional vesicles without any distinct surface features (Fig. 4–3, C).

Smooth Endoplasmic Reticulum

The SER has a variety of functions and is more prominent in certain cell types whose roles require an enhanced SER ability. Four common functions are the mobilization of glucose from glycogen, calcium storage, drug detoxification, and the synthesis of lipids. Glucose is stored as the polymer glycogen in close proximity to the SER, especially in the liver, which plays a key role in glucose homeostasis. The glucose-6-phosphate resulting from glycogen breakdown is dephosphorylated by glucose-6-phosphatase, an SER-bound enzyme that is primarily expressed in the liver and kidney. Type 1 glycogen storage disease (Von Gierke disease), one of about a dozen diseases that affect glycogen metabolism, is due to a genetic deficiency of glucose-6-phosphatase. Patients with this disease can store glycogen but cannot break it down to glucose and, with time glycogen accumulates, enlarging the liver. The disease causes chronic low blood sugar and abnormal growth and is frequently fatal.

The SER is a storage site for calcium within cells. Calcium is pumped into the SER by active transport and released in response to hormonal signals. This is particularly important in muscle cells where the SER is called the sarcoplasmic reticulum. Calcium is released in response to signaling pathways initiated by neurotransmitter binding to cell surface receptors.

Cytochrome P450s are a large family of enzymes resident in the membrane of the SER that use oxygen and nicotinamide adenine dinucleotide phosphate (NADPH) to hydroxylate a wide variety of substrates, including steroids and drugs. Hydroxylation often increases the solubility of hydrophobic drugs, facilitating clearance from the body, and selected cytochrome P450 enzymes are upregulated in response to different drugs. This upregulation can be large enough to cause dramatic expansion of the SER membrane. For example, chronic barbiturate use leads to expansion of the SER caused by induction of detoxifying cytochrome P450 enzymes. The increased inactivation of the drug requires larger barbiturate doses to achieve an effect, which is part of the addictive spiral in chronic users. Carcinogens, such as polycyclic aryl hydrocarbons, are also hydroxylated by SER-associated cytochrome P450 enzymes, which frequently enhance their carcinogenic activity.

Phospholipids, ceramide, and sterols are primarily synthesized in mammalian cells by enzymes in the ER, usually associated with the cytoplasmic leaflet of the SER. Exceptions to this include mitochondria that make selected phospholipids and peroxisomes that can synthesize cholesterol and some other lipids. The initial step in phospholipid synthesis is the condensation of two molecules of fatty acyl coenzyme A (CoA) with glycerol phosphate to make phosphatidic acid (Fig. 4–4). Each molecule of fatty acyl-CoA is added separately, enabling the cell to control the type of fatty acid esterified to the 1 and 2 positions of the glycerol. Free fatty acids in the cytosol are usually bound to a fatty acid–binding protein and are converted to the fatty acyl-CoA derivatives that are substrates for the acyltransferase enzymes in the cytosolic side of the membrane. A phosphatase removes the phosphate from phosphatidic acid to make diacylglycerol, and a polar head group, either cytidine diphosphoethanolamine (CDP-ethanolamine) or cytidine diphosphocholine (CDP-choline), is added (see Fig. 4–4).

Phospholipids are assembled in the cytoplasmic leaflet of the ER and must be translocated to the other half of the bilayer to generate a particular phospholipid distribution between the two halves of the bilayer. The spontaneous flipping of phospholipids from one-half of the bilayer to the other is extremely slow, and proteins termed *flippases* have evolved that catalyze the flipping of specific lipids. Phospholipids are often asymmetrically distributed in the bilayer; for example, phosphatidylcholine and sphingomyelin are predominantly in the lumenal face of the ER membrane, whereas phosphatidylethanolamine and phosphatidylserine are mainly in the cytosolic face. How this flipping process is regulated to achieve the diversity of lipid asymmetry observed in different membranes is discussed in Chapter 2.

Ceramide, the precursor of phosphosphingolipids and glycosphingolipids, is synthesized in the ER from serine and palmitoyl CoA. Phosphosphingolipids are also made in the ER. Glycosphingolipids, such as gangliosides, are made when ceramide reaches the Golgi complex and is glycosylated on the luminal face of the Golgi complex

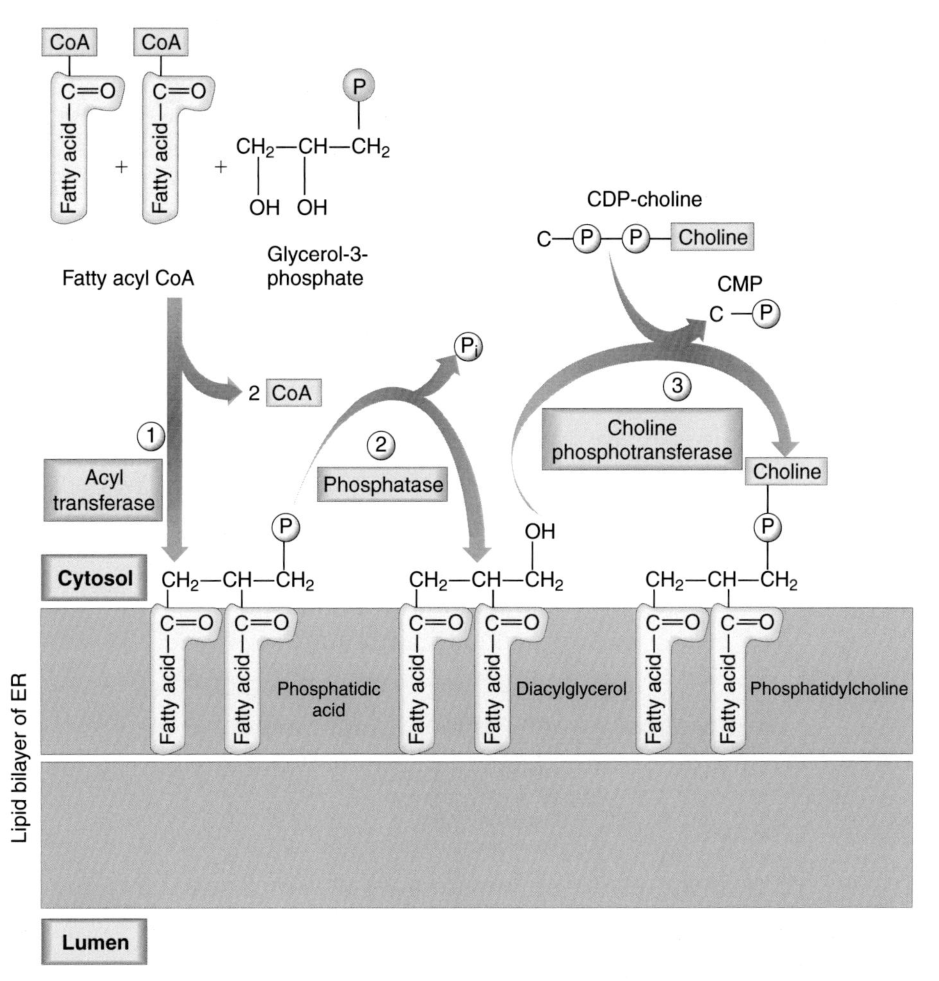

Figure 4–4. **Synthesis of phospholipids in the smooth endoplasmic reticulum.** Diagram presents the pathway for synthesis of phosphatidylcholine from fatty acyl coenzyme A (CoA), glycerol-3-phosphate, and cytidine diphosphocholine (CDP-choline).

by glycosyltransferases. Glycosphingolipids are found only in the lumenal face of intracellular membranes (equivalent to the extracellular face of the plasma membrane), suggesting there are no flippases for this type of lipid.

The committed step in cholesterol synthesis, the production of mevalonate, is catalyzed by 3-hydroxy-3-methlyglutaryl-CoA reductase (HMG-CoA reductase), an integral membrane protein of the SER. Other enzymes involved in the synthesis and modification of cholesterol are also ER residents. Although initially made on the cytosolic side of the SER, cholesterol is found in both halves of the bilayer, and evidence exists that cholesterol flippases catalyze its flipping.

Not only are many lipids asymmetrically arranged in the two halves of the bilayer, but also most membranes maintain a unique lipid composition. For example, the ER of mammalian cells is typically 50% or more phosphatidylcholine, with less than 10% each of sphingomyelin and cholesterol. In contrast the plasma membrane contains less than 25% phosphatidylcholine and more than 20% each of sphingomyelin and cholesterol. Thus, once a lipid is incorporated into the ER, it must not only be transported to other membranes, but also the characteristic compositions of the destination membranes must be maintained. While lipids can be transported between compartments in vesicle-mediated and nonvesicle-mediated processes, the mechanisms that control the lipid composition of specific membrane compartments remain unclear.

Rough Endoplasmic Reticulum

The RER is where secretory and membrane proteins (other than mitochondrial and peroxisomal proteins in mammalian cells) first pass through or enter a membrane. The RER contains machinery that, in cooperation with ribosomes, selects secretory and membrane protein from all the other proteins being synthesized in the cell and targets them to the RER. This process includes the binding of ribosomes to the RER, which gives the RER its defining rough appearance in electron micrographs (Fig. 4–3, B). Many other RER functions are related to the processing of secretory and membrane proteins. These include glycosylation, folding, quality control, and the degradation of proteins that do not pass quality-control standards.

The Translocation of Secretory Proteins Across the Rough Endoplasmic Reticulum Membrane Requires Signals

In the 1970s cell biologists noticed that mature versions of secreted proteins had a slightly smaller molecular weight than when the proteins were translated in a cell-free protein synthesis system containing ribosomes, tRNA, mRNA, and all the other factors necessary for protein synthesis in a test tube. This difference in molecular weight was traced to the presence of extra amino acids at the N-terminal end of secretory proteins that were cleaved off to form the mature protein during the secretory process. Evidence from electron microscopy and biochemical experiments showed that secretory proteins entered the lumen of the RER en route to the cell exterior, and it was proposed that these extra amino acids were the signal to select secretory proteins from all other proteins and somehow divert them to the ER. This was called the *signal hypothesis*, and the extra amino acids at the N-terminus are called the *signal peptide* or *signal sequence*. When signal sequences from different proteins were compared, several common features were found (Fig. 4–5).

The overall length of N-terminal signal peptides is usually 16–30 amino acids, and these peptides contain 1 or more positive amino acids at their N-terminus followed by a continuous core of hydrophobic residues.

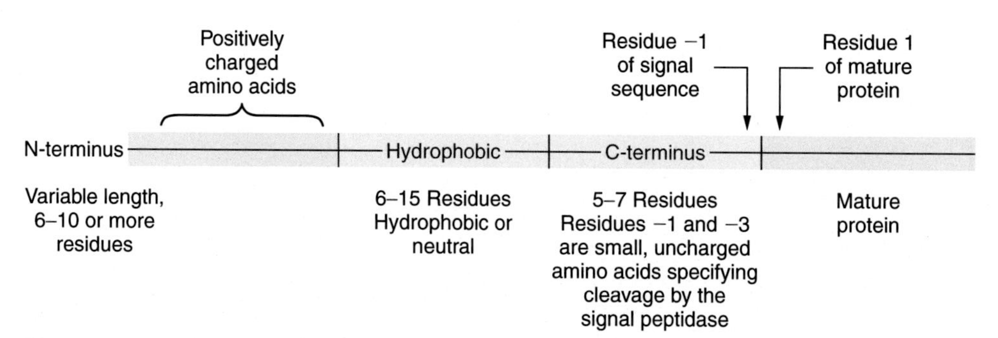

Figure 4–5. **Features of signal peptides.** Signal peptides typically have three functional regions. The N-terminus has a variable length and contains positively charged amino acids at the C-terminal side. The middle region has 6–15 hydrophobic or neutral amino acids. The C-terminal region has 5–7 residues, and positions −1 and −3 (where the first amino acid residue of the mature protein is +1) are small, uncharged amino acids that are important for proteolytic removal of the signal peptide by signal peptidase.

The hydrophobicity of this core sequence, and not a particular amino sequence, is what is necessary for the recognition of the signal peptide by the secretory machinery. The region after the hydrophobic sequence also contains amino acids that direct signal peptidase to remove the signal peptide from secretory proteins by proteolytic cleavage.

The signal hypothesis was tested in cell-free assays that reconstituted the transport of secretory proteins across the RER membrane. The reaction mixtures contained the mRNA for a secreted protein and all other components necessary to translate the mRNA, including ribosomes prepared from cytosol (i.e., the ribosomes were from the soluble pool, not those bound to the RER). In addition, it was possible to add microsomes, small intact vesicles that are derived from the RER when cells are homogenized. The experimental design is illustrated in Figure 4–6.

When the protein was made in the absence of microsomes, it contained the signal peptide and was not glycosylated. If microsomes were added after synthesis, the signal peptide was not cleaved, the protein remained outside of the microsomes, and no carbohydrate was added. The presence of the protein outside of the microsomes was demonstrated by adding proteases that digested the protein. However, when microsomes were present at the same time, the protein was being translated, the signal sequence was cleaved, the protein was glycosylated, and the completed protein was released inside the microsomes, deduced from the observation that it was protected from added proteases that could not enter the microsomes. When the microsomal membrane was dissolved with detergents, the protein inside was released and became protease sensitive.

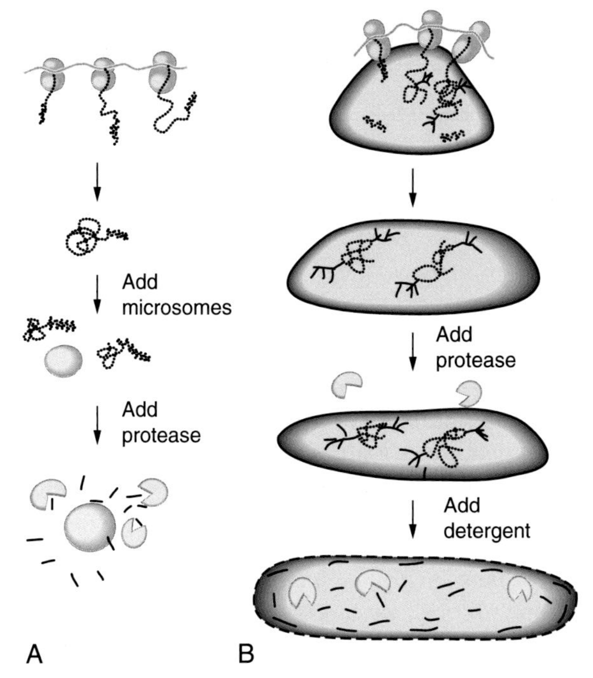

Figure 4–6. **Testing the signal hypothesis in cell-free protein synthesis assays.** (A) Messenger RNA (mRNA) for a secretory protein is translated in a cell-free system. If microsomes (vesicles derived from the RER) are added after translation, the translated protein does not enter the microsome, the signal peptide remains intact, and no carbohydrate is added to the protein. If a protease is added, the protein is digested because it is not protected inside the microsomes. (B) mRNA for a secretory protein is translated in the presence of microsomes. Ribosomes making the protein bind the microsomes, and the protein translocates to the interior of the microsome. Analysis of the proteins indicates that the signal peptide is cleaved and the protein is N-glycosylated. If protease is added to the mixture, the protein is not degraded because it is sheltered within the microsomes. However, if detergent is added to disrupt the membrane, the protease and the protein come into contact, and the protein is degraded. *(Adapted from Lodish H, Berk A, Matsudaira P, et al.* Molecular Cell Biology, *5th ed. New York: W. H. Freeman and Company, 2004.)*

The results from these experiments and others were incorporated into a model with the following features: (1) Secretory proteins are translocated across the ER membrane cotranslationally, while new amino acids are being polymerized into the protein (there are exceptions, especially in yeast). (2) The signal peptide is cleaved off during translocation, before synthesis is complete. (3) Certain types of carbohydrate are cotranslationally added to the growing polypeptide on the lumenal side of the RER. (4) Ribosomes from the soluble pool participate in synthesis and translocation of secreted proteins, indicating that there are not two different types of ribosomes, one for the synthesis of cytoplasmic proteins and a second for secreted proteins.

Because secretory proteins begin to cross the RER membrane before their synthesis is complete, there must be a mechanism to recruit ribosomal complexes engaged in making secretory proteins to the RER. Continued biochemical studies of the cell-free synthesis of secretory proteins identified a soluble factor that when added to the cell-free system arrested the elongation of secretory proteins, but not cytoplasmic proteins. Because the only known feature that distinguished ribosomes making secretory proteins from those making cytoplasmic proteins was the signal peptide on the secretory protein, the factor was called *signal recognition particle* (SRP). In eukaryotes, SRP is composed of six proteins and a 300 nucleotide 7S RNA molecule that associates with the ribosome as the signal peptide emerges from the large ribosomal subunit and arrests translation. Elongation arrest is relieved when the ribosome/SRP complex associates with the RER and translocation of the secretory protein into the ER proceeds. The discovery of SRP motivated the search for an SRP receptor on the RER. The receptor was found and is a heterodimer with one subunit spanning the ER membrane (Fig. 4–7). SRP binds the SRP receptor, which brings the SRP/ribosome complex in proximity of the *translocon*, or Sec61 complex, a protein pore in the ER membrane whose opening and closing is highly regulated. GTP hydrolysis by subunits of the SRP and the SRP receptor results in the release of the ribosome and its nascent polypeptide, which associates with the translocon. The interaction of the nascent

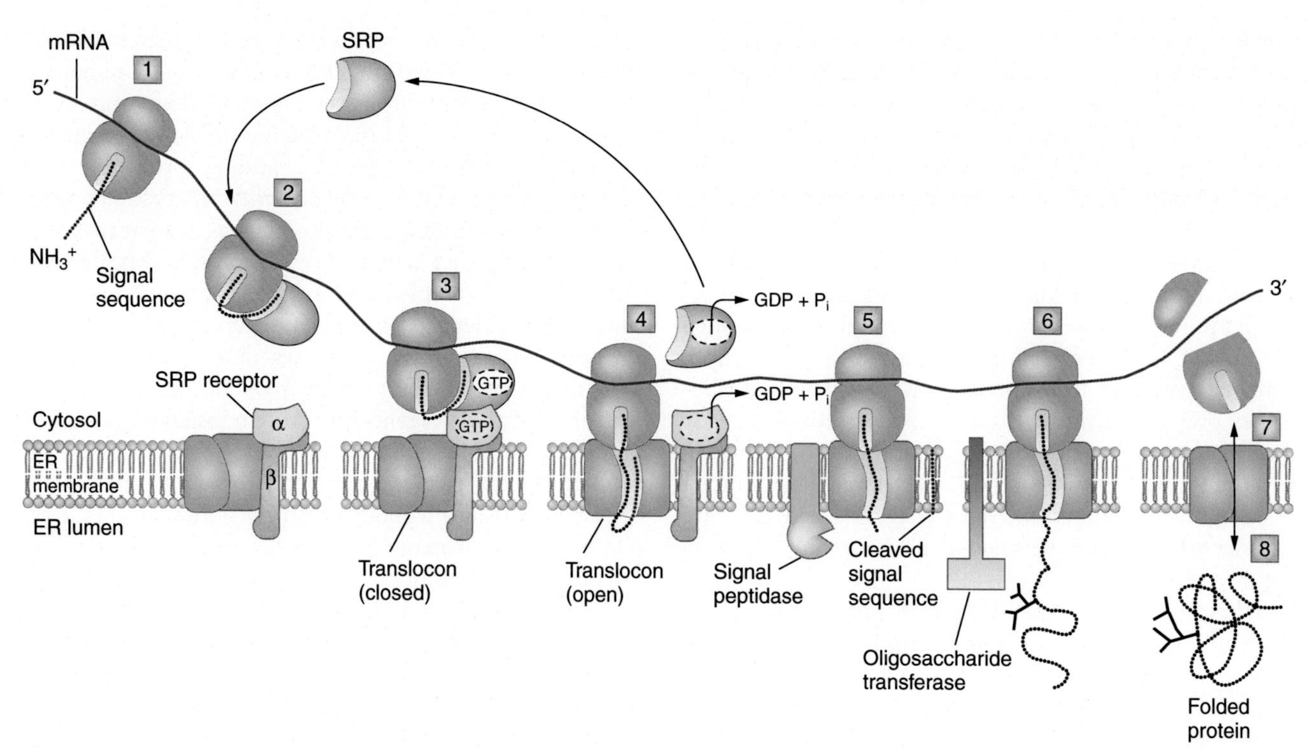

Figure 4–7. **Model for translocation of secretory proteins across the endoplasmic reticulum (ER) membrane.** When the signal peptide emerges from ribosomes, the signal recognition particle (SRP) binds to the hydrophobic signal peptide (in *red*). This transiently delays further elongation of the secretory protein until the SRP/ribosome complex binds to the SRP receptor on the rough endoplasmic reticulum (RER) via interactions between the SRP and the receptor. GTP hydrolysis dissociates SRP from the complex, which is now available to target another ribosome to the RER. Elongation of the protein resumes, and the peptide inserts into the translocon as a loop with the N-terminus of the signal peptide on the cytoplasmic side. As elongation proceeds, signal peptidase cleaves the signal sequence, and oligosaccharide transferase adds the preassembled N-linked oligosaccharide to Asn residues in the appropriate context. *(Adapted from Lodish H, Berk A, Matsudaira P, et al.* Molecular Cell Biology, *5th ed. New York: W. H. Freeman and Company, 2004.)*

polypeptide with the translocon stimulates its opening, and the continued synthesis of the nascent polypeptide results in its insertion into the aqueous channel of the translocon. This regulated gating of the translocon ensures that the ER membrane remains impermeable to ions, such as Ca^{2+}, which would leak out of the ER if the translocon was permanently in an open state. As discussed later in this chapter, the translocon is also gated laterally, which allows specific sequences in membrane proteins to move sideways out of the aqueous channel of the translocon and into the bilayer.

A model summarizing the translocation of secretory proteins through the ER membrane is depicted in Figure 4–7. Upon emerging from a ribosome in the cytosol, the signal peptide of the secretory protein interacts with SRP. Elongation of the nascent polypeptide is arrested until the ribosomal complex binds to the ER via the SRP receptor. The nascent chain is transferred to the channel of the translocon, and the signal peptide inserts into the translocon as a loop with the N-terminus on the cytoplasmic side. During this process, GTP is hydrolyzed to GDP, releasing SRP into the cytosol to participate in another round of signal sequence recognition. The signal peptide is removed by signal peptidase before protein synthesis is completed and is released laterally from the translocon into the bilayer where it is degraded. The continued synthesis of the polypeptide chain pushes the N-terminus of the cleaved protein further into the lumen of the RER. Proteins are also glycosylated during elongation by addition of a preformed oligosaccharide to certain asparagine residues, a process that is covered in more detail later in this chapter. The signal hypothesis and insights into how proteins are targeted to and across membranes have been so influential in cell biology and medicine that, in 1999, a major contributor in this area, Dr. Günter Blobel, was awarded the Nobel Prize in medicine and physiology for his work.

The Insertion of Proteins Into Membranes Requires Stop-Transfer Anchor Sequences

As explained in Chapter 2, single-pass integral membrane proteins are anchored to the bilayer by a membrane-spanning α-helix that contains predominantly hydrophobic or neutral amino acids. Single-pass membrane proteins may be oriented with their N-terminus on the lumenal side and their C-terminus on the cytoplasmic side of the membrane, or the reverse. How do the membrane-spanning domains get into the membrane? An extension of the signal hypothesis provides the answer. A protein containing an N-terminal signal peptide and a downstream stretch of hydrophobic residues is targeted to

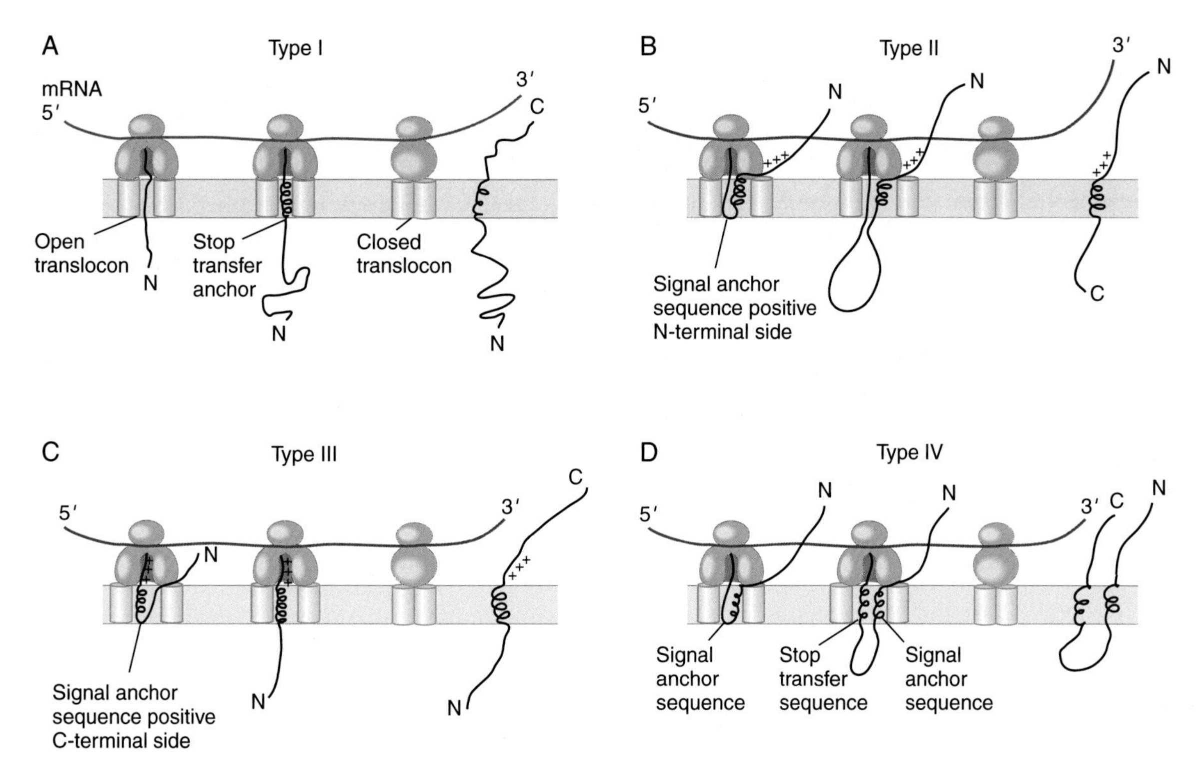

Figure 4–8. **Type I, II, III, and IV membrane proteins.** (A) Type I proteins contain a typical N-terminal hydrophobic signal peptide that directs ribosomes to the translocon in the rough endoplasmic reticulum (RER). Elongation of the protein results in the translocation of the N-terminus into the lumen of the RER. When a second hydrophobic region, termed a *stop-transfer anchor*, enters the translocon, translocation is halted, and the remainder of the protein is elongated on the cytosolic side of the membrane. The stop-transfer sequence causes the translocon to open laterally releasing the protein into the bilayer resulting in an integral membrane protein anchored in the membrane by the hydrophobic stop-transfer sequence. (B and C) Type II and III proteins do not contain an N-terminal signal sequence; rather an internal hydrophobic domain that will eventually span the membrane serves as both the signal sequence (that binds to signal recognition particle (SRP)) and the membrane anchor. (B) Type II proteins contain a cluster of positively charged amino acids on the N-terminal side of the anchor and insert into the translocon as a loop, and the N-terminus remains on the cytosolic side of the membrane. (C) Type III proteins have the cluster of positively charged amino acids on the C-terminal side of the signal anchor, which changes the orientation of the protein in the membrane. (D) Type IV proteins have multiple membrane-spanning domains that alternate as signal-anchor and stop-transfer sequences. Two membrane-spanning domains are shown, but the insertion mechanism can accommodate many more to produce proteins with 10 or more membrane-spanning domains.

the translocon by the mechanisms described earlier. As the protein translocates through the translocon, the downstream stretch of hydrophobic residues will eventually emerge from the ribosome and enter the translocon. If transfer through the translocon stopped at this point and the translocon opened laterally, the hydrophobic residues, which form a membrane-spanning α-helix, can move into and anchor the protein in the bilayer. Because translocation stops when the downstream hydrophobic sequence spans the translocon, this type of membrane-spanning sequence is called a stop-transfer anchor sequence. Cell-free studies on the synthesis of model membrane proteins in the presence of microsomes suggested this simple extension of the signal hypothesis was correct in producing single-pass membrane proteins that had their N-termini in the lumen of the ER and their C-termini in the cytosol. The two features of the protein that dictate it will be a single membrane-spanning protein with a lumenal N-terminus and a cytosolic C-terminus are an N-terminal cleavable signal peptide and an internal stop-transfer anchor sequence. Membrane proteins with these features are called type I integral membrane proteins (Fig. 4–8, A).

But what about proteins whose orientation is the opposite of type I proteins? This mystery was solved when it was realized that membrane-spanning domains could themselves act as signal peptides recognized by SRP, thus targeting a membrane protein to the RER by the same machinery that secretory proteins used. The only difference is that membrane-spanning domains do not contain the amino acid signal that directs cleavage by signal peptidase, so membrane-spanning domains are not cleaved. Integral membrane proteins of this type have no N-terminal signal peptide, but they do have an internal hydrophobic membrane-spanning domain that serves the purpose of the signal peptide, including one or more positively charged residues on the N-terminal side of the membrane-spanning domain. As the internal hydrophobic membrane-spanning domain emerges from the ribosome, SRP binds and targets the complex to the RER by associating with the SRP receptor. The positively charged N-terminal side of the membrane-spanning domain stays in the cytosol and the protein loops into the translocon, just as with proteins that contain a typical signal peptide. The result is that the N-terminus of the protein that has already emerged from the ribosome stays in the cytosol. As the protein is elongated, the C-terminus eventually emerges from the ribosome and passes completely through the translocon. The translocon opens laterally, and the protein moves into the bilayer with its N-terminus in the cytoplasm and C-terminus in the lumen of the ER. Integral membrane proteins that enter the membrane by this mechanism, termed *type II proteins*, have no N-terminal signal peptide, but they do have an internal hydrophobic membrane-spanning domain that has two functions: a signal peptide function and a membrane anchor function. Thus type II proteins are said to have a signal-anchor sequence (see Fig. 4–8, B).

Type III membrane-spanning proteins have a signal-anchor sequence but no N-terminal signal peptide, just like type II proteins; yet their N-terminus is lumenal, and their C-terminus is cytosolic, just the opposite of a type II membrane protein. The one structural difference researchers noticed was that the cluster of positive charges in type III membrane proteins was usually found adjacent to the C-terminal side of the signal anchor. By a mechanism that is not well understood, the side of the membrane anchor that is positively charged has a strong tendency to stay in the cytoplasm. Type III membrane proteins initially insert into the translocon as a loop such that the C-terminal side of the anchor stays in the cytoplasm and the N-terminal side goes through, threading the already completed N-terminal portion of the peptide through the membrane. As elongation proceeds the C-terminus of the protein is synthesized and remains in the cytosol (see Fig. 4–8, C).

Many proteins span the membrane multiple times, from 2 to more than 10 times, and are termed *multipass membrane proteins* or type IV integral membrane proteins. The insertion and orientation of these proteins can be understood as combinations of type I, II, and III mechanisms. For example, suppose a protein has two hydrophobic membrane-spanning domains and the first is a typical type II signal anchor with the positive charges adjacent to the N-terminal side of the hydrophobic sequence (see Fig. 4–8, D). The protein initially inserts just as a type II protein with the N-terminus in the cytosol, and elongation proceeds until the second stretch of hydrophobic residues reaches the translocon. It enters the pore and acts as a stop-transfer sequence; the translocon opens laterally releasing both membrane-spanning domains into the bilayer, and the protein has both its N- and C-terminal ends in the cytosol. If there were a third membrane-spanning domain in this protein, it would be equivalent to start-transfer anchor sequence in the sense that it would enter the translocon and serve as a membrane-spanning domain, although the ribosome does not detach from the RER and another SRP is not needed for synthesis to proceed. The orientation of all subsequent membrane-spanning domains in type IV proteins is determined by the first one, with all others continuously threading through the membrane, each with an opposite orientation. Thus, if the first domain crosses the membrane with its N-terminal side in cytosol and the C-terminal side in the lumen, the next membrane-spanning domain will have its N-terminal side in the lumen and C-terminal side in the cytosol, and so on. Given their amino acid sequences, a reasonable topology for most membrane proteins can be deduced. This is especially important in medicine given that the human genome sequence is available and the topology of a protein of unknown function can be predicted, offering clues to function. Deducing the topology is facilitated by using the primary amino acid sequence to prepare a hydropathy plot (Fig. 4–9).

In this plot, each amino acid is assigned a value proportional to its hydrophobicity. The values are summed over a window, usually 10–20 adjacent residues beginning with the amino terminus, and the sum is a point on the plot placed in the middle of the sequence window. The window is then moved by one residue toward the C-terminal end, and the hydropathy value is calculated again. The process is repeated with the window moving one residue each time until the C-terminus is reached. Positive hydropathic regions usually identify stretches of hydrophobic residues that are part of an N-terminal signal peptide or a membrane-spanning domain. Coupled with inspection of the charges on the N- or C-terminal side of the hydrophobic sequences, the topology of the protein can be predicted. For proteins with multiple membrane-spanning domains and complicated topologies, structural studies of the protein are necessary to confirm predicted topologies.

Some Membrane Proteins Are Anchored to the Bilayer by Covalently Attached Lipids

Several classes of proteins are covalently modified after synthesis to attach a hydrophobic lipid (either a fatty acid or a hydrophobic molecule such as geranyl or farnesyl groups) that inserts into the bilayer and serves as a membrane anchor. Many proteins in this class are synthesized in the cytosol; the lipid is covalently added, and the hydrophobic sequence inserts into the cytoplasmic side of a membrane. Proteins attached to membranes by this mechanism are highly relevant to medicine because many of them are protooncogenes that when mutated cause cancer. For example, mutations in the small GTP-binding protein Ras frequently cause cancer in humans, and active Ras has a covalently attached lipid that tethers it to the plasma membrane. Because the activity of Ras depends on the anchoring process, enzymes that catalyze the covalent addition of the lipid anchor have been studied intensely as potential targets for chemotherapy.

Another class of lipid-anchored membrane protein contains the complex lipid glycosylphosphatidylinositol

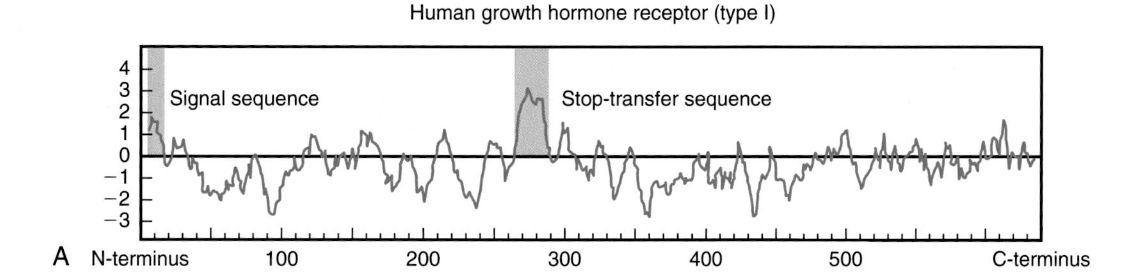

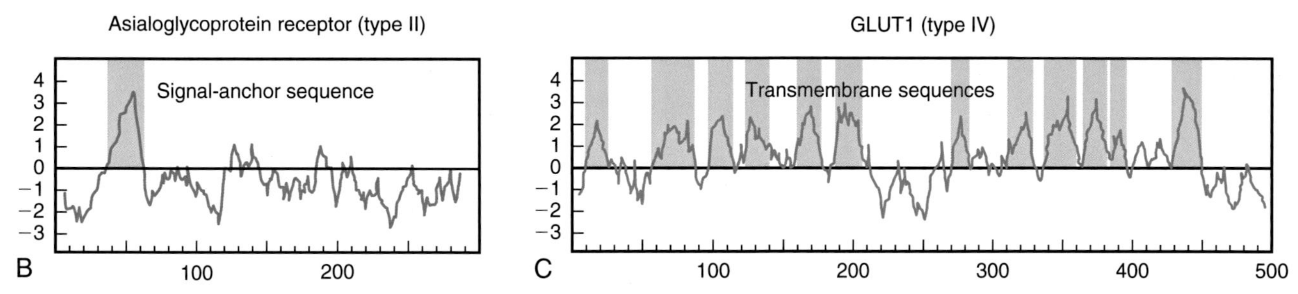

Figure 4–9. **Hydropathy profiles can identify likely topogenic sequences in integral membrane proteins.** Hydropathy profiles are generated by plotting the total hydrophobicity of each segment of 20 contiguous amino acids along the length of a protein. Positive values indicate relatively hydrophobic regions and negative values relatively polar regions of the protein. Probable topogenic sequences in (A) type I, (B) type II, and (C) type IV membrane-spanning proteins are highlighted in *blue*. The predicted topology of multipass (type IV) proteins, such as GLUT1 (C), must be verified using biochemical approaches. *(Adapted from Lodish H, Berk A, Matsudaira P, et al.* Molecular Cell Biology, *5th ed. New York: W. H. Freeman and Company, 2004.)*

(GPI). This lipid anchor is added to certain integral membrane proteins in the ER. The nascent protein is cleaved on the N-terminal side of its membrane-spanning anchor and covalently attached to GPI (Fig. 4–10). Because GPI-anchored proteins are on the lumenal side of the ER, they appear on the extracellular face of the cell after transport to the plasma membrane.

Glycosylation of Secretory and Membrane Proteins Begins in the Endoplasmic Reticulum

Many secretory proteins and integral membrane proteins are cotranslationally glycosylated by the transfer of a preassembled oligosaccharide unit to asparagine (Asn) residues of the proteins as they emerge from the translocon. This reaction is catalyzed by the enzyme oligosaccharide protein transferase (Fig. 4–11), and this type of glycosylation is called *N-linked* because the linkage is via the amide nitrogen of Asn. All glycosylated Asn residues appear in the tripeptide sequence asparagine-X-serine (or threonine) where X is any amino acid. This tripeptide sequence is necessary for glycosylation but is not sufficient, as every occurrence of this Asn-containing tripeptide is not glycosylated. The branched oligosaccharide (Fig. 4–11) is preassembled by one-at-a-time addition of nucleotide-activated monosaccharides onto a large hydrophobic polyisoprenoid lipid termed *dolichol*.

Three glucose residues and one mannose residue of this branched oligosaccharide are trimmed in the RER. The N-linked oligosaccharide is further trimmed in the Golgi complex, and other sugars are added. N-linked oligosaccharides on proteins appear to have complex and sometimes overlapping functions. One function is to stabilize protein structure, related, in part, to the affinity of the sugars for water so that glycosylated regions of a protein are highly water soluble. N-linked oligosaccharides are frequently recognized by other proteins, providing a molecular mechanism for protein-protein interactions. For example, certain cell-adhesion proteins on the cell surface have sugar-binding domains that help one cell adhere to another. In addition, the phosphorylation of mannose residues in the N-linked oligosaccharide in the Golgi complex provides a signal for targeting lysosomal enzymes to lysosomes (see later for a more detailed discussion).

Protein Folding and Quality Control in the Endoplasmic Reticulum

All the information for proteins to fold into their final three-dimensional structures is contained within their amino acid sequences, but the rate of folding is often too slow on a biological time scale. Hence, proteins have evolved that catalyze the folding of other proteins. One class of these folding catalysts, called *chaperones*, binds to nascent proteins as they emerge from free ribosomes in the cytoplasm to promote rapid folding. It is not surprising, therefore, that chaperones exist within the ER that assist in the folding of secretory and membrane proteins as they emerge from the translocon. Recent work has elucidated the quality-control machinery within the ER that facilitates protein folding, retains misfolded

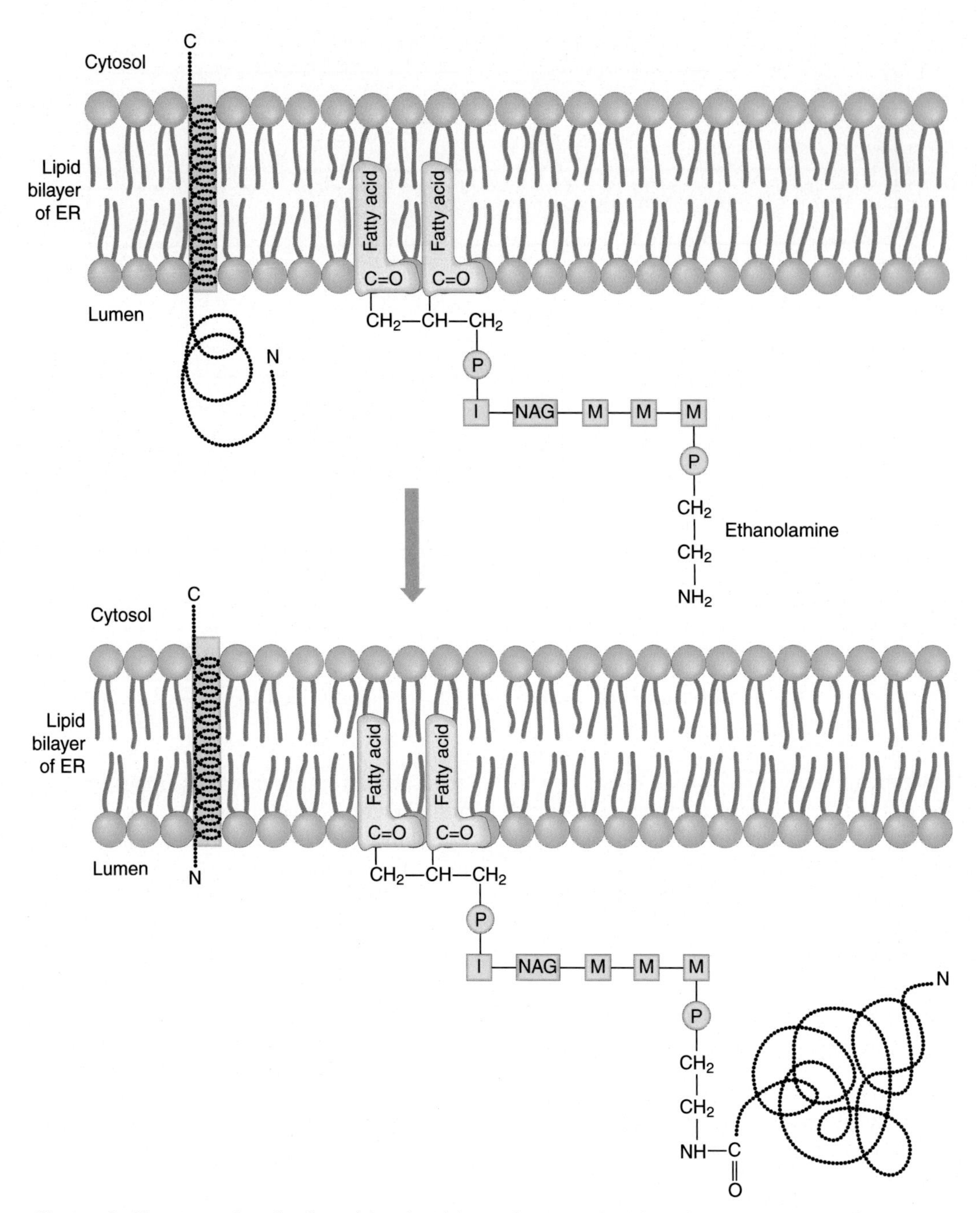

Figure 4–10. **Formation of a glycosylphosphatidylinositol (GPI)-anchored membrane protein.** An endoprotease in the lumen of the endoplasmic reticulum (ER) cleaves the proteins away from its C-terminal membrane-spanning domain. The new C-terminus of the protein is attached to the amine within the ethanolamine moiety of the GPI anchor.

proteins within the ER to prevent their transport to the Golgi apparatus, and disposes of proteins that cannot fold properly.

Abundant chaperones exist within the lumen of the ER that promote protein folding in different ways. BiP, which stands for heavy chain binding protein, was first discovered as a binding partner for the heavy chain of immunoglobulins and is now understood to associate in an ATP-dependent manner with hydrophobic peptide sequences of most proteins as they emerge from the translocon. The result is to shield hydrophobic residues and help them fold to the interior of the protein where they are protected from water. Secreted proteins and many membrane proteins that are on the exterior surface of the plasma membrane contain disulfide bonds between cysteine residues of the protein that help to stabilize proteins exposed to the variable environment outside the cell. To ensure that disulfide bonds are quickly and correctly formed by secreted proteins, another class of chaperones, called *protein disulfide isomerases*, promotes oxidation-reduction reactions that connect the sulfhydryl groups of cysteine residues into disulfide bonds.

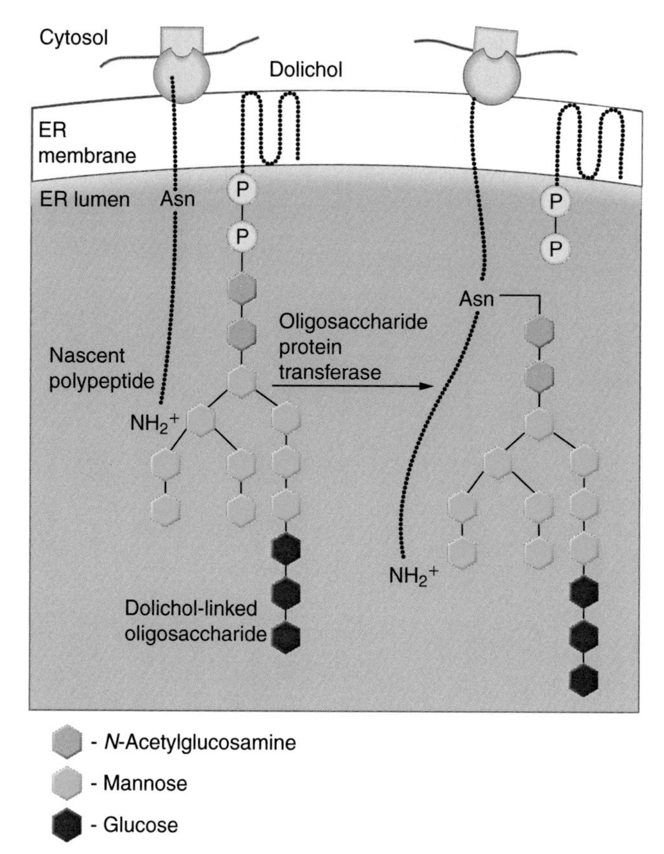

Figure 4–11. **N-linked glycosylation in the rough endoplasmic reticulum (RER).** In the early steps of N-linked glycosylation, the oligosaccharide is preassembled on a dolichol carrier, a large, aliphatic hydrocarbon terminating in a hydroxyl group that is phosphorylated. The oligosaccharide is assembled by one-at-a-time addition of *N*-acetylglucosamine, glucose, and mannose residues. Interestingly, most of this assembly occurs on the cytosolic side of the membrane, and the oligosaccharide is then flipped to the lumenal side of the membrane (these early steps are not shown). Once an Asn residue in the correct sequence context emerges from the translocon, the enzyme oligosaccharide protein transferase attaches the oligosaccharide to the protein.

As proteins emerge from the translocon, BiP binds to the nascent protein to assist in folding, and N-linked glycosylation occurs (Fig. 4–12). When synthesis is complete, the glycosylated proteins engage a second quality-control machinery that is dependent upon calnexin and calreticulin. Calnexin, a single-pass membrane protein, and calreticulin, a soluble protein, are lectins (sugar-binding proteins) that bind to N-linked oligosaccharides containing a single glucose residue. The role of calnexin/calreticulin in protein folding is illustrated in Figure 4–12. The first two glucose residues on the N-linked oligosaccharide are trimmed by glucosidases. Calnexin/calreticulin then binds to N-linked oligosaccharides containing a single glucose residue and in collaboration with protein disulfide isomerases further assist the protein in folding. Proteins released from calnexin/calreticulin that have achieved their native state undergo trimming of the final glucose residue by glucosidase and trimming of specific mannose residues that allows the proteins to be packaged into transport vesicles leaving the ER. However, proteins that are not properly folded following removal of the final glucose residue undergo reglucosylation and another round of calnexin-/calreticulin-assisted folding to achieve their native conformation.

Two other pathways come into play when, despite the best efforts of chaperones, a protein cannot fold properly. One pathway is ER-associated degradation (ERAD). This system identifies deglucosylated proteins that are improperly folded and targets them to a retrotranslocation channel, where the protein undergoes export to the cytosol (Fig. 4–12). During this retrotranslocation process, which is dependent upon a cytosolic ATPase of the AAA family, the misfolded proteins undergo deglycosylation and ubiquitination on the cytosolic side of the ER membrane. The ubiquitinated proteins are then targeted to the proteasome for degradation. Misfolded secretory proteins that do not contain N-linked sugar modifications are also delivered to the retrotranslocation channel where they are exported to the cytosol for proteasome-dependent degradation. If unfolded proteins become abundant in the lumen of the ER, a second pathway induces the synthesis of additional chaperones and increases the amount of proteins involved in ERAD to eliminate excess unfolded proteins. This pathway is called the unfolded protein response (UPR) and consists of a collection of mechanisms that upregulate genes encoding proteins involved in chaperone function and in ERAD.

Why has the cell devoted so many resources for quality control in the ER? Improperly folded proteins invariably expose hydrophobic amino acid sequences to water, which causes unfolded proteins to aggregate, thereby shielding the hydrophobic surfaces from water. These aggregates can become large and could recruit other proteins into them by inducing folded proteins to unfold; thus unfolded proteins could be catalysts that accelerate the unfolding of already folded proteins. If these aggregates left the ER, they could acutely interfere with the action of existing proteins at other locations in the cell. In addition, unfolded proteins present novel antigenic sites to the immune system that would elicit unwanted autoimmune reactions should the proteins be exposed on the cell exterior. The medical relevance of quality control in the ER is emphasized by the observation that there are more than 35 known diseases that are directly or indirectly related to improper protein folding in the ER. Among them is cystic fibrosis, which arises because of mutations in the CFTR chloride transporter. The most common mutation in CFTR is the deletion of a single residue (F508del mutation) in the protein that prevents its folding in the ER. The misfolded protein undergoes retrotranslocation to the cytosol and degradation. If proteins containing the F508del mutation fold properly, they undergo transport to the plasma membrane, where they function normally in chloride transport. Observations such as this have launched major efforts to develop drugs that enhance protein folding in the ER (Clinical Case 4–1).

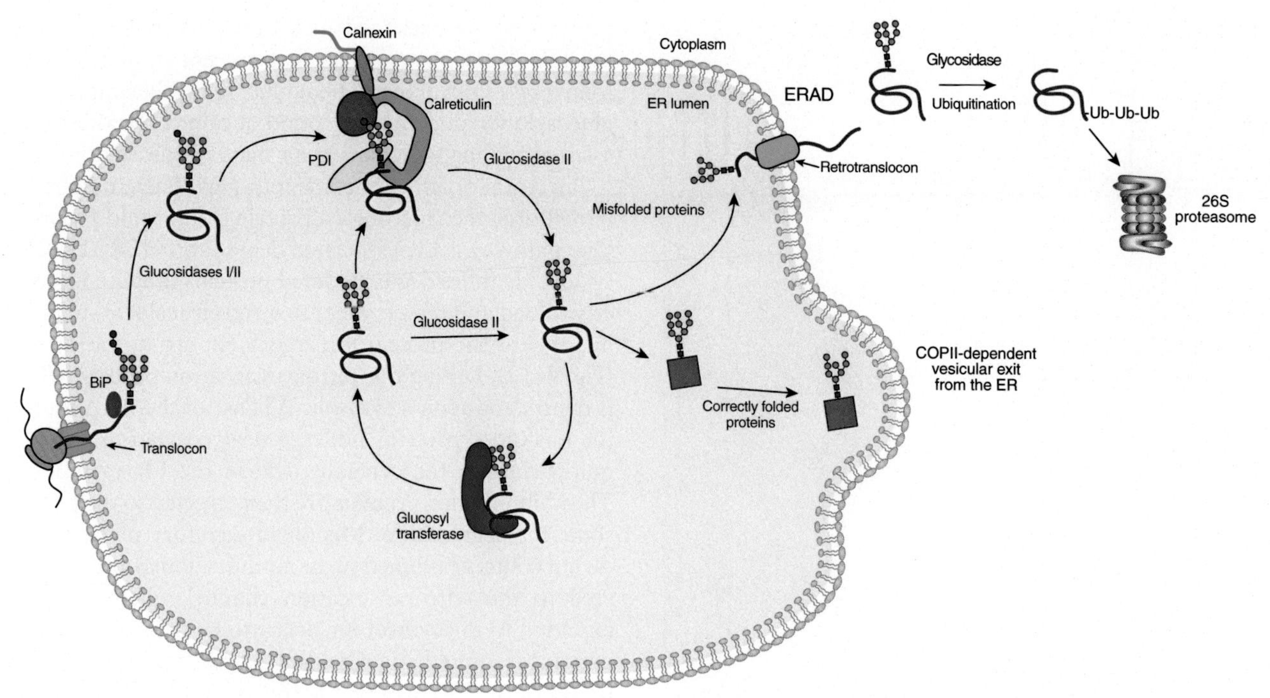

Figure 4–12. **Calnexin and quality control in the endoplasmic reticulum (ER).** As proteins emerge from the translocon, they cotranslationally acquire N-linked sugar modifications. In addition, BiP binds and assists in the folding of the nascent protein. The N-linked oligosaccharide of N-glycosylated proteins initially has three terminal glucose residues. Two of these glucose residues are removed by glucosidases. Calnexin and calreticulin then bind to N-linked oligosaccharides containing a single glucose residue and in collaboration with protein disulfide isomerase (PDI) further assist in the folding of the nascent protein. Correctly folded proteins are released, and the last glucose residue and a mannose residue are removed, and the protein exits the ER in vesicles formed in a COPII-dependent fashion (details described later in the chapter). However, if the protein is still misfolded, glucosyltransferase readds a single glucose residue, and the monoglucosylated protein undergoes another round of calnexin/calreticulin-assisted folding. Proteins that cannot achieve a stable folded conformation are shunted to the endoplasmic reticulum–associated degradation pathway (ERAD). The misfolded proteins are delivered to the retrotranslocation channel and are exported to the cytosol. The proteins are then deglycosylated and ubiquitinated. The ubiquitinated proteins are then targeted to the proteasome for degradation. *(Adapted from Adams et al.* Protein J *2019;3:317–329.)*

Clinical Case 4–1

Peter is a 13-year-old boy who was having a physical examination before changing from a public to a private school. He was very thin, appeared younger than his stated age, and appeared chronically ill. The initial examination showed that he had scattered rales throughout his lungs, an area of coarse rhonchi in the right middle lobe, and a questionable area of atelectasis in the left lower lung field. Prompted by these findings, the school physician elicited a more detailed history from Peter and his parents.

Peter had been a full-term normal-weight infant but shortly after birth was returned to the hospital because he was not passing any stools. Fortunately, with 3 days of intravenous hydration, he began to have normal gastrointestinal function and returned home. Though subsequently he grew slowly, his parents noted no particular problems until he began school at the age of 5. At that time, he developed his first episode of pneumonia, which required antibiotic therapy for *Haemophilus influenzae*. Since then, he has had repeated pneumonias two or three times a year. These are usually heralded by fits of heavy coughing producing thick green mucus. When cultured, his sputum shows variable organisms, and he has often responded to oral broad-spectrum antibiotics. Unfortunately the last two cultures have shown predominantly *Pseudomonas* variants.

In addition to his lung problems, Peter has recently developed right upper quadrant pain after eating, and an intravenous cholecystogram showed the presence of gallstones and probable cholecystitis. Furthermore, around the same time, he noted that his stools, although formed, were larger and frequently floated in the toilet bowl. With this information in hand, the physician took a small sample of Peter's axillary sweat secretions.

Cell Biology, Diagnosis, and Treatment of Cystic Fibrosis

Peter has cystic fibrosis, which results from a mutation in a protein called the cystic fibrosis transmembrane conductance regulator (CFTR). The CFTR is a chloride channel that contains 12 membrane-spanning domains encoded by a gene on chromosome 7. More than two-thirds of cystic fibrosis patients contain a deletion of a single phenylalanine residue at position 508 of CFTR. Loss of this single residue destabilizes the protein so that it folds slowly in the RER with the result that it is retained and degraded in the RER by the ERAD pathway described in this chapter. Interestingly, reducing the temperature of cultured cells expressing the defect or adding certain chemical chaperones to cells increases the amount of the CFTR that is correctly folded, and the folded mutant protein functions normally in chloride transport. This has launched a major research effort to discover drugs that facilitate CFTR folding.

CFTR regulates the transport of chloride (and water) across epithelial cell membranes. A multitude of physiologic defects occur when this protein fails to reach its intended membranes. These include inadequate hydration of mucus secretions in the lung, which renders the patient particularly susceptible to bacterial infections; inadequate hydration of an infant's stool, which can produce meconium ileus at birth or pathologic constipation thereafter; thickening of cervical mucus that can limit female fertility; and inadequate water and chloride secretion in the accessory gastrointestinal tract leading to gallstones and cholecystitis and destructive desiccation of the exocrine pancreas. Perhaps of less importance physiologically but most convenient diagnostically, it also leads to a marked increase in sweat chloride caused by the inability of the sweat gland epithelium to adequately resorb chloride.

Peter's sweat sample had a chloride of 110 mEq/L, which established the diagnosis. Medical treatments may delay the progressive nature of the disease. Pulmonary hygiene and specific antibiotic therapy will help his lungs temporarily. Oral lipase therapy may normalize his stools, and cholecystectomy may avert further biliary problems. In some cases lung transplant may be considered. Two currently available medications (ivacaftor and lumacaftor) can improve the function of particular mutant alleles, resulting in less buildup of mucus in the lungs and improvement in other symptoms. Unfortunately, progressive, permanent multiorgan damage ultimately occurs, and the average life expectancy for people with CF remains less than 40 years.

LEAVING THE ENDOPLASMIC RETICULUM: A PARADIGM FOR VESICULAR TRAFFIC

The ER, the Golgi apparatus, lysosomes, the plasma membrane, and various endosomes are all connected by pathways of vesicular traffic whereby membrane from one organelle is transferred to another. A vesicle forms at the originating membrane and moves to and fuses with the acceptor membrane, thus transferring selected lipids, proteins, and the lumenal contents of the vesicle to the destination organelle. All membranes involved in vesicle traffic must accurately select cargo to be loaded into the vesicle, bud the vesicle, target and fuse the vesicle to the correct membrane, and, finally, return to the originating membrane; those components dedicated to vesicle formation and fusion, so they can be reused for another round of vesicular traffic. The molecular mechanisms of vesicular trafficking differ in detail for different membranes, but mechanistic similarities have been identified. An overview of features common to most pathways of vesicle traffic is presented next, followed by specific details of ER to Golgi vesicle traffic.

Overview of Vesicle Budding, Targeting, and Fusion

Cargo Selection Requires a Specific Coat

Cargo selection is initiated by the assembly of a protein scaffold on the cytoplasmic side of the membrane where the vesicle will bud (Fig. 4–13). The scaffold is part of a structure called a *coat* that can be seen on membranes by electron microscopy, and different coat proteins exist for different organelles. The coat often contains two protein layers, with an inner layer closest to the membrane often called the *adaptor* that interacts with specific membrane-spanning proteins. The membrane-spanning proteins may be cargo themselves en route to another membrane compartment, or they may also be transmembrane receptor proteins with binding sites on the lumenal side of the membrane for binding to soluble proteins, thus concentrating the soluble proteins on the lumenal side of the membrane where a vesicle will form. Membrane protein cargoes have structural features recognized by adaptor proteins of the coat. These structural features, called *sorting signals*, not only can be stretches of amino acids common to several cargo types but also can be more subtle, depending on tertiary protein structure. Membrane proteins that bind to the coat adaptor on the cytoplasmic side and to a soluble protein on the lumenal side need an additional binding site for interacting with sorting signals on soluble proteins, and several of these signals have been defined.

How is the coat recruited to the cytoplasmic side of an organelle membrane? A common theme is to first recruit a small GTP-binding protein to the cytoplasmic surface of the membrane. Different membranes use different GTP-binding proteins, but recruitment is initiated by the exchange of GDP for GTP, catalyzed by a guanine nucleotide exchange factor (GEF) that is already associated with the membrane, which triggers binding of the GTP-binding protein to the membrane. The GTP-binding proteins contain a hydrophobic helix that is buried inside the protein when it is soluble in the cytosol, but which is displayed upon GTP binding so that it inserts into the membrane, anchoring the GTP-binding protein to the membrane. This anchored GTP-binding protein then interacts with coat components to start coat formation. Figure 4–13 illustrates a generic model of coat protein binding and cargo selection.

Vesicle Budding

To form a vesicle the parent membrane must deform with a high radius of curvature. In general, protein components of the coat, or other proteins that bind to the coat, are believed to associate with the membrane and cooperate to shape the vesicle. Once the vesicle has separated from the parent membrane, the coat protein complexes must dissociate, providing access to proteins on the cytoplasmic surface of the vesicle membrane that are needed for targeting

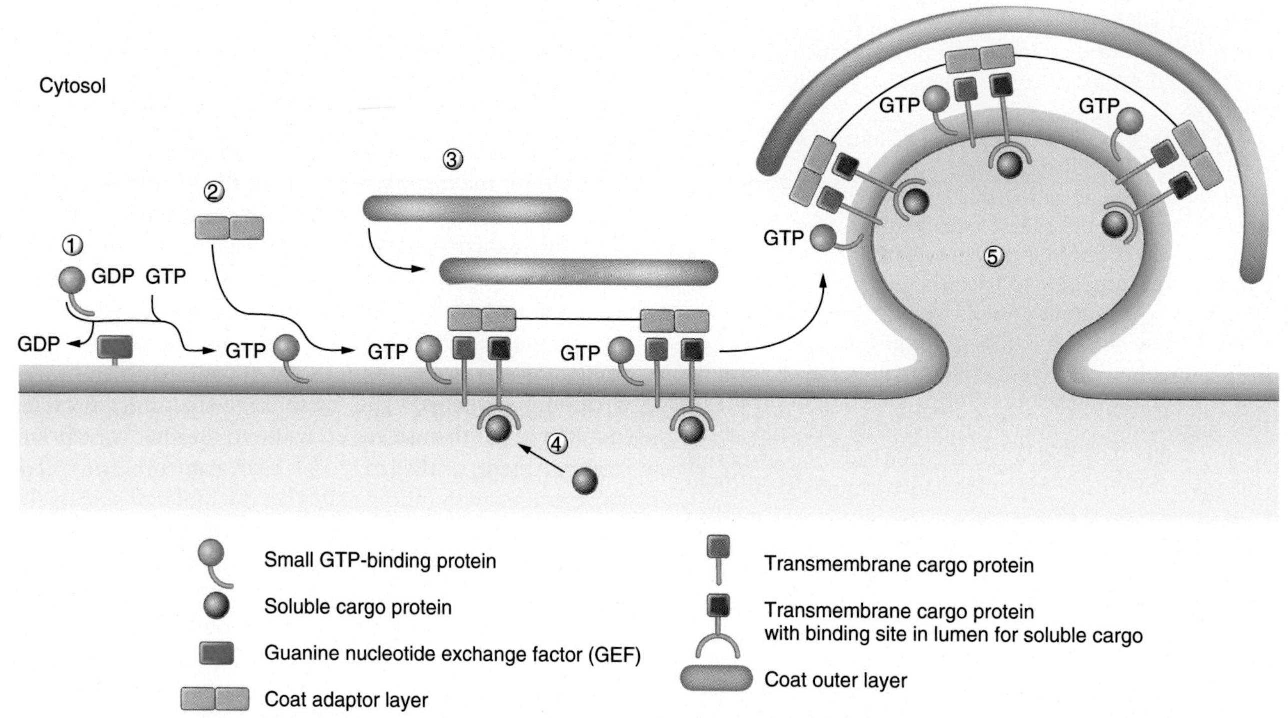

Figure 4–13. **Generic model showing the initial steps in vesicle coating, cargo recruitment, and vesicle budding.** (1) The exchange of GDP for GTP catalyzed by a guanine nucleotide exchange factor (GEF) stimulates the recruitment of a cytosolic small guanosine triphosphate (GTP)–binding protein to the membrane. GTP-binding exposes a hydrophobic segment on the GTP-binding protein that inserts into the membrane. (2, 3) The activated GTP-binding protein interacts with soluble coat components, recruiting them to the membrane. There are usually two parts to the coat, the adaptor layer and the outer layer. (4) The adaptor layer binds to transmembrane proteins that contain specific amino acid sequences that function as sorting signals. Some of the transmembrane proteins have domains on their lumenal side that bind to soluble lumenal proteins, thus coupling the soluble proteins to sites on the membrane where a vesicle will form. (5) The coat layers accumulate and contribute to deforming the planar bilayer into a vesicle. The vesicle eventually buds off and the coat dissociates upon hydrolysis of the GTP bound to the small GTP-binding protein (not shown).

to and fusion with destination membranes. In the best understood models, the hydrolysis of GTP by the GTP-binding protein that initiates coat recruitment is the signal for the coat to dissociate. Coat components and the GTP-binding protein in the GDP state are released to the cytosol, available for another round of vesicle formation.

Vesicle Targeting and Fusion

Once formed the vesicle moves to the destination membrane and fuses with it, combining the vesicle membrane with the target membrane and mixing the soluble contents of the vesicle with the lumenal contents of the destination organelle. Vesicle targeting and fusion must occur with great fidelity because delivery of the proteins and lipids to the wrong membrane could be lethal. There are three general levels to targeting and fusion: vesicle movement, vesicle tethering, and vesicle fusion. In small cells, such as yeast, vesicles are believed to move mainly by diffusion. However, in larger mammalian cells, vesicles often move on tracks of the cytoskeleton, such as microtubules. Proteins are known that connect vesicles to microtubules, and strong evidence has been found that motor proteins required for transport along microtubules drive directional vesicle movement (see discussion in Chapter 3).

Once near the target membrane, the initial interaction of the vesicle with the target membrane is a tethering event. Tethering involves an assembly of proteins that form an extended structure long enough to tether the vesicle to the membrane. Vesicle transport among different membranes uses different tethering factors, but most appear to be regulated, at least in part, by a class of small GTP-binding proteins called *Rab proteins.* Rab proteins use a typical GTP/GDP cycle that is controlled by Rab GTPase-activating proteins and GEFs. In the GDP state, Rab proteins are complexed with an escort protein in the cytosol, and when activated, they insert a covalently attached hydrophobic lipid anchor into their home membrane and are included in vesicles when they bud. More than 70 Rab proteins are encoded by mammalian cells, and a unique Rab protein appears to participate in each different vesicle-trafficking event, leading to the idea that Rab proteins contribute, at least in part, to the specificity of vesicle targeting. Distinct Rab proteins are also differentially expressed in different tissues. Several disease syndromes are known that result from defects in a Rab protein, a Rab escort protein, or a Rab regulatory protein. For example, the eye disease, choroideremia, is caused by a defective escort protein that normally binds to the GDP-bound form of Rab27a in the cytosol.

Vesicle fusion is mediated by integral membrane proteins called SNARES that are of two types, v-SNARES and t-SNARES ("v" stands for vesicle and "t" stands for target). Every vesicle when it buds incorporates v-SNARE proteins that have binding sites in their cytosolic domains for cognate t-SNARE proteins on the target membrane. Although there are variations on the theme, there is usually a single v-SNARE class in the vesicle and three different t-SNARES in the target membrane. When the SNARES interact, each contributes one amphiphilic α-helix to form an extremely stable coiled-coil structure where the four helices twist around one another, similar to the strands of a rope. The strength of this interaction is believed to bring the cytosolic face of the vesicle and the target membrane into close proximity facilitating fusion. A generic model of vesicle tethering and fusion is shown in Figure 4–14.

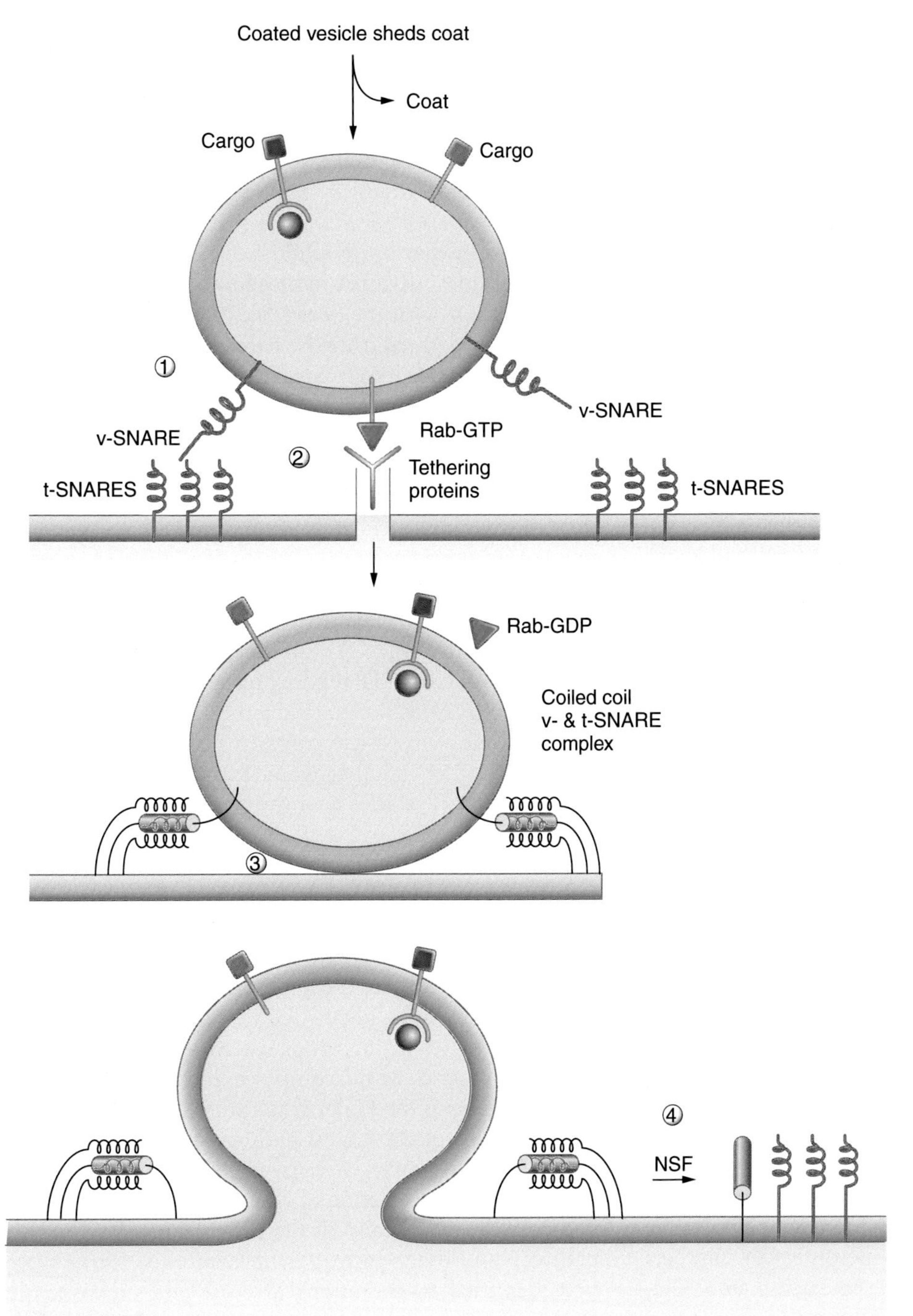

Figure 4–14. **Vesicle tethering and fusion.** (1) A coated vesicle sheds its coat and moves to the target membrane. In addition to cargo, the vesicle contains a v-SNARE that was recruited at the time of vesicle formation. A Rab protein in the active guanosine triphosphate (GTP)–bound form is also recruited to the cytosolic face of the vesicle. (2) The vesicle docks with the target membrane in a tethering step that involves the Rab-GTP and a tethering protein complex on the target membrane. (3) The v-SNARE on the vesicle interacts with three t-SNARES on the target membrane, bringing the cytosolic face of the two membranes close together so they can fuse. At some point in this process, the Rab protein hydrolyzes GTP to guanosine diphosphate (GDP) and dissociates into the cytosol. (4) After fusion is complete, *N*-ethylmaleimide sensitive factor (NSF), in association with other proteins, disrupts the v- and t-SNARE complexes freeing the v-SNARE, which is trafficked back to its home membrane so it can participate in another round of vesicle transport.

Evidence supporting this model of fusion includes studies where many different v- and t-SNARE combinations were incorporated into lipid vesicles and the ability of the vesicles to fuse was measured. Fusion activity was greatest for those v- and t-SNARE combinations that were known from other studies to be cognate pairs in living cells.

Recycling v-SNARES

After membranes fuse, there must be a mechanism for dissociating the stable v- and t-SNARE complexes, enabling the v-SNARES to recycle to their original membrane and support another round of vesicle fusion. SNARE complexes are disassembled by a protein called NSF, an ATPase of the AAA family. NSF was originally discovered as a soluble protein necessary for vesicle fusion in assays that reconstituted vesicle transport among Golgi membranes. NSF was sensitive to inhibition by *N*-ethylmaleimide, a reagent that binds to free thiol groups of cysteine residues in proteins, hence, the name *N*-ethylmaleimide sensitive factor. It is now thought that NSF is required to disassemble v- and t-SNARE complexes enabling additional rounds of vesicle targeting and fusion to occur.

Endoplasmic Reticulum to Golgi Vesicle Transport and COPII-Coated Vesicles

Vesicles that carry cargo from the ER to the Golgi complex form at specialized regions of the RER called the transitional ER. Transitional ER sites can be identified by electron microscopy and are widely dispersed throughout the RER membrane as islands that lack bound ribosomes. The coat protein used in ER to Golgi transport is called COPII, where COP stands for coat protein. As described in the generic model of vesicle formation, a small GTP-binding protein called Sar1 is used to recruit COPII to membranes. Sar1 is bound to GDP in the cytosol and is activated by the exchange of GTP for GDP, catalyzed by the membrane-bound GEF protein Sec12, in the transitional ER. A hydrophobic helix at the N-terminus of Sar1 is exposed upon binding GTP and inserts into the membrane as an anchor. Sar1-GTP binds to a dimeric protein complex containing proteins Sec23 and Sec24, the adaptor layer of the COPII coat protein complex. A second layer of the coat complex consisting of proteins Sec13/Sec31 subsequently binds to the Sar1/Sec23/Sec24 scaffold. Both Sar1 and Sec23/Sec24 interact with sorting signals in the cytoplasmic domains of membrane protein cargoes. Mammalian Sec24 contains two distinct cargo-binding sites that interact with specific amino acid sequences found in the cytoplasmic domains of cargo proteins. One site on Sec24 interacts with proteins that contain the sequence Asp-X-Glu, termed the *diacidic signal* (where X is any amino acid), while a second site on Sec24 binds to cargo proteins that have different sorting signals.

ER sorting signals can be subtle and not easily decoded. For example, the binding of certain cargo proteins to Sec23/Sec24 depends on the oligomerization of the proteins that contain multiple subunits. In other instances the cargo protein must bind to an escort protein to reveal the export signal. One class of escort protein is ERGIC 53, which interacts with mannose residues of the N-linked oligosaccharide on exported proteins. There is an autosomal recessive human syndrome in which mutated ERGIC 53 results in the failure of clotting factors V and VIII to efficiently enter COPII vesicles, resulting in a bleeding disorder. One concept emerging from the study of ER export sorting signals is that their presentation can depend on whether a protein is correctly folded; that is, quality control is an important parameter of whether a protein will display the correct signal to be included in a vesicle that leaves the ER.

In the COPII vesicle system, Sar1 initiates the membrane deformation events necessary for the vesicle to bud by inserting an amphiphilic helix into the membrane. The membrane curvature induced by Sar1 is stabilized by its association with the Sec23/Sec24 complex. Once the vesicle has separated from the parent membrane, the coat protein complexes dissociate, triggered by hydrolysis of GTP on Sar1-GTP, releasing the COPII coat components.

In mammalian cells, uncoated COPII vesicles aggregate together near ER transitional sites where they are formed and fused together to make a larger compartment that morphologically contains intermixed vesicles and tubules. Fusion of vesicles to others of a like kind is called *homotypic fusion*, and it also occurs in other examples of membrane traffic. The tethering step for this particular fusion event uses the GTP-binding protein Rab1, which when activated by GTP binding recruits a coiled-coil protein called p115 as part of the tether. Homotypic fusion requires that each vesicle contains both v-SNARES and t-SNARES. The compartment formed by homotypic fusion of COPII vesicles near the transitional ER is called the *vesicular tubular compartment* (VTC). To return SNARES to the transitional ER, it would be efficient if retrograde recycling vesicles emerged from the VTC while in close proximity to the transitional ER, and strong evidence exists that retrograde recycling of this type is mediated by another type of vesicle coated with a protein complex called COPI (COPI is discussed in more detail later in this chapter). After the VTC grows larger, it attaches to microtubules and moves en masse to the *cis* face of the Golgi apparatus.

The details of the COPII pathway have been worked out by a combination of yeast genetics, in vivo studies, and in vitro reconstitution of vesicle formation with purified components. Many of the proteins involved in this pathway and other pathways of membrane traffic were first identified by mutations in yeast and subsequently found in mammalian cells, an example of how work in nonmammalian organisms can lead to an understanding human cell biology and disease.

THE GOLGI APPARATUS

The structure and morphology of the Golgi apparatus is illustrated in Figure 4–15. The forming face, where VTCs from the ER arrive, is called the *cis* Golgi network (CGN). Secretory material arrives at the *cis* face and progresses through the medial and *trans* stacks to reach the *trans* Golgi network (TGN), where it is sorted and distributed to the next destination, either the plasma membrane or endosomes. Proteins passing through the Golgi complex undergo several types of covalent modification catalyzed by Golgi-resident enzymes. One main function of the Golgi apparatus is to modify N-linked oligosaccharides as they move through the Golgi stacks. In addition, some proteins acquire O-linked carbohydrates as they pass through the Golgi. In the TGN, the most distal compartment of the Golgi, some proteins are sulfated, and other proteins undergo controlled proteolytic events as part of their maturation process. The various modifications that occur during transit through the Golgi are described in the following section.

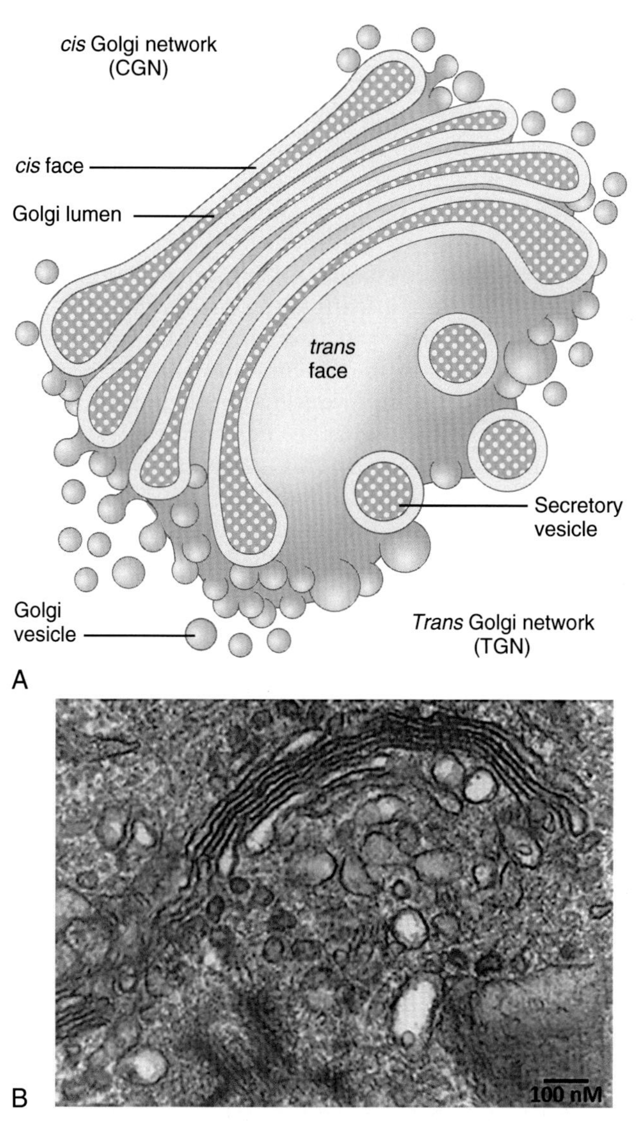

Figure 4–15. **Structure of the Golgi complex.** (A) The Golgi contains stacks of flattened cisternae. These cisternae have dilated edges from which small vesicles bud. The forming or *cis* face of the Golgi is the *cis* Golgi network (CGN), and it receives cargo from the endoplasmic reticulum. The exporting or *trans* face of the Golgi is the *trans* Golgi network (TGN), where vesicles and cargo depart for delivery to other membrane compartments. (B) Electron micrograph illustrating the flattened cisternae of the Golgi in a human leukocyte. *([B] Modified from https://en.wikipedia.org/wiki/File:Human_leukocyte,_showing_golgi_-_TEM.jpg.)*

Glycosylation and Covalent Modification of Proteins in the Golgi Apparatus

The packaging of nascent proteins with N-linked sugar modifications into ER export vesicles requires processing of the N-linked sugar (Fig. 4–12). Upon arrival in the Golgi apparatus, further trimming of the N-linked sugar occurs, and *N*-acetylglucosamine, galactose, fucose, and *N*-acetylneuraminic acid are added to the various branches of the oligosaccharide in the different Golgi subcompartments (Fig. 4–16). *N*-Acetylneuraminic acid

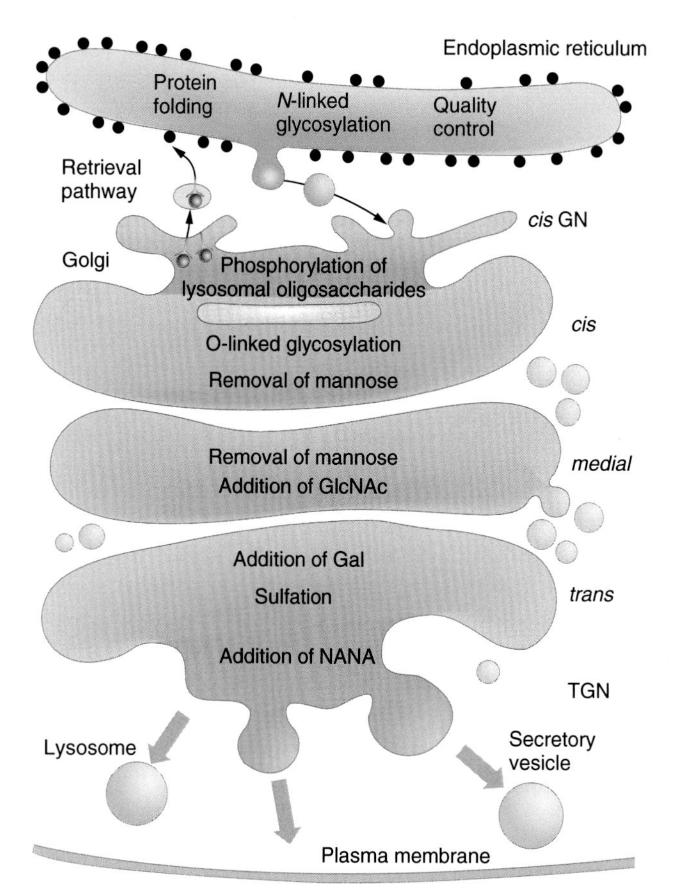

Figure 4–16. **Overview of compartmentalized functions in the secretory pathway.** Protein folding, N-linked glycosylation, and quality control begin in the rough endoplasmic reticulum (RER). Phosphorylation of mannose on lysosomal enzymes, O-linked glycosylation, and trimming of N-linked oligosaccharides begin in the *cis* Golgi network (CGN). Additional carbohydrate trimming and addition of monosaccharides to N- and O-linked oligosaccharides occurs in the medial and *trans* Golgi cisternae and in the *trans* Golgi network (TGN). Sulfation of proteins on tyrosine residues happens in the TGN.

(NANA) (also called sialic acid) is the terminal modification added to many N-linked sugars, and it is added in the TGN. A special N-linked carbohydrate modification occurs for lysosomal enzymes in the *cis* Golgi network, the phosphorylation of one or more mannose residues on position 6 (Fig. 4–16). Mannose 6-phosphate is later recognized by a receptor, the mannose 6-phosphate receptor that selects lysosomal enzymes from all the other soluble secretory proteins and diverts them to lysosomes (see later for a more detailed description).

A second type of carbohydrate modification also occurs in the Golgi complex, the addition of O-linked oligosaccharide chains to proteins (Fig. 4–16). O-linked carbohydrate is attached to proteins via the oxygen of the hydroxyl group of either serine or threonine, hence the term *O-linked.* O-linked oligosaccharides are built by one-at-a-time addition of UDP-activated monosaccharides, catalyzed by glycosyl transferase enzymes. Depending on the protein, there may be only a few or many O-linked carbohydrate residues added.

The sulfation of certain proteins on tyrosine residues occurs in the TGN. In addition, the proteolytic cleavage of some secreted proteins occurs in the TGN, or in secretory vesicles, and this cleavage is often necessary for the protein to achieve its fully activate state. Proteolytic processing of this type is common for secreted polypeptide hormones (e.g., insulin) and is a way of ensuring that the hormone will not be in its active form until it is ready for secretion from the cell.

Retrograde Transport Through the Golgi Complex

In the face of abundant membrane traffic from the ER to the Golgi and farther, retrograde vesicular transport from Golgi membranes to the ER is robust. One function of this retrograde traffic is to return SNARES involved in anterograde traffic back to the ER. Although perhaps not immediately obvious, another crucial reason for retrograde transport is to return resident ER proteins that have escaped back to the ER. This need arises because the volume of anterograde traffic from the ER to the Golgi can be so large that resident ER proteins are unintentionally included in COPII anterograde vesicles. These resident ER proteins would be lost unless captured by an efficient mechanism of retrograde transport to return them to the ER. Two types of proteins must be returned by this retrograde transport pathway. Soluble ER-resident proteins are included within the fluid volume of an anterograde COPII-coated vesicle when it buds, and ER-resident integral membrane proteins accidentally included in the membrane of a budding vesicle. Soluble and membrane proteins contain different retrograde sorting signals that ensure they are recognized and included within retrograde vesicles. Soluble proteins contain at their C-terminus the sequence KDEL (or a related sequence), whereas integral membrane proteins contain the sequence KKXX (or a related sequence—Table 4–1) at their C-terminus. These sorting signals were discovered by comparing the amino acid sequences of different soluble and membrane ER proteins. Their function as retrograde sorting signals was uncovered by grafting the signals at the DNA level onto other proteins that were not normally residents of the ER. The resulting chimeric proteins accumulated in the ER because they were so efficiently captured and returned to this compartment by this retrograde recycling pathway.

The mechanism of retrograde vesicular transport conforms to the generic model of vesicle transport described earlier in this chapter (see Figs. 4–13 and 4–14). The vesicle coat used by retrograde vesicles is called coat protein I (COPI), and it is composed of seven polypeptides termed the *coatomer*. Coatomer units are recruited from the cytosol to Golgi membranes by a GTP-binding protein called Arf1. Following activation by a GEF, Arf1 inserts into the Golgi membrane, initiating the binding of coatomer to form the COPI coat. Similar to COPII vesicles, there appear to be two layers to the COPI complex, an inner layer that interacts with sorting signals and an outer layer that may form a scaffold to stabilize the coat. The inner layer has sites that bind cargo. One site directly binds the KKXX motif, concentrating proteins containing this motif in budding vesicles. Lumenal proteins that contain the KDEL signal are recognized by a transmembrane protein called the KDEL receptor. This receptor has a lumenal domain that binds the KDEL signal and a cytoplasmic domain that binds to the COPI coat, thus concentrating lumenal proteins that have the KDEL signal to sites where COPI vesicles will bud. The coat dissociates after vesicle formation, and the vesicles fuse with a target membrane using tethering factors and SNARES. The two general target membranes for COPI-coated vesicles are the membrane of the preceding Golgi stack and the ER. Thus COPI-coated vesicles carry material in the retrograde direction from one stack to the next, *trans* to *cis*, and eventually from the *cis* stack of the Golgi to the ER. In addition to this COPI-dependent pathway, there is at least one COPI-independent retrograde transport pathway discovered because some medically important bacterial toxins (such as Shiga toxin made by pathogenic strains of *Escherichia coli*) are delivered to early compartments of the secretory pathway in a COPI-independent fashion.

Anterograde Transport Through the Golgi Complex

The two basic models for anterograde transport through the Golgi complex are the vesicle transport model and the cisternal maturation model (compared in Fig. 4–17). The vesicle transport model proposes that each Golgi membrane cisterna is a permanent structure that receives

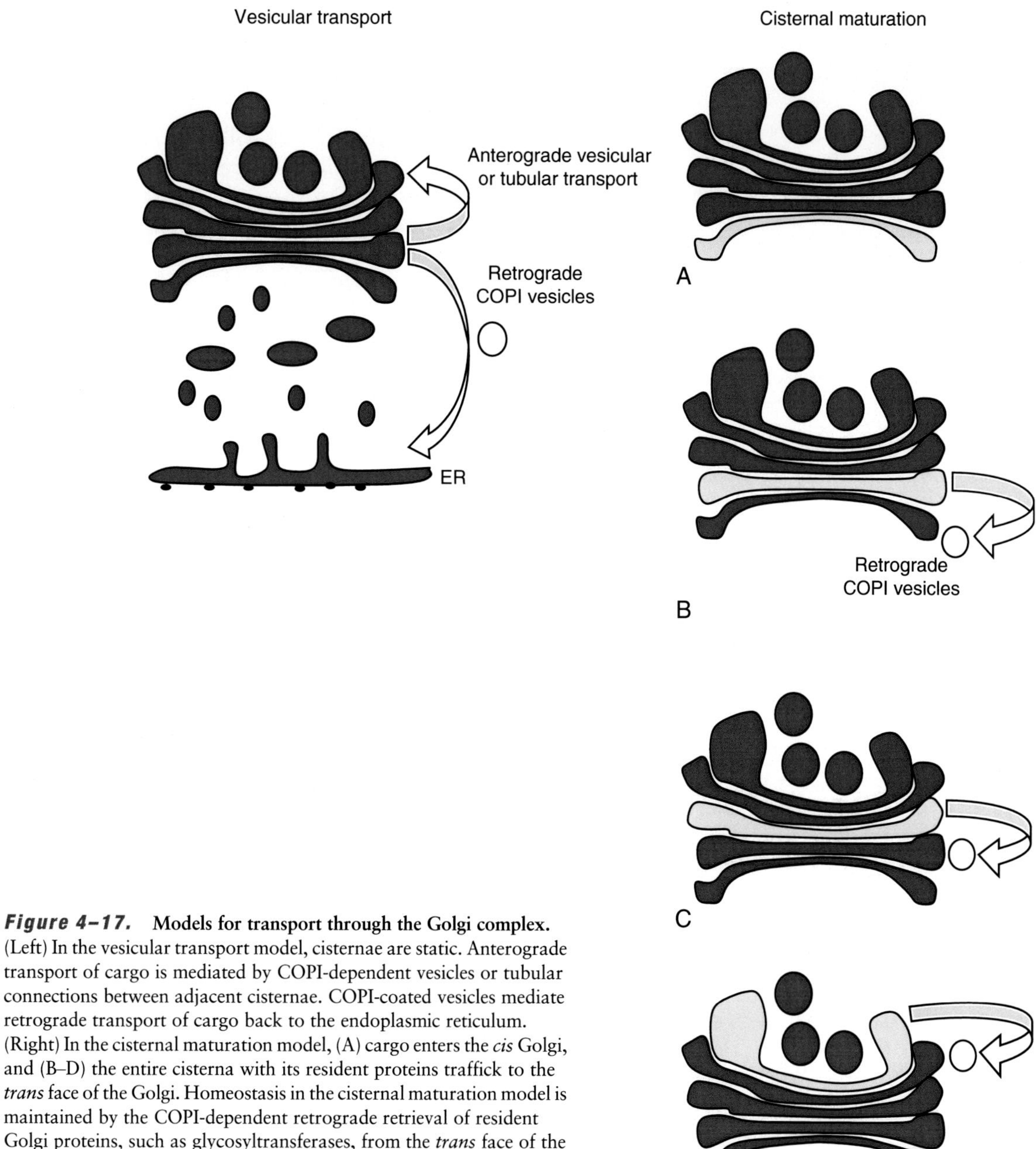

Figure 4–17. **Models for transport through the Golgi complex.** (Left) In the vesicular transport model, cisternae are static. Anterograde transport of cargo is mediated by COPI-dependent vesicles or tubular connections between adjacent cisternae. COPI-coated vesicles mediate retrograde transport of cargo back to the endoplasmic reticulum. (Right) In the cisternal maturation model, (A) cargo enters the *cis* Golgi, and (B–D) the entire cisterna with its resident proteins traffick to the *trans* face of the Golgi. Homeostasis in the cisternal maturation model is maintained by the COPI-dependent retrograde retrieval of resident Golgi proteins, such as glycosyltransferases, from the *trans* face of the organelle back to specific compartments in the Golgi where they function. *(Adapted from Martinez-Manarguez.* ISRN Cell Biol *2013;2013.)*

COPI-dependent vesicles carrying anterograde cargo from the adjacent cisterna on the *cis* side and then packages that cargo into new vesicles that deliver the cargo to the adjacent cisterna on the *trans* side. COPI-dependent retrograde transport operates in the reverse direction to recycle SNARES and other material to the preceding cisterna. In contrast the cisternal maturation model suggests that cargo enters the *cis* most cisterna of the Golgi and the entire cisterna with its resident proteins progresses through to the *trans* face of the organelle. Homeostasis in the cisternal maturation model is maintained by the COPI-dependent retrograde retrieval of resident Golgi proteins, including glycosyltransferases, from the *trans* face of the organelle back to specific compartments in the Golgi where they function.

Until the late 1990s the vesicle transport model was rarely challenged. However, experimental data indicated that COPI-coated vesicles primarily carried retrograde cargo. In addition, some cargo, such as fish scales and large collagen complexes, were too large to fit into small anterograde transport vesicles. These large cargos could, however, move through the Golgi complex within individual cisterna as the cisterna matured during its progression from the *cis* to the *trans* face. The study of cargo movement through the Golgi complex remains an active area of investigation, and it is possible that both of the

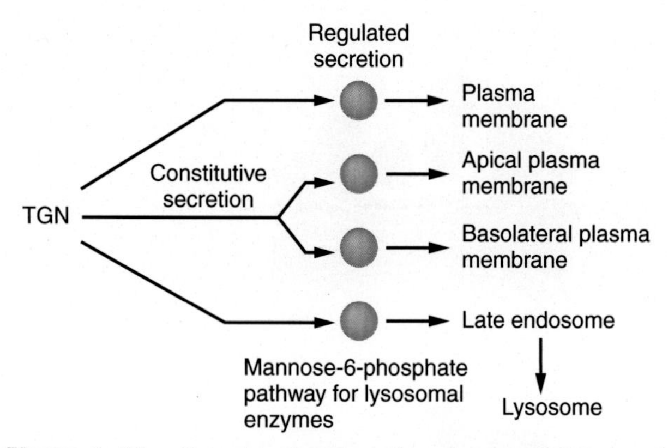

Figure 4–18. **Overview of transport from the *trans* Golgi network (TGN).** There are three main pathways for material leaving the TGN: regulated secretion of cargo destined for the plasma membrane, constitutive secretion of cargo destined for the plasma membrane, and the mannose 6-phosphate pathway for the delivery of lysosomal enzymes to late endosomes. Note that secretion to the plasma membrane in polarized cells involves targeting cargo and vesicles to either the apical or basolateral domains of the plasma membrane.

models described earlier contribute to the anterograde transport of cargo through this organelle.

Leaving the Golgi Complex

There are three general pathways of anterograde export from the TGN (summarized in Fig. 4–18): constitutive secretory vesicles that constantly ferry material from the TGN to the plasma membrane; regulated secretory vesicles that collect material destined for the plasma membrane but hold it until a signal for secretion triggers fusion with the plasma membrane; and vesicles that carry lysosomal enzymes to late endosomes or lysosomes.

Constitutive and Regulated Secretion

Vesicles constantly leave the TGN and fuse with cytosolic face of the plasma membrane, replenishing lipids and membrane proteins as a housekeeping function. Evidence suggests that multiple mechanisms contribute to cargo selection and vesicle budding in this constitutive transport pathway. Furthermore, there is evidence that both vesicles and membrane tubules are involved in delivering cargo from the TGN to the plasma membrane. This constitutive secretory pathway exhibits an additional level of complexity in polarized cells that possess compositionally distinct apical and basolateral plasma membranes. Although cargo destined for the apical and basolateral membranes of polarized epithelial cells possess distinct sorting signals, the coat proteins that direct the TGN budding of vesicles containing apical and basolateral cargo are not known. Following budding from the TGN, vesicles containing apical and basolateral cargo utilize unique complements of SNAREs to traverse distinct endosomal compartments prior to undergoing delivery to the plasma membrane. In hepatocytes, additional complexity is observed, as some proteins are first delivered to the basolateral membrane and then sorted into endocytic vesicles that carry the proteins from the basolateral membrane to the apical membrane, a process called *transcytosis*.

Regulated secretory vesicles collect proteins that are to be stored in vesicles until the cell receives the appropriate stimulus. The secretion of insulin by pancreatic β-cells is a well-characterized example of regulated secretion where insulin is retained in secretory vesicles until blood glucose levels are elevated. The increase in ATP that occurs in pancreatic β-cells when blood glucose levels are high triggers a series of events that eventually lead to the fusion of insulin-containing vesicles with the plasma membrane. The targeting of cargo to regulated secretory vesicles appears to be dependent upon the ability of the cargo to selectively aggregate at sites where the vesicles will form in the TGN. The aggregates can contain several different types of proteins that will become passengers in the same vesicle. Following budding from the TGN, cargo can continue to condense as the secretory vesicles mature.

Lysosomal Enzymes Are Targeted via a Mannose 6-Phosphate Signal

Lysosomes contain a wide variety of soluble hydrolytic enzymes capable of digesting most naturally occurring macromolecules. Lysosomal hydrolases contain a cleavable N-terminal signal peptide, and they are cotranslationally translocated into the lumen of the RER where they acquire N-linked oligosaccharides and fold prior to undergoing delivery to the *cis* Golgi. In the *cis* Golgi, mannose residues on the N-linked oligosaccharides of lysosomal enzymes are phosphorylated, and this modification provides a mechanism to select lysosomal enzymes from other secreted proteins and divert them to lysosomes. The discovery of the pathway is a great example of how the synergy between basic and clinical science can lead to advances in cell biology. Patients with I-cell disease (*inclusion body disease*) secrete a wide variety of lysosomal enzymes, rather than transporting them to lysosomes. Because so many different enzymes were affected, researchers realized that the origin of the disease was in a fundamental pathway used to target lysosomal hydrolytic enzymes to the correct subcellular organelle. When the secreted lysosomal enzymes from patients with I-cell disease were added to the media of normal cultured cells, the I-cell lysosomal enzymes were not taken up by the normal cells. However, when lysosomal enzymes from normal cells were added to the media of normal cells or I cells, the enzymes were taken up by both cell types. This provided a clue that there was a structural difference between lysosomal enzymes from healthy individuals and from patients with I-cell disease. The difference was shown to be the presence of phosphate on mannose residues of the N-linked sugar of normal lysosomal enzymes that were missing from the enzymes isolated from patients with I-cell disease.

Subsequent studies identified the mannose 6-phosphate receptor, a transmembrane protein that binds to lysosomal enzymes that contain the mannose 6-phosphate sorting signal at the TGN. This binding is strong at the pH found in the TGN (~6.4) but weak at the acidic pH found in lysosomes, providing an important clue to how the receptor binds and releases lysosomal enzymes in different compartments. The addition of phosphate to mannose residues occurs by a two-step process where *N*-acetylglucosamine phosphate is transferred from the UDP-activated monosaccharide to the 6th position of mannose by the enzyme *N*-acetylglucosamine phosphotransferase, followed by cleavage of the *N*-acetylglucosamine, leaving the phosphate group on the mannose (Fig. 4–19). In patients with I-cell disease, the phosphotransferase enzyme is defective, and as a result, many lysosomal hydrolases are secreted rather than undergoing delivery to the lysosome. There is no consensus amino acid sequence in lysosomal hydrolases that is involved in their recognition by the phosphotransferase in the *cis* Golgi suggesting that a common tertiary structure found in correctly folded lysosomal enzymes directs this modification to occur.

The pathway for transport of lysosomal enzymes from the TGN to lysosomes is illustrated in Figure 4–20. Lysosomal enzymes bearing the mannose 6-phosphate marker arrive together with other secretory proteins in the lumen of the TGN. At the slightly acidic pH found in the TGN, the mannose 6-phosphate receptor is able to bind lysosomal enzymes and recruit them into vesicles that bud from the TGN. Vesicle budding uses a coat protein complex called *clathrin* that has the prototypical features of coat proteins discussed earlier. The initial interaction of the clathrin coat with the membrane relies on the small GTP-binding protein Arf1. Recall that Arf1 also initiates the binding of COPI coats to Golgi membranes. Although we do not fully understand how a single Arf1 protein functions in the binding of different coats to different membranes, it is clear that the activation of Arf1 in different membrane compartments is dependent upon different GEFs. The recruitment of the clathrin coat to vesicles destined for lysosomes is also dependent upon the AP-1 adaptor, which contains one copy of four different subunits. AP-1 binds to the cytosolic domains of cargo proteins, including the mannose 6-phosphate receptor, thus coupling lysosomal enzymes to the coat machinery. The AP-1 adaptor then recruits *clathrin*, which is composed of one heavy and one light chain that assemble into a trimer containing three legs called a *triskelion*. Triskelions associate to form the clathrin coat (see Fig. 4–20). Once the clathrin-coated vesicle buds, the coat dissociates and the vesicle fuses to late endosomes using Rab and SNARE proteins unique to this type of vesicle.

The pH within **late endosomes** is acidic (~5.5), established by ATP-dependent proton pumps in the endosomal membrane. The affinity of the mannose 6-phosphate receptor for mannose 6-phosphate is low at the pH found in the endosome, dissociating lysosomal enzymes from the receptor. This is an important step in the targeting of lysosomal enzymes as it allows for the recovery of the unoccupied receptors by recycling from the endosome to the TGN, while the endosome containing the lysosomal enzyme fuses with the lysosome. Drugs that block the acidification of late endosomes prevent the delivery of lysosomal enzymes to lysosomes because the mannose 6-phosphate receptors do not dissociate from the bound lysosomal enzymes.

At least two other adaptor complexes, GGA and AP-3, function in the export of cargo from the TGN. GGA is a monomeric adapter that is involved in the clathrin-dependent transport of glycosylated cargo from the TGN to lysosomes. In addition, GGA appears to cooperate with AP-1 in directing the mannose 6-phosphate receptor into vesicles destined for lysosomes (Fig. 4–20). AP-3 is a multisubunit adaptor structurally related to AP-1 that may be necessary for the clathrin-dependent budding of secretory vesicles from the TGN. In addition, AP-3 is required for the delivery of some enzymes to lysosome-related organelles, including melanosomes. Hermansky-Pudlak syndrome type 2, a disease associated with pigmentation defects, immunodeficiency, and blood disorders, is caused by a defect in one subunit of the AP-3 adaptor.

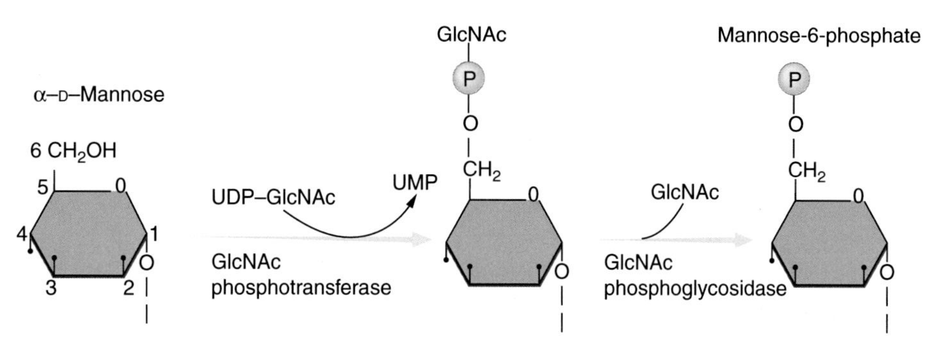

Figure 4–19. **Biosynthesis of mannose 6-phosphate on lysosomal enzymes.** The addition of mannose 6-phosphate occurs in two steps within the *cis* Golgi network. First, *N*-acetylglucosamine-phosphotransferase uses UDP-GlcNAc to add GlcNAc phosphate in a phosphodiester linkage to the 6th position of mannose residues on N-linked oligosaccharides of a lysosomal enzyme. Next a phosphoglycosidase removes the GlcNAc, leaving mannose phosphorylated on the 6 position.

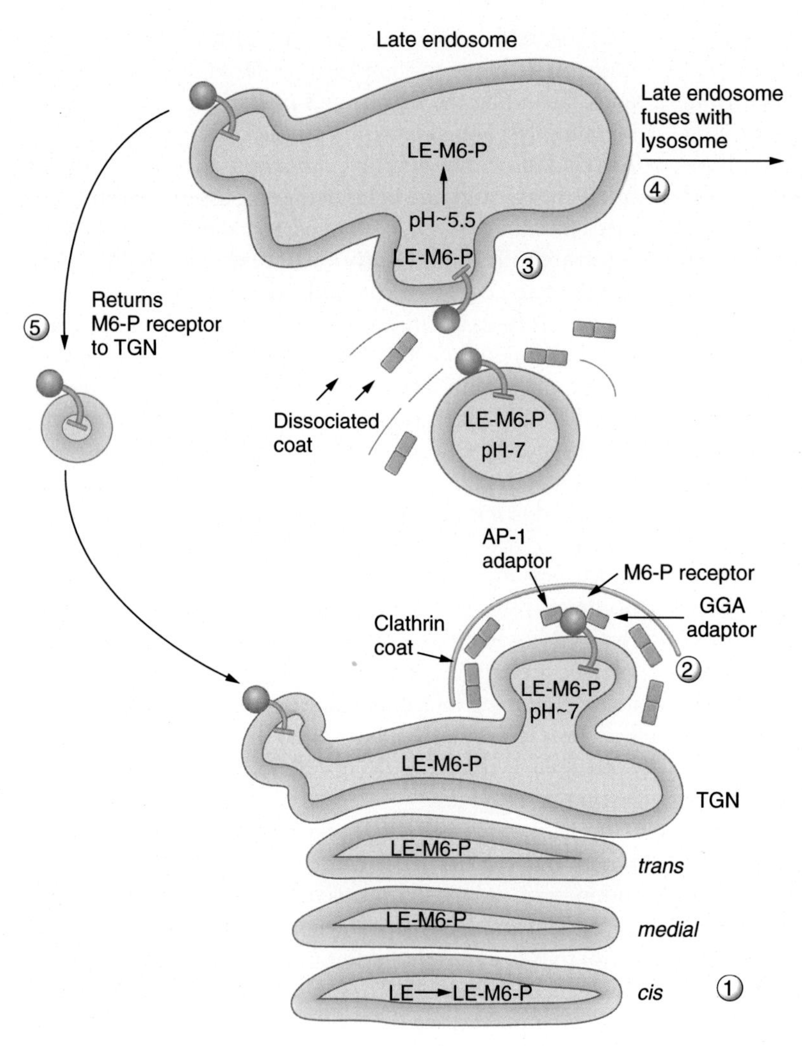

Figure 4–20. **Mannose 6-phosphate (M6-P) pathway for delivery of lysosomal enzymes to lysosomes.** (1) Lysosomal enzymes receive the M6-P marker in the *cis* Golgi network (CGN) and migrate through the Golgi cisternae with other secretory enzymes to the *trans* Golgi network (TGN). (2) The M6-P receptor in the TGN binds lysosomal enzymes via recognition of the M6-P modification, segregating lysosomal enzymes from other soluble secretory proteins. The clathrin/adaptor protein 1 (AP-1) adaptor coat is recruited by interactions with the small guanosine triphosphate (GTP)–binding protein Arf1 (not shown) and collects M6-P receptors with associated lysosomal enzyme cargo in the budding vesicles. The GGA adaptor is also involved in recruiting the clathrin coat to these budding vesicles. (3) After vesicle formation the coat dissociates, and the vesicles fuse with late endosomes where the low pH dissociates lysosomal enzymes from the M6-P receptor. (4) The lysosomal enzymes subsequently reach lysosomes when the late endosome fuses with existing lysosomes or matures into a lysosome. (5) The M6-P receptor is returned to the TGN in vesicles that bud from the late endosome.

ENDOCYTOSIS, ENDOSOMES, AND LYSOSOMES

Endocytosis is the general term describing the uptake of extracellular material through the internalization of vesicles formed at the plasma membrane. This process enables the entry of many physiologically important molecules into the cell, and it is often exploited by pathogens seeking access to the cell interior. There are different types of endocytosis, and some endocytic processes occur only in certain cell types. Phagocytosis refers to the formation of large ($\geq 1\,\mu m$), actin-dependent vesicles by macrophages and neutrophils that enable the cells to engulf large particles, such as invading bacteria. Following internalization from the cell surface, phagocytic vesicles (*phagosomes*) fuse with lysosomes, exposing ingested material to degradative lysosomal enzymes. Other types of endocytosis are divided into two general categories: clathrin-independent and clathrin-dependent endocytosis. Several types of clathrin-independent endocytosis exist. Macropinocytosis refers to a type of clathrin-independent and actin-dependent endocytosis in which extensions of the plasma membrane fuse to engulf a volume of extracellular fluid. A second type of clathrin-independent endocytosis involves the formation of vesicles at sites in the plasma membrane called *lipid rafts*, which are enriched in cholesterol and sphingolipids. The formation of endocytic vesicles at lipid rafts can be dependent upon a variety of proteins including caveolin and flotillin. Endocytic vesicles that are dependent upon caveolin for their formation are particularly abundant in endothelial cells, adipose cells, and fibroblasts.

Clathrin-Dependent Endocytosis

A typical mammalian cell internalizes about half of its plasma membrane surface area per hour, primarily by the formation of 50–100 nm clathrin-dependent endocytic vesicles. When these vesicles form, extracellular material is brought into the cell in two ways, either as a soluble component in the fluid that is inadvertently engulfed by the vesicle or as a substance that has bound to a receptor in the membrane where the vesicle forms.

The nonspecific internalization of fluid is called *pinocytosis* (cell drinking), whereas the internalization of receptor-bound material is called *receptor-mediated endocytosis*. Receptor-mediated endocytosis is a much more efficient process than pinocytosis because the material taken up is concentrated at sites in the plasma membrane where vesicles bud.

The clathrin coat that drives vesicle formation at the plasma membrane consists of an inner layer of adaptor protein 2 (AP-2) and an outer layer of clathrin. AP-2, which is structurally similar to the AP-1 adaptor, binds sequences in the cytosolic domains of transmembrane receptors, ensuring that they will be efficiently included in endocytic vesicles. At any given time the clathrin/AP-2 coat covers ~2% of the cytoplasmic surface of the plasma membrane, and these regions can be easily visualized by electron microscopy. Some cell surface receptors constitutively associate with these sites via their interaction with AP-2, whereas other receptors do not expose their AP-2 binding site and as a result do not associate with clathrin-coated regions of the membrane until a ligand binds to their extracellular domain.

After budding from the plasma membrane, the clathrin/AP-2 coat disassociates from the vesicle that either undergo homotypic fusion to form an early endosome or fuse with a preexisting early endosome. Early endosomes contain active proton-translocating ATPases that pump in protons from the cytosol, acidifying the endosome interior. The affinity of many receptor-ligand pairs is regulated by pH, and at the acidic pH found in the endosome, the receptor and ligand dissociate, releasing the ligand into the lumen of the endosome. This dissociation of ligand and receptor is a key event that allows the receptors to recycle back to the plasma membrane to initiate another round of endocytic uptake. Recycling begins from the early endosome and is believed to occur by the budding of vesicles from the endosome and their subsequent fusion with the plasma membrane. This remodeling of the early endosomal membrane occurs as it matures into a late endosome, which then undergoes fusion with lysosomes. In this way, ligands released from their receptors in the endosome are delivered to the lysosome and exposed to the degradative enzymes of this organelle. Figure 4–21 provides an overview of the clathrin-dependent endocytosis pathway. To illustrate the endocytic pathways, we next look in more detail at the receptor-mediated uptake of low-density lipoprotein (LDL).

Receptor-Mediated Endocytosis of Low-Density Lipoprotein

LDL is a lipoprotein particle in blood that contains an outer monolayer composed of phospholipids and the protein, apolipoprotein B-100 (apoB-100), which surrounds a core of cholesterol esters. LDL carries dietary cholesterol to cells and is taken up by receptor-mediated endocytosis. The LDL receptor is a single-pass transmembrane protein in the plasma membrane that has a binding site in its extracellular domain for apoB-100 and a sequence in its cytosolic domain that binds the AP-2 adaptor (Fig. 4–21). Studies of the human inherited disorder familial hypercholesterolemia identified mutations in the LDL receptor that contributed much to our understanding of receptor-mediated endocytosis and cholesterol homeostasis. Analysis of one class of LDL receptor mutants that failed to cluster in clathrin-coated pits resulted in the identification of the sorting signal (Asn-Pro-X-Tyr) in the cytosolic tail of the receptor that serves as a binding site for the AP-2 adaptor. This sorting signal also directs the clathrin-/AP-2–dependent uptake of other surface receptors.

Studies of LDL uptake in patients afflicted with familial hypercholesterolemia also contributed to the understanding of important features of cholesterol regulation. Either cells obtain cholesterol from the diet, delivered by the LDL particles, or they make cholesterol de novo. When dietary cholesterol is available and delivered to cells, they sense an elevated level of cholesterol and suppress cholesterol synthesis. However, in patients that have a defective LDL receptor, cells do not receive dietary cholesterol, even though it is present in the blood, and these individuals continue to synthesize cholesterol de novo. This inability to coordinate cholesterol synthesis with dietary availability greatly increases whole-body cholesterol and leads to premature atherosclerosis. Studies of familial hypercholesterolemia and the LDL system have made such important contributions to biology and medicine that the researchers most responsible for advances in the area, Drs Joseph Goldstein and Michael Brown, received the Nobel Prize in medicine and physiology in 1985.

Multivesicular Endosomes

Unlike the endocytic pathway utilized for the uptake of LDL, some ligands taken into the cell by receptor-mediated endocytosis are degraded in lysosomes, together with their receptors. For example, polypeptide hormones such as epidermal growth factor and its receptor are both degraded in lysosomes after receptor-mediated endocytosis. The destruction of both the hormone and its receptor transiently reduces the sensitivity of cells to certain hormone signaling pathways because the number of hormone receptors on the cell surface goes down. The degradation of epidermal growth factor receptor following its endocytic uptake is dependent upon the budding of vesicles derived from the endosomal membrane into the lumen of this organelle (Fig. 4–22). These endosomes, which contain small vesicles within their lumen, are called *multivesicular bodies*.

The internalization and degradation of certain cell surface receptors, such as the epidermal growth factor

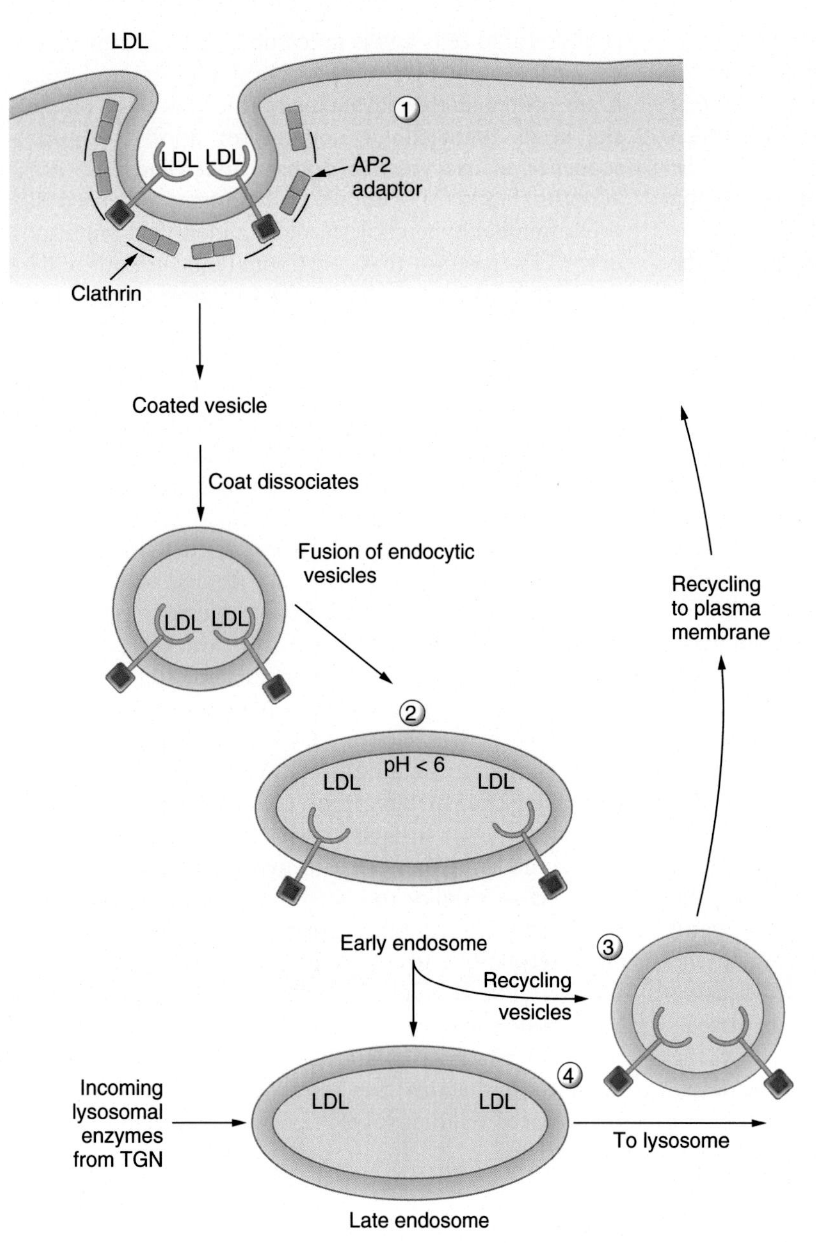

Figure 4–21. **Clathrin-dependent endocytosis of low-density lipoprotein (LDL).** (1) The LDL receptor has an extracellular domain that binds LDL and a cytosolic domain that binds to the adaptor protein 2 (AP-2) adaptor layer of the clathrin coat, and clathrin-coated vesicles containing LDL receptors with bound LDL bud from sites on the plasma membrane. (2) Following internalization the coat dissociates, and endocytic vesicles either fuse with themselves to form an early endosome or fuse with a preexisting early endosome. The pH within the early endosome triggers the dissociation of LDL from its receptor. (3) Recycling vesicles then bud from the early endosome (or a related structure called a *recycling endosome*, not shown here) carrying the unoccupied LDL receptors back to the plasma membrane. (4) LDL in the lumen of the early endosomes goes with the fluid to the late endosome and eventually reaches the lysosome where it is degraded, releasing its cargo of cholesterol.

receptor, are dependent upon the monoubiquitination of a lysine residue in the cytosolic domain of the receptor (Fig. 4–22). This ubiquitin moiety is recognized by the AP-2 adaptor complex and results in the recruitment of the receptor into clathrin-dependent endocytic vesicles. Following internalization the ubiquitinated receptors cluster in regions of the endosomal membrane that will undergo budding into the lumen of the endosome. This budding process is dependent upon multiple protein complexes, called ESCRTs, that select ubiquitinated proteins for inclusion in the budding vesicle and drive membrane deformation during the budding process. Other signals can also direct the incorporation of cargo into regions of the endosomal membrane that will undergo budding into the lumen of this organelle. Upon delivery to the lysosome, these intralumenal vesicles are degraded (Fig. 4–22). While degradation is the typical fate of these intralumenal vesicles, the fusion of multivesicular bodies with the plasma membrane releases intralumenal vesicles into the extracellular environment. The released vesicles are called exosomes, and they provide a mechanism for the delivery of material between cells.

Many viruses, including HIV, do not encode all the proteins needed for extracellular budding and instead commandeer a subset of the ESCRT machinery to complete the final steps of viral egress from the cell. Understanding the relationship between multivesicular body formation and the life cycle of pathogenic viruses has fueled work in the area because it may be possible to

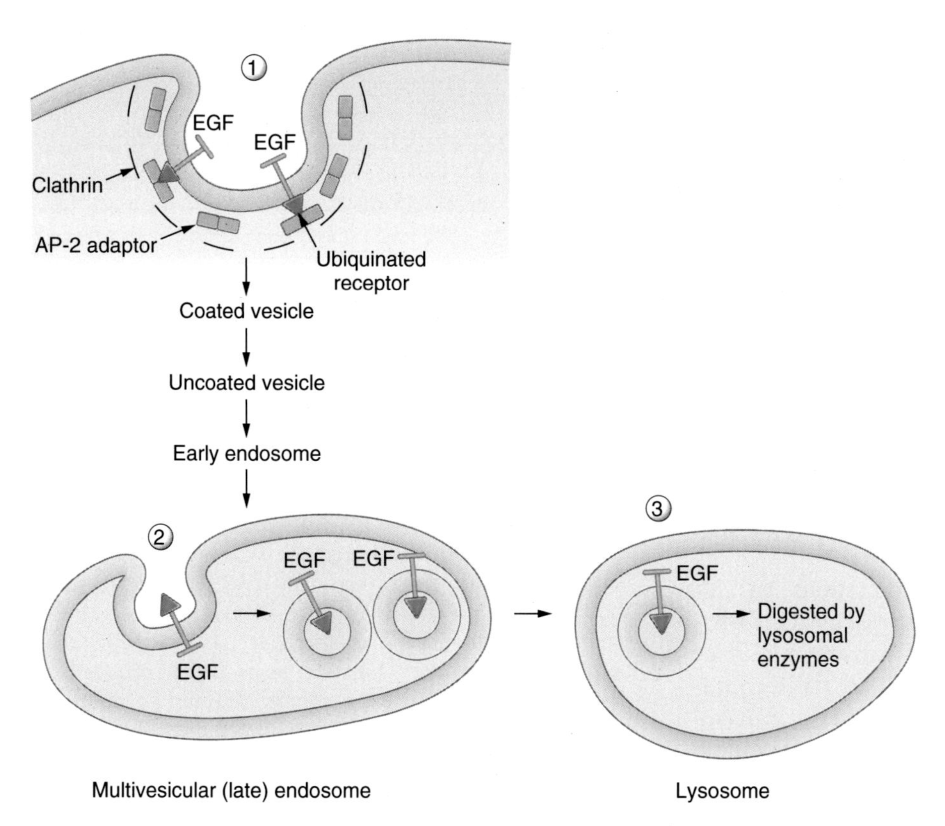

Figure 4–22. **Epidermal growth factor (EGF) endocytosis and the formation of multivesicular bodies.** (1) EGF binds to the EGF receptor, a transmembrane protein that is ubiquitinated on its cytosolic domain and that interacts with clathrin/adaptor protein 2 (AP-2) adaptors. After endocytosis and dissociation of the coat, the early endosome matures into a late endosome that can also become a multivesicular body. (2) The vesicles within multivesicular bodies are formed by invagination of the endosome membrane to form vesicles within a vesicle that are enriched in ubiquitinated receptors, such as the EGF receptor. (3) The small vesicles within the multivesicular body are substrates for lysosomal enzymes when the endosome fuses with lysosomes, subjecting the vesicle membrane, and eventually the vesicle contents, to degradation. In this way, membrane-bound material originally brought into the cell by endocytic vesicles, which includes transmembrane receptors, can be digested.

devise therapeutics that prevent virus budding, thus blocking transmission.

LYSOSOMES

Lysosomes were first discovered in biochemical assays as a collection of acid hydrolases that moved together as a large particle when cells were lysed and the contents analyzed by centrifugation. The large particles were biochemically characterized and predicted to be subcellular organelles before the development of electron microscopy techniques. Lysosomes were subsequently visualized in cells by electron microscopy, verifying the hypothesis that they were a membrane-bound subcellular organelle. For this work, Christian de Duve shared the 1974 Nobel Prize in Physiology and Medicine with Albert Claude and George Palade, a Nobel Prize that marks the founding of modern cell biology.

Lysosomes are membrane vesicles with a diameter of 0.2–0.6 μm, and they contain a wide variety of hydrolytic enzymes that have an acidic pH optimum. The pH (~4.5) within lysosomes is generated by proton-translocating ATPases in the lysosomal membrane that pump protons from the cytosol into the lysosome lumen. The hydrolases in lysosomes include proteases, nucleases, glycosidases, and lipases, and they are sufficient to degrade almost all naturally occurring macromolecules to their monomeric constituents. We have already seen how lysosomal enzymes are targeted to late endosomes via the mannose 6-phosphate marker. Before reaching lysosomes, most lysosomal enzymes are in an enzymatically inactive proenzyme form to control their hydrolytic capacity. When the proenzymes reach lysosomes, their processing by preexisting lysosomal proteases convert the proenzymes to their active state. The lysosomal membrane itself is resistant to the action of lumenal hydrolases because membrane glycoproteins that coat the lumenal side of the lysosomal membrane serve as a protective polysaccharide barrier.

Lysosomal digestion products, such as amino acids, sugars, or nucleotides, are transported across the lysosomal membrane to the cytosol where they join existing nutrient pools. Thus one function of lysosomes is to generate nutrients necessary for normal cellular metabolism. This was illustrated in the earlier discussion of how dietary cholesterol is delivered to cells via the uptake of LDL particles, which are degraded in lysosomes. The resulting cholesterol is released into the cytosol for reutilization. Lysosomes, however, have additional functions. They serve as a defense against microorganisms. Following their engulfment by phagocytosis, microorganisms are digested in the macrophage equivalent of a lysosome, the phagosome. Lysosomal enzymes can be released into the extracellular environment and participate in a variety of biological processes. For instance, lysosomal enzymes released by sperm during fertilization aid the sperm in penetrating the egg. Lysosomes also participate in the process of autophagy (self-eating) whereby the cell deliberately digests some of its own subcellular components. This not only occurs as part of the normal turnover process but also is enhanced in starved cells where cell components are recycled to provide nutrients. Autophagy begins when a cup-shaped membrane segment wraps around a subcellular structure to form a closed vacuole. Subcellular structures such as mitochondrial fragments can be visualized by electron microscopy inside of autophagic vacuoles that possess two membrane bilayers. Multiple membrane compartments appear to give rise to the enveloping membrane of the autophagic vacuole. Once formed the vacuole fuses with lysosomes, exposing the enveloped material to digestive enzymes. Defects in autophagy have been implicated in a variety of neurodegenerative diseases and tumorigenesis.

We have already seen how mutations in *N*-acetylglucosamine phosphotransferase results in the *lysosome storage disease*, I-cell disease. There are more than 40 other lysosome storage diseases that arise as a consequence of mutations in individual lysosomal hydrolases that prevent their folding, modification, or enzymatic activity. In these various diseases the accumulation of specific substrates in the lysosomes of affected individuals can have severe pathological consequences often resulting in death soon after birth. For instance, in Tay-Sachs disease, affected individuals accumulate the glycosphingolipid, GM2 ganglioside, in their lysosomes, particularly in nervous tissue where this ganglioside is prevalent. The disease is caused by a defective α-hexosaminidase enzyme that is required for the catabolism of this ganglioside. Children with severe Tay-Sachs disease rapidly develop mental retardation and paralysis and die within 3–4 years after birth. Once it was realized that the lysosomal storage diseases were caused by single enzyme defects, the idea of enzyme replacement therapy with functional enzymes developed. Applying this simple idea is not trivial because it requires that replacement enzymes injected into the patient be correctly targeted to lysosomes in affected cells and not targeted to unwanted sites. Early attempts at this therapy found that injected lysosomal enzymes were rapidly cleared from the blood by cells that destroyed the enzymes before therapeutic effects occurred. This clearance was traced to certain carbohydrate residues on the enzymes that were signals for uptake and destruction of the enzymes by some cells. Modifying the undesirable carbohydrates to avoid premature destruction of the enzymes aided the therapeutic value, and enzyme replacement therapy has been partially successful in ameliorating some symptoms of Gaucher's disease, where there is a defective glucocerebrosidase, and Fabry disease, caused by a defective α-galactosidase. Enzyme replacement therapy is another example of how basic cell biology and medicine together can lead to the development of novel strategies to treat human disease (Clinical Case 4–2).

Clinical Case 4–2

Porter is a 14-year-old girl brought in by her parents to the pediatrician because of progressive weakness. Porter says she has no trouble with her hands or feet but that her shoulder muscles and thigh muscles just "don't do what she wants them to." She has a hard time lifting heavy objects and cannot walk long distances. She also says that she often feels breathless and occasionally wakes up in the night short of breath. On examination, weakness was seen in a limb-girdle distribution. The pediatrician ordered a complete blood count and a metabolic panel including markers for muscle inflammation. Complete blood count and basic metabolic panel were normal, but creatinine kinase and LDH levels were elevated. The physician ordered a measurement of leukocyte alpha-glucosidase enzyme activity, which was undetectable. Electrocardiogram was normal. The pediatrician then referred the patient to a pediatric neurologist who performed an electromyogram that showed myopathic discharges associated with complex repetitive discharges. The specialist also referred the patient for pulmonary function tests, which revealed a decreased forced vital capacity (the amount of air that can be forcibly breathed out after a deep breath). A urine test for glucose tetrasaccharide was elevated. Molecular testing revealed two mutant alleles in the gene for alpha-glucosidase (GAA). The mutations are known to result in alpha-glucosidase proteins with reduced enzymatic activity. A pediatric cardiologist performed an echocardiogram that showed mild wall thickening but no decrease in ejection fraction.

Cell Biology, Diagnosis, and Treatment of Pompe Disease

Porter has Pompe disease, one of many lysosomal storage diseases that are caused by enzyme deficiencies within the lysosome resulting in accumulation of material. In the case of Pompe disease, the alpha-glucosidase enzyme, which is necessary for the breakdown of glycogen in lysosomes, is lacking. In the absence of this enzyme, lysosomes and autophagosomes within cells become enlarged and disrupt normal cellular function. The resultant tissue damage is particularly prevalent in muscle, including cardiac muscle in young children. The symptoms of Pompe disease reflect this tissue predilection, presenting with symptoms of hypotonia, weakness, respiratory dysfunction, and occasionally difficulty swallowing. Late-onset Pompe disease, which Porter has, is less severe than infantile-onset disease in which hypertrophic cardiomyopathy is common. In late-onset disease, heart function remains normal, and skeletal muscle symptoms predominate.

Diagnosis of Pompe disease is based on clinical presentation and blood analysis. Creatinine kinase is always elevated in early onset Pompe disease, and direct measurement of alpha-glucosidase can confirm the diagnosis. The alpha-glucosidase deficiency can be due to autosomal recessive mutations in the GAA gene that result in decreased levels of production of the protein. As such, germline molecular testing will reveal two pathogenic mutations in this gene.

Historically the infantile-onset form of Pompe disease was fatal early in life, and late-onset disease led to death in the third or fourth decade of life. However, the availability of synthetic alpha-glucosidase (alglucosidase alfa) has changed the course of the disease in clinical practice. The delivery of recombinant enzyme, known as enzyme replacement therapy (ERT), allows for correction of the most dangerous aspects of the disease. ERT, when initiated early, prevents the progressive hypertrophic cardiomyopathy that leads to heart failure in infantile-onset disease. This therapy can both correct and reverse the damage to the heart muscle. In late-onset Pompe disease, ERT improves skeletal muscle function and therefore improves respiratory capacity. This allows patients to continue to walk on their own and increases survival rates.

Porter was immediately started on intravenous alglucosidase alfa infusions every 2 weeks. Her breathing difficulty improved within the first few treatments. Six months later, she continues to have mild muscle weakness in her arms and proximal thighs but is able to walk farther than before. Because of the high risk of ultimately developing antibodies against the therapy, Porter knows that at some point, her muscle weakness may return and she awaits gene therapy trials designed to correct the mutant alleles. She has started a high-protein and low-carbohydrate diet and respiratory therapy as well.

The Ubiquitin-Proteasome System for Protein Degradation

As we mentioned earlier in the chapter, the posttranslational addition of ubiquitin to proteins can direct the ubiquitinated protein to the proteasome for degradation. The ubiquitin-proteasome system is the major nonlysosomal pathway for ATP-dependent protein degradation, and in 2004 Aaron Ciechanover, Avram Hershko, and Irwin Rose shared the Nobel Prize in Chemistry for their landmark discovery of this system. In the succeeding section, we will first look at the mechanism of protein ubiquitination and then the function of the proteasome in protein degradation.

Ubiquitin is an 8.6-kDa highly conserved polypeptide found both free and covalently conjugated to other proteins in all eukaryotic cells. Up to 14 different families of ubiquitin and ubiquitin-like proteins are known that differ in the amino acid sequence, but all have the same characteristic spatial structure. The posttranslational attachment of ubiquitin to target proteins is a dynamic process involving ubiquitin-conjugating enzymes and deubiquitination enzymes. As shown in Figure 4–23B, the conjugation of ubiquitin to target proteins is a multistep process.

A cascade of reactions catalyzed by several classes of enzymes is required to form an isopeptide bond between the C-terminal glycine of ubiquitin and a lysine ε-amino group of the target acceptor protein. Ubiquitin-activating enzyme (E1) uses ATP and forms a high-energy thioester bond between its active site cysteine and the C-terminal glycine of ubiquitin via the formation of ubiquitin adenylate, an activated ubiquitin intermediate. The activated ubiquitin is next transferred from E1 to an active site cysteine of the ubiquitin-conjugating enzyme (E2). The final step, the formation of an isopeptide linkage between ubiquitin and the target protein, is catalyzed by E3 ubiquitin-protein ligases. In general, E3 binds both E2 and the target protein, bringing them into close proximity. Ubiquitin is then transferred from E2 to the substrate either directly or via an E3-thioester intermediate. The specificity of substrate ubiquitination is attained through the hierarchy of the ubiquitination system. In mammalian cells, there are only two types of ubiquitin-activating E1 enzymes, approximately 30 ubiquitin-conjugating E2 enzymes, and more than 800 E3 ubiquitin-protein ligases. The E3 ligases are divided into two distinct families. E3 ligases containing a homologous to E6-AP carboxyl terminus (HECT) domain form thioester intermediates with ubiquitin prior to conjugating it to substrate, while family members containing a RING domain mediate the direct transfer of ubiquitin from E2 to the target protein. An additional conjugating factor, named E4, is associated with the formation of polyubiquitin chains.

For most known substrates, polyubiquitination leads to protein degradation by the 26S proteasomal complex. The polyubiquitin chain is recognized by the 19S regulatory complex of the proteasome, and the substrate protein is degraded by the 20S core complex. As shown in Figure 4–23B, the 20S complex has a barrel shape with four stacked rings. Initially the ubiquitinated protein associates with the 19S complex via interaction with ubiquitin receptors. Following binding, ubiquitin is removed by deubiquitinases, and the protein is unfolded by AAA ATPases of the 19S complex and inserted into the 20S core complex. The protein is then degraded into small peptides, which are subsequently degraded into amino acids by amino and carboxypeptidases within the cytoplasm.

The addition of ubiquitin to target proteins also serves as a signal for their uptake into phagosomes, which fuse with lysosomes where the target proteins are degraded. This process has been linked to the autophagosomal uptake and lysosomal degradation of old or damaged organelles. In addition, the addition of ubiquitin and ubiquitin-related proteins to cellular proteins regulates other cellular processes including translation, activation of transcription factors and kinases, activation of DNA repair, and regulation of membrane protein trafficking. As discussed in Chapters 9 and 10, the dysfunction of the ubiquitin-proteasome system plays an important role in several cancers and neurologic disorders.

MITOCHONDRIA

One of the main functions of mitochondria is to provide the cell with ATP generated by oxidative phosphorylation. Mitochondria are also involved in many other

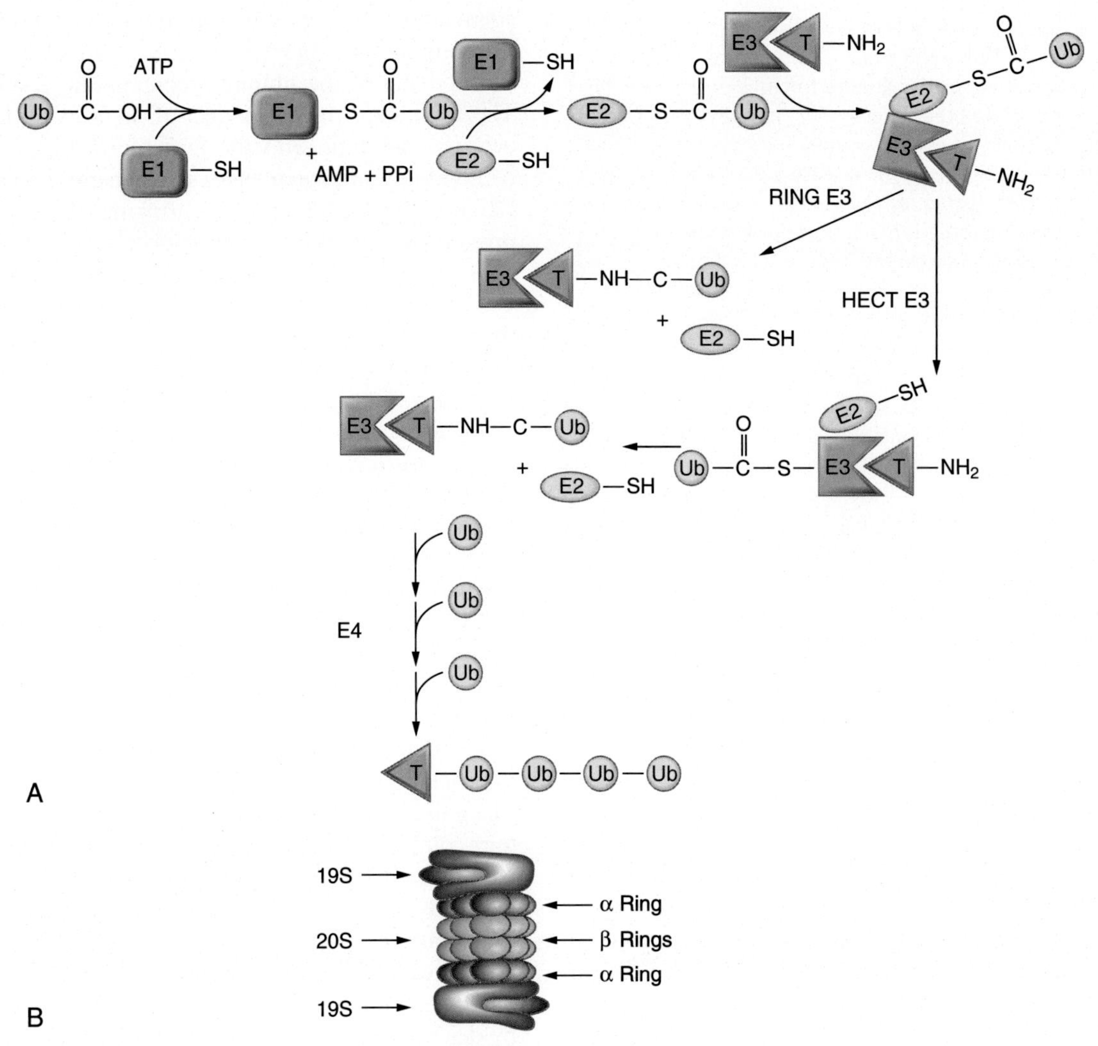

Figure 4–23. **The ubiquitination pathway.** (A) Free ubiquitin (Ub) is activated in an ATP-dependent reaction with the formation of a thioester intermediate between E1 and the C-terminus of ubiquitin. Ubiquitin is then transferred to an E2, again forming a thioester linkage. There are two classes of E3 enzymes. The RING domain E3s bind to the target (T) protein and directly transfer ubiquitin from the E2 to the target. HECT domain E3s form a thioester intermediate between ubiquitin and itself before transferring ubiquitin to the target. An E4 then adds a series of ubiquitins by linking through lysine 48 of one ubiquitin to the C-terminus of the next. (B) Structure of the proteasome.

metabolic functions, including heme biosynthesis, synthesis of iron/sulfur (Fe/S) clusters, steroid synthesis, metabolism of fatty acids, regulation of the cellular redox state, calcium homeostasis, amino acid metabolism, carbohydrate metabolism, and protein catabolism. In addition, mitochondria are the central regulators of programmed cell death (see Chapter 11).

Mitochondria are 1–2 µm in length and, when viewed in living cells with a fluorescent dye, such as MitoTracker Red or rhodamine 123, are highly dynamic organelles, changing shape, fusing, dividing, and moving. Mitochondria are thought to have arisen after engulfment of a bacterium by the progenitor of eukaryotic cells in which the bacterium established a symbiotic relationship with the progenitor cell (the *Endosymbiotic Theory*). Mammalian cells typically contain hundreds or thousands of mitochondria; the number of mitochondria per cell appears to depend on the metabolic requirements of that cell.

Mitochondria are surrounded by two membranes: an outer and an inner membrane. This creates two compartments: an intermembrane space between the inner and outer membranes and the matrix, which is the compartment enclosed by the inner membrane (see Fig. 4–24). The outer membrane contains a major integral protein, called *porin.* Mitochondrial porins, also known as voltage-dependent anion channels, are the most abundant proteins in the mitochondrial outer membrane and can form channels within the outer membrane through which molecules that are less than 5000 Da can pass freely. The inner membrane contains proteins involved in the respiratory chain, ATP production, and the transport of small molecules and ions across the inner membrane. The inner membrane is invaginated to form

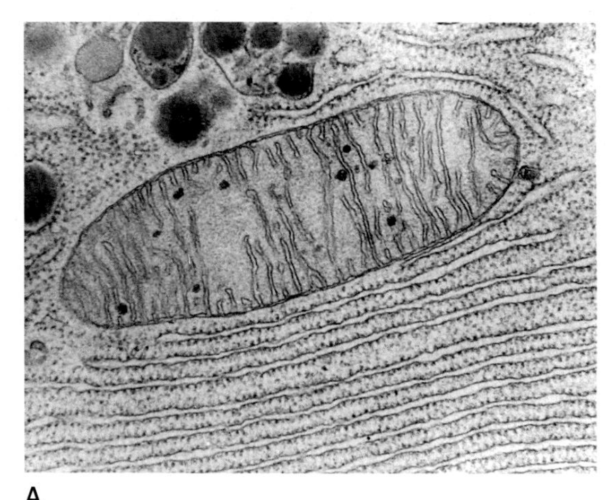

A

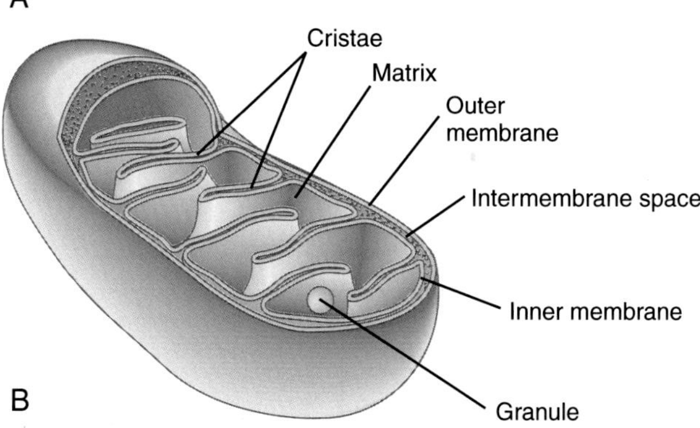

B

Figure 4–24. **Structure of a mitochondrion.** (A) Transmission electron micrograph of a mitochondrion in a human pancreatic acinar cell. Mitochondria are surrounded by two membranes, an outer and an inner membrane. The inner membrane has many invaginations, called *cristae*. Several granules can be seen in the matrix. (B) Diagrammatic presentation of mitochondrial compartments and membranes. *([A] Courtesy Keith R. Porter/Photo Researchers, Inc.)*

cristae (Fig. 4–24), which increase its surface area, thereby enhancing its ability to generate ATP. The inner mitochondrial membrane has a high concentration of the phospholipid, cardiolipin (diphosphatidylglycerol), which is essential for the optimal function of the numerous enzymes that are involved in mitochondrial energy metabolism. The matrix contains enzymes involved in the oxidation of pyruvate and fatty acids and most of the enzymes of the citric acid cycle. In addition, the mitochondrial genome and ribosomes for protein synthesis are located in the matrix. The intermembrane space contains a number of proteins, including cytochrome *c*, which plays a critical role both in electron transport and in programmed cell death.

ATP Production by Oxidative Phosphorylation

One of the main functions of mitochondria is the generation of ATP via oxidative phosphorylation. The steps involved in oxidative phosphorylation and the mechanism resulting in ATP production by the movement of electrons through the electron transport chain begin when glucose is converted to pyruvate in the cytosol by the glycolytic pathway (Fig. 4–25) and the two reduced nicotinamide adenine dinucleotide (NADH) molecules formed by glycolysis are reoxidized to NAD^+ by transfer of their electrons to complexes of the electron transport chain located in the mitochondria. Pyruvate also enters the mitochondrial matrix, where it is converted to acetyl-CoA, the key substrate for the citric acid cycle. Another source of mitochondrial acetyl-CoA is the oxidation of fatty acids. Free fatty acids, which are stored primarily in adipose tissue as triacylglycerol, can be released into the bloodstream. Free fatty acids can cross the plasma membrane of cells and are converted in the cytosol to fatty acyl-CoA, which can be transported into the mitochondrion. Within the mitochondrial matrix, free fatty acyl-CoA is broken down by a cycle of reactions (Fig. 4–26), which removes two carboxyl carbons and produces one acetyl-CoA molecule per cycle.

The acetyl-CoA formed by the oxidation of pyruvate and fatty acyl-CoA fuels the citric acid cycle, also known as the *Krebs cycle* or *tricarboxylic acid cycle* (Fig. 4–27). During the citric acid cycle, acetyl-CoA is oxidized to two molecules of CO_2 (which are released from the cell), and the released electrons are transferred to NAD and flavin adenine dinucleotide (FAD).

The 3 NADH and $FADH_2$ molecules formed by the citric acid cycle and the NADH molecules generated during glycolysis can now transfer electron pairs to acceptor molecules on the inner mitochondrial membrane, eventually leading to the reduction of oxygen and the formation of water. The detailed steps resulting in the generation of ATP via a process known as chemiosmotic coupling will not be described here.

Once ATP has been produced via oxidative phosphorylation, the mechanism for getting the newly synthesized ATP out of the mitochondria and into the cytosol is accomplished by the adenine nucleotide translocator, which couples the movement of ATP down its electrochemical gradient (from the matrix to the intermembrane space and into the cytosol) to the movement of ADP in the reverse direction. The synthesis of ATP within the mitochondrion also requires phosphate ions (P_i), so P_i must be imported from the cytosol. The transport of P_i is mediated by a phosphate transporter, which imports one phosphate ion together with one H^+ into the mitochondrial matrix. Figure 4–28 summarizes the generation of ATP in mitochondria.

Mitochondrial Genetic System

Mitochondria contain their own genetic system. The human mitochondrial genome is a circular, double-stranded DNA that is 16,569 base pairs in length. It contains 37 genes that encode 2 ribosomal RNAs (12S and 16S rRNA) and 22 transfer RNAs (tRNAs). It also encodes 13 proteins involved in electron transport and

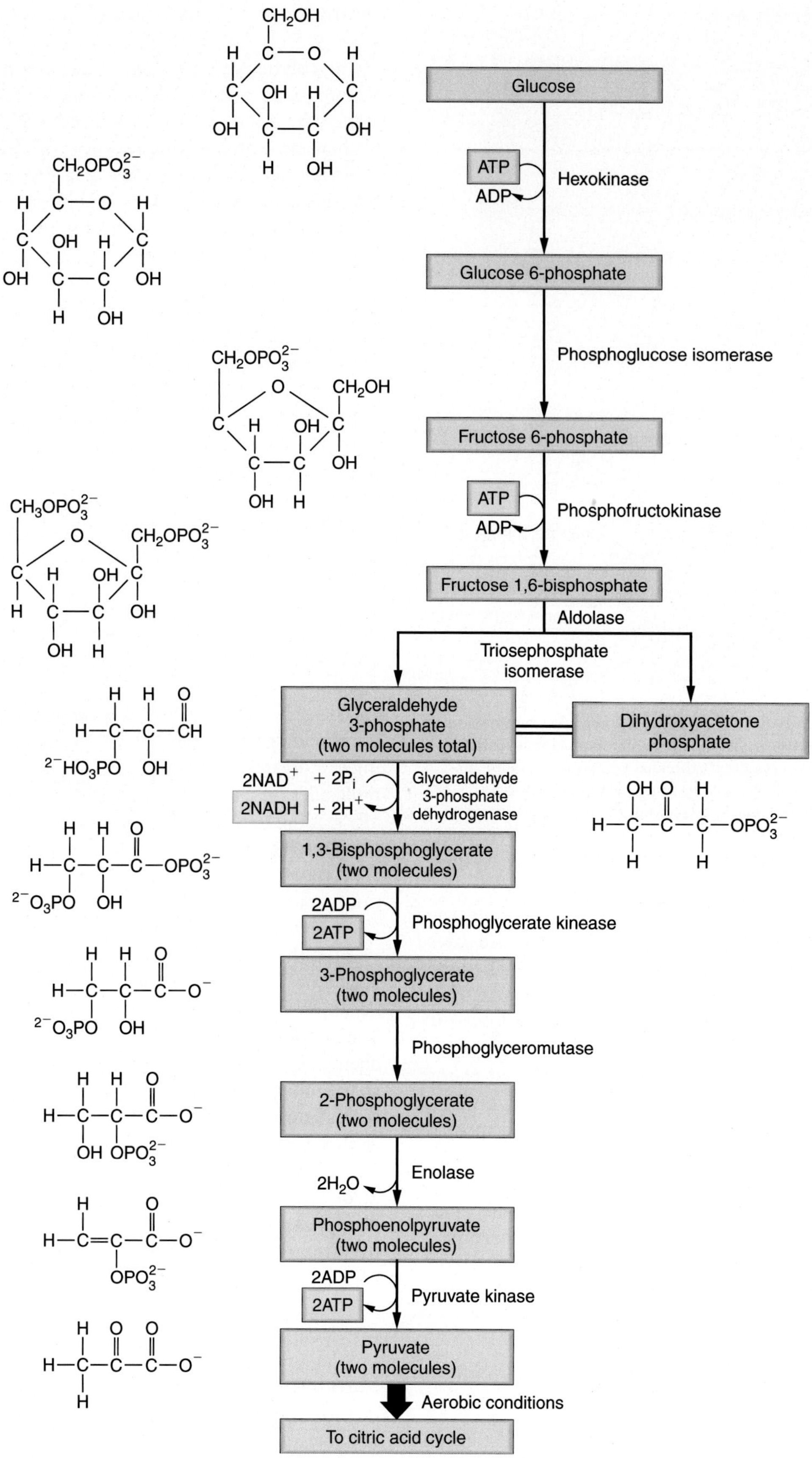

Glucose
ATP
ADP
Hexokinase
Glucose 6-phosphate
Phosphoglucose isomerase
Fructose 6-phosphate
ATP
ADP
Phosphofructokinase
Fructose 1,6-bisphosphate
Aldolase
Triosephosphate isomerase
Glyceraldehyde 3-phosphate (two molecules total)
Dihydroxyacetone phosphate
$2NAD^+ + 2P_i$
$2NADH + 2H^+$
Glyceraldehyde 3-phosphate dehydrogenase
1,3-Bisphosphoglycerate (two molecules)
2ADP
2ATP
Phosphoglycerate kinease
3-Phosphoglycerate (two molecules)
Phosphoglyceromutase
2-Phosphoglycerate (two molecules)
Enolase
$2H_2O$
Phosphoenolpyruvate (two molecules)
2ADP
2ATP
Pyruvate kinase
Pyruvate (two molecules)
Aerobic conditions
To citric acid cycle

Figure 4–25. Glycolytic pathway. The breakdown of glucose into pyruvate yields two molecules of NADH and two molecules of ATP.

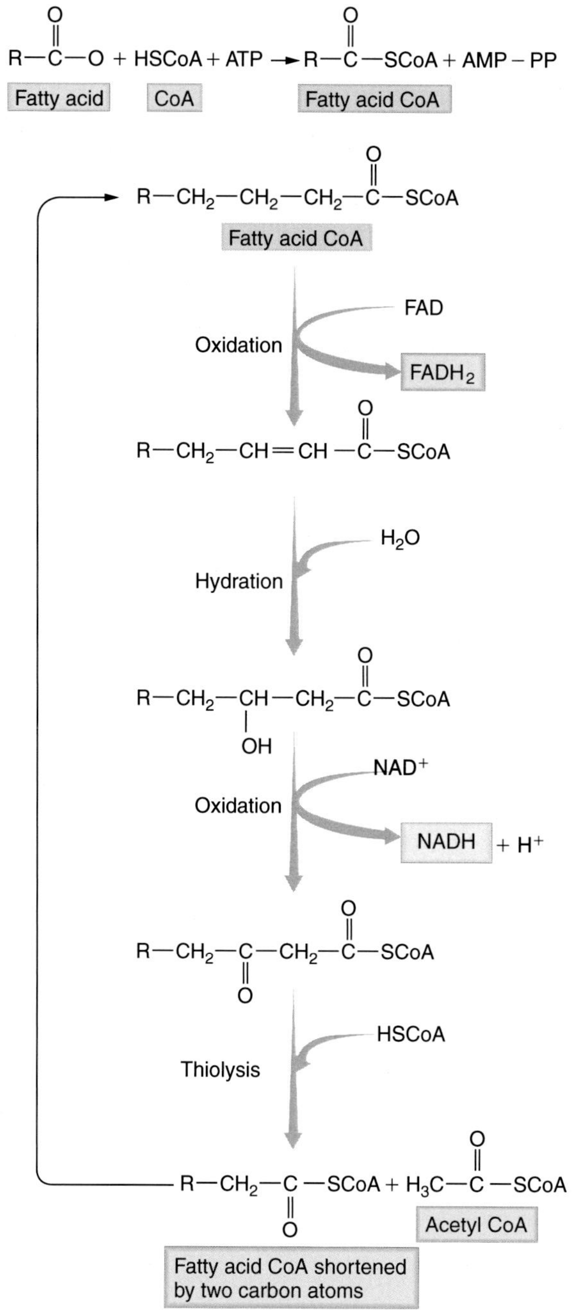

$R{-}C(=O){-}O + HSCoA + ATP \rightarrow R{-}C(=O){-}SCoA + AMP - PP$

Fatty acid | CoA | Fatty acid CoA

$R{-}CH_2{-}CH_2{-}CH_2{-}C(=O){-}SCoA$

Fatty acid CoA

Oxidation: FAD → $FADH_2$

$R{-}CH_2{-}CH{=}CH{-}C(=O){-}SCoA$

Hydration: H_2O

$R{-}CH_2{-}CH(OH){-}CH_2{-}C(=O){-}SCoA$

Oxidation: NAD^+ → NADH + H^+

$R{-}CH_2{-}C(=O){-}CH_2{-}C(=O){-}SCoA$

Thiolysis: HSCoA

$R{-}CH_2{-}C(=O){-}SCoA + H_3C{-}C(=O){-}SCoA$

Fatty acid CoA shortened by two carbon atoms | Acetyl CoA

Figure 4–26. Pathway for the oxidation of fatty acids in mitochondria.

oxidative phosphorylation. There are approximately 2–10 mitochondrial genomes per human mitochondrion and multiple mitochondria per cell.

There are several interesting features of the human mitochondrial genome. First, almost the entire genome is coding sequence. Mitochondrial protein synthesis requires only 22 mitochondrially encoded tRNAs, which leads to the second distinctive feature of the mitochondrial genome. Many of the mitochondrial tRNAs recognize any one of four nucleotides in the third position of codons. Therefore there is far greater "wobble" in the third codon position within mitochondrial mRNAs, leading to "two-out-of-three" pairing of tRNA. Lastly the human mitochondrial genetic code differs from the "universal" genetic code (Table 4–2). For example, UGA encodes a stop codon in the universal genetic code but codes for tryptophan in human mtDNA.

There are several important differences between the genetics of mtDNA and of nuclear genes. One key difference is that the human mitochondrial genome is maternally inherited. In addition, mtDNA is polyploid, and individual cells contain hundreds or thousands of copies of mtDNA. During cell division, mitochondria together with their genomes are distributed to daughter cells more or less randomly. In addition, the mitochondrial genome has a much higher mutation rate (~10-fold greater) than the nuclear genome.

Defects in Mitochondrial Function Can Cause Disease

As discussed earlier, mitochondria play a key role in energy production and also participate in a number of other biological functions. Defects in any of these mitochondrial processes can result in a mitochondrial disease. Because of their dual genetic control, mitochondrial diseases can be caused by mutations in either nuclear genes or mtDNA genes. Chapter 5 describes how specific mutations in mtDNA lead to a variety of disease states in humans.

Mitochondria Import Most of Their Proteins From the Cytosol

As discussed earlier, human mtDNA encodes only 13 proteins. However, analyses of the mitochondrial proteome indicate that mammalian mitochondria contain approximately 1500 unique proteins. Thus most mitochondrial proteins are encoded by nuclear genes and imported from the cytosol. Most mitochondrial proteins are synthesized on soluble ribosomes as precursors and imported into mitochondria posttranslationally. Mitochondrial precursor proteins contain specific targeting signals that direct them to the organelle. These mitochondrial targeting signals are diverse in nature and are recognized by receptor proteins located in the mitochondrial outer membrane. Once imported, mitochondrial proteins need to be sorted to different suborganellar destinations—outer membrane, inner membrane, intermembrane space, and matrix.

We first examine how proteins are imported from the cytosol into the mitochondrial matrix. This import pathway is summarized in Figure 4–29. Most mitochondrial matrix proteins are synthesized as a larger precursor with a cleavable amino-terminal mitochondrial targeting sequence (called a *presequence*). These presequences are approximately 20–40 amino acids in length with

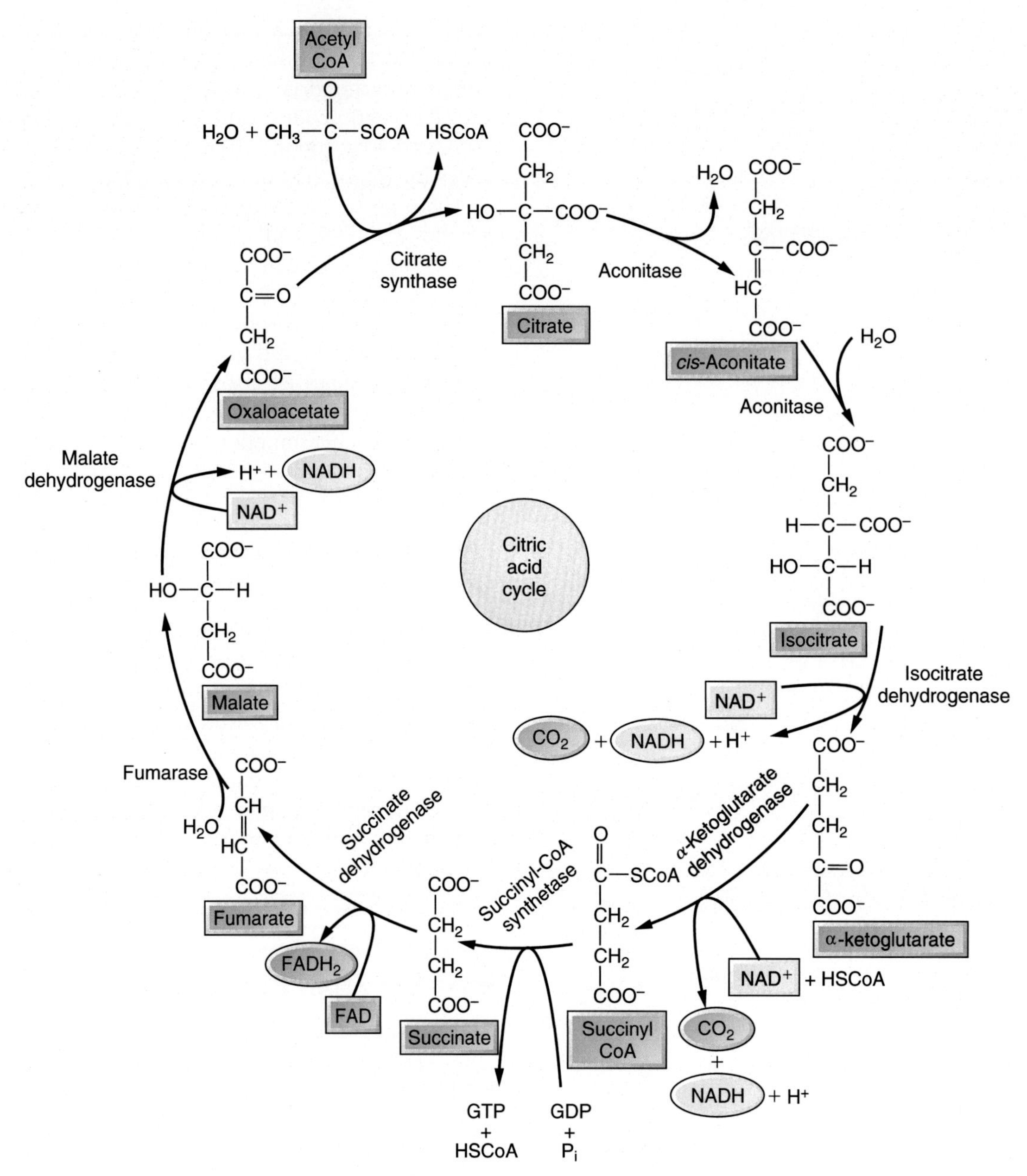

Figure 4–27. Citric acid cycle.

multiple positively charged residues and have the characteristic of forming an amphipathic α-helix where one side of the helix is nonpolar and the other side contains positively charged amino acids (Table 4–1). The precursor protein is maintained in an unfolded state by the attachment of cytosolic chaperone proteins, such as members of the Hsp70 family of heat shock proteins. The presequence first binds to receptors in the outer membrane of mitochondria. The precursor protein is then transferred to the general import protein Tom40, which is a member of the "translocase outer membrane" (TOM) complex family of proteins through which it translocates through the outer membrane. For the protein to be translocated across the membrane, cytosolic Hsp70 must be released from the protein; this step requires ATP hydrolysis. The precursor protein is then transferred to the presequence translocase of the inner membrane (TIM23 complex), which forms a channel across the inner membrane. Insertion into the channel formed by the Tim23 protein requires an electrochemical gradient across the inner mitochondrial membrane. The ATP-driven presequence translocase-associated motor (PAM) is needed to complete the translocation of the precursor protein into the matrix. The central component of the PAM import motor is the mitochondrial Hsp70 (mtHsp70) chaperone. As the precursor protein emerges in the matrix, there is binding of a mtHsp70 protein, and ATP hydrolysis in the matrix is used to translocate the precursor protein into the matrix. Once inside the matrix the presequence is

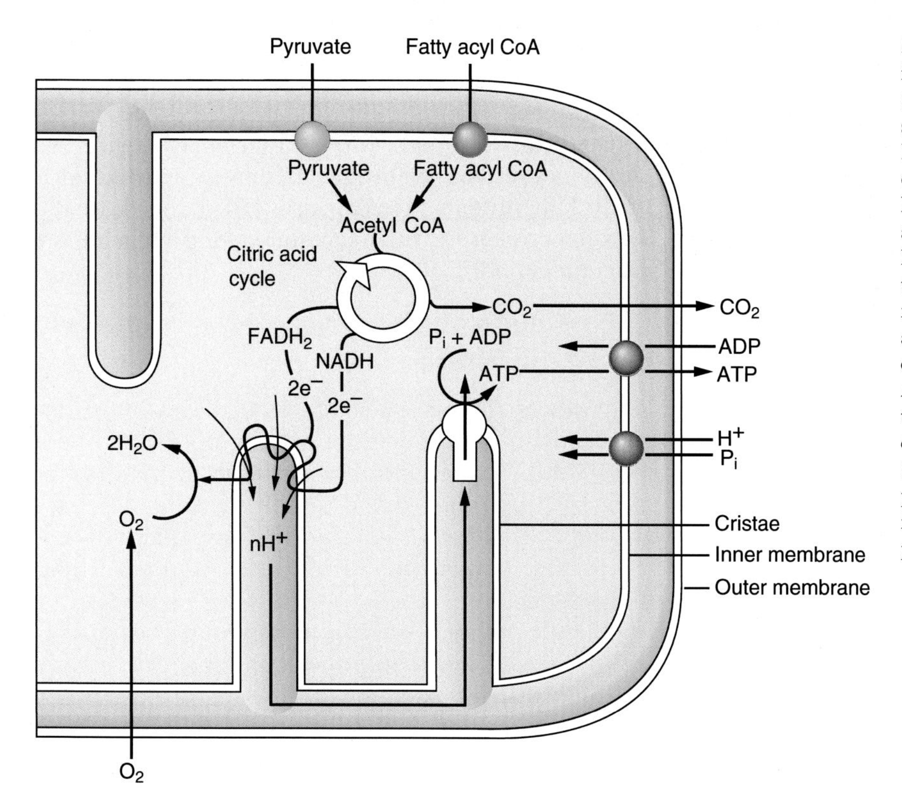

Figure 4–28. **Summary of mitochondrial function.** Pyruvate and fatty acyl coenzyme A (CoA) are transported into mitochondria by specific transporter proteins and are metabolized to acetyl-CoA. Acetyl-CoA is then metabolized by the citric acid cycle. NADH and $FADH_2$ are produced by the citric acid cycle, and electrons are transferred from NADH and $FADH_2$ to O_2 by a series of electron carriers in the inner mitochondrial membrane. A proton motive force is created by this electron transfer, and protons, moving back down their electrochemical gradient into the matrix, power the ATP synthase to produce ATP. ATP produced in the mitochondrial matrix is transported to the cytosol by the adenine nucleotide translocator (which exchanges ATP for ADP). Inorganic phosphate (P_i) is transported from the cytosol into the matrix of mitochondria by the phosphate transporter.

TABLE 4–2. Differences Between the Human Mitochondrial and the "Universal" Genetic Code

Codon	Universal Genetic Code	Mitochondrial Genetic Code
UGA	Stop	Trp
AUA	Ile	Met
AGA	Arg	Stop
AGG	Arg	Stop

cleaved by the matrix processing peptidase. Unfolded matrix proteins are next transferred to a mitochondrial chaperone of the Hsp60 family (mtHsp60), which, together with a cochaperone (mtHsp10), assists in folding the protein into its final stable conformation.

As noted earlier, some mitochondrial proteins are targeted to the outer membrane, inner membrane, or intermembrane space, rather than to the matrix. Mitochondrial precursor proteins are sorted to one of the submitochondrial compartments from the TOM complex. There are several modes for transport of precursor proteins to the inner mitochondrial membrane. Some proteins of the inner membrane are synthesized as a larger precursor with a presequence (similar to matrix proteins discussed earlier). These proteins use the presequence import pathway (TOM and TIM23 complexes). They are then inserted into the inner membrane from the TIM23 complex by a hydrophobic sorting signal located after the presequence and do not require the PAM import motor. Some proteins that are targeted to the inner membrane are hydrophobic proteins (such as the carrier proteins). These proteins do not contain presequences but have multiple internal mitochondrial import signals. These proteins are initially recognized by a different receptor, Tom70, in the mitochondrial outer membrane (Fig. 4–29). They are translocated across the outer membrane through Tom40. These proteins are then recognized by small soluble Tim proteins (Tim9 and Tim10 complexes) of the intermembrane space. The proteins are then inserted into the inner mitochondrial membrane by the carrier translocase of the inner membrane (TIM22 complex). This insertion requires a membrane potential across the inner mitochondrial membrane. Other proteins targeted to the inner membrane are first transported across the inner membrane into the mitochondrial matrix. After removal of the presequence, a second sorting signal is uncovered. This sorting signal targets the proteins to another translocase of the inner membrane, Oxa1, which inserts the proteins into the inner mitochondrial membrane. Oxa1 is also the translocase for some inner membrane proteins that are encoded by mtDNA and synthesized on mitochondrial ribosomes.

The TOM complex is sufficient for insertion of a subset of outer membrane proteins, with a relatively simple topology, such as single-pass proteins. However, β-barrel proteins of the outer membrane, such as porin and Tom40, are first imported via TOM into the intermembrane space. These precursors are then transferred with the help of small Tim proteins of the intermembrane space to another complex of the outer membrane, the SAM complex (sorting and assembly machinery).

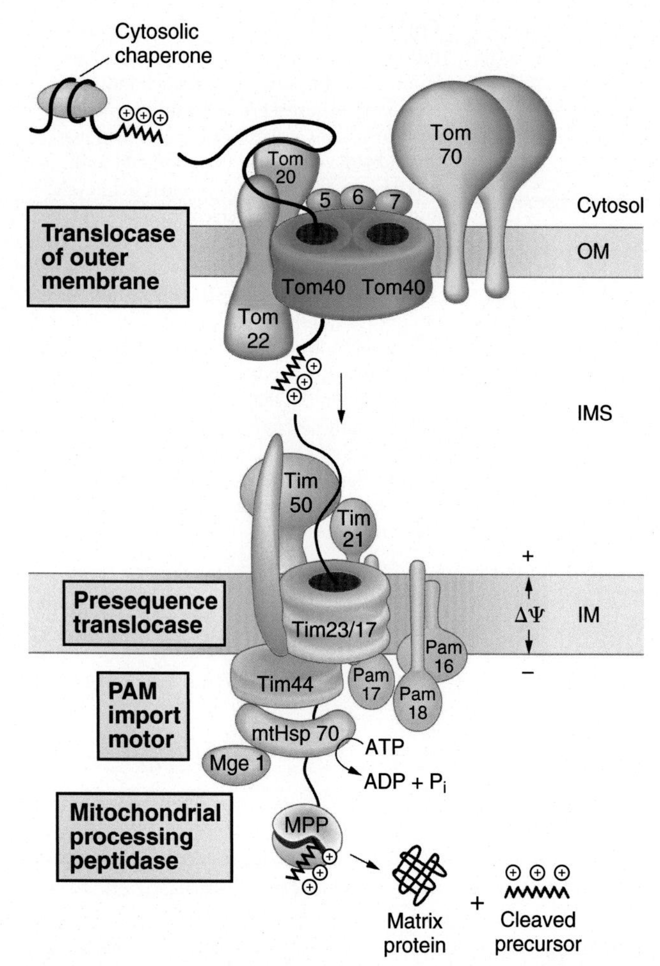

Figure 4–29. **Import of proteins into the mitochondrial matrix.** Proteins are targeted to the matrix of mitochondria by amino-terminal presequences that contain positively charged amino acids. Precursors are unfolded by cytosolic chaperones, such as heat shock protein 70 (Hsp70), before import into mitochondria. The presequence first binds to the receptors, Tom20 and Tom22, and is then transferred to the general import pore, Tom40, of the translocase of the outer membrane (TOM) complex. After passage through the OM, the presequence binds the intermembrane space (IMS) domain of the Tom22 receptor protein. The presequence-containing precursor then binds the Tim50 protein and is transferred to the presequence translocase of the inner membrane (IM; TIM23 complex). Membrane potential (ΔΨ) across the inner membrane is required for insertion of the precursor protein into the channel formed by the Tim23 protein. The presequence translocase-associated motor (PAM), which is driven by ATP, is required for completion of protein translocation into the matrix. The central constituent of this import motor is the mitochondrial Hsp70 chaperone (mtHsp70). mtHsp70 and its nucleotide exchange factor, Mge1, are transiently recruited to the presequence translocase by Tim44. Pam16 and Pam18 act as cochaperones. Once the precursor emerges in the matrix, it is bound by mtHsp70, and multiple rounds of ATP hydrolysis are required to translocate the protein across the inner membrane. In the matrix the presequence is cleaved by the mitochondrial processing peptidase (MPP). *(Adapted from Bohnert M, Pfanner N, van der Laan M. A dynamic machinery for mitochondrial precursor proteins.* FEBS Lett *2007;581:2802–2810, with permission.)*

Interestingly, defects in components of the mitochondrial import machinery can cause mitochondrial diseases. For example, the X-linked recessive neurodegenerative disorder Mohr-Tranebjaerg syndrome (deafness-dystonia syndrome) is due to a defect in a small Tim protein (Tim8a, also called the deafness-dystonia protein 1), which is required for protein import into mitochondria.

PEROXISOMES

Peroxisomes are small, ubiquitous organelles with a single membrane and a diameter ranging from 0.1 to 1 μm (Fig. 4–30). Peroxisomes participate in many different metabolic activities, including the oxidation of fatty acids, the breakdown of purines, the biosynthesis of cholesterol, the biosynthesis of bile acids, and ether lipid biosynthesis.

Peroxisomes are a heterogeneous group of organelles and contain a wide variety of enzymes that can catalyze the O_2-dependent oxidation of substrates (labeled RH_2) with the production of hydrogen peroxide. The reaction is as follows:

$$RH_2 + O_2 \rightarrow R + H_2O_2$$

Many different substrates are broken down by such oxidative reactions in peroxisomes, including uric acid, amino acids, purines, methanol, and fatty acids.

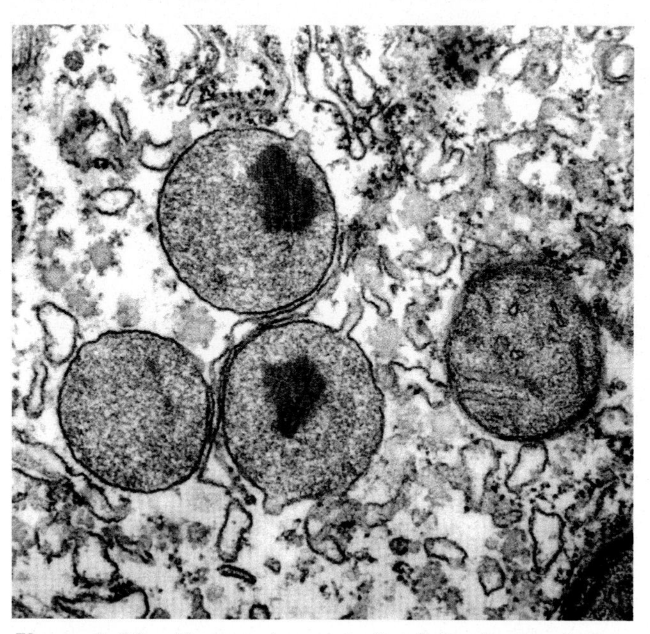

Figure 4–30. **Electron micrograph of a cell showing peroxisomes.** Transmission electron micrograph of a liver cell showing several peroxisomes. A urate oxidase crystal is evident within two of the peroxisomes. *(Courtesy Don W. Fawcett, Visuals Unlimited.)*

The hydrogen peroxide formed in this reaction is broken down by the enzyme, catalase, which is also present in peroxisomes, in one of two reactions. In one reaction, catalase uses the hydrogen peroxide produced to oxidize other substrates, such as alcohol, formaldehyde, nitrites, phenol, and formic acid. The general reaction is as follows:

$$H_2O_2 + RH_2 \rightarrow R + 2H_2O$$

This reaction is an important reaction in liver and kidney cells where peroxisomes detoxify various toxic substances. For example, approximately one-fourth of consumed alcohol is detoxified by this mechanism. In a second reaction, catalase can convert hydrogen peroxide to H_2O:

$$2H_2O_2 \rightarrow 2H_2O + O_2$$

A major function of peroxisomes is the β-oxidation of long and very-long-chain fatty acids. In humans, fatty acids are oxidized both in peroxisomes and in mitochondria (see Fig. 4–26). Short-, medium-, and most long-chain fatty acids are oxidized in the mitochondria, whereas peroxisomes oxidize very-long-chain fatty acids and some long-chain fatty acids.

In addition to oxidation reactions, peroxisomes are involved in the biosynthesis of some lipids. For example, in animal cells, dolichol and cholesterol are synthesized in peroxisomes and in the ER. Peroxisomes also contain enzymes for the synthesis of plasmalogens, a family of ether phospholipids. Plasmalogens are abundant lipids in the myelin sheaths that surround the axons of neurons.

The peroxisome does not have its own genome or ribosomes; therefore all of its proteins must be imported. Peroxisomes have two subcompartments, an internal matrix and an outer membrane. The majority of peroxisomal proteins are synthesized on free ribosomes and posttranslationally translocated across the peroxisomal membrane. Proteins are targeted to peroxisomes using specific peroxisomal targeting signals (PTS). Different targeting signals are used for peroxisomal matrix (PTS) versus peroxisomal membrane proteins (mPTS).

Two matrix PTS have been identified: a tripeptide located at the carboxy terminus (PTS1) and a 9-amino acid sequence close to the amino terminus (PTS2). The consensus sequence of the C-terminal PTS1 is S/A-K/R-L/M. Cytosolic soluble receptors recognize the PTS of peroxisomal matrix proteins and shuttle them to the peroxisome membrane. PTS1 is recognized by the receptor, Pex5p, and PTS2 by Pex7p. Current evidence suggests that the peroxisomal matrix protein enters the peroxisome together with its receptor. Once inside the peroxisome the receptor and matrix protein dissociate, and the receptor recycles to the cytosol. In contrast with import into mitochondria, proteins are imported into peroxisomes in a folded conformation.

Proteins are sorted to the peroxisomal membrane using a sorting mechanism that is distinct from the mechanism used by peroxisomal matrix proteins. At least one peroxisomal membrane protein, Pex3p, appears to be targeted to the peroxisome via the ER. Most new peroxisomes are formed by the growth and division of preexisting peroxisomes. However, recent evidence suggests that there might also be a de novo pathway for peroxisome biogenesis.

Numerous genetic disorders of peroxisome function have been identified. These can be divided into two categories. The first category includes disorders resulting from a defect in a single peroxisomal enzyme, such as X-linked adrenoleukodystrophy (X-ALD). In X-ALD, there is an accumulation of very-long-chain fatty acids in the brain and adrenal cortex because of a defect in a membrane protein that transports these fatty acids into the peroxisomes. Excess long-chain fatty acids in the brain destroy the myelin sheath surrounding nerve cells. The second category is a set of disorders that result from a deficiency in the biogenesis of the peroxisome and affects all the metabolic pathways of the peroxisome. These disorders are known as peroxisomal biogenesis disorders and include Zellweger's syndrome. Zellweger's syndrome is a fatal genetic disorder where the defect lies in an inability to import proteins into the peroxisome. Mutations in at least 12 different *PEX* genes that encode proteins important for peroxisomal protein import have now been identified in Zellweger's syndrome.

SUMMARY

In addition to the plasma membrane surrounding the cell, eukaryotic cells contain a variety of membrane-limited subcellular organelles that can be placed in one of two categories: those that are connected by pathways of vesicle-mediated membrane traffic and those that are not. Subcellular organelles that are connected by pathways of membrane traffic include (1) the ER, (2) the nucleus, (3) the Golgi apparatus, (4) various endosomes, and (5) lysosomes. The ER has the largest surface area of any subcellular organelle and is the site where most proteins and lipids are first inserted into a membrane. New lipids and proteins leave the ER for other sites by vesicle-mediated transport. The nucleus is bounded by two lipid bilayer membranes and contains dynamic pores through which DNA (mainly transfected plasmids), RNA, and proteins pass between the nucleoplasm and the cytoplasm. The nucleus is also connected to the ER and can be considered a specialized extension of the ER. The Golgi apparatus consists of membrane stacks involved in the modification and sorting of lipids and proteins. Membrane made in the ER passes through the Golgi apparatus en route to its final destination. Various endosomes are intermediates in the endocytic pathway that originates at the plasma membrane. Lysosomes contain enzymes capable of degrading most natural biopolymers. Two membrane-bound organelles that are for the most part not connected by pathways of vesicle-mediated transport are mitochondria and peroxisomes. Nuclear-encoded mitochondrial proteins and most peroxisomal proteins are synthesized on cytosolic ribosomes and imported posttranslationally. A major function of mitochondria is oxidative energy metabolism. Peroxisomes participate in a wide variety of biological reactions, including the detoxification of various compounds.

Suggested Readings

Aridora M, Hannan LA. Traffic jam: a compendium of human diseases that affect intracellular transport processes. *Traffic* 2000;1: 836–851.

Aridora M, Hannan LA. Traffic jams II: an update of diseases of intracellular transport. *Traffic* 2002;3:781–790.

Beal MF. Mitochondria take center stage in aging and neurodegeneration. *Ann Neurol* 2005;58:495–505.

Scriver CR, Sly WS, Childs B, et al. In: Scriver CR, Sly WS, editors. *The Metabolic and Molecular Bases of Inherited Disease*, 8th ed. New York: McGraw-Hill Professional, 2000.

Wanders RJA, Waterham HR. Peroxisomal disorders I: biochemistry and genetics of peroxisome biogenesis disorders. *Clin Genet* 2004;67:107–133.

Wickner W, Schekman R. Protein translocation across biological membranes. *Science* 2005;310:1452–1456.

Wiedemann N, Frazier AE, Pfanner N. The protein import machinery of mitochondria. *J Biol Chem* 2004;279:14473–14476.

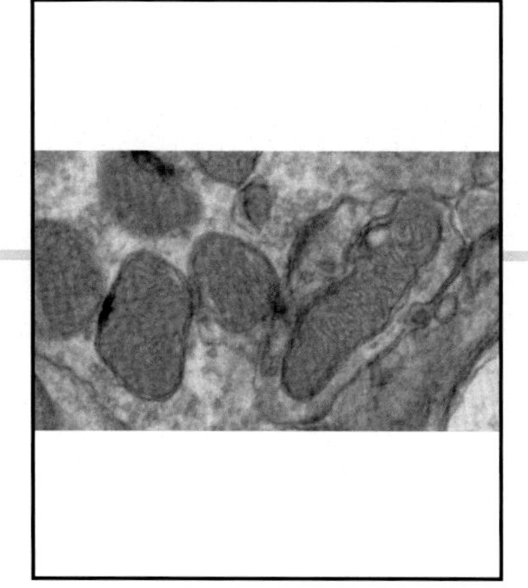

Chapter 5

Mitochondria and Diseases

As we saw in Chapter 4, the eukaryotic cell is a highly complex biological machine, composed of a variety of organelles with highly specialized functions that are required for maintaining normal, healthy cytophysiology. For example, the lysosome is a membrane-bound vesicle filled with hydrolytic enzymes responsible for the degradation and recycling of cellular components, while the nucleus encloses the cell's genetic material, providing separation from the cytoplasm and containing molecules that regulate gene expression. Because a compromise in the function of any of these specialized organelles can have devastating effects on cells, leading to disease or even death, cell biologists are dissecting and elucidating the functions of each organelle.

The mitochondrion is a particularly important and fascinating organelle. It appears to be a remnant of a once free-living bacterium that, at some point in the ancient past, entered into a symbiotic relationship with the single-celled ancestor of all eukaryotic organisms. The fundamental and mutually beneficial relationship between them appears to be quite simple: The mitochondrion provides energy in the form of ATP to the cell, while the cell provides nutrients and a stable environment that sustain the mitochondrion. Although energy production remains the major defining characteristic of the mitochondrion, its relationship with the cell has deepened and increased in complexity over evolutionary time, with the mitochondrion acquiring novel regulatory and biosynthetic roles within the cell.

Because of the mitochondrion's central role in powering the cell, conditions that disrupt mitochondrial function are devastating. Considered separately, each specific "primary" mitochondrial disease is rare. However, collectively, primary mitochondrial diseases are relatively common. In addition, a growing body of research has been implicating the mitochondrion in the pathogenesis of a variety of common diseases. Parkinson's disease (PD), for instance, has been associated with mutations in a gene required for the proper disposal and recycling of mitochondria. Other progressive degenerative diseases, including Alzheimer's disease (AD), are also thought to involve defects in mitochondrial function. In fact the aging process itself may be driven in large part by progressive genetic and cellular damage caused by the chemically reactive byproducts of mitochondrial metabolism. Additionally, mitochondrial dysfunction has been associated with the development of a large proportion of cancers, and cancer cells share, almost universally, the property of repressed mitochondrial ATP production in favor of glycolysis. Indeed, this reliance on glycolysis, which is thought to fuel the aggressive proliferative and metastatic properties of cancer cells, can provide a powerful oncological target wherein cancer cells may be "starved" selectively while leaving normal cells intact.

Goodman's Medical Cell Biology. https://doi.org/10.1016/B978-0-12-817927-7.00005-3

Finally, mitochondrial dysfunction is often a precursor and contributing factor to the development of diabetes, which is perhaps unsurprising given the organelle's central role in cellular energy utilization.

In this chapter, we describe the genetic, structural, and biochemical properties that make the mitochondrion uniquely important to the eukaryotic cell and how defects in mitochondrial function underlie a variety of human diseases. Most readers will have a passing familiarity with the mitochondrion as the "powerhouse" of the cell. Scientists and physicians interested in understanding and treating mitochondrial disease, however, need to be attentive to the more nuanced details of mitochondrial function. This chapter begins by introducing the reader to mitochondrial biology, before diving into how that biology impacts human disease and the treatment approaches researchers are developing for mitochondrial-related conditions.

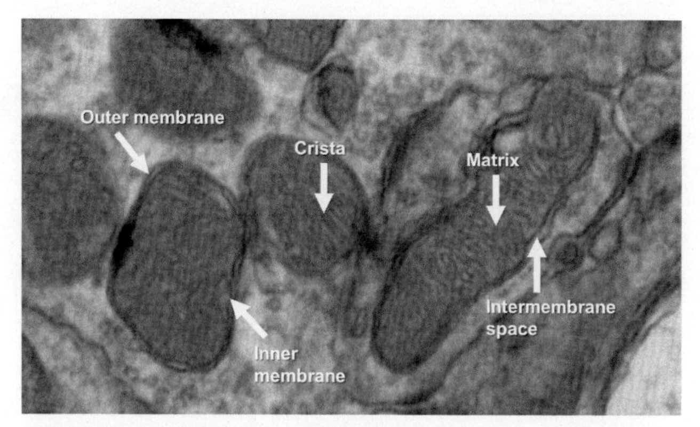

Figure 5–1. **Major mitochondrial structures.** The image shown was captured from a cell located in the cerebellum (a part of the brain), using a transmission electron microscope. The major mitochondrial structures have been indicated with *white arrows*. Note that the double-membrane structure of the mitochondrion creates two major spaces: an intermembrane space and a large central space referred to as the "matrix." Both compartments serve crucial functions in the generation of the chemical gradient that ultimately powers ATP production. To facilitate exchange between these two compartments, the inner mitochondrial membrane folds in on itself to form smaller subcompartments called "cristae," which serve to increase the surface area between the matrix and intermembrane space. *(Image courtesy of Li Yang.)*

THE FUNDAMENTALS OF MITOCHONDRIAL BIOLOGY

Mitochondrial Structure and Organization

Mitochondrial Membrane Structure

A defining feature of a eukaryotic organelle is the lipid bilayer that separates the inside of the organelle from the surrounding cytoplasm. Most organelles (e.g., the Golgi apparatus, endoplasmic reticulum, and lysosomes) possess a single lipid bilayer, similar to the plasma membrane. The mitochondrion, however, has an unusual and specialized double-membrane structure, consisting of an inner and an outer membrane with distinct lipid/protein compositions and functional properties. The possession of a similar double membrane in Gram-negative bacteria supports the view that this dual mitochondrial membrane is likely an evolutionary holdover from the free-living bacterial ancestors of mitochondria. However, much of the lipid composition of the outer mitochondrial membrane is clearly of a nonbacterial origin, leading many experts to argue that the outer membrane is actually of eukaryotic origin. Under this model the outer mitochondrial membrane is thought to be derived from the vesicle that originally surrounded the mitochondrial ancestor cell when it was engulfed by the primitive eukaryotic cell. This question remains a matter of some controversy within the field.

Whatever the origins may be for this membrane arrangement, the result is the creation of two major compartments—an intermembrane space and a centralized space called the "matrix"—with divergent biochemical and functional characteristics (Fig. 5–1). In addition, the inner membrane folds upon itself to create a unique structure of serial inward-folding ridges known as cristae, thereby creating a massive increase in surface area between the matrix and the inner mitochondrial membrane itself. Because this interface is the site of key biochemical reactions, an increase in its surface area enhances the metabolic ability of the mitochondrion dramatically. Cells derived from mitochondrial disease patients in which mitochondria are deficient in cristae may have severe reductions in mitochondrial function.

The complex membrane structure of the mitochondrion creates important logistical demands on the cell because a variety of biological components must be exchanged continuously between mitochondria and the cytoplasm of the host cell. While ATP molecules produced by mitochondria need to be exported to be utilized by the cell, most of the proteins that mediate ATP biosynthesis are translated in the cytoplasm and thus must be imported into mitochondria. Other aspects of mitochondrial biology, including mitochondrial DNA replication and mitochondrial fission/fusion, are also controlled by proteins made by the host cell. Import of proteins and metabolic precursors into the mitochondrion occurs through protein complexes in the outer and inner mitochondrial membranes that allow materials to be directed specifically to the intermembrane space, inner membrane, or matrix. Similar pathways also mediate export of molecules out to the cytoplasm. Import or export failures can disrupt local organelle functions and the broader cytophysiology of the host cell, causing or contributing to mitochondrial disease.

Mitochondrial Genome Structure

The mitochondrion occupies a somewhat unique position among organelles (shared only with the chloroplast in plants) in that it carries its own genome distinct from the nuclear genome. Mitochondrial genome mutations have been associated with many diseases and therefore must receive careful attention in the diagnosis and evaluation of a suspected mitochondrial condition.

Mitochondrial structure and function requires genes in both the nuclear genome and the mitochondrial genome. In humans the entire mitochondrial genome—commonly referred to as mtDNA—is just 16,569 base pairs long, tiny compared with the nearly 3 billion base pairs that comprise the nuclear genome. The human mtDNA contains only 13 protein-coding genes, 22 tRNA genes, and 2 rRNAs, all of which are encoded via polycistronic (multigene) transcripts (Fig. 5–2). Proteins encoded by the genes in the mtDNA are required for oxidative phosphorylation (i.e., the production of ATP). Over the course of mitochondrion evolution from free-living organism to cellular organelle, some components of its metabolism and related functions have been outsourced to the nuclear genome, leaving only those genes that must absolutely be produced within the organelle itself.

Generally speaking, each individual mitochondrion contains multiple copies of the mtDNA molecule. Thus each cell can contain hundreds (e.g., skin cells), thousands (e.g., muscle cells), or even hundreds of thousands (e.g., oocytes) of mtDNA molecules, depending on the energetic requirements of the cell type in question. Precise control of mtDNA copy number is a major concern for the cell, and the mechanisms by which this control is enforced constitute a major area of study. Key metabolic outputs of the electron transport chain (ETC) are thought to be major determinants of mtDNA copy number. Proteins with a role in mitochondrial transcription, such as transcription factor A, mitochondrial (TFAM), can also regulate mitochondrial replication efficiency. Mutations in key mtDNA replication proteins like TFAM can diminish mtDNA copy number, thereby compromising mitochondrial function by restricting its production of the RNA and protein molecules necessary for normal function. Collectively the class of diseases caused by mtDNA copy number insufficiency are referred to as mitochondrial DNA depletion syndrome, a classification that encompasses numerous severe (and often fatal) childhood mitochondrial disease states.

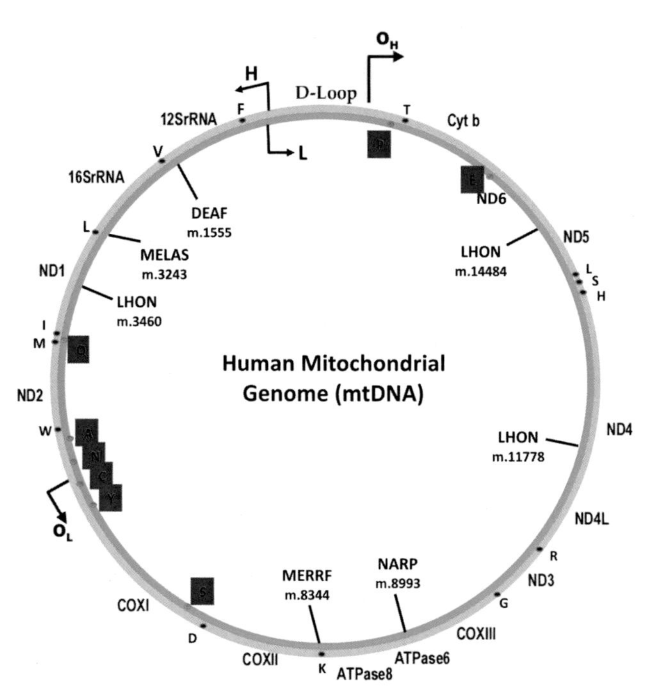

Figure 5–2. **Structure of the human mitochondrial genome.** Two distinct RNA molecules are transcribed from mtDNA: a "heavy strand" (indicated by the *orange line*) and a "light strand" (indicated by the *blue line*). Both strands contain multiple coding genes, tRNAs, and rRNAs transcribed as a single RNA from distinct origins of replication (O_H for the heavy strand and O_L for the light strand). The locations of the mitochondrial tRNAs are indicated by *red letters* on the light strand and with *black letters* on the heavy strand. The locations of common disease-causing variants are also indicated, based on the following abbreviations: *DEAF*, familial progressive sensorineural deafness; *LHON*, Leber's hereditary optic neuropathy; *MELAS*, mitochondrial encephalomyopathy, lactic acidosis, and stroke-like episodes; *MERRF*, myoclonic epilepsy with ragged red fibers; and *NARP*, neuropathy, ataxia, and retinitis pigmentosa.

Mitochondrial Inheritance

The inheritance of the mitochondrial genome deserves special comment here. Unlike the nuclear genome, which is inherited from both parents in equal measure, mtDNA is passed exclusively from mother to child via the oocyte. In fact, active measures appear to be taken in the early embryo to destroy paternal mitochondria and thus paternal mtDNA. The reasons for this maternal uniparental inheritance are unclear, but it is nearly ubiquitous in animals, with notable exceptions such as bivalves (which exhibit strict uniparental inheritance for both sexes, with males inheriting mtDNA from the father and females inheriting it from the mother).

Researchers are still in the process of trying to understand why paternal mtDNA is completely absent in nearly every case where it has been investigated. It has been hypothesized that the enormous energy demands placed on sperm mitochondria prior to fertilization lead to a high level of reactive oxygen species (ROS)-induced damage to the sperm's mtDNA. Hence, paternal mitochondrial elimination may filter out likely to be defective mtDNA from the offspring. Furthermore the ratio of sperm mitochondria to oocyte mitochondria in the oocyte is approximately 1:1000, putting paternal mtDNA at a severe disadvantage relative to maternal mtDNA even before active mechanisms of paternal mitochondrial elimination are taken into account. These caveats notwithstanding development of the ability to artificially induce paternal inheritance may be useful in reproductive medicine for cases in which a female patient carries a deleterious mutation in her mtDNA.

Owing to the maternal inheritance of mtDNA, mutations in mitochondrial genes follow a mother-to-child inheritance pattern rather than the Mendelian pattern of inheritance exhibited by most nuclear genes. Although both are maternally inherited, mitochondrial traits are distinct from X-linked traits in that they can affect male and female offspring equally, while most common X-linked conditions are X-linked recessive and tend to affect male offspring selectively. On a broader level the maternal lineage of mtDNA means that entire human populations can be classified into genetic groups based on the relatedness of their mtDNA sequences. These groups are referred to as "haplogroups" because the entire mtDNA sequence is inherited as a unit from one parent in a non-Mendelian manner; the same terminology is used for Y-chromosome DNA sequences passed on solely from the father. Distinct haplogroups possess different nucleotide variants in their mtDNA, some of which represent adaptations of particular populations to environmental conditions. For instance, haplogroups originating in colder climates tend to carry variants that sacrifice ATP production to create more body heat.

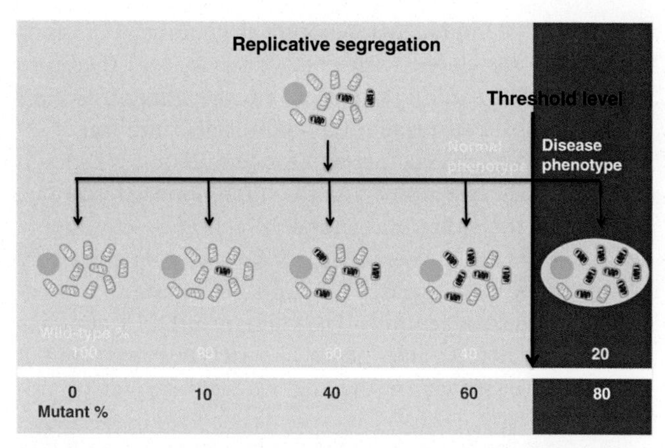

Figure 5–3. **Replicative segregation and heteroplasmy.** Diagram shows the range of possible outcomes when a parent cell carries a mutant mtDNA heteroplasmy (indicated in *red*). Despite the fact that the parent cell carries the mutation at a heteroplasmy well below the threshold for expressing the mutant phenotype, there is a possibility that some of its progeny will possess a heteroplasmy level above that threshold after several rounds of replication.

Heteroplasmy

Heteroplasmy describes the situation in which two or more mtDNA variants exist within the same cell. Heteroplasmies are often caused by de novo mutations occurring either in the germline or in the somatic tissues. In fact, heteroplasmy levels often vary even between the cells or somatic tissues of the same individual, leading to situations where only specific cells, tissues, or organs are affected by mitochondrial dysfunction. The accumulation of these mutations in somatic tissues over time may be a central factor in aging.

Instead of the discrete allele frequencies of 0%, 50%, or 100% observed for alleles of nuclear genes, allele frequencies for mtDNA loci have a continuous distribution ranging from 0% to 100% ("homoplasmy"). The threshold for the disease phenotypes can be anywhere along that distribution, depending on the gene and tissue in question. Thus, as we will see in the discussion of primary mitochondrial diseases later, low-mutation heteroplasmy is unlikely to cause a discernible phenotype, whereas high-mutation heteroplasmy is very likely to cause problems, especially when present in an energy-sensitive organ like the brain or muscles. The lack of paternal contribution precludes compensation for a defective allele inherited from the mother by a functional "wild-type" allele from the father. Furthermore, even if a heteroplasmy is subthreshold for disease presentation in the mother, the disease may occur in her offspring depending on how the mtDNA molecules are apportioned to her oocytes during gametogenesis (Fig. 5–3).

High Mutation Rate

The overall mutation rate in mtDNA is estimated to be about an order of magnitude higher than that in the nuclear genome of humans. Several factors contribute to this discrepancy. Most obviously, there is the increased risk of DNA damage presented by the ROS-generated mitochondrial environment compared with that in the less chemically reactive environment of the nucleus. Another major factor in the high mtDNA mutation rate is the mitochondrion-specific DNA polymerase gamma (POLG) enzyme, which is responsible for the replication of mitochondrial DNA. Experiments have shown that POLG is much more error prone than that the family B DNA polymerases utilized in the eukaryotic nucleus. Additionally, the mitochondrial genome is replicated at a much higher rate than the nuclear genome; mtDNA is replicated continually, even in nondividing cells like neurons in which the nuclear genome is not being replicated. Because each round of DNA replication represents an opportunity to generate mutations, mtDNA genes are at an elevated risk of mutation, even in terminally differentiated tissues.

mtDNA Sequencing

The accurate assessment of heteroplasmy levels and disease-causing mtDNA sequence variants is critical for the diagnosis of mitochondrial diseases. Inaccurate measurement may lead to an improper assessment of disease states or a failure to identify variants. Thus great care must be taken to utilize sensitive and accurate mtDNA analysis methods. Both requirements can be improved by next-generation sequencing (NGS) methods, which employ parallel sequencing approaches to provide thorough genome coverage. Typically the NGS approaches applied to mtDNA sequencing involve two steps: (1) long-range polymerase chain reaction amplification of the entire mitochondrial genome with mtDNA-specific primers and (2) deep sequencing of the products from step 1 with NGS technology, providing thousands of reads worth of sequence coverage for every position in the mitochondrial genome. This combined approach provides an efficient and cost-effective means of detecting mtDNA variants and heteroplasmies down to a frequency of around 2%. This level of sensitivity is more than adequate for most clinical purposes because the disease threshold for most mitochondrial mutations is well above this level, even in tissues susceptible to mitochondrial disease. A major caveat to this approach is the fact that different tissues can possess different levels of heteroplasmy within an individual, with different heteroplasmy threshold sensitivities. Multiple tissue samples from the same patient may be needed to get a better picture of the genetic underpinnings of the disease state in that individual.

Whole-genome sequencing (WGS) is used to assess the nuclear genome of patients suffering from diseases with an unidentified genetic cause. With the proper analytical approaches, mtDNA sequence data can be extracted selectively from a complete WGS dataset. The WGS approach carries the clear advantage of enabling one to analyze the nuclear genome and mtDNA simultaneously. As the cost of DNA sequencing continues to fall, WGS may become a routine part of clinical evaluation. Furthermore, mtDNA data in WGS should provide fertile ground for bioinformatic analysis and data mining to refine our understanding of mitochondrial biology and disease mechanisms (refer also to Chapter 14 for a broader discussion of the role of bioinformatics in modern precision medicine).

Mitochondrial Dynamics

Mitochondria possess the ability to undergo fission and fusion. Although the reasons for these processes are unclear, this process seems to be a key for the maintenance of proper organelle function. Mitochondrial fission-fusion may serve as a quality control mechanism to eliminate damaged or redundant mitochondria by way of autophagosome destruction. The reasons for fusion and fission events involving apparently healthy mitochondria are more obscure. Regardless, given the constant splitting and fusing of the mitochondrial biomass within a cell, it would be more accurate to view all of the mitochondria in a cell as a single organelle that can be combined or separated depending on the needs of the cell. Sometimes, meeting cellular needs may entail forming a single, conjoined network of mitochondria; at other times, cellular needs may require hundreds of smaller organelles. Abnormalities of mitochondrial fusion and fission are associated with many diseases.

Mitophagy

As we alluded to in the previous section, the autophagic destruction of dysfunctional or excess mitochondrial organelles is critical for cellular function. The uncontrolled destruction of mitochondria or permeabilization of the mitochondrial outer membrane can release a variety of damaging compounds into the cytosol, including protons from the intermembrane space, ROS, and proapoptotic factors such as cytochrome *c*. A mitochondrion with compromised metabolic function is a continual liability to the cell, if for no other reason than the increased amount of ROS produced by its compromised ETC. For these reasons the cell has mechanisms to detect compromised mitochondria and direct them to autophagosomes for destruction. Because of the mitochondrial specificity of this type of autophagy, it is often referred to as "mitophagy." Mitophagy represents a major factor to be considered in investigations of the pathogenesis of diseases with a suspected mitochondrial component.

Mitochondrial Function

Fundamentally the mitochondrion is a compartment designed to maintain a specialized set of chemical conditions for the efficient generation of ATP while keeping this highly acidic and caustic environment from damaging other components of the cell. To maintain this compartmentalization, proteins and chemical precursors must be trafficked into and out of mitochondria in a precisely controlled manner. Any compromise in the efficiency of these processes or any breach that allows leakage of the highly reactive byproducts of mitochondrial metabolism is a potential source of pathology. Furthermore, in addition to their primary role in ATP production, mitochondria provide several important ancillary functions to their host cells such as triggering apoptosis and biosynthesizing critical chemicals. Disruptions in these ancillary functions also underlie mitochondrial diseases, as will be discussed later in this chapter.

Glycolysis Versus Oxidative Phosphorylation

The most basic mechanism for generating energy within the cell involves the chemical breakdown of one molecule of glucose into two molecules of pyruvate. This reaction, called glycolysis, produces two molecules of ATP per glucose molecule split. Glycolysis is an ancient, evolutionarily conserved metabolic process, but it is relatively inefficient. The presence of mitochondria in cells provides a much more efficient pathway that can be utilized under aerobic conditions (i.e., conditions in which oxygen is available). Under these conditions the pyruvate produced by glycolysis can be imported into the mitochondrial matrix, where it is converted into acetyl-CoA. This acetyl-CoA can then enter the tricarboxylic acid cycle (a.k.a. citric acid cycle or Krebs cycle), a series of biochemical reactions that produce byproducts that are used by the mitochondrion to power its central biochemical process, oxidative phosphorylation. The latter will be described in more detail in the next section. The main concept to understand here is that these combined biochemical reactions of aerobic respiration produce several times more ATP than glycolysis, with as many 34 ATP molecules being produced by the ETC for every two pyruvate molecules produced by glycolysis.

When there is a shortage of oxygen, such as during intense exercise, cells resort to anaerobic respiration (glycolysis alone) to meet their energy demands temporarily. The pyruvate byproduct of glycolysis cannot undergo oxidative phosphorylation, but rather needs to be converted to lactic acid and exported out of the cell. Glycolysis is analogous to the alcoholic fermentation process that yeast uses to produce ethanol under anaerobic conditions; in fact the process is often referred to as lactic acid fermentation in humans. Extracellular lactate must be expelled from the body to prevent negative effects on surrounding tissues. It can build up rapidly should anaerobic conditions continue to prevail, especially in energy-intensive tissues such as the skeletal muscles. In fact, lactic acid buildup is the source of the muscle soreness that persists following intense physical activity. Not surprisingly, lactate buildup is also a common symptom of mitochondrial diseases that force cells to rely heavily on glycolysis to meet their energy needs.

Oxidative Phosphorylation

As we outlined earlier, the bulk of ATP production within the eukaryotic cell occurs in mitochondria, downstream of glycolysis. The mitochondrion accomplishes this feat via the aforementioned ETC, which is a series of protein complexes embedded in the inner mitochondrial membrane (Fig. 5–4). The chemical process used by the ETC to produce ATP is referred to as oxidative phosphorylation.

Mitochondrial oxidative phosphorylation is preceded by the generation of NADH and $FADH_2$ from acetyl-CoA by the tricarboxylic acid cycle. Acetyl-CoA may be derived from glycolysis or from the catabolism of fats and proteins, depending on what nutrients are available to the cell at the moment. Regardless the majority of the NAHD and $FADH_2$ produced by the tricarboxylic acid cycle are processed by Complexes I and II of the ETC, respectively, where they undergo a chemical reaction in which they donate electrons to these protein complexes. These donated electrons are passed to the next component of the ETC, known as Complex III, before being transferred to cytochrome *c* and then to Complex IV in turn. Complex IV then uses these electrons to split a molecule of oxygen into water.

One might be wondering what the purpose of all this electron shuffling may be and how exactly it is used to generate ATP. In the most direct sense, it is not. What actually happens is that the energy provided by the electron movement within the ETC is used to shuttle a large number of protons from the mitochondrial matrix into the intermembrane space. The resultant reservoir of protons produces an electrochemical gradient across the inner mitochondrial membrane that represents a massive store of potential chemical energy that the cell can harness to generate ATP. This harnessing is accomplished by the last component of the ETC, referred to as Complex V or ATP synthase, which acts as a channel through which protons in the intermembrane space can pass back into the matrix to relieve the electrochemical gradient. The ATP synthase complex couples the movement of these protons to drive another set of chemical reactions that convert ADP (the depleted form of ATP) back into ATP. This final enzymatic step links the movement of electrons in the ETC to the production of ATP to fuel the cell's energy needs.

Mitochondrial Membrane Potential

The production of ATP by the ETC requires continual maintenance of a reservoir of hydrogen ions in the intermembrane space and the strong difference in electrical potential it generates between the mitochondrial matrix and the intermembrane space. For this reason, one of

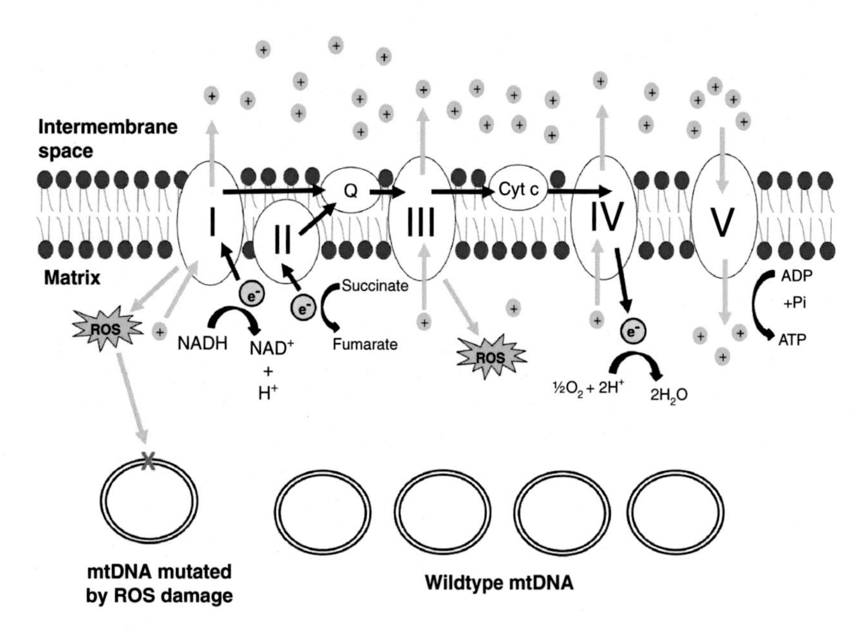

Figure 5–4. **Outline of the electron transport chain (ETC).** The conversion of NADH to NAD^+ at Complex I and succinate to fumarate at Complex II results in the transfer of electrons to each respective complex. These electrons are then transferred to each successive component of the ETC—where they power the movement of protons into the intermembrane space—before ultimately being used to generate water from oxygen molecules and protons. The protons that build up in the intermembrane space as a result of this process will eventually return to the mitochondrial matrix by moving through the ATP synthase complex (Complex V), which will use the energy created by this proton flux to generate ATP from ADP and inorganic phosphate (P_i). Reactive oxygen species (ROS) will be generated as a byproduct of this process, with Complex I and Complex III being major contributors. Some of these ROS byproducts will end up damaging the mitochondrial DNA (mtDNA) carried within the mitochondrial matrix, resulting in an accumulation of mtDNA mutations over time.

the strongest indicators of proper mitochondrial function is a negative charge difference between these compartments. Researchers use this charge, referred to as the mitochondrial membrane potential (ΔΨM), as a proxy for mitochondrial health. In fact, when certain types of cells are damaged, their mitochondria will initiate their own self-destruction by creating a pore in the inner membrane that allows protons to flow back into the matrix, eliminating the ΔΨM. This process is referred to as mitochondrial permeability transition.

Free Radicals

A major consequence of oxygen-based metabolism in mitochondria is the production of free radicals, such as hydroxyl radicals ($^{\bullet}OH$) and superoxide radicals ($^{\bullet}O^{-}_{2}$). These byproducts, referred to collectively as "reactive oxygen species" (ROS), are inevitable in any biochemical process involving oxygen and are even required in low concentrations for normal cell function. At high levels, chemical interactions of ROS with cellular proteins, lipids, and nucleic acids disrupt cellular functions (Fig. 5–4). For this reason, both the mitochondrion and the cell as a whole have evolved mechanisms to protect themselves from ROS-based damage, either by absorbing ROS, rendering them chemically inert, or coopting them as intracellular signaling molecules.

Mitochondria and Apoptosis

The ability to eliminate damaged or potentially malignant cells, which is a hallmark of multicellular life, helps to prevent faulty cells from compromising the health of the whole organism. This orderly form of cell death, referred to as apoptosis, is encoded at the cellular level in multicellular organisms, wherein each cell can trigger its own apoptosis when deemed necessary. The importance of this process is underscored by the fact that a loss in apoptotic ability occurs in cancer cells, allowing them to undergo uncontrolled proliferation (see Chapter 11 for a more detailed discussion of this issue).

In mammals the B-cell lymphoma 2 (Bcl-2) family proteins are master regulators of apoptosis, controlling the choice between continued cell survival and cell death. Although there are multiple pathways by which Bcl-2 influences this choice, the central mechanism by which it triggers cell death, referred to as the intrinsic pathway, acts through mitochondria. Bcl-2, which is localized to the mitochondrial outer membrane, can permeabilize the outer membrane in response to cellular stress signals, including signals originating from within the mitochondrion itself. This stress-induced permeabilization releases factors such as cytochrome *c* that activate a highly conserved set of cysteine proteases called caspases, which act as the ultimate mediators of apoptotic cell death. Owing to the mitochondrion's central position in the pathway, mitochondrial health and integrity have a powerful influence on the triggering of apoptosis. The fact that such a pathway has evolved to link mitochondria to the control of apoptosis is another indicator of the vital role of mitochondria in the basic functions of the eukaryotic cell.

Synthetic Functions

One of the more surprising functions of the mitochondrion is its role in critical biosynthetic pathways within the cell, many of which involve, directly or indirectly, the utilization of iron. For instance, both the early and late portions of the pathway for synthesizing the iron-containing compound heme—the oxygen-carrying component of hemoglobin—occur within mitochondria. Mitochondrial ferredoxin proteins are also required for the synthesis of iron-sulfur (Fe-S) clusters, which act as enzymatic cofactors in critical cell processes, such as DNA replication and repair. Fe-S clusters also interact with iron-regulatory protein 1 to convert it to its cytosolic aconitase form when intracellular iron levels are high. When intracellular iron is low, mitochondrial production of Fe-S clusters is reduced, and iron-regulatory protein 1 is converted into an RNA-binding protein that activates pathways that increase the cell's ability to take in and retain iron. Beyond these iron-based functions, the mitochondrial cytochrome P450 enzyme family proteins are necessary for the synthesis and metabolism of important steroid hormones, such as vitamin D. Through these diverse synthetic pathways, the mitochondrion modulates multiple regulatory processes, both at the cellular and organismal level.

MITOCHONDRIA AND DISEASE

Primary Mitochondrial Disease

Aerobic respiration is fundamental to the operation of the eukaryotic organism, allowing cellular specializations that would be difficult to obtain with anaerobic respiration alone. For that reason, any condition that disrupts oxidative phosphorylation—whether it be a mutation, lack of oxygen, or some other environmental factor—will have serious consequences for a eukaryotic organism. Humans being no exception, there are a host of mutations—both in the mtDNA and the nuclear genome—that cause disease primarily by compromising mitochondrial function.

Overall, mitochondrial disorders that are directly attributable to mtDNA mutations appear to be quite rare, affecting some 0.02% of the population according to widely accepted estimates. However, this prevalence represents a significant patient population that the medical community needs to be prepared to deal with in ways suited to the unique nature of these diseases. For instance, the issue of tissue-specific heteroplasmy must be considered at all times and on a case-by-case basis. If mitochondria are distributed asymmetrically to daughter cells

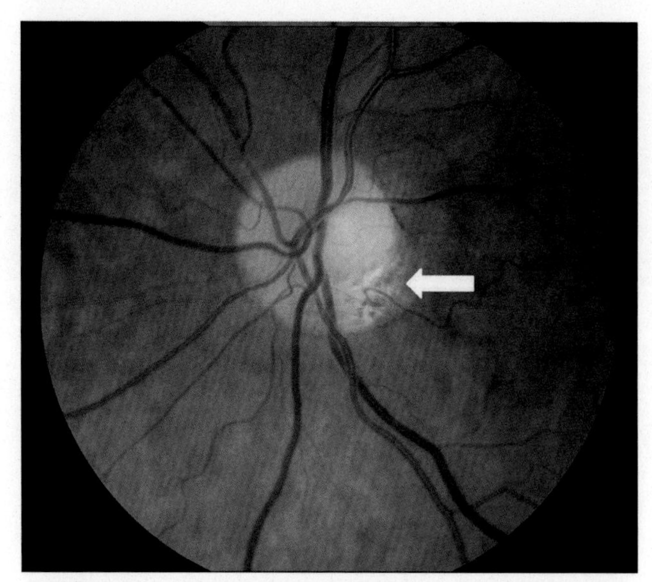

Figure 5–5. **Optic atrophy induced by mitochondrial disease.** The image shown is of the fundus (i.e., the back of the eye) from a patient carrying a mutation associated with both mitochondrial dysfunction and optic atrophy. The *white spot* in the center of the image is the optic disk (the point of exit for the optic nerve from the retina), while the rest of the image shows the retina proper, along with the blood vessels supplying the retina. Atrophy of the optic disk can be clearly observed along the edge of the optic disk *(white arrow)*.

during mitosis, each daughter cell forms a lineage at the level of the mtDNA. When this happens early in embryonic development, different somatic tissues will have differing levels of heteroplasmy. Furthermore, cells in high-energy demand tissues such as the brain, heart, and muscles are more sensitive to mitochondrial mutations than cells in low-energy demand tissues, such as adipose tissue. Optic atrophy, for instance, is a very common feature of mitochondrial disease, due to the high energy demands of retinal tissues in the eye (Fig. 5–5). An awareness of such common signs of mitochondrial dysfunction can be highly useful in identifying potential cases of mitochondrial disease.

Because most of the key mitochondrial proteins are encoded in the nuclear genome, it should be unsurprising that a significant portion of mitochondrial diseases is caused by mutations in nuclear genes. A substantial portion of these mutations affect ETC-related components. Nuclear mutations can have serious effects on other aspects of mitochondrial function, including mitochondrial dynamics and mtDNA copy number control. In the following sections, we describe some common mechanisms by which mitochondrial dysfunctions derived from mtDNA or nuclear DNA contribute to debilitating disease.

Defects in the Electron Transport Chain

Nuclear gene mutations affecting the ETC include mutations in the genes encoding subunits of Complex I and components of Complex II (e.g., *SDHA* and *SDHB*), Complex III (e.g., *BCS1L*), and Complex IV (e.g., *COX10* and *COX15*) and mutations in genes encoding the assembly factors needed to build the complexes. It is well documented that *MT-ATP6* mutations affect Complex V (ATP synthase); mutations affecting other factors involved in Complex V assembly, including the products of *ATP12* and *TMEM70*, can also be disease causing. Ataxia, motor problems, lactic acidosis, developmental delays, and peripheral neuropathy are common to this group of conditions. Several examples of common conditions associated with mutations in mtDNA and mitochondrial-related nuclear genes are elaborated in the succeeding text.

Leber's hereditary optic neuropathy (LHON). A hallmark of many mitochondrial diseases is a loss of visual acuity due to the neurodegenerative loss of retinal cells. LHON is a mitochondrially inherited disorder that causes vision loss through damage to the retinal ganglion cells that transmit visual information from the retina to the visual system of the brain. The pathogenic mechanism of the disease is a loss of ETC function caused by mutations to any of three genes that encode Complex I subunits: *MT-ND4*, *MT-ND1*, and *MT-ND6*. The most common genetic cause of LHON is a m.11778G>A mutation (i.e., a change from a guanine to an adenine nucleotide at position 11,778 in the mtDNA). This mutation is located in the coding region for *MT-ND4*.

In most cases the progression of LHON is rapid, with loss of central vision in both eyes. This bilateral visual loss is nearly always simultaneous. However, visual loss may start a few months earlier in one eye. Unlike more severe mitochondrial diseases, the onset of LHON symptoms is typically in the late teens or twenties. Usually, LHON does not produce any symptoms other than visual loss, although sometimes LHON may present with additional symptom such as heart blockage and irregular heartbeat. Curiously, symptom severity appears to be independent of the heteroplasmy level of the mutation, as most individuals carrying the homoplasmic version of the m.11778G>A allele never develop symptoms. Even more strangely, males with the m.11778G>A mutation develop LHON approximately 50% of the time, while females have only a 10% chance of developing symptoms. Studies examining LHON penetrance have shown that disease onset and severity appear to be modified by mutations in other genes, such as *YARS2* (a mitochondrial tyrosyl-tRNA synthetase encoded in the nuclear genome). In light of this fact, it is likely that the male bias in the occurrence of LHON is due to a modifier locus on the X-chromosome. More generally, this phenomenon demonstrates the fact that the phenotypic severity produced by disease-causing mtDNA mutations can be modified by mutations in nuclear genes. This kind of interaction between mtDNA mutations and nuclear mutations may help to explain the high variability in mitochondrial disease severity that has been observed, even within the same family.

Neuropathy, ataxia, and retinitis pigmentosa (NARP). NARP is a mitochondrial condition attributed to mtDNA mutations. As the name indicates, NARP affects the nervous system, including the visual system. Patients suffer progressive damage to light-sensing cells in the retina that leads to a gradual worsening of vision, culminating in legal blindness. Frequently, children affected by NARP exhibit developmental delays and various learning disabilities.

The mutations that cause NARP are found primarily in the gene encoding mitochondrial ATP synthase subunit 6 (*MT-ATP6*), indicating that this disease is the result of a loss of mitochondrial ATP production. The two most common mutant variants associated with NARP occur at the same position in the mitochondrial genome: m.8993T>G and m.8993T>C. While these two alleles produce slightly different effects on mitochondrial metabolism and function, the most important factor in disease presentation is heteroplasmy level from tissue to tissue. Typically, symptoms of the disease are observed only at a heteroplasmy level above 70%.

Leigh syndrome. One of the most common and severe forms of mitochondrial disease is Leigh syndrome (see Clinical Case 5–1), a neurological disorder that strikes

Clinical Case 5–1 Leigh Syndrome

William, a 2-year-old boy, was brought into the emergency room after experiencing a seizure. He walked normally starting at 12 months and had a vocabulary of around 30 words. About 2 months ago, he experienced a week-long febrile illness accompanied with runny nose, cough, and vomiting. He recovered fully from the illness but around a month ago started to become more hesitant with his walking, only taking a couple of steps at a time when previously he had always been on the move. He also became more irritable even after napping, and his appetite decreased notably. He has started talking less, and when he does speak, his words are less easily understood than previously. On examination, he is awake but doesn't engage well. His pupillary response is sluggish, and he has visible nystagmus (rapid rhythmic extraocular eye movements). His muscle tone is poor, and his deep tendon reflexes are diminished. A blood sample is taken and reveals normal blood counts and normal chemistry except for a bicarbonate level of 19 mEq/L (normal range 23–30) and high levels of lactic acid. Chest X-ray shows a mildly enlarged heart but no lung abnormalities. Electroencephalogram is nonspecific. MRI of the brain is performed and shows bilateral symmetric hyperintensities in the basal ganglia and in the substantia nigra and small cysts bilaterally in the cerebral cortex. With this finding a neurologist was consulted and ordered a proton magnetic resonance spectroscopy (MRS) that showed a lactate peak in the brain parenchyma. William's parents refused an invasive muscle biopsy but allowed for genetic testing. NGS testing of common OXPHOS genes revealed a mutation in NDUFS1.

Cell Biology, Diagnosis, and Treatment of Leigh Syndrome

William has Leigh syndrome, which in this case is caused by an autosomal recessive gene likely shared by his parents. Leigh syndrome, also referred to as subacute necrotizing encephalopathy, was described by Archibald Leigh in 1951, before the underlying physiology of the disease was understood. It was only in 1977 that it was determined that Leigh syndrome was directly associated with mitochondrial respiratory chain dysfunction. Dysfunction in one or more of the complexes that are responsible for oxidative phosphorylation results in the typical clinical findings. As oxidative phosphorylation dysfunction abounds, pyruvate (the end product of glycolysis) accumulates and is metabolized to lactic acid, which builds up in the tissues and causes damage. Additionally, reactive oxygen species create cellular instability. Cells that rely heavily on oxidative phosphorylation, such as muscle and neurons, are particularly vulnerable. Of note, mutations that cause dysfunction in the respiratory chain can be either nuclear (~80% of cases) or mitochondrial (~20%). Nuclear mutations are predominantly autosomal recessive, but X-linked patterns of inheritance have also been described.

Diagnosis of Leigh syndrome usually occurs within the first years of life after healthy full-term birth. Development is often normal until an inciting event that leads to metabolic challenges, such as an infection. Early symptoms include loss of previously achieved developmental milestones and muscle weakness or spasticity. Neurologic symptoms include abnormal extraocular movements (nystagmus), dizziness, difficulty swallowing or speaking, and seizures. Laboratory tests often reveal high lactic acid levels in the blood, and a directed test for pyruvate/lactate ratio is often high. Urine, blood, and cerebrospinal fluid may also show high levels of alanine, which is an additional breakdown product of pyruvate. Ultimately the diagnosis of Leigh syndrome is dependent on characteristic imaging findings. Brain magnetic resonance imaging (MRI) characteristically shows bilateral and symmetrical hyperintensities on T2-weighted imaging. They are often found in the basal ganglia or brain stem. Magnetic resonance spectroscopy (MRS) can be added on to the brain MRI to look for unusual peaks in lactate levels, which correlate well with other characteristic findings. As muscle tissue is often affected, muscle biopsy can provide diagnosis of respiratory chain dysfunction, but these tests are invasive and complicated. In recent years, with the advancements of next-generation sequencing tests, evaluation of the many genes associated with Leigh syndrome has become more efficient and less costly. The most common mutations are in Complex I proteins, as is the case of William.

Unfortunately, there is no curative option for Leigh syndrome, and most patients succumb to the disease within a couple of years. Mitochondrial cocktails have been proposed as a means to slow disease progression, but there has been no study to date that has shown any clear benefit of these agents. Supportive care is a mainstay of treatment. Genetic counseling is important for parents of patients with Leigh syndrome. In the future, gene therapy could potentially be a therapeutic modality.

primarily during infancy, though rare adult-onset cases have been observed. Typically, symptoms begin in infancy with vomiting and difficulty eating, leading to problems with growth and weight gain. The condition worsens rapidly, resulting in a progressive loss in neurological and motor function that can be lethal.

Like the other conditions discussed in this section, most of the genes associated with Leigh syndrome are involved in energy production. Because of this the symptoms of Leigh syndrome often manifest first during an event that strains the body's energy production, such as a viral infection. The majority of Leigh syndrome cases are caused by autosomal recessive mutations in one of over 75 nuclear genes, including NADH:ubiquinone oxidoreductase core subunit S1 (*NDUFS1*), mitochondrial isoleucyl-tRNA synthetase 2 (*IARS2*), and mitochondrial asparaginyl-tRNA synthetase (*NARS2*). However, approximately 20% of cases are caused by mutations in the mitochondrial genome, primarily in *MT-ATP6*. The careful reader will recall that mutant variants of *MT-ATP6* also cause NARP. For this reason, both diseases are sometimes considered to be a part of a continuum, with NARP at the less severe end of the spectrum.

Defects in Protein Translation

Mitochondrial encephalomyopathy, lactic acidosis, and stroke-like episodes (MELAS). Mutations in the mtDNA genes that encode ETC-related components cause disease primarily by reducing the efficiency of oxidative phosphorylation. One of the most common disease-causing mtDNA mutations, m3243A>G, occurs within a non-protein coding gene that produces a mitochondrial leucine tRNA. The loss of this tRNA reduces overall protein translation and favors the misincorporation of amino acids into proteins, leading indirectly to severe ETC deficits.

The m.3243A>G mutation is particularly troublesome and destructive. It causes the primary mitochondrial condition known as MELAS, which affects the brain, muscles, and other energy-intensive organs, leading to muscle weakness, seizures, severe headaches, and stroke-like episodes that may be misdiagnosed as epilepsy. Frequently, characteristic signs of brain damage, such as basal ganglia calcification and cerebellar atrophy, can be detected by CT scan (Fig. 5–6). Other symptoms such as muscle weakness, fatigue, and vomiting could be secondary to lactic acid buildup in the body. The onset of symptoms can occur during childhood or adulthood but nearly always before the age of 40. The prognosis for patients with MELAS is quite poor.

Myoclonic epilepsy with ragged red fibers (MERRF) is a common class of mitochondrial disease related to a translational defect. The major symptoms of MERRF are bouts of epilepsy, muscle twitching, and overall muscle weakness and spasticity. As suggested by the name, muscle biopsies of affected patients show an abnormally "ragged" appearance under the microscope due to the accumulation of defective mitochondria in a particular region of muscle fibers (Fig. 5–7). Symptom severity is highly variable, likely related to levels of heteroplasmy and additional modifiers carried in each patient's genetic background.

Mutations in several mitochondrial genes have been associated with MERRF, including the mitochondrial tRNAs *MT-TK*, *MT-TL1*, *MT-TH*, and *MT-TS1*. The vast majority of MERRF cases involve an m.8344A>G mutation in *MT-TK*, which encodes a lysine tRNA. Similar to MELAS the loss of this tRNA impedes protein production, which disrupts a variety of mitochondrial functions, including the all-important process of oxidative phosphorylation. MERFF patients have a poor prognosis, with many dying before the age of 40. However, some MERRF patients show a relatively limited set of symptoms and can survive well past this age.

Mitochondrial Depletion Syndrome

Mitochondrial depletion syndrome is a particularly interesting class of diseases that result from an abnormally low mtDNA copy number (see Clinical Case 5–2 for more details). Several nuclear genes involved in mtDNA synthesis have been found to be mutated in mitochondrial depletion syndromes (Fig. 5–8), including

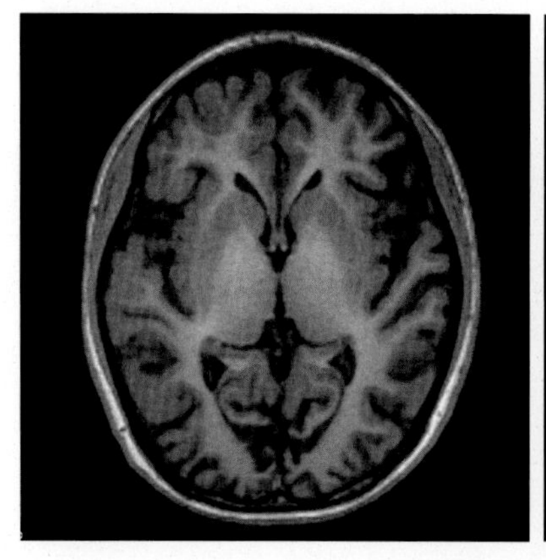
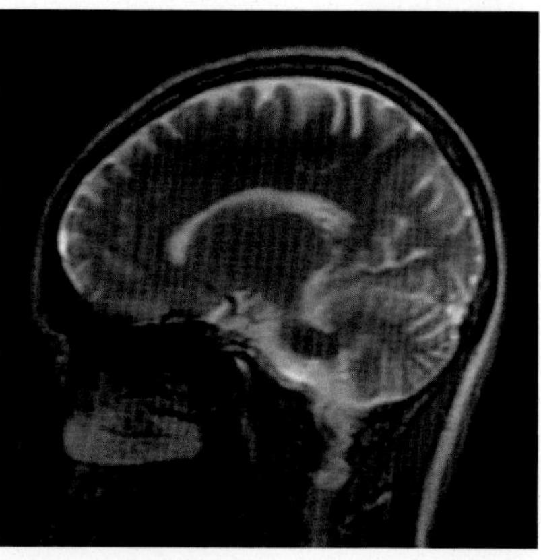

Figure 5–6. Structural changes in the brain due to mitochondrial disease. Images are from an MRI scan of a young patient carrying the MELAS mutation. Note the prominent sulci, particularly in the cerebellar and cerebral regions, that are indicative of atrophy caused by the metabolic defects of the MELAS syndrome.

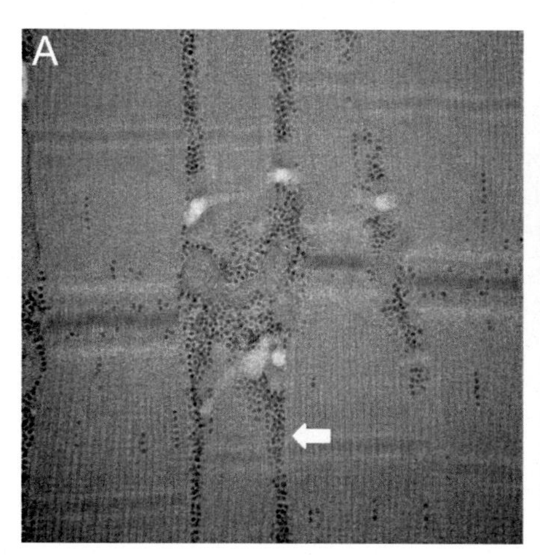

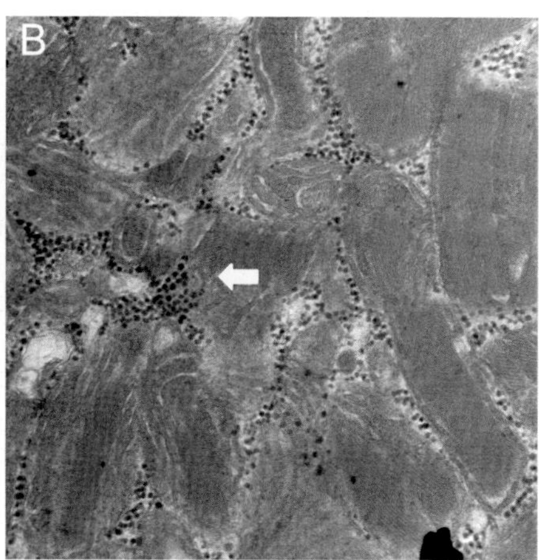

Figure 5–7. **Electron microscopy imaging of muscle fibers from a patient with mitochondrial myopathy.** Longitudinal (A) and cross-sectional (B) images from a patient carrying a deletion associated with Kearns-Sayre syndrome. Note the *dark*, granular structures indicated with the *white arrows*. These inclusions are aggregates of diseased mitochondria and are a characteristic signature of mitochondrial dysfunction within the mitochondria. Under Gömöri trichrome staining, these aggregates will take on a striking red appearance, leading to the name "ragged red fibers" to describe this histological condition.

Clinical Case 5–2 Mitochondrial Depletion Syndrome

Parker is a 20-year-old woman who presented to a neurologist over concerns about her balance and her school performance. She had been healthy until age 17 when she started noticing slight disturbances in her balance. Last year, she had a sudden episode of uncontrollable jerking movements of her right arm and upper body. One month later, this reoccurred and progressed to a generalized tonic-clonic seizure. She was treated with a benzodiazepine and valproic acid, and seizures have not recurred. However, she has since developed a tremor and worsening imbalance, and she states that she gets confused more often than previously. She has also been performing poorly in college, whereas she was getting good grades before. On examination, deep tendon reflexes were absent. She had occasional muscle jerking and ataxia that was worse with her eyes closed. Muscle strength was otherwise normal. Her eye movements were normal without nystagmus. Laboratory tests were notable for AST and ALT elevations and mildly elevated bilirubin. Serum glucose was normal. Serum creatine kinase level (a marker of muscle breakdown) was elevated. Nerve conduction studies were performed and showed sensory neuropathy and normal motor nerve responses. Electroencephalogram and brain MRI were normal. Lumber puncture was performed and showed an elevated protein level without any other abnormalities. Biopsy of the quadriceps muscle was performed and did not show any abnormalities. Whole-exome sequencing was performed using funds from a research grant and demonstrated homozygous mutation in the POLG gene (A467T). Parker was given a diagnosis of myoclonic epilepsy, myopathy, and sensory ataxia syndrome (MEMSA) and was started on dietary supplements. She was taken off valproic acid and started on lamotrigine.

Cell Biology, Diagnosis, and Treatment of Mitochondrial Depletion Syndromes

Parker has a MEMSA, which in this case is caused by a homozygous POLG mutation leading to a mitochondrial DNA depletion syndrome. Mitochondrial depletion syndromes are a heterogeneous group of disorders characterized by a reduction in the amount and copy numbers of mitochondrial DNA. As mitochondrial DNA encodes proteins essential for oxidative respiration, decreasing the amount of mitochondrial DNA leads to impairment of energy production in tissues and organs. These disorders are ultimately caused by defects in the nuclear DNA that encodes for various proteins that are responsible for maintenance, replication, or synthesis of mitochondrial DNA.

POLG encodes the catalytic subunit of DNA polymerase gamma, which is the DNA polymerase in mitochondria. Without functioning DNA polymerase G, the mitochondrial DNA is unable to replicate efficiently. Because mitochondrial DNA needs to replicate to make up for mtDNA turnover and cell division, the mutation leads to a decrease in mtDNA copy number. Mitochondrial DNA is required for the production of key subunits of the respiratory chain, and decrease in mtDNA copy number leads to insufficient amount of respiratory chain components to meet the energy needs of the tissues. Other nuclear genes leading to mitochondrial DNA depletion also exist (e.g., TK2, SUCLA2, and TYMP) and may present with different clinical pictures.

Mitochondrial depletion syndromes are generally very rare, may present in a wide variety of ways, and can be broadly characterized by the main organ system involved. Patients often have aberrations in any combination of their nervous system, muscular system, and gastrointestinal systems (generally liver). While most MDS present in early childhood, some can present in late adolescence or adulthood. POLG mutations have been associated with a wide spectrum of clinical presentations, including disorders of the eye muscles, seizure syndromes, and hepatic damage. Valproic acid is often prescribed for the treatment of seizures, but this must be undertaken with care, as it can also worsen hepatic symptoms and hasten progression to hepatic failure.

The differential diagnosis for mitochondrial DNA depletion syndromes is very broad and should be undertaken by a multidisciplinary team. With the rise in the availability of genetic testing using next-generation sequencing (NGS) techniques, querying the patient's genome has become more feasible in cases in which MDDS is suspected.

Continued

Unfortunately, there are no proven effective treatments for any MDS, and all are ultimately progressive. Most treatment is geared to symptom control and supportive care. In cases of hepatic failure, liver transplant has been attempted but is generally not successful due to the involvement of other organ systems. Wherever possible, valproic acid should be avoided, and adequate control of seizures should be achieved with other antiepileptics. Dietary supplements with antioxidants have been hypothesized as a mechanism to decrease oxidative stress in the mitochondria and therefore decrease mtDNA destruction, and a number of experimental therapies are undergoing clinical trials at the current time.

mitochondrion-specific DNA polymerase (*POLG*), and genes required for the synthesis of dNTPs such as thymidine kinase 2 (*TK2*) and deoxyguanosine kinase (*DGUOK*). The prognosis for patients with these diseases varies depending on the particular mutation and tissues involved. For instance, *TK2* mutations affect primarily the muscular system with a rapid progression, resulting in death in early childhood. Mutations in *DKUOK* and *POLG*, on the other hand, often target the liver.

Defects in Mitochondrial Dynamics

Genes with roles in mitochondrial fusion/fission dynamics are gaining attention as a source of mitochondrial disease. Mutations in such genes lead to abnormally fragmented or hyperfused mitochondria. For instance, mutations in *OPA1*, the product of which is expressed in the inner mitochondrial membrane and is required for mitochondrial fusion, lead to highly fragmented mitochondria and damage to optic nerves. Mutations in the solute carrier family gene *SLC25A46*, which plays a role in mitochondrial fission, cause hyperfused mitochondria. Hyperfused mitochondria not only compromise cristae and membrane structure but also are often difficult to traffic to synapses in elongated neurons, leading to optic nerve damage and neurodegenerative effects similar to those observed for mutations in mitochondrial fusion genes.

Aminoglycoside-Induced Hearing Loss

Aminoglycoside antibiotics, including well-known drugs such as neomycin and streptomycin, have been known for decades to lead to permanent hearing loss in some patients while leaving other patients unaffected. This variability in drug response has been linked to variants in two mtDNA genes, *MT-RNR1* and *MT-CO1*, with the m.1555A>G variant of *MT-RNR1* showing the strongest association with susceptibility. The reason for this drug-induced hearing loss appears to be related to the bacterial origins of mitochondria. It appears that, in affected individuals, aminoglycosides—which are designed to target bacterial ribosomes—also target ribosomes in the mitochondria of cochlear cells, killing the cells through the inhibition of mitochondrial translation.

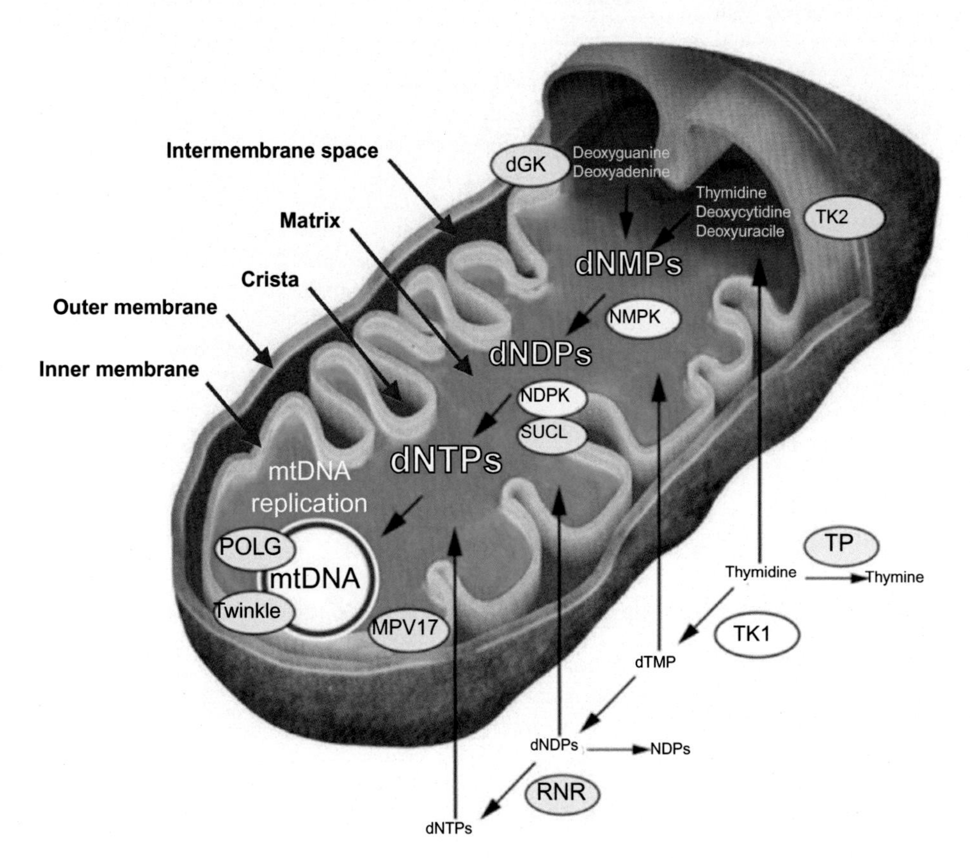

Figure 5–8. **Summary of genes and pathways involved in mitochondrial DNA depletion syndrome.** The major structures of the mitochondria have been marked with *red arrows*, while *black arrows* have been used to indicate the pathways involved in mitochondrial DNA depletion syndrome. Mutations causing mitochondrial DNA depletion can result from mutations in genes that affect the supply of mitochondrial dNTPs necessary to replicate mtDNA (e.g., *TK2* and *DGUOK*) or from mutations in genes physically required to replicate mtDNA such as the polymerase *POLG* and the helicase *TWINKLE*. Either class of defect creates a severe restriction in the ability of the patient's cells to replicate mtDNA over time, leading to a reduction in mtDNA content within the cells and a concurrent reduction in oxidative phosphorylation and mitochondrial metabolism. *(Modified from Brockhage R, et al.* J Genet Genomics *2018;45:333–335.)*

Variants that create this susceptibility, such as the m.1555A>G variant of *MT-RNR1*, alter the mitochondrial ribosome in such a way that makes it a stronger target for aminoglycoside drugs. This example suggests that the mitochondrial genome may underlie other instances of variable drug responses and should be considered in future pharmacogenomics studies.

Mitochondria and Common Diseases

There is likely to be a great deal of concordance between the mechanisms that cause rare primary mitochondrial diseases and more common diseases in which mitochondrial impairment plays a contributory role. Most obviously the progressive damage induced by mitochondrial ROS byproducts may underlie (at least in part) age-related cellular senescence—including the accumulation of mutations in somatic tissues and the buildup of abnormal protein aggregates, like those seen in AD. There is also a sizable portion of the population that carries heteroplasmic mutations at a very low frequency at the tissue or organ levels. Low-frequency heteroplasmies are usually assumed to be asymptomatic and ignored in genetic diagnoses. However, it cannot be ruled out that low-frequency heteroplasmies lead to dysfunction at the cellular levels. In addition, over the course of a lifetime, mutations may accumulate that accelerate aging or increase the risk of mitochondrial-involved diseases like diabetes or PD. Hence, it makes a great deal of sense to pursue studies of primary mitochondrial disorders because the results may provide important insights into how to prevent or slow down the progression of these diseases.

Somatic mtDNA Mutations and Aging

There is a well-established correlation between advancing age and a decline in mitochondrial function, which likely contributes to age-related senescence and geriatric disease. Decline in mitochondrial function is likely caused in large part by the gradual accumulation of somatic mtDNA mutations due to ROS damage and DNA replication errors. In turn, this accumulation of mtDNA mutations promotes ROS production and establishes a vicious cycle, accelerating the aging process. While a serious mtDNA mutation can be acquired early in life (thus leading to disease at a young age), for most people, it will take several decades to acquire one or more disease-causing mutations and have them reach a high enough level of heteroplasmy to cause serious health issues. This gradual accumulation of mutations and increase in heteroplasmy may help to explain the time-dependent decline in function that occurs with age.

Somatically acquired mtDNA mutations can occur anywhere in the mtDNA and may even include large deletions or duplications. Deletion mutations represent a particularly insidious risk because the reduced size of mtDNA molecules carrying large deletions (ΔmtDNAs) appears to give them a replicative advantage over normal mtDNA. This replicative advantage can drive the ΔmtDNA molecules to be represented at progressively higher frequencies over time, as has been observed in sequencing analysis studies of skeletal and cardiac muscle tissue cells. This progressive expansion in ΔmtDNAs may help to explain, perhaps in part, the loss in muscle tone and strength that occurs with aging.

Neurodegenerative Disorders and the Importance of Mitophagy

The recycling of malfunctioning or compromised mitochondria is crucial to the proper operation of the mitochondrial system. If damaged mitochondria are not removed in a timely manner, they disrupt other cellular components, not only by failing to provide sufficient ATP but also by generating ROS due to their compromised respiratory chain function. Thus the orderly destruction and recycling of such mitochondria is crucial for the maintenance of a healthy cell. Indeed, several diseases—particularly neurodegenerative diseases—appear to be caused by a loss of mitophagy.

One of the most important regulators of mitophagy is the PTEN-induced putative kinase 1 (PINK1)/Parkin pathway, wherein PINK1 phosphorylates Parkin, an E3 ubiquitin ligase. Normally, PINK1 is localized to the inner mitochondrial membrane but is translocated to the outer membrane in response to serious mitochondrial damage. This translocation of PINK1 causes Parkin to be recruited to the outer membrane, where it is activated by PINK1 phosphorylation. Parkin, in turn, ubiquitinates a variety of mitochondrial proteins, a chemical modification that marks the proteins for destruction by the ubiquitin proteasome system. Ultimately, this ubiquitination process directs the mitochondria to be subjected to mitophagy, an autophagy-based pathway of destruction.

As the reader may have intuited, the *PARKIN* gene was so named because of its association with the occurrence of Parkinson's disease (PD), perhaps the most well-studied common neurodegenerative disease with mitochondrial involvement. Mutations in *PARKIN* have been shown to lead to an autosomal recessive, juvenile-onset form of PD. Individuals with a family history of PD, especially early onset (<20 years old) PD, often carry mutant *PARKIN* alleles. Spontaneous *PARKIN* mutations are also often observed in isolated, nonfamilial cases of PD, further cementing the model that loss of Parkin function and mitophagy leads to PD. Mitochondrial dysfunction can also be observed in terms of abnormal mitochondrial structure, reduced function of ETC complexes, loss of mitochondrial membrane potential, and other mitochondrial defects in primary cell cultures and dopaminergic neuron samples from patient tissues.

PD is both a neurodegenerative disorder and a mitochondrial disease. The oxidative stress that results from mitochondrial dysfunction also appears to be a major contributor to the development of Alzheimer's disease (AD), the most common form of age-related dementia. This association seems to be a function of the neuronal aggregates of misfolded beta-amyloid and alpha-synuclein in AD interacting directly with the outer mitochondrial membrane, destabilizing it in the process. Such destabilization of the mitochondrial membrane may cause the release of cytochrome *c* and ROS, thus favoring neurodegeneration and cell death. Based on these observations and those described earlier for PD, it is tempting to suggest that mitochondria play a crucial role in most, if not all forms of neurodegenerative disease, though additional investigation is needed on this front.

Diabetes

Diabetes has already been linked to mutations in mitochondrial genes. For instance, the m.3243A > G mutation that causes MELAS has been linked with a superficially distinct form of diabetes known as maternally inherited diabetes and deafness (MIDD), which accounts for ~1% of diabetes diagnoses in the general population. The overt mitochondrial form of the disease and the "diabetes" form are merely different phenotypic manifestations of the same underlying genetic disorder. In addition, all primary mitochondrial diseases tend to have a high risk for diabetes comorbidity. Thus there may be some wisdom in viewing primary mitochondrial diseases and common diseases in which mitochondrial dysfunction plays a contributing role as points along a continuum.

Cancer and the Warburg Effect

Cancer cells appear to rely almost exclusively on anaerobic respiration (i.e., glycolysis) for their metabolism, eschewing oxidative phosphorylation entirely. This observation was first formalized by Otto Warburg in the mid-20th century and has been named the "Warburg effect" in his honor. It is likely that this metabolic change provides a number of advantages to the cancer cell, including a reduction in ROS levels from the ETC, so long as there is an ample supply of glucose in circulating blood and a sufficient blood supply to the tumor.

Defects in Biosynthesis

Mutations in genes involved in mitochondrial biosynthesis can also cause severe diseases. For instance, the steroidogenic acute regulatory protein is required for the transfer of cholesterol within mitochondria, a process required for steroid hormone biosynthesis. Thus mutations in this gene lead to a variety of severe congenital and/or developmental disorders, such as lipoid congenital adrenal hyperplasia and glucocorticoid deficiency 1.

Mutations that disrupt Fe-S cluster synthesis can also cause an abnormal intracellular buildup of iron, leading to high ROS and oxidative stress levels. Over time the accumulated effects of iron-induced toxicity can devastate afflicted tissues, damaging both nuclear and mitochondrial genomes and compromising a variety of cellular functions. Friedrich's ataxia is a well-studied iron-related pathology, being the most common form of genetically inherited ataxia. It is caused by the expansion of a trinucleotide repeat in *FXN*, the gene that encodes frataxin, with the age of onset and phenotypic severity being inversely correlated with the length of the expansion. Particularly interesting to the mitochondrial biologist, frataxin protein is localized to mitochondria, where it is required for the synthesis of both Fe-S clusters and heme, along with other critical functions. Thus a major consequence of losing frataxin activity is a loss of function for all Fe-S cluster proteins, along with an abnormal buildup of iron in mitochondria, which causes damage through oxidative stress. Both effects are likely to be major causes of the nerve damage underlying this progressive form of ataxia.

Mutations in another component of the Fe-S cluster synthetic pathway, ferredoxin reductase (encoded by *FDXR*), have also been found to cause cases of peripheral neuropathy and optic atrophy. The elucidation of these *FDXR*-associated diseases underscores the importance of proper regulation of iron homeostasis by mitochondrial proteins. Given these findings, it seems plausible that other genes involved in iron metabolism may be linked to mitochondrial disease.

TREATMENTS FOR MITOCHONDRIAL DISEASE

Current Treatment Strategies

Validated options for treating mitochondrial diseases are quite limited, and chemical, biological, or genetic manipulation of mitochondria is difficult. Treatments are often designed to reduce or ameliorate symptoms, rather than to cure underlying pathologies. The first and simplest intervention is to prescribe an increased exercise regimen, which can induce increased mitochondrial biogenesis and other metabolic changes that can increase mitochondrial function and efficiency. Unfortunately, fatigue, muscle weakness, and exercise intolerance are common features of mitochondrial disorders, limiting the physical activity of many patients. Furthermore, patients with mitochondrial diseases often have thermal regulation problems that make it difficult for them to partake in outdoor activities when the weather is too warm or too cold.

Careful monitoring of diet is another major consideration in treating mitochondrial disease symptoms. Fasting should be avoided, and small, frequent meals are better for some patients than a standard diet of three meals per day. Alcohol and tobacco use are strongly discouraged because they can damage mtDNA and increase oxidative stress, both of which can exacerbate the progression of mitochondrial diseases. Excess iron should also be avoided in these patients, due to its tendency to produce free radicals that can further compromise mitochondrial structures or damage mtDNA. High fat, ketogenic diets have been proposed as a possible intervention for mitochondrial disease, particularly in disorders involving otherwise untreatable bouts of epilepsy. However, there are conflicting data on this proposition, and more research is required to determine the circumstances where a ketogenic diet can be employed safely.

Even the most rigorous monitoring of diet and physical activity cannot resolve most cases of mitochondrial disease completely. For this reason, most patients afflicted with mitochondrial dysfunction are prescribed a "mitochondrial cocktail" of supplements and vitamins (e.g., riboflavin, thiamine, and L-carnitine) intended to reduce the severity of their symptoms. These mitochondrial cocktails can address mitochondrial impairments at multiple levels. Antioxidants such as vitamins E and C, for instance, help absorb excess free radicals produced by malfunctioning mitochondria, thereby reducing the collateral ROS-mediated damage to mtDNA, proteins, and lipids. Supplementation of coenzyme Q_{10} (CoQ_{10}), which serves as an electron carrier between Complex I and Complex II of the ETC, also appears to improve mitochondrial function by modulating ROS formation while restoring some lost ETC function. The amino acids L-arginine and L-citrulline have been shown to relieve the stroke-like symptoms of MELAS, likely by relieving patients' nitric oxide deficiencies, since L-arginine and L-citrulline are the substrates for nitric oxide synthase. Studies of the efficacy and risks of these supplements are ongoing. Ultimately, while mitochondrial cocktails can produce some clinical improvements in some patients, but not others, the ideal composition of supplements is unknown and will likely vary depending on the particular mutation and disease in question.

Future Treatment Strategies

Due to the limited efficacy of current treatments, there are ongoing efforts in the mitochondrial biology field to develop new treatments for mitochondrial diseases. Some of the more promising treatments currently being investigated as alternative interventions for mitochondrial disorders are described later.

Mitochondrial Replacement Therapy

The repair of mtDNA mutations in somatic tissues is a nontrivial undertaking, requiring the use of viral vectors or other radical (and sometimes risky) measures to deliver a nuclease or other repair agent to all of the cells. When possible, it would be much better to correct defective mtDNA in the newly fertilized zygote (i.e., germline therapy), where there is only a single cell that needs to be repaired. One particularly powerful way to prevent mitochondrial disease in the offspring of a mitochondrial mutation carrier mother is to remove her mitochondria from the embryo entirely. Although this may sound like a ludicrous proposition at first glance, it has been demonstrated in animal models and in in vitro fertilized human embryos. There are several mitochondrial replacement techniques, but all require the use of a healthy donor to provide oocytes with wild-type, nonmutant mtDNA that can then be combined with the recipient's nuclear genome.

The simplest and oldest mitochondrial replacement method is cytoplasmic transfer, wherein a portion of the cytoplasm from a donor oocyte is added to a recipient's oocyte to lower the percentage of mutant mtDNA in the latter oocyte. While crudely effective, this method leaves substantial mutant mtDNA in the embryo, which can lead to situations where the mutant heteroplasmy can return to disease-causing levels as the fetus/child develops or otherwise create conflicts with the nuclear genome. For this reason, two alternative mitochondrial replacement therapy (MRT) techniques have been developed to entirely replace the recipient's mutant mtDNA: oocyte spindle transfer and pronuclear transfer. Both methods involve replacing the native nuclear DNA from the donor oocyte with the nuclear genome of the patient carrying mtDNA mutation. In oocyte spindle transfer the replacement occurs prior to fertilization, while in pronuclear transfer it is performed after both the patient and donor embryos have been fertilized. The offspring that results from either of these techniques is referred to as a "three-parent baby" because such a child possesses DNA from three different individuals (nuclear DNA from the father and patient mother and mtDNA from the egg donor). The oocyte spindle transfer method has been employed successfully, producing a live birth in a case of maternal mitochondrial disease (Fig. 5–9). The resultant child is a healthy boy who remains free of disease symptoms at nearly 2 years of age.

Despite the apparent success of this recent attempt at MRT, there remains a great deal of controversy about the long-term safety and ethics of the technique. Concerns are related largely to the fact that, although MRT almost completely eliminates the mutant mtDNA, there is still a small amount of cytoplasm with mutant mtDNA carried over with the mother's nuclear material. The tiny amount of mutant mtDNA in those mitochondria, likely only a few percentage points of the total mtDNA content, should not create significant issues in most cases. However, there is a small chance that the mutant mtDNA could outcompete the wild-type mtDNA in its replicative ability, allowing the level of mutant mtDNA heteroplasmy to increase over time and restore the original disease. Alternatively the healthy mtDNA from the donor may have negative interactions with the nuclear genome of the mother, particularly if the donor mtDNA evolved in a widely divergent haplogroup. A growing body of data suggest that mitochondrial genomes evolve in a tight, symbiotic relationship with their host nuclear genome, and unanticipated issues may be created by

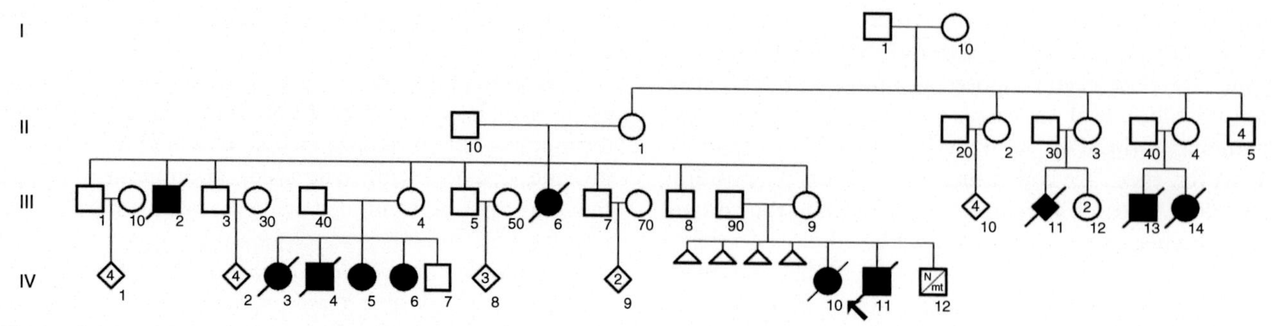

Figure 5–9. **Pedigree from a mitochondrial disease case treated by mitochondrial replacement therapy.** The mother referred for treatment in this case (labeled as "III-9") had previously experienced four miscarriages and two children who died early in childhood. Sequencing analysis revealed the mother to be a carrier of the Leigh syndrome mutation (mtDNA mutation 8993T > G) at 23.27%–33.65% heteroplasmy, depending on the tissue analyzed. Oocyte spindle transfer method was used to replace the defective mtDNA, resulting in the live birth of a health boy (IV-12). To separately indicate the status of the mitochondrial and nuclear genomes in this child, his box on the pedigree has been divided into two sections: "N" and "mt," representing the nuclear and mitochondrial genomes, respectively. *(Image originally from Zhang J, et al.* Reprod Biomed Online *2017;34(4):361–368.)*

mismatches. Thus careful experimentation in cell culture and animal models is needed to clarify if and under what circumstances residual mutant mtDNAs are likely to create serious issues. Those caveats notwithstanding the fact that the United Kingdom granted regulatory approval to MRT in 2016 are an encouraging sign for this burgeoning field of research.

Genome Editing

In many cases where reproductive assistance is required to treat mtDNA mutations, it is still a very complicated process technically and ethically. In addition, it relies on a favorable donor–patient match to prevent reemergence of the maternal mutation. Therefore additional genetic manipulation will likely need to be explored to repair the mutant allele directly. Recent revolutionary advancements in genome manipulation technologies provide a plausible mechanism for doing exactly that, although serious technical and ethical hurdles remain to be solved on this front.

Enzymes that enable selective editing of genome sequences have existed for decades in the form of restriction endonucleases (REs)—enzymes derived from bacteria or yeast for defense against exogenous viral DNA. REs target and cut specific DNA sequences—usually contiguous, palindromic sequences of 4–8 nucleotides—with exceptional specificity and efficiency. The protein sequences of REs can be altered to carry mitochondrial targeting signals, allowing them direct physical access to the mitochondrial genome when expressed transgenically by way of viral vectors or other means. Preliminary work in cell culture and animal models indicates that they can be used to target and cut mutant mtDNAs in vivo. The RE-linearized mutant mtDNA is subsequently destroyed by unrelated mitochondrial enzymes whose job it is to eliminate broken or damaged DNA, eliminating them from the organism while leaving wild-type DNA largely untouched. Assuming the appropriate viral vector and promoters are chosen, these mitochondrially localized REs should eliminate targeted alleles effectively with little chance of off-target damage to the rest of the mtDNA and no chance for nuclear genome damage owing to the mitochondrial localization of the REs.

Because the set of DNA sequences targeted by currently identified REs is, unfortunately, quite limited, most mutant mtDNA alleles will not have a cognate RE that can differentiate their sequences from the corresponding wild-type alleles. For these cases, custom nucleases will need to be designed to edit or remove the mutant mtDNA sequences. Fortunately, there are several options in the modern geneticist's toolbox for doing exactly that. For instance, transcription activator-like effector nucleases (TALENs) are a class of custom nucleases whose DNA-binding domain is composed of interchangeable repeats (transcription activator-like effectors or TALEs) that bind to a single, specific nucleotide. Practically, any nucleotide sequence can be targeted by mixing and matching these repeats in different arrangements. A similar technology is based on the use of zinc-finger nucleases, wherein DNA-binding zinc-finger repeats derived from zinc-finger proteins are combined with the RE *Fok*I. Both zinc-finger nucleases and TALENS are similar to REs with respect to their specificity and in that they can be fused to mitochondrial localization signals, but they have lower overall nuclease activity levels than REs. Regardless, both methods have been shown to eliminate mtDNA mutant alleles reliably in cultured cells and animal models, and both are promising and active lines of research.

For situations where neither REs, TALENs, nor zinc-finger nucleases can be employed to target a desired nucleotide sequence, there is hope that CRISPR-Cas9 genome editing may be used. Like REs, this system was originally derived from a prokaryotic viral defense system, but with a twist: namely, target sequences are determined by an RNA nucleic acid—called the guide RNA or gRNA—that cooperates with the Cas9 nuclease to determine where DNA cleavage will occur. Although a particular sequence (the PAM sequence) must be present on the guide RNA, this system allows for nearly total freedom in the choice of a target sequence. Unfortunately, although the Cas9 protein can be directed to the mitochondrial matrix with the addition of a protein localization signal, there is currently no simple method for directing gRNA molecules to the same location, creating a major technical challenge. Thus most attempts to employ CRISPR-Cas9 to degrade or alter mutant mtDNA sequences have failed. One proposed workaround would be to fuse the gRNA covalently to another RNA sequence that is imported into mitochondria; this approach has been productive in some experimental contexts. It remains to be determined whether the same strategy will work in vivo for editing or eliminating mutant mtDNA variants.

Ongoing Clinical Trials

There are a large number of clinical trials underway testing a wide variety of pharmacological treatments for mitochondrial disorders. Many of the treatments being tested are highly speculative, but a few have shown a great deal of success in clinical trials and are poised to form the basis of the next generation of mitochondrial drugs. Several promising drugs in development are described later.

Drugs to improve ETC functions. Of all of the compounds tested in the clinical trials to date, the antioxidant compound idebenone has shown the most promise at improving ETC activity. Idebenone can facilitate the transfer of electrons from Complex I to Complex III when those portions of the ETC are disrupted. Idebenone has been approved for the treatment of LHON in the European Union, though it has yet to be approved by the US FDA. A version of CoQ_{10}, which can facilitate ETC functions while absorbing excess ROS, has also been chemically altered to target the mitochondria with a covalently attached lipophilic moiety. This modified compound, called MitoQ, is the focus of a great deal of study aiming to demonstrate that its mitochondrial targeting leads to superior efficacy over normal CoQ_{10}.

Iron overload and chelation therapies. In many mitochondrial disorders the production of Fe-S clusters by the cytochrome P450 system is reduced or abolished. In addition to the loss of function for proteins that rely on Fe-S clusters for their activities, there is an ancillary issue created by the buildup of iron when these pathways fail. Because the redox properties of iron make it a powerful ROS generator when gathered in any significant amount, this iron buildup can itself wreak havoc on cells. Excess iron is receiving increasing attention as a pathogenic mechanism, particularly for PD. Rendering this iron chemically and biologically unavailable with chelators, such as deferiprone, represents a promising potential therapy approach. Several trials are underway to investigate the therapeutic potential of iron chelators in fighting neurodegenerative disorders.

Drugs that stabilize the mitochondrial membrane. One particularly promising pharmaceutical agent is the artificial tetrapeptide elamipretide. This drug, which is transported with high efficiency to the inner mitochondrial membrane, appears to have at least two mechanisms by which it helps to alleviate the negative effects of mitochondrial dysfunction. First, it interacts with and helps stabilize the structurally important mitochondrial membrane lipid cardiolipin, which is often destabilized in mitochondrial disorders. Secondly, elamipretide is chemically disposed to absorb free radicals and ROS. Thus, once localized to the inner mitochondrial membrane, elamipretide is perfectly positioned to absorb excess ROS produced by a dysfunctional ETC. Because of its multiple levels of function, elamipretide has shown promising across multiple studies and is currently in phase 3 trials to treat patients with primary mitochondrial myopathy. It is likely that its ultimate utility will extend far beyond primary mitochondrial myopathy to other mitochondrial disorders.

SUMMARY

In this chapter, we have provided the novice with a solid foundation in the biological operation of the mitochondrion. We have delineated the major categories of mitochondrial disease, namely, primary mitochondrial disorders and common diseases with mitochondrial involvement. We also summarized the current state of mitochondrial treatments and predicted likely future research directions. Our aim was to lead the reader to a deeper understanding of the critical issues in mitochondrial medicine, including the role of subtler forms of mitochondrial dysfunction in nominally "nonmitochondrial" diseases.

Suggested Readings

Giles RE, Blanc H, Cann HM, Wallace DC. Maternal inheritance of human mitochondrial DNA. *Proc Natl Acad Sci* USA. 1980;77(11):6715–6719.

Goto Y, Nonaka I, Horai S. A mutation in the tRNA(Leu)(UUR) gene associated with the MELAS subgroup of mitochondrial encephalomyopathies. *Nature* 1990;348(6302):651–653.

Li Z, Peng Y, Hufnagel RB, Hu YC, Zhao C, Queme LF, et al. Loss of SLC25A46 causes neurodegeneration by affecting mitochondrial dynamics and energy production in mice. *Hum Mol Genet* 2017;26(19):3776–3791.

Sheng ZH, Cai Q. Mitochondrial transport in neurons: impact on synaptic homeostasis and neurodegeneration. *Nat Rev Neurosci* 2012;13(2):77–93.

Taylor RW, Turnbull DM. Mitochondrial DNA mutations in human disease. *Nat Rev Genet* 2005;6(5):389–402.

Wallace DC. A mitochondrial paradigm of metabolic and degenerative diseases, aging, and cancer: a dawn for evolutionary medicine. *Annu Rev Genet* 2005;39:359–407.

Warburg O. On the origin of cancer cells. *Science* 1956;123(3191):309–314.

Wickner W, Schekman R. Protein translocation across biological membranes. *Science* 2005;310(5753):1452–1456.

Youle RJ, van der Bliek AM. Mitochondrial fission, fusion, and stress. *Science* 2012;337(6098):1062–1065.

Zhang J, Liu H, Luo S, Lu Z, Chavez-Badiola A, Liu Z, et al. Live birth derived from oocyte spindle transfer to prevent mitochondrial disease. *Reprod Biomed Online* 2017;34(4):361–368.

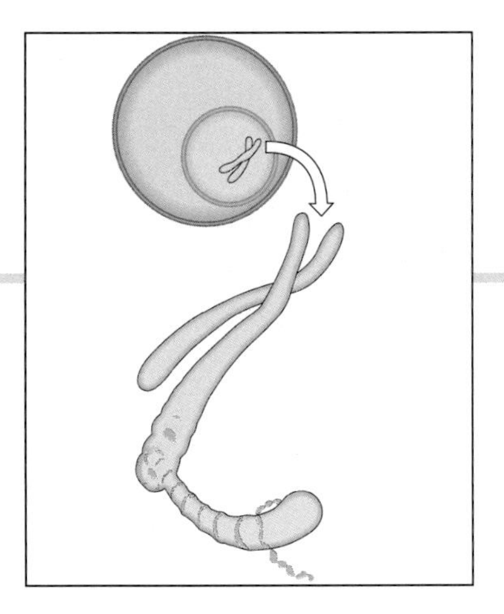

Chapter 6

Regulation of Gene Expression

CELL NUCLEUS

Nucleus Structure

The Nucleus Is Bounded by a Specialized Membrane Complex, the Nuclear Envelope

The most conspicuous organelle within a eukaryotic cell is the nucleus. It is the sequestering of nearly all the cellular DNA in the nucleus that marks the major difference between eukaryotic and prokaryotic cells. Nuclei are generally spherical and are bounded by a special membrane system, the **nuclear envelope**, which defines the nuclear compartment. The nuclear envelope is formed from two distinct lipid bilayers (Fig. 6–1).

The **inner nuclear membrane** is in close contact with a meshwork of intermediate filaments, the nuclear lamina, which provides support for this lipid bilayer. In addition, this membrane contains proteins that provide contact sites for chromosomes and nuclear ribonucleic acids (RNA), either directly or through proteins of the nuclear matrix. The **outer nuclear membrane** is contiguous with the membrane of the endoplasmic reticulum (ER). Ribosomes that are actively synthesizing transmembrane proteins are often observed associated with the outer nuclear membrane. The outer nuclear membrane is a specialized region of the ER. Proteins synthesized on ribosomes associated with the outer nuclear membrane are either destined for the inner or outer membranes or translocated across the membrane into the region between the inner and outer nuclear membranes, termed the **perinuclear space**.

Nuclear Pores Allow Communication Between the Nucleus and Cytosol

Inside the nucleus are all the components of the genetic apparatus. This includes **deoxyribonucleic acid (DNA)**, **ribonucleic acid (RNA)**, and nuclear proteins to organize and provide for nuclear function (Fig. 6–2).

The nuclear proteins, which include structural proteins of the matrix and lamina, RNA and DNA polymerases, and gene regulatory proteins, are synthesized in the cytoplasm and brought into the nucleus. Thus nuclear proteins must pass the double-membrane barrier of the nuclear envelope. The transport of materials to and from the nucleus is facilitated by "holes" in the nuclear envelope called **nuclear pores.**

The interior of the nucleus and the cytoplasm of the cell maintain contact or communication through the nuclear pores. In electron micrographs the pores appear as highly organized disklike structures surrounding a central hole or cavity. Nuclear pore structure is conferred by a set of protein granule subunits, which are arranged in an octagonal pattern and form the boundaries of the pore complex. These eight protein subunits consist of radial

Goodman's Medical Cell Biology. https://doi.org/10.1016/B978-0-12-817927-7.00006-5

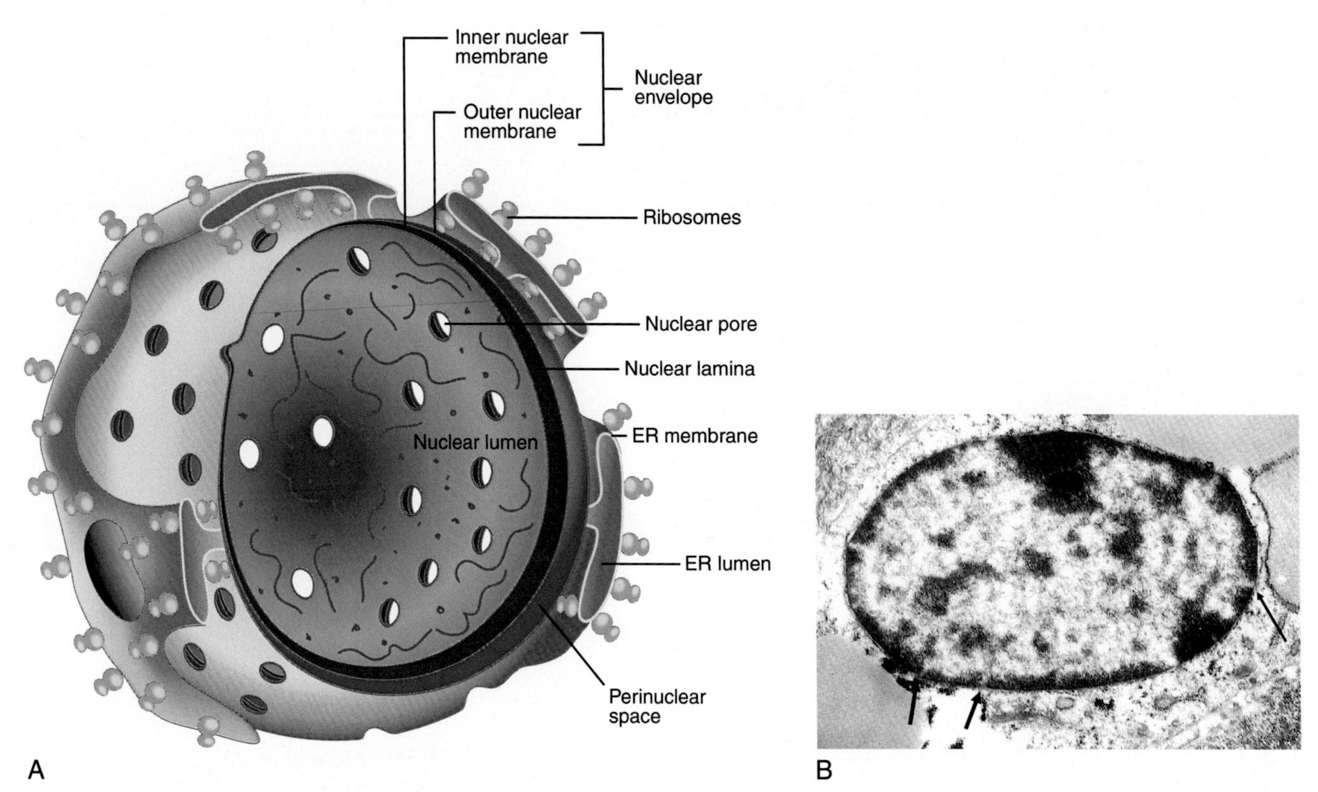

Figure 6–1. **Relation of the nuclear envelope with cellular structures.** (A) Diagram shows the double-membrane envelope that surrounds the nuclear compartment. The inner nuclear membrane is lined by the fibrous protein meshwork of the nuclear lamina. The outer nuclear membrane is contiguous with the membrane of the endoplasmic reticulum (ER). As illustrated the outer nuclear membrane often has ribosomes associated with it that are actively synthesizing proteins that first enter the region between the inner and outer nuclear membranes, the perinuclear space, which is contiguous with the lumen of the ER. The double membrane of the nuclear envelope is perforated with holes or channels of the nuclear pores. (B) Electron micrograph of a nucleus from a luteal cell. *Thick arrowheads* denote the inner and outer nuclear membranes of the nuclear envelope, which contains the nuclear pores *(thin arrow)*. *(Courtesy of Dr. W. Zimmer, Texas A&M College of Medicine.)*

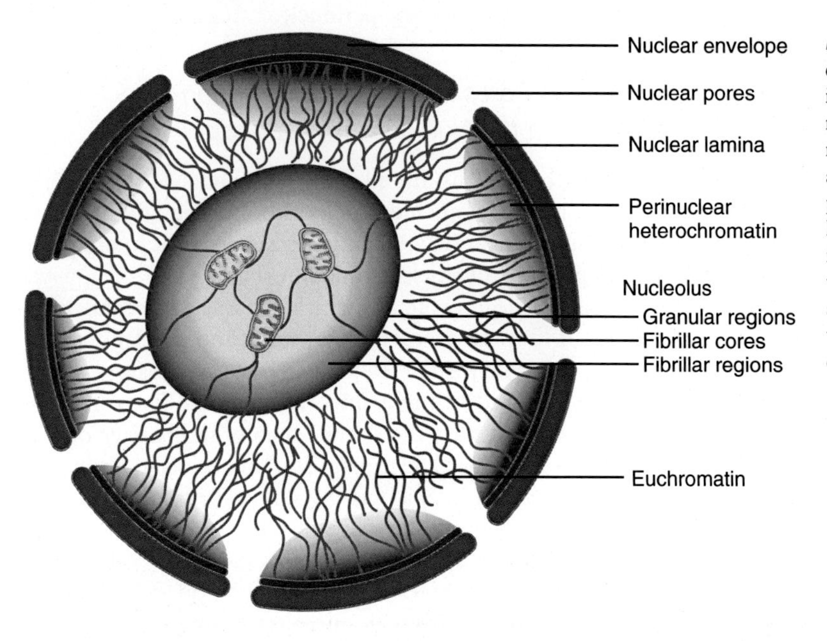

Figure 6–2. **Schematic model of a typical eukaryotic interphase nucleus.** The organization of the internal nuclear compartment is shown. The inner nuclear membrane is in direct contact with the protein network of the nuclear lamina, which is often associated with a highly condensed DNA protein complex, referred to as the perinucleolar heterochromatin. Most of the nuclear compartment is filled with noncondensed DNA-protein complexes, the euchromatin. The most obvious structure within an interphase nucleus is the dark-staining nucleolus, which can be further divided into fibrillar and granular components.

arm segments joined together by a set of proteins referred to as **spokes**, which traverse the pore membrane. Sophisticated electron microscopy and genetic ablation studies have indicated the presence of transporter subunits that line the central channel of the pore and, as detailed later, forms the barrier between the nucleus and cytoplasm. On the cytoplasmic and nucleoplasmic faces of the outer and inner nuclear membranes are ring structures formed from eight bipartite subunits that are in apparent contact with the spoke proteins and the membrane phospholipids. These are important to the nuclear pore complex, not only for overall structure but also because they anchor filaments that extend into the cytoplasm and nuclear compartments. These nuclear pore ring-attached filaments allow for direct coupling of the nuclear compartment with the cell cytoplasm through interaction with cytoskeletal and nucleoskeletal filaments. In addition, the nuclear pore ring filaments appear to participate in the recognition of molecules that need to be transported through the nuclear pore. Thus the nuclear pore contains an intricate composition of proteins that form this important structure (Fig. 6–3).

The pore complex penetrates the double membrane of the nuclear envelope, bringing together the lipid bilayers of the inner and outer nuclear membranes at the boundaries of each pore. Although this would appear to allow exchange of components (i.e., proteins and phospholipids) between these two membranes, evidence now indicates that these two membranes remain chemically distinct. Therefore the protein components of the nuclear pore complex must provide a barrier preventing bulk exchange between these two membranes.

Measurements of nuclear pore complexes have demonstrated that they are highly organized structures with an outside diameter of approximately 100 nm and an internal channel 9–10 nm in diameter. The properties of transport through the nuclear pores have been addressed by injection of radiolabeled compounds into the cytosol and examination of the rate of their appearance into the nucleus. Such experiments have demonstrated that the nuclear pores are freely permeable to ions and small molecules, including proteins with a diameter smaller than 9 nm (equivalent to 60 kDa or less relative molecular mass [M_r]). Nonnuclear proteins larger than 9 nm in diameter (greater than 60 kDa) are excluded from nuclear transit. However, nuclear resident proteins that are synthesized in the cytoplasm and are larger than 60 kDa are readily transported into the nucleus, indicating that there must be mechanisms for the selective transport of the molecules across the nuclear envelope.

Selective transport of large molecules and complexes across the nuclear envelope occurs through the nuclear pore by a receptor-mediated process. The key experiments illustrating this concept have made use of the protein nucleoplasmin, a 165-kDA M_r pentameric protein found in high concentrations in frog oocyte nuclei. When purified nucleoplasmin is injected into the cytosol, it accumulates into the nucleus at a rate greater than can be explained by simple diffusion, showing that the protein was concentrated into the nucleus by a selective uptake mechanism. Furthermore, by separating the nucleoplasmin protein into fragments, it was shown that the ability to selectively import the entire 165-kDA protein was conferred by a small domain of the protein.

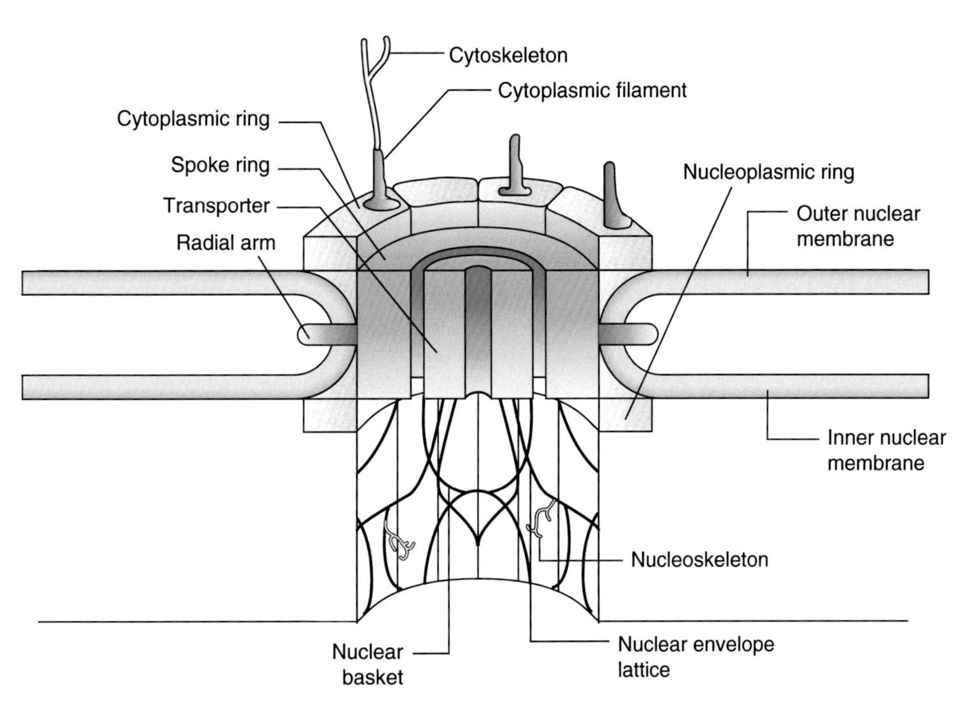

Figure 6–3. **Diagram of nuclear pore complex that allows communication of the cell cytoplasm with the internal nuclear compartment.** The diagram shows the proteins that comprise the ~100-nm, octagon disk-shaped nuclear pore complex. The pore complex is anchored into the nuclear envelope (outer nuclear membrane and inner nuclear membrane) by the radial arms and spoke rings. Subunits that make up the transporter are just inside the spoke ring and form the aqueous channel of the nuclear pore complex. Adjacent to these structures are the cytoplasmic and nucleoplasmic rings that anchor the connections of cytoplasmic and nucleoplasmic filaments (basket). It is through these filament structures that potential connections to the cytoskeleton *(outside)* and nuclear matrix *(inside)* are made, adding a physical connection between the nuclear compartment and the rest of the cell. *(Modified from Goldberg M, Allen T.* Curr Opin Cell Biol *1995;7:301–307, by permission.)*

That is, the domain contained amino acids forming a signal that marked the nucleoplasmin as a nuclear resident protein. Such signals are called **nuclear localization sequences/signals** (NLS).

In addition, the nucleoplasmin nuclear localization signal could be linked to other proteins and thus mark them for import into the nuclear compartment; this is true even for proteins that are normally never found in the nucleus. The selective transport of material into the nucleus occurs only when energy-generating molecules [guanosine triphosphate (GTP)] are present. These early studies conclusively showed that nucleoplasmin contains a domain that functions as a signal sequence for nuclear localization of the protein, and this transport is an energy-dependent, specific transport through the nuclear pores.

The mechanism of nuclear localization signal-mediated protein import appears to encompass a variety of proteins, both soluble in the cytoplasm and located on the nuclear pore complex, that work in a multistep pathway. First a protein that has a nuclear transport signal, or NLS, binds a receptor complex. This is a multisubunit receptor, soluble within the cell cytoplasm, which functions to dock the protein to be transported with filaments extending from the cytoplasmic ring of the nuclear pore complex. After the docking, more proteins associate with the complex; most important are GTPases and their activating proteins, which provide the energy of the protein translocation by the hydrolysis of GTP. Not all of the proteins that form the translocation complex go through the pore to the nucleoplasm, and those factors that do are apparently recycled by some mechanism to the cell cytoplasm for further rounds of nuclear transport. A model for nuclear import, based on current data, is shown in Figure 6–4.

One mechanism for recycling factors is to use them to export materials out of the nucleus into the cytoplasm. The export cycle (see Fig. 6–4) is similar to import, except that cargo destined for export would contain a signal marking it for transport to the cytoplasm. Such cargo exported out of the nucleus would be messenger RNA (mRNA), ribosomal RNA (rRNA), and ribosomes and perhaps proteins that shuttle to the cytoplasm. These nucleic acid and proteins associate with factors containing an export signal, generally a short domain of leucine-rich, hydrophobic amino acids. Although many factors that participate in this process have been examined, there are likely many more that have not yet been identified.

Nuclear transport signal sequences have been identified from a variety of nuclear-targeted proteins. Table 6–1 lists nuclear resident protein for which an import signal sequence has been identified. Each identified sequence is a small region of the total protein (about four to eight amino acids in length), and most are basic; however, no apparent consensus of sequences exists, either in primary structure or in any location within the different proteins. The same nuclear pore complexes are responsible for the transport of RNA out of the nucleus to the cytosol. Transport out of the nucleus is selective and requires a receptor in the nuclear pore complex that binds with the exported protein-RNA complex. It is now clear that the complexes to be exported contain an export signal sequence. Thus it would appear likely that the nuclear pore complex contains multiple receptor sites that recognize a variety of signal sequences on protein complexes to be transported across the nuclear envelope. Whether this is a property of a single- or multiple-subunit protein within the pore complex remains to be elucidated.

In summary the nuclear pore is an important channel of communication between the interior of the nuclear compartment and the cytoplasm of the cell. It displays properties of a molecular filter, in that ions and small molecules are freely permeable through the aqueous channel, whereas larger protein complexes are selectively transported through the pore. This selective transport is highly specific because of the presence of amino acid signal sequences within the protein complexes that cause them to bind with receptor-like proteins on the pore complexes. Then an energy-dependent process is responsible for the translocation across the nuclear envelope.

The Structure of the Nucleus Is Determined by Proteins of the Nuclear Lamina and the Nuclear Matrix

Lining the inner surface of the nuclear envelope in interphase cells is a protein meshwork, the nuclear lamina. This meshwork forms an electron-dense layer 30–100 nm in thickness that provides connections between the inner nuclear membrane and perinuclear chromatin. When examined by electron microscopy, the lamina appears as a square latticework built from filaments that are about 10 nm in diameter. These filaments are classified as intermediate filaments and are composed of three extrinsic membrane proteins, called **lamins A, B,** and **C**, which have M_r of 60–70 kDA. The mRNA for lamins A and C are formed from alternately spliced transcripts of the same gene and encode identical proteins, except that lamin A contains a COOH-terminal extension of 133 amino acids. Lamin B is encoded from an mRNA that is synthesized by a gene distinct from the lamin A/C gene.

Isolated lamins have a rodlike structure approximately 52 nm in length and a globular head domain. Similar to other intermediate filament proteins, the formation of the long filaments is mediated by the globular head domain. Lamin B is different from lamins A and C in that it is posttranslationally modified by the addition of an isoprenyl group, which allows membrane lipid attachment. The inner nuclear membrane contains a receptor molecule of about 58,000 M_r that binds specifically to lamin B. Multiple receptor proteins

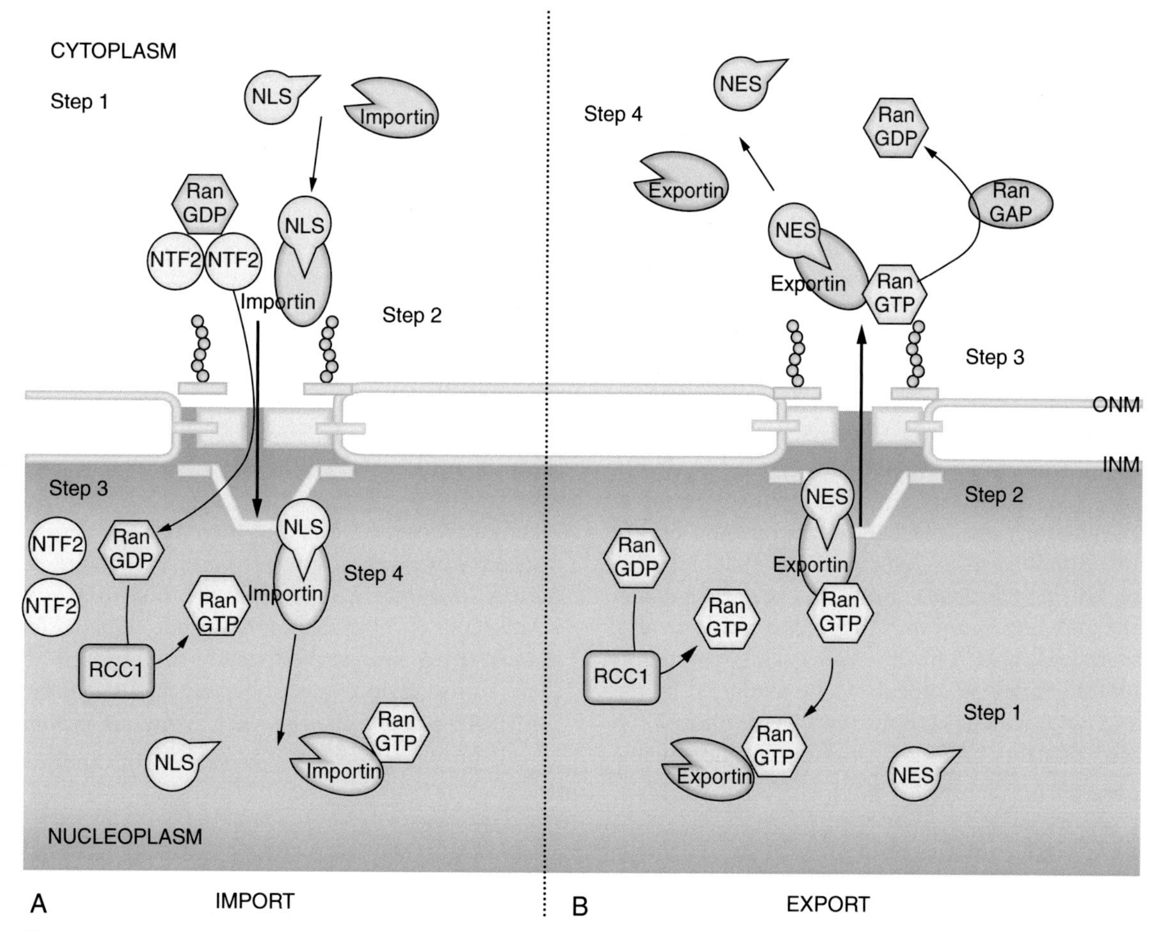

Figure 6–4. **Diagram of the steps necessary for import and export through the nuclear pore complex.** (A) Proteins to be transported to the nucleus demonstrate its nuclear localization sequence (NLS), which is recognized by a cytoplasmic receptor molecule (importin) forming a soluble complex in the cytoplasm, called *step 1.* This binding/recognition leads to a docking of the complex with fibers emanating from the cytoplasmic ring of the pore complex (step 2). This docking recruits accessory proteins, the Ran-GTPases and NTF2, to promote interaction between the transported complex and the control transporter channel and the Ran-GTPases that catalyze the hydrolysis of GTP to provide for transport (step 3). The hydrolysis of energy, GTP to GDP, allows for movement through the pore (step 4). (B) Export from the nucleus follows a similar plan. The exported cargo demonstrates a nuclear export signal recognized by a receptor called *exportin* (step 1), after which the complex becomes associated with the pore complex (step 2). This docking recruits the energy-generating Ran-GTPases to provide the energy (step 3) to export the cargo into the cytoplasm of the cell (step 4). Thus import and export through the nuclear pores follow similar pathways.

TABLE 6–1. Nuclear Import Signal Sequences Derived From Various Nuclear Resident Proteins

Protein	Location of Signal Sequence	Amino Acids of Signal Sequence
SV40-large T antigen	Internal: residues 126–132	Pro-Lys-Lys-Lys-Arg-Lys-Val
Influenza virus nucleoprotein	COOH-terminus: residues 336–345	Ala-Ala-Phe-Glu-Asp-Leu-Arg-Val-Leu-Ser
Yeast mat α 2	NH_2-terminus: residues 3–7	Lys-Ile-Pro-Ile-Lys
Yeast ribosomal protein L3	NH_2-terminus: residues 18–24	Pro-Arg-Lys-Arg

Data from Dingwall C, Laskey R. *Annu Rev Cell Biol* 1986;2:367–390; Lyons RH, Ferguson B, Rosenberg M. *Mol Cell Biol* 1987;7:2451–2456; Newmeyer DD, Forbes DJ. *Cell* 1988;52:641–653; Christophe D, et al. *Cell Signal* 2000;12:337–341.

are on the inner nuclear membrane, such as the nesprin protein family, which facilitates binding of the nuclear lamina to the membrane. The nesprin proteins also contain actin-binding domains and thus may be transducers of signals from the cytoskeleton to the nuclear skeleton. Lamins A and C then interact with lamin B, which mediates interactions with the lamina and chromatin. Therefore, in interphase cells, all three lamin proteins are found adjacent to the inner nuclear membrane, forming the nuclear lamina complex.

As the chromatin condenses in the prophase stage of mitosis, there is an apparent disappearance of the nuclear membranes and the nuclear lamina. Figure 6–5 summarizes a model to explain the participation of the nuclear lamins in the breakdown and reformation of the nuclear envelope during the cell cycle.

Examination of cells by electron microscopy shows that during prophase the nuclear membrane fragments into smaller vesicles that remain associated with the ER. Lamin B is found tightly coupled with these vesicles, whereas lamins A and C are depolymerized and are found throughout the cell. These depolymerization and subsequent breakdown of nuclear membrane are thought to be mediated by phosphorylation of the lamins by a lamin kinase, p34/cdc2, and other downstream kinases. When the lamins are phosphorylated, they depolymerize, and the nuclear membrane breaks down into small vesicles, whereas the chromosomes condense. During telophase of the cell cycle, the nuclear membrane and associated structures reassemble around the separated daughter chromosomes. This reassembly of the nuclear envelope appears to be mediated by the lamins and is coincident with the removal of phosphates from these proteins. The reformation of the nuclear membrane closely follows the decondensation of the daughter chromosomes. As the chromatin becomes dispersed, it apparently induces the dephosphorylation of the lamins, allowing them to polymerize, which in turn causes the small vesicles associated with lamin B to fuse and form a normal interphase nuclear membrane. Although the interactions that lead to reassembly of the nuclear membrane are not yet clearly defined, it is thought that the phosphatase responsible for removal of phosphates from the lamins is tightly associated with the chromatin, possibly a component of the internal nuclear structure called the **nuclear matrix.**

Although defined structurally and biochemically, the function of the nuclear matrix is unclear. Perhaps, its most obvious role would be to provide organization and structure to the internal nuclear compartment. Newly replicated DNA and the enzymatic components necessary for DNA synthesis are associated with the matrix, suggesting a role in the organization of the DNA replication machinery. Recent evidence indicates that actively transcribed genes and the products of their transcription [e.g., heterogeneous nuclear RNA (hnRNA)] are enriched in nuclear matrix preparations. Localization of RNA transcripts in the nucleus using

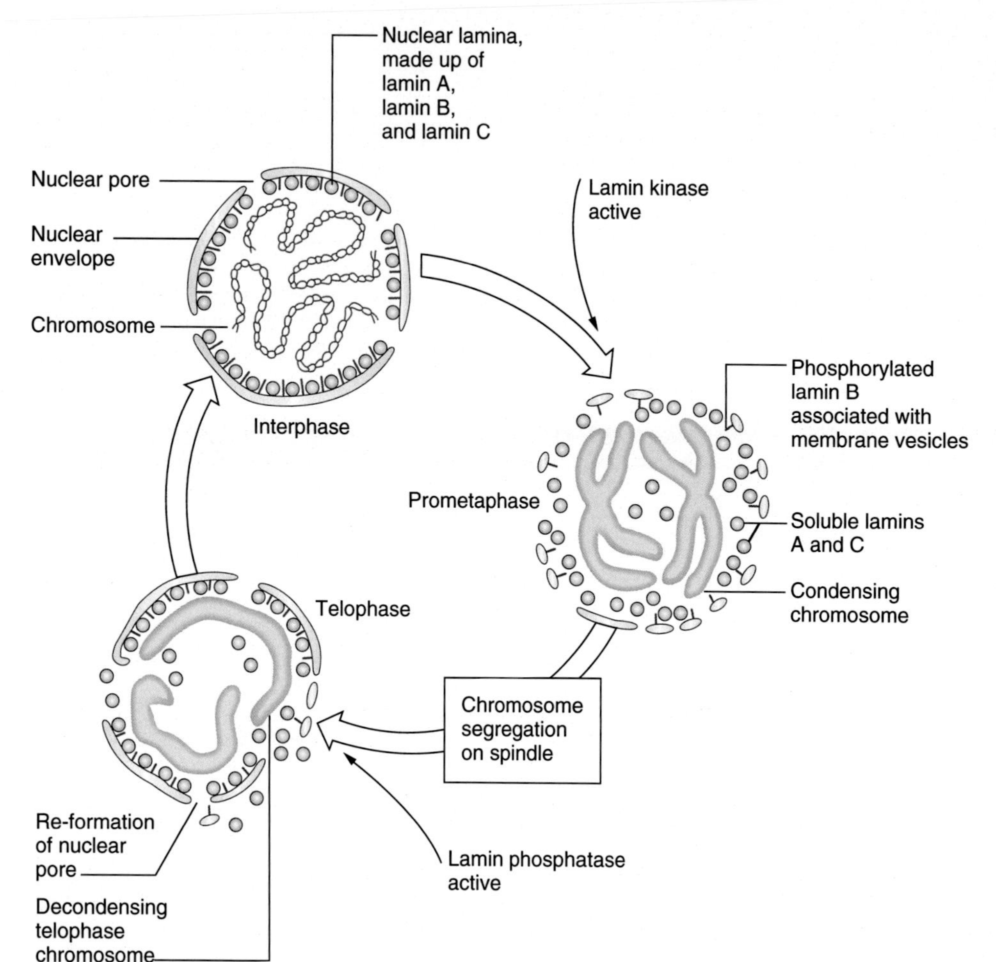

Figure 6–5. **Correlation of lamin protein phosphorylation and nuclear envelope structure during mitosis.** As cells proceed from interphase to prophase of the cell cycle, there is a condensation of the chromosomes and a breakdown of the nuclear envelope. Nuclear envelope breakdown is concomitant with the activation of lamin kinase, which phosphorylates the lamin proteins (A, B, and C) and causes the depolymerization of the nuclear lamina matrix. Lamin B remains associated with remnant membrane vesicles when phosphorylated, whereas lamins A and C are dispersed within the prometaphase cell. Coincident with the decondensation of chromosomes in daughter cells *(telophase)*, the phosphates are removed from the lamin proteins by an activated lamin phosphatase. This allows the polymerization of the nuclear lamina, using the membrane-bound lamin B as a nucleation site and the formation of a nuclear envelope.

fluorescent-labeled nucleic acid probes has shown that the RNA follows tracks in the nuclear compartment, with a more intense fluorescent signal seen near the nuclear borders. Thus, after transcription, RNA does not diffuse within the nucleoplasm, but is possibly bound to nuclear matrix fibers as they are spliced to form mature RNA bound for the cytoplasm. These experiments support the concept that the nuclear matrix plays an important role in the organization of the nuclear compartment. In addition, many components of the actin cytoskeleton, including actin itself, have been recently observed in the nuclear compartment. This emerging view of the nuclear compartment is one of a highly organized compartment, and this organization is fundamental to efficient operation of nuclear function. We must also remember that receptors on the nuclear envelope contact the cellular cytoskeleton, the nuclear lamina, and likely the nuclear matrix. Thus cellular organization extends inside important cellular organelles/compartments and aids in the efficient functioning of the cell, such as providing vital links from outside the cell through the cytoplasm and to the nuclear compartment. This allows the cell to adapt quickly to a changing environment and survive many of the challenges it must confront.

We have discussed how the nuclear and cytosolic compartments are in contact through an elaborate set of interconnecting filaments (Fig. 6–3). Recent experiment has defined biochemically a set of proteins that interconnect the nuclear and cytoplasmic compartments throughout the nuclear membrane boundary. There have been a number of proteins residing in the outer nuclear membrane and appear to be important in various nuclear functions, including the reformation of the nuclear envelope during the cell cycle as described earlier. Using the *Caenorhabditis elegans* model, researchers have documented that an outer nuclear membrane protein, called Anc-1, contacts cytoplasmic cytoskeletal elements and participates in the actin-dependent nuclear positioning in the cell. Using genetics, it was demonstrated that nuclear positioning in the cell is also dependent upon a second protein, called Unc-84, which is resident in the inner nuclear membrane. Recent experiments demonstrated two important points: First the Anc-1 and Unc-84 proteins contacted each other in the space between the inner and outer nuclear membranes, which suggests that they are both necessary for actin-based nuclear tethering of the nuclear compartment, and second the Unc-84 protein was shown to make specific contacts with lamin proteins lining the inner nuclear membrane. Thus it is now recognized that there are connections between the inner nuclear and cytoplasmic compartments via connections of proteins that collectively are called the **linker of nucleoskeleton and cytoskeleton** (LINC) protein complex.

The LINC protein complex provides an important mechanism for the cell to adjust to forces that change the cellular environment. It has been well established that cells under mechanical stress will often remodel their actin cytoskeleton (discussed in Chapter 3). The rearrangement of the actin fibers in the cell is transduced into the nuclear compartment via the changing interactions of the remodeled actin cytoskeleton and the LINC. It has been recently shown that disruption of LINC-actin interactions will result in an aberrant differentiation of mesenchymal stem cells. Thus appropriate control of events in the nuclear compartment can be influenced by signals acting upon elements of the cytoskeleton. These interactions may become important as targets for therapeutics in disease states.

Nuclear Function

The Genome of the Cell Is Sequestered in the Nuclear Compartment

The nucleus contains almost all the genetic information of the cell in the form of DNA. DNA is composed of four nucleotides: Two are purines that have a double-ring structure (adenine and guanine), and two are pyrimidines that have a single-ring structure (thymine and cytosine). The basic structure of DNA, derived in 1953 by Watson and Crick, is that of two polynucleotide chains that are held together by hydrogen bonds between adenine and thymine (A-T base pairing) and guanosine and cytosine (G-C base pairing). The two chains are antiparallel or complementary and are coiled into a double helix approximately 2 nm in diameter. The nucleotides are arranged in a nonrandom fashion, such that the genetic information is contained in the specific linear arrangement of bases. The information is stored in "words" consisting of three nucleotides, termed the **codons**, of the genetic code. A model of DNA as genetic information and its relation to chromosomes stored in a eukaryotic nucleus is presented in Figure 6–6.

In eukaryotic cells, each DNA molecule is packaged into linearly arranged units, termed the **chromosomes**, and the total genetic information stored within the chromosomes is referred to as the **genome** of the organism. The human genome contains about 3×10^9 nucleotide pairs that are packaged into 24 separate chromosomes (22 autosomes and 2 different sex-determinant chromosomes). In diploid somatic cells, there are two copies of each chromosome present—one inherited from the mother and one from the father, except for the sex chromosomes in male individuals, in which the Y chromosome is from the father and X chromosome from the mother. Thus a diploid human cell contains 46 chromosomes and approximately 6×10^9 nucleotide pairs of DNA. For these chromosomes to remain as discrete functional units, they must possess the ability to replicate, separate into daughter cells at mitosis, and maintain their integrity between cell generations. Experiments examining chromosomal architecture and function have defined

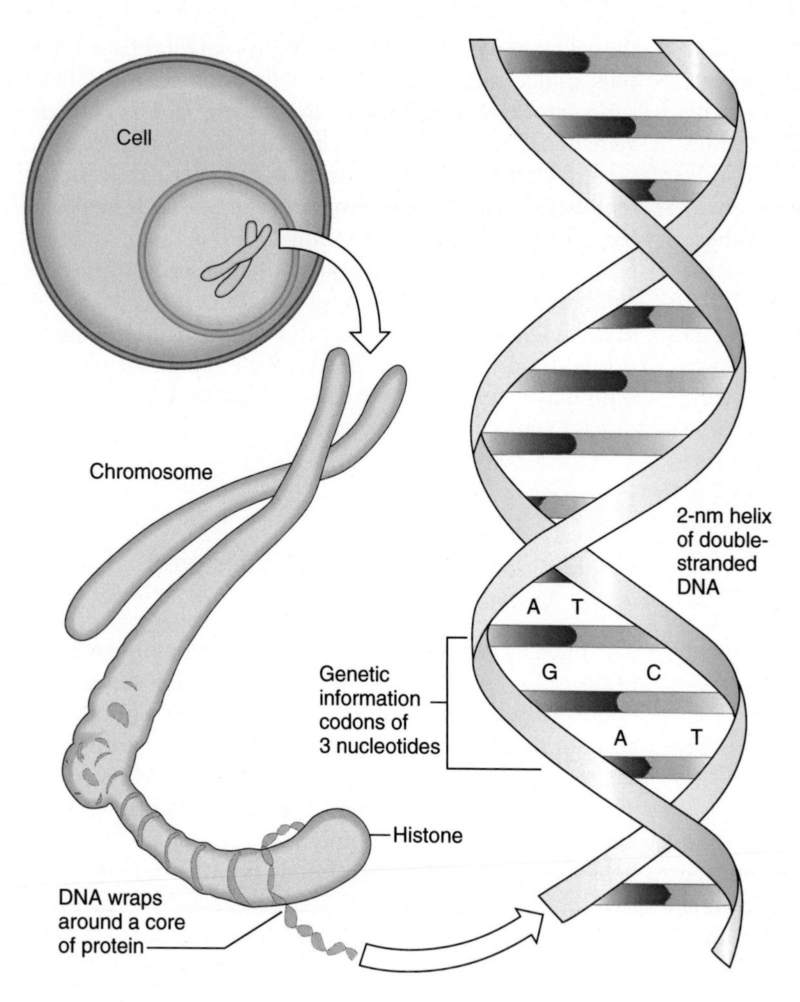

Figure 6–6. **Relation of molecular details of the genetic code stored in DNA and chromosomes within the nucleus of the cell.** At the right is a model of DNA in which two antiparallel strands (one is 5′ to 3′ top to bottom and the other is the reverse) are held together by pairing of nucleotide bases. It is the strict arrangements of the nucleotides, in codons of three bases, that are the storage of genetic information. As indicated the 2-nm helix of DNA is associated with protein, forming the chromatin fibers of individual chromosomes that are housed in the nucleus of eukaryotic cells.

three domains or elements necessary for the maintenance and propagation of individual chromosomal units (Fig. 6–7).

To replicate, the DNA contains specific regions that function as the focal point for the initiation of DNA synthesis, termed the **DNA replication origin**. Each chromosome contains many replication origins dispersed throughout its length, which become activated in an asynchronous fashion during the S phase of the cell cycle. These are specific nucleotide sequences at which DNA synthesis begins, although not all of the origins are active at the same time. This suggests that there must be heterogeneity among these sequences, allowing the ordered replication of the chromosome. DNA replication and repair are important processes for cellular maintenance and in disease (see later for more detailed discussion).

A second sequence element that is responsible for attachment of the chromosome to the mitotic spindle during M phase of the cell cycle is termed the **centromere**. Each chromosome contains one centromere region in which the DNA interacts with a complex set of proteins, forming a structure called the **kinetochore**. This structure is responsible for the segregation of the chromosome into daughter cells on cell division.

The third sequence element that is required for maintenance of chromosomal structure is located at termini of the linear chromosome and is called the **telomere**. This is a specialized sequence that defines the ends of the linear chromosome and is built from sequence repeats enriched in guanosine and cytosine bases. These sequences are replicated by telomerase, which folds the guanosine-rich DNA strand to form a special structure that protects the end of the chromosome. When the telomere is not replicated correctly, there can be a shortening of the chromosome. This shortening has been correlated with diseases such as cancer. Moreover, chromosomal shortening is also found in the aged. Thus researchers are making a concerted effort to find compounds to enhance telomere maintenance.

DNA REPLICATION AND REPAIR ARE CRITICAL NUCLEAR FUNCTIONS

Replication of DNA Occurs During the Synthetic (S) Phase of the Cell Cycle

The ability of cells to divide and multiply is critical for normal growth and development of the organism. In addition, there are tissues in adult organisms that

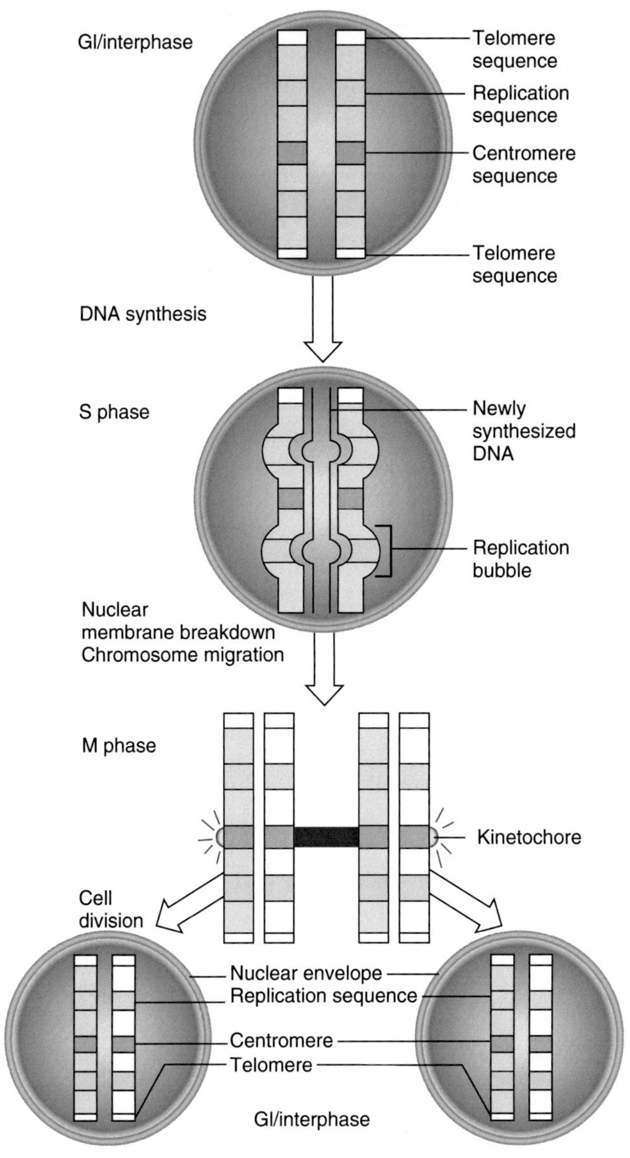

Figure 6–7. **Sequence elements of chromosomes that are necessary for maintenance of structure and propagation.** Three sequence elements are needed to maintain chromosomes as individual units in nuclei of eukaryotic cells. First is the telomere sequence, which caps the ends of the chromosomes, keeping degradative enzymes from attacking the units during interphase or G phase of the cell cycle. For duplication, chromosomes contain numerous origins of replication that serve as points of initiation for DNA synthesis during the synthetic (S) phase of the cell cycle. After S phase the nuclear envelope breaks down, and the chromosomes are segregated into the daughter cells by use of the kinetochore, which forms at the region of chromosomal constriction called the centromere DNA element.

undergo constant, rapid cell renewal, such as blood and digestive tract epithelium, which indicates that cell multiplication is a vital process. Cellular growth occurs through a highly regulated process called the **cell cycle,** which results in the creation of two equal "daughter cells." During the cell cycle, each component of the cell must be replicated so that the resultant cells can provide their function to the organism. As discussed earlier, one of the important processes is the replication of the cell's genome and maintenance of the chromosome. This chapter discusses the mechanisms of DNA replication and repair; the regulation of the cell cycle is examined in Chapter 10.

DNA replication occurs in the part of the cell cycle called the S phase, so named because it is the synthesis of DNA that is ongoing during this time. DNA synthesis begins at specific locations along the chromosome called the **replication origins.** These are distinct sequences; however, there does not appear to be a singular origin sequence in higher eukaryotes. In contrast, bacteria, viruses, and yeast have very well-defined nucleotide sequences that form their replication origins. Replication origins in human cells are sequence enriched in A-T base pairs, likely because it takes less energy to pull apart A-T pairs. Thus the separation of the DNA strands so they can be copied would be facilitated at origins containing A-T bases.

The replication origins are recognized by a group of protein factors that form the **initiation complex.** It is interesting that not all replication origins are initiated at one time; indeed, there are groups of replication origins that initiate DNA synthesis at different times during S phase. In general, genes that are actively transcribed are replicated during the first part of the S phase and DNA that is not actively transcribed is replicated later. This suggests two concepts that are actively being pursued experimentally: (1) Transcription and replication may share factors that link the two processes, and (2) because no two cell types express the exact same genes, there may be heterogeneity in the replication initiation factors that help govern chromosomal replication. Heterogeneous complexes exist for initiating replication and for recognizing damaged DNA for repair, and mutations found in these factors form the basis of disease processes.

Once the origin has been identified by the initiation complex factors, the parental DNA strands are separated, and each is used as a template for the synthesis of new DNA. This synthesis is catalyzed by the enzyme DNA polymerase. In eukaryotes, four DNA polymerases have been recognized, termed polymerases α, β, γ, and δ. DNA polymerase α is important for replication, whereas DNA polymerase β primarily functions in DNA repair. Polymerase γ is enriched in mitochondria and is involved with replication and maintaining the mitochondrial genome. DNA polymerase δ plays a role in both replication and repair, primarily elongating DNA strands.

All DNA polymerases copy the DNA template from **3′ to 5′** and produce a newly synthesized strand in the **5′ to 3′** direction. Because polymerases all work in a singular direction, the replication of DNA must involve the actions of many proteins. DNA replication occurs from the origin in a bidirectional fashion. This means it begins at a specific location (as indicated in Fig. 6–8); then the process moves simultaneously in both directions.

Because there are many origins on the eukaryotic chromosome, the bidirectional movement ensures replication of the entire chromosome is accomplished in a

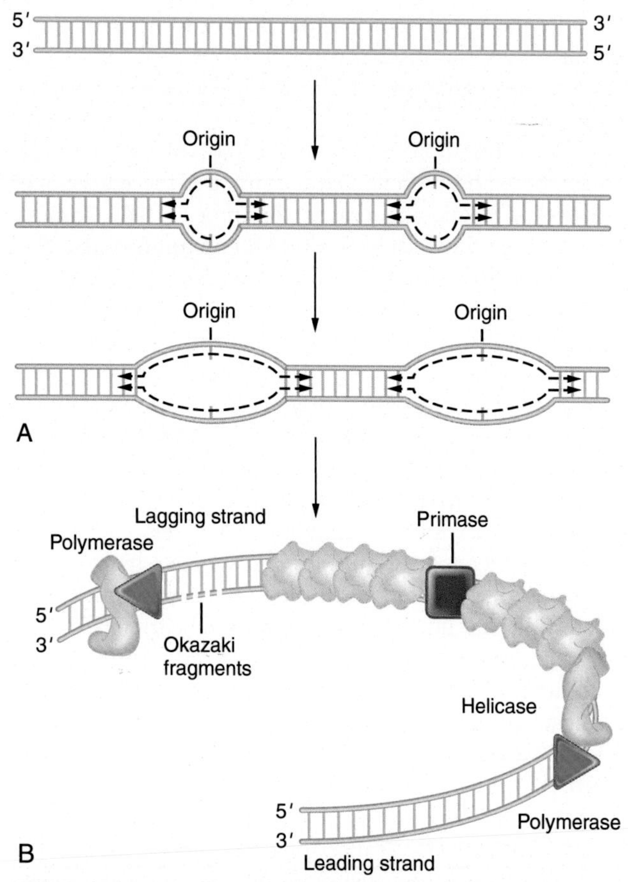

Figure 6–8. **Schematic of events leading to chromosome replication.** (A) The major events of DNA replication are shown, beginning with two separate origins of replication. After recognition of the replication origins, the synthesis of DNA proceeds in both directions from the initiation complexes and proceeds down the DNA molecule. Once synthesis complexes meet, the DNA strands are ligated or joined together, forming a daughter DNA strand composed of one parental strand and one newly synthesized strand. (B) Some of the proteins that are responsible for DNA synthesis from a replication origin. Because DNA polymerases synthesize DNA only from 5′ to 3′, there are two sets of reactions occurring as synthesis occurs away from a replicating origin. On one side the leading strand, there is synthesis, 5′ to 3′, using a singular polymerase complex. On the other strand, there is the participation of many proteins. This includes single-strand binding proteins keeping the DNA strands apart, a primase activity to "prime" DNA synthesis using an RNA strand, and the synthesis from these RNA strands creating the smaller DNA fragments called the *Okazaki fragments*. Once synthesized, these smaller fragments are ligated together, forming the newly synthesized DNA strand.

relatively short time frame. As shown in Figure 6–8, once synthesized, each daughter molecule contains one intact parental strand and one newly synthesized strand, which are joined by appropriate base pairing interactions. Thus the resultant chromosomes are replicated by what is called a **semiconservative process.** The "bubble" of DNA formed at each origin is termed the **replication fork.** The replication fork is where active synthesis is ongoing and requires the coordinated efforts of multiple proteins. A class of protein called helicases unwinds the parental DNA strands, which are then kept separated by the actions of single-strand binding proteins and a second group of proteins called **topoisomerases.** With these protein complexes the DNA strands are maintained separately, and each is then copied by the DNA polymerases.

DNA is synthesized in both directions from the replication fork, and the simplest mechanism to accomplish this would be binding a polymerase to each strand and synthesizing the DNA strands. This would require the synthesis of DNA in two directions: 5′ to 3′ and 3′ to 5′. However, DNA polymerases synthesize DNA only in the 5′ to 3′ direction; thus mechanisms to synthesize the genome must take the strandedness of DNA synthesis into account. Using radioactive DNA precursors, researchers showed that the synthesis of the new DNA strands near the origin was asymmetric and not equal.

Moreover, these experiments showed that near the replication fork there were short pieces of DNA about 1–200 bases in length. The asymmetry of DNA synthesis is created by the continuous synthesis 5′ to 3′ of a template strand, called the **leading strand,** and the synthesis of short fragments, called **Okazaki fragments,** on the other strand, which is referred to as the lagging strand. Because the Okazaki fragments are also synthesized 5′ to 3′, the direction of nucleotide addition is opposite to the overall growth of the DNA strand. After the synthesis of the short Okazaki fragments, they are joined or ligated together to form a continuous strand of new DNA.

DNA Repair Is a Critical Process of Cell Survival

DNA replication occurs with few errors. This is likely enhanced by having polymerases that synthesize DNA in a singular direction and by having cellular mechanisms that efficiently detect and repair errors. In addition to errors of synthesis, cellular DNA is constantly being challenged by outside forces, such as ultraviolet light, chemicals, and products of cellular reactions such as oxygen radicals. In general the repair of damaged DNA follows three steps: (1) recognition of the damaged or altered DNA, (2) assembling the proteins needed to repair the damage, and (3) repairing the DNA (Fig. 6–9).

It is critical that damaged DNA is repaired accurately, because the propagation of the errors will lead to detrimental effects for the organism. Because there are many ways that DNA can be damaged, several proteins can accurately detect the damage and signal to assemble appropriate repair complexes. It is therefore not surprising that mutants of these detection systems can be a root cause for diseases. Table 6–2 shows a partial list of diseases whose primary cause can be linked to mutants in DNA surveillance complexes (Clinical Case 6–1).

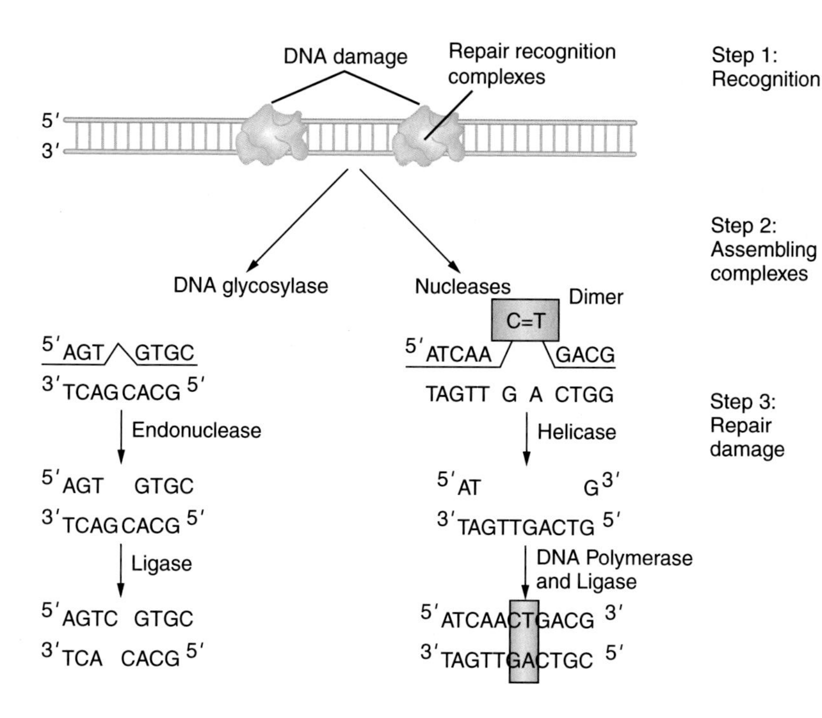

Figure 6–9. **Major events of DNA repair processes in eukaryotic cells.** The major steps in the DNA repair pathway are (1) recognition of the DNA to be repaired, (2) assembling of the appropriate repair complexes, and (3) repair of the DNA. The recognition step requires constant surveillance of the DNA by complexes that can identify mutant or damaged DNA. This means that there are many components involved, which is likely why so many diseases involve mutations in these protein complexes. Recruiting the appropriate repair complexes requires the decision to use either the base excision or nucleotide excision repair pathway. Finally the actual repair requires the polymerases and ligases to "patch" the double-stranded DNA molecule.

TABLE 6–2. Diseases Associated With DNA Repair Defects

Disease	Defect/Name
Skin cancer, ultraviolet light sensitivity, neurologic defects	Nucleotide excision repair/xeroderma pigmentosum (XP)
Colon cancer	Mismatch repair
Leukemia, lymphoma, genome instability	ATM protein kinase/ ataxia-telangiectasia (AT)
Breast cancer	Homologous recombination repair network/BRCA-2
Premature aging, genome instability	DNA helicase/Werner syndrome
Stunted growth, genome instability	DNA helicase/Bloom syndrome
Congenital abnormalities, leukemia, genome instability, cancer at various sites	DNA intrastrand repair/Fanconi anemia

Data from Cleaver JC, Mitchell DL. *Cancer Medicine,* vol. 1. 2003; Mitchell JR, et al. *Can Opin Cell Biol* 2003;15:232–240; Yang Y, et al. *J Neurosci* 2005;25:2522–2529; Proietti DSL, et al. *DNA Repair* 2002;1:209–223; Wood RD, et al. *Science* 2001;291:1284–1289; Sancur A, et al. *Annu Rev Biochem* 2004;73:39–85.

Clinical Case 6–1

Raymond Tanaka is a 9-year-old Japanese American boy who recently moved with his family from Seattle to Los Angeles. After a few months in LA, his mother took him to a dermatologist because even minimal exposure to the Southern California sun seems to cause severe sunburns that don't go away, even after weeks. He also complains of a sore on his tongue that has not healed for a couple of months. His mother mentions that Raymond, though active indoors, rarely played outside in Seattle, especially when the sun was out because he said it irritated him. His mother has always been very vigilant about covering his skin and head and ensures he wears sunscreen whenever going outside. On physical exam, Raymond is a bright, alert 9 years old. His conjunctivae are injected with bloodshot areas overlying a somewhat milky appearance, though his vision is 20/20. His skin is extensively freckled, and he has large confluent areas of leathery, dark patches on his face and arms. He has an irregular weeping lesion of his left wrist and a "punched-out" appearing lesion at the tip of his tongue.

On questioning the mother stated that Raymond has two brothers and a sister, all of whom have never had freckling or sun sensitivity, and that no one in either of their extended families has anything like the freckling that Raymond exhibits.

Despite the boy's protests the dermatologist took three small punch biopsies of his wrist lesions and a thin slice biopsy from his tongue. Two of Raymond's skin biopsies showed actinic keratoses, but the third showed a basal cell carcinoma that was fully contained within the margins of the biopsy. The tongue sample showed an invasive squamous cell carcinoma with local invasion beyond the biopsy site. This required subsequent surgical excision and lip reconstruction.

Multigene testing was performed and revealed a homozygous mutation in the XPA gene.

Cell Biology, Diagnosis, and Treatment of Xeroderma Pigmentosum

Raymond has xeroderma pigmentosum. This autosomal recessive hereditary disorder usually develops in early childhood and is associated with a plethora of sun-induced premalignant and malignant changes in the skin. Affected individuals are about 1000 times more likely to develop

Continued

skin cancers than individuals without the disorder. Actinic keratoses and basal cell carcinomas are common. Squamous cell carcinomas occur less frequently but regularly, and there is a 5% lifetime risk for melanoma for patients with xeroderma, which is significantly higher than in the general population. Corneal clouding occurs in many and probably represents the same epithelial damage in the cornea as in the skin. The disease is relatively common in Japan (1/20,000) but much less common in North American and European populations. Males and females are affected equally.

Patients with xeroderma pigmentosum are exquisitely sensitive to the sun because of inherited inadequate DNA repair mechanisms. Ultraviolet light, which is present in sunlight, generally causes mutations in skin DNA in all individuals. Healthy individuals, however, have intact repair mechanisms, which can remove mutated sections of DNA and replace them with the normal base pairs. To do this, there are specific proteins, called endonucleases, that are serve to excise and replace these mutations. In patients with xeroderma pigmentosum, important endonucleases are either absent or dysfunctional, and therefore routine DNA damage to skin cells cannot be repaired. Exposure to ultraviolet light, especially of the UVB type, which penetrates the epidermis, can produce reactive cyclobutanol dimers and other 6,4-photoproducts that react to form dysfunctional adducts with pyrimidines in skin cell DNA. If the damaged DNA segments involve and inhibit tumor suppression genes such as p53, unbridled proliferation and tumor genesis may occur in the affected cells.

Sun protection and avoidance are critical in patients with xeroderma pigmentosum and can be very difficult to achieve. Patients must constantly use sunscreen and must wear ultraviolet protective goggles at all times. They should be kept away from windows while indoors and fully covered when outdoors. Bare fluorescent light bulbs should be avoided as they emit small amounts of UV light. Vitamin D deficiency may occur due to lack of sunlight, and supplementation is often required. Surgery remains the treatment of choice for skin cancers, though, due to the large number of cancers over a lifetime, excision margins should be relatively narrow as cosmesis ultimately becomes an important issue. Actinic keratosis can be managed with cryotherapy, while photodynamic therapy may be appropriate for superficial skin cancers or to shrink cancers before surgery. To prevent the emergence of skin cancers, systemic retinoids given at high doses can be used. Retinoids likely modulate cell proliferation, differentiation, and apoptosis and therefore decrease the carcinogenic potential of UV-affected cells. Topical chemotherapy such as imiquimod or 5-fluorouracil can be used intermittently (every 3–6 months) indefinitely. Research is ongoing as to the role of immunotherapies, such as PD-1 inhibitors, in both treating and preventing future squamous cell cancers in individuals with XP. Research is also being performed on T4 endonuclease V, a bacterial DNA repair enzyme that can temporarily repair DNA in vitro. Genetic counseling is required, and siblings should be evaluated for being carriers. Metastatic skin cancer is the most common cause of death and often occurs in the fourth decade of life.

After recognition of the DNA lesion, damaged DNA is then repaired by one of two mechanisms. One pathway, called **base excision repair**, relies on the lesioned base to be removed by an enzyme called **DNA glycosylase**, followed by excision of damaged DNA. The second repair pathway, called **nucleotide excision**, depends on a small patch of DNA surrounding the damage being removed as a unit. Subsequent to removing the damage, both pathways then resynthesize the DNA and ligate the repaired DNA into place, completing the repair. Efficient repair systems allow the stability of DNA, ensuring the fidelity of stored information and enhancing the survival of the organism.

We have discussed previously the concept of linking the nuclear compartment with the cytoplasmic compartment of the cell via the LINC protein complex. There is a nucleoskeletal complex of proteins that seem to be responsible for organizing the nucleus in a way that facilitates nuclear functions, including DNA repair. It has been shown that the structural protein α-spectrin (specifically the nonerythroid αSpIIΣ1 isoform) is present in the nuclear compartment and participates in organizing complexes during DNA damage for repair. This is evident in a variety of disease states including agents that induce interstrand cross-links (ICL) caused by carcinogens such as mitomycin C (MMC) and 8-methoxypsoralen (8-MOP). Additionally, αSpIIΣ1 has been shown to localize and organize enzyme complexes at sites of DNA repair in disease states such as Fanconi anemia. It recruits the enzyme complexes necessary to facilitate the repair of nuclear DNA by direct interactions with the repair proteins such as XPF, which creates incisions at sites of DNA ICLs. Additionally, the nuclear lamins are found to participate in the DNA repair process, and mutants of lamin B are found to have reduced repair capacity. Finally, it has been recently demonstrated that other proteins originally considered as strictly cytoplasm resident are localized in the nuclear complexes. These experiments make it very clear that the nuclear compartment is very much organized via cytoskeletal components and this organization is critical for appropriate nuclear functions like gene transcription and regulation and DNA repair. This will be a fruitful area for future research into the ability to develop therapeutics for diseases.

DNA Is Packaged by Nuclear Proteins to Form the Nuclear Chromatin

The DNA within the cell nucleus is associated with a variety of nuclear proteins, and the DNA protein complex is referred to as **chromatin**. Proteins associated with the DNA can be divided into two general categories: the histone and the nonhistone chromosomal proteins. Nonhistone proteins are a heterogeneous class of polypeptides that includes structural proteins [the high-mobility group of proteins (HMG)], regulatory proteins (those that appear to have a direct role in gene regulation; e.g.,

Fos and *Myc*), and enzymes needed for nuclear function (RNA polymerases and DNA polymerases). The histones are found only in eukaryotic cells and are by far the most abundant proteins present in the nucleus. Histones are relatively small proteins that are rich in positively charged amino acids (arginine and lysine), which gives them an overall strongly positive charge (basic) that enables them to bind tightly with the negatively charged (acidic) DNA molecules.

There are five types of histones, designated H1, H2A, H2B, H3, and H4. Four of the histones, H2A, H2B, H3, and H4, are termed the **nucleosomal histones**, because they are responsible for the formation of the inner core of a DNA-protein complex called the **nucleosome**. The nucleosome is the basic unit of chromatin fiber and gives chromatin the bead-on-a-string appearance in electron micrographs. Examination of the structure of histone-DNA chromatin complexes has relied on the digestion of the chromatin with nonspecific nucleases (Fig. 6–10).

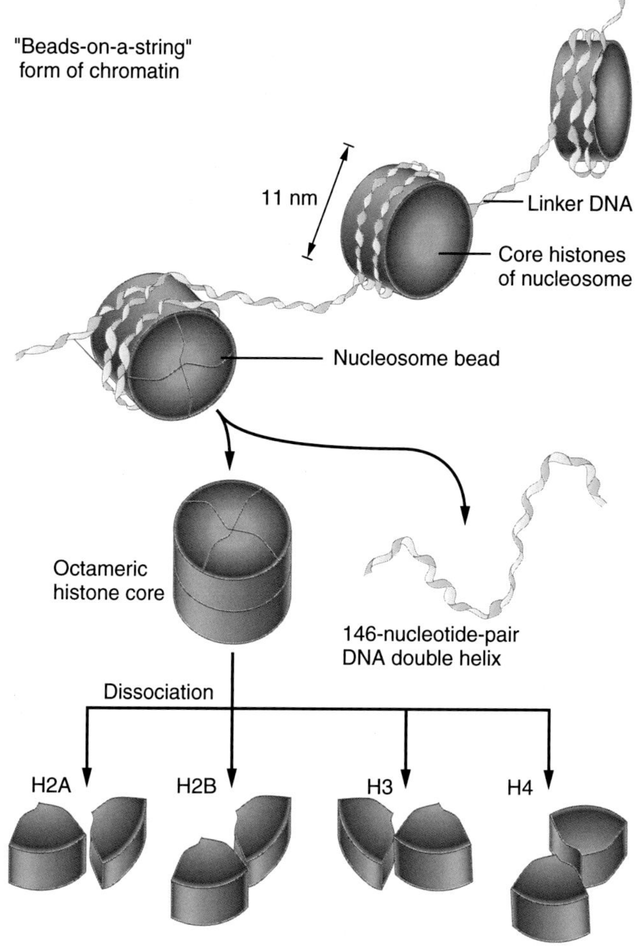

Figure 6–10. **Outline of experiment examining chromatin repeating structure.** The digestion of chromatin with nonspecific nucleases, such as DNase I, results in the release of repetitive units termed the *nucleosome*. This represents the "beads" of the chromatin fiber. Analysis of the components of the nucleosome demonstrates that the bead is made from DNA, a repeating size of 146 base pairs, wrapped around a core of protein. The protein core is made from two molecules each of the core or nucleosomal histones: H2A, H2B, H3, and H4.

These studies have shown that the basic structure of chromatin can be resolved into a repeating unit, called the **nucleosomal bead**. Each nucleosome bead is formed from an octamer of proteins containing two copies of each of the H2A, H2B, H3, and H4 histones, around which is wrapped about 150 nucleotide pairs of DNA. This is the amount of DNA that will make two complete turns around the octamer core of nucleosomal histones, forming a chromatin fiber that is approximately 11 nm in diameter. Because it contains the simplest arrangement of DNA and protein, the 11-nm chromatin fiber is considered the basic unit of chromatin packaging in the nucleus. However, only a small portion of the DNA is found packaged as an 11-nm fiber in an interphase cell and is probably limited to those regions of DNA that are actively transcribing gene sequences. When nuclei are treated gently and examined by electron microscopy, most of the chromatin is found in a fiber that measures 30 nm in diameter. This 30-nm chromatin fiber is thought to represent the packaging of the nucleosomes by the remaining histone, H1. One model that accounts for the formation of the 30-nm chromatin fiber is the cooperative binding of H1 molecules to nucleosomal DNA. Each histone H1 molecule binds through the central region of the molecule to a unique site on the nucleosome and extends to contact sites on adjacent nucleosomes (Fig. 6–11). This cooperative binding would compact the nucleosomes such that they are pulled together into regularly repeating arrays, forming the 30-nm chromatin fiber.

Chromatin inside an interphase eukaryotic cell nucleus in interphase has been divided into two classes, based on its state of condensation. Chromatin that is highly condensed and considered to be transcriptionally inactive is referred to as **heterochromatin**. In electron micrographs of interphase nuclei, the heterochromatin is generally concentrated in a band around the periphery of the nucleus and around the nucleolus. The amount of heterochromatin present in the nucleus is correlated with the transcriptional activity of the cell. That is, little heterochromatin is present in transcriptionally active cells, whereas nuclei of mature spermatozoa, a transcriptionally inactive cell, contain predominantly highly condensed chromatin. In a typical eukaryotic cell, about 90% of the chromatin is thought to be transcriptionally inactive. This amount of inactive chromatin is much more than can be accounted for as the highly condensed heterochromatin. Therefore heterochromatin is thought to be a special class of inert chromatin that may have specialized functions. For example, the DNA near the centromere region is composed of repetitive DNA, and these sequences appear to constitute a major portion of the heterochromatin DNA. The remaining 10% of chromatin that is transcriptionally active is found in a more extended, dispersed conformation and is called **euchromatin**. Euchromatin is responsible for providing the RNA molecules that exit the nucleus and encode the proteins of the cell type.

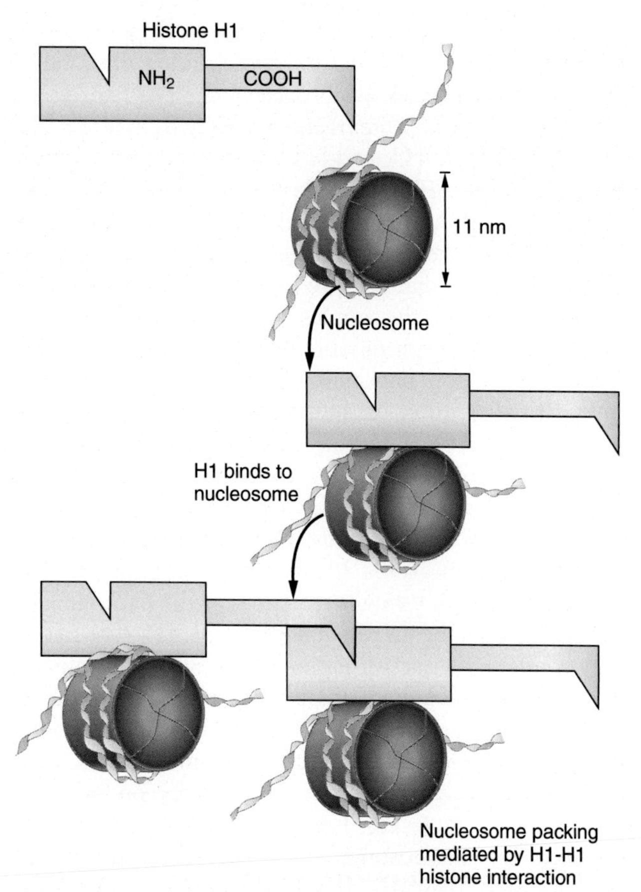

Figure 6–11. **Model for histone H1 packaging of chromatin into a 30-nm fiber.** A histone H1 molecule contains two distinct domains: a globular NH2-terminal domain and a COOH-terminal "arm" segment. In the presence of nucleosomes, the H1 molecule binds to a specific region of the nucleosome through its globular domain with the COOH-arm segment able to reach out to subsequent H1-containing nucleosomes. The COOH-terminal domain of histone H1 is then able to interact with specific sites on the adjoining H1 nucleosome by cooperative H1-H1 protein interactions.

Chromosomes are visible as distinct units in the light microscope when the chromatin is extensively condensed at mitosis. As a 30-nm fiber, the chromatin could not account for the degree of DNA condensation in metaphase chromosomes. Consequently, higher-order packaging units are required to achieve this state. From studies examining the appearance of specialized chromosomes, such as the lampbrush chromosome found in frog oocytes, it is thought that regions of the chromosome are present as extended loops of the 30-nm chromatin fiber held together at the base of the loop by a specific protein-DNA complex. The model of chromatin condensation presented in Figure 6–12 shows that, to account for the size of the typical human chromosome (~1.4 μm) in its most condensed state, the extended loop structures of the chromatin must be condensed again, possibly by drawing in the loop domains to form a tightly wound helical formation. Thus, to achieve the compaction necessary to fit the $\sim 10^7$ base pairs of DNA in the individual chromosomes into the ~1.4-μm chromosome seen at metaphase, there must be at least four orders of packaging in the DNA above the 2-nm double-helical chain of the DNA molecule.

The human genome contains 46 chromosomes in each diploid cell—1 pair of the sex-determining chromosomes and 22 pairs of autosomal chromosomes. Cytologic methods, involving staining of fully condensed metaphase chromosomes with various stains or dyes, have been useful for the identification of individual chromosomes. For example, staining metaphase chromosomes with the Giemsa reagent results in a characteristic pattern of bands on each chromosomal unit, termed **G banding**. Once the chromosomes have been stained, they can be examined under the microscope. The display of the chromosomes prepared in such a manner is referred to as the karyotype of the organism. Figure 6–13 shows an example of karyotype analysis, which is a Giemsa-staining pattern of the metaphase chromosomes of a normal human female (46,XX karyotype).

With the aid of these methods, it has been possible to correlate a variety of human syndromes with abnormalities in chromosome number. An example of Down syndrome, which is characterized by the presence of an additional chromosome 21, is shown in Figure 6–13. Other abnormalities that result in the loss or movement of a particular region of an individual chromosome in the genome (e.g., **cri du chat** syndrome, which results from the loss of a portion of the small arm of chromosome 5) can also be identified by these technologies. Thus karyotype examination provides a powerful tool for the recognition of chromosomal abnormalities associated with particular genetic diseases, and it is a particularly useful technique for prenatal diagnosis of such disorders (Clinical Case 6–2).

Clinical Case 6–2

Nathan Rubenstein is a 16-year-old boy who is referred to a pediatric hematologist because he is sufficiently anemic to be symptomatically dyspneic with minimal effort. The referring doctor found a hemoglobin level of 7.2 g, a white count of 2200, and a platelet count of 76,000. He has had no bleeding or infectious episodes. Nathan's mother is alarmed because one of his older brothers died after a rapid course of acute myelogenous leukemia several years ago. Three other brothers and a sister are in excellent health.

On examination, Nathan is a pale, frail "funny-looking kid" with short forearms and thumbs, who seems younger and shorter than his stated age. He is afebrile, but his pulse is elevated at 108, as are his respirations, at 23. He is in mild respiratory distress after climbing one flight of stairs to the office. The physician notes that Nathan has several "café au lait" spots on his shoulders and back. The rest of

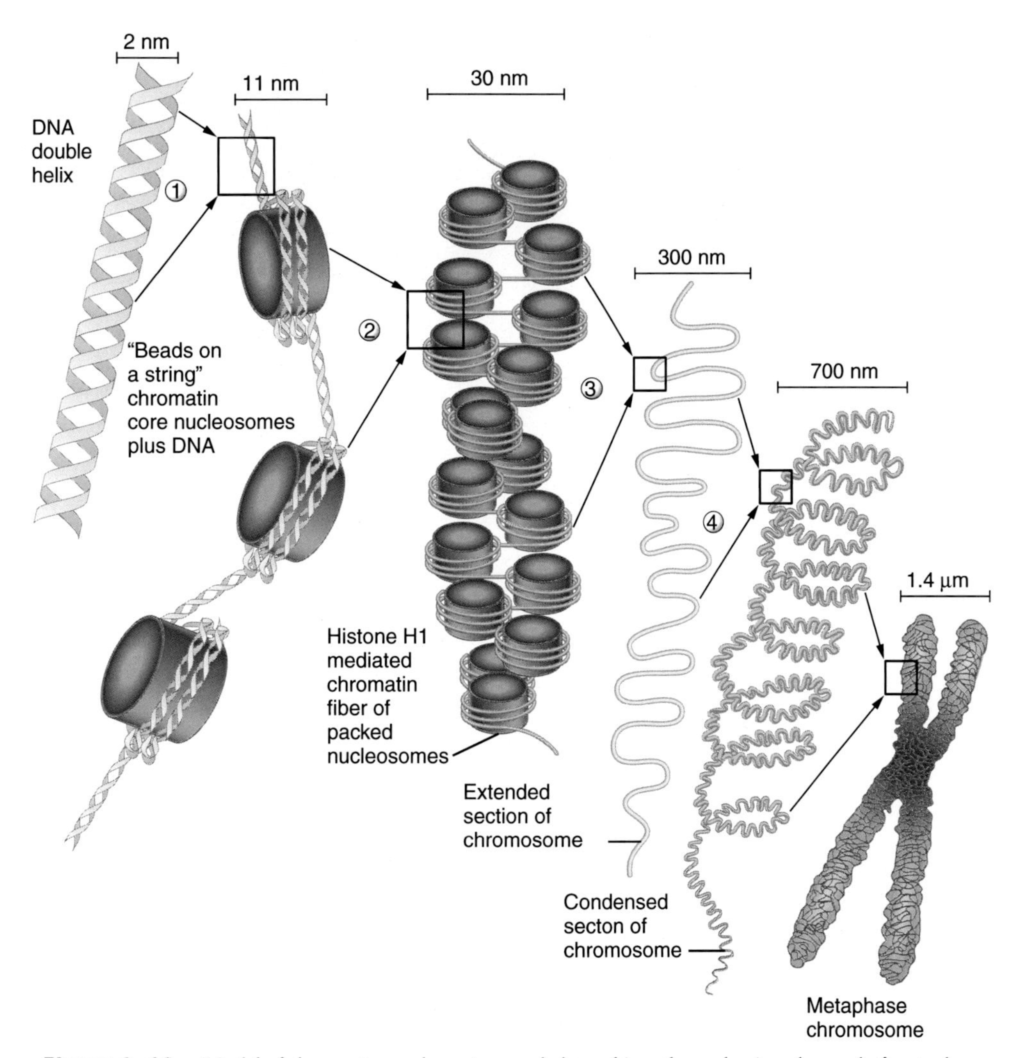

Figure 6–12. **Model of chromatin condensation needed to achieve the packaging observed of metaphase chromosomes.** The first-order packaging involves the formation of the 11-nm chromatin fiber by association of the DNA helix with the core nucleosome proteins. To form the 30-nm fiber, there is the cooperative binding of histone H1, molecules pulling the nucleosomes into close apposition. The 30-nm fiber is representative of a looped section of the chromosome, and this folding results in a 10-fold packing unit of ~300 nm. The looped domains are thought to be arranged in a secondary loop, folding the chromatin into a 700-nm structure; however, the interactions that result in this packaging are not well defined.

the examination is unremarkable, including the absence of any palpable organs or masses in his abdomen.

The hematologist draws some blood for laboratory studies and obtains a bone marrow core biopsy and aspiration. He also orders a skeletal bone series of radiographs. In addition, he asks the mother to bring in Nathan's four remaining healthy siblings for tissue-typing studies and to return for laboratory results in a week. He tells Mrs. Rubenstein that Nathan is profoundly anemic and suggests that he avoid any significant exercise for the next week. He tells her, however, that he does not want to transfuse him at this time because he does not want to sensitize him to different blood groups.

Nathan's blood smear shows red cell macrocytosis, reduced white cells with 30% polys, and reduced but large-sized platelets. The marrow examination results show substantial hypoplasia of all three formed elements, with an increase in marrow fat to 85%. The red cell series shows early megaloblastic changes; the white cell series shows a shift to the left, with increased promyelocytes and occasional "Pelger Huet-like" peculiarly lobed anomalous nuclei in the more mature forms. The megakaryocytes show a small number of binucleate forms. The hematologist reads this picture as marrow hyperplasia with substantial trilineage dysplasia. He also asks the laboratory to perform an assay of the DNA stability on the lymphocytes from Nathan's blood sample. After incubation with a mild DNA cross-linking reagent, they find a 36-fold increase in the incidence of obvious chromosome breaks in comparison with a concurrent control. The bone X-ray films show only vestigial radii and short-thumb metacarpals bilaterally.

Continued

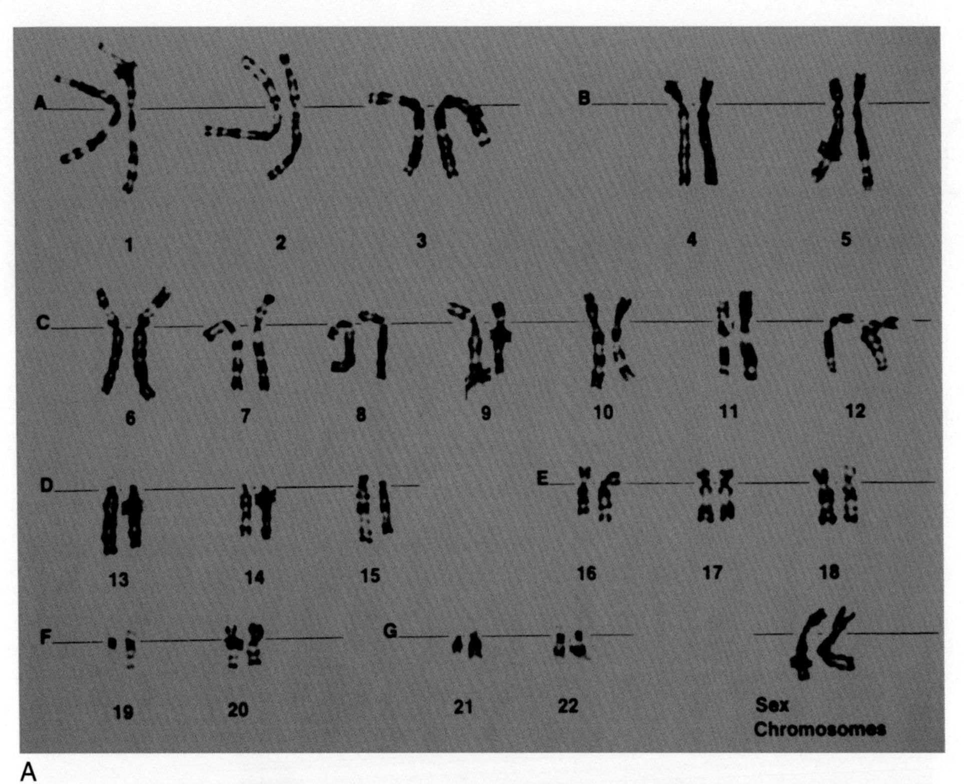

A

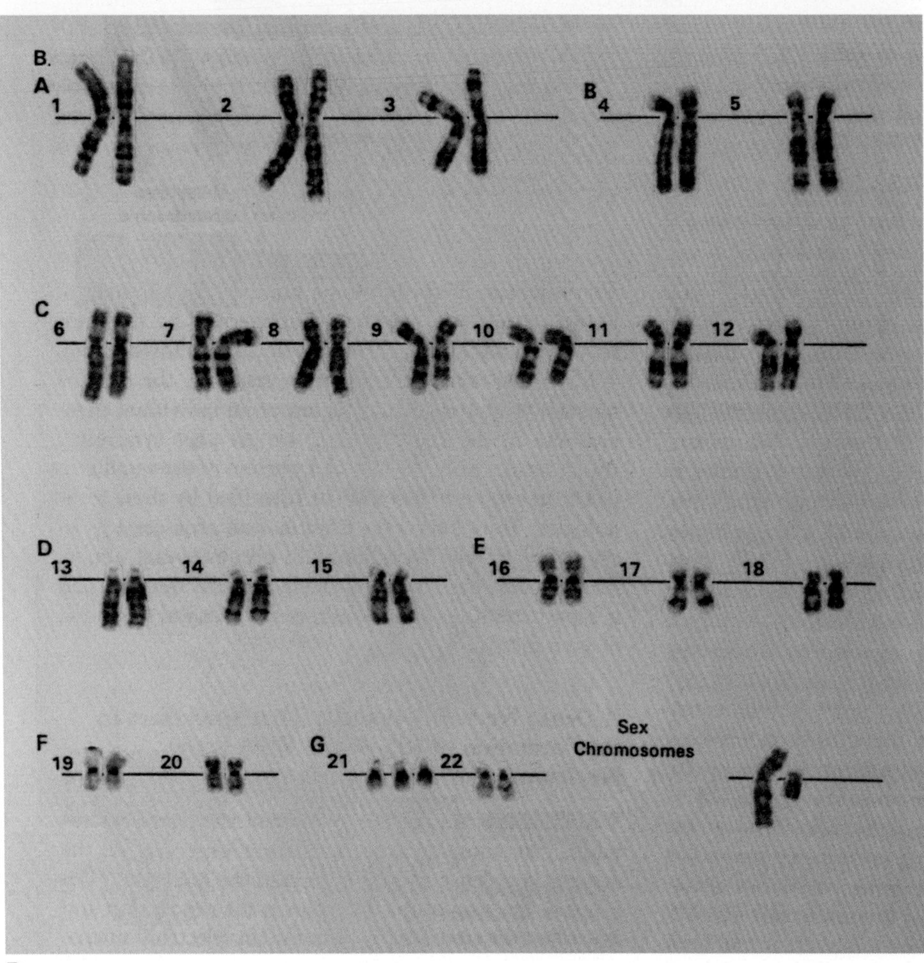

B

Figure 6–13. **Giemsa-staining pattern of human metaphase chromosomes.** (A) An example of G-binding karyotype analysis. Shown are the aligned metaphase chromosomes of a normal female human, as illustrated by the 46,XX karyotype. (B) Example of a karyotype from an individual with Down syndrome is shown with an extra chromosome 21. *(Courtesy Dr. Cathy Tuck-Muller, Department of Medical Genetics, University of South Alabama, College of Medicine.)*

Cell Biology, Diagnosis, and Treatment of Fanconi Anemia

On the strength of these findings, the hematologist makes a diagnosis of Fanconi hypoplastic anemia. On the return visit, to discuss the patient's laboratory findings and the tissue matching of his siblings, the physician explains that untreated, Nathan has a poor prognosis, with a high likelihood of death from marrow failure, leukemia, or some form of gastrointestinal cancer before the age of 25. He believes, however, that he can avert much of this risk with a peripheral blood stem cell-bone marrow transplant.

Fortunately, one of Nathan's older brothers is a perfect tissue-type match, and Nathan undergoes a successful marrow transplant using mobilized stem cells from him 1 month later. After the return of adequate marrow function in 4 months, Nathan is no longer dyspneic, his hemoglobin level is 12.5 g, and his white cell and platelet counts are within normal limits. Nathan understands that he has avoided many of the serious effects of his genetic disorder but that he must remain vigilant for early signs of cancer for the rest of his life.

The Nucleolus Is a Dense Nuclear Organelle That Specializes in the Formation of rRNA

When interphase cells are examined under the microscope, the most prominent feature observed in the nucleus is a dense structure, termed the **nucleolus**. The size and shape of the nucleolus are dependent on its activity. In cells that are actively synthesizing large amounts of proteins, the nucleolus may occupy up to 25% of the total nuclear volume, whereas in dormant cells it may be hardly visible. Examinations of cells from different physiological states have shown that the observed differences in nucleolar size are primarily due to differences in cellular status. Cells that are active in protein synthesis contain more of the maturing ribosomal precursor particles in their nucleus. This increased nucleolar size probably reflects the time necessary to assemble the rRNA with proteins of the ribosomal subunits, because electron micrographs of cells containing large nucleoli demonstrate an increase in the number of active ribosomal genes and an increase in the apparent rate that each gene is transcribed. The size and shape or number of visible nucleolar centers in the nucleus can be used in pathologic examinations as a determinant of the cell.

In general the nucleolus is visible only in interphase cells. Concomitant with the condensation of chromosomes as the cell approaches mitosis, the nucleolus is observed to decrease in size and then disappears as RNA synthesis stops. In humans the rRNA genes represent clusters of DNA segments located near the tip of five different chromosomes (chromosomes 13, 14, 15, 21, and 22); thus there are 10 different ribosomal gene loci in diploid somatic cells. Following mitosis, rRNA synthesis is restarted initially on small nucleoli located at the 10 ribosomal gene loci, which are often referred to as **nucleolar-organizing regions** (NORs). These small NORs usually are not observed separately because of the rapid infusion of the NORs that forms the larger characteristic interphase nucleolus.

In interphase cells the nucleolus is responsible for the synthesis of rRNA and the production of mature ribosomal subunits by complexing the rRNA with appropriate proteins. The completed subunits are transported to the cytoplasm through the nuclear pores, where they provide the machinery for translation.

REGULATION OF GENE EXPRESSION

Genomics and Proteomics

The Study of Gene Expression Has Been Facilitated by Recombinant DNA Technology

A major focus of modern cell biology is to understand how a cell works in molecular detail. Although classic biochemical approaches have made possible the purification and examination of many cellular components, until recently, the only way to investigate the informational content of the cellular genome was the examination of the phenotype of mutant organisms to deduce gene function. This approach remains an important investigative and diagnostic tool (karyotyping analyses); however, investigators now possess the ability to directly examine a specific gene and how it functions in normal and pathologic circumstances through the application of techniques, referred to as **recombinant DNA technology**. Although new technical advances occur rapidly in this field of study, the key techniques that constitute the basis of recombinant DNA technology are as follows:

1. The cleavage of DNA at specific locations by restriction endonuclease, which facilitates the identification and manipulation of individual gene sequences.
2. The propagation of eukaryotic DNA fragments in bacterial cells by gene cloning, which allows the isolation of large quantities of a specific DNA.
3. The determination of the order of nucleotides contained in a purified DNA by DNA sequencing, which allows the examination of gene structure and the amino acid sequence it encodes.
4. The direct amplification of a DNA sequence by the polymerase chain reaction (PCR), which enhances the ability to quickly examine specific regions of the genome for genetic defects.

Because recombinant DNA techniques are becoming increasingly important as clinical diagnostic tools, we devote an initial section to explaining these techniques, after which we examine the concepts of gene expression, leading to a discussion of the importance of these concepts toward establishing viable genetic therapies.

Restriction Nucleases: Enzymes That Cleave DNA at Specific Nucleotide Sequences

One of the most important developments of recombinant DNA technology was the discovery of enzymes that catalyze the double-stranded cleavage of DNA at specific nucleotide sequences, the restriction endonucleases. This discovery came from an understanding of the defense mechanism used by bacteria to protect themselves from foreign DNA molecules carried into the cell. The genome of a bacteria contains a host-specific pattern of DNA methylation, and when a DNA that does not contain this pattern is encountered by the cell (e.g., carried in by bacteriophage), it is degraded, thereby protecting the bacteria from the foreign DNA. The first of these enzymes was purified from *Escherichia coli*. Remarkably, this enzyme catalyzed the double-stranded cleavage of DNA within a specific short nucleotide sequence. Hundreds of enzymes capable of cleaving DNA at specific nucleotide sequences have been isolated from different species of bacteria, providing powerful tools for the characterization of DNA molecules.

With these enzymes the DNA isolated from a particular cell can be cleaved into a series of discrete fragments called **restriction fragments**. The size of DNA fragments produced by an enzyme digest can be analyzed by resolving the cleaved DNA on an electrophoretic gel. Thus, by examining the sizes of restriction fragments produced from a specific gene region after treatment with combinations of different restriction endonucleases, a map of that region can be drawn that shows the location of each restriction site relative to adjacent restriction sites (Fig. 6–14).

Because restriction endonucleases cleave DNA at positions of specific nucleotide sequences, a restriction map reflects the arrangement of these sequences within a given fragment of DNA. This is useful to characterize similarities and differences between isolated, homogenous DNA fragments (e.g., cloned DNA). The DNA fragments containing a specific sequence can be identified within the thousands of fragments produced when a population of DNA molecules, such as the total genome of an organism, is cleaved with restriction enzymes by using a technique referred to as **Southern hybridization** (Fig. 6–15).

Southern hybridization is a powerful way to examine the organization of specific genetic loci among individual members of a family or population. A difference in restriction maps between two individuals is called a **restriction fragment length polymorphism (RFLP)**. This analysis has become an important technique to identify loci that are close to, or contain, a defective gene associated with a genetic disease (Fig. 6–16). For example, the molecular basis for Duchenne's muscular dystrophy (the dystrophin gene) and cystic fibrosis (the cystic fibrosis transporter gene) was elucidated using RFLP technology.

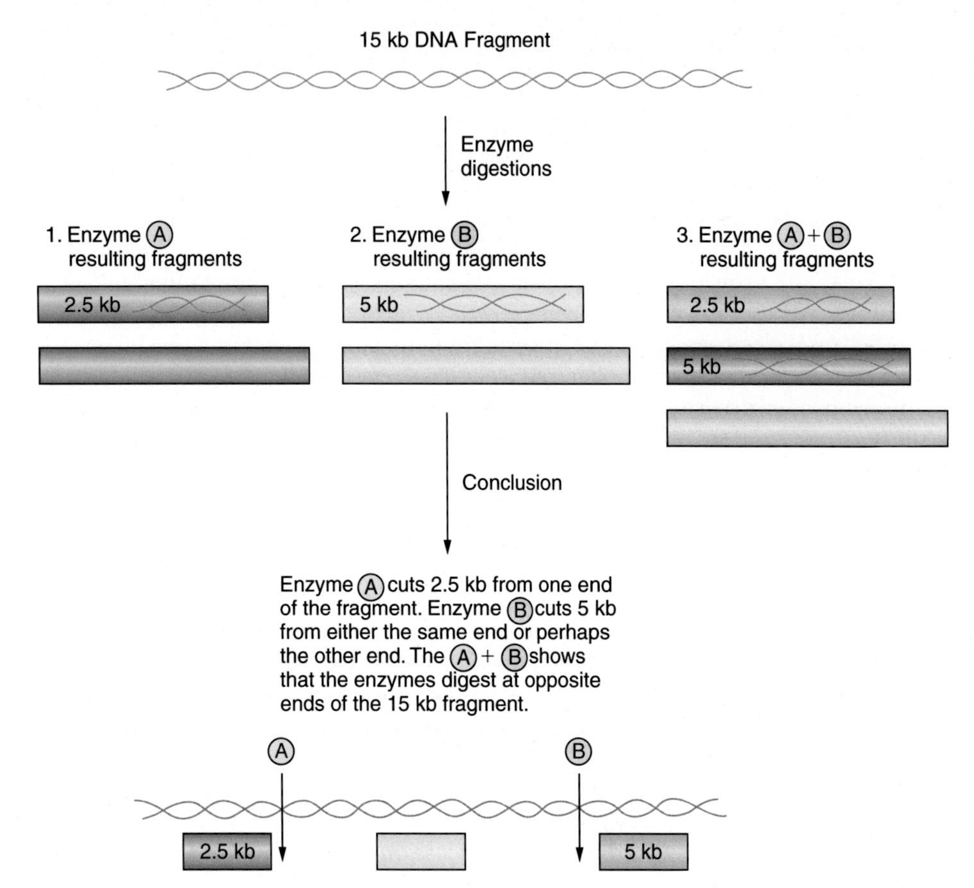

Figure 6–14. **Example of restriction mapping.** Restriction enzymes are useful tools to characterize segments of DNA. Shown is an experiment demonstrating how the cleavage sites for different restriction endonucleases are positioned relative to each other to create a restriction map of a DNA fragment. kb is an abbreviation that means 1000 nucleotides or 1000 nucleotide pairs.

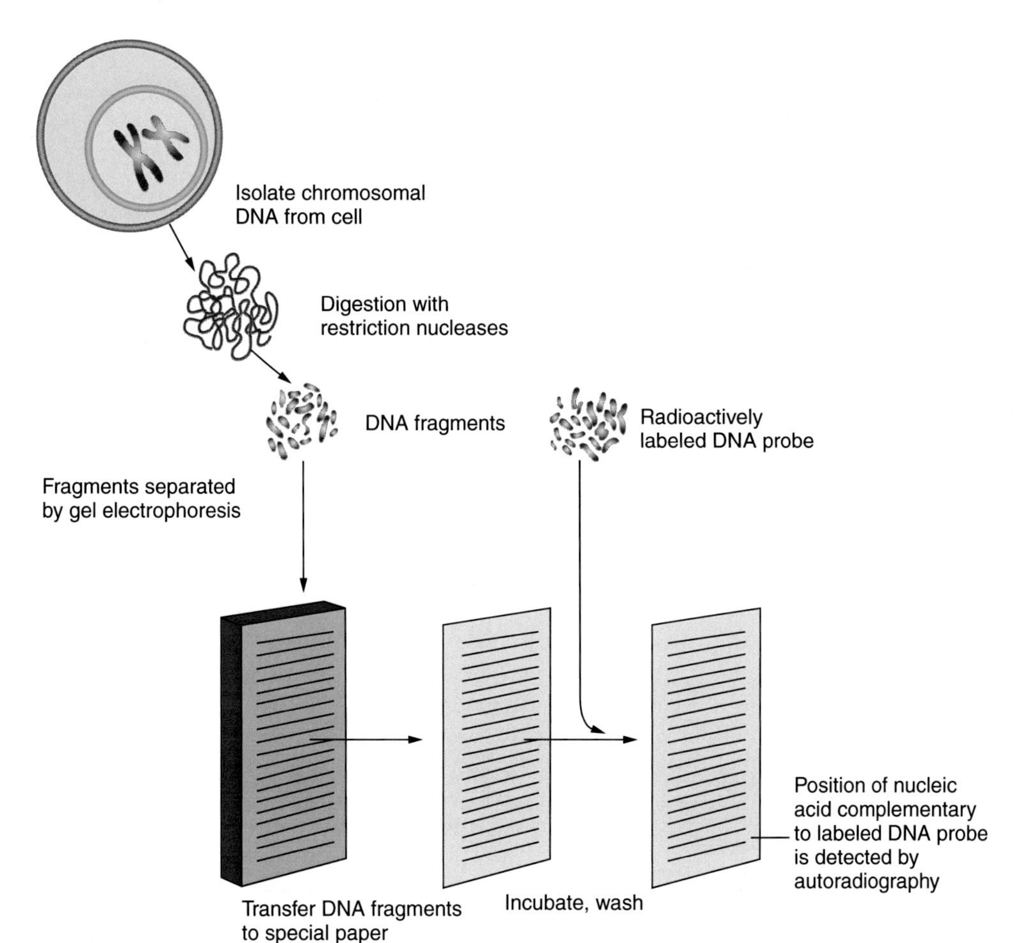

Figure 6–15. **Southern blotting analysis.** The size or migration of specific DNA fragments within a mixture of DNA can be examined by the Southern hybridization technique. After treatment with restriction endonucleases, the DNA fragments are resolved on electrophoretic gels, and the fragments contained within the gel are then transferred to a nitrocellulose paper by blotting. The paper is incubated with a radioactive DNA fragment under conditions that permit this DNA probe to bind with complementary molecules on the paper sheet (hybridization). After hybridization the sheet is washed free of nonspecific probe binding, and the immobilized molecules complementary to the probe are visualized as radioactive bands on X-ray films placed next to the nitrocellulose paper.

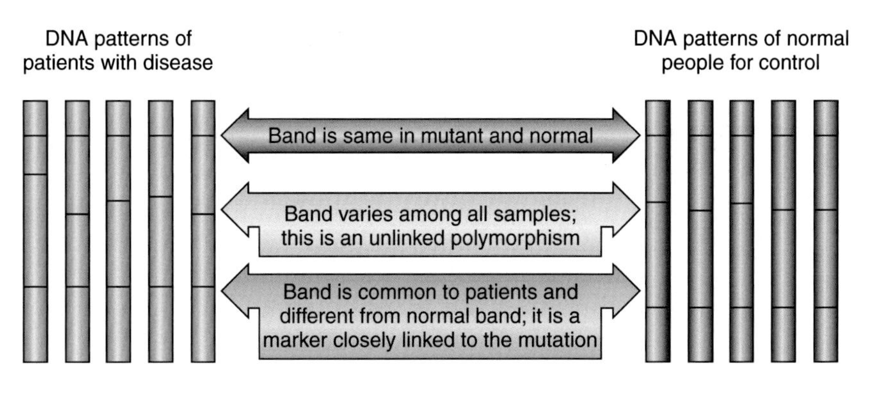

Figure 6–16. **An example of restriction fragment length polymorphism analysis.** Southern blotting examination of the chromosomal DNA from patients who have a disease trait and comparison with the DNA from unaffected individuals has been instrumental in elucidating the molecular basis for many genetic disorders. *(Modified from Lewin B.* Genes, *4th ed. New York: Oxford University Press, 1991, by permission.)*

Gene Cloning Can Produce Large Quantities of Any DNA Sequence

In 1973 Boyer and Cohen recognized that DNA molecules from any source are capable of being joined covalently by DNA ligase, an enzyme that links DNA molecules together. In their pioneering experiments, Cohen and Boyer linked a fragment of eukaryotic DNA to a DNA molecule isolated from bacteria that could direct its own replication in bacterial cells. This hybrid DNA formed in vitro is called a **recombinant DNA molecule** because it is formed from the end-to-end joining of two different DNA. Moreover, because the bacterial DNA fragment replicates when the bacterium grows, the eukaryotic DNA fragment of the recombinant molecule is also replicated.

This experiment is an example of **DNA cloning.** Successful cloning requires two basic elements. The first is a suitable host bacterial strain. Most, if not all, bacteria contain a restriction-modification system. However, many bacterial strains have been modified or engineered by classic genetic selection such that they no longer contain this defense mechanism. Because these bacterial strains can undergo genetic transformation (the uptake of DNA) with a foreign DNA molecule with the DNA becoming resident within the recipient cell, they provide an excellent host for cloning experiments. The second element is the segment of bacterial DNA capable of directing replication in these modified bacterial cells. These DNA are commonly referred to as DNA vectors and can be used to "carry" the foreign DNA linked to them in the bacterial cell (Fig. 6–17).

Cloning has been helpful in isolating specific eukaryotic DNA for analysis. Two basic types of clones have been used: complementary DNA (cDNA) clones, which are copies of a specific mRNA, and genomic clones, which are pieces of the genome spliced with a DNA segment capable of replication in bacteria. Because the bacterial segment contains an origin of replication that directs replication of the recombinant DNA molecule (e.g., the bacterial vector and the foreign DNA linked to it) as the bacteria grow, there is an unlimited supply of the target DNA for further analyses.

The Primary Structure of a Gene Can Be Rapidly Determined by DNA Sequencing

Many human diseases are the consequence of single-base changes in genes, causing abnormal proteins to be formed, or perhaps abnormal function of the gene. Sickle cell anemia, for example, is the result of a single-base change that ultimately replaces a glutamic acid at residue 6 with a valine in the β-chain of the hemoglobin molecule. Other diseases of hemoglobin caused by single-nucleotide changes exist; however, the changed base is not within the protein-coding sequence of the gene. For example, a particular class of disease in which hemoglobin is not produced (the β-thalassemias) is caused by single-nucleotide change of the β-globin gene that creates improper processing of the pre-mRNA to the mature mRNA capable of directing the synthesis of the protein. The basis of these thalassemic diseases remained a mystery until the genetic material from affected individuals was cloned and the primary structure of the genes was determined by the DNA sequencing.

DNA sequencing is simply the determination of the order of the nucleotides in a particular DNA fragment. The initial two methods for determining DNA sequence were Maxam and Gilbert chemical degradation and the dideoxynucleotide enzyme-based synthesis of DNA. The first method is based on the chemistry of the DNA molecule. The second depends on an enzyme that synthesizes DNA from a single-stranded template and is referred to as the dideoxynucleotide chain termination method. Both methods are reliable and have been used by individual research laboratories to determine sequence and are generally displayed on special electrophoretic gels that allow separation of DNA fragments that differ by a single DNA nucleotide (Fig. 6–18).

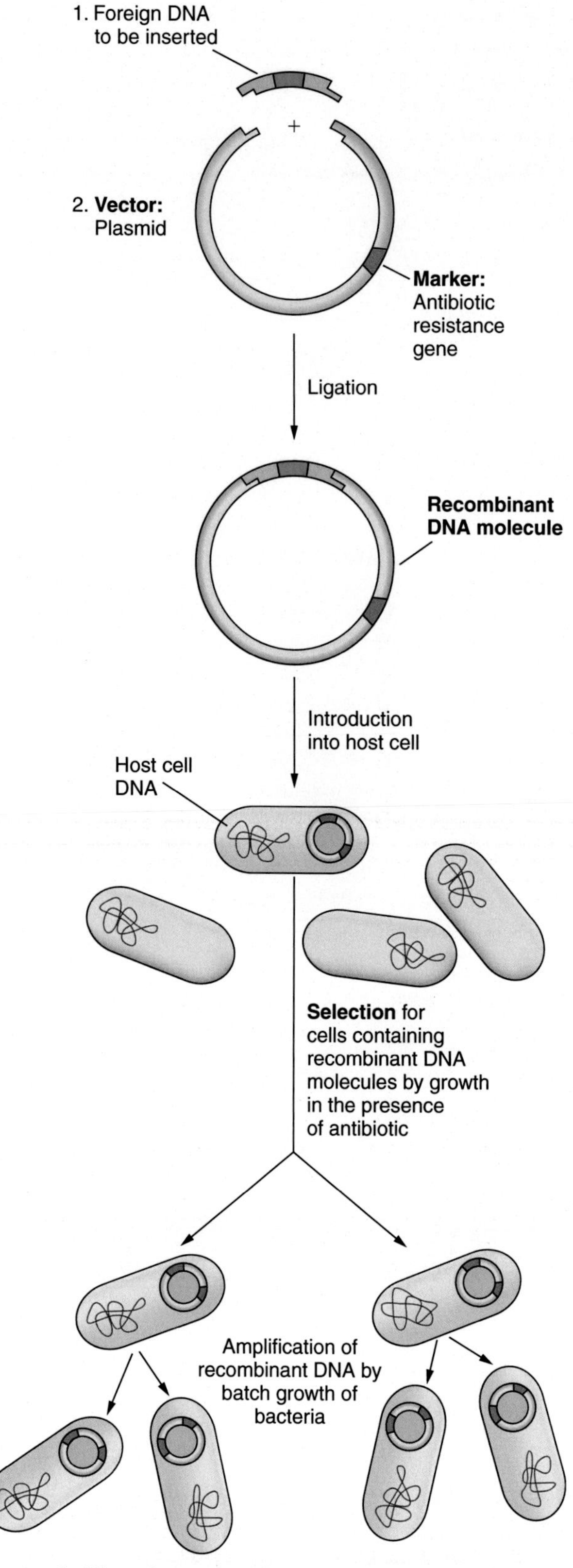

Figure 6–17. **Cloning DNA fragments.** Two DNA fragments can be covalently joined together, forming a recombinant DNA molecule. If one of the DNAs contains a bacterial origin of replication and a vector (such as a plasmid, as shown here), then the products of the in vitro ligation reaction can be placed into host bacteria. Plasmid vector molecules also contain genes that confer resistance to antibiotics, allowing the selection of bacteria that take up the recombinant molecule.

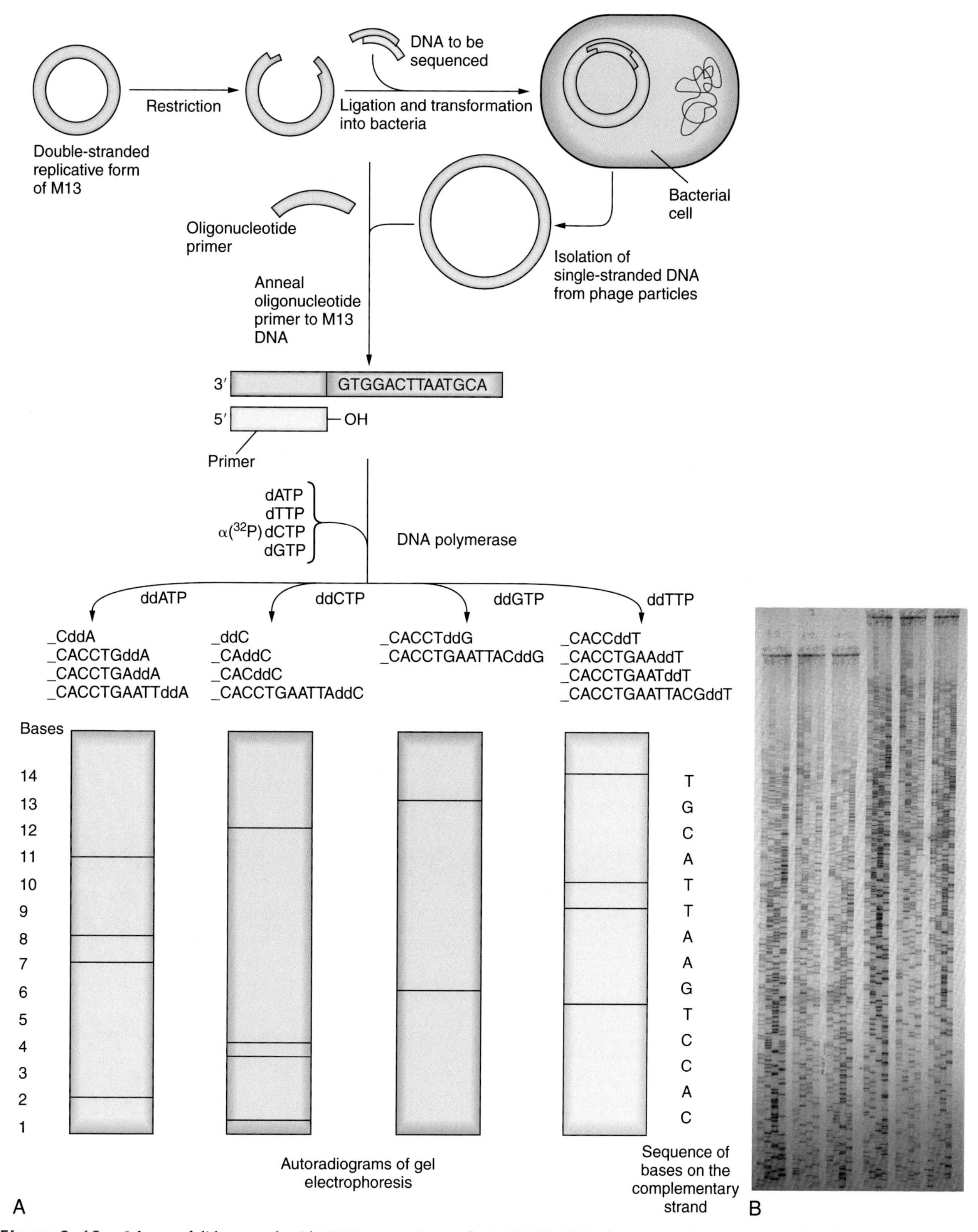

Figure 6–18. **Scheme of dideoxynucleotide DNA sequencing analysis.** (A) The DNA fragment to be sequenced is cloned into a vector that is capable of synthesizing a single-stranded version of the recombinant DNA molecule. The single-stranded DNA is then used as a template for in vitro DNA synthesis in the presence of a dideoxynucleotide. The resultant DNA fragments are resolved on an electrophoretic gel, which allows determination of the DNA sequence. (B) Example autoradiogram from a dideoxynucleotide chain termination sequence experiment. *(Zimmer, W. Unpublished.)*

It is now more common to determine the sequence of a protein by deducing the amino acids from a nucleotide sequence; there are only four bases in nucleic acids and 20 different amino acids, making DNA a chemically simple molecule. Moreover the importance of knowing and understanding the primary structure of a gene as related to human disease can be inferred from the recent effort leading to the determination of the complete sequence of the 10^9 nucleotides of the human genome. New DNA sequencing technologies, termed second-generation technologies that have been developed since 2005, has accelerated the rate of discovery and the application of genome science to medical practice. Major advances have been demonstrated in the speed and cost of sequencing entire genomes of individuals or of cellular samples (e.g., individual tumor cells). While several techniques have been described, the most often used techniques are the Illumina and Ion Torrent DNA sequencing techniques. Both rely upon construction of short fragment random DNA libraries representing the genome by PCR, and the small fragments are then affixed to a solid support. The Illumina sequencing techniques relies on a modification of the dideoxynucleotide sequence method (Fig. 6–18). Genome DNA is fractionated and affixed to the short DNA strands on the slides, and then, synthesized DNA strands in the presence of specific fluorescent nucleotide molecules are then read in the sequence reactions. In Ion Torrent sequencing the short DNA fragments are affixed to microscopic beads, and the DNA-bead complex is amplified to create a library of individual beads each with a unique sequence. These individual beads capture genome sequences and the sequence determined by a pH change in the well when the appropriate nucleotide is added to the growing chain via a DNA polymerase reaction. In either case, these new second-generation sequencing techniques are allowing the simultaneous sequencing of many genomes and have put DNA science into the forefront as a diagnostic tool for medical practice. Indeed, by applying the same techniques, genomes from a variety of species have been sequenced in their entirety, which has resulted in an emerging technology termed *bioinformatics*. This is essentially a way to analyze and/or use the volumes of data generated by today's automated technologies. Bioinformatics will be important in the future to help determine why certain individuals tolerate certain drugs better than others. As discussed later in this chapter and in more detail in Chapter 14, the newer technologies of genomics and proteomics will eventually lead to "individualized or personalized medicine."

Specific Regions of the Genome Can Be Amplified With the PCR

The techniques discussed in this chapter have revolutionized our understanding of how cells work. However, they are somewhat time consuming. In 1987 a novel technology was introduced that allows amplification of a nucleic acid sequence without the need to clone it. This technique, called *PCR*, does require knowledge of the sequences that surround the region to be amplified. Synthetic oligonucleotides complementary to these sequences are used as primers in a series of reactions that use a special thermostable DNA polymerase isolated from a bacterial species that lives at high temperatures (Fig. 6–19).

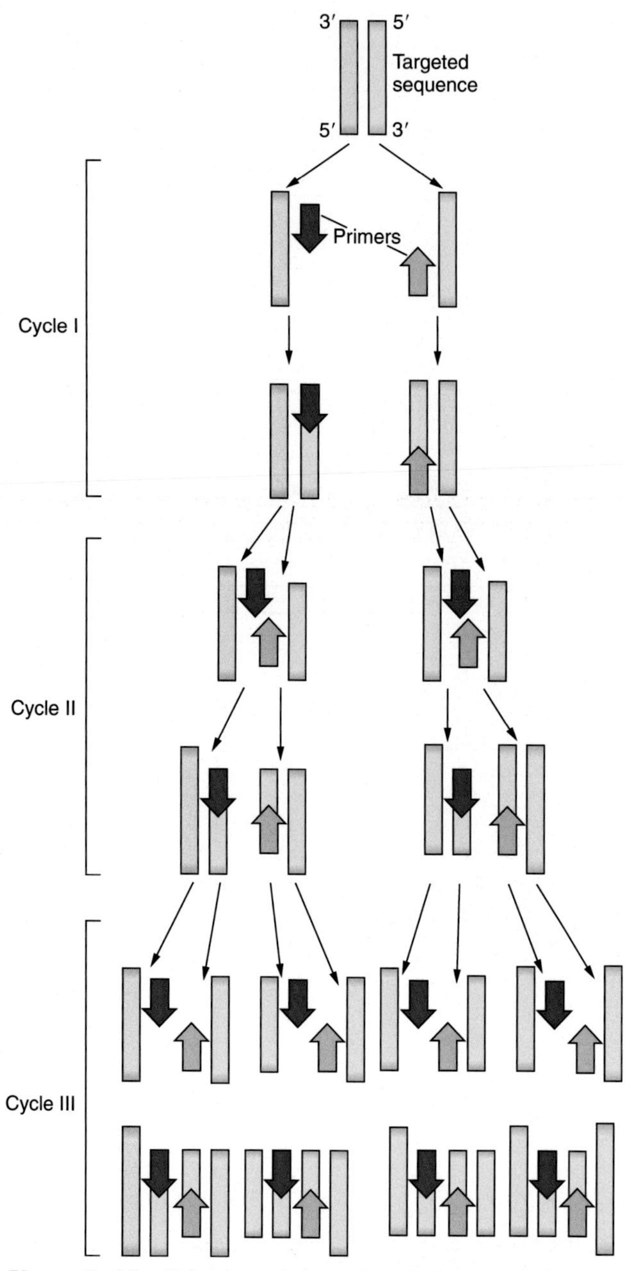

Figure 6–19. **Polymerase chain reaction (PCR).** Amplification of a DNA fragment through three cycles of a PCR experiment. In each cycle the DNA strands are denatured, permitting annealing of synthetic oligonucleotides and the synthesis of the complementary DNA strand. Notice that there is an exponential increase in the amount of the target DNA at the end of each cycle, resulting in an amplification of the DNA segment.

The reactions involve a cycle of steps that first denatures the DNA duplex at high temperature and then allows binding of the oligonucleotides by cooling to a lower temperature and extension of the oligonucleotide primers by DNA polymerase. Because the reagents are not inactivated by high temperatures, they can be added to a single tube, and the cycles of denaturation, annealing, and extension can be repeated multiple times. Each cycle increases the concentration of the duplex DNA bound by the oligonucleotide primers such that microgram quantities of DNA are isolated from the nucleus of a single cell (a specific sequence is amplified exponentially; e.g., 30 cycles would represent 228, or 27 millionfold amplification). Moreover, this can be accomplished in hours, compared with weeks or months needed for conventional cloning techniques.

Although PCR is a relatively new technique, it has become an important diagnostic tool. Because of its speed and sensitivity, PCR is becoming a vital technique for the clinical diagnosis of infectious diseases such as acquired immune deficiency syndrome (AIDS). The PCR-based technologies can detect the presence of the viral genome much earlier than can antibody-based tests that require months (and even years) of infection to detect the presence of a viral protein. In addition, PCR tests have been developed for prenatal diagnosis of a variety of genetic diseases. Finally, because PCR is such a sensitive technique, it is possible to analyze DNA from a tissue sample as small as a human hair. Thus PCR is an ideal technology for use in forensic medicine.

Bioinformatics: Genomics and Proteomics Offer Potential for Personalized Medicine

Comparative Genomics Allow the Generation of New Therapies

With the automation of DNA sequencing techniques has come the need to organize and evaluate vast amounts of data. This has opened a new area of science called **bioinformatics.** Essentially, bioinformatics is the application of computer science to give an understanding of increasing loads of data generated from automated experiments. Indeed, complete sequences are now available for many species, from bacteria to humans, and the emphasis is to use these data. Direct comparisons of genomes have shown that many human genes are similar to genes found in other mammals (e.g., 98% similarity human/chimpanzee and 90% similarity human/mouse), and this similarity extends to other organisms. Thus, by identifying gene regions of high similarity, scientists can begin to predict features of genes that are critical for the function of a protein. Extending these findings, these studies will give greater information to the understanding of human gene structure and function, thereby allowing the development of better strategies to combat common diseases. In addition to gene segments, the DNA surrounding genes has been found to be similar among species. These analyses will promote finding genomic features conserved among disparate species and thus provide important information concerning signals that regulate gene expression. This will lead to new, innovative therapies for human genetic diseases. For example, certain primates share ~98% genomic identity with humans, yet they are not susceptible to certain human syndromes such as malaria. Thus, from an understanding of why primates do not suffer from these diseases, scientists will be able to generate better therapies for certain human disease states.

Another technique that has developed from the human genome sequencing project is the ability to examine the expression of every gene or potential gene in cells or tissues. This technique is called **microarray analysis** (Fig. 6–20), and it involves isolating RNA from cells and testing the whole mixture by hybridization with special slides or chips that contain small spots of nucleic acid, each spot representing a specific gene. Two components that allow this technique to be useful are the ability to make short nucleotide sequences representing each gene as derived from the complete genome sequence information and the ability to make picoliter spots of these DNA sequences on special glass plates. By hybridizing the total RNA content of a cell to the glass slides or gene chip, scientists can accurately determine which genes of the entire genome are active in the cells of interest and which genes are not active in these cells. In addition, newer technologies are available that can link or label RNA populations with different fluorescent tags and use them in microarray analyses. Thus one can isolate RNA from cell populations, such as from a specific tumor and the normal cells; label them with different fluorescent dyes; and then mix the labeled RNA to hybridize to the same microarray gene chip. By incorporating sensitive color scanning and detection equipment, it is now possible to examine whole-genome expression patterns *and* simultaneously know how the patterns change in cancer (tumor) cells from the same individual patient. This has revolutionized medicine because physicians can order microarray analysis of a patient's tumor cells, or cell populations affected by specific drug treatments, and quickly determine whether specific metabolic pathways are altered. Thus, although currently experimental, the gene chip microarray holds great promise to move the practice of medicine to include individualized therapeutics.

A direct application of the bioinformatic analyses of genomic sequences has come in the ability to predict how an individual patient will respond to certain drugs. It is well known that, although most drugs can be generally effective (i.e., therapeutic), a patient may not respond to the therapy. Often the nonresponsive patient can be grouped with other nonresponsive patients by some criteria, suggesting that there may be a genetic component to the ability of an individual to respond to drug

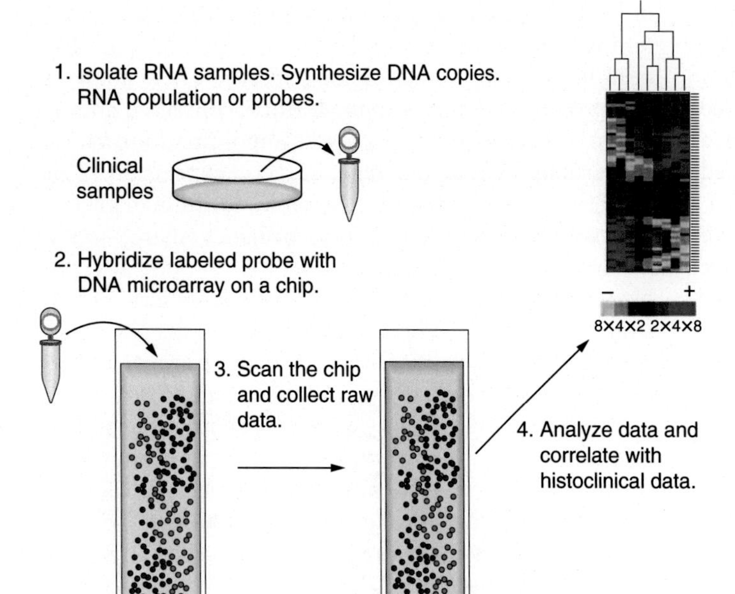

Figure 6–20. **Outline of a microarray analysis.** An outline of expression profiling of RNA samples from cells, called *microarray analysis*. The initial step (1) requires that RNA samples are isolate from the cells or tissues to be examined. These may be cell populations that are derived from a tumor and normal cells, from cells before and after a drug therapy, or from other clinical samples. The RNA populations are then made into DNA copies, or cDNA, at which time fluorescent dyes are incorporated into the copied DNA. This creates a DNA probe for hybridization to the gene chip (2). This is a special slide that contains small spots of nucleic acids representing every gene in the human genome. Once the hybridization is performed, the slide is scanned to collect the expression data (3), and the data can then be analyzed (4) by bioinformatics computer systems. The expressed RNA can determine whether whole pathways (e.g., gluconeogenic pathway or signaling cascades) are activated by treatment of the cells, and this information can be correlated with clinical data to affect the therapeutic course for the patient.

TABLE 6–3. Pharmacogenetics of the Cytochrome P450 System: Examples of Clinical Importance

		% of Natural Dose[a]		
Disease	**Affected Enzyme**	**Ultrarapid Metabolism**	**Poor Metabolism**	**Examples**
Depression	cyp2C9[b]	–	–	Bipolar disorder and valproate
	cyp2C19	–	40	Selective serotonin reuptake inhibitors
	cpy2D6	200	30	Side effects of tricyclic antidepressants
Psychosis	cpy2D6	160	30	Haloperidol and Parkinsonian side effects, oversedation, and thioridazine
Ulcers	cyp2C19	–	20	Proton pump inhibitors and pH/gastrin
Cancer	cyp2B6	–	–	Cyclophosphamide metabolism
	cyp2D6	250	60	Nonresponse to antiemetic drugs
Cardiovascular	cyp2C9	–	30	Warfarin dosing and blood pressure response
Pain	cyp2D6	–	–	Codeine dosing and nonresponders
Epilepsy	cyp2C9	–	–	Phenytoin dosing and unwanted side effects

[a]Represents the percentage change of dose for normal population.
[b]cyp = cytochrome P450.
Modified from Katzung T. *Pharmacology*. New York: McGraw-Hill, 2002.

therapies. An example can be seen in patients with clinical depression, where 20%–40% of the patients do not respond to drug therapy. Moreover a significant fraction of these patients exhibit resistance to asthmatic, ulcer, and antihypertensive therapies. Comparing the genetics of these individuals has demonstrated variations in the enzyme systems that are responsible for drug metabolism, the cytochrome P450 system.

The study of heredity and responsiveness to drug therapy is referred to as **pharmacogenetics**, and the application of bioinformatics to these analyses is called **pharmacogenomics.** As shown in Table 6–3, numerous variations of the cytochrome P450 enzyme system lead to an individual's ability to respond to specific drug therapies.

Similar pharmacogenetic analyses are in process for other classes of metabolic enzymes. Thus, in the future, medicine will be individualized based on the genetic makeup of the patient, which will allow for more efficient delivery of drug therapies. Further coverage of precision medicine can be found in Chapter 14.

The Next Level of Biological Complexity Is the Study of Protein Structure and Function or Proteomics

Now that the entire human genome has been sequenced, a central challenge is to understand the structure and function for all the proteins encoded by the DNA. Proteins perform most of the biological functions within the cell; thus, to complete an understanding of how the cell works, you must know which proteins are expressed in the cell and how these interact with each other to allow cellular function. The entire protein complement expressed by a cell is referred to as the proteome of the cell. Proteomics encompasses a broader definition, which includes protein sequence, posttranslational modifications, protein copy number, protein network formation, protein structure and function, and the localization of proteins within the cell confines. Proteomics includes techniques discussed in Chapter 1, namely, protein purification, protein separation, and gel electrophoresis; however, with the recent application of sophisticated mass spectroscopy, a new level of information concerning the cellular proteome is now possible.

Two basic mass spectrometric techniques are used in proteomic analyses, as well as variations of these methods. Both methods examine proteins derived from a cell or tissue that (as illustrated in Fig. 6–21) has been initially digested with proteases to liberate manageable, smaller peptide fragments. Peptides are placed into mass spectrometers that can ionize the peptides into a gas phase and then detect the mass-to-charge ratios of the individual peptides that are created. In the technique referred to as MALDI (matrix-assisted laser desorption/ionization), the peptides are mixed with a small organic molecule with a chromophore that absorbs light of a specific wavelength. The sample and matrix are then dried, and the crystal lattice is heated with a beam of light from a laser. The matrix absorbs photons and transfers energy to the peptides that are released as ions. A time-of-flight instrument then measures the individual peptide mass-to-charge ratio. This mass fingerprinting analysis gives a highly individualized result that can provide specific identification of the peptide when compared with databases of peptide mass information. In electron spray ionization the peptides are in an acidic solution and passed through a needle at high voltage. The resulting positively charged ions enter the first component of a tandem mass spectrometer yielding a primary spectrum (see Fig. 6–21). The mass ion can be isolated and fractured at peptide bonds via energetic collision with particular gasses, and the resultant fragments analyzed by a tandem mass spectrometer, a process called **MS/MS analysis.** The fragmentation pattern is diagnostic of the peptide composition, and examining the fragmentation pattern against databases can lead to the sequence of the original peptide fragment. In addition, because the MS/MS is extremely sensitive, alterations to a particular residue of the peptide can be determined. That is, posttranslational events are also easily determined using the MS/MS technique. Thus the peptide fragments give information of the "proteome" or proteins expressed in particular cellular populations and can also derive information of the status of a specific protein within the analyzed cellular lysate. This can be important for determining the status of cells undergoing changes in response to disease or drug treatments and can provide information regarding molecular control of cellular processes.

The application of automated data handling (bioinformatics) to compare the calculated peptide mass against databases can determine the origin of a peptide and then identify a protein by its "peptide footprint." The combination of increasingly automated techniques and the improvements in data handling have made proteomics and MS/MS a powerful technique for drug discovery and diagnostics in the clinical setting. For example, several companies are now in trials to develop diagnostic tools for identifying cancer stages using proteomic scanning of body fluids (blood and urine). In addition, it is possible to monitor the efficacy of a treatment by examining markers in body fluid, using high-throughput proteomic techniques. Comparing the proteome in a disease state versus control, or in a disease state plus or minus a specific treatment, is called protein profiling. Therefore treatments will become more individualized with the utilization of genomics and proteomics, thus making strides toward the promise of personalized medicine to improve health care and change the landscape of health care delivery.

Transgenic Mice Offer Unique Models of Genetic Diseases

Medical science has prospered dramatically when there is the ability to model a disease process in animals. The mouse has presented an excellent model for much of this work, in large part because of the tremendous knowledge base of mouse genetics. Many experiments mapping human genes and diseases have relied on somatic cell techniques using mouse cells. The finding of rodent genes that can complement the function of the human counterparts and the similarities observed for many genes at the primary structure (DNA sequences) and chromosome locals between mice and humans lend support for the use of mice as human disease models.

It has been demonstrated in a variety of eukaryotic systems that linear DNA fragments introduced to a cell will be ligated together to form tandem arrays (i.e., DNA molecules joined end to end) and can be integrated into the genome at random positions. This provides a technique to test the function of a gene within the confines of the recipient cell. If the gene-modified chromosome can enter the germline cells, it can then be passed on to progeny. These animals, if maintained, contain permanently

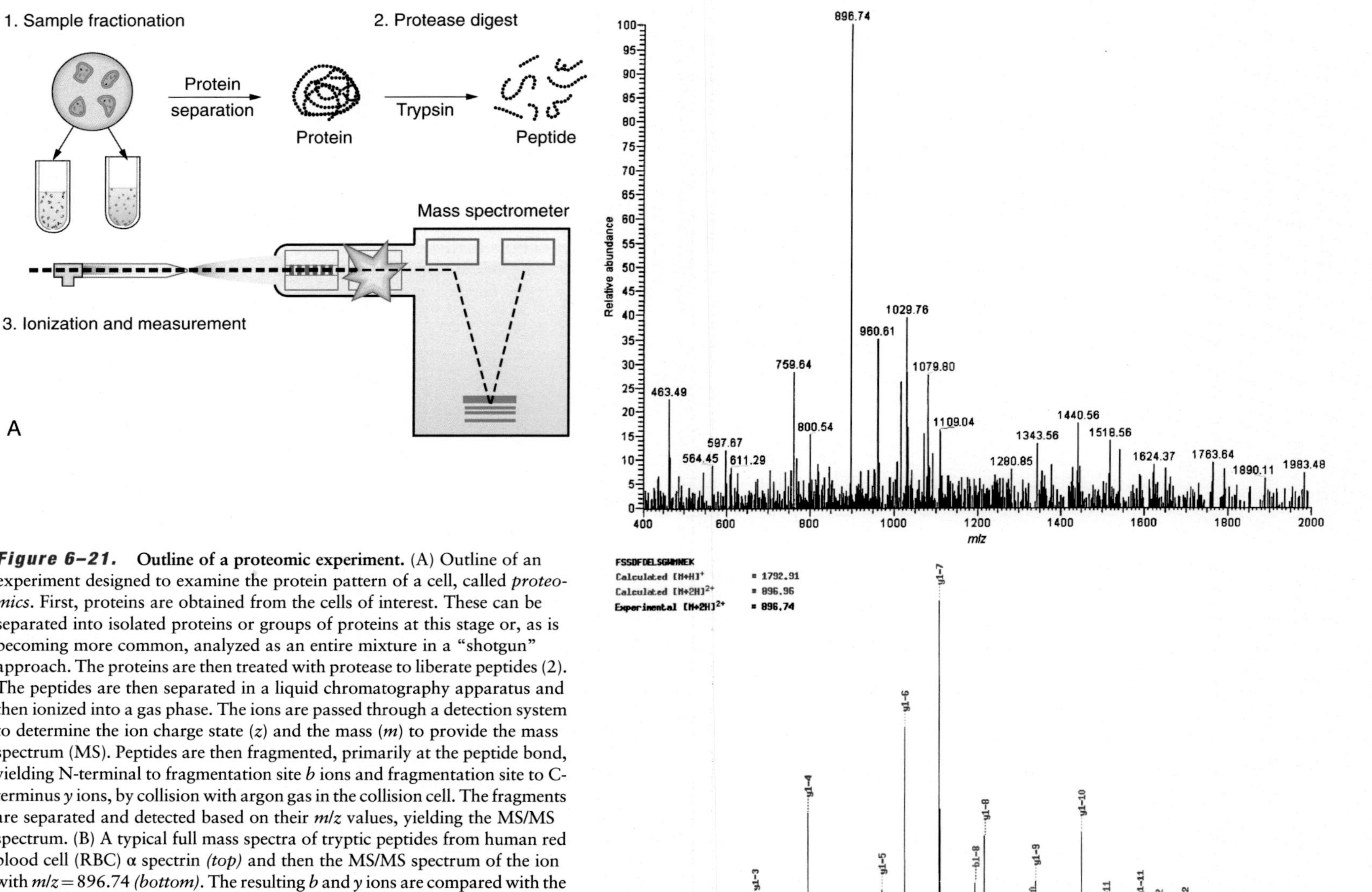

Figure 6–21. **Outline of a proteomic experiment.** (A) Outline of an experiment designed to examine the protein pattern of a cell, called *proteomics*. First, proteins are obtained from the cells of interest. These can be separated into isolated proteins or groups of proteins at this stage or, as is becoming more common, analyzed as an entire mixture in a "shotgun" approach. The proteins are then treated with protease to liberate peptides (2). The peptides are then separated in a liquid chromatography apparatus and then ionized into a gas phase. The ions are passed through a detection system to determine the ion charge state (z) and the mass (m) to provide the mass spectrum (MS). Peptides are then fragmented, primarily at the peptide bond, yielding N-terminal to fragmentation site b ions and fragmentation site to C-terminus y ions, by collision with argon gas in the collision cell. The fragments are separated and detected based on their m/z values, yielding the MS/MS spectrum. (B) A typical full mass spectra of tryptic peptides from human red blood cell (RBC) α spectrin *(top)* and then the MS/MS spectrum of the ion with $m/z = 896.74$ *(bottom)*. The resulting b and y ions are compared with the known molecular mass of each amino acid to generate the peptide sequence. In this case the peptides sequence in FSSDFDELSFWMNEK can only be generated from RBC α spectrin. This sequence analysis provides unequivocal analysis of the peptide and the parental protein from which it was derived. *([A] Modified from Aebersold R, Mann M. Mass spectrometry-based proteomics.* Nature *2003;422:198–207, and [B] MS and MS spectra provided by S. Goodman.)*

Figure 6–22. **Nkx3.2-deficient mice have altered limb growth.** Two-month-old mouse that does not express the regulatory molecule Nkx3.2. These mice do not have normal gastrointestinal development, and as illustrated, they exhibit altered growth of the forelimbs. This altered growth was not predicted, but the phenotyping of this mouse strain has shown that multiple organ systems can be affected through a single molecule. Furthermore, in annals of pediatric genetics and development, cases describe linking birth defects in certain bone growth and gastrointestinal problems. Thus this mutant mouse has helped to describe the molecular basis for defects in seemingly unrelated organ systems and provides an important model of human disease. *(Courtesy Dr. Monique Stanfel.)*

altered genomes and are said to be transgenic animals with the foreign DNA referred to as **transgenes.**

An example of a gene-targeting mouse experiment is shown in Figure 6–22. In this mouse, deleting a particular segment from the mouse genome using homologous recombination turned off a gene encoding a factor important in the regulation of genes in visceral smooth-muscle cells. The gene, called *Nkx3.2*, indeed was found to be important for proper development of gastrointestinal tract smooth muscle because these mice demonstrate altered stomach and intestinal morphologies. Another interesting and unexpected result from this experiment (see Fig. 6–22) is that Nkx3.2 also exerts effects on certain bone structures such that Nkx3.2-deleted mice have altered limb growth and "kinky" tails. This occurs from an altered regulation of growth and development of only certain bone structures, such as the forelimbs shown in the Nkx3.2-deficient mouse. This result indicated that multiple mesenchyme-derived structures could be linked through a single regulatory molecule and has provided a clue into an understanding of human birth defects that exhibited anomalies in seemingly unrelated organ systems.

Many transgenic mice have been formed by the introduction of a human gene that houses altered information leading to disease, allowing the function of the human gene to be examined. Furthermore the recent ability to add DNA that integrates at a specific location as a gene within the mouse genome (e.g., gene targeting) has allowed the ability to generate transgenic mice with altered expression of the mouse gene or to generate a mouse that expresses an altered gene product, that is, altered at the information storage level of the genome. These types of experiments have provided powerful knowledge of human disease states, from the ability to model the disease, and have enhanced our understanding of genetics so that gene-based therapies for human diseases are now within reach.

Gene Expression: The Transfer of Information From DNA to Protein

According to standard definitions the word **gene** is defined as "a complex molecule associated with the chromosomes and acting as a unit or in various biochemically determined combinations in the transmission of specific hereditary characteristics." The word **expression** is defined as an "outward indication or manifestation of some feeling, condition, or quality." Thus the phrase **gene expression** is an outward indication or manifestation of the complex molecules associated with the chromosomes.

What is a measure of this outward manifestation? The most obvious answer is that the appearance of a functional protein within a given cell would be the outward indication of the expression of a gene. The aggregate functional capabilities of the collective set of proteins expressed in a particular cell are what specify the biochemical and phenotypical properties of the cell. For example, skeletal muscle is composed of elongated, multinucleated cells that are able to contract when stimulated by neuronal input because of the collective set of proteins expressed in these cells. This does not mean that all the proteins found in a skeletal muscle cell are expressed exclusively in this cell type; in fact, many of these proteins are expressed in a variety of cells. However, there are proteins unique to skeletal muscle, implying that there must be mechanisms that stringently govern their appearance. The appearance of proteins in a cell occurs by a transfer of the information housed within the genetic material to the machinery responsible for protein synthesis. Thus the mechanisms that govern the expression of proteins must, in some way, act on this information transfer.

The Basic Steps of Information Transfer Are Transcription and Translation

The process of information transfer from DNA to protein involves two major steps: transcription and translation (Fig. 6–23). Transcription occurs in the nucleus and is the synthesis of a single-stranded RNA copy of the information stored in the double-stranded DNA molecule. Synthesis of the RNA is catalyzed by the enzyme RNA polymerase, and the synthesis of RNA from DNA is asymmetric; that is, only one strand of the DNA is transcribed to make the RNA copy. In eukaryotes, three types of RNA molecules are transcribed: transfer RNA

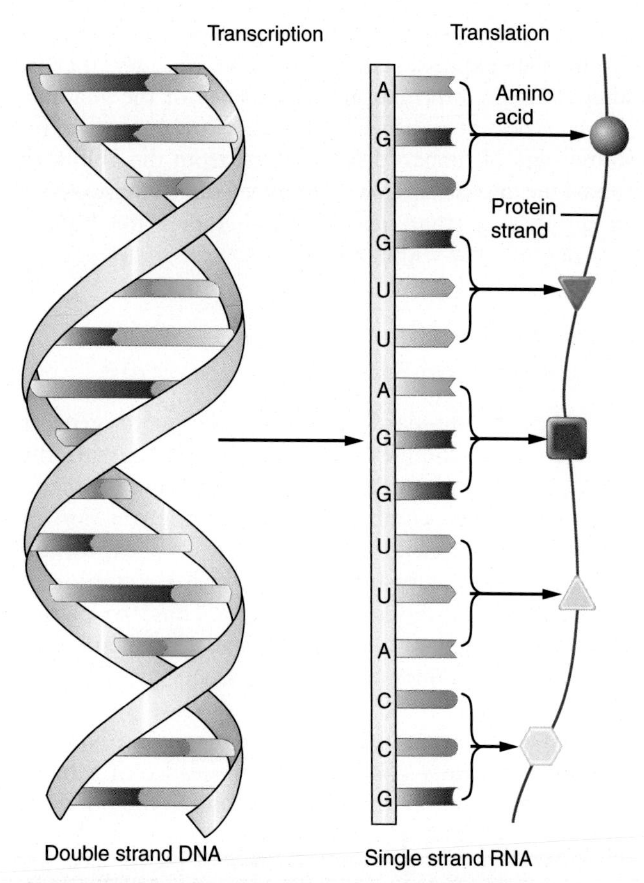

Figure 6–23. **Two major steps are involved in information transfer from DNA to protein.** The first step involves the synthesis of a single-stranded RNA molecule from a double-stranded DNA template. This process, catalyzed by the enzyme RNA polymerase, is called *transcription*. The product of transcription by eukaryotic RNA polymerase II is an RNA that directs the synthesis of a protein molecule by the translation machinery, the *ribosome*. The genetic code of three nucleotide units, the codons, stored in the double-stranded DNA, are passed to the single-stranded RNA molecules (the messenger RNA), which are the "words" that specify a particular amino acid building block of the protein sequence.

(tRNA), rRNA, and mRNA, each of which is synthesized by a different RNA polymerase. rRNA (transcribed by RNA polymerase I) and tRNA (transcribed by RNA polymerase II) are often considered to be structural RNA because they make up the integral components of the translational machinery, the ribosome. The RNA molecules that are synthesized by RNA polymerase II have a special role; they carry the information needed to code a protein sequence from the DNA to the ribosome. The information is stored in this RNA as discrete "words" made up of three nucleotides, referred to as **codons.** Because this RNA provides the link between the genome and the protein synthesis machinery, it is called **messenger RNA (mRNA).**

Regardless of the RNA that is synthesized, the process of transcription begins with binding of RNA polymerase to specific sequences of DNA in or near the gene, called **promoter DNA elements.** Synthesis of the RNA then occurs by the progressive addition of ribonucleotides to form an RNA chain with the polarity of synthesis being 5′ to 3′. Therefore transcription is a multistep process, and the interruption or enhancement at any of the steps necessary for transcription would be important factors for the regulation of expression. Transcriptional controls are thought to be paramount regulatory mechanisms because the synthesis of RNA from the DNA template constitutes the primary step of gene expression.

The second basic step of information transfer is the translation of codons encrypted in the nucleic acid of the mRNA into a chain of amino acids. Protein synthesis is conducted by the ribosomes, and in eukaryotic cells, this synthesis occurs in the cytoplasmic compartment of the cell. Ribosomes are large complexes of RNA and protein composed of one large subunit and one small subunit that come together to form the protein synthesis machinery. The basic function of the small subunit is binding the mRNA and tRNA, whereas the large subunit is required for catalyzing the peptide bonds of the growing protein chain. Thus the synthesis of a protein requires all three of the RNA classes synthesized in the nucleus working in concert in the cytoplasm.

Translation occurs through a series of distinct steps (Fig. 6–24). The first step is binding of the small ribosomal subunit with an mRNA molecule. The small ribosomal subunit is prepared for binding of mRNA by its association with proteins, referred to as **initiation factors,** and its binding of special tRNA, called the **initiator tRNA,** which contains a methionine amino acid residue. The binding of the small ribosomal subunit occurs at or near the 5′ end of the mRNA, after which the activated ribosomal subunit scans the messenger for an AUG initiation codon for methionine. This results in the formation of an initiation complex.

After the formation of the initiation complex, the large ribosomal subunit binds to the small-subunit mRNA structure, permitting progressive "reading" of subsequent codons. The reading of mRNA involves the recognition of the three-nucleotide codon, with recruitment and binding of the appropriate tRNA-amino acid complex to the ribosome. Once the tRNA is bound, the amino acid it is carrying is linked to the growing protein chain by peptide bond formation, catalyzed by the large ribosomal subunit. The formation of the peptide bond causes a release of the tRNA from the previous cycle, a shift of the tRNA polypeptide chain within the ribosomal complex, and the subsequent reading of the next codon sequence. This cycling continues until the ribosome encounters a codon that specifies the end of the protein, called the **termination** or **stop codon,** after which the mature protein is released from the ribosome. The ribosome-mRNA complex dissociates into separate components, which, after the appropriate charging of the small ribosomal subunit, can reassemble to make more protein.

Because mRNA is composed of four nucleotides (adenosine, cytosine, guanosine, and uridine), there are

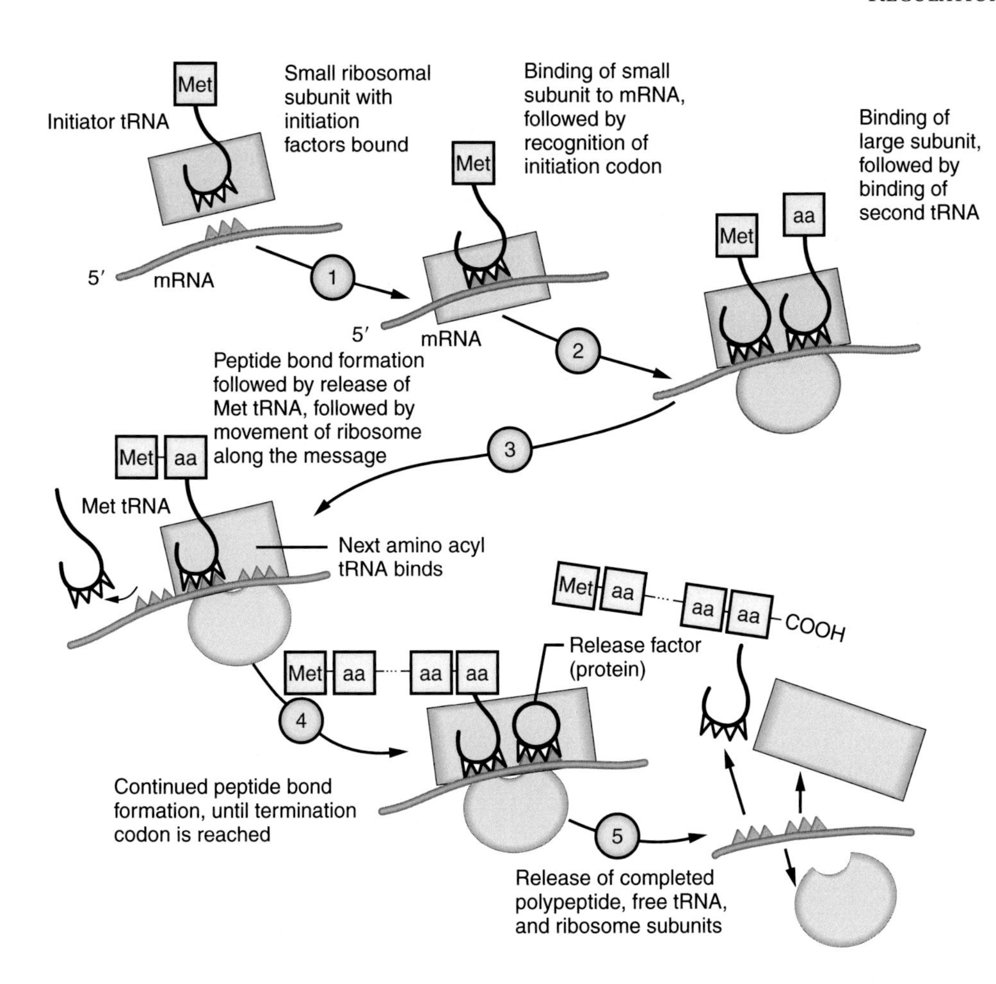

Figure 6–24. **Scheme of events for protein translation from a messenger RNA (mRNA) molecule.** The process of translation begins with binding of activated small ribosomal subunits with the mRNA. On identification of an initiation codon (AUG), the large ribosomal subunit binds to form the translation machinery that recruits tRNA specified by the three nucleotide codons of the mRNA. This enables the ribosome to join the amino acids of the polypeptide chain. When a termination codon is encountered, the protein chain is released, and the translation machinery breaks up, which can then reform a translation complex on reactivation of the small ribosomal subunit. *(Modified from Widnell C, Pfenninger K.* Essential Cell Biology. *Baltimore: Williams & Wilkins, 1990, by permission.)*

64 possible combinations available to form the three nucleotide codon sequences. Three of these nucleotide combinations, UAA, UAG, and UGA, do not specify an amino acid, but rather instruct the ribosomal machinery to end the protein synthesis; they are referred to as **stop codons.** This leaves 61 possible combinations to encode only 20 amino acids; the amino acids are specified by binding of complementary nucleotides carried in specific tRNA molecules. Two amino acids, methionine and tryptophan, are encoded by a single codon, and the other 18 amino acids are encoded by multiple codons. Thus the genetic code is said to be degenerate, meaning that most amino acids are specified by more than one triplet sequence.

In prokaryotic cells, mRNA is available to the translational machinery as soon as it is transcribed. Translation is often observed to begin before the transcription process is finished. This is not true in eukaryotes, in which there is a separation of the transcriptional and translational machinery. In eukaryotic cells, transcription occurs in the nucleus, and the transcribed product, called the **primary RNA transcript,** is often larger than the final, mature mRNA. The primary RNA transcript is modified or processed to form the mature RNA, which is then transported from the nucleus to the cytoplasm to participate in protein synthesis. Thus gene expression in eukaryotes is more complex than that of prokaryotes; it requires additional steps, and each step in the process provides a potential point of regulation.

Each Cell Type of Multicellular Organisms Contains a Complete Complement of Genes

Two central observations indicate that the entire blueprint, or genetic plan, for an organism is contained within the nuclei of all different cell types. First, examination of the DNA content of different cell types within an organism demonstrated that all somatic cells have approximately equal amounts of DNA, and this DNA content was twice that found in the gametic cells. For example, physical, chemical, and kinetic experiments showed that a liver cell contained the same DNA content as a brain cell. This is referred to as the **constancy of DNA.** However, it is difficult to imagine that cells that differ so dramatically in morphologic structure and function do not suffer some type of irreversible change in genetic material (e.g., loss of nonessential genes) during the progressive specialization of the cells during development. Because only a fraction of the total genetic material is expressed in each cell, it was reasoned that a loss of genetic material might be so minimal that it had escaped detection when examining gross DNA content. This was

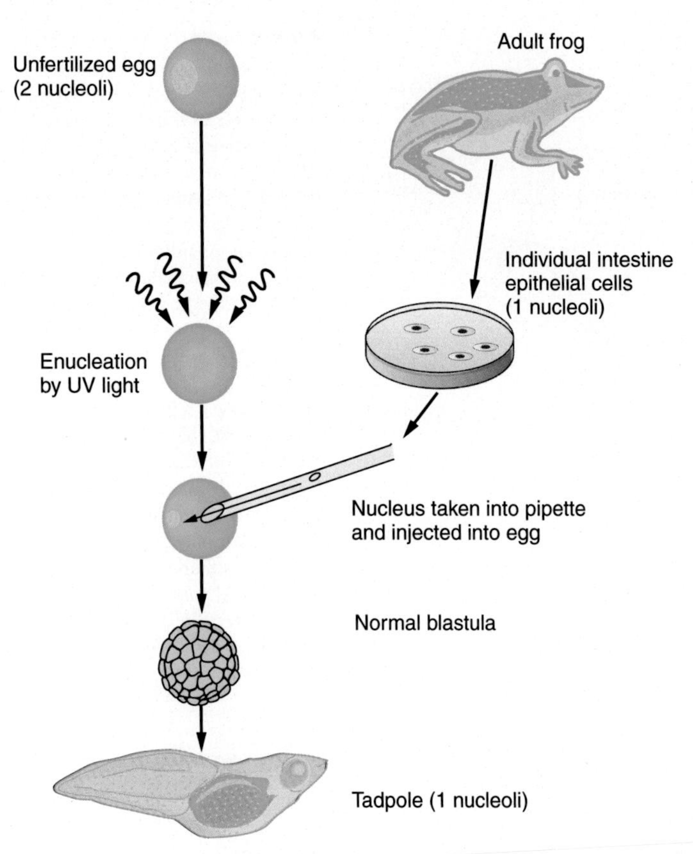

Figure 6–25. **Nuclear transplantation experiments to examine the constancy of DNA in eukaryotic cells.** Diagram outlines an experiment that examined the capacity of nuclei from an adult differentiated tissue to express specific gene sequences. Nuclei from the intestinal epithelium of an adult frog were injected into an oocyte that had had its genetic material inactivated by ultraviolet (UV) light. Genetic markers for the two kinds of nuclei were derived from the difference in the number of nucleoli expressed in the cells: The donor nucleus had one nucleolus, and the acceptor oocyte had two nucleoli. The injected nuclei have the capacity to express all the genes necessary to make a tadpole and subsequently an adult frog, demonstrating that differentiated cells contain an entire complement of genes to specify the many differentiated cells of an adult organism. *(Modified from Gurdon JB.* Sci Am *1968;219:24–35, by permission.)*

shown to be incorrect by an elegant set of experiments by Gurdon and colleagues, who examined differentiation in frogs. The seminal experiment from these studies is diagrammed in Figure 6–25.

When a nucleus of a fully differentiated frog cell (e.g., intestinal epithelium cells) was injected into a frog egg from which nuclei had been removed, the injected donor nucleus was capable of programming the recipient egg to produce normal, viable tadpoles. Because a tadpole contains the full range of differentiated cell types found in adult frogs, it was concluded that the nucleus of the original differentiated donor cell contained all the necessary information to specify the frog's many different cell types. No irreversible loss of important DNA sequences was evident in the differentiated cell nucleus; the development of an organism requires the expression of a specific set of sequences at the appropriate time. These seminal findings have been extended with recent experiments cloning a number of mammalian species, including sheep, cows, cats, and mice. These experiments will ultimately expand medical therapeutics as better techniques become available to replace diseased organs with newly constructed systems, perhaps grown in culture systems, derived from individual genomes. Such therapies called **regenerative medicine are discussed in Chapter 13.** *Stem Cell Biology and Regenerative Medicine* have the potential to cure or treat a vast number of diseases; however, a convergence of science and ethics is necessary to determine how this potential is used.

The Molecular Definition of a Gene as a "Unit" of Information

The primary function of the genome is to produce RNA molecules, and only specific regions of the DNA sequence are copied or transcribed into a corresponding RNA nucleotide sequence that can function to encode a protein (mRNA) or as structural molecules (tRNA and rRNA). Therefore every segment of the DNA molecule that produces a functional RNA would constitute a gene.

Until recently, genes were defined primarily by their abilities to confer a biochemical or phenotypic trait on a cell. The application of molecular cloning technologies has allowed a refinement of the definition of a gene based on the structural organization of the DNA nucleotide sequence. One of the most notable findings from these analyses is that many of the DNA sequences transcribed by RNA polymerase II produce functional mRNA that represent more nucleotides than are found in the cytoplasmic messenger.

One of the first eukaryotic genes in which extra DNA sequences were identified was the chicken ovalbumin gene (Fig. 6–26). The extra length of nucleotides consists of long stretches of noncoding DNA that interrupt the segments of informational DNA. The informational or coding sequences are called **exons**, and the interrupting stretches of noncoding DNA are referred to as **introns**. Thus the noninformational segments of the RNA molecule as synthesized from the DNA (called the **RNA**

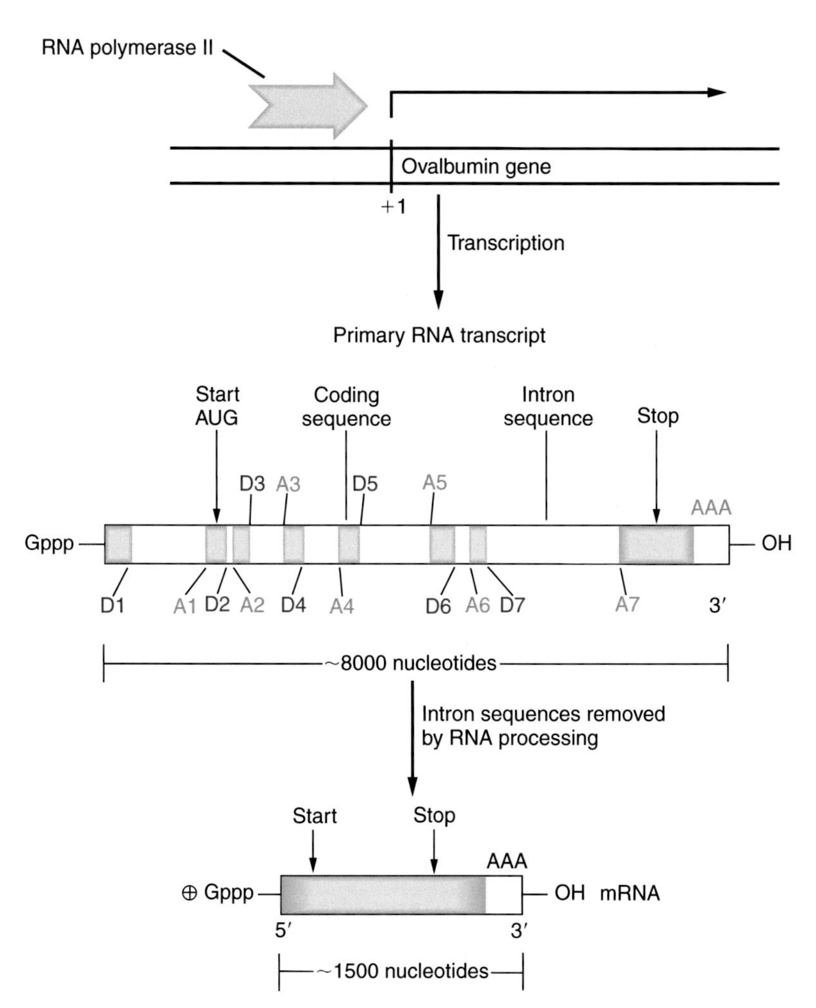

Figure 6–26. **The chicken ovalbumin gene makes RNA that is larger than the translated messenger RNA (mRNA).** Transcription of the ovalbumin gene leads to the formation of RNA that is larger than the mRNA found in the cytoplasm for translation. This larger RNA, called the *primary RNA transcript*, contains segments of noninformational content—introns—that are found in the double-stranded DNA of the gene. The intron segments are removed, and the segments containing informational content are joined together at specific sites, called *donor* (D) and *acceptor* (A) sequences, to form the mature mRNA by a mechanism of RNA processing.

primary transcript) must be removed and segments and informational content joined together to form the mature mRNA. These events occur in the nucleus and are termed **RNA splicing**. Therefore, in eukaryotes, a gene does not necessarily reflect, nucleotide for nucleotide, the functional RNA that it encodes.

A second refinement of the definition of a gene has evolved from a correlation of structural organization of the DNA sequence with the ability to visualize regions of chromatin supporting active transcription. Electron micrographs of transcribed DNA segments (genes) demonstrate the transcribed segment as an expanded, bead-on-a-string region of chromatin (e.g., ~11-nm chromatin fiber) on which RNA polymerase molecules appear as globular particles, with a single RNA molecule trailing the polymerase particle. Active RNA polymerase II molecules are often observed to be single units, indicating an infrequent transcription of many gene segments.

However, on occasion, many polymerase particles and associated transcripts are observed on a single gene, indicative of high-frequency transcription. In that situation the length of the associated RNA molecules is observed to increase progressively in the direction of transcription. Micrographs of such gene regions demonstrate a characteristic "Christmas tree" pattern. More importantly, these experiments demonstrated that transcription of a DNA segment begins and ends at discrete sites, defining a transcription unit (Fig. 6–27).

Because transcription begins with the binding of a polymerase molecule to specific DNA segments, an expanded definition of a gene as a unit of transcription must then take into account the segments of DNA that are associated with the transcribed segments. These associated DNA regions are said to direct or promote transcription. Comparisons of DNA sequences from several genes show that certain of these elements are conserved in sequence and position relative to the transcribed sequences.

Because nuclei of somatic cells from multicellular organisms contain a complete complement of genetic material, the drastic physiologic and biochemical differences among differentiated cell types must arise by the specific activation of individual genes, allowing their expression within a given cell. This activation process includes sequences that are copied into an RNA molecule and adjacent DNA sequences that are required for appropriate transcription, causing transcription units. Moreover the primary transcripts synthesized from

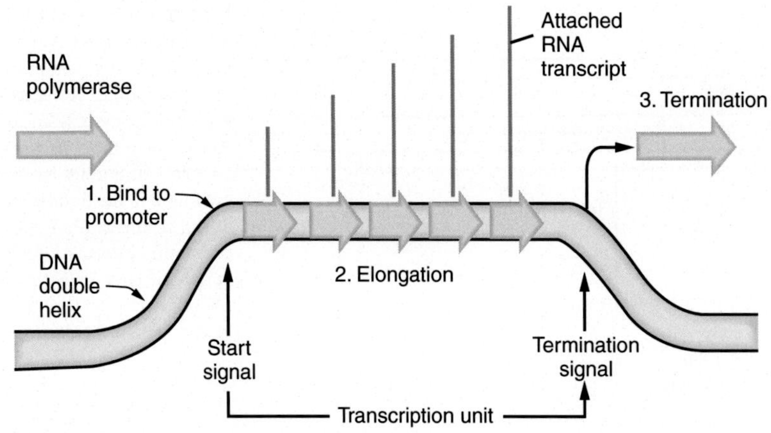

Figure 6–27. **Model of a transcription unit, derived from electron microscopy experiments, visualizing transcribed DNA.** A transcribed segment of DNA appears as an unfolded 11-nm chromatin fiber that is thought to be looped from the chromosome. Transcription begins with the binding of an RNA polymerase to a promoter DNA element that specifies the starting place for polymerization of the RNA molecule. The growing RNA chain remains linked to the polymerase as transcription proceeds, forming the characteristic "Christmas tree" pattern observed in the electron microscope. Transcription terminates by the release of the primary RNA transcript, which undergoes RNA-processing reactions. Thus the transcription unit is defined by discrete start and termination sites of the transcription process. *(Modified from Alberts B, Bray D, Lewis J, et al.* Molecular Biology of the Cell, *2nd ed. New York: Garland Publishing, 1989, by permission.)*

eukaryotic genes contain informational segments (exons) and noninformation segments (introns) and must be significantly modified to form a functional RNA sequence.

Accordingly the pathway of information transfer from DNA to protein in eukaryotic cells involves a complex set of steps; altering the pathway at any of these steps can be a point of regulatory control for gene expression. As shown in Figure 6–28, a cell may exert control of the proteins it expresses by the following steps:

1. Transcription control specifying when and how often a gene sequence is copied into RNA.
2. Processing of the primary transcript altering the modifications of the synthesized RNA molecule to form a functional mRNA.
3. RNA transport selection of which mRNA is exported from the nucleus to the cytoplasm.
4. mRNA stability selectively degrading the mRNA molecules in the cytoplasm.
5. Translation control selection of which mRNA in the cytoplasm is translated.
6. Protein posttranslational control activating, inactivating, or compartmentalizing specific polypeptide chains after they have been translated.

Although there is evidence to show that each of these steps may function as points of gene expression control, the expression of any given gene is governed by the collective set of interactions along the pathway of information transfer.

Transcriptional Control Requires Two Basic Steps: Activation and Modulation of Gene Sequences

The control of gene transcription occurs at two levels: activation, the conversion of compacted chromatin to an extended structure, and modulation, the fine-tuning of transcription mediated by DNA-binding proteins called **transcription factors** (Fig. 6–29).

When chromatin is in its most compacted state, such as at metaphase of the cell cycle, there is little or no RNA transcription. This inhibition of transcription occurs because the DNA is packaged so tightly to form the metaphase chromosomes that it becomes inaccessible to the nonhistone proteins responsible for transcription (RNA polymerases, transcription factors, and others). To be transcribed a gene sequence must first be made available to the RNA polymerases and other proteins (activated state) so that the regulatory proteins can provide their functions of influencing the rate that a gene is transcribed (modulation).

In electron micrographs of transcription units (a DNA segment that is being transcribed), the chromatin appears as an 11-nm fiber. Nucleosomes are present; however, the transcribed segment of DNA is in an extended fiber arrangement of chromatin. These observations correlate well with experiments showing that transcriptionally active gene segments are arranged in a chromatin structure that is biochemically distinct from inactive genes. For example, if a gene is expressed in skeletal muscle

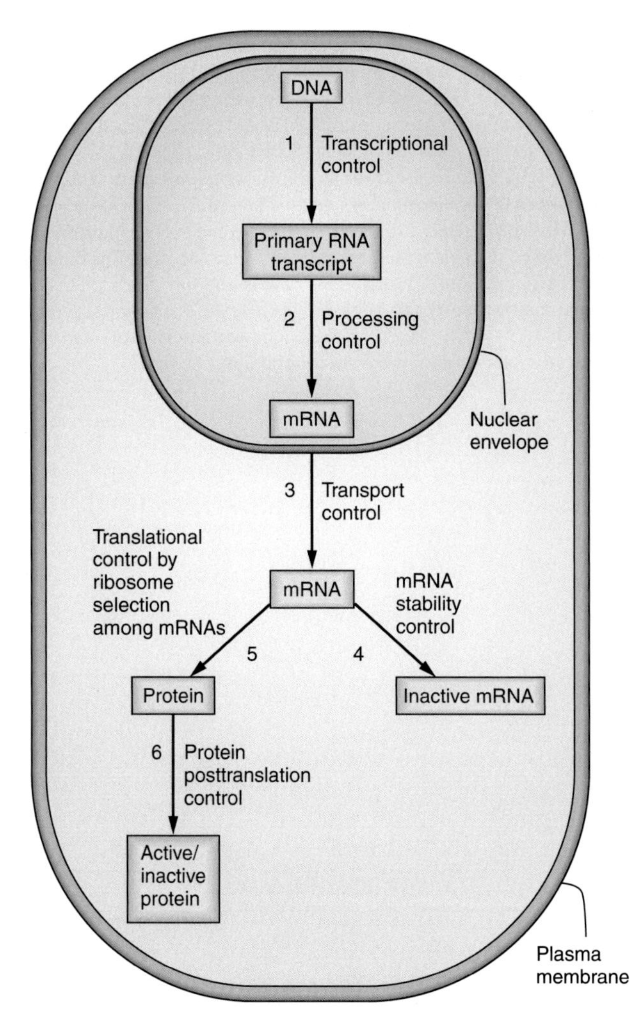

Figure 6–28. **Six steps of information transfer in eukaryotes that constitute potential regulatory points of gene expression.** Three steps of potential regulation occur in the nucleus of the cell: (1) transcription, (2) RNA processing, and (3) transport of the messenger RNA (mRNA) from the nucleus to the cytoplasm. Once in the cytoplasm, RNA is subjected to degradation: (4) mRNA stability control or selective translation and (5) translation control by the pool of ribosomes in the cytoplasm. The translated protein may be modified (6) to form an active protein, to inactivate the protein, or to compartmentalize the protein by posttranslational controls.

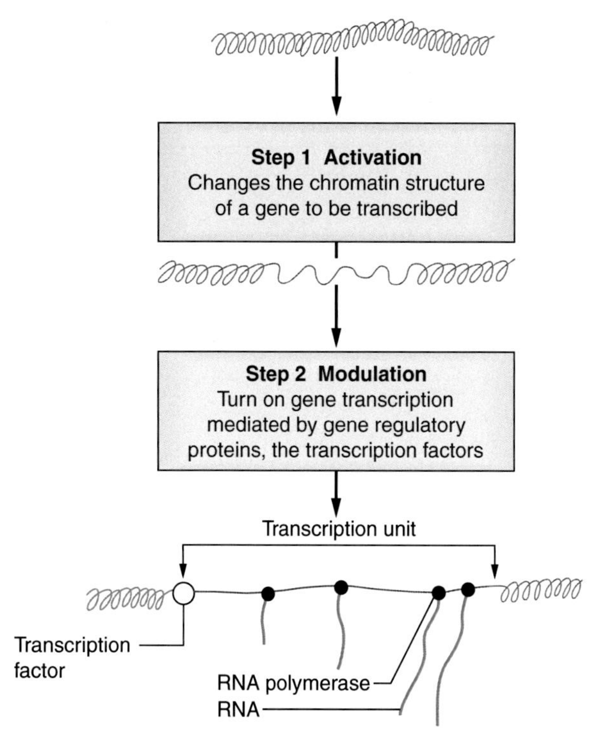

Figure 6–29. **Steps involved in the transcriptional control of gene sequences.** The transcriptional control of gene expression occurs in two discrete stages. The first step, *activation*, involves changes in the structure of chromatin containing the DNA sequences to be transcribed. The molecular interactions that cause this change in chromatin structure are unknown, but they result in the formation of a region of DNA-protein complex that is biochemically distinct from that of inactive (condensed) chromatin. The second step of transcriptional control, *modulation*, involves the binding of gene regulatory proteins to the activated DNA sequences that operate to fine-tune the transcription of a specific gene.

but not in liver cells, the gene sequence is more accessible to probes, such as nucleases, in nuclei isolated from skeletal muscle than in those isolated from liver cells. Chromatin in this accessible state is referred to as **active chromatin**, and its nucleosomes are thought to be altered in such a way that their packing is less condensed.

The mechanisms that form active chromatin are not understood. It is thought that active chromatin acquires an extended loop structure that emerges from the surrounding highly condensed chromatin. This model implies that there must be some way to recognize the gene sequences that are to be expressed, such as the possibility that they are converted by a specific protein-DNA interaction to an active, less-condensed form. Evidence now indicates that the conversion of chromatin from an inactive to an active form requires DNA synthesis and that the DNA near active gene sequence contains fewer modified bases, in particular, methylated cytosine residues. In addition, there is less histone H1 and more nonhistone proteins associated with active chromatin. It is unknown whether these changes cause the activation of gene sequences, or whether they occur subsequent to the conversion of the chromatin to an active state.

Eukaryotic cells contain a variety of sequence-specific DNA-binding proteins that function to modulate gene expression by turning transcription on or off. Collectively, these proteins are known as **gene regulatory proteins**. These proteins contain structural domains that can "read" the DNA, allowing their binding to specific sequences. DNA-binding domains are conserved, allowing the broad classification of regulatory proteins as helix-loop-helix, homeodomain, zinc finger, or leucine zipper proteins. Gene regulatory proteins are generally present in few copies in the individual cells and perform their function by binding to a specific DNA nucleotide sequence. The DNA sequences recognized by these proteins can be classified into two broad categories: the core or basal promoter sequences and the enhancer sequences (Fig. 6–30).

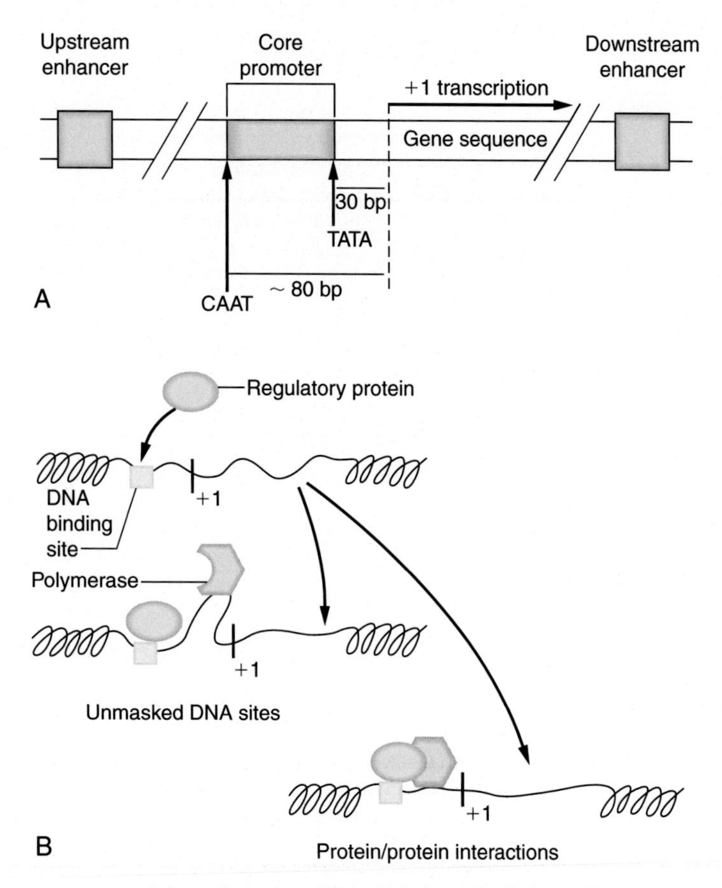

Figure 6–30. **DNA segments that can modulate transcription by binding gene regulatory proteins.** Several DNA sequences have the potential to bind gene regulatory proteins and are important in the modulation of transcription. (A) These can be divided into two categories: core promoter elements that are conserved A-T-rich DNA sequences required for basal transcription activity and enhancer DNA elements that may be placed 5′ (upstream) or 3′ (downstream) relative to the gene. (B) Two general mechanisms can be evoked to explain the action of DNA-binding sites: The binding of a gene regulatory protein may change the surrounding DNA, such that it is more favorable for binding the transcriptional machinery (polymerases), or the bound regulatory protein may directly interact with the transcriptional machinery.

Core promoter sequences are generally located close to the transcribed portion of DNA and function to specify the exact point of RNA chain initiation. These sequences are often enriched in adenine and thymidine bases located approximately 20–30 (called **TATA sequence**) and 70–80 bases (called **CAAT sequence**) to the 5′ side of the transcribed DNA segment. Because they appear to function in all cell types, these sequences are thought to promote basal transcriptional activity. The second class of DNA elements, the enhancer sequences, are regulatory DNA sequences that activate or enhance transcription from a core promoter, with RNA synthesis beginning at the site specified by the core promoter, and appear to function regardless of their sequence polarity and location relative to the transcribed DNA. The enhancer sequences are variable and capable of providing their function even when located at relatively long distances away from a transcribed gene. In some cases, enhancer sequences have been found in DNA that follows (to the 3′ side) the transcribed gene. Each of the different enhancer elements appears to bind a specific, distinct protein factor, allowing the specificity of its function to be based on the appearance or absence of the protein factor within the nucleus of a given cell. RNA polymerase catalyzes the synthesis of RNA from a DNA template at a rate of about 30–40 nucleotides per second. Because the rate of synthesis is constant, the absolute rate that a gene is transcribed is effectively regulated by the number of polymerases that are synthesizing RNA from the gene sequence. Thus the principal role of the gene regulatory proteins is to modulate the number of polymerase molecules actively synthesizing RNA from a given segment of DNA. That is, an increase in transcription requires an increase in active RNA polymerase molecules, and reduction in transcription is accomplished by fewer active RNA polymerase molecules. There are two ways that binding of a protein to a specific DNA sequence can affect gene transcription (see Fig. 6–30).

First, binding of a regulatory protein may affect the conformation of surrounding DNA. The binding may unmask sequences, allowing an optimized presentation of polymerase-binding sequences to increase transcription, or conversely the binding may present a block to transcription by tightening DNA conformation or by occupation of polymerase-binding elements or both.

Second a regulatory factor may interact directly with the polymerase. The binding of a protein to a specific sequence near the gene could create a complex that attracts polymerase molecules by an optimal binding complex formed between the polymerase and the regulatory factor. Recent studies have shown that some proteins do not bind DNA directly, but rather influence transcription through binding to a regulatory factor and perhaps facilitating interactions with RNA polymerase. Experiments have shown that both mechanisms are used (often in combination) and support the conclusion that the major function of regulatory factors is the recruitment of RNA polymerase molecules to a gene locus.

A way to have tissue- and cell-specific transcription of a gene would be simply to have a single regulatory region for each gene. Transcription could then be governed by a single protein-DNA interaction. However, most eukaryotic genes contain multiple DNA elements that are able to bind to different regulatory factors. The synthetic rate of a gene is then governed by the sum-total effects derived from multiple protein-DNA interactions. The best example of this regulatory scheme is the expression of the β-globin gene. The DNA surrounding the chicken β-globin gene contains 13 distinct sites (seven that are 5′ and six that are 3′ to the gene) capable of binding eight different protein factors (Fig. 6–31).

In early development the chromatin-containing β-globin gene is converted to an active state, and nine of the binding sites for regulatory factors are occupied, even though the gene is not transcribed. Later, all but one of the sites have bound their regulatory factors allowing transcription, after which the change in binding of proteins at the 5′ region of the gene again inhibits the transcriptional capacity of the β-globin gene. Even though this represents a singular example, most eukaryotic genes appear to be subject to transcriptional regulation derived from the sum of activities of multiple regulatory proteins. The cellular specificity of gene transcription is thereby governed by a delicate balance of regulatory protein activities.

A class of regulatory factors has been defined that is able to regulate multiple genes specifying a particular cell type (Fig. 6–32). MyoD1, a nuclear protein found in skeletal muscle, is an example of this class of master regulatory proteins. The introduction of DNA containing the *MyoD1* gene into cultured fibroblasts (cells that never express muscle-specific gene products) can convert the

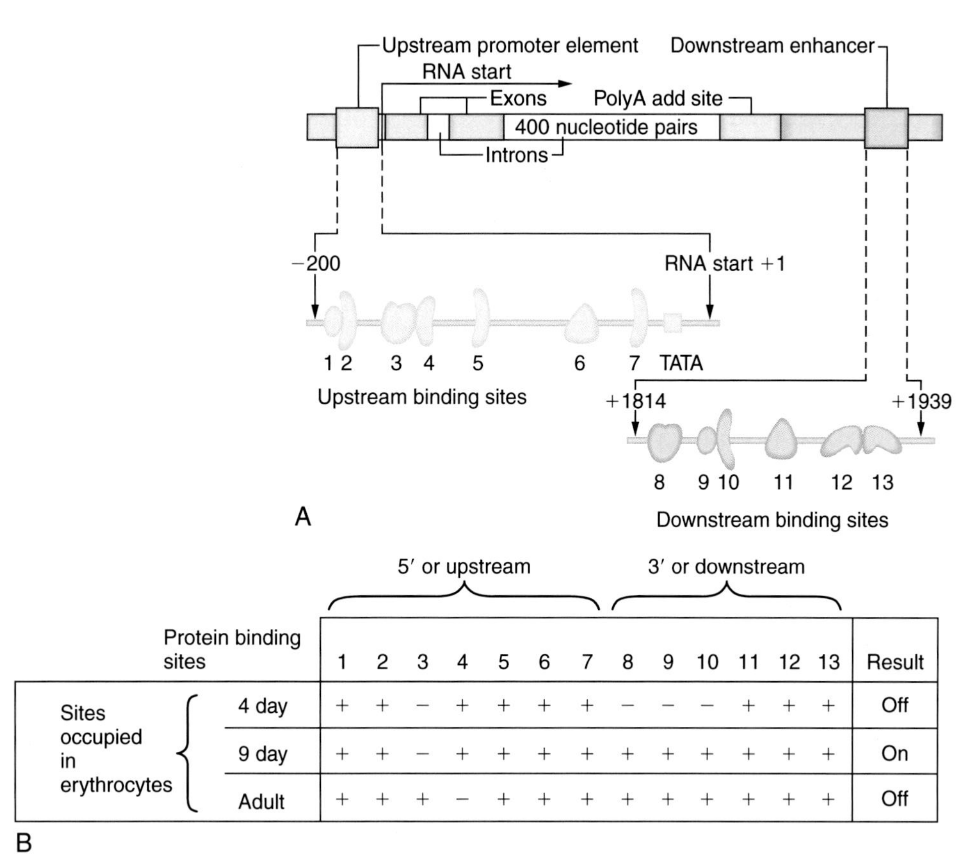

Protein binding sites		1	2	3	4	5	6	7	8	9	10	11	12	13	Result
Sites occupied in erythrocytes	4 day	+	+	−	+	+	+	+	−	−	−	+	+	+	Off
	9 day	+	+	−	+	+	+	+	+	+	+	+	+	+	On
	Adult	+	+	+	−	+	+	+	+	+	+	+	+	+	Off

Figure 6–31. Site of gene regulatory protein binding that controls the expression of the chicken β-globin during development. (A) Diagram of the chicken β-globin gene with the known binding sites of the 13 different gene regulatory proteins. Notice that some of the regulatory proteins (shown by the different shapes) have two binding sites (e.g., sites 3 and 8) and that some binding sites are close to each other—sites 1 and 2, 3 and 4, 9 and 10, and 12 and 13—to promote protein-protein interactions between the two regulatory factors. (B) The occupation of regulatory protein-binding sites during development is shown by the *plus* (+) occupied and *minus* (−) unoccupied sites. As indicated by this chart, the difference in the β-globin gene being on (9 days) and off (4 days and adult) is due to a balance of the activities of multiple gene regulatory proteins that may exhibit binding to different DNA sequences near the gene. *(Modified from Alberts B, Bray D, Lewis J, et al.* Molecular Biology of the Cell, *2nd ed. New York: Garland Publishing, 1989, by permission.)*

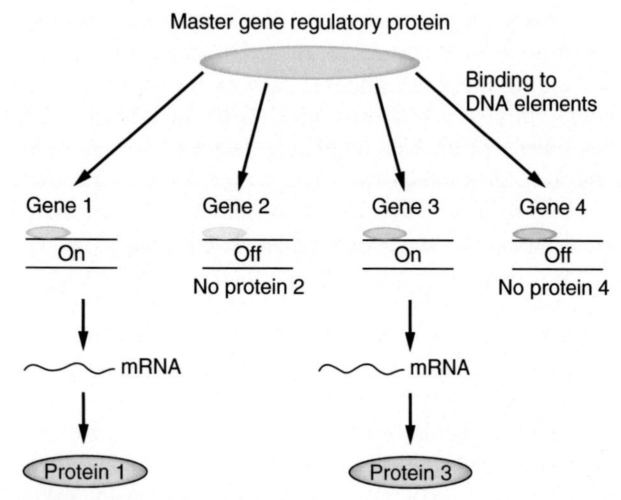

Figure 6–32. **Scheme of the activity of a master gene regulatory protein.** The expression of a protein, such as MyoD1, that has the ability to regulate the expression of multiple genes—some positive, some negative—could lead to cellular specialization. The activities of proteins that are subject to regulation by the master gene protein can be a biochemical trait for a differentiated cell (actin and myosin in skeletal muscle) or can be a protein that is required in the nucleus to regulate specific genes of the differentiated cell.

fibroblast to muscle cells, phenotypically and biochemically. The exact process by which MyoD1 is able to convert cells to a skeletal muscle phenotype is unknown; however, this protein binds to the regulatory region of several (but not all) skeletal muscle-specific genes. That this protein does not bind to the regulatory regions of all skeletal muscle genes implies that it must trigger other regulatory events required for the expression of the muscle phenotype. Furthermore the regulation of specific gene sequences must be governed by a combination of regulatory-binding activity. Regardless of the mode of action, the transcriptional regulation leading to cellular specialization is subject to a coordinate activation of gene sequences, with the subsequent modulation by the binding of sequence-specific regulatory protein factors.

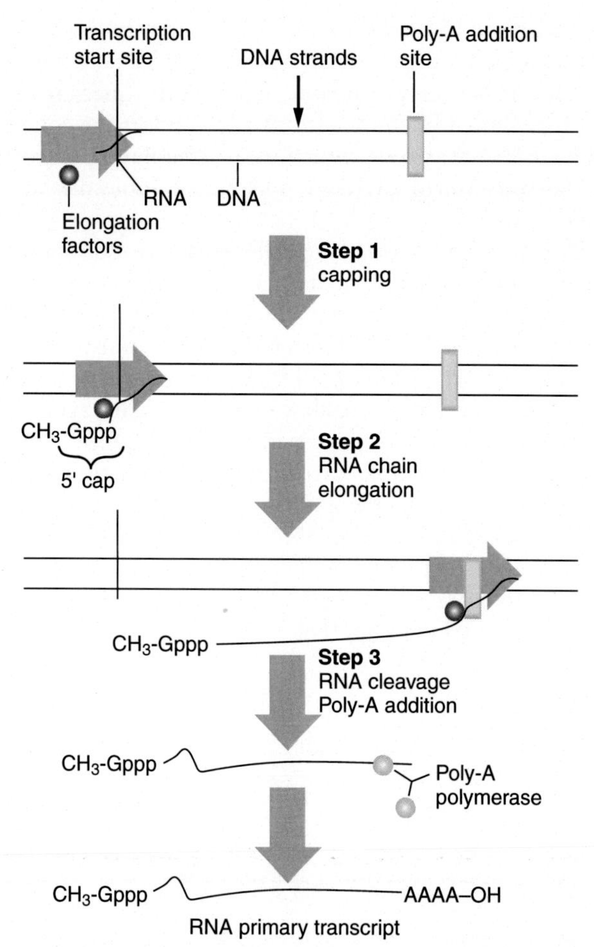

Figure 6–33. **Synthesis of a primary RNA transcript involves two modifications of the RNA strand.** Almost immediately after RNA synthesis is initiated, the 5′ end of the RNA is capped by a guanosine residue *(step 1)*, which protects it from degradation during the elongation of the RNA chain *(step 2)*. On reaching the signal sequence for the addition of the 3′ polyadenylic acid (poly-A) tail, the RNA is cleaved *(step 3)*, allowing the poly-A polymerase to add multiple adenosines to the 3′ end of the RNA. This RNA is the primary transcript and is ready for splicing of intron segments to form the mRNA. Although the steps involved in transcriptional termination are not well defined, one model is that the polymerase is altered in its activity and continues to synthesize RNA, but this synthesis is not productive because the RNA is degraded.

Primary Transcripts Are Modified to Form Mature mRNA

The primary transcripts of RNA synthesized by RNA polymerase II (mRNA) are modified in the nucleus by three distinct reactions: the addition of a 5′ cap, the addition of a polyadenylic acid (poly-A) tail, and the excision of the noninformational intron segments. These modifications are required to form a mature RNA capable of supporting the translation of a protein, and the entire set of events is called **RNA processing**. The 5′ end of the mRNA (the end that is synthesized first during transcription) is capped by the addition of a methylated guanosine nucleotide. The addition of the 5′ cap is the first modification of the mRNA primary transcript and occurs almost immediately with the onset of transcription (Fig. 6–33).

The formation of the cap involves the condensation of the triphosphate moiety of a GTP molecule with the diphosphate group of the nucleotide at the 5′ end of the initial transcript. The enzymes responsible for the capping reaction(s) are thought to reside within the subunit structure of RNA polymerase II. The addition of the 5′ cap structure is critical for mRNA to be translated in the cytoplasm and appears to be needed to protect the growing RNA chain from degradation in the nucleus.

The second modification of an mRNA transcript occurs at its most 3′ end, the addition of a poly-A tail. The 3′ end of most polymerase II transcripts is not defined by the termination of transcription, but by the specific cleavage of the RNA molecule and the addition of adenosine residues to the cleaved molecule by a separate polymerase, poly-A polymerase. The signal for cleavage is the appearance of the sequence AAUAAA in the growing RNA chain, with the actual cleavage occurring about 10–30 nucleotides away from this signal sequence. Immediately on cleavage, poly-A polymerase adds 100–500 residues of adenylic acid to the 3′ end of the cleaved RNA molecule. The RNA polymerase II appears to continue transcription well beyond the cleavage site, with the subsequent RNA being rapidly degraded, presumably because they lack 5′ cap structure. The exact functions of the poly-A tail are not well defined; however, experimental evidence suggests that it plays an important role in the export of mature mRNA from the nucleus to the cytoplasm. In addition, it may serve a regulatory function, in that some genes contain multiple sites for poly-A addition.

After modifications of the 5′ and 3′ ends of the primary transcript, the noninformational intron segments are removed, and the coding exon sequences are joined together by RNA splicing. The specificity of exon joining is conferred by the presence of signal sequences marking the beginning (called the 5′ *donor site*) and the end (called the 3′ *acceptor site*) of the intron segment (Fig. 6–34).

These signal sequences are highly conserved (they are approximately the same in all known intron segments), and as might be predicted, alterations in these sequences lead to aberrant mRNA molecules. For example, a group of genetic diseases, collectively called the **β-thalassemia syndromes** (characterized by the abnormally low expression of hemoglobin), are directly attributable to single-base changes in the genome at splice junctions of the β-globin gene that disrupt the appropriate joining of exon segments. Therefore splicing reactions must occur with exquisite precision to ensure that a functional RNA molecule is formed.

The excision of an intron segment from RNA is conducted by a ribonucleoprotein complex called the **spliceosome.** The spliceosome is formed from a set of undefined proteins complexed with a series of small RNA molecules

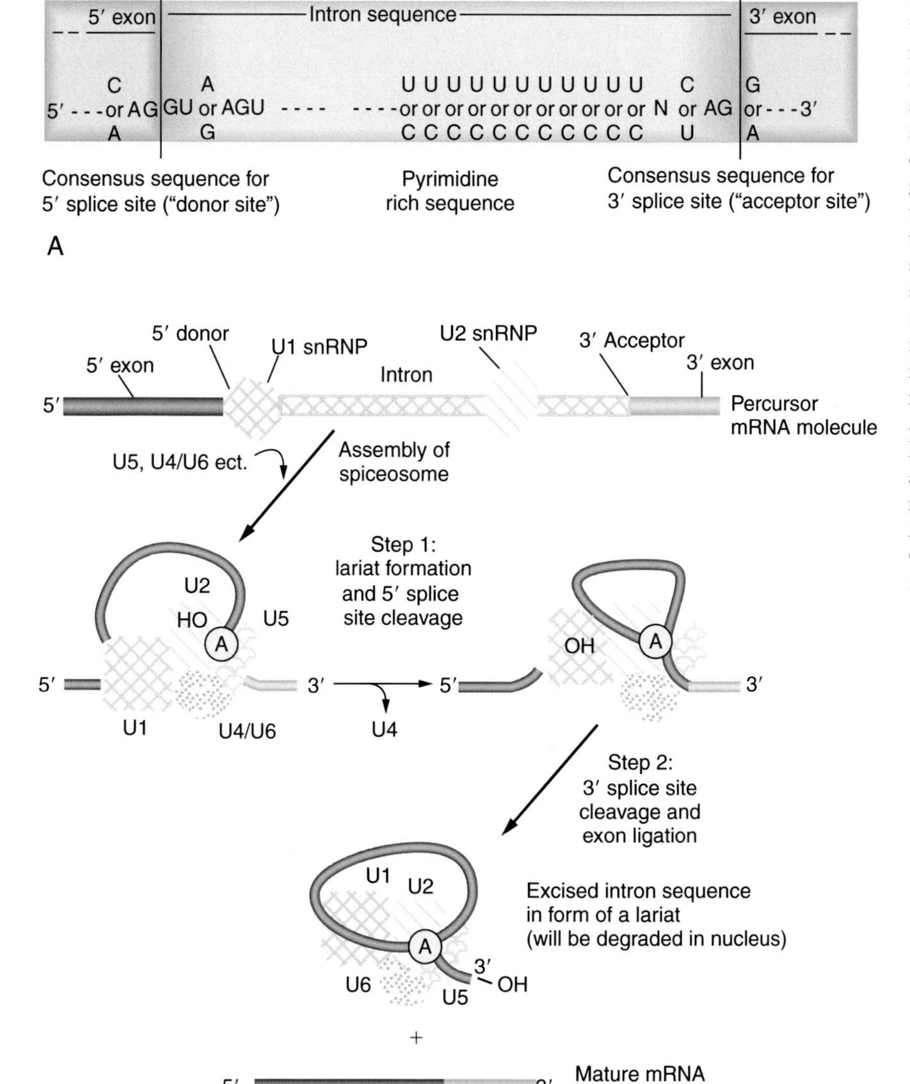

Figure 6–34. **Mechanism of RNA splicing to form mature messenger RNA (mRNA) molecules.** (A) RNA splicing occurs at discrete locations that are marked by conserved sequences. The consensus sequences for RNA splicing, listed here, have been determined by comparison of many eukaryotic polymerase II gene sequences. The most conserved nucleotides *(shaded regions)* mark the boundaries of the intron sequence. (B) The mechanics of RNA splicing involve the recognition of signal sequences by the U1 (5′ donor) and U2 (polypyrimidine sequence), which leads to the formation of the spliceosome [a combination of many small nuclear ribonucleoprotein (snRNP) molecules]. Once the spliceosome is formed, the 5′ donor is cleaved by the formation of an RNA lariat, the 5′ donor is then ligated with the 3′ acceptor, and the spliced intron is degraded into the nucleus. *(Modified from Alberts B, Bray D, Lewis J, et al.* Molecular Biology of the Cell, *2nd ed. New York: Garland Publishing, 1989, by permission.)*

referred to as *U1* to *U12*. The splicing reaction occurs in steps that include (1) recognition of consensus 5′ donor and 3′ acceptor sequences, (2) cleavage of the 5′ splice site and formation of a looped RNA structure termed the **lariat**, and (3) the cleavage of the 3′ place site and subsequent ligation of the RNA molecule. The exact role of individual components of a spliceosome is currently being studied; however, it is known that the excision of introns requires the energy of ATP hydrolysis. The excised intron is degraded almost immediately after its release from the primary RNA transcript.

Although it would be logical that the splicing of RNA proceeds by the removal of the most 5′ intron to the last or 3′ intron, experimental evidence has demonstrated that the removal of introns from any given transcript follows a preferred path, often beginning with introns internal to the transcript. This appears to be an inefficient mechanism, and initially, it was viewed with skepticism; however, the recent discovery that a single gene may express multiple different proteins by the selected joining of exon sequences to form different mRNA has shed some new light on the pathways of intron removal. For example, a single gene encoding the protein troponin T can produce at least 10 distinct forms of the molecule by simply joining different combinations of encoded exon segments. This variability can be influenced by cell type or by factors extrinsic to a cell, enabling the expression of protein isoforms needed to compensate for alterations in cellular metabolism. The ability of certain genes to form multiple proteins by joining different exon segments in the primary transcript is called **alternate splicing** and has caused a reexamination of the concept of "one gene, one protein."

RNA Transport to the Cytoplasm Occurs via the Nuclear Pore Complex

After the modifications of the primary transcript, the functional mRNA must transverse the nuclear envelope to the cytoplasm to direct the synthesis of a protein. Although this step is important, it is probably the least understood stop of the gene expression pathway. It is clear that transport of mRNA to the cytoplasm requires that RNA passes the nuclear envelope through the nuclear pores, presumably by an active transport mechanism that requires the recognition of the RNA, or a protein bound to it, by a receptor molecule either within the nucleoplasm or directly associated with the pore complex (Fig. 6–35).

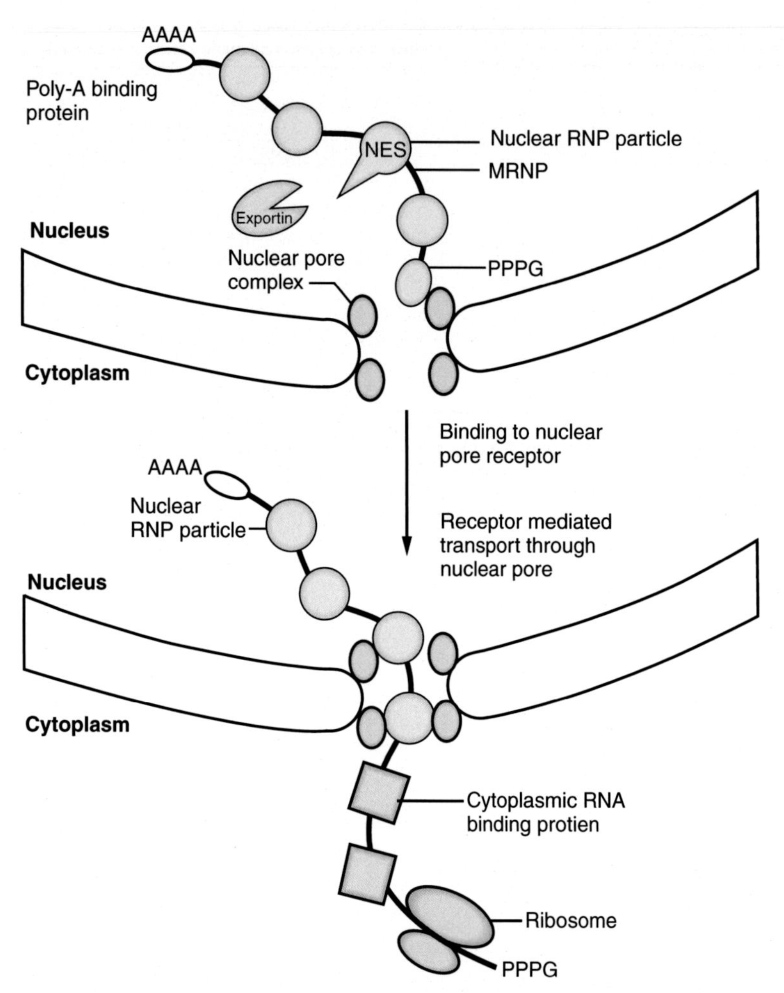

Figure 6–35. **Potential mechanism of messenger RNA (mRNA) transport through the nuclear pore complex.** mRNA that is ready for transport to the cytoplasm is bound by a variety of proteins, including a polyadenylic acid (poly-A)-binding protein and ribonuclear protein (RNP) particles. These proteins protect the mRNA from degradation and perhaps bind with a receptor molecule on the nuclear pore complex. Once bound the nuclear pore receptor may facilitate the transport of the mRNA to the cytoplasm, similar to the mechanism of protein import to the nuclear compartment.

Experiments have shown that primary transcripts that cannot be appropriately processed are retained in the nucleus and will be degraded; they are not allowed to transport into the cytoplasm until all processing steps are completed. Therefore there must be components within the nucleus that cause selective retention of RNA molecules. This selective retention might be operative on a broader scale, allowing the possibility of nuclear components serving as a filtering mechanism in the determination of which RNA is transported. Currently the mechanisms that allow the transport of RNA are largely unclear, although without this transport an RNA cannot complete the path of information transfer. It is clear that this involves receptors and an energy-dependent transport system through the pores, which is similar to protein import processes (see Fig. 6–4). Thus it is a necessary step that has potential regulatory control of gene expression.

RNA in the Cytoplasm Is Subject to Degradation

Once RNA reaches the cell's cytoplasm, it is subject to degradation by nuclease components resident in the cytoplasm; that is, RNA and other cellular components are continuously being replenished by a balance of their synthesis and degradation. In eukaryotic cells, mRNA is degraded at different selective rates. A measure of degradative rate for a particular mRNA is called its **half-life** (the period it takes to degrade an RNA population to half its initial concentration), and RNA with longer half-life measurements are said to be more stable. For example, β-globin mRNA has a half-life longer than 10 h, whereas RNA encoding the growth factors called *fos* and *myc* have measured half-lives in the same cells of approximately 30 min. Therefore the β-globin RNA is more stable than are *fos* and *myc*, and these experiments demonstrate the ability of selective mRNA degradation within the cell. In addition, this example is indicative of how selective degradation of mRNA might control expression. Simply stated a β-globin mRNA, by virtue of its longer resident time in the cytoplasm, can direct the synthesis of more protein because it is in contact with the synthetic machinery (the ribosomes) for a longer period than are *fos* and *myc* RNA.

The stability of mRNA can be influenced by extracellular signals. The primary response of cells to steroid hormones is an increased transcription rate of selective genes. However, the hormone also can influence the expression of these gene products by increasing the stability of their mRNA in the cytoplasm of the cell. Certain signals may cause the selective degradation of RNA, leading to less protein being expressed. For example, the addition of iron to cells decreases the stability of the mRNA that encodes the ion-scavenging transferrin receptor (Fig. 6–36).

The altered stability of the mRNA is mediated by a specific nucleotide sequence within the 3′-nontranslated region of the molecule. This region of the transferring receptor mRNA is bound by an iron-sensitive receptor protein that, when resident, protects the RNA from degradation. In the presence of excess iron, the receptor is dislodged from the 3′-nontranslated binding site, and the RNA is rapidly degraded, preventing the synthesis of the transferring receptor protein.

As suggested by experiments examining transferrin mRNA stability, the selective degradation of much mRNA is controlled, at least in part, by specific nucleotide sequences within the 3′-nontranslated region of the RNA molecule. This concept was demonstrated by genetic engineering, mixing, and matching specific regions from RNA displaying different stabilities, such as the experiments shown in Figure 6–29. When the 3′ noncoding segment of a stable mRNA, such as globin, is substituted for the analogous region of a nonstable growth factor mRNA (e.g., *fos*), the resultant growth factor mRNA displays a stability similar to that of the globin RNA. That is, the engineered growth factor RNA becomes more stable solely because of its new 3′-nontranslated segment. Similarly, when the 3′-terminus of histone RNA, an RNA that shows selective stabilization during the DNA S phase of the cell cycle, is placed onto a globin mRNA, the globin mRNA then acquires the cell cycle-dependent degradation characteristics of the histone mRNA. The conclusion drawn from these mix-and-match experiments is that the specific ability of RNA is governed, in part, by sequences resident within the 3′ noncoding portion of the RNA molecule. However, the cellular components responsible for this selective degradative process are not well established.

Gene Expression Can Be Controlled by Selective Translation of mRNA

The second basic step of gene expression is the translation of mRNA into protein. In eukaryotes the translation of proteins occurs in the cytoplasm; however, not all mRNA is translated on arrival to the cytoplasm. The RNA molecules in the cytoplasm, as in the nucleus, are constantly associated with proteins, some of which may function to regulate translation. Most of the defined mechanisms that regulate translation operate to repress protein synthesis (negative translational controls), although some evidence has been reported (derived from studies examining viral RNA) that positive or enhanced translation of certain RNA may be operative. Translational controls are important in many fertilized eggs, which must rapidly switch from making the proteins required for maintenance of a quiescent oocyte to making proteins required for cell division and growth. These eggs have been stored as RNA-protein complexes maternal mRNA that is not translated until the egg is fertilized.

Another important negative translational control has been demonstrated for the expression of iron storage protein ferritin. Ferritin mRNA in the cytoplasm shifts from an inactive RNA-protein complex to a translationally

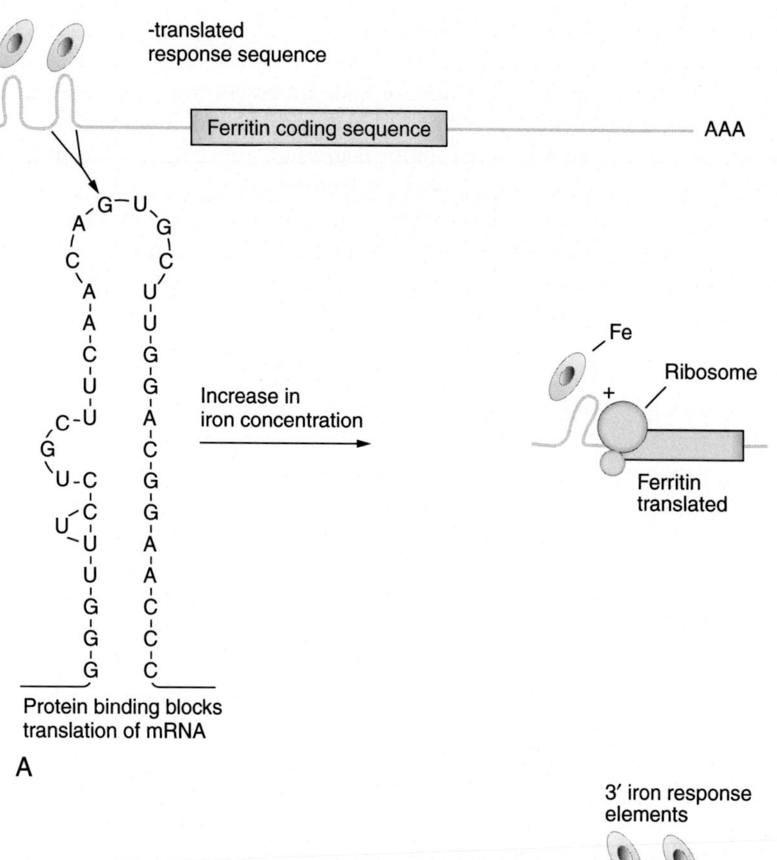

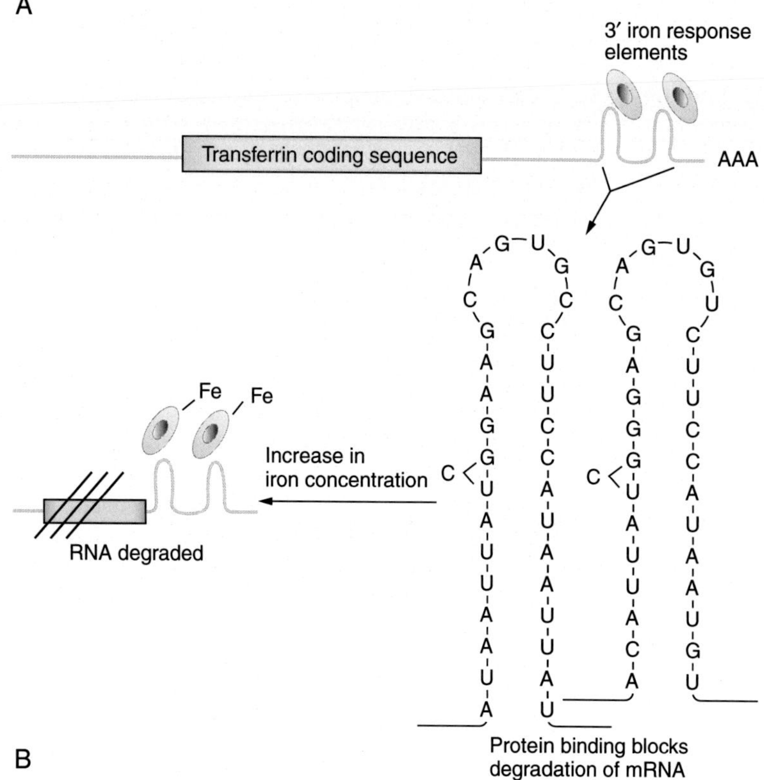

Figure 6–36. **Iron metabolism in cells is regulated by modulation of ferritin translation and destabilization of transferring messenger RNA (mRNA).** An iron-sensitive receptor protein is able to bind to specific sequences of the ferritin and transferring mRNA. (A) The ferritin sequence is located in the 5′-nontranslated region of the mRNA, and the bound protein blocks the translation of this mRNA. (B) The transferring mRNA contains similar binding sequences, and the binding of the same iron-sensitive receptor protein to these sequences stabilizes this mRNA, allowing more protein to be translated. An increase in iron concentration in cells is sensed by the receptor, and on binding the excess iron, the conformation of the protein changes so that it no longer is bound to the mRNA. The result is a release of the translational block of ferritin mRNA and a destabilization of the transferring mRNA. Thus the regulation of intracellular iron concentration is accomplished quickly by posttranscriptional regulation of gene expression.

active polyribosome on increase of intracellular iron concentration. The block of ferritin mRNA translation is mediated by a 30- to 40-nucleotide segment of the RNA at the 5′ leader (5′-nontranslated region) segment of the molecule (see Fig. 6–36). This segment of the RNA binds a repressor protein that blocks the ability of ribosomes to form active complexes on the mRNA. This repressor, called the **iron response molecule**, is the same protein that is bound to the 3′-nontranslated segment of the transferring receptor mRNA discussed in the previous section. Thus this iron response protein allows exquisite control of intracellular iron metabolism by increasing the degradation of the mRNA encoding an iron-salvaging protein, the transferring receptor, and simultaneously releasing the translational block of the iron-binding protein, ferritin. This provides rapid, sensitive controls of gene expression without affecting the synthetic rates (transcription) of these mRNA.

Modification of Protein Posttranslationally Can Affect the Expression of an Active, Functional Molecule

Once a protein has been synthesized by the ribosomal complexes in the cytoplasm, the functional capabilities of the protein often are not realized until the protein has been modified. Although these mechanisms, collectively referred to as *posttranslational modifications*, are not often thought of as gene expression controls, it is recognized that the manifestation of gene expression is not complete until a protein is performing its function within the cell. An example of posttranslational modifications occurs in the formation of a functional insulin protein. Insulin is a secreted polypeptide hormone that is synthesized on ribosomes associated with the ER. It is synthesized as a single polypeptide referred to as a *preproinsulin*. The prefix *pre-* refers to a signal peptide sequence that directs the translocation of the proinsulin molecule across the ER membrane. This presequence is immediately removed by protease cleavage, the first posttranslational modification. The proinsulin molecule then folds such that the NH_2- and COOH-terminal ends are held in close proximity by sulfhydryl bonds. The proinsulin is inactive until a second cleavage event removes the connective or C-peptide, leaving two chains formed from the NH_2- and COOH-terminal domains of the proinsulin peptide. Thus multiple posttranslational events are required to form an active insulin hormone. Moreover the sequestering of the hormone into secretory vesicles can be viewed as a regulatory event, in that some proteins are required at specific cellular locations for them to exert the functional properties. Any number of events or modifications alter the activity of a protein (e.g., phosphorylation, methylation, and glycosylation) or may play a role in the selective degradation of proteins (the addition of a ubiquitin molecule) that can constitute posttranslational controls of gene expression.

Structural RNA (tRNA and rRNA) Are Subjected to Regulatory Mechanisms

The preceding sections have limited discussions to RNA synthesized by polymerase II, the mRNA. However, it is important to point out that the structural class of RNA—rRNA and tRNA—is also subject to regulation. RNA polymerase III transcribes the tRNA and a class of small RNA, referred to as 5S RNA, each of which is made as precursor molecules that are subsequently modified to form functional molecules. Moreover the transcription of these RNA is facilitated by the specific binding of proteins to the promoters of these genes, one of which has been purified and is called *transcription factor (TF) IIIC*. Similarly, RNA polymerase I transcribes rRNA, a process facilitated by the binding of factors. For example, there is a protein called TFID that regulates the transcription of the 45S rRNA precursor, which is modified in the nucleolus to form 18S (small) and 28S (large) rRNA. Because each of the RNA (mRNA, rRNA, and tRNA) must work in concert to provide the cell with functional proteins, it is clear that regulatory events for each RNA class are important for the overall transfer of information from DNA to protein.

Gene Expression Can Be Regulated by Noncoding RNAs

In the preceding sections, we have discussed the "central dogma" of information transfer from genome to protein (Fig. 6–28). This flow of information can be regulated by changes at any of the multiple steps in the process. We have also introduced the concept that RNA molecules other than the mRNA can also exert a regulation of the information transfer. These are called the noncoding RNAs. Recent evidences have shown the noncoding RNAs make a large portion of the total RNA found in the cell, often 100–1000 times greater in content than the mRNA. Noncoding RNAs can be divided into three major categories: structural RNAs, short noncoding RNAs, and a long noncoding RNAs (LncRNAs). We discuss the structural RNAs in the previous section and will discuss here the short noncoding RNA and LncRNA classes.

The class of short noncoding RNAs include small interfering RNAs (SiRNAs), micro-RNAs (MiRNAs), and small nucleolar RNAs (SnoRNAs). The SiRNA an MiRNA classes are short RNAs, about 22 bases in length, that are formed from precursor RNAs synthesized by RNA polymerase II. The precursor RNAs vary greatly in length, with the segments that eventually become the mature 22 base noncoding RNA adopting a structural conformation that is recognized by an RNAse III endonuclease system in the cytoplasm called DICER. Both SiRNA and MiRNA had been found to be important in gene regulation initially during development, but more recently, they have been shown to influence cellular responses to several stimuli. They exert regulatory effects by finding homologous sequences in mRNA and binding the sequences via base pairing with the mRNA. This marks the messenger for degradation. Thus one mechanism of the small noncoding RNAs affecting gene expression is by eliminating the mRNA from the cell. The MiRNA also have been found to bind nuclear RNAs interrupting the transcription of the target messenger precursor.

There are at least two therapeutic uses for the SiRNA and MiRNA populations. First, because they target specific RNAs, there are small molecules designed to interrupt this activity, thus preventing the downregulation of the targeted gene. It is also possible to synthesize analogs of the SiRNA or MiRNA to be targeted to the cells and specifically downregulates the expression genes that might be detrimental to cellular metabolism. An example of small noncoding RNA used in therapeutics is the microRNA MiR-122. MiR-122 is highly expressed in hepatocytes in the liver and has been shown to have a critical function in liver development. It is also implicated as an essential host factor for hepatitis C infection. In the liver, one function for MiR-122 is an inhibitor of translation for enzymes involved in cholesterol regulation. In animal models, molecules that inactivate MIR-122 have been demonstrated to allow cholesterol metabolism enzymes to be translated, and this expression reduces the plasma cholesterol. This example suggests this as a potential therapeutic mechanism for patients with elevated cholesterol. These results implicate the small noncoding RNA is up as potential targets for therapeutics in many disease states.

The last class of small noncoding RNAs, the SnoRNA, are derived from intron sequences of transcribed mRNA genes. As the intron segments of the mRNA are spliced together, those containing SnoRNA are retained in the nucleus and are recognized by A nuclear resident 3′ and 5′ ribonuclease complexes, possibly by structure, and then are trimmed to form the mature SnoRNA species. These are heterogeneous in size and can vary from 60 to 300 nucleotides in length. Mature SnoRNAs are resident primarily in the nucleolus of the cell and participates in the maturation of rRNAs. Although some SnoRNAs have been found associated with ribosomal machinery, to date, their function is in translation and is not well defined. However, continued research may find the SnoRNAs as therapeutic targets.

The most abundant and heterologous class of noncoding RNA is the LncRNAs. Advances in sequencing technologies have shown that a large portion of the genome is transcribed into a RNA, whereas only about 2% of these transcripts are translated into protein. Of the transcribed RNAs that do not encode proteins, the noncoding RNA, those that are greater than 200 nucleotides in length are designated as the LncRNAs. With this designation greater than 8000, LncRNA have been discovered from the human genome, and such large numbers have made it difficult to understand their cellular function. Most of the LncRNAs are transcribed by polymerase II, the mRNA polymerase. They are subject to nuclear metabolism and emerge as RNA possessing 5′ caps and polyadenylated 3′ ends. In short, they are assembled essentially as the mRNAs, but by definition, they do not encode proteins. Generally the biological functions of the LncRNA are thought to be very diverse. Two main qualities for the LncRNA have emerge: RNA molecules acting as a scaffold to build complexes coordinating the functions of multiple proteins and secondly acting as guide sequence bringing together LncRNA bound proteins to a particular cellular location. Both functions demonstrate LncRNA as a facilitator of cellular processes. The latter activity occurs through a recognition of sequences, DNA or RNA, homologous with a segment of the LncRNA and binding the sequences through base pairing. This base pairing-driven binding brings the proteins bound to the RNA to a location that they can now provide a cellular function. This can be a range of functions including gene transcription (increasing or decreasing transcription) and mRNA translation into protein. The latter has also been shown to be affected by the LncRNA base pairing inhibition of translation by simply blocking the protein synthetic machinery. Thus the LncRNA is a diverse group of RNA molecules that affect gene expression at multiple steps; however, no direct therapeutic methods have been yet designed for these molecules. Clearly the LncRNA represent a class of molecules for the design of a targeted therapeutics in the future.

The recent research on noncoding RNA has caused a reevaluation of the central dogma of gene expression as illustrated in Figure 6–37. The pathway from gene to protein remains firm; however, changes in the central dogma must be incorporated to accommodate the role of noncoding RNA that is playing in this pathway. First, some of the noncoding RNA are transcribed and processed just as the mRNA yet provide functions as an RNA molecule not a protein. Additionally, some noncoding RNAs are produced from processes initially thought to generate only "waste" molecules. These molecules large and small actively participate at a variety levels in the regulation of gene expression that causes a reexamination of the process. The scheme of informational transfer from the genome to protein is now somewhat more complicated due to the workings of the noncoding RNA molecule. Understanding their roles more fully will allow the development of target therapeutics for the many disease states.

GENETIC THERAPY

There Are Many Obstacles to the Development of Effective Gene Therapies

With the development of recombinant DNA technology has come the promise, or potential, of it dramatically improving the practice of medicine. Indeed, as advances have been realized, with various aspects of DNA technologies, there are parallel applications of new methods in the clinical management of patients. RFLP analysis is commonly used to aid in the diagnosis of specific inherited diseases. The sensitivity of the PCR technique makes it useful not only for the diagnosis of inherited diseases and for latent viral diseases (e.g., AIDS) but also for forensic medicine. An ability to understand the roles of a specific gene product in the pathogenesis of human disease has allowed precise and effective clinical

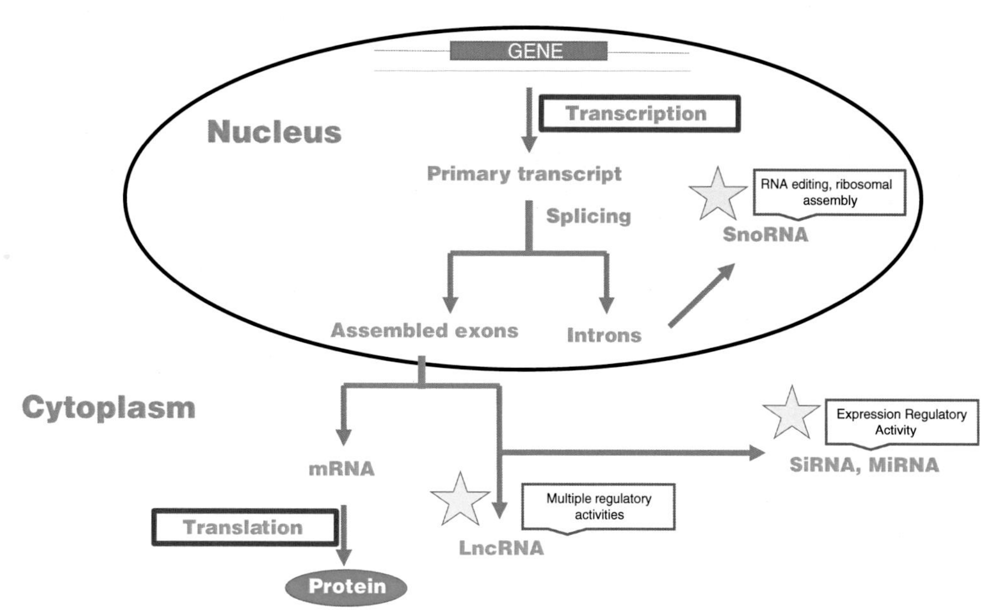

Figure 6–37. **Expanding the gene regulation dogma.** As detailed in Fig. 6–28, the central dogma for gene expression includes the two major steps of transcription and translation to transfer information stored in the genome, the gene, to direct production of a functional protein. The path from DNA to protein remains intact; however, there are a number of RNA molecules derived from mechanisms in the nucleus from splicing of the primary transcript(s) into mRNA, the SnoRNA; mechanisms in the cytoplasm that cleave RNAs to smaller molecules, the SiRNA and MiRNAs; or RNAs that appear in the cell cytoplasm that do not encode proteins, the LncRNA, which are able to modify or influence the gene regulation pathway. Thus the influence of these molecules at various steps in the pathway indicates that we must expand the dogma to include them in the new paradigm of gene regulation.

intervention. Moreover, recombinant DNA technologies have led to the development of new therapeutic products made possible by the ability to engineer the overexpression of genetic material in bacteria and eukaryotic cells. Although recombinant DNA technologies have made significant inroads toward diagnosis and management of human disease, the treatment of human disease through transfer of genetic material to a patient (e.g., gene therapy) is not yet commonplace in medical practice.

In general a variety of questions must be addressed as a prelude to the implementation of an effective gene-based therapy. First the gene that is the root cause of the genetic disease must be well studied. This would include its identification with cloning and sequencing and an understanding of its regulation and the function of the expressed gene product within the cell. Second the gene must be delivered to the appropriate cell(s) and maintained stably expressed at an appropriate level within the cell. For example, the targeting to and expression of a normal β-globin molecule in a neuron of the brain would not provide an effective therapy for sickle cell anemia. Finally the expression of the gene product must be able to correct or reverse the disease process. This is important, particularly for somatic cell therapies (discussed later), when the expression of an appropriate gene product will be able to affect a curative response; that is, it will reverse the disease phenotype. Few inherited diseases are understood at the level of complexity outlined earlier. However, the daily contribution to our knowledge of the molecular bases underlying inherited diseases will soon bring genetic therapies to the forefront. Two considerations for successful gene-based therapies are a suitable model system for study and strategy for gene replacement.

Many Strategies Are Available for Gene-Based Therapies

The essence of a successful genetic therapy is the availability of a strategy for gene replacement. There are basically two types of gene replacement therapies: altering of germline cells and altering of somatic cells. Whereas the fixing of germline cells could potentially cure the disease, many ethical issues must be addressed before this is a viable alternative. The best opportunity for gene therapy is the ability to alter somatic cell expression of the diseased gene. As stated earlier the disease must be studied to the greatest of molecular detail before a strategy can be devised. Perhaps the best examples to date and those for which strategies have been devised are sickle cell anemia, adenosine deaminase deficiency, cystic fibrosis, Duchenne muscular dystrophy, and familial hypercholesterolemia. In each case the gene deficiency is known, its expression pattern studied, and the function of the expressed product defined.

To effect a successful gene-based therapy, one must be able to target the gene to the appropriate cell. This can be done in two ways, ex vivo (the cells are outside

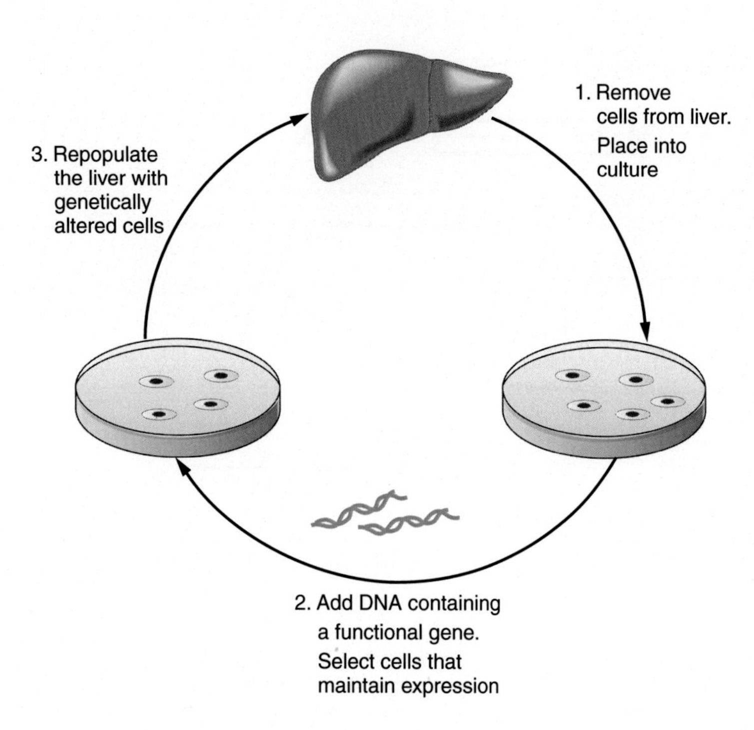

Figure 6–38. **A potential ex vivo method for genetic-based therapy of a liver disease.** Hepatocytes are first isolated from the liver and placed into culture dishes with an appropriate media, such that they remain as hepatocytes. The DNA carrying an altered gene is added to the cells—either by physical means or facilitated by virus vectors—and the cells that received the gene are selected for the culture. The cultured cells containing the altered gene are then added back to the original liver tissue. It is necessary for the cells to have either a selected advantage or the ability to repopulate the liver tissue, so that the genetically altered cells will be effective in their therapeutic nature.

organisms) or in vivo (the cells remain within the organism). Many protocols rely on ex vivo techniques because of the ability to control the cellular environment during the delivery of the genetic molecule. A potential ex vivo strategy for the genetic intervention of familial hypercholesterolemia is outlined in Figure 6–38.

The potential therapeutic DNA must be delivered to the cell in a way that it becomes part of the recipient cell's genome. There are essentially two methods for accomplishing this: physically adding the DNA and allowing it to integrate or using a virus to facilitate DNA uptake. The viral approach is most used because of the ability to have DNA integrate in higher numbers of cells; however, the virus (retrovirus or DNA virus are both used) can cause immunologic problems later. A third method for incorporating DNA into the genome and potentially the most selective is to target a specific genome sequence/locus using the recently developed CRISPR/CAS 9 system. This system is based upon the bacterial CRISPR System, which in bacteria is a defense against viral invasion. The system relies upon a RNA that specifically recognizes a DNA sequence and binds to the specific locus via base pairing. The DNA recognized by the CRISPR molecule is cleaved by proteins complexed with the CRISPR guide RNA, called CAS 9 protein. The guide RNA must specifically bind to a genome sequence to bring the cleavage machinery into the singular genome locus. Adapting the guide RNA by substituting a human genome sequence for the base pairing domain of the CRISPR machinery allows this system to be adapted for altering the human genome in a very specific location of interest. This is advantageous because it allows the change in specific spots of the human genome, for example, the disease-causing gene sequence. When applied to stem cells, this system will allow the therapeutic intervention to genetic diseases. In any case the cells containing the altered gene must be replaced into the organism.

There are as many strategies as there are genetic diseases, and no one approach will be applicable to all diseases. For example, it is possible to stably express protein for longer periods in skeletal muscle cells. Thus, if the gene product is used outside the cell (e.g., α1-antitrypsin is a secreted molecule made in the liver, but its absence leads to lung disease), it could be possible to use the skeletal muscle as a synthetic protein "factory." As the efficiency of genetic-based therapies becomes greater and our knowledge base expands, using better molecular techniques and more expansion model systems, there is great promise in the ability to treat human disease with a specifically designed DNA molecule.

SUMMARY

The central regulator of the function of a cell is the nucleus. The nucleus houses most of the DNA within the cell, and it has the ability to use this stored information to allow the cell to develop and survive. The genome of the organism is acted on by proteins within the nucleus that allow the cell to identify the specific gene segments needed and to express these genes in an appropriate fashion. There are several steps to this process, and regulation of gene expression can occur at any or all of these steps. Misexpression of genes can be the root cause of disease processes. Thus the cell has efficient ways to detect and fix genomic segments when they become altered. Mutants in specific enzyme systems can affect how the organism responds to outside influences. For example, mutations in the cytochrome P450 genes expressed in liver can cause an individual to be a poor metabolizer of a specific drug, leading to complications if the drug is prescribed. With the advent of whole-genome sequencing and new technologies of proteomics, medicine is now beginning a new path toward individualized therapeutics. This means that, by understanding the genetic makeup of an individual, the medical team can give therapeutics in an individualized, optimized format. This will lead to major changes in medical delivery because unwanted side effects of therapeutics can be somewhat eliminated. Thus a continued understanding of genetics and how genetics works will facilitate medicine in the future.

Suggested Readings

Asbersold R, Mann M. Mass spectrometry-based proteomics. *Nature* 2003;422:198–207.

Crisp M, Liu Q, Roux K, Rattner JB, Shanahan C, Burke B, Stahl PD, Hodzic D. Coupling of the nucleus and cytoplasm: role of the LINC complex. *J Cell Biol* 2006;172:41–53.

Dean DA, Strong DD, Zimmer WE. Nuclear entry of nonviral vectors. *Gene Ther* 2005;(11):881–890.

Dewar JC, Hall IP. Personalized prescribing for asthma: is pharmacogenetics the answer? *J Pharm Pharmacol* 2003;55:279–289.

Dietel M, Sers C. Personalized medicine and development of target therapies: the upcoming challenge for diagnostic molecular pathology. A review. *Virchows Arch* 2006;(6):744–755.

Hunter CV, Tiley LS, Sang HM. Developments in transgenic technology: applications for medicine. *Trends Mol Med* 2005;11:293–298.

Ingelman-Sundberg M. Pharmacogenetics of cytochrome P450 and its application in drug therapy: the past, present, future. *Trends Pharmacol Sci* 2004;16:337–342.

Kozarova A, Petrinac S, Ali A, Hudson JW. Array of informatics: applications in modern research. *J Proteome Res* 2006;5:1051.

Lambert MW. The functional importance of lamins, actin, myosin, spectrin, and the LINC complex in DNA repair. *Exp Biol Med* 2019;1–25.

Qadir MI, Bukhat S, Rasul S, Manzoor H, Manzoor M. RNA therapeutics: identification of novel targets leading to drug discovery. *J Cell Biochem* 2019;1–32.

Tsongalis GJ, Silverman LM. Molecular diagnostics: a historical perspective. *Clin Chim Acta* 2006;2:350–355.

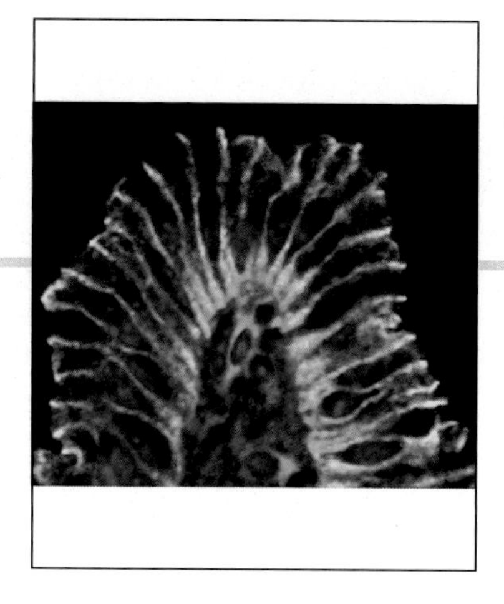

Chapter 7

Cell Adhesion and the Extracellular Matrix

Multicellular organisms exist because cells can bind to each other. The process by which cells attach to a cell or a substance is called *cell adhesion*. Cells may adhere directly to each other, a process called **cell-cell adhesion**. Cells also bind to extracellular components that provide a structural framework for cell binding. These extracellular components are collectively called the *extracellular matrix* (ECM). The binding of cells to the ECM is termed **cell-matrix** or **cell-substratum adhesion**.

Cell adhesion and the ECM are, together, crucial for the development and maintenance of tissue structure and function. Abnormalities of the matrix or cell adhesion dynamics compromise tissue functions and cause human diseases. These abnormalities may be specific such as a mutation in the gene encoding an essential molecule involved in leukocyte adhesion or an autoimmune attack on a specific cell adhesion molecule as in the disease called pemphigus. Alternatively, they may be part of an accumulation of changes that cause cancer, where the loss of adhesion promotes metastasis or spread of cancer cells to other tissues.

Cell adhesion is not only a static process but also a highly dynamic process involving cell signaling-mediated regulation of assembly and disassembly of adhesion structures under physiologic and pathophysiologic conditions. For example, cell migration during the development of an embryo involves coordinated changes in cell-cell adhesion and cell-substratum adhesion. Another example involves the cells of the epidermis, called keratinocytes. These cells are tightly bound together to provide the barrier properties of the skin that protect humans against fluid loss, infection, and the wear and tear of everyday activities. However, epidermal keratinocytes are constantly being lost from the outer surface and replaced from the basal layer of the multilayered epidermis. These cells are renewed approximately once every 28 days, depending on location in the body. Therefore there is a constant upward migration of keratinocytes, and this involves changes in both cell-cell and cell-substratum adhesion. Other cells, such as blood platelets, must circulate freely in the blood and thus must be nonadhesive. However, when wounding occurs, they must become adhesive rapidly to participate in hemostasis, that is, the coagulation of the blood that prevents bleeding. Cells must possess mechanisms for modulating their adhesiveness. The "inside-out" signaling that arises intracellularly modifies the stickiness of the cell surface, thus leading to changes in cellular functions such as cell migration.

On the other hand the "outside-in" signals provide cells with information about their environment by changes in cell adhesion. Such signals are essential regulators of cellular functions such as gene activity and differentiation, cell proliferation, and programmed cell

Goodman's Medical Cell Biology. https://doi.org/10.1016/B978-0-12-817927-7.00007-7

death or apoptosis. A dynamic relationship exists between the cells and ECM. Cells produce the matrix and also the enzymes that degrade it, thus regulating its composition and turnover. A prime function of the ECM is to provide tissues with shape, strength, and elasticity. It also acts as a substratum for cell adhesion. However, the matrix also regulates cellular activity. It is an essential reservoir of growth factors, which may activate cellular receptors. Many of the "outside-in" cell adhesion signals are generated by specific interaction between cells and the matrix.

To understand cell adhesion, one must know about the molecules that mediate cell adhesion (cell adhesion molecules, CAMs), organization of these molecules into multiprotein adhesion complexes, and regulation of protein-protein interactions by cell signaling pathways. The domain structures of CAMs determine their precise organization into specific cell-cell and cell-substratum junctions. Modifications of CAMs, such as phosphorylation and dephosphorylation, by cell signaling molecules, regulate the assembly and disassembly of adhesion complexes affecting cell behavior and function. This chapter summarizes the current knowledge of structural features of CAMs, their interactions, and regulation in health and disease.

CELL ADHESION

Most Cell Adhesion Molecules Belong to One of Four Gene Families

Most of CAMs belong to one of four gene families, the cadherins, immunoglobulin superfamily CAMs (IgCAMs), selectins, and integrins (Fig. 7–1). Proteoglycans may be considered as another class of CAMs. CAMs can also be classified as calcium-dependent and calcium-independent CAMs. Adhesive properties of cadherins and selectins are calcium dependent, whereas integrins and IgCAMs associations are independent of calcium.

CAMs are typically single-pass transmembrane proteins with an extracellular domain that participates in adhesion, a transmembrane domain that anchors the protein in the cell membrane, and a cytoplasmic domain that mediates attachment to the cytoskeleton. Cytoplasmic domains first recruit multifunctional adapter proteins, which in turn connect CAMs to the cytoskeleton directly or indirectly. The cytoplasmic domain is also the region that interacts with intracellular signaling molecules and thus is involved in the regulation of cell adhesion by signal transduction. Almost invariably the N-terminal part of the protein forms the extracellular domain, and the C-terminal part forms the intracellular domain.

The adhesive binding of adhesion molecules is either homophilic or heterophilic. **Homophilic** adhesion means the adhesion is mediated by the interaction between the extracellular domains of the same type of CAM molecules. On the other hand the **heterophilic** binding refers to the adhesion mediated by the interaction of the extracellular domain of a CAM with the extracellular domain of a different type of molecule in another cell or ECM (Fig. 7–1). Cell-cell adhesion is commonly (but not always) homophilic. But cell-matrix adhesion is always heterophilic. Cell adhesion may also be homotypic or heterotypic. **Homotypic** interaction refers to adhesion between the same type of cells such as epithelial cells, whereas **heterotypic** cell adhesion is between different

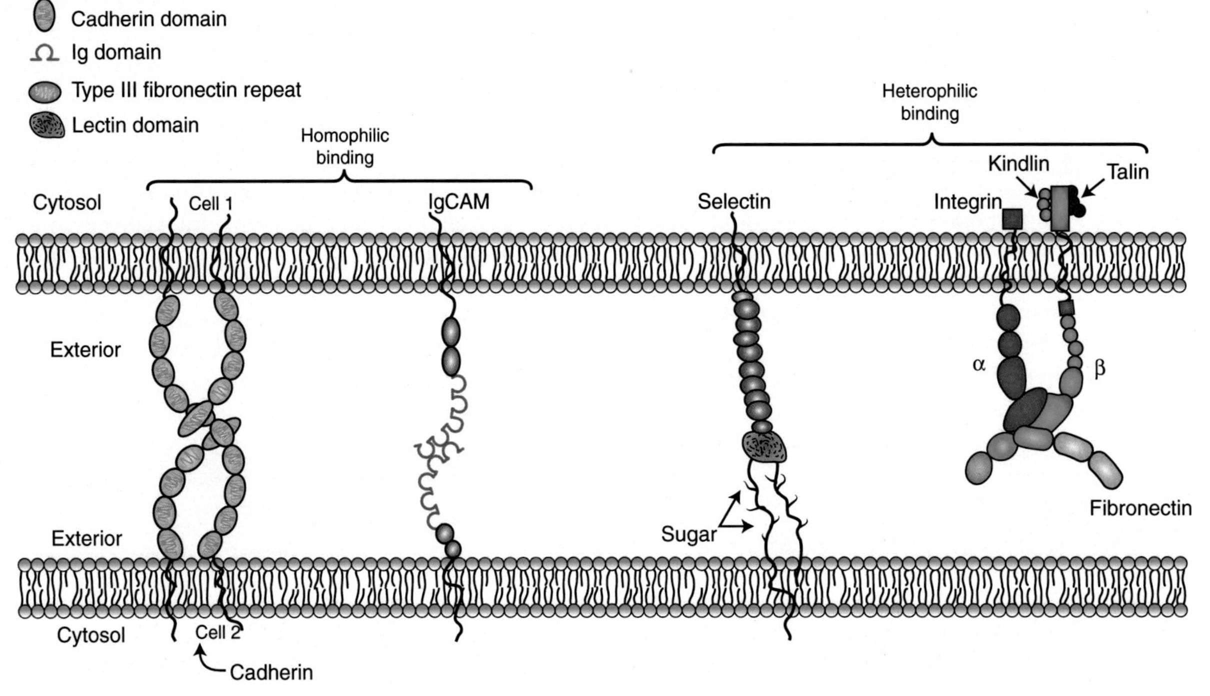

Figure 7–1. Types of cell adhesion molecules.

types of cells, such as that between leukocytes and endothelial cells or between cells and ECM. The CAMs may be distributed in the membrane as **diffuse** or **clustered** structures. Clustering of CAMs such as integrins and cadherins may occur as a result of certain physiological stimuli. The signaling event may occur due to the formation of a cluster of CAMs, thus affecting cellular functions. Therefore, stimulated clustering is a crucial functional module. The clustering and cell adhesion involve two types of intermolecular interactions. The interaction between molecules form clusters in an individual cell is called ***cis*** interaction, whereas the interactions between CAMs in adjacent cells are called ***trans*** interaction. Both *cis* and *trans* interactions are involved in the formation of adhesive junctional complexes such as zonula occludens or tight junctions and zonula adherens or adherens junctions. The adhesion may either be **tight** as in tight junctions between epithelial cells or **loose** as in the interaction of leukocytes with the endothelial cells.

Cadherins Are Calcium-Dependent Cell-Cell Adhesion Molecules

The cadherins are members of a superfamily of calcium-dependent cell-cell adhesion molecules. They are classified as Type I (e.g., E-cadherin and N-cadherin) and Type II (e.g., cadherin-6 and 11) classic cadherins, desmosomal cadherins (e.g., desmoglein and desmocollin), and protocadherins (α, β, and γ). Type I cadherins include E-cadherin (epithelial), N-cadherin (neural), P-cadherin (placental), and R-cadherin (retinal), named depending on tissue source. **E-cadherin,** the most well-characterized cadherin, is widely expressed in epithelial cells. The extracellular domain of cadherins is uniquely composed of 5–34 repeats of barrel-shaped structures named EC domain, although most cadherins contain only 5–6 EC domains. The extracellular domain of E-cadherin consists of 6 EC domains that bear calcium-binding sites. In the presence of calcium, the extracellular domains of dimeric E-cadherin bind to the extracellular domain of E-cadherin dimer in the adjacent cell forming cell-cell adhesion (see Fig. 7–1). The removal of calcium causes the EC domain to collapse and adhesion to be lost. Adhesive binding by most cadherins, including E-cadherin, is preferentially homophilic and homotypic, although exceptions exist.

The intracellular domain of E-cadherin binds to catenins such as **α-catenin; β-catenin;** and **p120-catenin,** which is also known as δ-catenin (see Fig. 7–12). The intracellular domain binds to p120-catenin and β-catenin. P120-catenin stabilizes the interaction between E-cadherin and β-catenin, whereas β-catenin binds to α-catenin (the actin-binding protein) linking the cadherin-catenin complex to the actin cytoskeleton. These protein-protein interactions are crucial for cell-cell adhesion and the formation of adherens junctions. An experimental mutation of the cytoplasmic domain of E-cadherin that prevents catenin binding abolishes cell-cell adhesion. Homophilic binding by cadherins is involved in the dynamic structure of adherens junctions in epithelial tissues. E-cadherin interactions also contribute to tissue segregation during embryonic development, particularly in cranial neural crest (CNC) formation. During CNC formation the neural crest cells undergo a series of processes, including epithelial-to-mesenchymal transition (EMT) and collective cell migration. In the premigratory stage, neural crest cells express high levels of E-cadherin. At the onset of migration, neural crest cells show increased expression of N-cadherin and a decrease in E-cadherin expression leading to EMT and cell migration. However, the remnant of E-cadherin is required for the collective cell migration, a process in which cells migrate as a sheet of interconnected cells, which is discussed further in later sections of this chapter. E-cadherin is essentially required for cell-cell adhesion during the collective cell migration during the formation of the neural crest.

E-cadherin-based cell-cell adhesion contributes to contact inhibition of cell proliferation. Therefore, mutation or downregulation of E-cadherin plays a role in the development of certain types of cancer, such as gastric and breast cancer. Loss of E-cadherin-based cell-cell contact also induces cancer cell EMT leading to tumor metastasis. Similarly, defects in desmosomal cadherins lead to many types of human diseases. Desmosomes and desmosomal cadherins play a central role in maintaining the integrity of solid tissues. Therefore disruption of desmosomes due to inherited, infectious, or autoimmune etiology leads to many pathophysiologic conditions such as cardiomyopathies, epidermolysis bullosa, and pemphigus. Pemphigus vulgaris is discussed in detail with a specific clinical case later in this chapter (Clinical Case 7–1).

The Immunoglobulin Family Contains Many Important Cell Adhesion Molecules

In this family of CAMs, the extracellular domains are characterized by the presence of at least one but usually multiple subdomains that resemble the structure of the basic subdomain of antibody molecules or immunoglobulins (see Fig. 7–1). The structure of these Ig subdomains is stabilized by disulfide bonds rather than by calcium ions, so these molecules participate in calcium-independent adhesion. Adhesion may be homophilic (see Fig. 7–1) or heterophilic but usually between the same type of cells (homotypic).

The Ig family is large and diverse, probably because the basic structure of the Ig subdomain is versatile and readily adaptable to a variety of different binding functions. Three major groups of IgCAMs are classified based on their tissue origin. IgCAMs of the nervous system includes the **neural cell adhesion molecule (NCAM)**, **L1**, and **SynCAM**, which are involved in neuronal guidance and fasciculation or bundling during development and regeneration of the nervous system. NCAM, L1, and SynCAM in neurons stabilize the ultrastructure of synapses and play a role in neurotransmitter release and neuronal plasticity. In epithelial cells the Ig family molecule **nectin** is closely associated with E-cadherin. Junction adhesion molecules (JAMs) are another group of IgCAMs. **JAM2** is localized in tight junctions of epithelial and endothelial cells. The vascular IgCAMs include lymphocyte function-related antigen 2 and 3 (LFA-2 and LFA-3), **intercellular adhesion molecule (ICAM)**, and the **vascular cell adhesion molecule (VCAM)**. LFA-2 and LFA-3 are located in T cells, whereas ICAM and VCAM are located in endothelial cells. The heterophilic binding between LFA-2 and LFA-3 contributes to cell-cell adhesion of T cells. ICAM and VCAM play an important role in the adhesion of leukocytes to endothelial cells during the inflammatory response by heterophilic binding between the endothelial ICAM and VCAM with integrins on the leukocytes.

The cytoplasmic domains of this diverse group of adhesion molecules are involved in the interaction with the actin cytoskeleton and spectrin membrane skeleton. For example, NCAM has several alternative COOH-termini that arise through differential **splicing of messenger RNA** derived from a single gene. This generates a series of different molecules, a soluble form with no membrane anchor and three membrane-anchored forms of various sizes, 120, 140, and 180 kDa. The two largest forms of molecules have transmembrane and cytoplasmic domains, but the 120-kDa form has neither and hence anchored to the cell membrane by a glycosylphosphatidylinositol linkage. This variation in structure offers the opportunity for alternative functional regulation. The cytoplasmic domains of the 140-kDa form bind to the cytoskeletal protein α-actinin and the 180-kDa form to α-actinin, actin, and spectrin. Another family of the actin-binding protein, ezrin, radixin, and moesin (the ERM proteins), interacts with the cytoplasmic domain of L1. The cytoplasmic domain of nectin binds to afadin that links the adhesion molecule to the actin cytoskeleton. ICAM interacts with the actin cytoskeleton and transduces outside-in signals that mediate cytoskeletal reorganization.

Dysfunctional NCAM and other neuronal IgCAMs are associated with the pathogenesis of many neurodegenerative diseases such as Alzheimer's disease, schizophrenia, and depression. The role of ICAM and VCAM in inflammatory responses has been well characterized. ICAM-I is constitutively expressed in endothelial cells at low levels. However, its expression is upregulated in response to inflammatory stimuli such as TNF-α, IL-1β, and IFN-γ. Anti-TNF-α antibody treatment effectively reduces ICAM-1 expression. ICAM-I expression is upregulated in the intestinal mucosa of patients with inflammatory bowel disease (IBD) that includes Crohn's disease and ulcerative colitis. It plays an important role in leukocyte extravasation that involves three important processes—rolling, adhesion, and migration. ICAM-I is required for adhesion, and therefore it is an attractive therapeutic target. Indeed, ICAM-I antisense oligonucleotide (also known as **alicaforsen**) is being considered for treating IBD patients. Whereas the efficacy of alicaforsen in the treatment of Crohn's disease has not been encouraging, the phase II clinical trials have produced promising results in the treatment of ulcerative colitis patients.

Selectins Are Carbohydrate-Binding Adhesion Receptors

The selectins, as indicated by their name, participate exclusively in heterophilic cell-cell binding because they have a **lectin**-like domain at their NH_2 terminus that binds to specific carbohydrate residues on the opposite cell surface (see Fig. 7–1). Selectins have a major physiologic role in initiating the adhesion of leukocytes and platelets during the inflammatory and hemostatic responses. Deregulation of selectin functions is associated with thrombus formation and involved in the pathogenesis of inflammatory diseases. Therefore selectins are targets for the development of therapeutics to treat sickle cell disease (SCD), acute myocardial infarction, and malignancies.

There are three members of selectins (named on the basis of cellular expression), **P-selectin**, **E-selectin**, and **L-selectin** localized on the surface of platelets, endothelial cells, and leukocytes, respectively (Table 7–1). The exception is that P-selectin is also expressed in activated endothelial cells. Selectins are transmembrane proteins with a long extracellular domain (that binds to carbohydrate moiety of their ligands), a transmembrane domain, and a short intracellular domain (binds to cytoskeleton).

TABLE 7–1. Characteristics of Various Selectins

	Cellular Expression	Protein Ligands	Rolling Velocity of Leukocytes
P-selectin (GMP-140, PADGEM, CD62P)	Activated endothelium and platelets	PSGL-1 CD24	Slow
E-selectin (ELAM-1, CD62E)	Activated endothelium	PSGL-1 ESL-1 L-selectin Podocalyxin	Slow
L-selectin (MEL-14, CD62L)	Constitutive expression on leukocytes	PSGL-1 GlyCAM-1 MAdCAM-1 CD-34 Podocalyxin	Fast

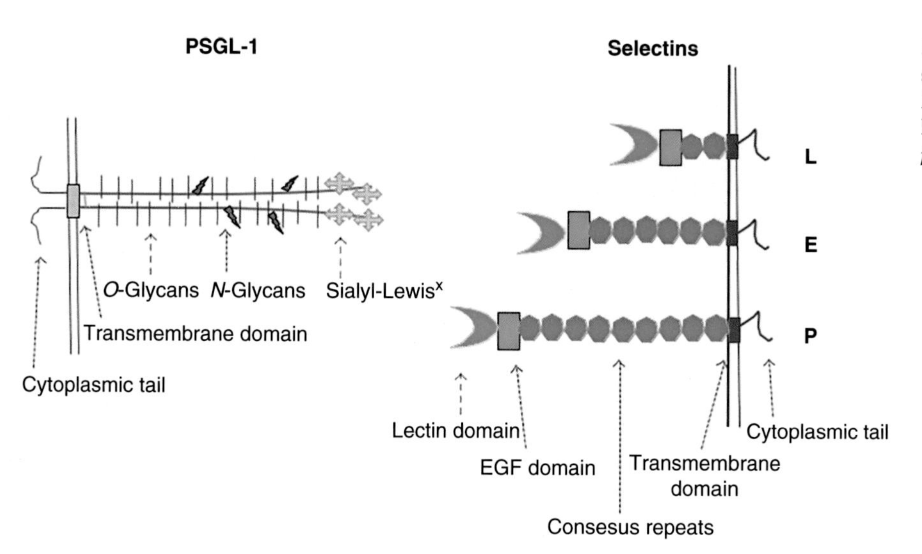

Figure 7–2. **Structural features of selectins and PSGL-1.** *(Adapted from Kappelmayer, Nagy.* Biomed Res Int *2017,* https://doi.org/10.1155/2017/6138145, *by permission with added colors.)*

The extracellular domains of all three selectins are composed of a lectin domain, epidermal growth factor (EGF) domain, and consensus repeats. Three selectins differ in the number of consensuses repeats: The longest P-selectin has nine consensus repeats, the medium E-selectin has six consensus repeats, and the shortest L-selectin has only two consensus repeats (Fig. 7–2).

The adhesive binding of selectins is calcium dependent. Selectins bind weakly to small sialyl, fucosylated oligosaccharides. The prototype selectin ligand is a tetrasaccharide, sialyl Lewis X (sLe^x). They also bind with high affinity to glycans in proteoglycans and glycoproteins. The best characterized ligand or a counterreceptor is a mucin-like glycoprotein, P-selectin glycoprotein ligand-1 (PSGL-1), expressed on most lymphocytes. PSGL-1 binds with high affinity to all three selectins. There are other ligands that bind to selectins with varying selectivity, such as ESL1 and podocalyxin (Table 7–1). E-selectins are the most permissive in ligand binding, whereas P-selectin and L-selectin bind preferably to ligands with sulfated tyrosine residues.

Selectins and PSGL-1 play crucial role in leukocyte recruitment during inflammation. Endothelial P-selectin interacts with leukocyte PSGL-1 (see later). Deregulation of this process leads to pathological states. For instance, patients with a mutation in the gene encoding the fucose transporter cannot form normally glycosylated PSGL-1, which cannot bind to any selectins. These patients exhibit defective leukocyte infiltration and suffer from bacterial infections. P-selectin has been considered a risk factor for recurrent thromboembolism. Soluble P-selectin has been shown to be elevated in patients with cardiovascular disorders. In experimental models, anti-P-selectin antibodies seem to prevent pulmonary arterial fibrin accumulation after blunt thoracic trauma.

Chronic inflammation is an important event in the pathogenesis of SCD. Chronic inflammatory vasculopathy in SCD involves the recruitment of leukocytes and thrombus formation, leading to ischemic tissue damage. Therefore selectins are targets for the development of therapeutics for the treatment of SCD. Evidence indicates that selectins and their ligands are involved in cancer progression. Cancer cells express selectin ligands (mucins) that facilitate their binding to leukocytes and endothelial cells, leading to extravasation during tumor metastasis. Additionally, cancer cells themselves rarely express selectins, thus helping in seeding in distant metastases.

Integrins Are Dimeric Receptors for Cell-Cell and Cell-Matrix Adhesion

As the name indicates, integrins play a role in the integration of cells into tissues. Integrins function as cell surface receptors that interact with components of the extracellular matrix. Such cell-matrix adhesion provides mechanical adhesion of cells in the tissues. Integrins also play a role in cell-cell adhesion in lymphoid cells. Tight adhesion through integrins enables leukocytes to transmigrate through the vascular endothelium (extravasation) during inflammation. Cancer cells take advantage of this extravasation process during tumor metastasis. Another important function of integrins is signal transduction. Interaction of integrins with intracellular and extracellular proteins triggers signaling pathways in cells that regulate a variety of cellular functions, such as cell proliferation, differentiation, motility, and cell survival. Integrins are mechanochemical sensors. Binding of specific components of ECM to integrins initiate cell survival signals, an "outside-in" signal. Loss of such interactions results in loss of survival signal leading to detachment-induced apoptosis, a process termed "anoikis." Gene mutations in cancer cells make them resistant to anoikis, thus enabling metastasis. Interaction of intracellular proteins with integrin initiates "inside-out" signals that regulate cellular functions, such as cell migration. Again, cancer cells manipulate this signaling mechanism during tumor metastasis.

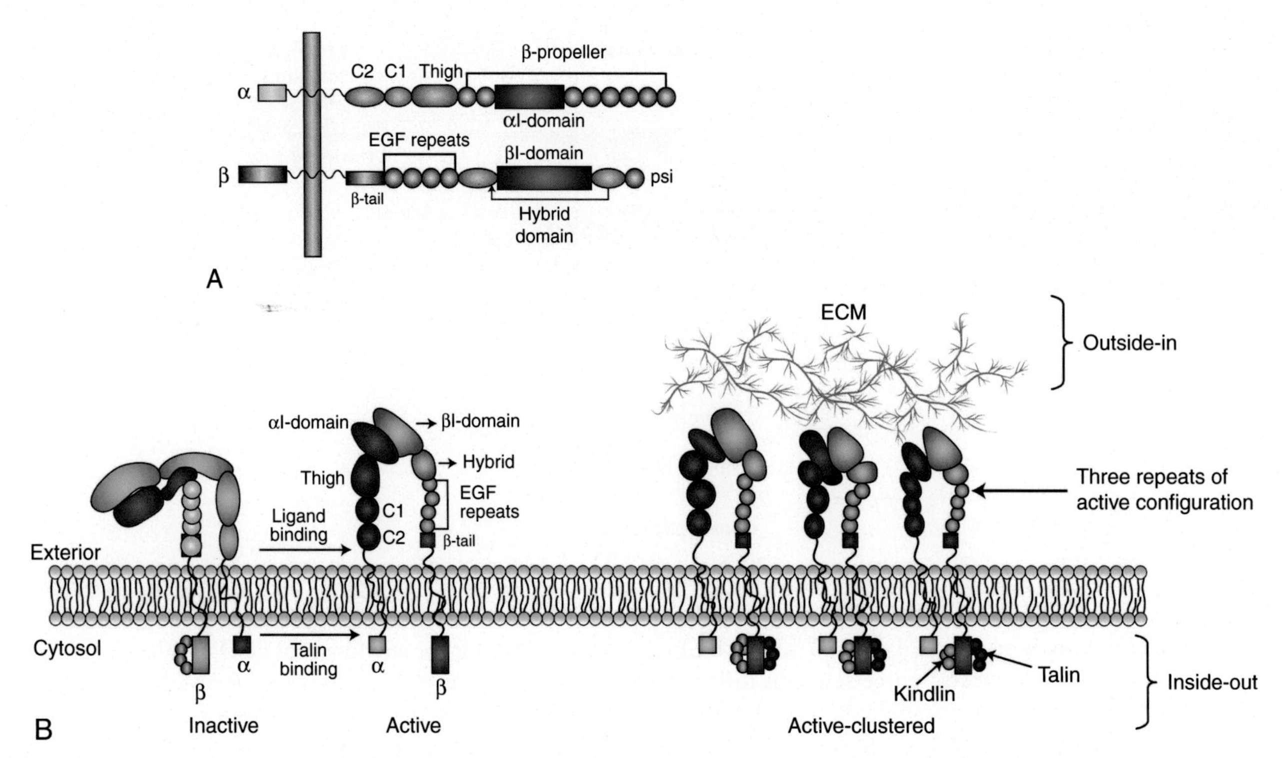

Figure 7–3. **Integrin activation and clustering.** (A) Domain structure of integrin subunits, (B) activation and clustering of integrins.

Unlike other adhesion receptors, integrins are heterodimeric transmembrane proteins that consist of one α-subunit and one β-subunit (Fig. 7–3). Eighteen different α-subunits and eight different β-subunits pair to form at least 24 different heterodimers in vertebrate tissues with distinct cellular functions. On the basis of subunits involved, integrins may be classified into different subfamilies. Thus α_1-integrin may associate with one of eight different β-subunits to generate a series of predominantly matrix receptors of differing ligand specificity. In contrast the α_2-integrins are a group of cell-cell adhesion receptors of lymphoid cells with three alternative β-subunits.

Integrin subunits consist of an extracellular domain, a transmembrane domain, and an intracellular domain (Fig. 7–3, A). The extracellular domain of α-subunit consists of the seven-bladed β-propeller head domain, a thigh domain, two calf domains, and a large αI-domain with metal ion-dependent ligand-binding site. The I-domain consisting of about 200 amino acids is inserted between blades 2 and 3 of the β-propeller. The extracellular domain of the β-subunit consists of a PSI (plexin-semaphorin-integrin) domain, a hybrid domain, and a βI-domain (homologous to αI-domain) that are inserted into the hybrid domain. The β-subunit also has four epidermal growth factor (EGF) repeats and a β-tail domain. The intracellular domains of integrins are short and lack subdomain structure. αI-domain and βI-domain of some integrins exhibit quite specific binding properties, whereas others are promiscuous. Thus $\alpha_5\beta_1$ binds to the arginine-glycine-aspartic acid (RGD in single-letter amino acid code) tripeptide sequence of the ECM protein fibronectin, but $\alpha_v\beta_3$ binds to several matrix components including vitronectin, fibronectin, fibrinogen, von Willebrand factor (vWF), thrombospondin, and osteopontin. An interesting example is $\alpha_4\beta_1$, which can bind not only to a specific domain of fibronectin but also to the Ig family adhesion receptor VCAM on endothelial cells. Additionally, individual cell types usually express multiple integrins. For example, blood platelets express predominantly $\alpha_{IIb}\beta_3$ (GPIIb-IIIa), which binds to fibrinogen, fibronectin, vWF, and vitronectin, and lesser amounts of $\alpha_v\beta_3$, $\alpha_5\beta_1$, $\alpha_6\beta_1$ (collagen), and $\alpha_6\beta_1$ (laminin) (see Fig. 7–25).

Recent studies have shown that EILDV and REDV sequences in ECM proteins also interact with integrins in an isoform-specific manner. For instance, $\alpha_5\beta_1$ interacts with EILDV and REDV sequences in fibronectin. Each of these specific interactions is suggested to trigger a unique signaling cascade regulating distinct cellular functions such as cell migration, proliferation, differentiation, and survival. The ligands for cell-cell interactions are transmembrane proteins such as syndecans, tetraspanins, and CD47. The intracellular domains interact with the proteins such as talin and kindlin, which in turn interact with the actin cytoskeleton via multiple cytoskeletal binding proteins, such as α-actinin, vinculin, paxillin, filamin, and tensin. Intracellular protein complexes are in dynamic interactions with the intracellular signaling proteins that play crucial roles in both inside-out and outside-in signal transduction.

As indicated earlier the αI-domain of α-subunit is inserted between blades 2 and 3 of β-propeller, and the βI-domain of β-subunit is inserted into the hybrid domain and thus unavailable for ligand binding, or the integrin is in an inactive state. The extracellular domains of integrins can switch between a low-affinity inactive state and a high-affinity active state (Fig. 7–3, B). The inside-out signals involving talin and kindlin binding to the intracellular domain of β-subunit induce conformational change to an active state, thus enabling I-domains to bind to their specific ligands. This mechanism is important in regulating cell behavior (see later) and is called **integrin activation.** The active state is dependent on the divalent cations Mn^{2+} or Mg^{2+}, and the inactive state is stabilized by calcium. During the inactive state, α- and β-subunits are held close to each other by calcium. During activation the release of calcium-binding results in the separation of α- and β-subunits. Ligand binding to partially active integrins can also induce some conformational changes and reinforces the active state. Activation of integrins by either intracellular signaling or binding to ECM induces the clustering of integrins on the plasma membranes.

An important property of cell adhesion is that small tight modular clusters of CAMs such as integrins or cadherins form loose aggregates contributing to cell-cell and cell-matrix adhesions. Recent advances in the development of nanoscale superresolution microscopy have revealed more details of integrin clustering in response to cellular activities and extracellular stimuli. The advantages of clustering receptors include enhancing avidity and specificity of receptor-ligand interactions, regulation of adhesion strength and dynamics, and spatial segregation of signals. Integrins play an essential role in the assembly and disassembly of focal contacts where migrating cells form multiple attachments to cell-matrix or focal adhesions. Focal adhesions are formed by the interaction of the extracellular domain of integrins with the matrix ligands and the cytoplasmic domain with the actin cytoskeleton. The interaction with the cytoskeleton is regulated by the phosphorylation-mediated assembly and disassembly of focal adhesion complex that includes integrins, paxillin, focal adhesion kinase, and other actin-binding proteins. Figure 7–4 shows a superresolution microscopic image revealing the nanoscale clustering of paxillin in migrating mouse embryonic fibroblasts. Cooperative signaling might independently occur in one cluster of integrins regulating a specific cellular function. Therefore the clustering of integrins and other CAMs is an essential mechanism for dynamic regulation of different cellular functions.

Integrin activities are involved in the pathogenesis of many human diseases, including cancer, IBD, and multiple sclerosis. Uncontrolled proliferation, altered stress response to favor cell survival, and invasion or metastasis are some of the hallmarks of cancer. Integrins play crucial roles in all these properties of cancer cells. A variety of integrins are overexpressed in cancer cells such as melanoma, breast cancer, prostate cancer, pancreatic cancer, and lung cancer. Deregulated integrin functions, either due to altered signals that activate integrins or the signals downstream to integrin activation, cause uncontrolled proliferation of cancer cells and enable their survival under hostile microenvironments. Integrins play an essential role in cell migration, particularly in leukocyte rolling and extravasation in inflammatory responses. This property of integrins is adapted by cancer cells in metastasis and invasion. Normal or deregulated functions of various integrins are also associated with the pathogenesis of other diseases such as IBD, rheumatoid arthritis, multiple sclerosis, and cardiovascular diseases. Therefore integrins are important targets for drug development for many diseases. Several drugs targeting specific integrins have been developed recently for cancer and other diseases. For instance, Cilengitide (RGD pentapeptide of $\alpha_v\beta_5$) has progressed to phase III clinical trials for the treatment of recurrent glioblastoma. Volociximab, a chimeric human IgG4 antibody inhibitor of

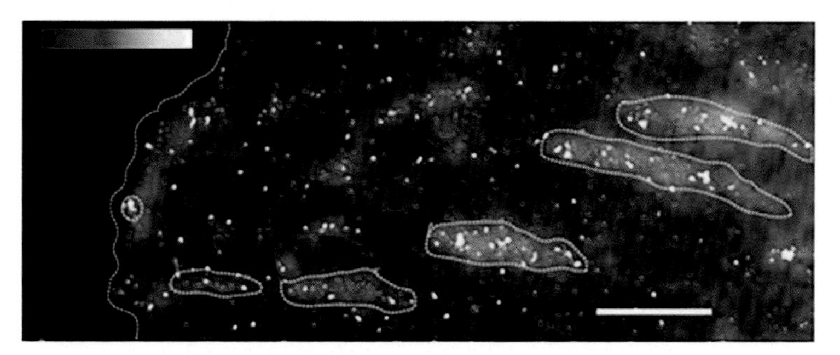

Figure 7–4. **Hierarchical organization of integrin clusters in migrating cell.** Image of a mouse embryonic fibroblast spread on fibronectin-coated substrate for 30 min. dSTORM imaging of endogenous paxillin conjugated (labeled with paxillin antibody, BD biosciences) to antibody tagged with AlexaFlour-647 (antimouse secondary antibody conjugated to AlexaFlour-647). Gray scale shows the TIRF image, whereas false color image shows the reconstructed superresolution image. Color code (top left corner, from left to right) indicates increasing intensity of molecules. Encircled in *dotted lines* from left- to right-cell edge, nascent adhesions, maturing adhesions, mature adhesions. Scale bar 2 μm. *(Adapted from Changede, Sheetz.* BioEssays *2016;39:1, 1600123, by permission.)*

$\alpha_5\beta_1$-integrin, is in phase II clinical trial for metastatic melanoma. Other integrin-based therapeutic candidates include natalizumab (anti-α4-integrin antibody) for multiple sclerosis and Crohn's disease that targets lymphocyte infiltration.

INTERCELLULAR JUNCTIONS

Organized clustering of CAMs in the cell surface form specialized adhesive structures called *cell junctions*. These junctions serve very specific functions in addition to cell-cell or cell-substratum adhesion. The special functions include sealing the spaces between cells, regulating cell polarity, and participation in direct cell-cell communication.

Electron microscopic analyses of cell junctions have revealed the distinct ultrastructure of different junctional complexes at different locations of the cell. On the basis of their ultrastructural morphology and functions, three types of junctions have been identified at the apicolateral borders of simple epithelial cells, such as those of the intestinal mucosa, bronchial epithelium of lung, and tubules in the kidney (Fig. 7–5). From apical to basal the junctional complexes consist of **tight junction** or zonula occludens, **adherens junction** or zonula adherens, and **desmosome** or macula adherens. Tight junction serves the barrier function and fence function of the epithelial tissues, while adherens junction is important for the development and maintenance of the polarity of epithelial cells. The primary function of desmosome is cell-cell adhesion and stabilization of tissue structure. The fourth type of intercellular junction is the **gap junction**, which plays a major role in intercellular communication.

At the cell-substratum interface of various epithelial cells, adhesion of cell base to the matrix is maintained by another type of cell junctions called **hemidesmosomes.** The structure of hemidesmosomes resembles half desmosomes but differs from desmosomes in molecular composition. Moving cells make focal contacts with the cell substratum by junctional complexes called **focal adhesions.** Both hemidesmosomes and focal adhesions play roles in cell-substratum adhesion and stabilizing the cells at its location. The structure and function of these junctions are considered in more detail in the following section.

Tight Junctions Regulate Paracellular Permeability and Cell Polarity

A monolayer of epithelial cells lines the intestinal mucosa. Similarly the epithelial lining is present in airways (bronchiole and alveoli), renal tubules, and glandular ducts. An important function of epithelial tissues is the barrier function that prevents diffusion of macromolecules across the epithelium. In the intestinal mucosa, epithelium prevents diffusion of toxins, allergens, and pathogens from the intestinal lumen into the mucosa and beyond. This barrier function is enabled by tight junctions that seal the paracellular space as a circumferential belt around the cell at the apicolateral surface. The tight junction is formed and maintained by a multiprotein complex. Tight junctions in different epithelia share common structures and types of proteins involved, except for differences in specific protein composition and regulatory mechanisms that control tight junction dynamics. Tight junction biology has been most studied in the intestinal epithelium, and hence, this chapter focuses on the intestinal epithelium.

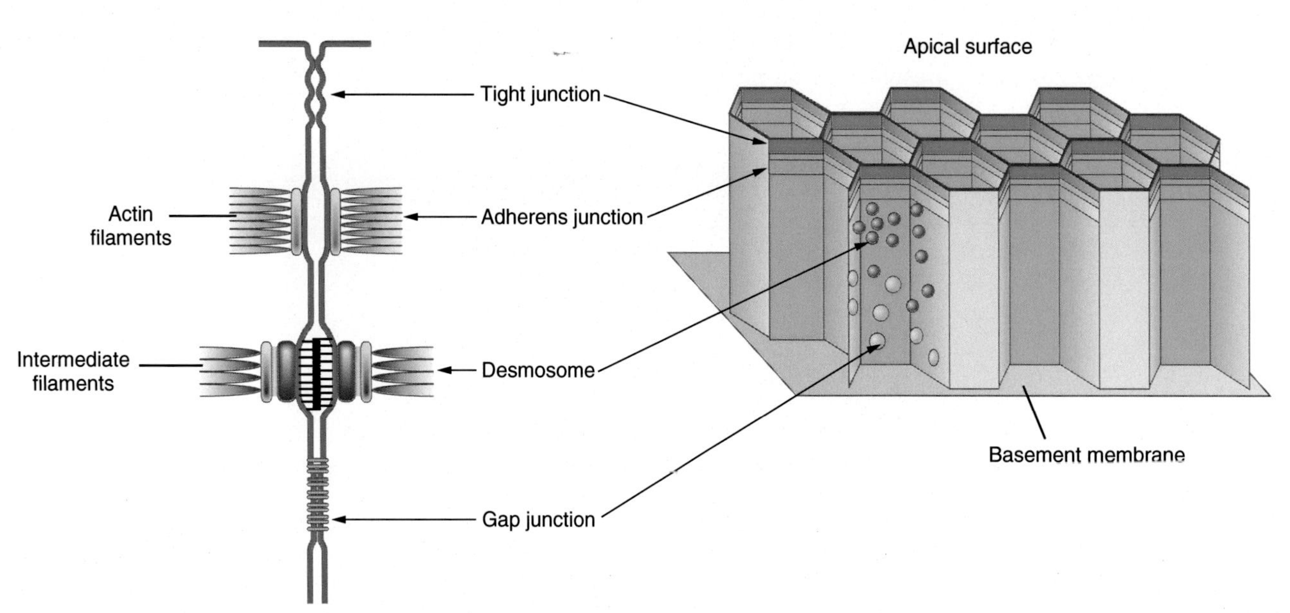

Figure 7–5. **Intercellular junctions and the junctional complex.** Simple epithelial cells have a junctional complex at their apicolateral borders. The components are the tight junction, the adherens junction, and the desmosome. The tight and adherens junctions are zonular, extending right around the cells, whereas desmosomes are punctate. Desmosomes are also present beneath the junctional complex, as is another punctate junction, the gap junction.

Tight junctions can be visualized by electron microscopy and immunofluorescence microscopy, but not visible by light microscopy. In transmission electron microscopy the tight junction can be visualized as electron-dense regions at the apicolateral region with tightly occluded adjacent cell plasma membranes and contact points called **kisses** (Fig. 7–6). The freeze-fracture electron microscopy that involves splitting cell membranes between the inner and outer leaflets in frozen tissues revealed tight junctions as anastomosing strands of beads forming a zone or belt around the apicolateral margin of the cell (Fig. 7–6). The beads represent molecules of transmembrane proteins of tight junctions. The strand numbers and the depth of the tight junction structure are inversely proportional to the epithelial permeability to macromolecules. The intestinal epithelial barrier function is diminished in many diseases, such as celiac disease. The freeze-fracture electron microscopy of the jejunal epithelium in control and celiac disease patients indicate that the strand numbers and the depth of tight junctions are significantly low in celiac disease (Fig. 7–7) that corresponds to higher intestinal permeability in patients with celiac disease.

The composition of tight junction strands was debated until 1986 when the first tight junction specific protein zonula occludens-1 (**ZO-1**) was discovered, followed by identification of another tight junction specific protein

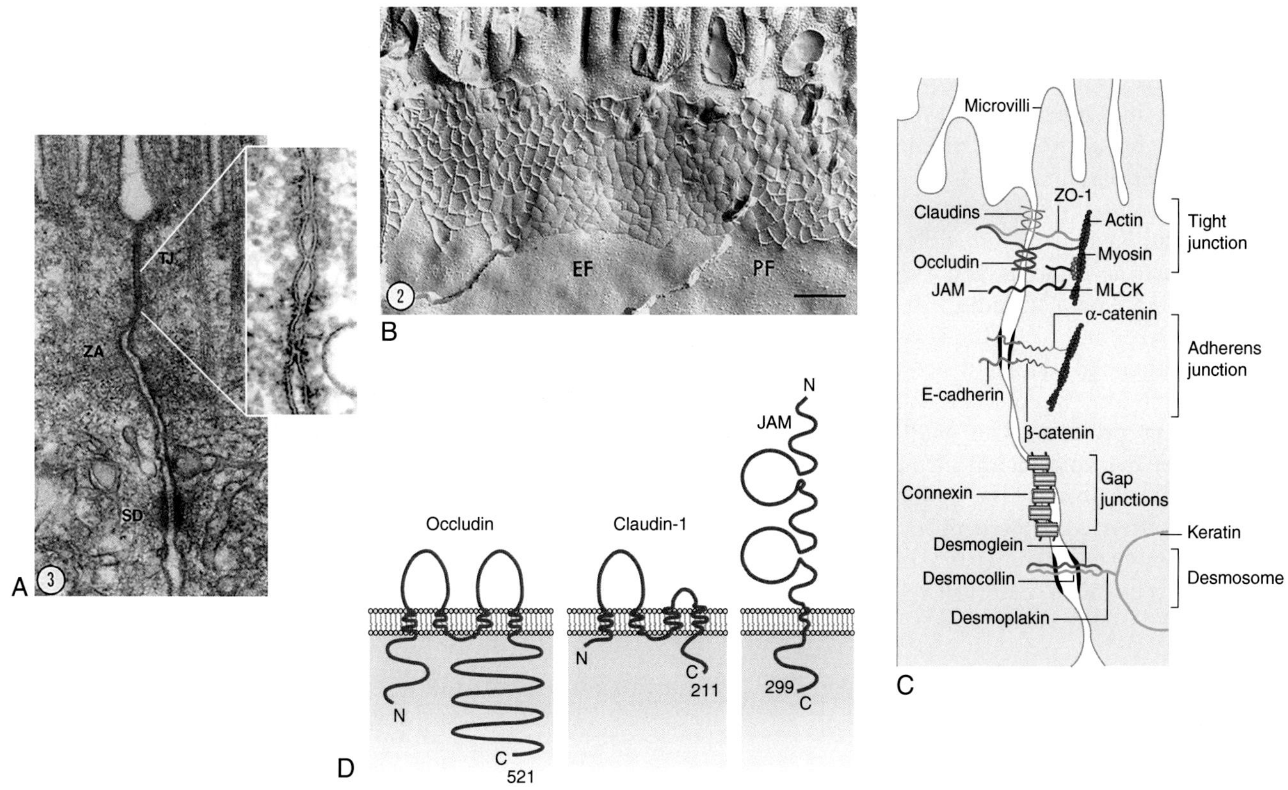

Figure 7–6. **Tight junctions are anastomosing strands of multiprotein complexes.** (A) The close relationship between the organization of the junctional complex, and that of the terminal web in the small intestine is illustrated in this thin-section electron micrograph. In the cytoplasm at the level of the tight junction (TJ), the core microfilaments appear as compact bundles enmeshed in a matrix of 70 Å filaments. Mats of 70 Å filaments line the membranes of the zonula adherens (ZA) or intermediate junction, and the bundles of core microfilaments became more diffuse in the adherens zone. Numerous tonofilaments course through the plaque of the spot desmosome (SD); some of these tonofilaments enter the adherens zone, while others form the basal zone. 80,000 ×. The inset image is a higher magnification of TJ region showing the "kisses." (B) An evenly cross-linked tight junction network found between the absorptive cells of the small intestine of a *Xenopus laevis*, stage 57. Short ridges or grooves on the protoplasmic (PF) or extracellular (EF) faces, respectively, intersect at acute angles to form a series of adjoining, similarly shaped polygons, which display no predominant orientation relative to the cell surface. Bar denotes 0.25 μm in the electron micrographs unless marked otherwise. 60,000 ×. (C) Line drawing of the apical junctional complex of an intestinal epithelial cell. Tight junction proteins include claudins, occluding, junctional adhesion molecule (JAM), and zonula occludens-1 (ZO-1), whereas E-cadherin, α-catenin, and β-catenin interact to form the adherens junction. Myosin light chain kinase (MLCK) is associated with the perijunctional actomyosin ring. Gap junctions are tubelike connections between adjacent cells formed by assembly of connexins. Desmosomes are formed by interactions between desmoglein, desmocollin, desmoplakin, and keratin filaments. (D) Occludin has four transmembrane domains with two extracellular loops. The first loop is characterized by a high content (~60%) of glycine and tyrosine residues. Claudin-1 also has four transmembrane domains but shows no sequence similarity to occludin. Note that the cytoplasmic tail of claudin-1 is shorter than that of occludin. Junctional adhesion molecule (JAM) has a single transmembrane domain, and its extracellular portion bears two immunoglobulin-like loops that are formed by disulfide bonds. *([A] Adapted from Hull, Staelin.* J Cell Biol *1979;81:67–82, by permission; Tsukita, Furose, Itoh.* Nat Rev *2001;2:285, by permission; [B] Adapted from Hull, Staelin.* J Cell Biol *1976;68:688–704, by permission; [C] From Turner JR. Intestinal mucosal barrier function in health and disease.* Nat Rev Immunol *2009;9:799–809, by permission; [D] Adapted from Tsukita, Furose, Itoh.* Nat Rev *2001;2:285, by permission.)*

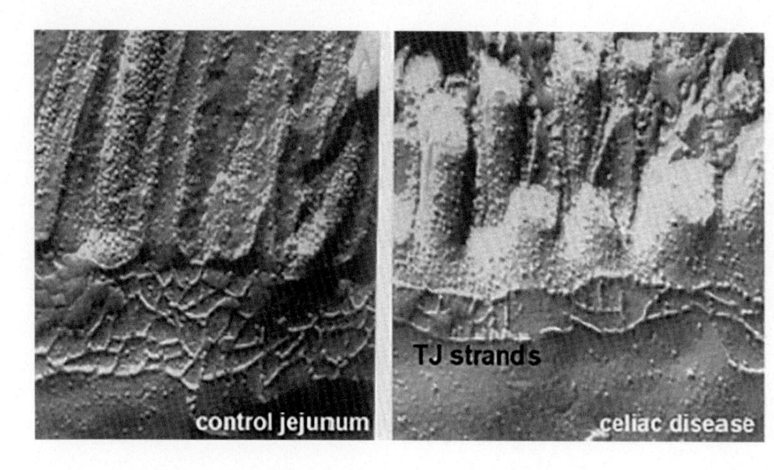

Figure 7–7. **Disrupted tight junction strands in intestinal disease.** Freeze-fracture electron microscopy of epithelial tight junctions from the jejunum of control subject and acute celiac disease patient. The jejunum from celiac disease patient shows significant reduction in the number or horizontally oriented strands and the depth of the tight junctional strand network. Additionally, strand discontinuities appeared in celiac disease. *(Adapted from Rosenthal, Barmeyer, Schulzke.* Tissue Barriers *2015;3:1–2, e977176, by permission.)*

called **cingulin.** The theory of protein composition of the tight junction was further supported by the discovery of the first known transmembrane protein of tight junction called **occludin.** Other occludin-related proteins associated with tight junction are **tricellulin** (MarvelD2) and MarvelD3. Tricellulin is localized predominantly at the tricellular junctions of the epithelial monolayer. Subsequent studies have identified additional transmembrane proteins such as **claudins** and JAM. The intracellular domains of occludin, claudins, and tricellulin bind to ZO-1, ZO-2, and ZO-3. ZO proteins are PDZ domain-containing adapter proteins that interact with many plaque-forming proteins, such as cingulin and actin-binding proteins. This multiprotein complex is anchored at the apicolateral membrane by binding to the actin-myosin belt in the apical region of epithelial cells. Interaction with the actin cytoskeleton is crucial for the assembly and maintenance of tight junctions. Disruption of the actin cytoskeleton by cytochalasin-D results in disruption of tight junction and loss of barrier function. The cellular localization of tight junctions and its integrity can be visualized by immunofluorescence staining of tight junction proteins and confocal microscopy (Fig. 7–8, A). Such techniques show a chicken-wire pattern of distribution of these proteins on the apical end of epithelial cells. Disruption of tight junctions, for example, by proinflammatory cytokines show a redistribution of tight junction proteins from the intercellular junctions into cytoplasmic compartments (Fig. 7–8, B).

Although occludin was initially considered as the primary protein of tight junction, the development of occludin knockout mice in the year 2000 indicated that tight junction could be formed in intestinal and other epithelia in the absence of occludin. This striking observation concluded that occludin is not required for the assembly of the tight junction, which led to a myriad of subsequent studies to understand the role of claudins in tight junction assembly. Although tight junction assembly occurs in the

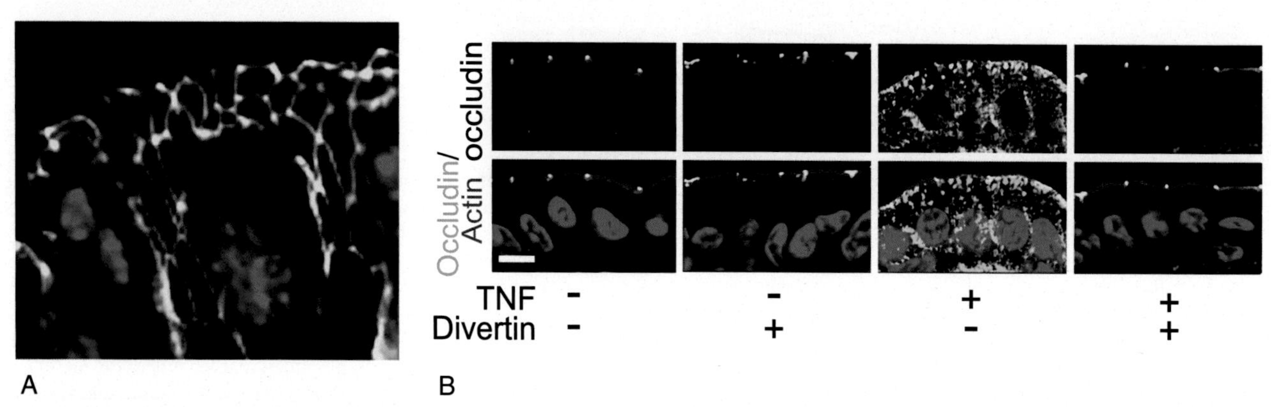

Figure 7–8. **Fluorescence labeling of epithelial tight junctions.** (A) Frozen section of mouse colon was stained for occludin *(green)* and ZO-1 *(red)* by immunofluorescence method. Nucleus was stained by a DNA-binding fluorescent dye *(blue)*. Confocal fluorescence microscopic image shows organization of tight junctions on the apical ends of colonic epithelial cells. (B) TNF-α induces redistribution of occludin from the junctions to cytoplasm. Mice were injected with vehicle or TNF-α. Cryosections of jejunum were stained for occludin *(green)* by immunofluorescence method. Filamentous actin *(red)* was stained by binding to fluorescently labeled phalloidin, and nucleus *(blue)* was stained by binding to fluorescent dye. The images show the TNF-α disrupts tight junctions. In one group of mice, the jejunum was perfused with divertin (MLCK inhibitor), which shows that divertin blocks TNF-α-mediated disruption of tight junctions. *([A] From Rao et al. Unpublished; [B] Adapted from Graham et al.* Nat Med *2019,* https://doi.org/10.1038/s41591-019-0393-7, *by permission.)*

absence of occludin, it is widely distributed in the tight junction in different epithelia under normal physiologic conditions, which led few laboratories to continue to investigate the function of occludin in tight junctions. Several such studies indicated that occludin might be required for cellular functions such as tight junction dynamics, apoptosis, and cell migration.

Occludin is a tetraspanin with four transmembrane domains, two extracellular loops, and one small intracellular loop. C-terminal and N-terminal domains extend into the cytoplasm (Fig. 7–6, D). The extracellular loops form homophilic interaction with extracellular loops of occludin in the adjacent cells. The intracellular domain binds to ZO-1, ZO-2, and ZO-3. The intracellular domain also interacts with various cell signaling molecules such as protein kinases, protein phosphatases, and Rab GTPases. The intracellular domain of occludin is heavily phosphorylated on Ser, Thr, and Tyr residues that are localized within a sequence of 10 residues. Reversible phosphorylation of occludin regulates its assembly into the tight junction. Whereas phosphorylation on Thr appears to be required for assembly and/or maintenance of occludin in the tight junction, the phosphorylation of Tyr residues is associated with the disassembly of occludin and disruption of tight junctions. A recent study characterized this phosphorylation hot spot in the intracellular domain of occludin as "occludin regulatory motif or ORM." Absence of occludin or deletion of ORM attenuates the dynamic properties of tight junctions and blocks collective cell migration in the intestinal and renal tubular epithelium.

Claudins (23–25 kDa MW) are also tetraspanins but are smaller than occludin (59–66 kDa MW). Twenty-seven isoforms of claudin genes have been identified in human cells. However, their expression varies with tissues and species. The C-terminal domains of claudins also interact with ZO proteins and the actin cytoskeleton. The occludin extracellular loop interactions between cells clearly form the barrier function. The claudin extracellular loop interactions form pores that are selective for certain ions. Claudin-2 forms cation-selective pores that are highly permeable to sodium, whereas Claudin-16 forms pores that are selective to magnesium and calcium. Claudin-1 and claudin-3 appear to play a role in the formation of barrier function. However, the barrier function of claudins is poorly understood.

One of the major roles of tight junctions is **the "barrier" function** or **"gate" function**. Tight junction occludes the space between the adjacent cells forming a barrier to diffusion of macromolecules and certain ions across the epithelium (Fig. 7–9, A). This means that an epithelial cell layer can act as a barrier separating biological compartments and that the epithelial cells can regulate what passes from one compartment to another. If the electrical resistance between the apical and basal compartments across the epithelium with well-formed tight junctions is measured, it is recorded much greater than

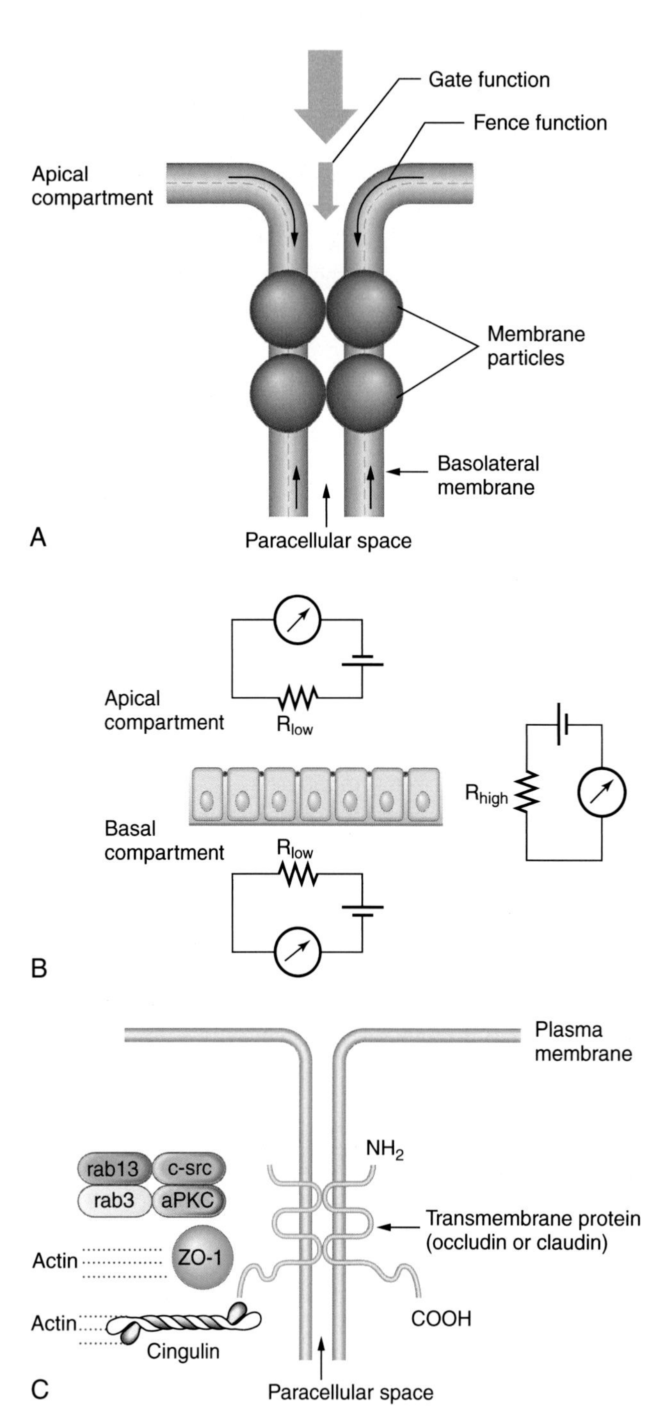

Figure 7–9. **Structure and function of tight junctions.** (A) The "gate" function of the tight junction refers to its property of regulating the permeability of the paracellular channels, and the "fence" function maintains the separation of molecules in the apical and basolateral cell membranes. (B) The low permeability of tight junctions causes high electrical resistance across the epithelial cell layer between the apical and basal compartments. (C) Some of the molecules that contribute to the structure and function of the tight junction are shown.

that of the media on either compartment. This is referred to as **transepithelial electrical resistance (TER)**. TER is caused by impedance to the diffusion of ions across the epithelium due to tightly occluded intercellular spaces by tight junctions (see Fig. 7–9, B). Ions that carry electric current can move much more easily within the apical or basal media than they can across the epithelium. If tight junctions are disrupted experimentally, for example, by hydrogen peroxide, osmotic stress, or proinflammatory cytokines, the TER is decreased. Therefore TER is an indicator of epithelial permeability to ions and solutes.

The tight junction barrier function has been studied extensively in the epithelia of the intestine, kidney tubules, and airways and in the endothelia that line blood vessels, particularly in the blood-brain barrier. In the gut, disruption of tight junction and increase in epithelial permeability are associated with the pathogenesis of numerous gastrointestinal diseases such as IBD, irritable bowel syndrome, celiac disease, and infectious enterocolitis. Disruption of tight junctions in the intestinal epithelium leads to the absorption of bacterial toxins such as lipopolysaccharides (LPS) into the portal circulation, liver, and subsequently into the systemic circulation causing a condition called endotoxemia. Endotoxemia may trigger or promote many systemic diseases, including obesity, diabetes, and cardiovascular diseases. Disruption of bronchial and alveolar epithelial tight junctions is associated with allergic rhinitis and asthma in the airway. Similarly, renal tubule epithelial barrier defect is involved in many renal diseases.

Two distinct components of barrier function are the **pore pathway** and the **leak pathway**. The pore pathway of tight junction permeability represents permeability to ions, with selectivity to different ions. These pores are formed by the interactions between extracellular loops of different claudin isoforms. For example, pores formed by claudin-2 are selective to sodium, whereas those created by claudin-16 are selective to magnesium and calcium. The precise functions of pore pathways are unclear. The claudin-2 expression is increased in many diseases such as IBD; however, its role in the pathogenesis of the disease is not understood. A genetic disorder called *familial hypomagnesemia* caused by mutation of the claudin-16 gene demonstrates the important role of claudin-16 in magnesium reabsorption in the kidney tubule. Hypomagnesemia and hypercalcemia in this disease are associated with nephrocalcinosis. However, the magnesium pore theory is too simplistic. The in vitro studies and claudin-16 knockout mice suggested that a defect in sodium permeability is the primary cause that leads to hypomagnesemia. Another disease associated with claudin expression is claudin-14 mutation-induced deafness. Although the evidence is circumstantial, a defect in K^+ permeability across the lining of intracochlear space is suggested.

Unlike the pore pathway the permeability of molecules in the leak pathway is independent of their size and the charges they carry. A study analyzing the permeability to noncharged solutes like graded polyethylene glycol (PEG) distinguished the leak pathway from the pore pathway. PEG molecules with a molecular radius of 4 Å shows a size-dependent apparent permeability. This pathway shows ionic charge discrimination, the selectivity, and the extent of which varies in different epithelia. As described earlier the charge selectivity is determined by the type of claudin involved, and the extent of selectivity is determined by the level of expression. On the other hand the leak pathway impedes the diffusion of PEG molecules with a molecular radius greater than 4 Å. However, above 4 Å, the pathway shows no size or charge discriminations. It is, therefore, important to recognize that the pore pathway (electric resistance) and the leak pathway (permeability to markers) are not directly correlated. The disruption of the leak pathway is necessary to induce increased permeability to LPS and to result in endotoxemia. In fact, oxidative stress, osmotic stress, mechanical stress, and proinflammatory cytokines have been shown to disrupt the leak pathway under pathophysiologic conditions. Cytokines can disrupt the leak pathway without affecting the pore pathway. The pore pathway can be altered by changes in the expression of different claudins, whereas the leak pathway is highly sensitive to the disruption of the actin cytoskeleton. Disruption of leak pathway and barrier dysfunction by proinflammatory cytokines and oxidative stress is associated with disruption of the actin cytoskeleton in different epithelia. Tumor necrosis factor-α (TNF-α) and interleukin-13 (IL-13), the two major inflammatory mediators in IBD, disrupt the tight junction barrier in the intestinal epithelium in vitro. Interestingly, TNF-α enhances the leak pathway of the barrier, whereas IL-13 activates the pore pathway. TNF-α disrupts tight junction architecture to increase macromolecular permeability by a myosin light chain kinase (MLCK)-dependent mechanism (Fig. 7–8, B). IL-13 increases the expression of claudin-2 in the intestinal epithelium and increases cation permeability. Oxidative stress is another important inflammatory mediator involved in the pathogenesis of various human diseases. Hydrogen peroxide-mediated oxidative stress disrupts tight junction barrier function that is associated with remodeling of the actin cytoskeleton. Tight junction disruption by oxidative stress is associated with tyrosine phosphorylation and threonine dephosphorylation of occludin and likely other tight junction proteins. Inhibitors of tyrosine kinase and protein threonine phosphatases (PP2A and PP1) block oxidative stress-mediated disruption of tight junction and permeability to macromolecular markers.

A second vital role of tight junctions is called the **"fence" function** (see Fig. 7–9, A). Epithelial cells are polarized both morphologically and functionally. The surface of epithelial cells that are attached to the tissue substratum is the basal surface, and the surface that faces the open space (intestinal, tubular, and ductal lumen) is referred to as apical surface. Most epithelial cells are cuboidal or cylindrical in shape. The tight junction separates the plasma membrane into the apical and basolateral surface. The morphologically apical surface is different from the basolateral surface because apical membranes are folded into microvilli. Functionally the functional proteins present in the apical membrane are distinct from those in the basolateral membrane. This **polarity** is crucial for the specific functions of different epithelia. Lateral diffusion of membrane molecules is extremely rapid. However, the mixing of components of the apical membrane with those of basolateral membrane and vice versa is prevented by the tight junctions. This property of tight junction is called the fence function and therefore helps maintain cell polarity. Disruption of tight junction leads to loss of cell polarity, and the loss of cell polarity is a hallmark feature of epithelial cancer cells (carcinoma cells) that have become poorly differentiated.

Adherens Junctions Are Important for Cell-Cell Adhesion and Cell Signaling

Situated immediately beneath the tight junction in the apical junctional complex is the adherens junction. Adherens junctions are characterized by parallel cell membranes separated by an intercellular space approximately 20 nm in width, a cytoplasmic plaque of low electron density, and association with microfilaments of the **actin cytoskeleton** (see Fig. 7–5). In simple epithelial cells the adherens junction is zonular (hence the name zonula adherens), extending around the entire circumference of the cells accompanied by a ring of actin filaments. Figure 7–10 shows immunofluorescence localization of two adherens junction proteins, E-cadherin, and β-catenin in the intestinal epithelium. This z-section image shows the colocalization and zonular organization of these adherens junction proteins. In other cell types, adherens junctions take various forms, such as the extensive **fasciae adherentes** of cardiac muscle and the smaller, punctate junctions formed by fibroblasts and other migratory cells.

The primary function of the adherens junction is cell-cell adhesion. In epithelia where adherens junctions and desmosomes are present, both contribute to the cohesiveness of the tissue. In simple epithelia where desmosomes are relatively few, adherens junctions probably make the major adhesive contribution, whereas, in stratified epithelia where desmosomes are abundant, adherens junctions likely make a minor contribution. Irrespective of their relative abundance, the adherens junctions and desmosomes are interdependent. Initial contacts between cells are made by fine processes called *filopodia* (singular, filopodium) that interdigitate and adhere by means of small adherens junctions forming an "adhesion zipper" (Fig. 7–11, A). Contacts are then stabilized by the

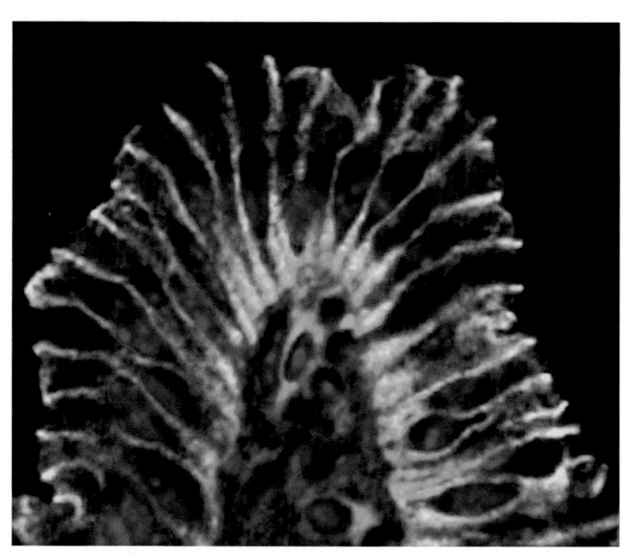

Figure 7–10. **Fluorescence labeling of epithelial adherens junctions.** Frozen section of mouse ileum was stained for E-cadherin *(green)* and β-catenin *(red)* by immunofluorescence method. Nucleus was stained by a DNA-binding fluorescent dye *(blue)*. Confocal fluorescence microscopic image shows organization of adherens junctions along the lateral membranes of colonic epithelial cells. *(Adapted from Shukla, Meena, et al.* Am J Physiol (GI&Liv) *2016;310: G705, by permission.)*

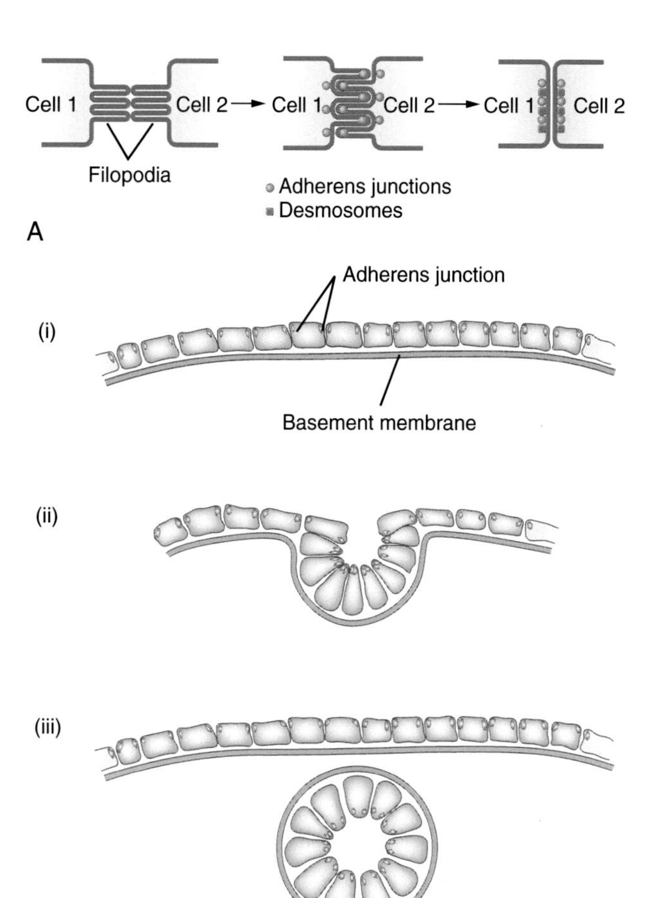

Figure 7–11. **Functions of the adherens junction.** (A) Initial cell contact is made by filopodia. These interdigitate and punctate adherens junctions are formed to constitute an adhesion zipper. The adhesion is then expanded and stabilized by the formation of desmosomes. (B) Bending of epithelial cell sheets during embryonic development is accomplished by contraction of the actin filament ring underlying the apicolateral adherens junctions. This acts like a purse-string to narrow the apices of the cells. This process can result in tube formation as in generation of the neural tube.

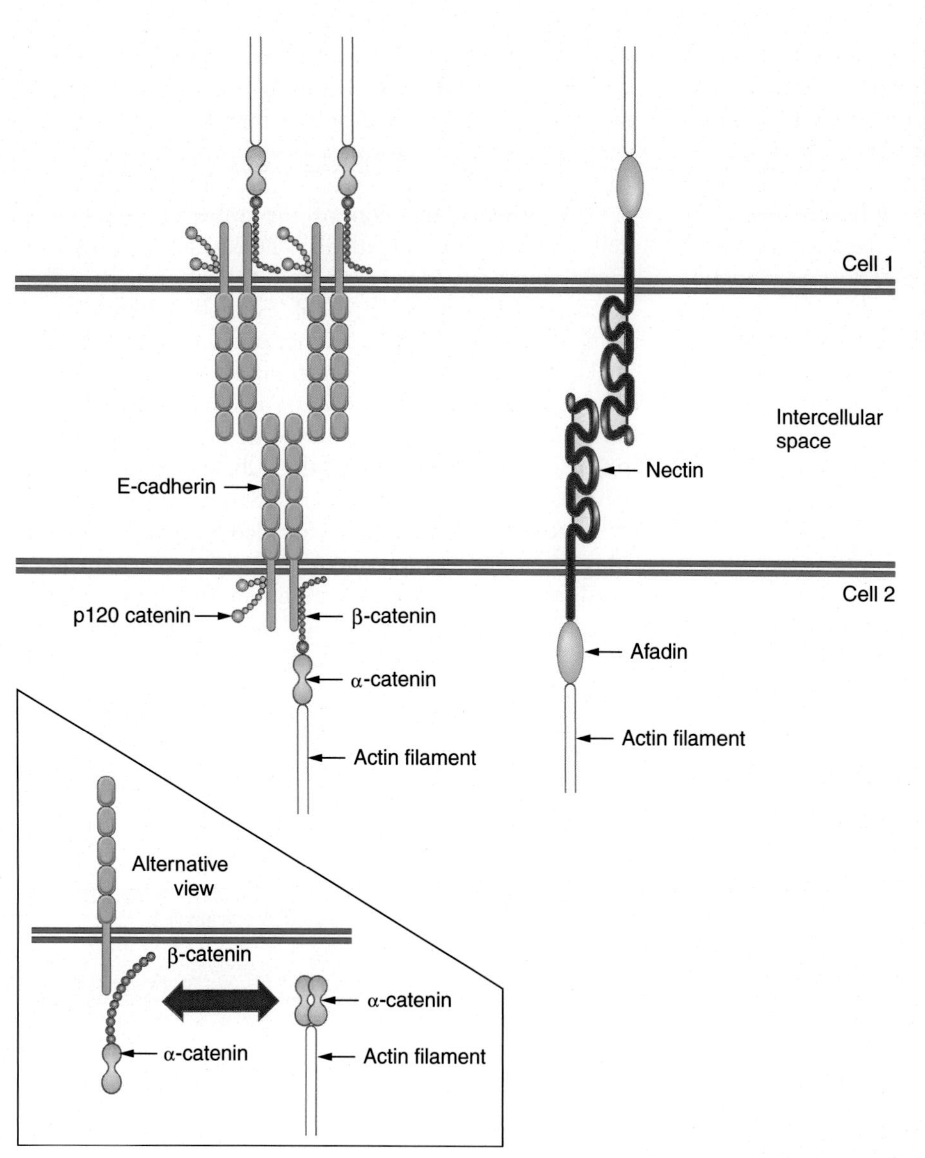

Figure 7–12. **Molecular structure of the adherens junction.** Two types of adhesion molecule are involved, E-cadherin and the immunoglobulin (Ig) family protein nectin. The cytoplasmic domain of E-cadherin binds β-catenin, which, in turn, binds α-catenin. This complex may link E-cadherin to the cytoskeleton, but an alternative view (inset) suggests that the whole complex cannot form simultaneously. Nectin is linked to actin by afadin.

formation of desmosomes. Further maturation of adherens junctions is dependent on stabilization by desmosomes. When desmosomal adhesion in mice is experimentally compromised or malformed due to inherited mutations in humans, adherens junctions do not stabilize, and the epidermis falls apart through the loss of keratinocyte adhesion. Contraction of the actin ring that underlies the adherens junctions can cause morphogenetic shape change of the cell sheet, for example, in rolling of the **neural plate** to form the neural tube, the precursor of the central nervous system, during embryonic development. Such contractility is probably of broader importance but has not been thoroughly investigated (see Fig. 7–11, B). Similarly the adherens junctions are likely to stabilize tight junctions in simple epithelia.

In epithelia the first adhesion molecule of adherens junction is E-cadherin, though this is replaced by other cadherins in other tissues (Fig. 7–12). As discussed earlier the extracellular domain of cadherins form homophilic interactions with cadherins of adjacent cells, and this interaction requires calcium. Incubation of simple epithelium in calcium-free medium leads to disruption of adherens junctions and loss of epithelial structure; cells acquire rounded shape. If the calcium is restored in the medium, adherens junctions assemble, and cells organize into the epithelium. The cytoplasmic partners of E-cadherin are p120-catenin, α-catenin, γ-catenin (plakoglobin), and β-catenin. The cadherin-catenin complex interacts with the actin-binding proteins such as **α-actinin.** This protein complex is anchored to the actin cytoskeleton, particularly to the actin-myosin ring. The α-actinin links cadherin-catenin complex with the actin cytoskeleton. This concept was challenged by the observation that only free α-actinin, but not the α-actinin bound to cadherin-catenin complex, binds to F-actin. This raised the question of whether there are additional partners involved in linking cadherin-catenin-actinin complex with F-actin. Evidence indicates that adherens junction complexes also interact with other actin-binding proteins such as **formin** and **vinculin**. The interaction of cadherin-catenin complexes with the actin cytoskeleton is essential for the assembly and maintenance of adherens junctions. The deletion of α-catenin leads to the disruption of adherens junctions. Additionally, disruption of the actin cytoskeleton by toxins such as cytochalasin-D results in the disruption of adherens junctions.

The intracellular protein complex of adherens junctions interacts with many signaling molecules such as protein kinases and protein phosphatases, indicating that adherens junctions are also dynamically regulated by intracellular signaling pathways. P120-catenin interacts with p190RhoGAP that regulates Rho/Rac activity and E-cadherin endocytosis and turnover. E-cadherin is heavily phosphorylated on serine and tyrosine residues that are required for interaction with catenins. E-cadherin and β-catenin are phosphorylated on tyrosine residues by Src family kinases such as c-Src and c-Fyn. Tyrosine phosphorylation of β-catenin on Y654 prevents its interaction with E-cadherin. In the intact adherens junctions, β-catenin is maintained in the dephosphorylated state by a protein tyrosine phosphatase, PTP1B, that binds to E-cadherin at a site adjacent to the β-catenin binding site. Inhibition of PTP1B, for example, by reactive oxygen species such as hydrogen peroxide or alcohol metabolites such as acetaldehyde inhibits PTP1B leading to tyrosine phosphorylation of β-catenin and disruption of adherens junctions.

The Ig family molecules, **nectins**, are colocalized with cadherins at the adherens junctions. Unlike cadherins the adhesions by nectins is independent of calcium. The C-terminal PDZ binding motif interacts with AF6/afadin. Afadin binds to α-catenin and interacts with the actin cytoskeleton. Both afadin and α-catenin binds to F-actin. Binding to α-catenin also suggests that the nectin-afadin complex may interact with the cadherin-catenin complex. The extracellular domains of nectins interact with nectins in adjacent cells either in a homophilic or heterophilic way. The precise functions of nectin-based cell adhesion are unclear. However, during the early formation of cell-cell contacts, the nectins aggregate at the contacts first, followed by cadherins. The heterotypic interactions of nectins may help recruit cadherins to the cell-cell borders. This is more common in synaptic connections.

In addition to its crucial role in adherens junction, β-catenin is an essential signaling molecule regulating cellular proliferation (Fig. 7–13). Any β-catenin that is free in the cell cytoplasm usually is proteolytically degraded. Free β-catenin forms a part of a soluble multi-protein complex that includes APC (adenomatosis polyposis coli), axin, casein kinase-1α, and GSK-3β. The kinases phosphorylate β-catenin, releasing it from the complex. The phosphorylated β-catenin gets recognized by a specific E3 ligase (βTrCP), which catalyzes the ubiquitination and rapid degradation by proteasome complex. Activation of the **Wnt signaling pathway** by binding of Wnt to their cell surface receptor, Frizzled, blocks β-catenin degradation by disrupting its interaction with the APC-axin complex and preventing β-catenin phosphorylation. APC, the tumor suppressor, participates in the phosphodestruction of β-catenin by

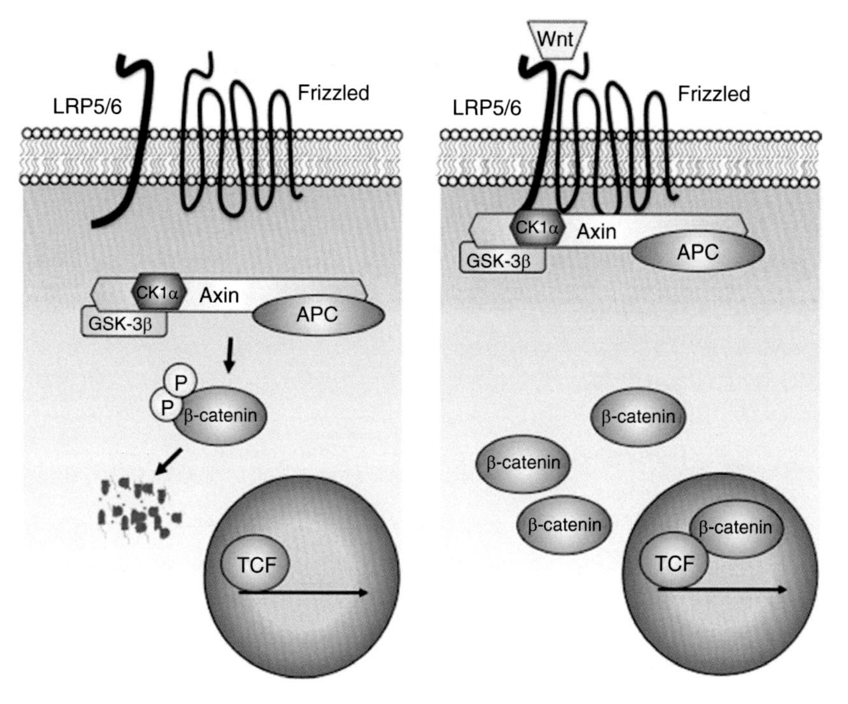

Figure 7–13. **Wnt signaling pathway.** In the absence of Wnt (left), cytosolic β-catenin is continually phosphorylated by casein kinase-1α (CK1) and glycogen synthase kinase-3β (GSK3β) within an APC-axin scaffold complex. This phosphorylation allows β-catenin to be ubiquitylated and rapidly degraded by proteasome. During Wnt activation (right), GSK3β activity is inhibited directly by Lrp5/6, which allows β-catenin to accumulate, enter the nucleus, interact with LEF/TCF family members, and promote transcription. *(Adapted from McEwen, Escobar, Gottardi.* Subcell Biochem *2012;60:171.* https://doi.org/10.1007/978-94-007-4186-7_8, *by permission.)*

antagonizing β-catenin dephosphorylation by phosphatases. Thus Wnt activation increases the level of free, intact β-catenin, which enters the nucleus and functions as a transcription factor. In the nucleus, β-catenin forms a complex with members of the **TCF/LEF** (T-cell factor/leukemia enhancer factor) family, activating transcription of a number of genes that affect cell proliferation and transformation. The Wnt signaling pathway is vital in embryonic development, in regular tissue maintenance, and when abnormally activated in cancer. For a complete description of Wnt and other signaling pathways see Chapter 9.

Desmosomes Maintain Tissue Integrity

The third and most basal component of the junctional complex is the desmosome (see Fig. 7–5). Unlike the zonula occludens and the zonula adherens, the desmosome is a spotlike or punctate junction that occupies a roughly circular region of 0.5 μm or less in diameter on the cell surface. However, desmosomes are not confined to the junctional complex and commonly have a wider distribution at cell-cell interfaces. Its principal function is to maintain tissue integrity by providing strong intercellular adhesion and acting as a link between the cytoskeletons of adjacent cells (Fig. 7–14); many tissues have such structural scaffolding extending through their cells. Cytoskeletal intermediate filaments (e.g., cytokeratin) provide the scaffolding poles, and desmosomes provide the couplings between them. The **desmosome-intermediate filament complex** is particularly well developed in tissues such as epidermis that are subject to constant shear stress and abrasion and is fundamentally important in maintaining tissue structure.

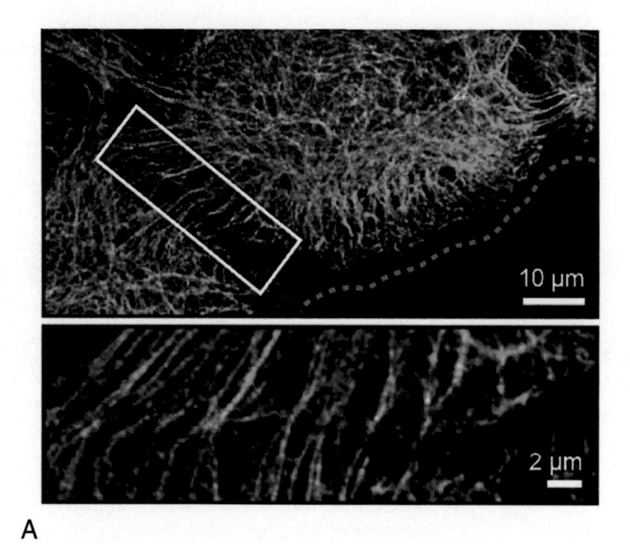

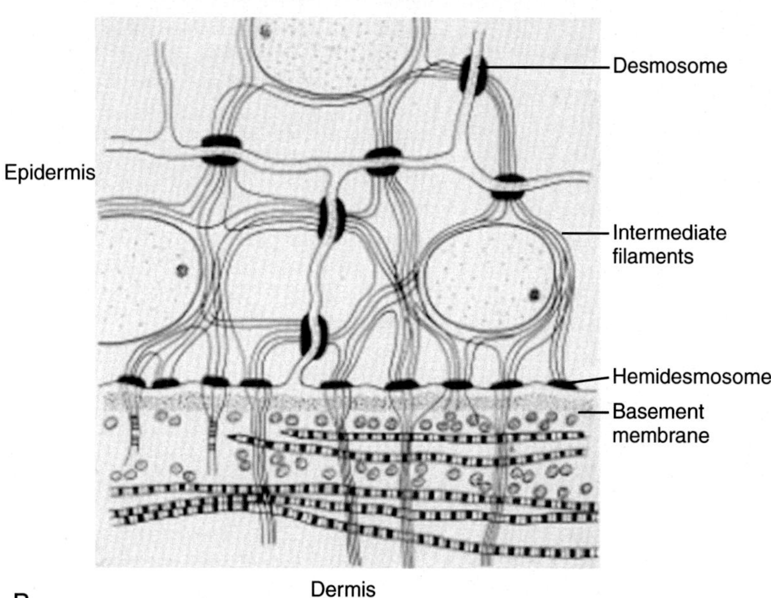

Figure 7–14. **The desmosome-intermediate filament complex strengthens epithelia, particularly the epidermis.** (A) Fluorescence micrograph showing desmosome-intermediate filament complex in cultured epithelial cells. Desmosomes stained by anti-desmoplakin antibody (*magenta*), intermediate filament stained by anti-keratin antibody (*green*), and nuclei (*blue*). (B) Diagram showing the desmosome-intermediate-hemidesmosome complex in the basal epidermis. *([A] From Moch M, Schwarz N, Windoffer R, Leube RE. Keratin-desmosome scaffold: pivotal role of desmosomes for keratin network morphogenesis.* Cell Mol Life Sci *2019;77:543–558, by permission; [B] modified from Ellison JE, Garrod DR.* J Cell Sci *1984;72:163–172, with permission.)*

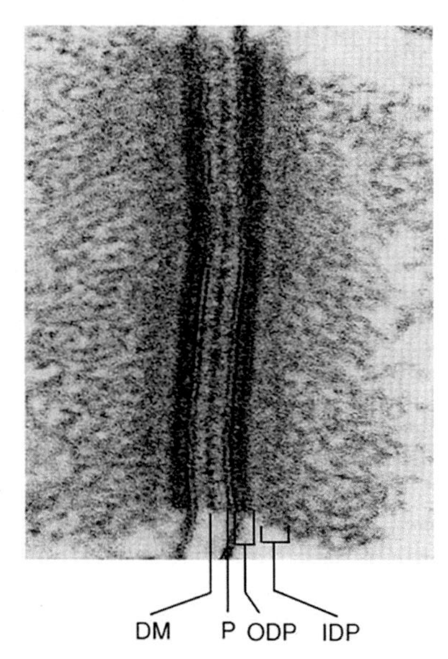

Figure 7–15. **Electron micrograph of a desmosome from bovine tongue.** *IDP*, inner dense plaque; *ODP*, outer dense plaque; and *P*, plasma membrane. The desmosome is about 0.5 μm wide. *(From Yin T, Green K. Regulation of Desmosome Assembly and Adhesion.* Semin Cell Dev Biol *2004;18:665–667, by permission.)*

Electron microscopy shows that the cytoplasmic face of the desmosome consists of a dense plaque that is joined to a bundle of intermediate filaments (Fig. 7–15). The intercellular space is more than 30 nm wide and is characterized by the presence of midline with branches extending between it and the plasma membrane of the adhering cells. This structure probably represents a highly organized arrangement of the adhesive material, and this organization may, in turn, explain why desmosomes are so strongly adhesive.

Desmosomes have two types of adhesion molecules, the **desmocollin** and **desmoglein**. These are representatives of the cadherin family and known as the desmosomal cadherins (Fig. 7–16). Their extracellular domains from the midline structure and their cytoplasmic domains lie in the dense plaques where they bind three other molecules, **plakoglobin, plakophilin,** and **desmoplakin,** that provide the link to the cytoskeleton. Plakoglobin and plakophilin are related to β-catenin. Desmoplakin belongs to the plakin family of cytoskeletal proteins. Human diseases involving desmosomes are rare but include both autoimmune and inherited conditions. Such conditions can result in abnormalities of skin or cardiomyopathy. An autoimmune disease that affects desmosomes, pemphigus, is considered in Clinical Case 7–1. Although they

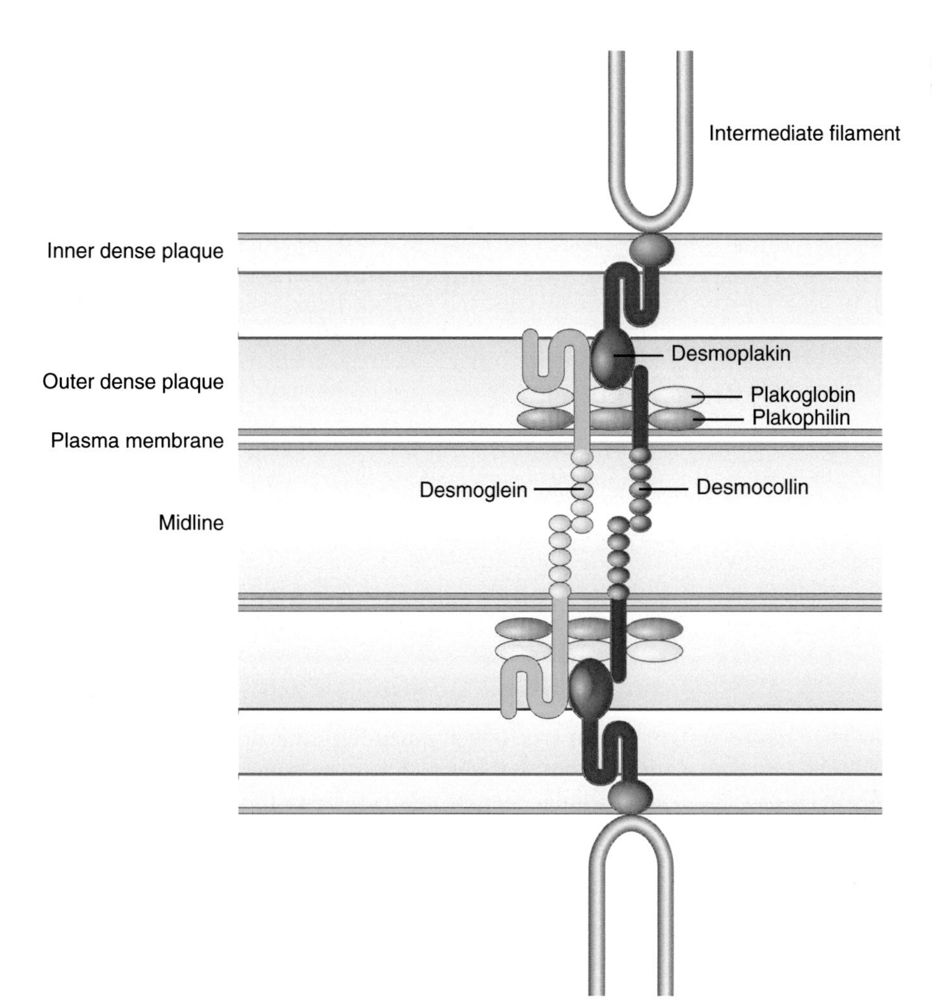

Figure 7–16. **Molecular composition of the desmosome.**

have an essential structural role in tissues, desmosomal adhesion has to be dynamic and reversible. It has to permit the upward movement of cells (stratification) required for renewal of epithelia such as epidermis, and it has to release cells for movement when this is required, for example, during the closure or reepithelialization of wounds.

Clinical Case 7–1

Joanne is a 58-year-old Caucasian woman with a history of Type II diabetes who comes to see a dentist for the development of repeated painful sores on the top of her mouth (palate). These sores had been relapsing and recurring over the last year or so but have recently become so severe that it has been difficult to eat due to the pain. She has lost about 30 pounds in the last 4 months. She mentions that, over the last couple of years, she has gotten what she feels are severe blistering sunburns, despite not spending significant time in the sun. She states that the blistering seems to be getting worse, and she notes that she has a couple of blistering burns on her back at the present time. Her internist started her on acyclovir for the mouth sores, suggesting a diagnosis of recurrent herpes simplex. The physician also recommends petroleum gel and hydrocortisone cream for her burns.

On examination the dentist is able to appreciate significant erosions on both the buccal and palatine mucosa, which are exquisitely tender on the exam. He notes that Joanne has several erosive lesions on her forehead and two large flaccid blistering lesions on the back of her neck. One of these blisters ruptures as the patient prods it to point it out to the dentist. The dentist attempts to wipe away the clear fluid that emerges, but the skin around the ruptured blister sloughs off with a mild circular rubbing motion. The dentist obtains a buccal scraping for microbiology testing and empirically prescribes nystatin mouth wash for presumed candida. He also recommends she see a dermatologist for her "burns."

The following week, Joanne is seen by a dermatologist and reports multiple additional blistering lesions on her torso and no improvement in her mouth pain. The dermatologist notes multiple flaccid blisters on both face and torso exam, but the skins surrounding the blisters are not especially red. Again, slight rubbing of the edges of the blisters results in skin sloughing. Punch biopsy of the edge of one of the blisters and a second biopsy right by the edge of the lesion is performed. Routine histology shows multiple detached keratinocytes in the suprabasal region, as well as a basement membrane zone with cells that resemble a "row of tombstones." Direct immunofluorescence of the perilesional skin shows the intercellular deposition of IgG. Blood work is performed, and ELISA demonstrates IgG antibodies to desmoglein-1 and desmoglein-3.

Cell Biology, Diagnosis, and Treatment of Pemphigus Vulgaris

Pemphigus vulgaris is a rare chronic blistering disease that generally impacts middle-aged adults (age 40–60). Symptoms usually start in oral mucosa first, but cutaneous blisters may soon follow. The typical skin findings are flaccid blisters, rather than tense blisters. Pemphigus vulgaris almost always involves the oral mucosa early on and presents with oral ulcers, which can be painful enough to limit adequate nutrition. Other mucosal areas may be involved as well. In the early stages the disease is often misdiagnosed as herpes or candidiasis.

Pemphigus vulgaris is caused by autoantibodies formed against desmogleins. Desmogleins are the adhesive molecules that make up desmosomes and are crucially important for cell-to-cell adhesion. The autoantibodies target the membrane surface epitopes of these cadherin-type molecules and either directly block their adhesive function or induce their proteolytic dissolution. Loss of the desmosomes results in the inability of skin cells (keratinocytes) to adhere to one another, disrupting the barrier that intact skin provides. This causes skin layers to separate, causing blisters. Because the blistering is specifically at the epidermal/dermal junction, rather than within the dermis itself, the blisters are flaccid rather than taut.

On histology the basal keratinocytes remain still attached to the basement membrane, but not to one other, leading to a characteristic appearance called "tombstoning." The fluid that accumulates between the keratinocytes and the basal layer forms the blister and results in a positive Nikolsky sign, which is when slight rubbing of the skin results in exfoliation of the outermost layer. Direct immunofluorescence of the perilesional area, as well as serology tests, usually reveals the causative factor, IgG directed against desmogleins 1 and 3.

If left untreated, pemphigus vulgaris is always progressive and usually fatal. Steroids, however, cause a significant improvement in symptoms for most patients. The doses of steroids used are usually somewhat high (1–1.5 mg/kg/day of prednisone). Once the disease is controlled, steroids can be tapered very slowly to the lowest level at which no lesions recur. Long-term use of steroids can cause significant side effects, so other immunomodulatory agents may be used for maintenance as steroids are tapered. Azathioprine or mycophenolate mofetil is most commonly used for this purpose. More intensive antiimmune treatments may be given for disease refractory to initial treatment, such as rituximab, IVIG, plasmapheresis, or cyclophosphamide. With these treatment strategies the mortality of pemphigus vulgaris has been substantially reduced.

Gap Junctions Are Channels for Cell-Cell Communication

A fourth type of cell junction that is not part of the junctional complex but has a more general distribution on intercellular membranes is the gap junction. These junctions were identified as characteristic gaps (<2 nm in width) in the intercellular junctions when examined by electron microscopy. The principal function of the gap function is **cell-cell communication.** Gap junctions isolated from tissues and seen in plain view rather than in section by electron microscopy have the appearance of rafts of circular particles each about 7 nm in diameter and each with a dot at its center. Each particle is called a **connexon**, and each dot represents the end of a channel that passes through the middle of the connexon (Fig. 7–17). It is this channel that provides the pathway for intercellular communication. Gap junctional communication is an important integrator of cell function in both excitable tissues such as nerves and cardiac muscle and nonexcitable tissues such as epithelia. In cardiac muscle, gap junctions provide the route through which electrical impulses are propagated between the muscle fibers and thus are important in coordinating the heartbeat. The muscle fibers are said to be electrically coupled. In epithelia, they function in **metabolic cooperation** in which small metabolites and signaling molecules pass between cells. The latter function is vital during embryonic development, as well as in adult tissues. In some instance neurons are connected by gap junctions known as an **electrical synapse**. Transmission through these channels are thousands of fold faster than chemical synapses. Cluster of gap junctions in the enteric neurons and intestinal smooth-muscle cells are called **nexuses** that play an important role in intestinal peristalsis.

Communication is possible because the central channels of the connexons provide an aqueous link between the cytoplasm of adjacent cells through which small soluble molecules can pass. The size restriction of the channels is about 1 kDa so that molecules such as inorganic ions and small sugars, peptides and small signaling molecules such as cyclic AMP, and inositol trisphosphate can penetrate but molecules greater than 2 kDa cannot.

Connexons are composed of 20–60-kDa proteins called **connexins**. These are a family of proteins: 20 connexin genes have been identified in humans (Fig. 7–17, A). Connexin has four transmembrane domains, similar to but not related to the tight junction proteins occludin and claudin. NH_2- and COOH-termini of connexin are intracellular, so the transmembrane domains are joined by one intracellular and two extracellular loops. Six connexin molecules form a hexamer that constitutes the connexon or hemichannel (see Fig. 7–17, B). By docking of the connexin extracellular loops with those of a connexon on the surface of another cell, gap junctional communication is established (see Fig. 7–17, C). Gap junctions are homotypic or heterotypic. Some cells have homotypic channels, whereas many cells have heterotypic channels. In heterotypic channels, different isoforms of connexins are involved, and connexin isoforms determine the permeability characteristics. For example, connexin-43 homotypic channel is 100-fold more permeable to ATP and ADP than connexin-32 channel.

While the extracellular domain interacts with the similar domains from the adjacent cells, the intracellular domains bind to various proteins such as α-catenin, β-catenin, ZO-1, ZO-2, plakoglobin, desmoplakin, and plakophilin-2. The multiprotein complex is anchored to the actin cytoskeleton. The connexon channel can be regulated by intracellular signals to adopt an open or closed configuration so that intercellular communications can, in turn, be regulated. Some of the signals that are known to regulate gap junction function is intracellular pH, calcium, and reversible phosphorylation of connexins. Intercellular communication through gap junctions can be demonstrated by passing an electric current between cells through intracellular microelectrodes (see Fig. 7–17, D) or by injecting a low-molecular weight fluorescent dye into one cell and detecting its spread to adjacent cells by fluorescence light microscopy.

Connexin gene mutations are associated with a variety of human diseases including cardiovascular anomalies and cataract. For example, mutation of one family member, connexin $43\alpha_1$, is linked to a syndrome called *oculodentodigital dysplasia* that may involve developmental abnormalities of the face, eyes, limbs, and teeth. Mutation of connexin-26 leads to neurosensory deafness. Mutations of connexins 43, 46, and 50 are associated with cataract and heart malfunction.

Hemidesmosomes Maintain Cell-Matrix Adhesion

Cell-matrix adhesion of certain epithelia, especially epidermis, is mediated by specialized junctions called *hemidesmosomes*. So named because they resemble half desmosomes by electron microscopy, hemidesmosomes are responsible for strong binding between the basal surface of the epithelial cells and the underlying basement membrane and intracellularly providing a link to the intermediate filament cytoskeleton.

Hemidesmosomes have dense cytoplasmic plaques that interact with the cytoskeleton (Fig. 7–18). Within the basement membrane, fine filaments called *anchoring filaments* appear to link to the outer surface of the plasma membrane opposite to the plaque. The anchoring filaments, in turn, connect to anchoring fibrils that extend from the basement membrane into the underlying collagenous matrix. Thus hemidesmosomes appear to provide the link in a contiguous series of filaments that extends from the cell cytoplasm, through the basement membrane, and into the matrix beneath.

Any ultrastructural resemblance to half desmosomes disappears when the molecular composition of hemidesmosomes is analyzed (Fig. 7–19). The major adhesion molecule of hemidesmosomes is $\alpha_6\beta_4$ **integrin**. Also present is a type II membrane protein (NH_2-terminal cytoplasmic and COOH-terminal extracellular) called **BP180**. (BP represents bullous pemphigoid, an autoimmune blistering disorder in which the autoantibodies target this 180-kDa protein.) Within the plaque, two molecules are involved in linking to cytokeratin, **BP230** and **plectin**, both members of the plakin family and related to desmoplakin. Outside the membrane the anchoring filaments

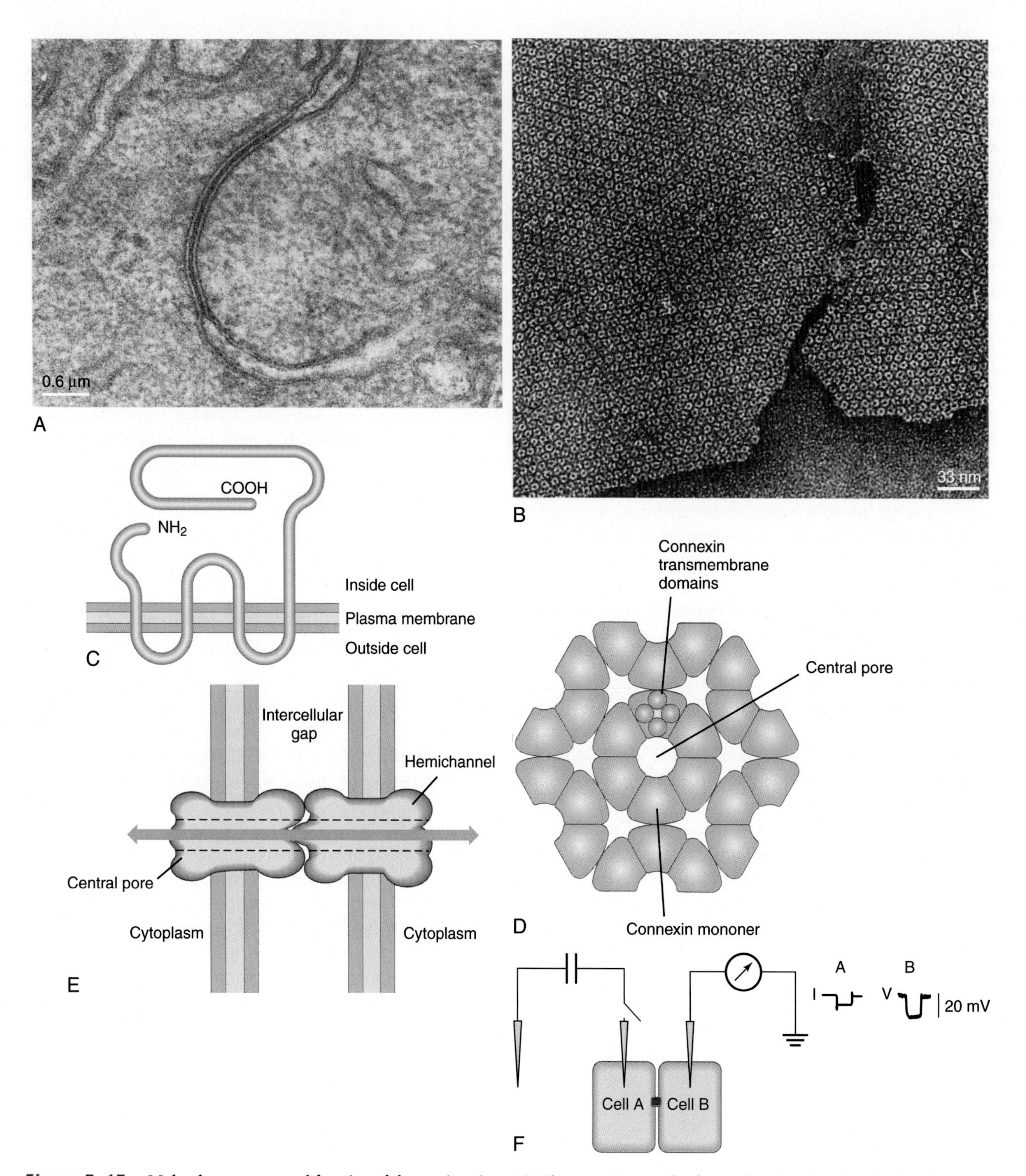

Figure 7–17. **Molecular structure and function of the gap junctions.** (A) Electron micrograph of a gap junction showing the close (2 nm) approach of the cell membranes. Scale bar = 0.6 μm. (B) Electron micrograph of surface view of an isolated gap junction showing connexons with central pores. Scale bar = 33 nm. (C) Connexin has four transmembrane domains. (D) Six connexin molecules form the channel of the connexon. (E) Docking of two hemichannels between adjacent cells establishes intercellular communication. (F) Adjacent cells are coupled electrically by gap junctions. *([A] and [B] From Gilula NB. Gap junctional contact between cells. In: Edelman GM, Thiery J-P, eds.* The Cell in Contact: Adhesions and Junctions as Morphogenetic Determinants. *New York: John Wiley & Sons, 1985, pp. 395–405, by permission.)*

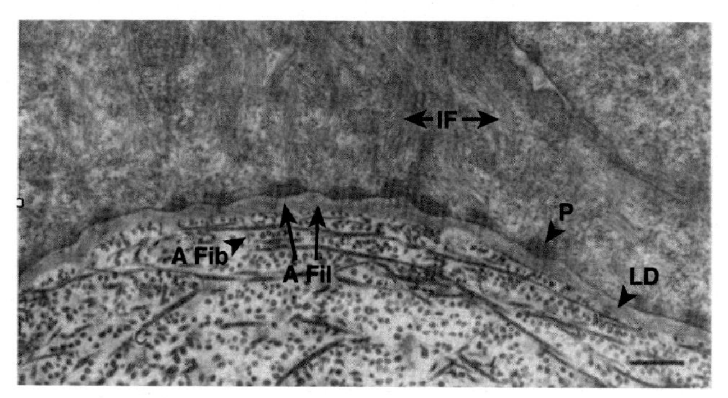

Figure 7–18. **Structure of the hemidesmosome as observed by electron microscopy.** *A Fib*, anchoring fibril; *A Fil*, anchoring filaments; *IF*, intermediate filaments; *LD*, lamina densa of basement membrane; and *P*, plaque. Scale bar = 0.4 μm. *(From Ellison JE, Garrod DR.* J Cell Sci *1984;72:163–172, by permission.)*

appear to be composed of a member of the laminin family of ECM proteins, laminin 5, and form the substrate for $\alpha_6\beta_4$ integrin binding. The anchoring fibrils are composed of collagen type VII, a specialized member of the collagen family.

The continuity of structure formed by cytokeratin filaments, hemidesmosomes, **anchoring filaments,** and **anchoring fibrils** literally anchors the epidermis to the dermis. A variety of genetic diseases affect this **dermal-epidermal junction.** Mutations in the genes for specific **cytokeratins,** $\alpha_6\beta_4$ **integrin, laminin 5,** or **collagen VII** give rise to various forms of **epidermolysis bullosa (EB),** a group of blistering diseases that result in differing degrees of epidermal detachment that vary in severity from extreme debilitation to neonatal lethality. (EB

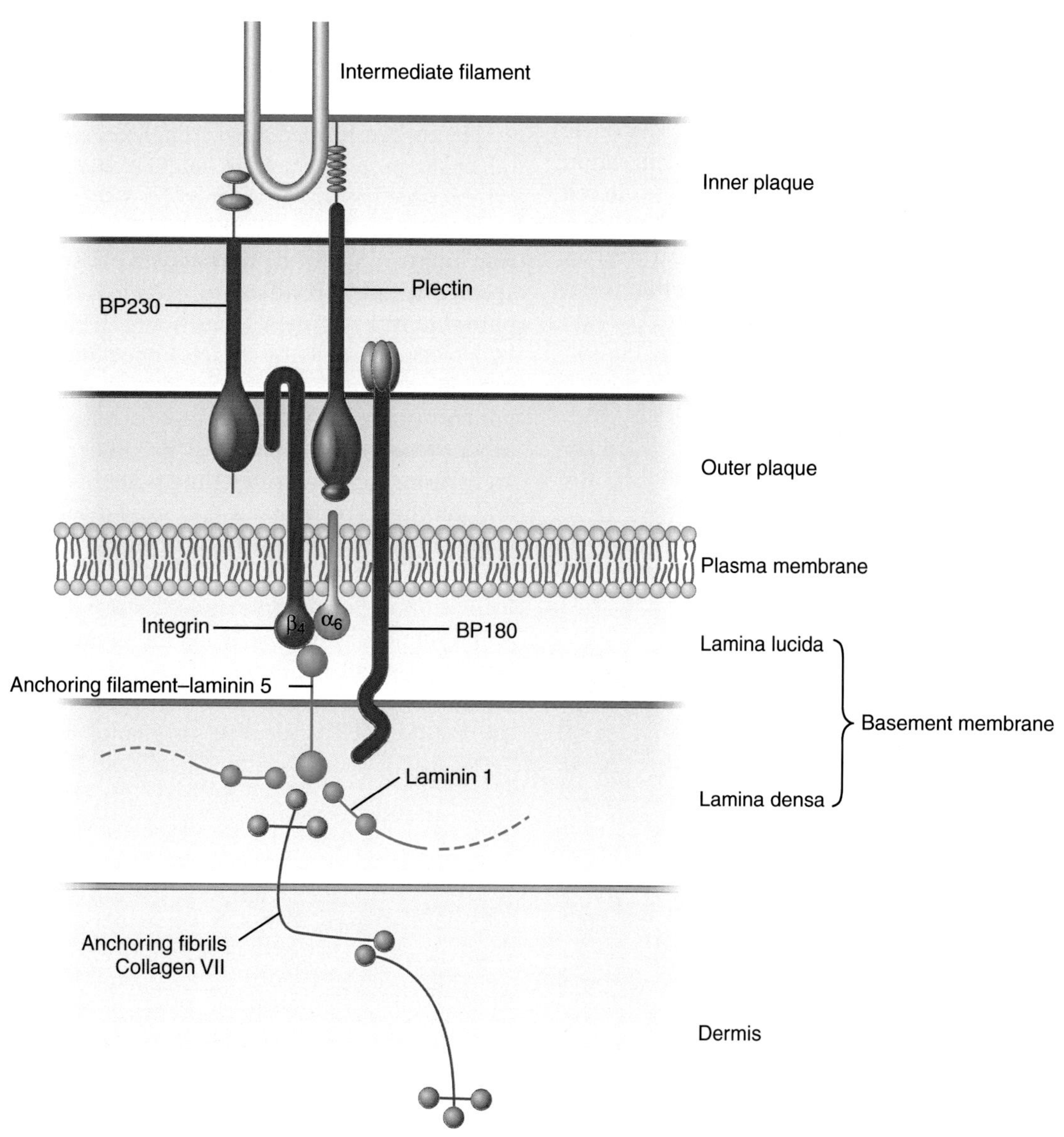

Figure 7–19. **Molecular components of the hemidesmosome.**

simplex involves keratin filaments and causes epidermal blistering; EB junctional form involves $\alpha_6\beta_4$ integrin or laminin 5 and is lethal in early infancy; EB dystrophica involves anchoring fibers and causes epidermal blistering that leads to syndactyly.) Like desmosomes, hemidesmosomes need not only to provide strong adhesion but also to relinquish this when required. For example, replenishment of cells in the epidermis occurs in the basal layer. As necessary, cells need to move up from the basal layers to join the upper epidermal cell layers. To do this, cells must lose their hemidesmosomal adhesions to relinquish contact with the basement membrane, and this must be done in a regulated fashion. Similarly, during cell migration, to close epidermal wounds, cells must lose their hemidesmosomal adhesions and reacquire them when wound closure is complete.

Focal Contacts Are Adhesions Formed With the Substratum

Formation of cell contacts is important during wound healing and single cell migration such as leukocyte extravasation or tumor cell metastasis. Cell migration and formation of focal contacts can be well studied in cell culture. When cells are plated on glass or plastic in tissue culture, they adhere to the surface and spread over it, often resembling the shape of a fried egg with thin edges and a bulky nucleus somewhere near the center. They do not actually adhere directly to the glass or plastic, but rather to a thin layer of ECM molecules that adsorb to the surface. Principal among ECM molecules is fibronectin, which is abundant in the serum component of tissue culture medium and is usually also secreted by the cells (Fig. 7–20).

Adhesion to this adsorbed layer is mediated principally by integrins, with $\alpha_5\beta_1$ **integrin** being the principal fibronectin receptor. To function effectively, integrins must cluster together on the cell surface. At the edges of spreading cells, integrins cluster to form small structures (<1 μm) called **focal complexes.** As cell spreading progresses, these complexes evolve into slightly larger, elongated structures called **focal adhesions** or focal contacts (Fig. 7–21); these are effectively adhesive junctions formed between the cell and the substratum. The underside of the cell is not flat. Focal contacts are the regions of closest association (about 15 nm) between the cell membrane and the substratum. In the cytoplasm, focal contacts are associated with the actin cytoskeleton. In well-spread cells, actin filaments are bundled into **stress fibers** that originate toward the center of the cell and terminate at focal contacts.

As well as being points of adhesion, focal contacts are important sites of signal transduction. Signals from outside the cell regulate cell function, whereas signals originating within the cell regulate cell adhesion. The cytoplasmic domains of the integrin subunits may recruit up to 50 different types of structural and signaling molecules to participate in adhesive linker and signal transduction functions.

The earliest formed focal complexes recruit cytoskeletal linker proteins such as vinculin and paxillin to the cytoplasmic face, accompanied by a signaling molecule, **focal adhesion kinase (FAK)** (Fig. 7–22). FAK is a **tyrosine kinase**, an enzyme that adds a phosphate group to specific tyrosine residues in protein substrates. **Phosphorylation** by protein kinases and dephosphorylation by protein phosphatases are important regulators of protein activity and function. Phosphorylation of components of the early focal complex cause recruitment of other structural components; the linker proteins **talin** and **tensin**; some with signaling potential such as **zyxin**; and another tyrosine kinase, c-**Src**. Further changes can result in either maturation or turnover of the complex. Regulation of the actin cytoskeleton also determines the formation of focal contacts and stress fibers. Key signaling molecules involved in this process are the Rho family of **small GTPases**. These molecules are active when they have GTP bound to them but inactive when the bound GTP is cleaved to guanosine diphosphate

Figure 7–20. **Fluorescence micrograph showing fibronectin produced by cells in culture.** *(From Mattey DL, Garrod DR.* J Cell Sci *1984;67:171–188, by permission.)*

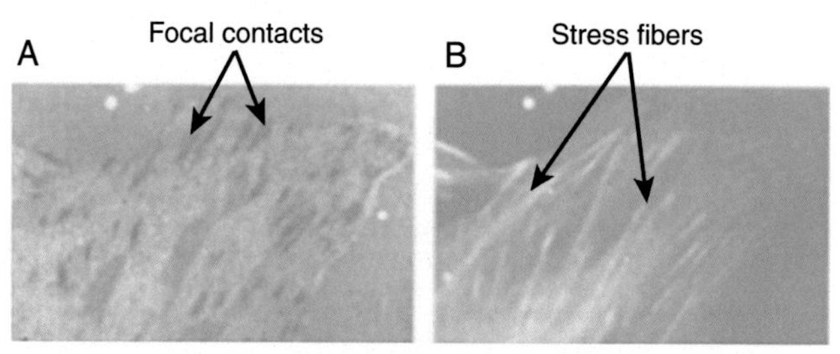

Figure 7–21. **Focal contacts and stress fibers of cells in culture.** (A) Micrograph of cells under interference reflection microscopy showing focal contacts. (B) Fluorescence micrograph showing actin stress fibers. Note how each stress fiber terminates at a focal contact. *(From Morgan J, Garrod DR.* J Cell Sci *1984;66:133–145, by permission.)*

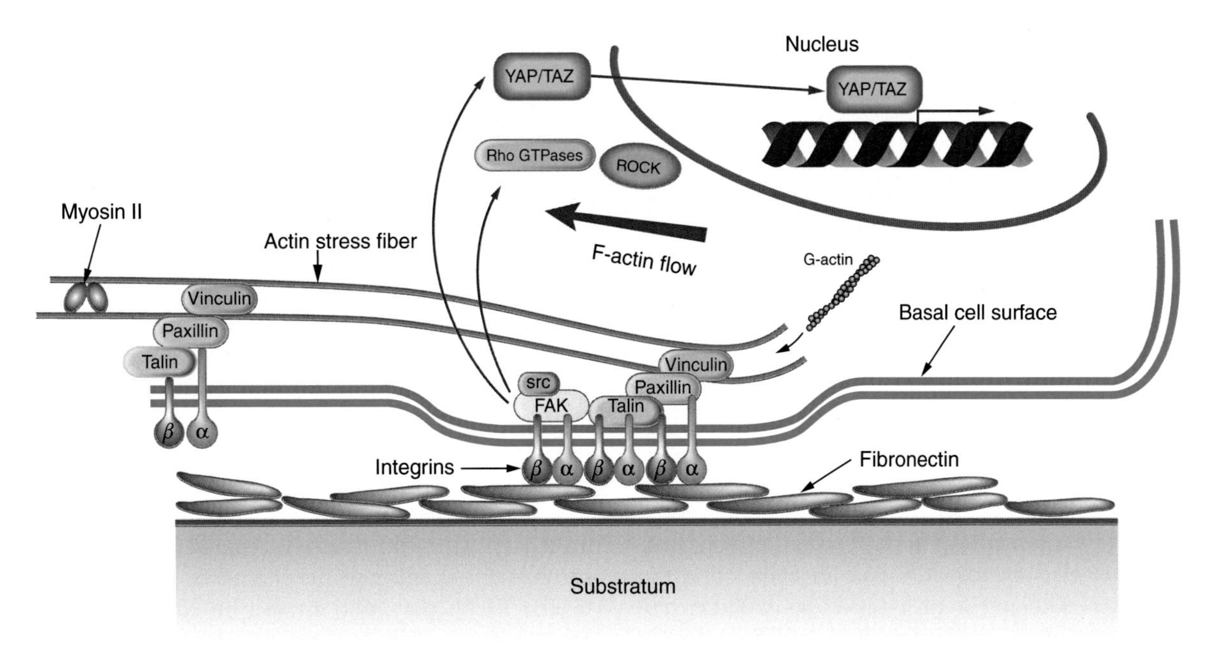

Figure 7–22. **Molecular composition of focal contact and mechanotransduction.** ECM stiffness increases the loading rate of focal adhesions (ECM-integrin-actin cytoskeleton machinery), while compliant ECM reduces its loading. The extracellular integrin head domain binds the ECM, and the intracellular tail of integrins binds the F-actin through adaptor proteins, such as talin and vinculin. Actin protein polymerization and extension of F-actin at the leading edge of a cell lead to retrograde flow of the actin filaments and directional migration, which is also facilitated by myosin II-mediated pulling forces in the cells. Besides cell motility, mechanotransduction also regulates cell response to ECM mechanical property by regulating GTPases, Rho/Rac activity, and nuclear translocation of transcriptional factor YAP/TAZ.

(GDP). Important members of this family are *cdc42*, ***Rac***, and ***Rho***. *Rho* promotes the formation of focal contacts and stress fibers, whereas *Rac* causes the formation of large, thin **lamellipodia** with small adhesions localized to the extreme edge. *Cdc42* promotes the formation of **filopodia,** narrow processes that extend from the cell surface that are supported internally by actin filament bundles and make long adhesions with the substratum. Downregulation of either *Rho*, *Rac*, or *CDC42* expression blocks cell migration, indicating that all these signaling molecules are required for cell migration.

Much is known about focal contacts and their regulation because they are relatively easy to study in culture. The dynamic nature of cell adhesion and the complexity of its regulation are important for controlling cell motility in vivo and respond to the extracellular environmental cues as discussed in the next section. Some cells such as leukocytes migrate substantially as part of their normal function. To do this, they must dynamically regulate their adhesions and cytoskeleton. Other cells such as epithelial cells are normally less motile, but when they develop into cancer cells, they invade the surrounding matrix and spread or **metastasize** to other parts of the body. These processes involve loss of cell-cell adhesion at the primary site and dynamic regulation of adhesion during migration. Thus it is important to investigate these processes to understand both normal and abnormal cell behavior.

ROLE OF CELL ADHESION IN TISSUE FUNCTION

To this point the chapter has dealt with the nuts and bolts of cell adhesion—the adhesion molecules and intercellular junctions. It is now appropriate to consider how cell adhesion is involved in tissue function and cell behavior.

Focal Contact and Mechanotransduction

Cells not only communicate with the chemical signaling molecules within the ECM, such as adhesive receptor ligand proteins, glycoproteins, proteoglycans, and growth factors, but also respond to the mechanical properties of the ECM, such as its stiffness and compliance, and to the forces transmitted through the ECM. The mechanisms that orchestrate this cellular response to the mechanical signals from the environment are called **mechanotransduction,** which converts a mechanical stimulus into a biochemical response and changes cell growth, shape, and migration. ECM-residing cells continually assess and respond to the structural integrity of the ECM to maintain the homeostasis of tissues and organs. The process of cells assessing the mechanics of the ECM, known as **mechanosensing,** is mediated through integrins and the actomyosin cytoskeleton. Focal contact discussed earlier is an integral component of these mechanosensing complexes. In response to the mechanosensing signaling inputs, cells mobilize a **mechanoregulation** process, which includes alteration of cell behaviors and the removal, rearrangement, or repair of the ECM, to preserve their overall form and function.

Mechanotransduction can be triggered by numerous mechanical stimuli, such as shear stress from the blood flow or stretching and compression forces from muscle activity. The mechanical properties of ECM largely depend on elastic fibers, fibrillar collagens, glycosaminoglycans, and proteoglycans. ECM stiffness increases the transmission of the mechanical forces, which regulates the loading rate of the ECM-integrin-actin cytoskeleton machinery in the form of focal adhesions. Rigid ECM increases mechanosensing complex (ECM-integrin-cytoskeleton) loading, while compliant ECM reduces the loading rate of this complex, due to decreased backward movement of the filamentous actin (F-actin) polymerization. Integrins play an essential role in connecting the ECM with intracellular F-actin cytoskeletons. The extracellular integrin head domain binds the ECM, and the intracellular tail of integrins binds the actin cytoskeleton through adaptor proteins. Talin and vinculin are the main components linking integrins to F-actin. Monomeric actin proteins (G-actin) polymerize at the leading edge of a cell, which generate F-actin. The continuous polymerization and extension of F-actin fibers results in a force by retrograde flow of the actin filaments. The increased engagement of these filaments with integrins, facilitated by talin and vinculin, on stiff ECM slows down the retrograde flow. As a result, cells push forward the cell membrane to accommodate the F-actin extension, leading to cell protrusion and migration, which is also facilitated by myosin II-mediated pulling forces in the cells (Fig. 7–22).

The ECM-integrin-cytoskeleton complexes are highly dynamic structures, and the adhesion proteins rapidly join and leave the mechanosensing complex. Both talin and vinculin undergo activation processes in response to mechanical forces and interaction with actin in living cells. The activation of these proteins and consequent engagement of F-actin with integrin-ECM slow down the retrograde flow of actin filaments. Activation of talin and vinculin also reduces their turnover. The stabilized talin-vinculin complexes further maintain the active conformation of integrins. Engagement of talin and vinculin with the retrograde flow of F-actin regulates the recruitment and release of other mechanotransduction signaling proteins such as FAK and paxillin, thereby regulating the molecular stoichiometry of mechanosensing complexes and orchestrating downstream signaling events. Recruitment of FAK and paxillin further activates the GTPases Rho and Rac, which regulate contraction or actin polymerization, respectively. These feedback mechanisms enable the mechanosensing complexes to respond to and reshape the actin cytoskeleton.

The F-actin cytoskeleton plays a critical role in mechanotransduction by linking mechanosensing complexes to other cytoskeletal systems and the nucleus. Cellular response to ECM mechanical properties and forces also involves alteration of gene transcription. Transcription factors YAP and TAZ have been identified as mechanotransducers, which shuttle between the cytoplasm and the nucleus in an ECM stiffness-dependent manner. YAP/TAZ can be activated in response to stiff ECM, decreased cell-cell contact, disturbed flow shear force, and stretching of a confluent cell monolayer. A rigid ECM promotes nuclear translocation of YAP/TAZ that modulate gene expression in response to mechanical signals from the ECM. Mechanical force-dependent conformational change of mechanosensing complexes increases actin polymerization and F-actin filaments contractility, which promotes YAP/TAZ activation.

Dysregulated mechanotransduction has been associated with multiple diseases. The defect in cell sensing mechanical signals and the change of physical properties of the ECM play major roles in the pathogenesis of atherosclerosis, fibrosis, cardiac hypertrophy, muscular dystrophy, and cancer. The age-related stiffening of elastic arteries and skin wrinkling are primarily associated with the loss of elastic fiber integrity in the ECM. The blood vessels are constantly exposed to mechanical forces from blood flow shear stress and blood pressure. Abnormal activation of YAP/TAZ by the excessive mechanical forces in endothelial cells promotes cell proliferation and expression of adhesive molecules recruiting monocytes, which are the critical initiating events during the development of atherosclerosis. Dysregulated remodeling and excessive deposition of ECM by fibroblasts activated by mechanical forces for extended period increase ECM rigidity and trigger tissue stiffening, which could lead to cellular dysfunction and eventually organ failure. These pathological ECM alterations are the key drivers of fibrotic diseases, such as liver cirrhosis; pulmonary, kidney, and cardiac fibrosis; and systemic sclerosis. In tumor microenvironment, abnormal tissue architecture, stromal cell accumulation, chronic inflammation, ECM remodeling by deregulated metalloproteinases, and increased ECM rigidity accompanied with augmented mechanical forces and pressure all contribute to the malignant characteristics of cancer. Mechanical stimuli from the aberrant tumor microenvironment drive hyperactivation of YAP/TAZ, which lead to uncontrolled cell proliferation, enhanced cell survival, and increased cancer cell migration and metastasis. Additionally, the aberrant YAP/TAZ activation in tumor microenvironment may dedifferentiate neoplastic cells into cancer stem cells/cancer progenitor cells, when considering the critical roles of YAP/TAZ in enhancing survival and preventing differentiation of embryonic stem cells and progenitor cells. Meanwhile, increased cancer-associated fibroblasts (CAFs) in tumor microenvironment increase the stiffness of ECM and restructure cell composition within tumor niches, thereby promoting tumor cell migration, invasion, and angiogenesis.

Junctions Maintain Epithelial Barrier Function and Polarity

More than 200 different cell types exist in the human body, and perhaps surprisingly, about 65% of these are epithelial cells; they are the components of the cell sheets that line body surfaces and cavities. **Epithelia** provide functional and physical separation between biological compartments within the body and often also have a protective or barrier role. A unique feature of the epithelia is that, on one surface, they are in direct contact with the so-called external environment such as gut epithelia and airway bronchial epithelia. The epidermis protects the human body from water loss and from the entry of environmental pathogens and toxins and resists the minor abrasion and shear stress to which it is constantly subjected. The airway epithelium separates inhaled air from tissue fluid, not only providing an absorptive surface but also maintaining the cleanliness of the airway and providing a protective barrier against the entry of airborne pathogens and allergens. The intestinal mucosa separates the gut contents from tissue fluid, performing a digestive function and selective absorption of digested products and preventing the entrance of allergens, toxins, and pathogens.

To perform these functions, all epithelia must be polarized; that is, their apical surfaces are structurally and functionally different from their basal surfaces (Fig. 7–23). In stratified epithelia such as the epidermis, the basal layer is concerned with attachment to the underlying matrix and producing new cells from its population of **stem cells** to replenish those lost from the outer surface. By contrast the outer layers are dead or dying but, in the process of progressing upward, have developed tough, impenetrable properties essential for barrier function. In simple epithelia such as the intestinal mucosa, polarity of structure and function lies within each cell. Cells of the small intestine have an apical surface specialized for absorption and a basolateral surface that has distinct properties for transferring absorbed molecules to the circulation and tissues. Cells in these epithelia are also continuously renewed without loss of epithelial barrier function. Intestinal epithelia are folded into pitlike structures that are the base called crypts and fingerlike projections called villi (villus is singular). Stem cells at the bottom of crypts proliferate and differentiate into different types of intestinal epithelial cells that migrate upward and shed into the intestinal lumen at the tip of the villi. Epithelial layers, such as endothelium and the renal tubule epithelium, are derived from mesenchymal cells, which have high motility but without apicobasal polarity. During development the mesenchymal cells can change their morphology by forming new cell-cell adhesions, a process called epithelial-to-mesenchymal transition (EMT), thereby establishing new apicobasal polarity and eventually forming a functional epithelial sheet.

Cell adhesion is central to epithelial structure and function because it maintains the integrity of the cell layers and enables the polarity to be established and maintained. Different cell-cell adhesion receptors frequently cooperate to guide the localization of cell polarity protein complexes and promote their local activation. Adherens junctions and desmosomes are important for establishment and maintenance of epithelial structure, whereas tight junctions provide the barrier function of the epithelium. In a simple epithelium the cell-cell adhesion junctions are on the lateral surface, and integrin-based adhesions are located on the basal surface. It is just as important that the apical surface be nonadhesive; otherwise the opposite faces of the intestines would stick together, and the lumen would be occluded.

Supracellular Polarity Is Maintained During Collective Cell Migration

Polarization at the supracellular level has an essential role in epithelial cell migration as a group of cells attached to each other, also known as **collective migration**; an example is wound healing. During collective cell migration, individual cell within a multicellular cohort moves interdependently, which is coordinated through stable or transient cell-cell contacts. While migrating, cells within the cohort maintain contacts, which is a central regulator of directional movement of the group. Cadherin-based adherens junctions are the major intercellular adhesions between cells of the migrating cell cohorts. Supracellular polarization of collectively migrating cohorts supports efficient movement (Fig. 7–24). Highly specialized cells, so-called **leader** or **tip cells**, locate at the front of the cell group. These cells direct the migration direction by detecting guidance cues and remodeling the surrounding ECM.

Leader cells normally present prominent pseudopodia aimed at forward direction and show intracellular front-rear polarization, which make them morphologically distinguishable from follower cells. The cadherin-based adherens junctions are absent at the front of leader cells, and this asymmetric distribution of adherens junctions is sufficient for the front-rear polarization of the cytoskeleton and the cytoplasm of lead cells. A variety of mechanisms, such as tension, regulate cell-cell and cell-ECM adhesions, thereby promoting front-rear polarization and directional migration. Besides the cadherin-based adherens junctions, tight junction, and gap junction complexes between cells also contribute to the polarization of leader cells.

Both inner follower cells and peripheral follower cells lack the asymmetric distribution of adherens junctions and rely on the direction cues derived from leader cells for collective migration. The movement of follower cells is also mediated by cadherin-dependent directional

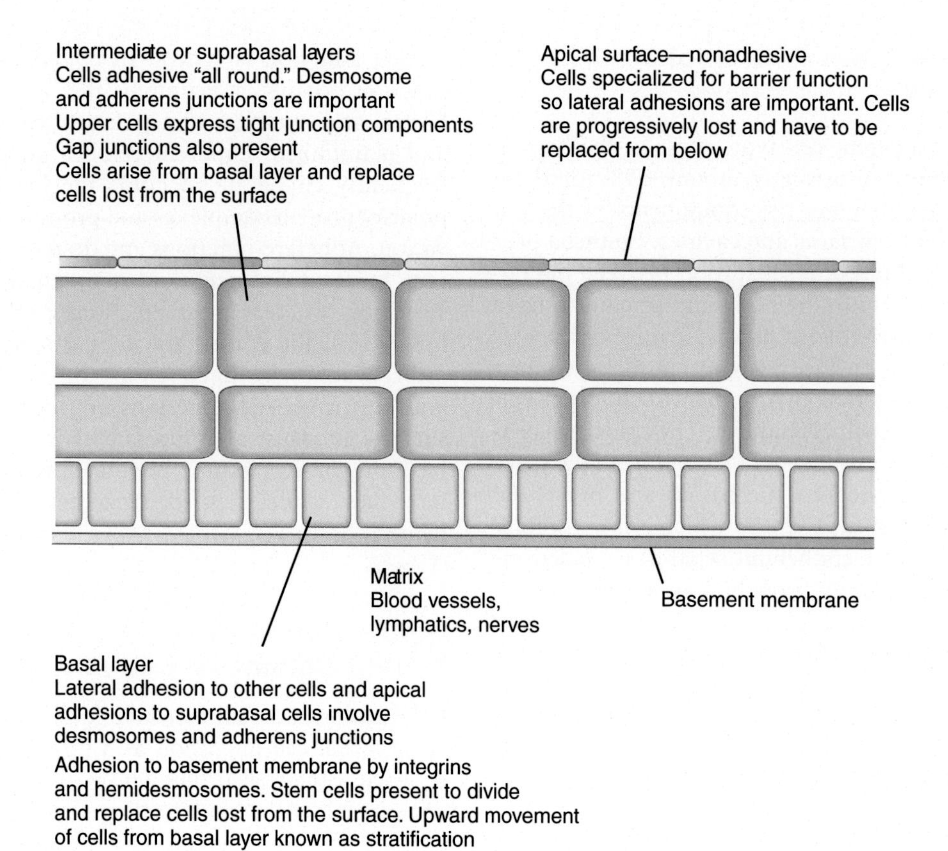

Apical surface—nonadhesive
Specialized for absorption, secretion, or bearing cilia
Lumen
Compartment 1
Polarity
Cell layer
Basement membrane
Submucosal connective tissue, blood vessels, lymphatics, nerves, etc.
Compartment 2
Basal surface
Integrin-based adhesion to basement membrane
Lateral surface—adhesive
Tight junction—occlude paracellular channels
Adherens junction } Adhesion
Desmosome
Gap junction—intercellular communication
B

Figure 7–23. **Cell adhesion in epithelia.** (A) Stratified and (B) simple epithelium showing distribution of adhesive properties.

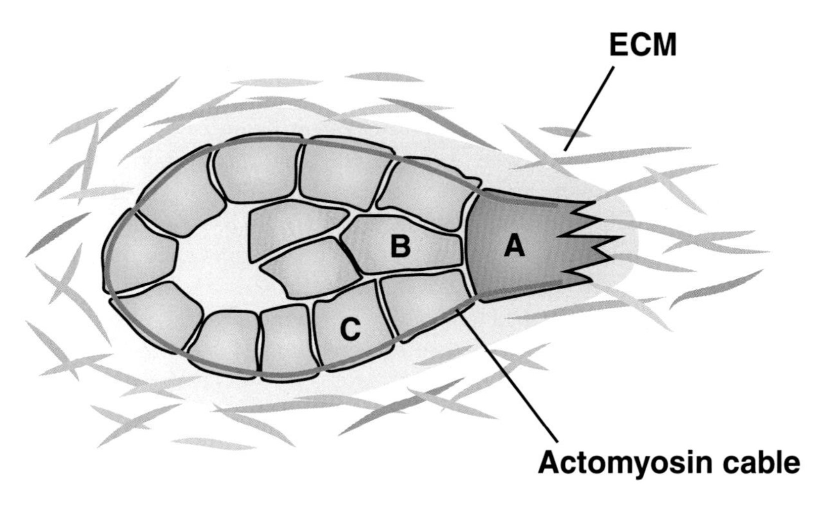

Figure 7–24. **Model of collective migration.** Leader cells (A, one or several cells at the front of the migrating group) determine the direction of migration by forming protrusions or a leading edge in response to environmental cues. The front-rear polarity of leader cells was determined by asymmetric environment of lateral and rear cell-cell adhesions and ECM contact at the front. Inner follower cells (B, behind leader cells inside of the cluster) and peripheral follower cells (C, locate at the outside of the cluster) also contribute to the front-rear polarity of the leader cells and directional movement of the group. Supracellular actomyosin cables formed in peripheral follower cells at the outer surface of the cluster maintain cohesion of the migrating group.

migration. Within the multicellular cohort, cell-cell and cell-ECM contacts provide integrated forces as long-range directional cues to control front-rear polarization of follower cells. Intriguingly, adherens junctions between leader cells and follower cells also provide an effective polarizing cue for leader cells. Thus the follower cells are not only following but also actively steering the directional movement of the leader cells. In such a teamwork the efficacy of collective cell migration is achieved by activity of cohesive groups as a sum of the entire cell cohort.

The collective migration is predominant during embryonic development and tissue homeostasis and drives tumor cell invasion. In many cancers, clusters of tumor cells can detach from the primary tumor and initiate **metastasis**. Moreover, tumor cell clusters are more efficient for initiating metastasis than single cells when examined in vitro and in vivo. Evidence from cancer patients and animal models supports that collective cell invasion mediates metastasis in various epithelial cancers including prostate, pancreatic, lung, colorectal, and breast cancer. RNA sequencing of circulating tumor cells from patients identified that plakoglobin (a component of desmosomes and adherens junctions) and keratin14 (enriched in desmosomes) are enriched in circulating tumor cell clusters and increase their metastatic potential.

Leukocytes Must Adhere and Migrate to Combat Infection and Injury

In contrast with epithelial cells that need to be in a constant adhesive state, other cell types need to be constitutively nonadhesive but need to increase their adhesiveness when this is functionally required. Leukocytes are one major example. Mostly, they circulate freely in the blood, showing no tendency to attach either to each other, to other blood cells, or to the endothelial cells that line blood vessels. However, when the need arises because of tissue damage or infection, they must leave the blood and congregate at the appropriate site to combat the problem. This is the so-called **inflammatory response**. It involves changes in adhesive properties both for leukocytes and the **endothelial cells** lining the small blood vessels near to the site of injury.

The inflammatory response is initiated by the release of diffusible molecules and inflammatory mediators, from the injured tissue, or by complement activation. These cause the rapid cell surface expression of P-selectin, stored inside the endothelial cells in vacuoles called **Weibel-Palade bodies**, and a slower increase in the surface expression of E-selectin and integrin ligands (ICAM-1 and VCAM-1). The newly exposed selectins bind to carbohydrates on the surfaces of circulating leukocytes, causing them to become loosely attached to the endothelial cells. This initial attachment is such that the leukocytes slow their free-flow movement and roll along the endothelial cell surface under the force of blood flow.

Initial adhesion and inflammatory mediators trigger a response from the leukocytes that results in firm adhesion to the endothelial cells. This involves an "inside-out" signal that activates the normally nonfunctional integrin dimers on the leukocyte surface, enabling them to adhere firmly to the endothelial cells by binding to the Ig family adhesion molecule ICAM. The leukocyte adhesion is further strengthened by binding of leukocytes to CXCL1, CXCL2, or CXCL8 on the apical surface of endothelial layer. Although this adhesion is firm, it permits cell migration for the next phase of the process, extravasation. Here, leukocytes migrate between the endothelial cells into the matrix of the tissue. Extravasation involves loosening of the junctional contacts of the endothelial cells in response to **inflammatory mediators. Pericytes** localized along with endothelium provide more permissive entry point for leukocytes invasion of endothelial layer, which may depend on pericyte-promoted upregulation of adhesion molecules and chemokines. In response to inflammatory signals, endothelial cells and pericytes increase expression of cytokine CXCL1 and facilitate leukocytes such as neutrophil crawling inside the blood vessels and entering the compartment between endothelium and pericytes adjacent to blood vessels. Meanwhile, CXCL2 released from neutrophils binds to its receptor located at endothelial junctions, leading to the directional neutrophil emigration.

After extravasation the leukocytes migrate to the site of injury or infection within the tissue guided by **chemotaxis** toward tissue-released diffusible molecules called **chemokines**. Leukocytes recognize chemokines by specific cell surface G protein-coupled receptors (GPCRs). Various GPCRs on leukocytes cell surface could guide leukocyte trafficking and homing, along with activating integrins. Many chemokine GPCR receptors display promiscuous ligand specificity that allows a given chemokine to stimulate different responses depending onto which receptor and where the chemokine binds on the cell surface. The mechanism of migration is not completely understood, but probably involves the type of dynamic regulation of adhesion and the cytoskeleton that is referred to earlier in the discussion of focal contacts.

The inflammatory response is a defensive mechanism aimed at returning homeostasis after injury or localized infection. It is immensely important when it operates in a regulated fashion. However, it can be overactive, in which case it results in the tissue damage of inflammatory diseases such as arthritis. Conversely, rare human mutations in β_2 integrin gene cause a disease called **leukocyte adhesion deficiency (LAD)**. Defect in E-selectin or integrin activation cascade can also cause LAD. Patients who have this disease cannot make **pus**, an accumulation of white blood cells, and are susceptible to death from overwhelming infection due to defective B-cell and T-cell immunity. The patients with severe LAD usually succumb to infectious diseases before 1 year of age.

Platelets Adhere to Form Blood Clots

Blood platelets are the smallest, anucleate blood cells that normally circulate freely in the blood vessels, which need to become adhesive rapidly to assist in the formation of **blood clots** at sites of blood vessel damage. They are present in enormous numbers, $\sim 1.5–4.0 \times 10^{11}$ per liter of blood and have an 8–10 day turnover period. Platelets are maintained in a resting state in the absence of activation signals, which relies on inhibiting molecules, such as NO and prostacyclin (PGI2), released from endothelial cells. Once activated, platelets can adhere to matrix components, including collagen that is exposed by blood vessel damage. **Von Willebrand factor** (vWF) is a matrix protein that is released from Weibel-Palade bodies of endothelial cells and platelet α-granules, and fibrin is a protein that is generated from plasma to form the blood clot (Fig. 7–25). Platelets can also aggregate together. Platelet activation can be triggered by a variety of soluble agonists including **thrombin**, adenosine diphosphate (ADP), and collagen or peptides derived from it. Platelet adhesion involves a number of cell surface adhesion receptors that have a complex nomenclature that originates from a time before integrins and other adhesion molecules were recognized. The three major platelet adhesion receptors are **GPIIb-IIIa** ($\alpha_{IIb}\beta_3$-integrin), **GPIa-IIa** ($\alpha_2\beta_1$-integrin), and **GPIb-IX-V**, a nonintegrin adhesion complex consisting of four gene products. GPIIb-IIIa is a promiscuous integrin that exhibits binding to the matrix molecules fibrinogen, vWF, fibronectin, vitronectin, and thrombospondin. GPIa-IIa is the major receptor for collagen and GPIb-IX-V for insoluble vWF. Platelet activation leads to activation of these receptors, release of vWF, and consequent adhesive binding. Platelet activation is further amplified by the activation of G protein-mediated signaling pathways through GPCRs (e.g., thromboxane-a2 or TxA2 receptor), PAR3, PAR4, and PAR1 (thrombin receptors), which further increases the formation and release of platelet agonists, such as PAF, thrombin, ADP, TxA2, and epinephrine. These positive-feedback reactions thus amplify the initial signals and reinforce the rapid activation and recruitment of platelets for plaque growth. Balanced platelet activation between activating and inhibitory signals is also critical for proper function of platelet coagulation, whereas disruption of this balance may cause thrombosis and bleeding disorders. To avoid overactivation of the platelets, activation of endogenous inhibitory mechanisms and ectodomain shedding of receptors on the surface of platelets both play essential roles in this balancing act.

Platelet adhesion is subject to a number of human genetic diseases that involve mutations in the genes for the adhesion proteins, including von Willebrand disease (see Clinical Case 7–2), **Bernard-Soulier syndrome** (involving GPIb-IX-V), and **Glanzmann thrombasthenia** (involving GPIIb-IIIa). Platelet adhesion can also be triggered inappropriately so that platelets adhere to endothelial cells in much the same way as leukocytes during the inflammatory response. This causes the development of **atherosclerotic lesions** (see Fig. 7–40) and **atherothrombosis**, which are enormously important health problems.

Embryonic Development Involves Many Adhesion-Dependent Events

Cell adhesion has a crucial function throughout embryonic development. The first morphogenetic event in mammalian development is **compaction**, in which the loosely attached cells or blastomeres of the eight-cell embryo "zip up" their adhesions to become tightly bound together (Fig. 7–26, A [i, ii]). This event involves the adhesion molecule E-cadherin and other adherens junction components. Also, at this stage, tight junctions are beginning to form between the cells. By the time a hollow ball of cells, the **blastocyst**, is formed, the first

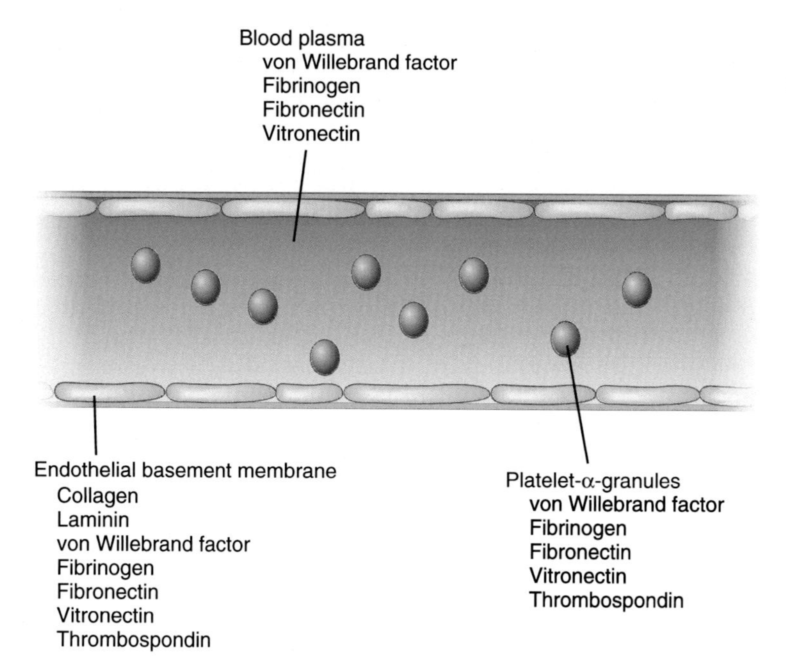

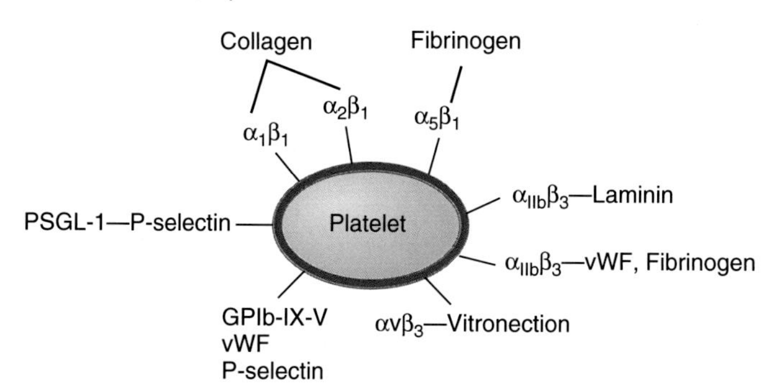

Figure 7–25. **Adhesion of blood platelets.** (A) The adhesive environment. Diagram summarizing the molecular adhesive components in blood plasma, platelets, and the endothelial cell basement membrane. (B) The adhesion receptors of platelets and their ligands. *PSGL-1*, P-selectin glycoprotein ligand-1.

epithelium, the **trophectoderm** that will generate the **placenta**, has a full complement of tight junctions, adherens junctions, and desmosomes, and this is still before implantation (see Fig. 7–26, B). Shortly after implantation the process known as **gastrulation** is initiated. This is the event (some say the most important event in our lives!) that generates the three-layered organization of the embryo, with **ectoderm** on the outside, **endoderm** on the inside, and **mesoderm** in between (see Fig. 7–26, C). Ectoderm forms the epidermis and nervous system, mesoderm the muscles and bones, and endoderm of the gut and some related organs. Gastrulation involves a massive (in embryonic terms) amount of cell movement. This generates the correct embryonic shape and positions the cell layers with respect to each other. These events are crucially dependent on cell adhesion, with molecules such as E-cadherin playing a key role.

Formation of the neural tube, the forerunner of the central nervous system, involves expression of different cadherin molecules (see Fig. 7–26, D). From the junction between the neural tube and the ectoderm arises a migrating cell population called the **neural crest** or ectomesenchyme (Fig. 7–27). This forms mainly nerves and bones in the head and parts of the autonomic and peripheral nervous systems in the trunk. Migration and precise positioning of neural crest cells are regulated by a series of changes in the expression of adhesion molecules, especially cadherins, fibronectin, and integrins. These initiate and guide migration and then cause the cells to stop and aggregate when they have reached the correct position.

Development of the nervous system involves multiple examples of directed cell migration and the extension of nerve fibers. During each of these events, the migrating cells must reach the correct target. For example, motor neurons must reach and form synapses called *motor*

Clinical Case 7–2

Jessica is a 29-year-old pregnant woman admitted to the hospital at 39 weeks of pregnancy after experiencing contractions. This is Jessica's first pregnancy. Her past medical history was notable for recurrent nosebleeds as a child and menstrual periods prior to pregnancy that could last up to 6 days. Family history was unremarkable. Her exam was normal, and cervical dilation was at 3 cm. At intake, her laboratories were notable for a hemoglobin of 9.8 g/dL (normal >11 g/dL) and an activated partial thromboplastin time (aPTT) of 40 s (normal 25–35). Childbirth proceeded without complications, but approximately 4 h following delivery, she had a rapid onset of gross high-volume vaginal bleeding that soaked through three pads in less than 30 min. She complained of lightheadedness and her blood pressure dropped to 85/50 from 120/80 previously. She was immediately started on high-volume normal saline, and stat labs (including a blood typing) was performed. CBC showed hemoglobin of 6.5 (normal >11) but normal platelet levels. aPTT had increased to 55 s with normal prothrombin time (PT). Factor VIII levels were low at 35 IU/dL. She was transfused two units of packed red blood cells and factor VIII concentrate.

Cell Biology, Diagnosis, and Treatment of von Willebrand Disease

Jessica has von Willebrand disease (VWD), the most common human-inherited bleeding disorder, occurring in up to 1% of all individuals. In Jessica's case, inheritance was likely autosomal recessive, explaining lack of symptoms in close family members. VWD results from a defect in von Willebrand factor (VWF), a glycoprotein necessary for platelet adhesion to the endothelium (internal lining of blood vessels) after vascular injury. VWF is a complex large protein that helps form adhesive bridges between platelets and the vascular endothelium. This adhesive bridge, made with this large, "sticky" protein, is necessary to initiate hemostasis and coagulation. Without adequate levels of functional VWD, the first step of developing a normal clot after bleeding injury, known as primary hemostasis, cannot occur.

VWF is most necessary in small blood vessels such as capillaries because of the increased shear stresses seen in these vessels. As a result the clinical symptoms of VWD are usually seen in tissues with extensive small vessels, such as the skin, the gastrointestinal tract, and the female reproductive tract.

VWD can occur as a result of a decrease in amount of VWF (Type 1), inactive or qualitatively impaired VWF (Type 2), or complete absence of VWF (Type 3). The main symptom of VWD is abnormal bleeding. The most common type is Type 1, and bleeding symptoms are often mild, though this can range between minimally symptomatic to routinely life threatening depending on the type and penetrance. Patients can experience occasional nosebleeds or bleeding gums with minimal trauma. Women may experience heavier than normal menstrual periods. The disease is often diagnosed only after a major bleeding event, which, like in Jessica, may occur during childbirth.

Patients with a history of mucosal bleeding, heavy or prolonged menstrual periods, or bleeding complications during childbirth should be worked up for a bleeding disorder such as von Willebrand. Laboratory tests often show prolonged bleeding times with normal coagulation parameters (prothrobin time (PT) and activated partial thromboplastin time (aPTT)) and normal platelet numbers. Complete blood counts should be performed in all patients to ensure that bleeding has not depleted hemoglobin to life-threatening levels and to evaluate the need for blood transfusion. Evaluation of VWF levels can be performed indirectly using a number of tests including a glycoprotein-binding assay, a collagen-binding assay, ristocetin cofactor activity, or ristocetin-induced platelet agglutination assay. VWF binds to Factor VIII to stabilize the factor in plasma, so decreases in VWF may lead to decreased levels of Factor VIII as well, which could mildly increase the aPTT.

Treatment of VWD is usually either supportive or preventative. In women with heavy menstrual periods, estrogen-containing contraceptives can decrease amount and duration of bleeding. Patients with nasal or oral bleeding can use topical thrombin therapy. Patients with minor bleeding or undergoing minor surgical procedures can often be managed by desmopressin, which stimulates release of VWF from endothelial cells. While this therapy may be effective in Type 1 disease (decreased VWF), it has no effect in Type 3 disease (complete absence of VWF) but may be minimally effective or even harmful in Type 2 disease (qualitative deficiency of VWF). Patients can administer this medication via intranasal spray, and in patients who have shown good responses in the past, minor bleeding can be managed by the patient in an outpatient setting.

In patients experiencing major hemorrhage or scheduled for major surgery, direct replacement of VWF is often necessary. This may entail blood or plasma products that contain VWF in high concentration, including intermediate purity Factor VIII and purified VWF. Recombinant VWF was approved for use by the FDA in 2015 as well. In severe VWD, prophylactic use of long-term VWF has been shown in small trials to decrease bleeding episodes.

Childbirth in women with VWD is riskier than in other adults, with up to 20% of women experiencing miscarriage or postpartum hemorrhage in some studies. Desmopressin is contraindicated prior to delivery due to induction of contractions. Recommendations are to maintain VWF and Factor VIII levels of 50 UI/dL during delivery and postpartum using factor replacement. Jessica recovered from her bleeding episode with blood transfusion and Factor VIII replacement. She was able to undergo two more healthy deliveries with monitoring of factor levels and factor replacement when levels decreased. Between pregnancies, she takes estrogen-containing contraceptives that have controlled bleeding in her menstrual periods.

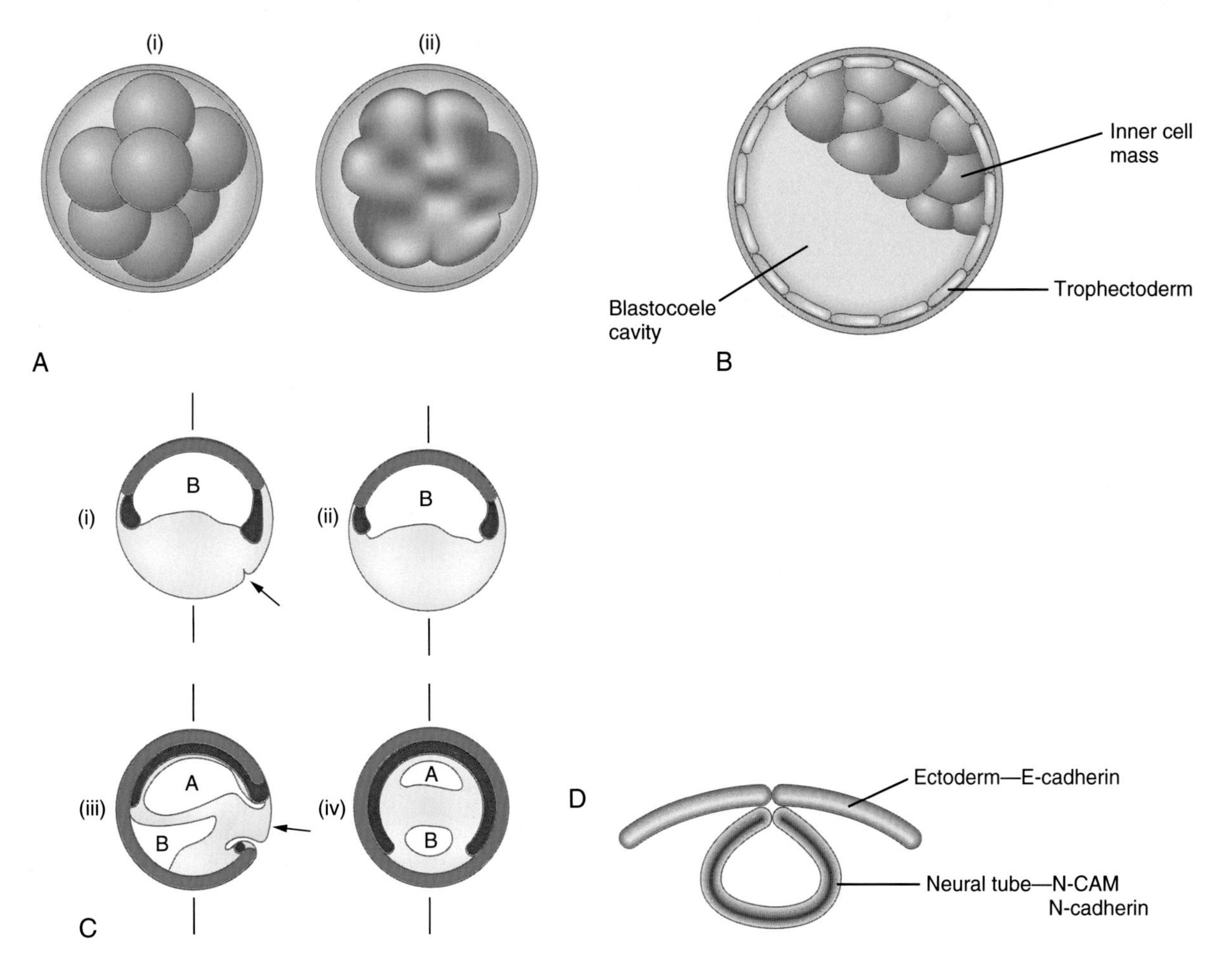

Figure 7–26. **Cell adhesion in embryonic development.** (A) Compaction in the mammalian embryo. (i) At the early eight-cell stage, the embryonic cells, the blastomeres, are loosely attached. (ii) Without further cell division, they "zip up" their adhesive contacts. (B) At the blastocyst stage the first epithelium, the trophectoderm, is formed. This contains the inner cell mass and the fluid-filled blastocoel cavity. (C) Gastrulation is shown in the amphibian embryo because it is easier to visualize than in the mammal where gastrulation is more complex because of the presence of extraembryonic tissue that forms the placenta. (i) Longitudinal and (ii) transverse sections of the early gastrula. Cell invagination is just beginning with the formation of the blastopore lip *(arrow)*. (iii, iv) Comparable sections of the late gastrula where invagination is almost complete though the blastopore *(arrow)* is not quite closed. Ectoderm *(blue)*, mesoderm *(red)*, and endoderm *(yellow)*. *A*, archenteron, the future gut cavity; *B*, blastocoel. *Vertical lines* indicate the relationship between the sections. (D) Formation of the neural tube, the future central nervous system, involves expression of different adhesion molecules.

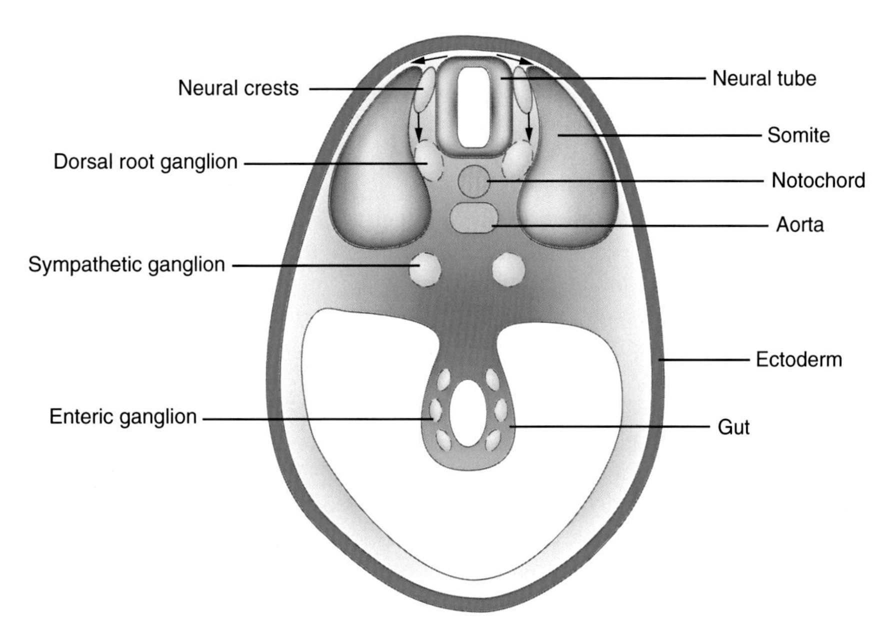

Figure 7–27. **Neural crest and its derivatives in the trunk.** *Arrows* indicate paths of migration of neural crest cells. The ventral pathway gives rise to ganglia and their associated nerves. The dorsal pathway gives rise to pigment cells of the skin. Cell adhesion plays a crucial role in guiding the migration and final positioning.

end plates with the appropriate skeletal muscles. Similarly, sensory nerve fibers must form a precisely mapped series of connections with the appropriate part of the brain; for example, those from the eye with the contralateral **optic tectum** in chicks and frogs or the **visual cortex** in mammals. Specific adhesions involving Ig and cadherin family members play important roles in guiding this complex wiring process. The cell polarity determined by cell adhesions is essential for the normal development of various organs. Moreover, loss of cell polarity due to disrupted cell adhesions could increase cell proliferation and lead to tumorigenesis.

SIGNAL TRANSMISSION BY CELL ADHESION RECEPTORS

Transduction of signals by or involving CAMs has been alluded to several times earlier. To form an integral component of a tissue, cells need to monitor their environment and to respond accordingly to the signals they receive. They have a variety of surface receptors that receive signals from diffusible molecules such as **growth factors** and chemokines. However, they also receive signals from insoluble components of their environment, that is, from the ECM and from other cells. Transduction of these signals is an essential second function of adhesion molecules or adhesion receptors. Modern research has reported many examples of signaling by specific adhesion molecules. Some important examples are considered in the following sections.

Contact Inhibition of Locomotion and Proliferation of the Cells

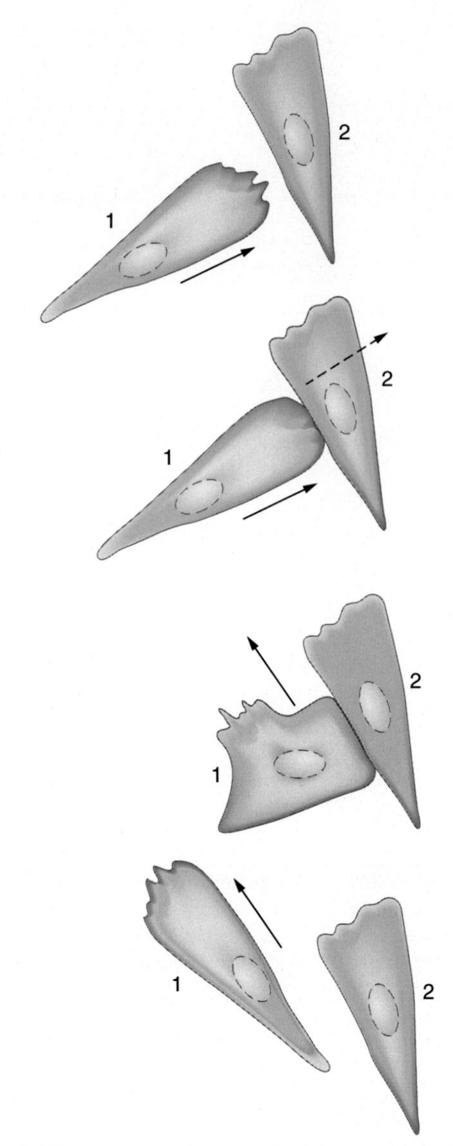

Figure 7–28. **Contact inhibition of cell movement.** Cell 1 moving in the direction of the *arrow* makes contact with and adheres to cell 2 (shown completely stationary for the purpose of illustration). Movement in the initial direction is inhibited. Cell 1 forms a new leading lamella or lamellipodium and moves off in a different direction.

An important example of adhesion signaling that was studied before any adhesion molecules had been characterized is known as contact inhibition (Fig. 7–28). Thus, when a cell such as a fibroblast moving over the surface of a substratum in tissue culture encounters another cell, its leading lamellipodium first forms an adhesive contact with the surface of the other cell. The newly formed cellular adhesion reduces the retrograde flow of F-actin in the surrounding area of the adhesion, resulting in decreased cell velocity and motility. The mechanical constraints yielded from cell-cell contact also inhibit mitosis of the contacted cells and reduce cell division. The cell-cell contacts suppress growth factor-mediated intracellular signaling (e.g., ERK and Akt), leading to cell arrest at G0/G1 phase. In addition, the mechanical tension also reduces YAP/TAZ activity in contacted cells, which is a key regulator of contact inhibition. Contact-inhibited movement in the initial direction eventually drives the cell to move off in another direction. The net result of these behaviors is that cells do not move over each other in culture but instead remain as a monolayer on the substratum. In general, cells in confluent monolayers, that is, continuous layers of cells in contact with each other, though not completely static, do not move around very much. However, if the monolayer is experimentally wounded to produce a free edge, cells at that edge begin moving into the wound and restore the monolayer cell sheet. Importantly, contact inhibition ensures that an organ within an organism maintain proper size during the tissue regeneration after injury. Early excitement about the phenomenon of contact inhibition arose because many types of transformed cells were found not to be subject to it. The loss of contact inhibition in both locomotion and proliferation is considered a hallmark of cancer, which involves defective adherens junction formation in part due to loss of E-cadherin in many malignant cells. Transformed cells were found to be able to move freely over the surfaces of other cells and continue cell growth thus apparently mimicking the **invasive behavior** of tumor cells.

Cell Growth and Cell Survival Are Adhesion Dependent

Cell-substratum adhesion is an important regulator of cell division. To proliferate, cells need to attach and spread on the substratum (Fig. 7–29). Cells that are well spread proliferate faster than those for which spreading is restricted. This aspect of cell regulation is referred to as **anchorage dependence**. This is a feature of normal or untransformed cells. By contrast, transformed cells (i.e., cells that are able to produce tumors) are commonly anchorage independent. Thus they can proliferate with little or no contact with the ECM and also when suspended in soft agar. Anchorage-independent growth in culture mimics the ability of tumor cells in vivo to grow in abnormal situations such as when they have become detached from the basement membrane or even when in suspension in circulation and ascites fluid in the peritoneal cavity. A signaling pathway involved in regulation of cell proliferation by ECM adhesion involves integrins and a cytoplasmic protein kinase called Erk. When cells are substratum attached, Erk can be activated and enter the cell nucleus to regulate proliferation, but if cells are held in suspension, Erk remains cytoplasmic, and proliferation does not occur. Regulation of Erk activity appears to depend on the actin cytoskeleton and the ability of integrin adhesion to regulate assembly into stress fibers. Stiffness of the adhesion interface also regulates cell cycle and cell growth, which is mediated through small GTPase Rac1. Activation of Rac1 increases the level of cell cycle regulator cyclin D, thereby promoting cell cycle progression and proliferation.

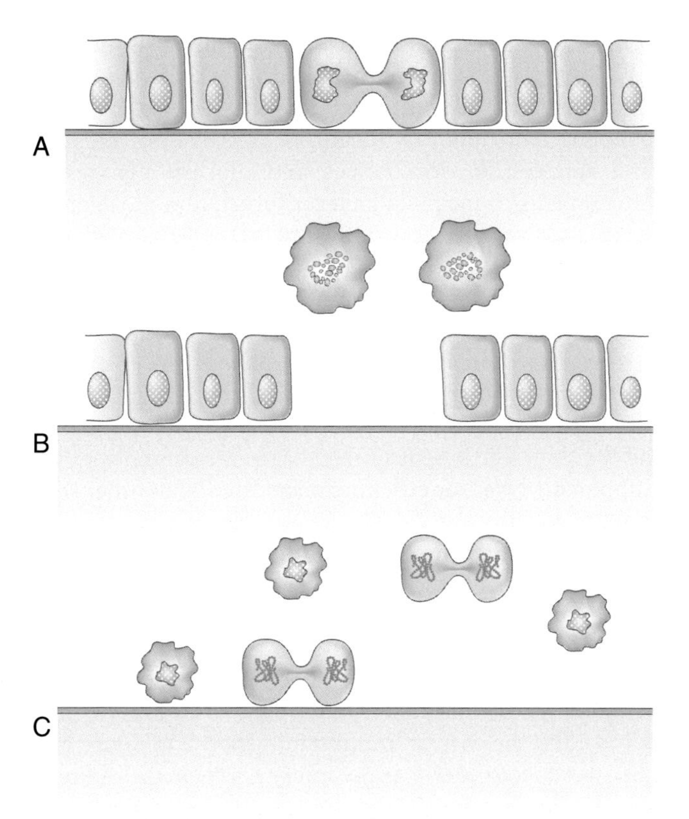

Figure 7–29. **Cell attachment, division, and death.** (A) Normal cells are anchorage dependent requiring attachment to the substratum to proliferate. (B) Cells that become detached from the substratum undergo anoikis, a form of programmed cell death or apoptosis. (C) Many tumor cells both survive and proliferate in suspension.

Conversely, cells that are released from substratum contact tend to undergo **programmed cell death** or **apoptosis** (see Fig. 7–29). This particular form of cell death that is triggered by release from the substratum has been named *anoikis*. In response to the cell-ECM detachment, proapoptotic proteins Bid and Bam (members of BH3-only proteins) are activated that promotes the oligomerization of Bcl-2 family proteins Bax-Bak inside the outer mitochondria membrane. The oligomerized Bax-Bak creates channels on outer mitochondria membrane, resulting in increased mitochondrial permeabilization, cytochrome *c* release, and activation of intrinsic apoptosis pathway. Additionally, the detachment of cells from ECM also upregulates cell membrane death receptor Fas and its ligand, supporting a critical role of extrinsic apoptosis pathway in anoikis execution. Another property of many transformed or tumor cells is the ability to survive in abnormal situations, for example, not to be susceptible to **anoikis**. Normal cells require survival signals from cell-substratum adhesion to avoid anoikis; many tumor cells do not require such signals.

An example of the regulation of cell survival and cell death in normal tissue function comes from the **mammary gland** (Fig. 7–30). During pregnancy the mammary gland enlarges because of an elaboration of ECM and growth of the mammary epithelium so that the latter can produce milk. When suckling is over, the process is reversed. Enzymes called **matrix metalloproteinases (MMPs)** degrade the matrix. As a consequence, many epithelial cells have no matrix to which to adhere. They therefore lose their integrin-mediated **survival signals** and undergo apoptosis. In tumor cells the loss of anchorage dependence in cell proliferation and ability to escape anoikis play instrumental role in tumor invasion and metastasis. Loosening of the surrounding ECM by MMPs secreted from tumor cells and tumor-associated stromal cells further enhances detachment of tumor cells from the primary tumors and their distant invasion.

Cell Adhesion Regulates Cell Differentiation

The mammary gland also provides a well-characterized example of the regulation of cell differentiation by cell adhesion (see Fig. 7–30). The primary function of the mammary gland is to produce milk during **lactation**. It has been shown that activation of the genes that code for milk proteins is dependent on a combination of two signals, one a diffusible signal from the hormone **prolactin** and the other an adhesive signal mediated by β-integrins. Thus, if epithelial cells from lactating mammary glands are cultured on collagen in the presence of

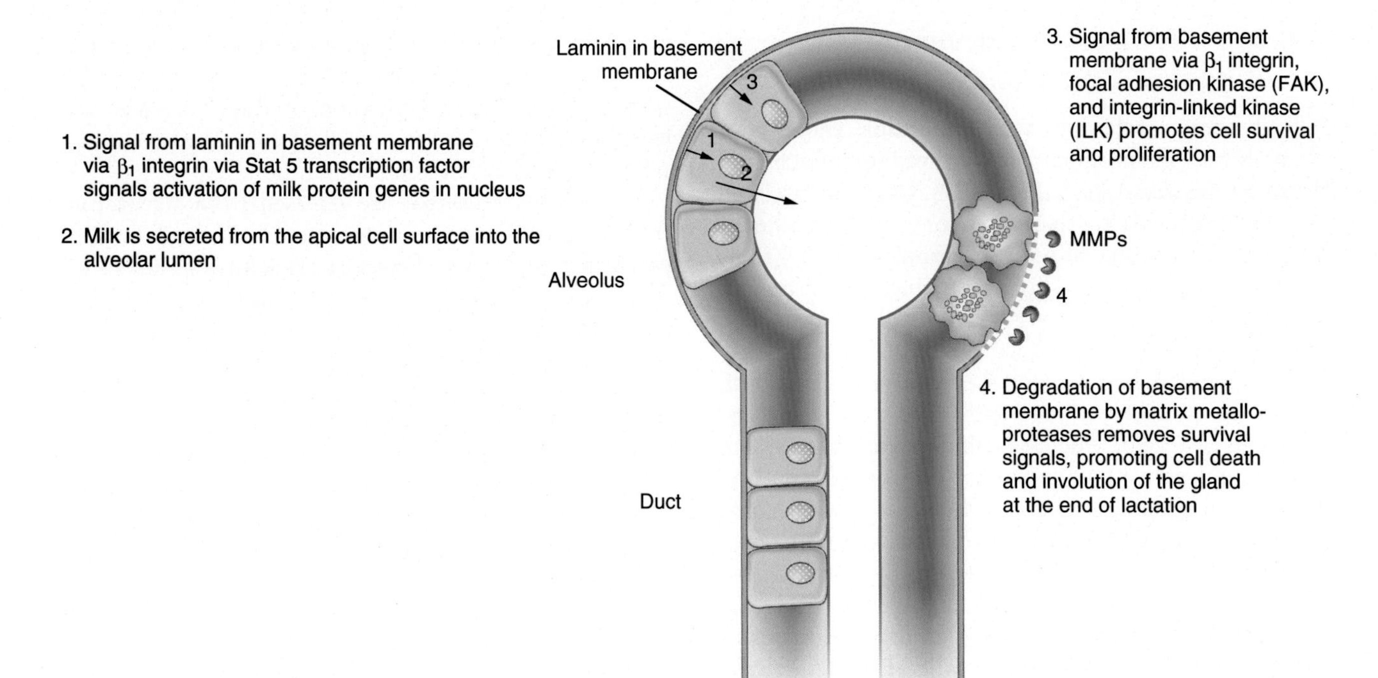

Figure 7–30. **Mammary gland.** A good example of the regulation of cell survival and differentiation by cell adhesion signaling.

prolactin, they survive but do not produce milk proteins; however, if laminin is the culture substratum, milk proteins are produced. Many other examples also exist of the regulation of gene expression and cell differentiation by signals originating through cell contact and adhesion. The surrounding environment of stem cells, also known as stem cell **niche**, has an essential role in regulating stem cell homeostasis, self-renewal, and differentiation. ECM stiffness, shearing force, hydrostatic pressure, and mechanical stress all can trigger stem cells/progenitor cells differentiation into functional cell types in response to the specific environmental cue. Integrins play a critical role in stem cell differentiation. Integrin A2 (CD49b) is a cell surface marker for mammary gland luminal progenitor cells, mesenchymal stem cells within bone marrow, and those derived from umbilical cord blood. Integrin A6 (CD49f) and CD44, a receptor for hyaluronic acid, are recognized as markers for cancer stem cells from multiple types of cancer. The cross talk between cellular contacts with ECM and stem cell fate will be discussed in detail in the later sections.

EXTRACELLULAR MATRIX

All tissues consist of two components, a cellular component and an extracellular component. The latter comprises a variety of specialized structures that constitute the ECM. Molecular components of this matrix are secreted and to some extent assembled by the cells of the tissue.

The amount of ECM varies enormously between different tissues. Thus, in **bone** and **cartilage** and in the **dermis** of the skin, the bulk of the tissue is composed of matrix. In contrast, in epithelia and muscles, most of the tissue is cellular, the matrix being confined to a basement membrane or basal lamina that surrounds or underlies the cellular component. Composition and amount of ECM differ according to the function of the tissue. Bone is calcified for strength and consists largely of ECM, enabling it to fulfill its functions of providing strength and support for soft tissues and of carrying muscle attachments to facilitate its lever function in movement. Cartilage also consists mainly of ECM, but it has very different properties from bone because it needs to provide articulation in joints, while at the same time needing to resist compression and provide a cushioning effect between hard bones. The dermis connects the epidermis to the underlying tissues and needs to provide great strength and elasticity to dissipate the stresses impinging on the skin. The basement membrane is essentially a thin supporting layer for cell attachment, but it has other specialized functions in tissues such as the kidney.

In adult organisms the majority of extracellular matrices exhibit slow turnover; that is, they are permanent or semipermanent in nature. They do, however, need to retain the capacity to respond to changes such as injury, for example, in the healing of fractures or wounds. Another type of matrix, the blood clot, needs to form rapidly and in the correct location in response to injury but then needs to disperse as the injury is repaired. Modulation of the ECM is also important in angiogenesis, the generation of new blood vessels in response to injury or tumor growth. The role of the matrix is not exclusively structural; it also regulates the level of hydration and the pH of the local environment. ECM provides the basis for signals transmitted to cells by adhesion receptors that bind to its components, and it acts as a reservoir for growth factors that also bind reversibly to its constituents. The major components of the ECM include two classes of macromolecules, fibrous proteins (e.g., collagens and elastin) and glycoproteins (e.g., fibronectin, proteoglycans, and laminin), which are discussed in more detail in the following section.

Collagen Is the Most Abundant Protein in the Extracellular Matrix

Collagen is secreted mainly by fibroblasts and accounts for up to 30% of the total proteins in human bodies. Rather than a single protein, collagen comprises a family of 28 genetically distinct proteins. Collagens are the principal structural elements of all connective tissues. They are characterized by the presence of a repeated sequence of three amino acids—a **tripeptide; glycine-X-Y,** where X and Y are commonly proline; or **hydroxyproline.** (Hydroxyproline is proline that has been post translationally modified by addition of a hydroxyl group.) All collagens are trimers in which at least some and often most of the protein chains are involved in forming a triple helix. The Gly-X-Y tripeptide plays a key role in the triple-helix structure.

The different family members can be divided into groups according to the structures that they form. These groups are fibril-forming or **fibrillar collagens,** fibril-associated collagens, network-forming collagens, anchoring fibrils, transmembrane collagens, **basement membrane collagens,** and others. By far the most abundant group is the fibrillar collagens, which constitute about 90% of total collagen. Of these, fibrils composed of collagen types I and V form the structural framework of bone, and collagen types II and XI contribute to the fibrillar matrix of articular cartilage. The structure of these collagens gives them great tensile strength and torsional stability, which are essential properties for these tissues. The flexible triple helices of type IV collagen form a meshwork in basement membranes. Types IX, XII, and XIV are fibril-associated collagens that associate as simple molecules with collagen fibrils formed by other collagens. Collagen types XIII, XVII, XXIII, and XXV are expressed on cell surface as receptors, which affect cell adhesion. Type XVII collagen, which is discussed earlier as BP180, an adhesion molecule of hemidesmosomes, has its collagenous domain within the basement membrane and is a transmembrane protein with its noncollagenous domain extending into the hemidesmosomal plaque.

Different collagens have different structures and properties because they are composed of different trimeric combinations of protein chains (Fig. 7–31). They may be either homotrimers composed of three identical chains or heterotrimers composed of two or three different chains. Type II and III collagens are examples of homotrimers and types I and IV of heterotrimers. The chains are called α-chains; the formula or molecular composition for type II collagen is [α1 (II)]3 and for type I collagen [α1 (I)]2 α2 (I). Each different α chain is encoded by a different gene.

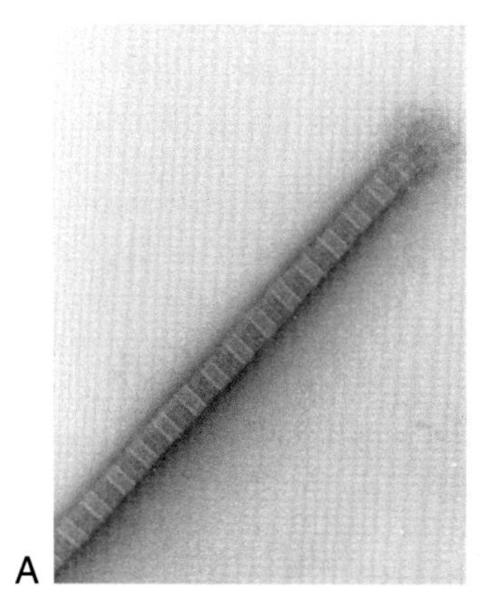

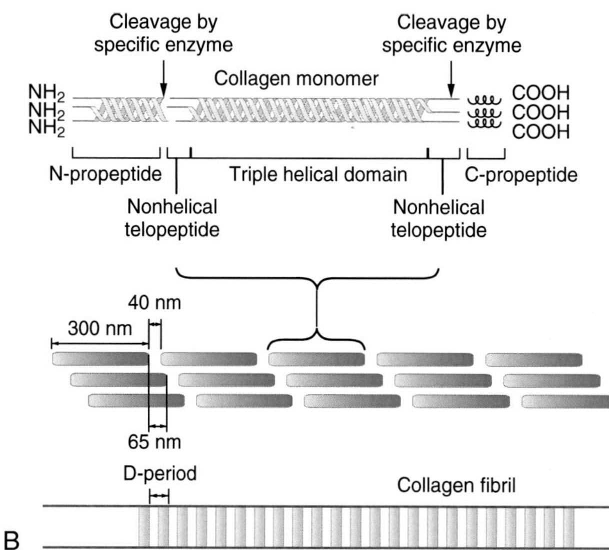

Figure 7–31. **Molecular structure of fibrillar collagen.** (A) Electron micrograph showing the banded appearance of a type I collagen fiber from tendon. (Courtesy Drs. Helen Graham and Karl Kadler.) (B) Diagram showing the fiber assembly. The N- and C-propeptides of procollagen are cleaved to give a collagen monomer that is triple-helical with nonhelical telopeptides at each end. Monomers assemble in a regular manner to cause the banded collagen fibril.

Characteristic of all collagens is the presence of a triple helix formed by mutual coiling of the three α-chains. The so-called collagenous domains that form this triple-helical substructure possess the (Gly-X-Y)$_n$ repeat amino acid sequence, which is prerequisite for α-helix formation. The collagenous domain may embrace the majority of the molecule as in the fibrillar collagens such as type I, or it may be limited to part of the molecule as in specialized collagens such as type XVII. The presence of hydroxyproline at either the X or Y position in many of the tripeptide repeats is essential for the stability of the helix by enabling the formation of intermolecular hydrogen bonds. In fibril-forming collagens the triple-helical domain is 300 nm (about 1000 amino acids) in length (see Fig. 7–31). Many of these triple-helical molecules (rather confusingly called monomers) assemble into fibrils through interaction of their nonhelical end domains and the side chains of amino acids that are exposed on the surfaces of the helices. In the fibril the ends of the helical monomers are separated by a distance of 40 nm, and a continuous row of monomers is staggered by about 27 nm with respect to the adjacent row, causing a periodicity (called the D-period) of about 67 nm. This staggered arrangement accounts for the characteristic **banded appearance of fibrillar collagens** viewed by light and electron microscopy.

It would be inappropriate here to enter into more detail about the variety of collagen substructure. Type IV collagen is considered further later in this chapter in the discussion of basement membranes. The structures formed by collagen fibrils are truly remarkable. For example, **tendons** consist of large-diameter collagen fibrils of indeterminate length that are strictly parallel to each other and that enable the tendons to withstand repeated application of tension. Formation of these long parallel fibril bundles depends partly on self-assembly of the collagen molecules and partly on the activity of the cells that produce them.

Synthesis of collagen protein occurs in the **endoplasmic reticulum** of fibroblasts (Fig. 7–32). Here, posttranslational modifications such as the hydroxylation of proline residues and the addition of carbohydrate chains occur. Here also the protein chains are assembled into a triple helix called **procollagen.** In procollagen the protein chains are longer at the NH_2 and COOH-termini than those of mature collagen. These extra bits are called N- and C-propeptides. The C-propeptide initiates the formation of the triple helix, which then progresses toward the N terminus. Procollagen then proceeds through the **Golgi apparatus** where it is packaged in the ***trans*** **Golgi network** (TGN) into vesicles called **Golgi to plasma membrane carriers** (GPCs). During the formation of GPCs, specific enzymes cleave the N- and C-propeptides, the essential step that initiates **self-assembly** of banded collagen fibrils. The nascent fibrils then grow in length and number within the GPCs. A GPC containing several fibrils then joins the cell surface forming a structure called a **fibropositor** that extrudes the fibrils from the cell where it joins in a regular parallel array with other fibrils to form a bundle. The parallel alignment of fibrils in the tendon is therefore generated by the tendon fibroblasts. This process occurs only during fetal development. Thereafter, tendons grow by addition of further fibrils to the parallel array.

Mutations in collagen genes result in a variety of human diseases including **chondroplasia, osteogenesis imperfecta,** Alport syndrome, Ehlers-Danlos syndrome, and dystrophic EB, and other collagen abnormalities contribute to osteoarthritis and osteoporosis. During wound healing, remodeling of collagen needs to occur. Certain members of the matrix metalloproteinase family of enzymes are involved in the necessary degradation of collagen required for this process. These enzymes are produced by a variety of cell types including fibroblasts; inflammatory cells such as granulocytes; hypertrophic chondrocytes, osteoblasts, and osteoclasts that are involved in the remodeling of cartilage and bone. Increased collagen cross-linking and deposition during tumorigenesis stiffens ECM, disrupts tissue morphogenesis, and

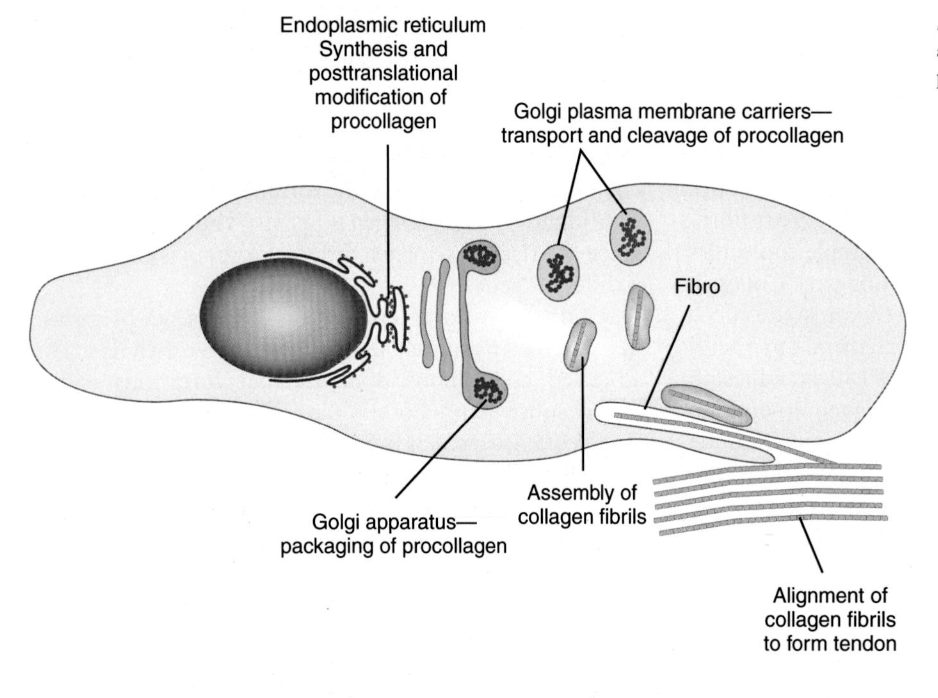

Figure 7–32. **How cells synthesize and assemble collagen fibrils into the parallel array that constitutes a tendon.**

promotes signaling activated by cell surface receptors binding to collagens, which enhance progression of tumors.

Glycosaminoglycans and Proteoglycans Absorb Water and Resist Compression

Other major bulk constituents of the ECM are long carbohydrate chains called **glycosaminoglycans (GAGs)**. GAGs are usually linked to proteins to form proteoglycans. GAGs consist of repeating disaccharides linked into long, unbranched chains. One of the sugars in the repeating unit is an **amino sugar**, *N*-acetylglucosamine, and the other is **uronic acid**, either glucuronic or iduronic acid (Fig. 7–33). GAGs are strongly negatively charged because most of the sugars bear **carboxylic acid groups**, and in **chondroitin sulfate, dermatan sulfate, heparan sulfate**, and **keratan sulfate**, the amino sugars are commonly **sulfated**. These long carbohydrate chains have two important properties that underlie their major role in tissue structure and function. First, unlike protein chains, carbohydrate chains do not fold into compact units. Second, their negative charge attracts cations such as Na^+ that are **osmotically active** and thus attract large amounts of **water**. These properties mean that GAGs fill large volumes of space and are able to resist compressive forces such as the huge pressures that are exerted on cartilage in joints.

Hyaluronic acid (HA), a nonsulfated GAG, consists of up to 25,000 disaccharide units and is widely distributed in tissues. HA is synthesized on the cell surface membrane, instead of at the Golgi apparatus, in a protein-free form. HA molecules can reach molecular weights of several million daltons, and a single molecule, swollen with water, can occupy a space of 10^7 nm^3. HA is a lubricant in **joints** and facilitates cell migration in embryonic development and wound healing.

Proteoglycans consist of sulfated GAGs covalently linked to a polypeptide chain, the **core protein** (Fig. 7–34). They vary enormously in size and carbohydrate composition. The largest can consist of up to 95% carbohydrate by weight, and **aggrecan**, a major constituent of cartilage, has a molecular weight of about 3 million Da. At the other extreme, **decorin** has a molecular weight of 40 kDa and a single carbohydrate chain. Already an enormous molecule in its own right, aggrecans in cartilage form giant complexes that have molecular weights of the order of 100 million Da and occupy a space of 5×10^{16} nm^3. An aggrecan aggregate consists of a central core HA molecule with many aggrecan molecules joined to it laterally by means of linker proteins; the entire substructure resembles a bottle brush under

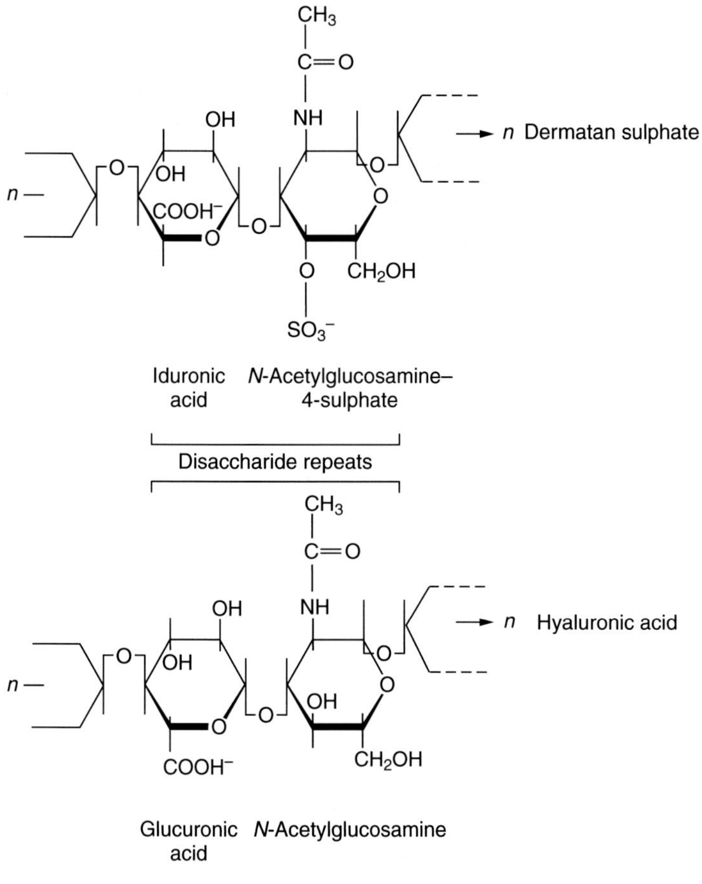

Figure 7–33. The repeating carbohydrate units that constitute sulfated and nonsulfated glycosaminoglycans.

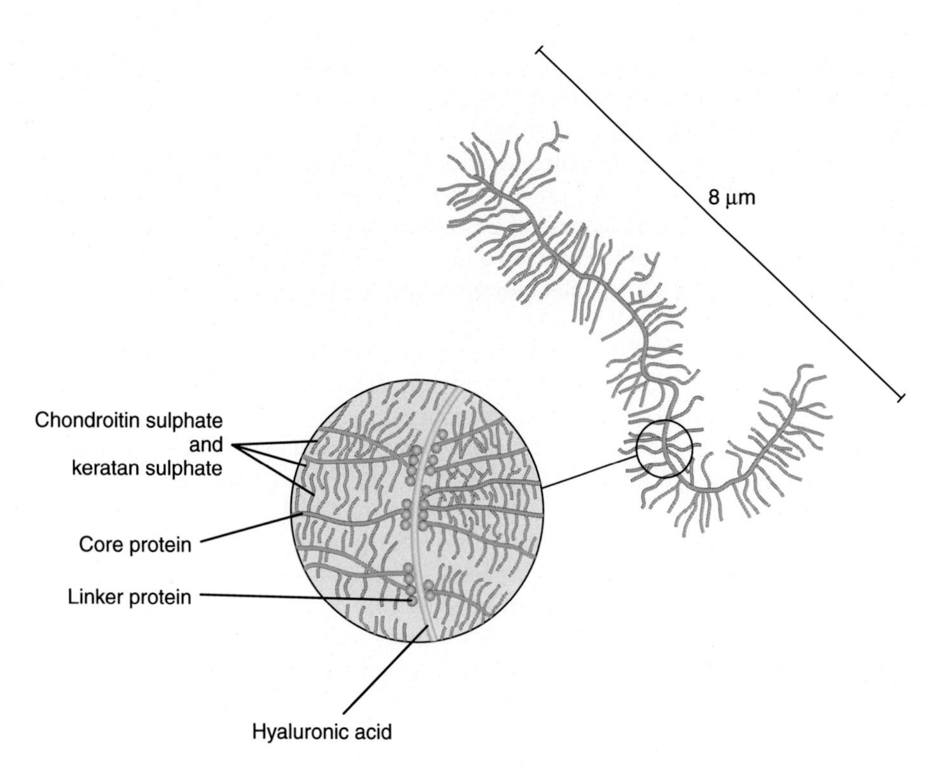

Figure 7–34. **Structure of a large proteoglycan, aggrecan, from cartilage.**

the electron microscope. Proteoglycans have several additional functions apart from their space-filling and mechanical properties. For example, by binding to growth factors such as **fibroblast growth factors** (FGFs) or **transforming growth factor-α** and chemokines, they can regulate the activity/availability of these diffusible signaling molecules. The G3 domain of **versican** harbors EGF-like repeats, which can activate epidermal growth factor receptor (EGFR) signaling and promote cell growth and migration. In contrast, decorin can directly bind to transforming growth factor-β (TGF-β) and inhibit cell growth. Decorin is also involved in the formation of collagen fibers through its ability to bind to collagen. Another proteoglycan called *perlecan* is a critical component of the basement membrane of the kidney where its properties contribute to the filtration of plasma. Some proteoglycans are transmembrane proteins rather than components of the ECM. For example, **syndecans** are integral membrane proteoglycans that contribute to the adhesive properties of focal contacts.

Elastin and Fibrillin Provide Tissue Elasticity

In addition to resisting tensile, torsional, and comprehensive forces, tissues require considerable elasticity, the ability to return to normal shape after being disturbed. This property is particularly important in the skin, lungs, and blood vessels. **Tissue elasticity** resides largely in a network of elastic fibers that are interwoven with collagen fibers. The principal components of **elastic fibers** are the proteins elastin and fibrillin (Fig. 7–35). Elastin is the major component and constitutes up to 50% by weight of large arteries. It contains a series of **hydrophobic domains** that are responsible for its elastic properties and α-helical linker segments, rich in lysine, that are involved in forming cross-links to adjacent molecules. The resulting ECM complex consists of a network that confers on the fibrils five times the extensibility of an elastic band of the same size.

Elastic fibers are covered with a sheath of 10-nm diameter **microfibrils** composed of the protein fibrillin. These

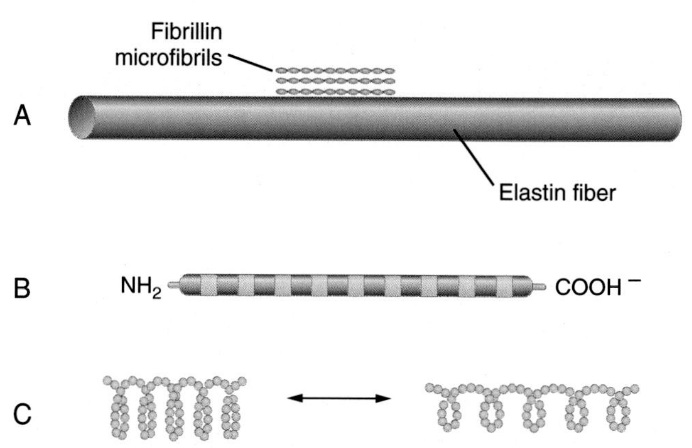

Figure 7–35. **Elastic fibers.** (A) Elastic fibers are composed of a central core of elastin surrounded by microfibrils of fibrillin. The whole is cross-linked by γ-glutamyl-lysine bonds. (B) Elastin molecules consist of tandem repeats of hydrophilic *(purple)* and hydrophobic *(pink)* domains. (C) The hydrophobic domains are responsible for elasticity.

are important for fiber assembly. Mutations in the fibrillin gene cause a human hereditary disease called **Marfan syndrome** in which the integrity of elastic fibers is compromised leading to a rupture of the **aorta** in severe cases. Mutations in the elastin gene cause narrowing of major arteries. On the contrary, elastin overproduction in vascular walls could contribute to the pathogenesis of atherosclerosis. Elastolytic enzymes, such as aspartic protease and MMPs, degrade elastic fibers, which release elastic peptide fragments, such as Gly-Val-Ala-Pro-Gly (VGVAPG). VGVAPG peptides serve as chemotaxin to recruit monocytes and fibroblasts during development of vascular diseases and cancers, resulting in vascular intimal thickening and tumor progression.

Fibronectin Is Important for Cell Adhesion

Fibronectin is the best studied of many noncollagenous ECM proteins that play a role in regulating cell adhesion and cell behavior. There was considerable excitement when it was first discovered because it was found to be substantially less abundant in cultures of certain tumor cells than those of normal cells, suggesting that it might contribute to the lower adhesiveness and metastatic properties of tumors. Meanwhile, fibronectin matrix production is also significantly increased around tumor vasculature that may contribute to tumor progression.

Fibronectin is important in embryonic development where it provides a substratum for guiding gastrulation movements and the migration of neural crest cells. As well as being a component of the ECM, a soluble form of fibronectin is abundant in blood plasma, where it is believed to contribute to blood clotting, wound healing, and **phagocytosis**.

Fibronectin is a dimer that consists of two similar or identical protein chains, each about 200-kDa molecular weight, that are linked together near their COOH-termini by two disulfide bonds (Fig. 7–36). The major structural elements of these chains are three modules of barrel-like repeating units. Distributed along the chain are various sites for interaction with other molecules including domains for heparin, collagen, cell binding, and self-association. The major cell-binding site consists of a tripeptide sequence (Arg-Gly-Asp or RGD in single-letter amino acid code) that is present on an exposed loop extending from one of the type III repeats. This is the site for binding $\alpha_5\beta_1$-integrin, the principal cellular fibronectin receptor. Syndecan is another cell surface receptor family that bind to fibronectin and functions as coreceptors with integrins to enhance integrin-mediated cell migration and intracellular signaling. RGD sequences have subsequently been discovered in other matrix protein, for example, the blood clot protein fibrinogen. Snakes produce an RGD-containing protein, disintegrin, in their venom to prevent blood clotting, and drugs based on RGD peptides have been developed as anticlotting agents.

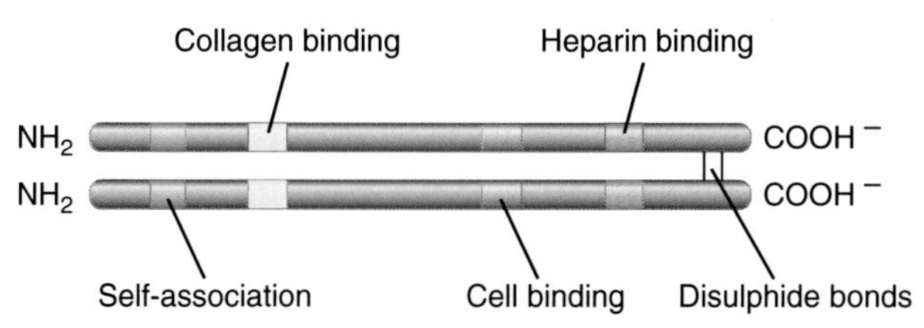

Figure 7–36. Fibronectin dimer showing various binding sites.

Laminin Is a Key Component of Basement Membranes

An important component of basement membranes is the trimeric protein laminin. Laminin consists of three different protein chains—α, β, and γ, the products of different genes. Each is actually a family of genes, and the various α, β, and γ chains (five different α-chains, four β-chains, and three γ-chains) can be combined to give a variety of laminins. So far, 16 different laminin heterotrimers have been identified.

The classical laminin molecule is **laminin-1**. Alternatively, it is termed laminin-111 as it is composed of α1β1γ1 trimeric chains. Laminin-1 consists of an α-chain of about 400 kDa and β- and γ-chains of about 200 kDa each (Fig. 7–37). These chains together form a cross-shaped molecule. Globular domains are at the NH_2-terminal regions of all three chains and at the COOH terminus of the α-chain. The COOH regions of the β- and γ-chains form a coiled-coil α-helical domain that associates with a rodlike region of the α-chain. The NH_2-terminal globular domains of the cross-like structure contain domains for self-association enabling laminin to form a network that is the basis of basement membrane structure. The COOH-terminal globular domain contains a site for cell binding, for example, through $\alpha_6\beta_1$-integrin. Other forms of laminin, such as laminin-5, are composed of three NH_2 terminally truncated chains that do not form a cross but that nevertheless bind to other basement membrane components to form a substratum for hemidesmosomal adhesion.

A number of congenital diseases have been linked with mutations in laminin chains. **Congenital muscular dystrophy type 1A** is caused by mutations in the laminin α2-chain. Mutations in the laminin β2-chain lead to **Pierson syndrome**, which is associated with renal failure and loss of vision. Mutations in all chains of laminin 332 (α3, β3, and γ2) have been found in patients with **junctional epidermolysis bullosa**, whose symptoms include skin fragility and blistering in response to mild mechanical trauma, resulting from the separation of the epidermis and the underlying dermis. Laminin-332 also plays a role in tumorigenesis, which is associated with the activation of PI3K and RAC1 downstream laminin-322 interaction with basement membrane components and cell surface receptors, such as integrins, syndecans, and EGFR.

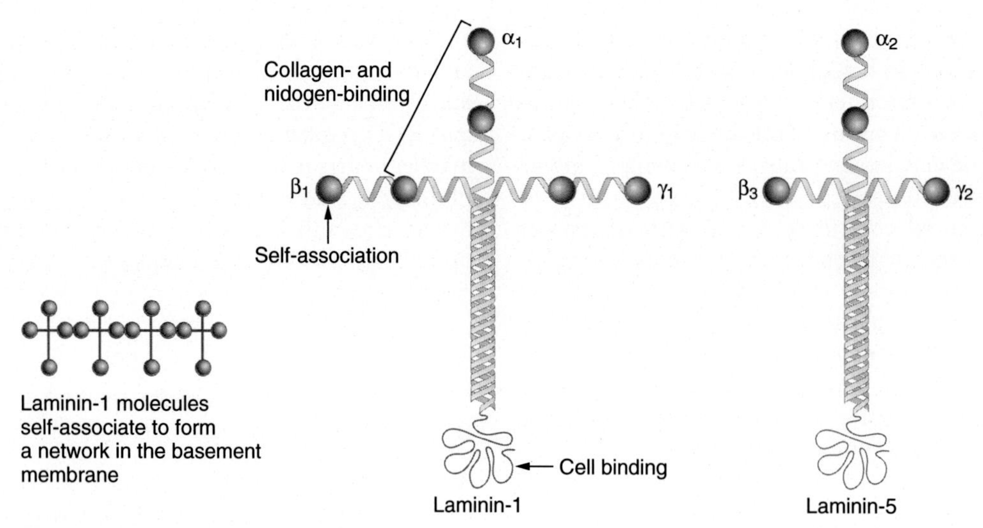

Figure 7–37. Structure of laminin.

Basement Membranes Are Thin Matrix Layers Specialized for Cell Attachment

Basement membranes are thin (50–100 nm), continuous layers of ECM that underlie epithelial and endothelial cell sheets and surround muscle cells, fat cells, and Schwann cells. They appear early during development, functioning to segregate tissues as macromolecular filters, form a substratum for cell attachment, and link cells sheets to the underlying connective tissue. The electron microscope shows two components to basement membranes, a clear or electron lucent layer next to the basal surfaces of the cells and a dark or electron-dense layer beneath this. These layers are called the **lamina lucida** and the **lamina densa**, respectively (see Fig. 7–19).

Basement membranes are highly cross-linked complexes of several proteins and proteoglycans (Fig. 7–38). Constitutive components are **type IV and VII collagen**, laminin, a protein called **nidogen/entactin**, and heparan sulfate proteoglycan. In total, about 50 basement membrane proteins have been identified, of which collagens, especially type IV, constitute 50% of all basement membranes. The basement membranes of different tissues have specific properties in addition to the general requirement for cell attachment. Specificity is conferred by different isoforms of type IV collagen, laminin, and heparan sulfate proteoglycan. Thus 7 different type IV collagens and 12

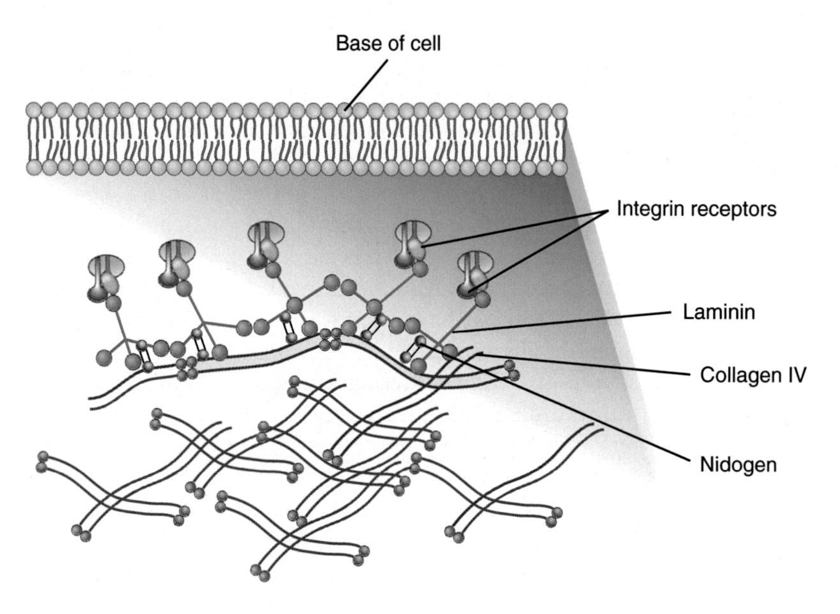

Figure 7–38. Molecular composition of the basement membrane.

different laminins are known. Such specificity is important for regulating the varied functions of different tissues and organs. Of the major basement membrane components, laminin and type IV collagen have the ability to self-assemble into sheetlike structures, whereas the other components do not. Studies on basement membrane assembly by cultured cells indicate that laminin first forms a network that associates with the cell surfaces through integrin adhesion receptors (especially β_1) and dystroglycan, a transmembrane proteoglycan. Type IV collagen forms an independent but associated network, the interaction between the two being facilitated by nidogen/entactin. This complex then forms a scaffolding for the binding of other basement membrane constituents. Type VII collagen then further stabilizes this structure by interacting with type IV collagen, thereby promoting the formation of the basement membrane. The highly cross-linked basement membrane is an efficient barrier to both cellular and molecular traffic between distinct tissue layers. Meanwhile the structure of the basement membrane also enables it to be permissive in certain pathophysiological conditions, such as immune cells infiltration during inflammation.

Basement membranes are the targets of a number of human diseases. Mutations in the gene for the α_5-chain of type IV collagen are associated with **Alport syndrome**, a disease that involves nephritis and deafness. Junctional EB, a severe blistering disease of the skin that results in early infantile death, is associated with mutations in the genes that encode the laminin-5 chains, whereas dystrophic EB, a severely debilitating blistering disease that results in syndactyly, is due to mutations in collagen VII genes and consequent absence of anchoring fibrils. Autoantibodies to type IV collagen α_3-chain, which is present in the glomerular basement membrane of the kidney, are associated with **Goodpasture syndrome**.

A key component of **tumor growth** is **angiogenesis**, the elaboration of new blood vessels. Growing tumors cannot exceed a few millimeters in diameter without acquiring a new blood supply, or they will die of anoxia. Tumor cells produce growth factors that promote angiogenesis. The basement membrane inhibits the proliferation and migration of endothelial cells, thus preventing them from branching out to produce new vessels. During tumor growth, inflammatory and stromal cells within the matrix near the tumor produce matrix metalloproteinases that degrade the vascular basement membrane, thus enabling the endothelial cells to proliferate, migrate, and form new blood vessels to supply the tumor. Research on this process offers hope that inhibition of angiogenesis may be used to prevent tumor growth.

Fibrin Forms the Matrix of Blood Clots and Assembles Rapidly When Needed

A major ECM component of blood clots is a protein called *fibrin* that forms an **elastic network** to which cells and other ECM components bind. Polymerization of fibrin to form the network occurs when its precursor molecule **fibrinogen**, present in substantial quantities (2–4 g/L) in blood plasma, is cleaved by the enzyme **thrombin**. Fibrinogen molecules are elongated structures 45 nm in length that consist of two sets of Aα-, Bβ-, and γ-chains linked by disulfide bonds (Fig. 7–39). Each molecule consists of two outer D domains linked to a central E domain by a coiled-coil segment. Thrombin cleaves a small fragment called *fibrinopeptide A* from the Aα-chains to initiate polymerization. This involves formation of double-standard fibrils by end-to-middle interaction of the D and E

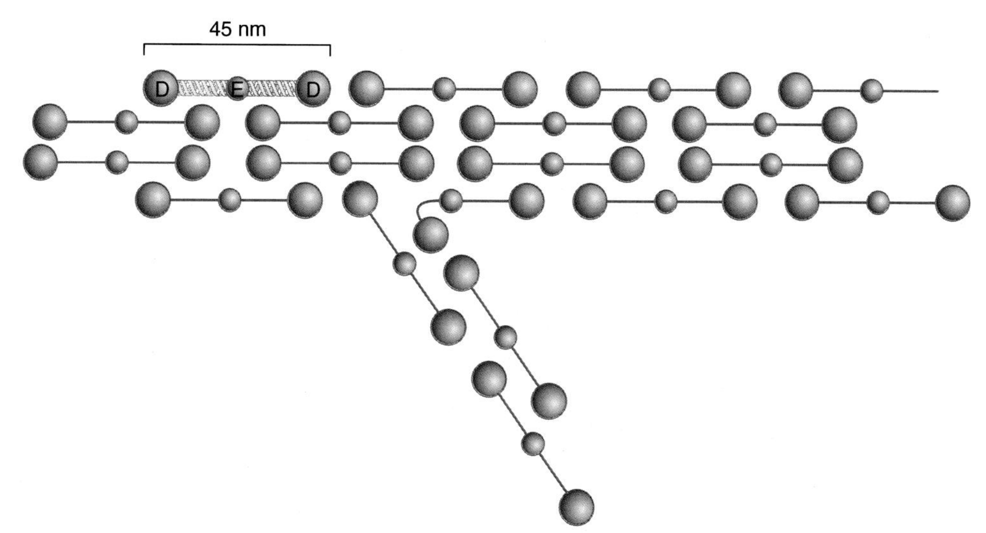

Figure 7–39. Structure of the fibrin molecule and the branching fibers that it forms in blood clots.

domains, as well as lateral and branching fibril associations to form the network of the clot. The network is stabilized by covalent cross-linking through intermolecular **E-(γ-glutamyl) lysine bonds** by the action of an enzyme called *factor XIII* or plasma **transglutaminase**. The cross-linked clot has substantial elasticity and can recover its original form after extension by up to 1.8 times its length. Inhibiting mechanisms exist for both thrombin generation and factor XIII activity so that the clotting process can be regulated.

Fibrin has binding interactions with a variety of extracellular components including the ECM components fibronectin and heparin, the growth factors FGF-2 and vascular endothelial growth factor, and the cytokine **interleukin-1**. It also has binding sites for a number of CAMs including **vascular endothelial (VE)-cadherin, ICAM-1**, the platelet integrin $\alpha_{IIb}\beta_3$, and the leukocyte integrin $\alpha_m\beta_2$ (Mac-1). These are important for promoting angiogenesis, incorporating platelets into the developing **thrombus**, and recruiting **monocytes** and **neutrophils**, respectively. Fibrin and its degradation products could also bind to very-low density lipoprotein receptor (VLDLR) on endothelial surface, which further promotes leukocyte transmigration.

In addition to needing to form quickly when required, clots need to disperse when their function is no longer required or if they form inappropriately. The dispersal process called *fibrinolysis* is mediated by an enzyme called **plasmin** that cleaves fibrin. Plasmin is activated by cleavage of a precursor protein **plasminogen,** through the action of either **tissue-type plasminogen activator or urokinase-type plasminogen activator**, which binds to fibrin and promotes its breakdown.

von Willebrand Factor in Blood Clotting and Angiogenesis

vWF plays a key role in the major response of platelets to vascular injury by mediating the initiation and progression of thrombus formation (Fig. 7–40). Blood flow produced substantial **shear forces** at the blood vessel wall,

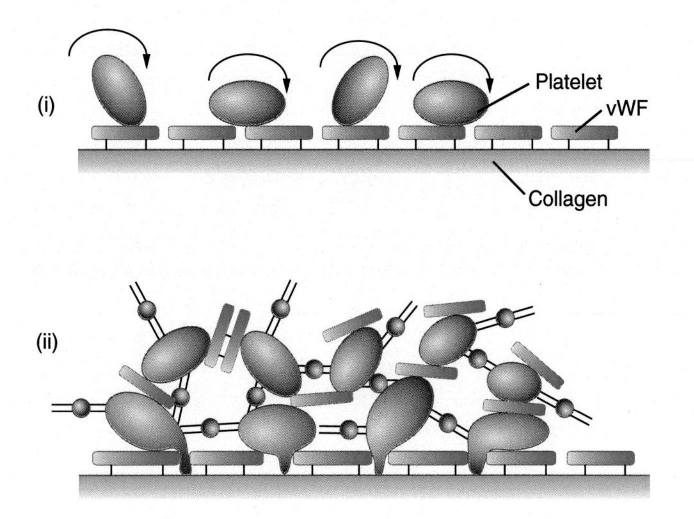

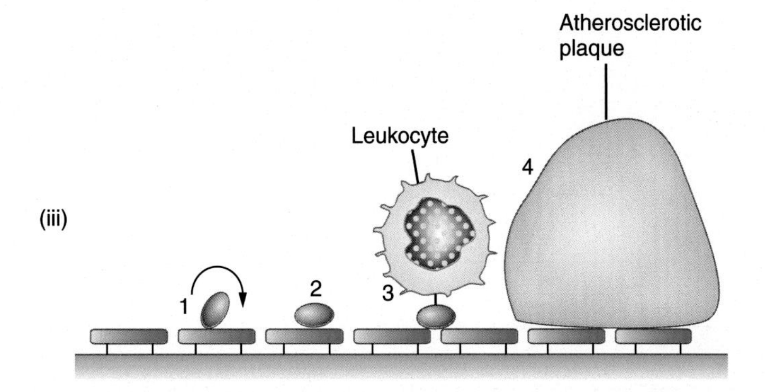

Figure 7–40. **Normal and abnormal adhesion of blood platelets.** (i) von Willebrand factor (vWF) *(blue)* attaches to exposed endothelial basement membrane collagen *(green)*, and platelets adhere loosely, rolling under the force of blood flow. (ii) Platelets form stable adhesions to collagen and more plates attach by aggregation mediated by vWF and fibrinogen *(brown)*. The thrombus then continues to grow. (iii) Platelets may adhere to the surfaces of endothelial cells initiating the formation of a thrombus by recruitment of other plates and leukocytes and, eventually, development of atherosclerotic plaque.

and these forces oppose cell adhesion. vWF forms a bridge between collagen in the vessel wall and blood platelets sufficient to enable cell adhesion to develop.

Mature vWF is a multimeric protein that consists of a variable number of identical subunits linked together by disulfide bonds. Each precursor subunit is a protein chain of 2050 amino acids with a substantial number of carbohydrate chains bound to it. These subunits become linked together by disulfide bonds between their COOH-termini to form dimers of about 500-kDa molecular weight. Further assembly requires cleavage of a propeptide from the NH_2 terminus by an enzyme **furin.** The NH_2 termini can then also become linked together by disulfide bonds. Such linking may proceed until multimers exceeding 10,000 kDa are formed. Electron microscopy shows the largest such multimers to be up to 1300 nm in length and 200–300 nm in cross section. vWF multimers are synthesized intracellularly and stored in Weibel-Palade bodies in endothelial cells and α-granules in platelets and megakaryocytes (large cells that give rise to platelets). Some vWF is **secreted constitutively** by endothelial cells, giving a residual plasma concentration of this protein. **Regulated secretion** by endothelial cells and platelets occurs in response to vascular injury. The vWF monomer has binding sites for collagen and the platelet adhesion receptors GPIb (part of the GPIb-IX-V complex) and GPIIbβ_3, as well as for collagen and the blood-clotting protein factor VIII. Thus the multimeric complexes are literally strings bristling with binding sites. The larger the multimers, the more effective they are at promoting thrombus formation.

Initial platelet adhesion is mediated by the binding of GPIb on the platelet to vWF, which is, in turn, bound to collagen. This attachment is easily broken and does not result in firm adhesion. Instead, platelets exhibiting this type of adhesion roll along the surface under the force of blood flow. It is possible that binding between selectin molecules on the platelet surface and the carbohydrate chains on vWF also contribute to initial adhesion.

GPIb-vWF interaction generates an intracellular signal within the platelets that involve changes in intracellular Ca^{2+} concentration and the signaling enzyme **protein kinase C**. The function of these signals is to bring about activation of the platelet integrin GPIIbβ_3 that then mediates firm adhesion to vWF.

Because it has multiple binding sites for platelet adhesion molecules, vWF can also mediate platelet aggregation by bridging between them. Platelets also adhere to fibrinogen and fibrin, other important participants in the clotting process. It appears that vWF and fibrin have complementary roles in thrombus formation. Thus vWF mediates rapid thrombus formation at high shear rates in the absence of fibrinogen, but the thrombi are unstable. In the presence of fibrinogen, thrombus development is slower but more stable. Thus patients having congenital defects in either vWF (von Willebrand disease) or fibrinogen have clotting disorders.

As seen with fibrin the process of thrombus formation by vWF requires regulation. This is done extracellularly by a plasma enzyme called **ADAMTS13**, which cleaves vWF multimers to restrict their size and possibly to prevent excessive thrombus formation. The normal function of ADAMTS13 is clearly important because mutations in the gene for this enzyme cause a disease called **chronic relapsing thrombocytopenic purpura.** It is important to understand the mechanisms involved in thrombus because it is likely that their abnormal function cause thrombocytic diseases such as **stroke, coronary thrombosis, phlebitis**, and **phlebothrombosis.**

Some patients with von Willebrand disease also have symptom of gastrointestinal vascular malformations, which leads to digestive tract bleeding in these patients. As a key regulator of hemostasis, VWF predominantly inhibits blood vessel formation. Hyperactivated VEGFR-2 signaling found in vWF-deficient cells indicates that vWF may control angiogenesis by maintaining VEGF signaling at physiological levels, thereby preventing formation of unstable, fragile, and leaky vessels due to dysregulated VEGFR-2 signaling.

Stem Cell Interaction With ECM Is Critical for Cell Stemness and Plasticity

Stem cells are a cell population who has extensive ability to self-renew and to differentiate daughter cells into specialized cell types with distinct functions. Embryonic stem cells can differentiate into all cell types that makes up the body, while adult stem cells, such as epithelial stem cells, intestine stem cells, hematopoietic stem cells, and mesenchymal stem cells (MSCs), have more defined differentiation potential. Stem cell renewal and differentiation are regulated by their local microenvironment, known as **niche**, in which ECM is a key component. Interactions between stem cells and ECM influence the stem cell migration, proliferation, survival, and differentiation. A complete discussion of stem cells and regenerative medicine can be found in Chapter 13. Here, we focus on stem cell-ECM interactions.

ECM stiffness and topography have a role in controlling stem cell fate. The spatial distribution of the ECM components participate in determining division axis orientation of epithelial stem cells, by remodeling the actin cytoskeleton. A stem cell with elongated shape is less likely to initiate differentiation than if it is in a circular island shape. Spreading MSCs tends to differentiate into osteoclasts and chondrocytes, while MSC rounding favors adipogenesis. Moreover, increased stiffness of ECM surrounding MSCs promotes bone differentiation, whereas compliant ECM in MSC niche enhances adipocyte differentiation. Mechanical forces, such as fluid flow stress, have been shown to promote more rapid differentiation of MSCs into chondrocytes. Mammary epithelial stem cells predominantly differentiate into basal epithelial cells when ECM stiffness increases, but aging decreases their mechanosensitivity and differentiation potential.

Stem cell-ECM interactions are largely mediated by integrins. Integrin interactions with the ECM, along with signals initiated by mechanical force and other ECM receptors within the niche, are critical for balancing stem cell self-renewal and differentiation. Binding of integrin to basement membrane constituents (e.g., laminin, fibronectin, and collagen IV) promotes asymmetric cell division in many stem cell types. The integrin α6/ITGA6/CD49f, which binds to laminin, is expressed in multiple adult stem cells types and is recognized as a marker for different cancer stem cells. Integrin-mediated mechanotransduction senses biophysical cues from ECM and modulates cellular responses by altering cytoskeleton and changing gene transcription. Rho-ROCK signaling activation in response to stiff ECM promotes osteogenesis in MSCs and favors neural stem cell differentiation into astrocytes over neuronal differentiation. In addition to the integrins and laminin, other critical ECM components mediating cell-ECM interaction, such as CD44 and Robo4, have critical roles for the hematopoietic stem cells homing and settling in the niche during transplantation.

The transcriptional changes, orchestrated by transcription factors YAP/TAZ and myocardin-related transcription factors (MRTFs)/serum response factor (SRF), have been shown to have increasing importance in controlling the cell stemness and plasticity. Decreased G-actin/F-actin ratio due to enhanced F-actin assembly releases cytoplasm-sequestered MRTF for nuclear translocation, where it acts as cofactor of SRF for regulating transcription. MRTF/SRF signaling is required for hematopoietic stem cells to colonize in bone marrow during development. YAP is activated in response to tension of the actomyosin cytoskeleton in a Rho GTPase-dependent manner. YAP/TAZ activation contributes to stemness maintenance in adult stem cells, such as intestinal stem cells and neural stem cells, and in certain cancer stemlike cells. Activated YAP/TAZ could also support the stem cell niche by promoting of laminin-511 matrix deposition, which depends on increased transcription of laminin α5-subunit by YAP/TAZ.

ECM in Pathogenesis and Treatment of Human Diseases

Given the crucial role of the ECM in maintaining tissue homeostasis, dysregulation of ECM has been linked with a variety of pathological conditions. Mutations in ECM component genes, such as *COL1A1* (collagen I α-chain), could affect the architecture of the ECM, which leads to bone formation defects. Increased ECM breakdown by hyperactivation of ECM-destruction proteinases, such as MMPs and ADAMs, decreases contractility of cardiomyocytes and destructs cartilage in the bone, resulting in cardiomyopathy and osteoarthritis, respectively. On the contrary, overproduction and excessive deposition of ECM lead to fibrosis. Fibrotic ECM can stimulate fibroblasts to further increase ECM production and exacerbate fibrosis. Severe fibrosis causes organ failure and increases risk of cancer development, such as those in the liver, lung, and breast.

Balanced ECM is essential for upholding tissue architecture and cell polarity, which regulate cell proliferation and apoptosis and restrain malignant tumor progression. Various stromal cells within the tumor microenvironment, such as cancer-associated fibroblasts, endothelial cells, immune cells, and tumor cells themselves, play critical roles in the formation of tumor ECM that promotes tumorigenesis. ECM stiffening, caused by overexpression of ECM component genes (e.g., collagen IV) and increased cross-linking, promotes tumor cell survival and facilitates metastatic tumor cell growth in the metastatic niche. Increased ECM stiffness promotes β1-integrin clustering, focal adhesion formation, and PI3K signaling through mechanotransduction, which drive invasion and tumor progression. In tumors from breast cancer patients, increased expression of protease inhibitors correlates with good prognosis, whereas high levels of MMPs are associated with poor prognosis and increased recurrence. ECM also modulates angiogenesis to promote tumor growth. Besides promoting pancreatic cancer cell survival and proliferation, tenascin C within ECM also promotes angiogenesis and enhances blood vessel permeability. Regulating immune and cancer cell migration is another crucial mechanism by which cancer cell ECM promotes tumor progression. Immune cells interact with ECM through integrin-mediated adhesion that regulates the recruitment and activation of immune cells in the tumor microenvironment. The composition and architecture of the ECM fibers also controls migration of immune cells and tumor cells. Loose ECM areas facilitate cell motility, whereas dense ECM areas impede migration. Confined ECM geometries also control the cells ability to deform and promotes bleb migration patterns of tumor cells at the edge of tumor mass. Additionally, ECM also serves as a reservoir for growth factors, such as VEGF, PDGF, and TGF-β, whose release due to ECM remolding further modulates tumor growth and progression.

Since ECM is substantially modified in various diseases, including atherosclerosis, autoimmune, inflammatory diseases, and cancer, directly targeting disease-associated ECM molecules is an attractive therapeutic strategy. Interference with the hyaluronic acid/CD44 signaling pathway and using CD44 or hyaluronic acid as cytotoxic drug targeting strategies have been examined in numerous targeting therapeutic approaches for cancer treatment. ECM is also a useful therapeutic target for improving drug delivery and efficacy, considering its ability to regulate drug transport and delivery. One strategy is to normalize vasculature through modulating ECM to improve anticancer drug distribution.

SUMMARY

Cell adhesion and the ECM are fundamental to the normal structure and function of human tissues. Adhesion is mediated by molecules that mostly belong to one of four families of adhesion receptors, the cadherins, the Ig family, the selectins, and the integrins. Adhesion receptors are commonly clustered in cell junctions; desmosomes and adherens junctions mediate cell-cell adhesion, whereas hemidesmosomes and focal contacts mediate cell-matrix adhesion. Intercellular junctions include tight junctions that regulate the paracellular permeability across the epithelium and determination of cell polarity and gap junctions that facilitate intercellular communication. In addition to participating in cell adhesion, adhesion receptors transduce signals that regulate many aspects of cell behavior including movement, proliferation, differentiation, and survival. Cell adhesion is a dynamic process that is particularly evident where nonadhesive cells rapidly become adhesive, for example, leukocytes in inflammation and platelets in blood clotting.

The ECM has many components of which the most abundant are fibrillar collagens. These provide the strength of tendons and the dermis and form the basis of bone and cartilage. Much of the bulk of tissues resides in GAGs and proteoglycans, negatively charged polymers that absorb water and resist compression especially in cartilage. Tissue elasticity is dependent on elastic fibers composed of elastin and fibrillin. The cells of many tissues, such as epithelia, adhere to basement membranes, the principal components of which are laminin and type IV collagen. Basement membranes of different tissues have specific properties dependent on differences in molecular composition. Most extracellular matrices are semipermanent in nature, but blood clots form rapidly in response to injury. Key components of the clot matrix are vWF and fibrin, both of which provide a substratum for platelet adhesion. All types of cell adhesions and the corresponding cell functions are regulated by intracellular signaling. Dysregulation of different cell adhesions by genetic, chemical, or environmental factors leads to pathophysiologic conditions and is associated many major human diseases.

Suggested Readings

Anderson JM, Van Itallie CM. Physiology and function of the tight junction. *Cold Spring Harb Perspect Biol* 2009;1:a002584, https://doi.org/10.1101/cshperspect.a002584.

Bonnans C, Chou J, Werb Z. Remodeling the extracellular matrix in development and disease. *Nat Rev Mol Cell Biol* 2014;15:786–801.

Buckley A, Turner JR. Cell biology of tight junction barrier regulation and mucosal disease. *Cold Spring Harb Perspect Biol* 2018;10(1) https://doi.org/10.1101/cshperspect.a029314.

Cooper J, Giancotti FG. Integrin signaling in cancer: mechanotransduction, stemness, epithelial plasticity, and therapeutic resistance. *Cancer Cell* 2019;35:347.

Kappelmayer J, Nagy Jr. B. The interaction of selectins and PSGL-1 as a key component in thrombus formation and cancer progression. *BioMed Res Int* 2017; https://doi.org/10.1155/2017/6138145. Article ID 6138145.

Mège RM, Ishiyama N. Integration of cadherin adhesion and cytoskeleton at adherens junctions. *Cold Spring Harb Perspect Biol* 2017;9: a028738 https://doi.org/10.1101/cshperspect.a028738.

Mouw JK, Ou G, Weaver VM. Extracellular matrix assembly: a multiscale deconstruction. *Nat Rev Mol Cell Biol* 2014;15:771–785.

Multhaupt HA, Leitinger B, Gullberg D, Couchman JR. Extracellular matrix component signaling in cancer. *Adv Drug Deliv Rev* 2016;97:28–40.

Panciera T, Azzolin L, Cordenonsi M, Piccolo S. Mechanobiology of YAP and TAZ in physiology and disease. *Nat Rev Mol Cell Biol* 2017;18:758–770.

Skerrett IM, Williams JB. A structural and functional comparison of gap junction channels composed of connexins and innexins. *Dev Neurobiol* 2017;77(5):522. https://doi.org/10.1002/dneu.22447.

Spindler V, Waschke J. Pemphigus—a disease of desmosome dysfunction caused by multiple mechanisms. *Front Immunol* 2018;9:136.

Venhuizen JH, Zegers MM. Making heads or tails of it: cell–cell adhesion in cellular and supracellular polarity in collective migration. *Cold Spring Harb Perspect Biol* 2017;9: pii: a027854.

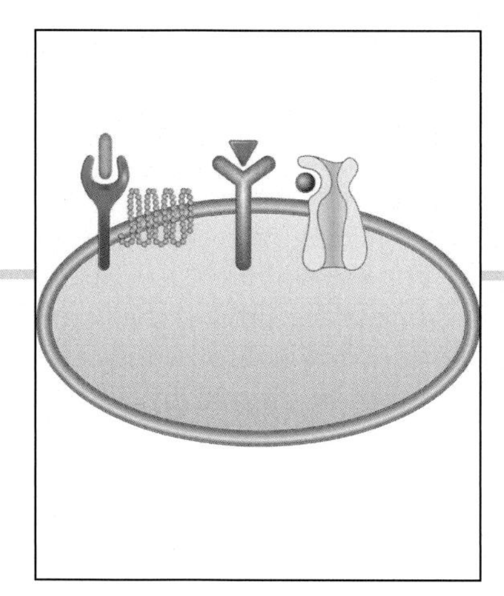

Chapter 8

General Modes of Intercellular Signaling

INTERCELLULAR SIGNALING MOLECULES ACT AS LIGANDS

Regardless of chemical structure, all cell-to-cell signaling molecules act as **ligands** and initiate biological responses in target cells by binding to specific **receptors** (Fig. 8–1). Molecules that are too large or too hydrophilic to cross the plasma membrane use receptors at the plasma membrane to relay the signal to the inside of the target cell. Plasma membrane receptors contain an extracellular ligand-binding domain, a transmembrane domain, and an intracellular domain that initiates the intracellular events that lead to a biological response. These plasma membrane receptors can be ion channels, enzymes, or linked to enzymes. Smaller hydrophobic ligands diffuse across the plasma membrane and activate receptors located inside the target cell. These intracellular receptors are most often transcription factors that initiate changes in gene transcription. Intercellular signaling ligands and receptors are expressed at low levels and are not amenable to isolation and characterization using traditional biochemical approaches. Recent advances in molecular biology and pharmacology have circumvented these issues and significantly increased the rate at which intercellular ligands and receptors are identified/characterized. This information has provided new insights regarding the complex cellular and molecular events that cells use to respond to their ever-changing environment.

CELLS EXHIBIT DIFFERENTIAL RESPONSES TO SIGNALING MOLECULES

Cells must be able to selectively respond to some signals and at the same time disregard other signals. Differential responses can be attributed to variations in the combination of ligands, receptors, and/or intracellular signaling pathways involved. If the ligand is absent or present in reduced quantities, there will be no response. Even if the ligand is present in high concentrations, there will be no response if the target cell does not express the appropriate receptor. Furthermore the same ligand does not necessarily produce the same biological effect in all cells. For example, acetylcholine regulates contraction in the heart and skeletal muscle but regulates secretion in the salivary gland (Fig. 8–2).

Ligand interaction with a different receptor subtype or coupling of the same receptor to different intracellular signaling pathways can generate differential responses. Finally, it is the sum of the cell's response to all of the

Goodman's Medical Cell Biology. https://doi.org/10.1016/B978-0-12-817927-7.00008-9

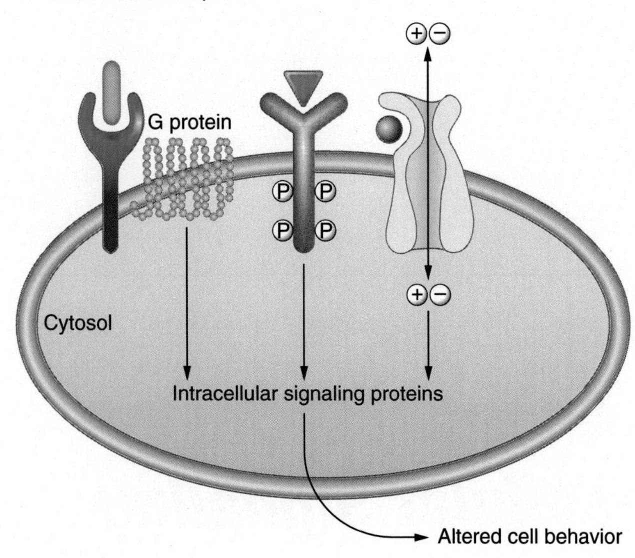

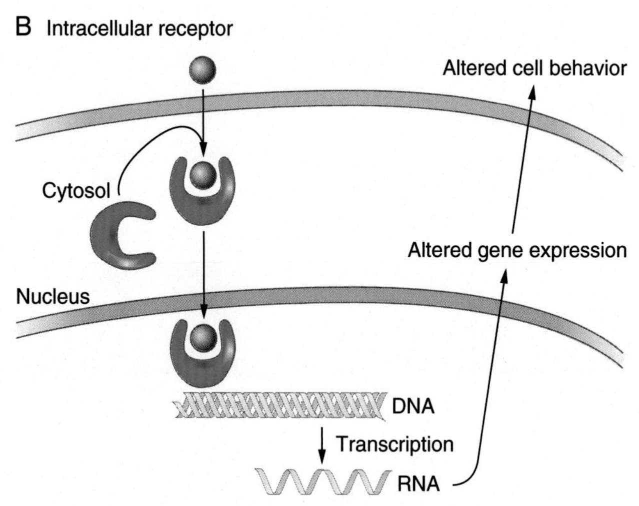

Figure 8–1. **Intercellular signaling molecules are ligands and exert their effects via interaction with specific target cell receptors.** Signaling molecules that are large and hydrophilic cannot enter the target cell by diffusion and exert their effects via interaction with cell surface receptors (A). These receptors can be ion channels, G-protein coupled, or enzyme coupled. Receptor-mediated changes in downstream second messenger pathways alter cell behavior. Small/hydrophobic signaling molecules readily diffuse into the target cell and interact with receptors located inside the cell (B). These ligand-receptor complexes then bind to regulatory regions in DNA and promote the transcription of new gene products that alter cell behavior.

individual ligands it is exposed to that determines what alteration in cell behavior occur. Highly divergent responses can be observed with only subtle changes in the combination of ligands or in the physiological state of the target cell or in both.

INTERCELLULAR SIGNALING MOLECULES ACT VIA MULTIPLE MECHANISMS

As our understanding of the events involved in intercellular signaling has increased, the distinctions between classes of intercellular signaling molecules have become blurred. Traditionally, signaling molecules have been classified as endocrine, paracrine, autocrine, or juxtacrine, depending on the distance over which they act and the source of the ligand (Fig. 8–3).

However, individual signaling molecules can act via multiple mechanisms. For example, epidermal growth factor (EGF) is a transmembrane protein that can bind to and signal a neighboring cell via direct contact (juxtacrine signaling). It can also be cleaved by a protease, released into the circulation, and act as a hormone (endocrine signaling). Epinephrine functions as a neurotransmitter (paracrine signaling) and a systemic hormone (endocrine signaling). This broad diversity in ligand structure/biochemistry and ligand synthesis/distribution/metabolism, as well as target cell receptors, allows only a hundred intercellular signaling molecules to generate an unlimited number of signals. In the following sections, examples of clinically relevant intercellular signaling pathways are used to present the unique cell biological properties of the different classes of intercellular signaling molecules.

HORMONES

Hormones allow organisms to coordinate the diverse activities of cells over long distances. Specialized endocrine glands (pituitary, thyroid, parathyroid, pancreas, adrenal glands, and gonads) and other organs secrete/release a long list of chemically diverse hormones (Table 8–1). These hormones enter the circulation and act on target cells located throughout the body. Because hormones distribute throughout the body, they can induce simultaneous changes in many different organs. However, the long distances that they travel to reach their target cells limit their actions to the regulation of physiological processes—reproduction, growth and development, and metabolism—with a timescale of minutes to years. Hormones can be grouped into two distinct classes: small lipophilic molecules that interact with **intracellular receptors** and hydrophilic molecules that interact with **cell surface receptors.** The distinctive features of each class are described in the following section. Finally the hypothalamic-pituitary axis is used to illustrate the interplay between these two classes of hormones and the complex positive and negative feedback mechanisms that fine-tune hormone function.

Lipophilic Hormones Activate Cytosolic Receptors

The small lipophilic hormones control a diverse list of biological processes. Sex hormones such as progesterone, estradiol, and testosterone are produced by the gonads and regulate sexual differentiation and function. Corticosteroids are synthesized by the adrenal gland and divided

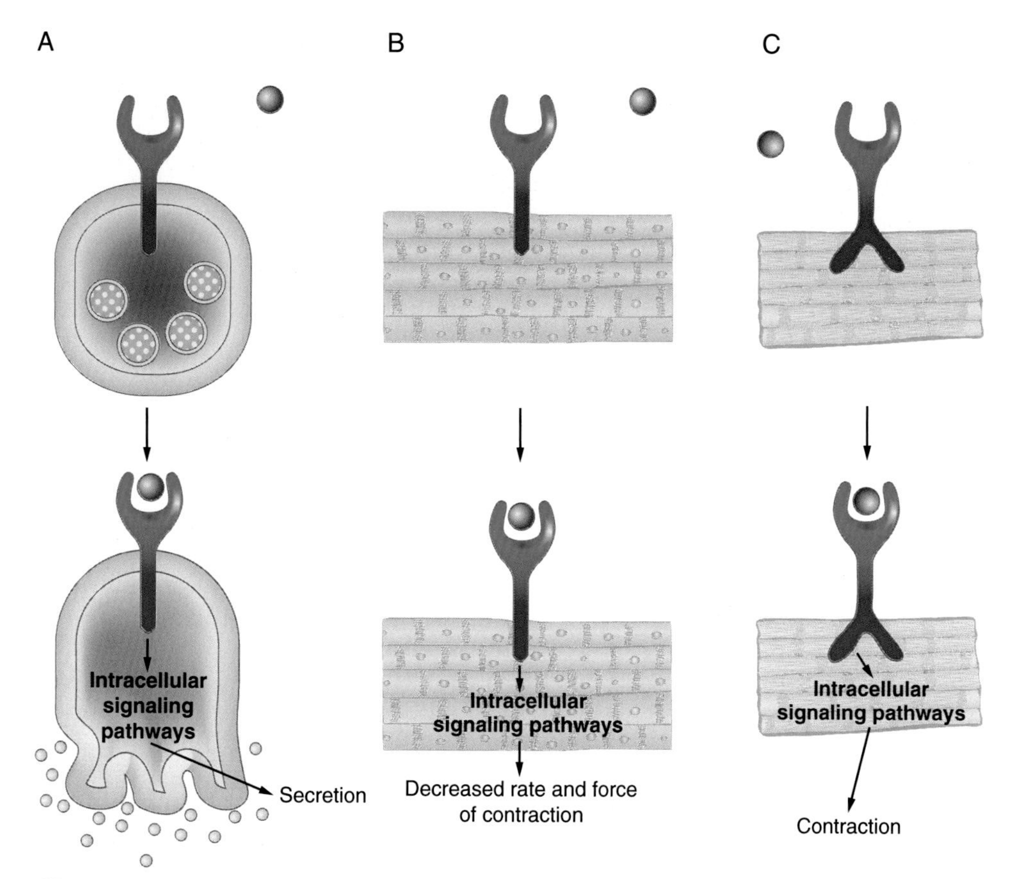

Figure 8–2. **Signaling molecules are versatile and induce differential responses.** In the salivary gland (A), acetylcholine activates a muscarinic receptor subtype, resulting in secretion. In the heart muscle (B), acetylcholine activation of the same muscarinic receptor subtype has a different biological effect, decreased rate, and force of contraction. The differential effects on cell behavior are due to the coupling of the muscarinic receptor to different intracellular signaling pathways in the two cell types. In skeletal muscle (C), acetylcholine activates a different receptor subtype, the nicotinic receptor, resulting in depolarization of the muscle cell and contraction.

into two groups based on function: Glucocorticoids increase glucose production in many different cell types, and mineral corticoids regulate salt and water balance in the kidney. Thyroxine is synthesized in the thyroid gland and regulates metabolism in virtually every organ. Vitamin D_3 regulates Ca^{2+} metabolism and bone growth. Retinoic acids and other retinoids play important roles in development. All lipophilic hormones freely diffuse across the lipid bilayer, interact with cytosolic receptors, and alter the expression of specific genes. It is the particular genes that are activated/inactivated that determine the ultimate effect of the hormone on the target cell/tissue.

The time frame via which lipophilic hormones act is determined by their synthesis and metabolism. All steroids are synthesized from cholesterol and have similar chemical backbones. Steroid-producing cells store only a small supply of hormone precursor; they do not store the mature active hormone. When stimulated, cells convert the precursor to the active hormone, which diffuses across the plasma membrane and enters the circulation. This process can take several hours to several days. Because their solubility in the aqueous environment is poor, steroids are tightly bound to carrier proteins in the bloodstream. This dramatically slows their degradation and allows steroid hormones to remain in the circulation for hours to days. Thus, once a steroid-hormone response develops, it can persist for a prolonged period.

Although structurally related to the steroid hormones, retinoids are synthesized from retinol (vitamin A), not cholesterol. Retinol is found in high concentrations in the liver and in the bloodstream, where it is complexed with serum-binding proteins. Retinol diffuses across the plasma membrane and forms a complex with a cytosolic retinol-binding protein. Cytosolic retinol is converted to retinoic acid by a series of dehydrogenases. The newly synthesized retinoic acid exits the cell by diffusion and acts on neighboring cells. Retinoic acid is unique in that it can remain in the cytosol and signal within the synthesizing cell.

Thyroid hormone synthesis is unusual and complex. The thyroid gland stores large amounts of thyroid hormone as amino acid residues on a large, multimeric,

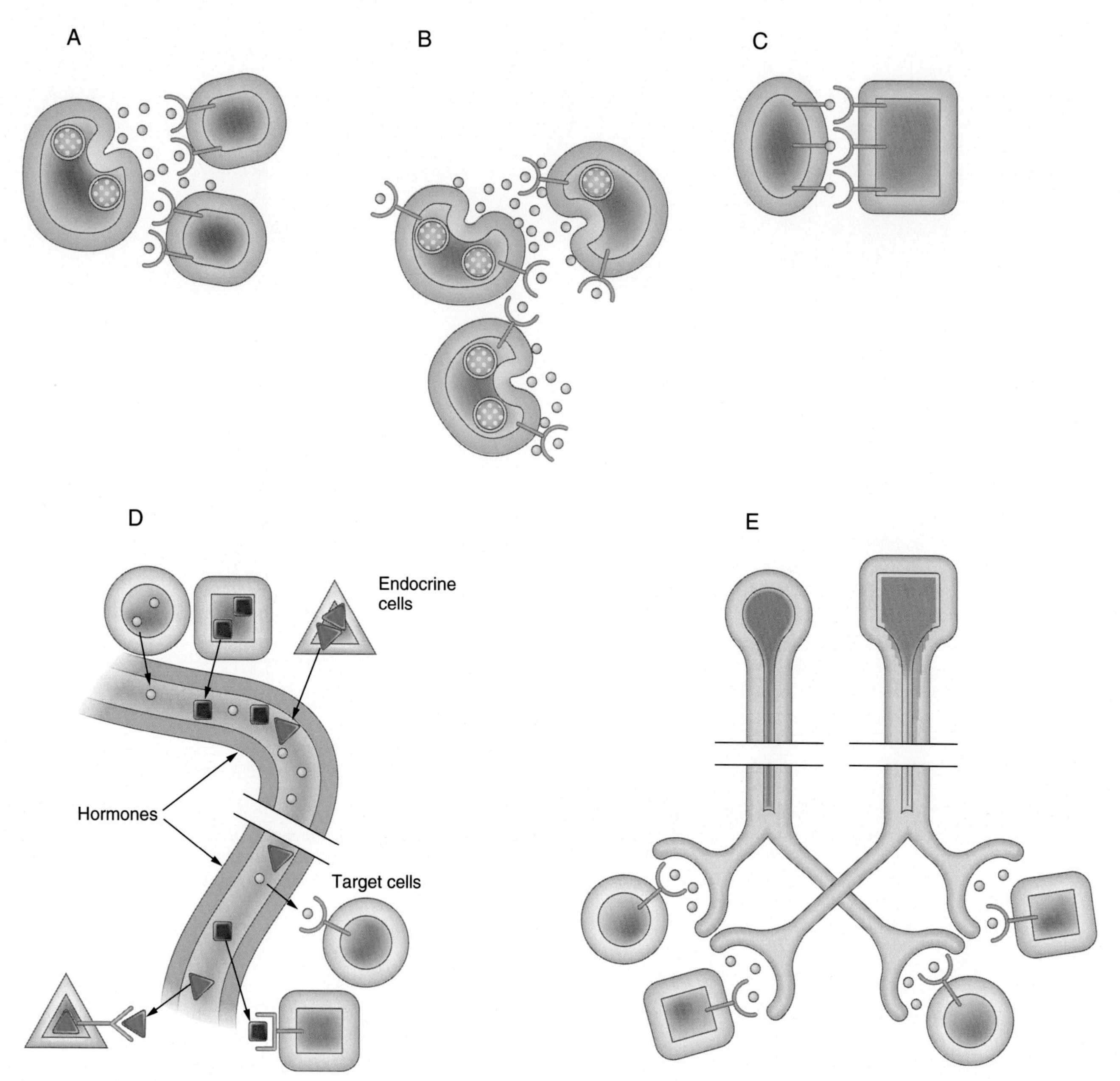

Figure 8–3. **General schemes of intercellular signaling.** Cell-to-cell signaling can occur over short (A–C) or long distances (D, E). In paracrine signaling (A), chemicals are released into the extracellular environment and exert their effects on neighboring target cells that express the appropriate receptor. In autocrine signaling (B), the cell that synthesizes/releases the signaling molecule is also the target cell. In juxtacrine signaling (C), the signaling molecule remains attached to the plasma membrane and interacts with receptors on adjacent target cells. In endocrine signaling (D), hormones are released into the circulation and distribute throughout the body but alter only the behavior of cells that express the appropriate receptor. Synaptic signaling (E) is a specialized form of paracrine signaling that occurs over long distances because the signal is transmitted along neuronal cell processes that can span the entire length of the organism. Specificity in synaptic signaling is generated by the formation of synaptic contacts and not the signaling molecule/neurotransmitter.

extracellular glycoprotein called **thyroglobulin.** Thyroglobulin is endocytosed at the apical membrane and appears as colloidal droplets within the cell. These droplets fuse with lysosomes where the thyroglobulin is degraded and thyroxin (T_4) and T_3 (3,5,3′-triiodothyronine) are liberated. This process is extremely inefficient and can take hours to days. The liberation of two to five T_4 molecules requires the complete breakdown of 1 thyroglobulin molecule—approximately 5500 amino acid residues and 300 carbohydrate residues. The T_3 and T_4 then diffuse across the basement membrane and enter the circulation. Because their solubility in the aqueous environment of the blood is low, T_3 and T_4 are tightly bound to carrier proteins. This inhibits their degradation and allows responses to persist for hours to days.

TABLE 8–1. Major Biological Activities of Representative Hormones

Hormone	Site of Origin	Major Biological Activity
Proteins/polypeptides		
Insulin	Pancreatic β cells	Carbohydrate utilization
Growth hormone–releasing hormone (GHRH)	Hypothalamus	Stimulate growth hormone secretion
Growth hormone	Anterior pituitary	General stimulation of growth
Luteinizing hormone (LH)	Anterior pituitary	Stimulate LH secretion
Parahormone	Parathyroid	Increased bone resorption
Follicle-stimulating hormone (FSH)	Anterior pituitary	Stimulate ovarian follicle growth and spermatogenesis
Thyroid-stimulating hormone (TSH)	Anterior pituitary	Stimulate thyroid hormone secretion
Erythropoietin	Kidney	Increase red blood cell production
Prolactin	Anterior pituitary	Stimulate milk production
Glucagon	Pancreas	Stimulate glucose synthesis
Insulin-like growth factor-1	Liver	Stimulate bone and muscle growth
Small peptides		
Somatostatin	Hypothalamus	Inhibit growth hormone release from anterior pituitary
TSH-releasing hormone (TRH)	Hypothalamus	Stimulate TSH release from anterior pituitary
LH-releasing hormone (LRH)	Hypothalamus	Stimulate LH secretion from anterior pituitary
Vasopressin/antidiuretic hormone (ADH)	Posterior pituitary	Elevate blood pressure, increase kidney water resorption
Oxytocin	Posterior pituitary	Stimulate smooth-muscle contraction
Amino acids		
Norepinephrine	Adrenal medulla	Increase blood pressure and heart rate
Dopamine	Hypothalamus	Inhibit prolactin secretion
Lipophilic hormones		
Estradiol	Ovary, placenta	Develop/maintain secondary male sex characteristics
Cortisol	Adrenal cortex	Metabolism, suppress inflammatory reactions
Progesterone	Ovary, placenta	Prepare uterus for pregnancy, maintain pregnancy
Testosterone	Testis	Develop/maintain secondary male sex characteristics
Thyroxine	Thyroid	Increase metabolic activity in many cells
Retinoic acid	Diet	Epithelial cell differentiation

Receptors for Lipophilic Hormones Are Members of the Nuclear Receptor Superfamily

Although the receptors for lipophilic hormones are not identical, they are evolutionarily related and belong to a large superfamily called the **nuclear receptor superfamily.** This superfamily also includes receptors that are activated by intracellular metabolites. Family members that have been identified by only DNA sequencing and for which ligands have not been identified are referred to as **orphan receptors**. Members of the nuclear receptor family contain related domains for ligand binding, DNA binding, and transcriptional activation.

Lipophilic hormone receptors are located in the cytosol, nucleus, or both, where are they are complexed with other proteins (Fig. 8–4). After ligand binding the complex dissociates and the receptor dimerizes, undergoes phosphorylation, binds DNA, and induces the transcription of specific target genes. Inactive thyroid hormone receptors are located in the nucleus and complexed with DNA in a conformation that inhibits transcription (see Fig. 8–4). The ligand receptor complex remains associated with the DNA and undergoes a conformational change that results in activation of transcription. When the ligand dissociates, the receptor is dephosphorylated and returns to its inactive state/location. Initially, these receptors directly activate the transcription of a small number of specific genes. These primary phase gene products activate other genes to produce a delayed secondary response. Although the activity of these receptors has become synonymous with transcription, increasing evidence exists that some effects of small hydrophobic hormones may be due to direct regulation of cellular processes without genomic effects.

Peptide Hormones Activate Membrane-Bound Receptors

Peptide hormones range in size from a simple tripeptide (thyrotropin-releasing hormone (TRH)) to a 198-amino acid protein (prolactin), to a glycosylated multisubunit oligomer: human chorionic gonadotropin (HCG). Because these agents mediate rapid responses to the environment, they are stored in secretory vesicles adjacent to the plasma membrane and are available for immediate release. The synthesis and release of peptide hormones occurs via a regulated exocytotic pathway (see Chapter 4). The environmental signals that trigger peptide hormone release also stimulate synthesis of peptide hormones to ensure that the released hormone is replaced. Released hormones are in the blood for only a few seconds or minutes before

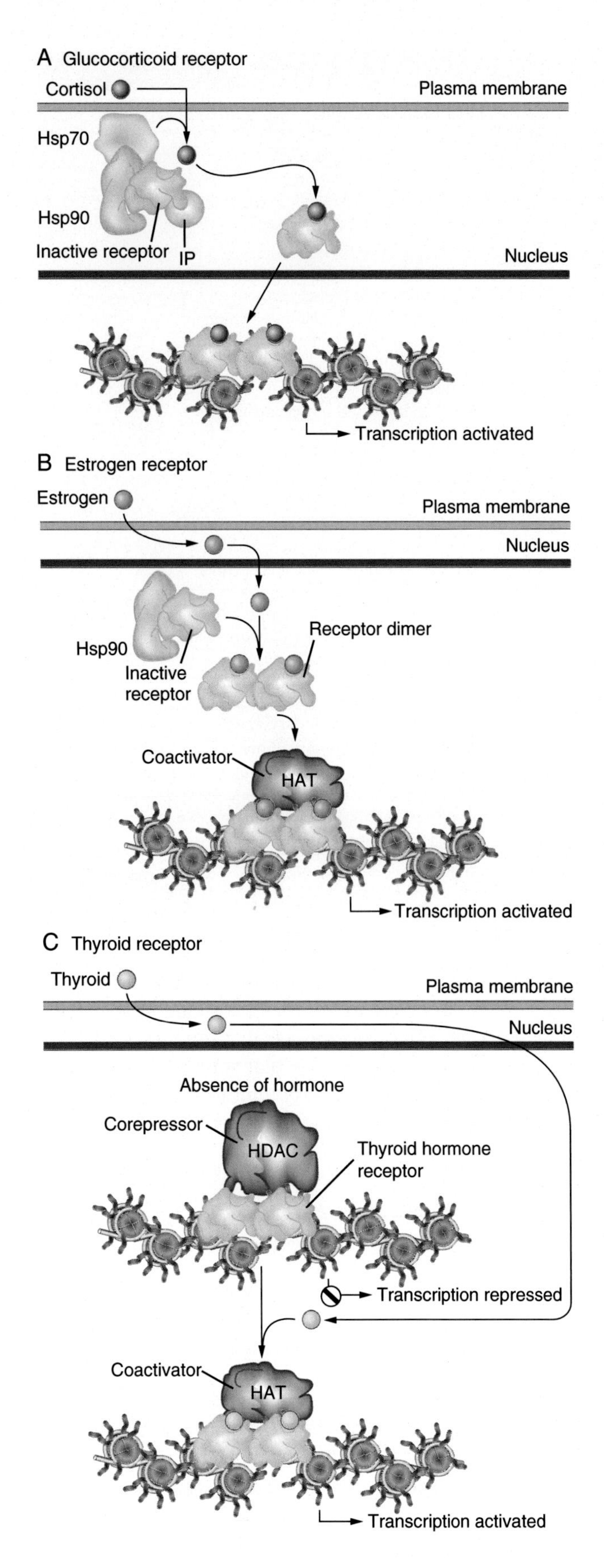

Figure 8–4. **Gene regulation by members of the nuclear receptor superfamily.** (A) The inactive glucocorticoid receptor is located in the cytosol in a complex that contains heat shock protein (Hsp) 70, Hsp90, and immunophilin (IP). Cortisol binding to the receptor displaces accessory proteins, and the activated ligand-receptor complex translocates to the nucleus where it activates target genes. (B) The inactive estrogen receptor Hsp90 complex is located in the nucleus. Estrogen binding displaces Hsp90 and the activated receptors dimerize, bind to DNA, associate with the coactivator histone acetyltransferase (HAT), and activate target genes. (C) The thyroid hormone receptor binds DNA in the presence and absence of ligand. In the absence of ligand, the receptor is complexed with the corepressor histone deacetylase (HDAC), which prevents gene transcription. In the presence of hormone, the ligand-receptor complex binds the coactivator HAT and activates target gene expression.

they are degraded by blood/tissue proteases or taken up into cells. Water-soluble hormones cannot diffuse across the plasma membrane and exert their effects by binding to receptors on the surface of the target cell. The signal is carried to the inside of the cytoplasmic region of the receptor and involves the generation of a second messenger (see Chapter 9). Some receptors activate a cascade of phosphorylation events, and others activate G-proteins. In contrast with lipophilic hormones, the effects of peptide hormones are almost immediate and usually persist for only a short period. One exception is growth hormone (GH), which can produce long-lasting and even irreversible changes as a result of downstream changes in gene transcription (Clinical Case 8–1).

Clinical Case 8–1 ***Type I Diabetes***

Carolyn, a 12-year-old girl, comes to her physician for a routine physical before starting the 7th grade. She has no specific complaints, but her mother notes that she seems constantly thirsty over the summer and has been more tired than usual recently.

On physical examination, Carolyn is a thin but well-developed girl who appears her stated age. She is showing normal secondary sex characteristics. Her blood pressure and pulse are normal. Her HEENT (head, ear, eyes, nose, throat) examination is normal, including her eye grounds. Her heart and lungs are clear, and her abdominal examination is normal. However, in the middle of the examination, she excuses herself to urinate and get a drink of water. Her neurologic examination is normal. The doctor sends a blood sample to the blood laboratory next door and asks for a urine sample.

Laboratory tests were notable for blood sugar of 360 mg/100 cc, sodium of 124 mg/dL, and potassium of 5.2. Her urine has a low specific gravity and shows 4+ sugar and 2+ ketones but no protein or white blood cells. The physician checks a Hemoglobin A1c, which returns at 8.5%. Thyroid-stimulating hormone (TSH) level is 9.8.

On questioning the mother the physician determines that there is no family history of premature heart disease, cancer, or diabetes.

Cell Biology, Diagnosis, and Treatment of Type 1 Diabetes Mellitus

Carolyn has Type 1 diabetes mellitus. She is symptomatic now from hyperglycemia, glucose-induced osmotic diuresis, and probable dehydration. This so-called juvenile form of diabetes mellitus usually occurs in younger patients. It is not unusual for it to start with the stress of puberty, and it is not always associated with a family history though diabetes in a first-degree relative increases the likelihood of developing the disease. Type 1 diabetes mellitus is caused by an "autoimmune" destruction of the insulin-secreting cells of the pancreas by a β-cell–specific antibody. Lack of insulin production leads to decreased circulating levels of insulin. Insulin interacts with cell surface proteins to ensure glucose uptake by peripheral target tissues. Lack of circulating insulin leads to build up of glucose in the bloodstream, which in turn can lead to both short- and long-term toxic effects. Short-term effects include ketosis, dehydration, weight loss, and if untreated and severe acidosis and death. Long-term effects include retinal disease, neurologic and infectious sequelae, and increased cardiovascular risk. It is also often associated with other autoimmune phenomena that can target other endocrine tissues. In this case the high level of TSH likely reflects an underlying Hashimoto's thyroiditis, which results from autoimmune destruction of thyroid hormone and resultant increase in pituitary production of TSH. While Carolyn's fatigue is likely due to her newly diagnosed diabetes, hypothyroidism may also play a role.

Carolyn is an intelligent young lady with a supportive family. It is likely that she will understand the need for careful glycemic control to avoid the extremes of hypoglycemia and ketoacidosis. Her physician has already told her about self-monitoring of blood glucose levels and the potential utility of a portable insulin pump. Furthermore, initiation of instruction in dietary basics will start her on lifelong attention to her personal glucose metabolism.

As she gets used to the sudden discovery of her chronic disease, her physician will help her to understand some of its cardiovascular, renal, and ocular complications. He will also encourage her to recognize that with good glycemic control, which she can watch by following her HbA1c (glycosylated hemoglobin) levels, she may be able to delay or avert those complications.

The Hypothalamic-Pituitary Axis

The synthesis of classical pituitary hormones is controlled by a complex integrated feedback loop that involves hypothalamic neurons and peripheral endocrine glands (Fig. 8–5). These complex **positive and negative feedback** loops ensure that pituitary hormone secretion is in tune with all aspects of the organism's environment.

The classical pituitary contains five different cell types, each of which secretes a particular biologically important hormone(s): Somatotropes secrete GH, lactotropes secrete prolactin, corticotrophs secrete adrenocorticotropic hormone (ACTH), thyrotropes secrete TSH, and gonadotrophs secrete follicle-stimulating hormone (FSH) and luteinizing hormone (LH). The pituitary also synthesizes many nonclassical hormones including growth factors, cytokines, and neurotransmitters. The secretion of the classical pituitary hormones is regulated by neurons in the hypothalamus that receive inputs from target tissues, hormonal feedback, and stimuli from other brain areas. These neurons fire at regular intervals,

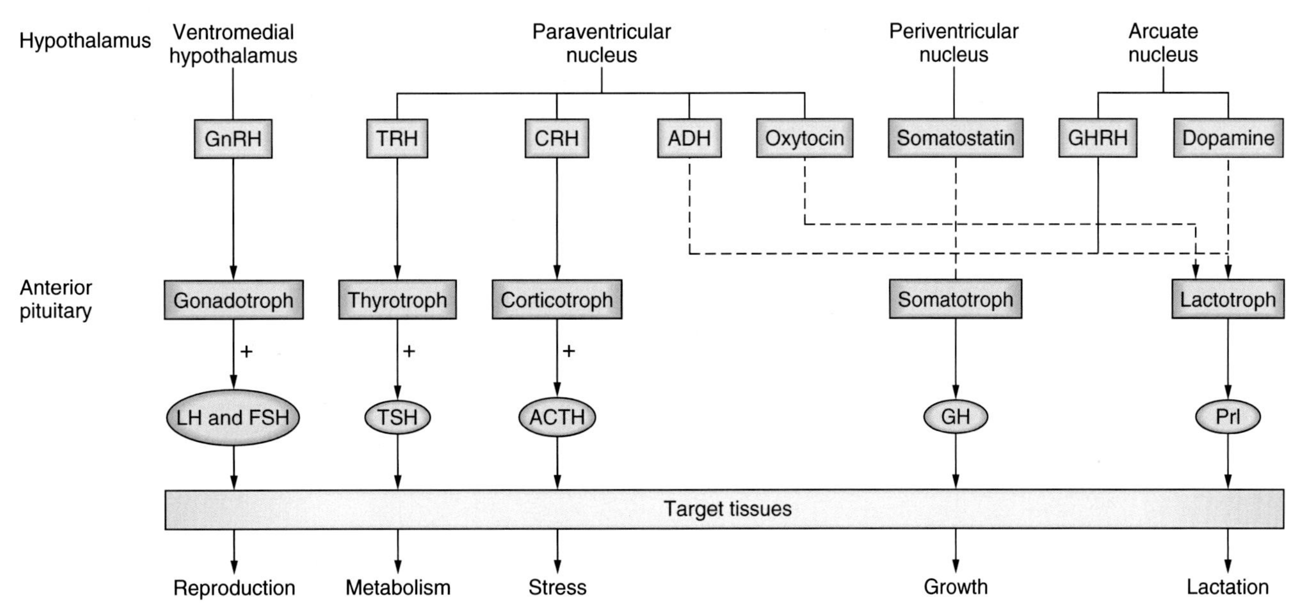

Figure 8–5. **Hypophysiotropic control of pituitary hormones.** Neuronal cells in various nuclei/regions of the hypothalamus release signaling molecules into the bloodstream, rather than synapses. These signaling molecules *(blue)* travel through the bloodstream to the pituitary where they stimulate *(solid lines)* or inhibit *(dashed lines)* the release of one or more pituitary hormones *(purple)*.

generating the pulsatile release of hypothalamic peptide hormones that is essential for proper endocrine system function. Because these agents are released into the bloodstream, specifically the superior or inferior hypophyseal artery, they are classified as hormones. Hypothalamic hormones travel through the portal circulation and activate receptors on specific pituitary cells to stimulate or inhibit the release of one or more pituitary hormones. The pituitary hormones then act on target tissues, such as the gonads, and stimulate the release of still other hormones, both peptides and steroids. GH secretion is an excellent example of the various factors that impact pituitary hormone secretion. GH levels must be coordinately regulated with metabolic fuel availability during prenatal and postnatal growth spurts and again during puberty (Fig. 8–6).

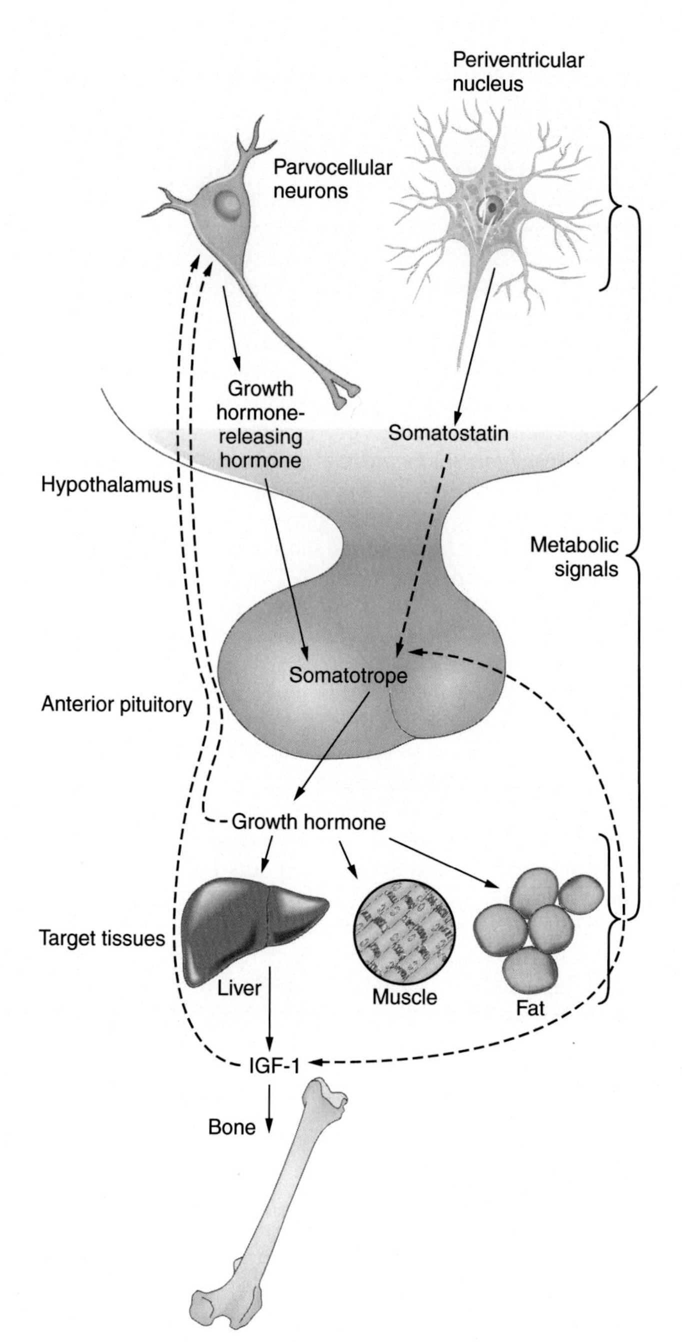

Figure 8–6. **Regulation of GH release.** GH release from the pituitary is stimulated by growth hormone–releasing hormone (GHRH), which peaks at night, and inhibited by somatostatin, which reaches high levels during the day. GH then stimulates the growth of muscle and adipocytes and the release of IGF-1 from the liver. IGF-1 then stimulates bone growth. If there are adequate metabolic fuels, metabolic signals from the peripheral target tissues will act on the hypothalamus and stimulate GH release. GH and IGF-1 provide negative feedback and inhibit the release of GHRH from the hypothalamus. IGF-1 also inhibits GH release from somatotropes in the pituitary.

GROWTH FACTORS

Growth factors represent a large number of polypeptides that play an important role in cell growth and survival from conception to death. All cells synthesize one or more growth factors (Table 8–2). In most cases the original names of growth factors do not reflect currently known biological activities. Although their usual physiologic effects are paracrine, some growth factors also act over longer distances. Thus it is not uncommon for some growth factors to be classified as hormones and for hormones that regulate cell growth to be classified as growth factors. Growth factors act in one of three ways: **Mitogens** stimulate cell proliferation, **trophic factors** promote growth, and **survival factors** inhibit apoptosis. Many growth factors are **pleiotropic**—that is, they have multiple effects within the same cell or elicit different responses in different cell types or both. Depending on the environment of the cell, growth factors can promote cell growth at one time and inhibit it at another time. Abnormalities in growth factor signaling are the basis for many types of cancers (see Chapter 10). Therapeutic agents for the treatment of neurodegenerative disorders, the side effects from chemotherapy and viral infections, and the harvesting of stem cells and regenerative medicine (see Chapter 13) target growth factor-signaling pathways.

Nerve Growth Factor

Nerve growth factor (NGF) was discovered in the 1950s and was the first growth factor to be characterized. NGF and other members of the neurotrophin family (see Table 8–2) regulate the development and survival of neurons. The main distinction among neurotrophin family members is their site(s) of synthesis and the target cells on which they act. During nervous system development, 50% or more of the neurons in the brain, spinal cord, and peripheral nervous system routinely die. This excess of neurons ensures that all postsynaptic cells will be innervated. NGF secreted from the future postsynaptic cells binds to NGF receptors on the growth cones of the closest approaching axons. NGF receptor activation alters gene expression; in particular, it downregulates genes that promote programmed cell death (see Chapter 11) and upregulates genes that promote survival and neurite extension. Neurons that do not receive NGF will eventually die. Only those neurons that receive

TABLE 8–2. Major Growth Factor Families

Signaling Molecule	Source	Major Biological Activity
Neurotrophins		
Nerve growth factor (NGF)	Brain, heart, spleen	Neuronal differentiation and survival
Brain-derived neurotrophic factor (BDNF)	Brain and heart	Neuronal differentiation and survival
Neurotrophin 3 (NT-3)	Brain, heart, kidney, liver, thymus	Neuronal differentiation and survival
Epidermal growth factor (EGF) family		
EGF	Salivary gland	Cell proliferation
Transforming growth factor-α (TGF-α)	Many cells/tissues	Cell proliferation
Fibroblast growth factor (FGF) family		
Fibroblast growth factors (22 total)	Many cells/tissues	Mitogenic
Transforming growth factor (TGF)-β family		
TGF-β	Ubiquitous	Inhibits proliferation
Inhibins/activins	Gonads and hypothalamus	Inhibits follicle-stimulating hormone (FSH) secretion
Bone morphogenetic proteins (>30 total)	Many cells and tissues	Osteogenesis Establishment of embryonic axis
Platelet-derived growth factor (PDGF) family		
PDGF	Platelets	Tissue repair
Vascular endothelial growth factor (VEGF)	Neural tissue Vascular smooth muscle	Endothelial cell proliferation ↑ Vascular permeability
Hematopoietic growth factors		
Erythropoietin	Kidney	↑ Red blood cell production
Colony-stimulating factors (CSFs)	Endothelium, T cells, fibroblasts, macrophages	↑ Red blood cell production
Thrombopoietin	Liver	↑ Platelet production
Insulin-like growth factor (IGF) family		
IGF-1	Many cells/tissues	Mitogenic, trophic, survival
IGF-2	Many cells/tissues	Fetal growth
Tumor necrosis factor (TNF) family		
TNF-α and TNF-β	Macrophages, natural killer cells, T lymphocytes	Tumor regression
Interferons (types I and II)	**Helper T cells**	**Antiviral activity**
Interleukins (29 total)	**T, B, and mast cells predominantly**	**Proliferation and differentiation of T and B lymphocytes**

NGF will survive and innervate the postsynaptic cell. There has been great interest in using NGF, and possibly other neurotrophins, to minimize neuronal cell death in neurodegenerative disorders such as Alzheimer's disease, Parkinson's disease, Huntington's disease, multiple sclerosis, encephalomyelitis, diabetic neuropathies, and spinal cord injury.

Growth Factor Families

All growth factors are classified into families based on their amino acid sequences and the receptors that they activate (see Table 8–2). Considerable variation in the size and biological activities of the growth factor families exists. The two main members of the EGF family are EGF and transforming growth factor-α (TGF-α). EGF and TGF-α interact with the same receptor and act as mitogens in a large number of tissues. The two members of the platelet-derived growth factor (PDGF) family, PDGF and vascular endothelial growth factor (VEGF), are important in tissue repair after injury. The fibroblast growth factor (FGF) family is one of the largest and contains 22 mitogenic growth factors. Members of the FGF family play a central role in angiogenesis, the formation of new blood vessels. The transforming growth factor-β (TGF-β) family is also large, and the biological activities of its members are quite diverse. At low concentrations, TGF-β stimulates growth; at high concentrations, it inhibits growth. The bone morphogenetic proteins promote osteogenesis and establishment of the early embryonic axis. The inhibins/activins act as classical hormones to inhibit FSH secretion and as growth factors to regulate formation of the embryonic notochord, somites, and neural tube. Another family that overlaps with hormones is the insulin growth factor (IGF) family. Although insulin and the IGFs (IGF-1 and IGF-2) have similar structures, their biological activities are quite different: Insulin promotes anabolic activity and has no mitogenic activity, whereas the IGFs are mitogenic, trophic, and survival factors. Two other clinically important growth factor families are the interferon family, whose members exhibit antiviral activity and are used to treat hepatitis C, and the tumor necrosis factor (TNF) family, whose members act as tumor regressors.

Growth Factor Synthesis and Release

Virtually, all cells synthesize polypeptide growth factors and secrete them by a "classical" regulated exocytosis (see Chapter 4). The major exception is members of the hematopoietic growth factor family that are not stored but rapidly synthesized when needed. The mechanisms responsible for the release/processing of growth factors are diverse. Most members of the FGF family have a classical leader sequence that ensures efficient secretion from cells via classical regulated exocytotic pathway. However, FGF-1 is released by a nonclassical release mechanism under stress conditions. Some FGF members accumulate in the cytosol and nucleus, where they are complexed with other proteins. The two members of the EGF family, EGF and TGF-α, are synthesized as membrane-bound precursors that are cleaved to yield smaller soluble peptides. Both the precursor and the soluble peptide have biological activity. In the mammary gland, preferential expression of the precursor form occurs. PDGF can also be retained on the plasma membrane via electrostatic interactions—it does not have a membrane-spanning domain. IGF-1 and IGF-2 are found in the serum associated with IGF-binding proteins. This sequestration serves several functions: it prolongs the half-life of the growth factor, forms a reservoir of growth factor, and inhibits growth factor activity by preventing interaction with its receptor. The major mechanism for termination of growth factor signaling is receptor-mediated endocytosis and lysosomal degradation (see Chapter 4).

Growth Factor Receptors Are Enzyme-Linked Receptors

All growth factor receptors are membrane-bound enzyme-linked receptors. Like other membrane receptors, they contain three domains: an extracellular ligand (growth factor) binding domain, a transmembrane domain, and a cytoplasmic domain that acts as an enzyme or forms a complex with another protein that acts as an enzyme. The majority of growth factor receptors are receptor tyrosine kinases. Growth factor binding leads to phosphorylation of tyrosine residues on a number of intracellular signaling molecules, and these molecules transmit the signal to the inside of the cell. The activation of the FGF receptor (FGFR), a tyrosine kinase, is detailed later in this chapter. However, not all growth factor receptors are tyrosine kinase receptors. TGF-β activates receptor serine-threonine kinases that phosphorylate the SMAD protein transcription factor, resulting in downstream changes in gene transcription. Erythropoietin and the cytokines signal through the Janus kinase pathway. GH, prolactin, and colony-stimulating factors (CSFs) signal through the JAK-STAT pathway. These intracellular signaling pathways are detailed in Chapter 9.

Growth Factors Are Paracrine and Autocrine Signalers

Growth factors use a variety of different signaling modes to exert their biological effects. Most growth factors signal in a paracrine fashion; that is, the growth factor is synthesized in one cell and exerts its effects on a neighboring cell. However, growth factors can also signal in an autocrine fashion. One example is the proliferation of helper T cells during an immune response. Antigen-presenting macrophages secrete interleukin-1 (IL-1), which binds to IL-1 receptors on resting helper T cells, converting them to activated helper T cells. These activated cells synthesize and secrete interleukin-2 (IL-2) and express IL-2 receptors. The secreted IL-2 binds to the IL-2 receptor on the synthesizing cell and induces proliferation. If T cells do not proliferate, then an immune response will not be mounted. Another example is PDGF, which is best known for its paracrine actions on smooth-muscle cells and fibroblasts during wound healing. Formation of the placenta requires the rapid clonal proliferation of cytotrophoblasts. This proliferation occurs as a result of secretion of PDGF by the cytotrophoblasts. Because these cells also express a PDGF receptor, they respond to the PDGF and proliferate.

Some Growth Factors Can Act Over Long Distances

The ability of growth factors to signal over long distances has permitted their use as therapeutic agents. In fact the administration of CSFs to cancer patients undergoing chemotherapy has become standard practice. Although the source is different, the exogenous CSFs act just like the endogenous CSF and mobilize the stem cell populations needed to replenish the rapidly dividing hematopoietic cell populations destroyed by chemotherapy. This same approach is used to ensure that the harvested marrow used in bone marrow transplantation surgery contains high numbers of stem cells. More recently, granulocyte-colony stimulating factor (G-CSF) has been shown to reduce infarct size in animal models of stroke by crossing the blood-brain barrier and interacting with its receptors in the brain.

Some Growth Factors Interact With Extracellular Matrix Components

A variety of growth factors and cytokines have been reported to bind specific carbohydrate moieties. The interaction of FGF with carbohydrates has been studied extensively (Fig. 8–7). All four FGFRs are tyrosine kinase receptors. In the presence of the extracellular matrix proteins heparin sulfate (HS) and glycosaminoglycans (see Chapter 7), a rigid FGF/FGFR/HS (2: 2: 2) dimer is

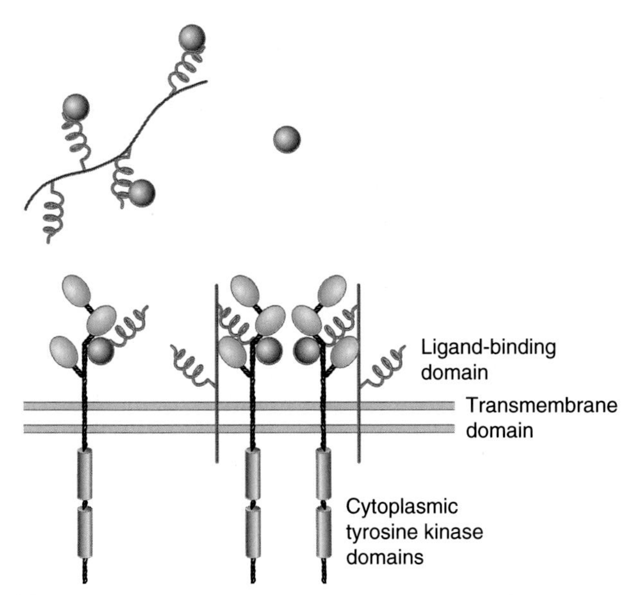

Figure 8–7. **Extracellular matrix components modulate FGF activity.** FGF interacts with the heparin sulfate side chains found on proteoglycans in the extracellular matrix and bound to the plasma membrane. These proteoglycans act as low-affinity receptors for FGF, sequester the FGF, and prevent its degradation. FGF interaction with HS is required for high-affinity interaction with its receptor FGFR. The formation of an FGF/FGFR/heparan sulfate (HS) (2: 2: 2) dimer induces the transphosphorylation of the cytoplasmic receptor tyrosine kinase domains that activates FGF signaling and alters cell behavior.

formed. This dimer then activates the cytoplasmic domains, and transphosphorylation of the receptors occurs. Once phosphorylated the receptor activates downstream signal transduction cascades. In addition to activating the receptor, interactions between growth factors and extracellular matrix components may provide a mechanism for storing and concentrating these particular growth factors. In addition to FGF, TNF-α and IL-2 also recognize carbohydrate sequences, and this recognition modulates their biological activity. However, this is not a general feature of all growth factor receptors—EFG receptor (EGFR) dimerization requires the binding of EGF only.

HISTAMINE

Histamine mediates a diverse list of processes including immediate hypersensitivity reactions, allergic responses, gastric acid secretion, bronchoconstriction, and neurotransmitter release. Although all tissues contain histamine, the greatest histamine concentrations are found in the skin, bronchial mucosa, and intestinal mucosa. Endogenous histamine is released into the local environment and acts in a paracrine fashion to influence the activity of neighboring cells. Venoms, bacteria, and plants are sources of clinically important exogenous histamine. Many of the popular over-the-counter drugs that are used to treat allergic reactions (Benadryl and Dramamine) and ulcers/heartburn (Tagamet) antagonize histamine signaling.

Endogenous histamine is synthesized locally and in most organ systems is stored in secretory granules. Cytosolic histidine is converted to histamine by the enzyme histidine decarboxylase. Histamine is transported into secretory granules, where it is complexed with heparin or chondroitin sulfate proteoglycans and stored until the cell is activated. In peripheral tissues, mast cells and basophils are the primary sources of histamine. A variety of endogenous and exogenous compounds stimulate the exocytotic release of histamine. Some drugs, venoms, high-molecular-weight proteins, and basic compounds (X-ray radiocontrast media) alter the mast cell membrane and induce the nonexocytotic release of histamine. In addition, thermal and mechanical stresses (scratching) displace histamine. Once a cell releases its histamine stores, it may take weeks for the supply to reach normal levels again. Alternatively, some cells in the immune system (platelets, monocytes/macrophages, neutrophils, and T and B cells) synthesize large amounts of histamine that are released immediately. Histamine signaling is terminated by its metabolism to inactive forms and/or uptake into cells by specific transporters.

Histamine Receptor Subtypes

Four different receptors, designated H_1, H_2, H_3, and H_4, mediate the actions of histamine (Table 8–3). These G protein–coupled receptors use Ca^{2+} or cyclic adenosine monophosphate (cAMP) second messenger systems or both (see Chapter 9). The H_1 and H_2 receptors are the main targets for antihistamine therapies and were first described in the mid-1960s. The H_3 and H_4 receptors have only recently been characterized. H_1 receptors are expressed on smooth muscle and endothelial cells and are responsible for many of the symptoms of allergic disease and anaphylaxis. H_2 receptors are found in the gastrointestinal tract and are the main mediators of gastric acid secretion. H_3 receptors are expressed on histamine-containing neurons and act as presynaptic autoreceptors that mediate feedback inhibition of histamine release and synthesis. H_4 receptors are preferentially expressed on hematopoietic and immunocompetent cells and are responsible for mast/eosinophil chemotaxis and recruitment. Recent studies have demonstrated that the ineffectiveness of H_1 receptor antagonists in treating asthma is attributable to their inability to block H_4 receptor–mediated recruitment of mast cells, basophils, and eosinophils to the lung. Additional characterization of histamine receptor subtypes will be needed to delineate the cellular events that mediate histamine signaling.

TABLE 8–3. Histamine Receptor Subtypes

Receptor	Signaling Mechanism	Distribution	Major Biological Activity
H_1	$G_{q/11}$	Smooth muscle Endothelial cells CNS	Bronchoconstriction and vasodilation Sleep and wake cycles
H_2	G_s	Gastric parietal cells Cardiac muscle Mast cells CNS	Gastric acid secretion Smooth-muscle relaxation Cognition
H_3	$G_{i/o}$	CNS	Neurotransmitter release, obesity, movement disorders, narcolepsy, schizophrenia, Parkinson's disease
H_4	$G_{i/o}$	Mast cells Hematopoietic cells	Chemotaxis, allergy, asthma

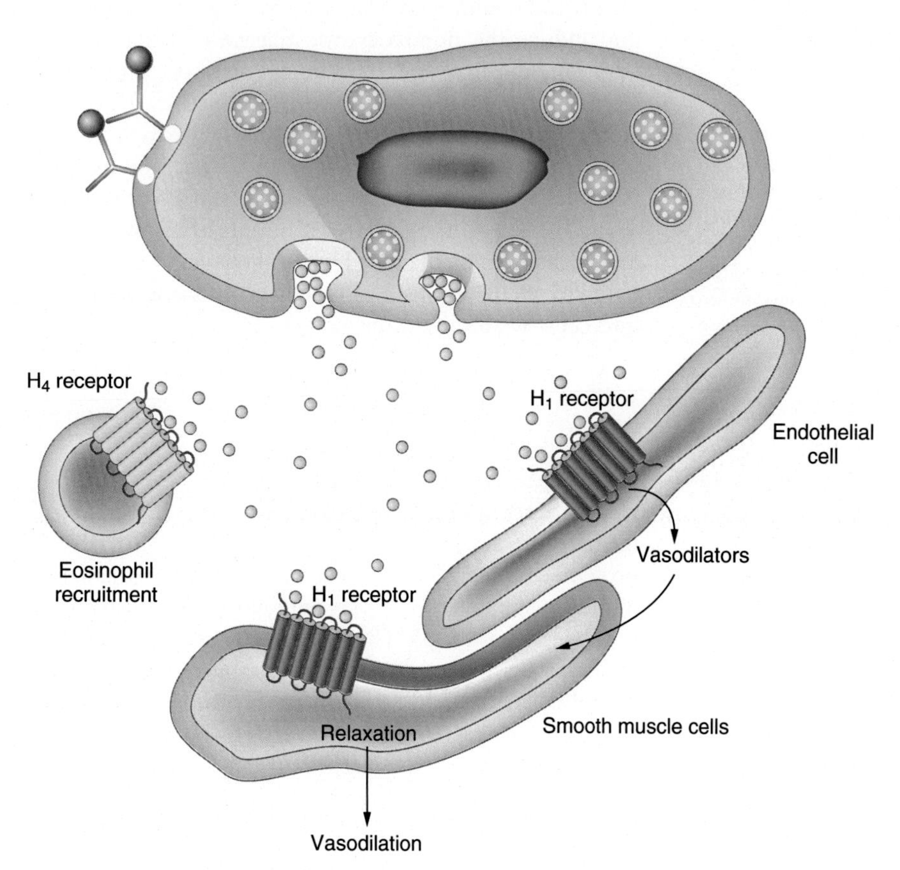

Figure 8–8. **Mast cell degranulation/histamine signaling.** Mast cells and eosinophils in the blood contain large numbers of secretory granules that are filled with histamine. In the case of an allergic reaction, binding of the allergen to an IgE antibody immobilized on the mast cell triggers the release of histamine-containing granules. The released histamine binds endothelial and smooth-muscle H_1 receptors to induce vasodilation. If concentrations are high enough, the released histamine acts on eosinophil H_4 receptors to induce their migration into the affected area. These eosinophils can then exacerbate the reaction by releasing additional histamine.

Mast Cell Histamine Release and the Allergic Reaction

One the best characterized actions of histamine is its role in allergic reactions (Fig. 8–8). Initial exposure to an antigen, such as bee venom protein or a food product, activates B lymphocytes to produce antibodies that recognize various regions of the antigen. Some of these antibodies will be IgE molecules. Mast cells express Fc receptors that bind these IgE antibodies, and the resulting IgE-Fc receptor complex forms a receptor for the antigen. When a second antigen exposure occurs, the new antigen will bind to and dimerize these IgE-Fc receptor complexes. Dimerization activates a cascade of intracellular signaling events that increase intracellular Ca^{2+} levels and trigger **degranulation,** the exocytosis of histamine-containing secretory granules. The released histamine then binds to H_1 receptors on the endothelial cells, resulting in increased capillary permeability. White blood cells and eosinophils migrate into the area to neutralize the antigen and repair tissue damage. Activation of endothelial cell H_1 receptors also stimulates production of local vasodilator substances, including nitric oxide (NO). Histamine activation of smooth-muscle cell H_1 receptors also contributes to vasodilation. In tissues where local mast cell degranulation has occurred, the histamine concentration will be sufficient to activate H_4

receptors on nearby eosinophils, and these cells will migrate into the site of the allergic reaction. In the event that these responses are exaggerated, a potentially fatal anaphylactic reaction can occur. Prior sensitization is not required for histamine release from mast cells. In fact the "allergic reaction" associated with some drugs can be attributed to their ability to activate mast cell degranulation without prior desensitization.

GASES: NITRIC OXIDE AND CARBON MONOXIDE

Carbon monoxide (CO) and NO were once considered to be only toxic pollutants, but they are now recognized as important intercellular signaling molecules. NO is a major paracrine signaling factor in the nervous, immune, and circulatory systems. In the circulatory system, both CO and NO mediate blood vessel dilation. In fact, NO is the final common mediator for many endogenous and exogenous smooth-muscle relaxants. When nitroglycerin is administered to patients with angina, it is rapidly converted to NO in the bloodstream; the resulting NO enters the coronary vasculature, causing vessel dilation and increased blood flow. In septic shock, cell wall material released from gram-negative bacteria triggers the release of NO from macrophages, resulting in widespread dilation of blood vessels and a dramatic, sometimes fatal, decrease in blood pressure. Drugs for the treatment of erectile dysfunction inhibit the downstream second messenger for NO, guanylyl cyclase.

The vasodilator effects of NO have been well characterized (Fig. 8–9). The enzyme NO synthase converts L-arginine to citrulline and liberates NO. Cells and tissues express three NO synthase (NOS) isozymes: Neuronal NOS (nNOS; NOS-1) was discovered in the brain, inducible NOS (iNOS; NOS-2) was originally purified from macrophages, and endothelial NOS (eNOS; NOS-3) was discovered in endothelial cells. These enzymes are the major site for regulating NO-mediated signaling. eNOS is constitutively expressed in endothelial cells and is responsible for the basal production of NO that is needed to maintain normal vascular tone. NO readily

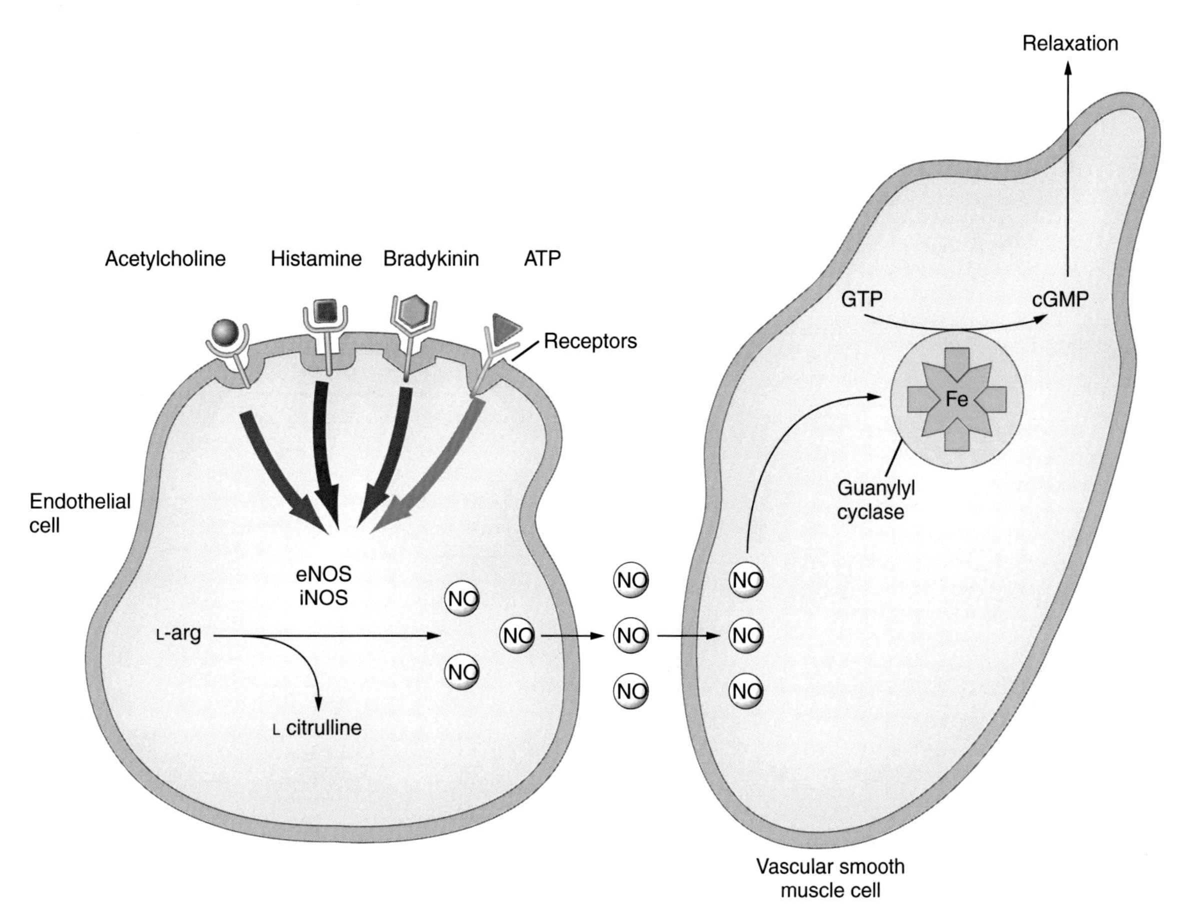

Figure 8–9. **NO-mediated endothelial cell relaxation.** eNOS produces NO in endothelial cells in response to a variety of intercellular signaling molecules including acetylcholine, histamine, bradykinin, and adenosine triphosphate (ATP). NO diffuses out of the endothelial cell and into the vascular smooth-muscle cell, where it binds to the heme (iron) moiety in guanylate cyclase. This increases cGMP levels and activates the cGMP-dependent kinases that ultimately produce smooth relaxation/vasodilation. During stress, iNOS will be activated, and NO levels will increase.

diffuses out of the cell. Because NO has a half-life of only a few seconds, it acts only on neighboring smooth-muscle cells. NO binding to a heme group (iron) in the active site of the enzyme guanylyl cyclase activates the enzyme and increases intracellular cyclic guanosine monophosphate (cGMP) levels. This, in turn, activates the cGMP-dependent protein kinases responsible for the phosphorylation/activation of proteins that cause smooth-muscle relaxation. Under stress conditions a new form of NOS, iNOS, is expressed. iNOS is essentially unregulated and once expressed can maintain high NO levels for prolonged periods, resulting in clinically significant hypotension.

The intercellular mediator functions of CO have only recently been described. Significant amounts of cellular CO are produced as an elimination product during microsomal heme oxygenase catalyzed heme degradation. Heme oxygenases are found in all tissues, and CO is generated in all cells. CO is relatively inert and reacts only with iron-containing compounds. CO activates the same soluble guanylate cyclase enzyme activated by NO but has a much lower potency than NO. In some instances CO may antagonize NO signaling. CO has also been implicated in oxygen sensing, oxygen-dependent changes in gene expression, and neuronal signaling.

EICOSANOIDS

The eicosanoids are a family of oxygenated derivatives of 20-carbon polyunsaturated fatty acids that includes **prostaglandins, thromboxanes, leukotrienes, endocannabinoids,** and **eicosanoids.** Because they are rapidly broken down, eicosanoids are limited to autocrine and paracrine signaling. Alterations in eicosanoid signaling are associated with failure of the ductus arteriosus to close at birth, platelet aggregation, inflammatory/immune responses, bronchoconstriction, and spontaneous abortion. The popular over-the-counter nonsteroidal anti-inflammatory drugs (NSAIDs) aspirin, acetaminophen (Tylenol), and ibuprofen (Advil) target eicosanoid-mediated signaling pathways.

Cells synthesize eicosanoids on demand and in response to physical, chemical, and hormonal stimuli. All eicosanoids are synthesized from a common precursor, arachidonic acid (Fig. 8–10). Phospholipase A

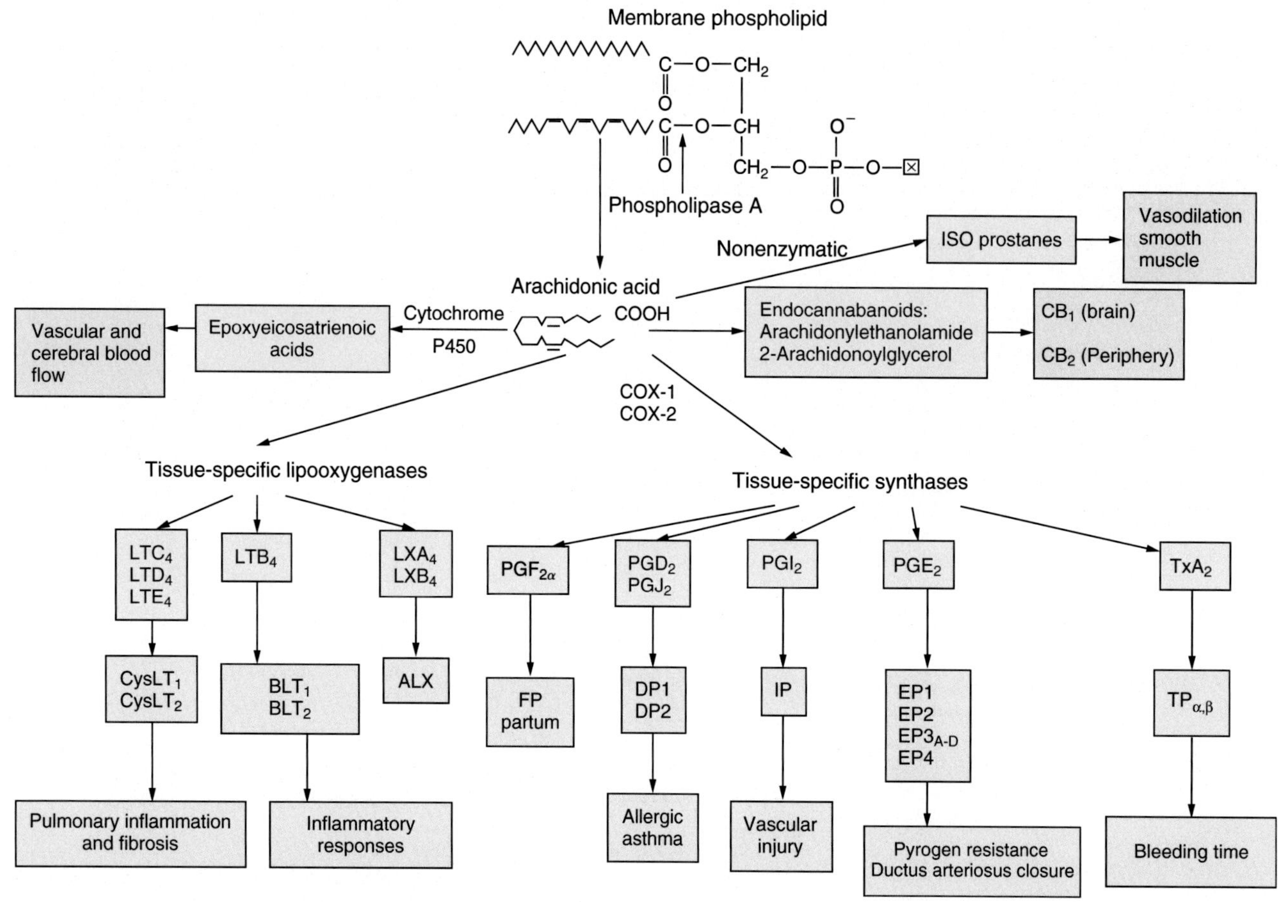

Figure 8–10. **Eicosanoid-mediated intercellular signaling.** Arachidonic acid, liberated from phospholipase A hydrolysis of plasma membrane phospholipids, is the precursor for all members of the eicosanoid family. The major signaling metabolites *(blue)*, their receptors *(yellow)*, and biological responses *(yellow)* are listed.

hydrolysis of phospholipids located on the cytoplasmic surface of the plasma membrane releases arachidonic acid. In unstimulated cells, arachidonic acid is incorporated back into the membrane. In stimulated cells, several different pathways convert arachidonic acid to eicosanoids. There are two **cyclooxygenases (COX)**, COX-1 and COX-2, that convert arachidonic acid to prostaglandins. Both enzymes contribute to unstimulated and stimulated prostaglandin synthesis and are targets for NSAIDs. There are five human **lipoxygenases** that convert arachidonic acid leukotrienes. Endocannabinoids, the newest members of the family, are synthesized primarily in the nervous system and are immediately released into the synaptic cleft—they are not stored in synaptic vesicles. The enzymes that convert arachidonic acid to endocannabinoids have not been characterized. Epoxyeicosatrienoic acids, important regulators of renal and cardiovascular function, are synthesized in endothelial cells by the cytochrome P450 enzymes. Eicosanoids are unique in that they are produced by a nonenzymatic free radical-based attack on arachidonic acid. Once synthesized, eicosanoids are transported out of the cell. Specific transporters have been identified for prostacyclin and thromboxane. Although the molecular mechanisms involved in terminating endogenous eicosanoid signaling have not been delineated, some signaling is terminated by uptake into target cells followed by enzymatic degradation.

Eicosanoids exert their effects by binding to a diverse list of G protein–coupled cell surface receptors (see Fig. 8–10). In the case of prostaglandins, seven different subclasses of receptors have been identified: EP_1, EP_2, EP_{3A-D}, and EP_4. The EP_3 receptor participates in the fever response, and the EP_4 receptor participates in closure of the ductus arteriosus. The endocannabinoids are endogenous ligands for the cannabinoid receptors CB_1 and CB_2. These are the same receptors for the psychoactive ingredients in leaves of the *Cannabis sativa,* marijuana and hashish. CB_1 is preferentially expressed in the nervous system and is responsible for the "high" produced by marijuana. Endocannabinoid signaling in the brain is associated with control of movement, learning and memory, pain perception, appetite, body temperature, and emesis. CB_2 is expressed primarily in the immune system. Despite their recent identification a number of CB_1 and CB_2 agonists are already in clinical trials for the treatment of pain and multiple sclerosis. The wide variety of differential responses to only small changes in the concentration/complement of prostaglandins is due to that cells express multiple eicosanoid receptors. In addition, some members of the eicosanoids family can interact with the same receptor. For example, eicosanoids alter vascular smooth-muscle behavior via interaction with one of the thromboxane receptors.

NEUROTRANSMITTERS

A **neurotransmitter** is defined as a chemical that is released from a stimulated presynaptic neuron, binds to the membrane of a postsynaptic target cell, and induces a response (inhibitory or excitatory) in the target cell. Some neurotransmitters can act as both neurotransmitters and hormones. For example, the neurotransmitter epinephrine is also produced by the adrenal gland and signals glycogen breakdown in the cell. Defects in neurotransmission are associated with neurologic and psychiatric disorders. A basic understanding of synaptic transmission is essential for understanding the pharmacological basis of therapeutic and recreational psychoactive drugs.

Electrical and Chemical Synapses

In the nervous system, information travels one cell to the next via specialized contact sites known as **synapses** (Fig. 8–11). Synapses that use chemicals to transmit information are called **chemical synapses.** Chemical synapses are common and are restricted to the nervous system. **Electrical synapses** are relatively rare and are found in neuronal and nonneuronal cells. In an electrical synapse, membrane depolarization in presynaptic cells passes directly to the postsynaptic cells via **gap junctions.** Gap junctions are bidirectional and allow ionic current to pass in either direction. Signaling across electrical synapses occurs almost instantaneously and is fail-safe: an action potential (AP) in the presynaptic neuron will always produce an AP in the postsynaptic cell. In invertebrates, electrical synapses mediate escape reflexes and allow animals to quickly retreat from danger. In humans, electrical synapses permit all of the cells within an organ to function as one large unit or **syncytium.** Electrical synapses make the synchronized beating of the heart, peristaltic movement in the gut, coordinated growth/maturation of the developing nervous system, and synchronized activity of neighboring neurons in specialized regions of the adult brain possible.

Chemical synapses are a specialized form of paracrine signaling that allows for fast and precise delivery of signals across large distances. In the central nervous system (CNS), chemical synapses carry signals between neurons. In the peripheral nervous system, chemical synapses carry information from nerves to myocytes and gland cells. In a chemical synapse the presynaptic cell converts the electrical signal into a chemical signal. This chemical signal is transmitted to the target cell where it is converted back into an electrical signal. Unlike growth factors and hormones, it takes only a millisecond or less for chemicals to transverse the 20- to 50-nm distance across the synaptic cleft. Unlike the all-or-none electrical synapse, chemical synapses are capable of receiving and summing inputs from multiple sources: a process that is essential for learning and memory and other higher brain functions.

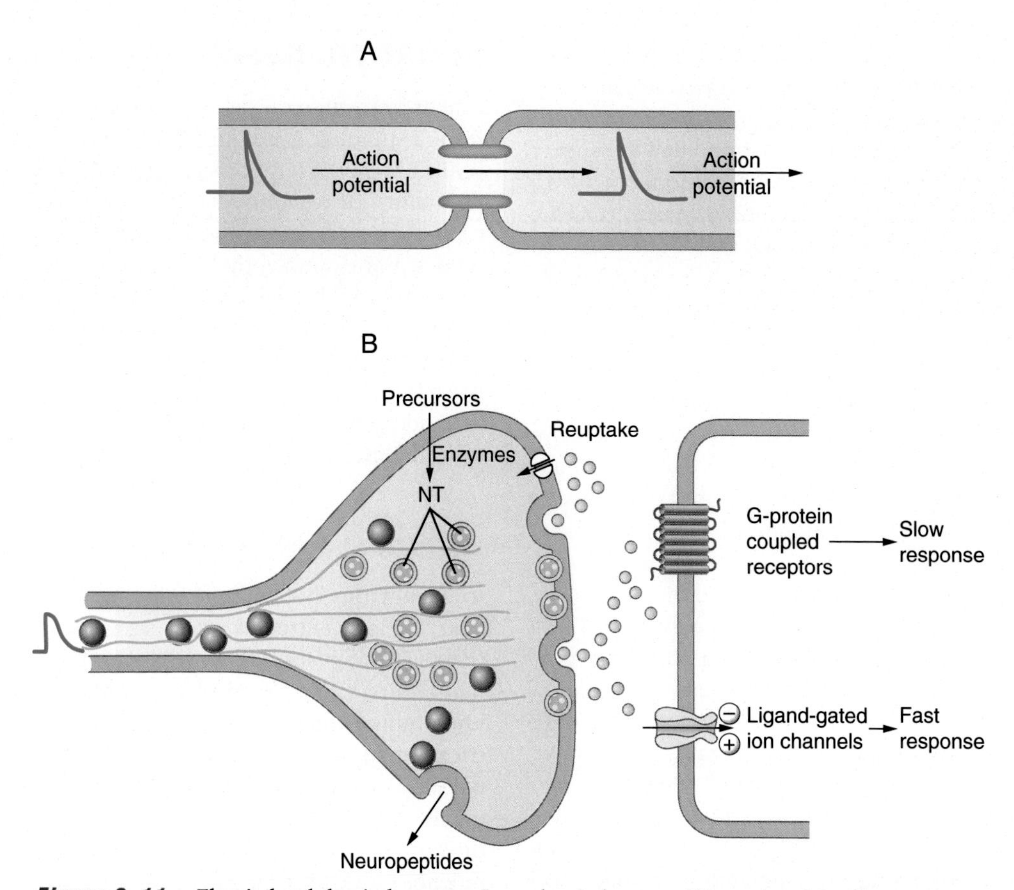

Figure 8–11. **Electrical and chemical synapses.** In an electrical synapse (A), gap junctions allow current to flow directly from the presynaptic cell to the postsynaptic cell. In a chemical synapse (B), the presynaptic cell converts the current into a chemical signal. Neurotransmitters *(yellow)*, located in synaptic vesicles, and larger neuropeptides, located in dense core granules, form the chemical signal. The neurotransmitters are released into the synaptic cleft and interact with ligand-mediated G protein–coupled receptors or ligand-mediated ion channel receptors on the postsynaptic cell. Neuropeptide *(red)* release occurs adjacent to the synaptic cleft. An action potential will be generated in the postsynaptic cell only if these chemicals produce a sufficient depolarization.

Prototypical Chemical Synapse: The Neuromuscular Junction

The neuromuscular junction is formed between axon terminals of cranial or spinal motor neurons and skeletal-muscle cells (Fig. 8–12). Before reaching the muscle the axon divides into multiple synaptic varicosities or synaptic boutons. These synaptic boutons are ensheathed by the Schwann cell plasma membrane but unmyelinated and allow a single axon to innervate hundreds of muscle fibers within a muscle. Every synaptic bouton, or terminal axon end, is in close apposition with a junctional fold, or invagination, of the muscle membrane. Both the presynaptic and postsynaptic sides of the neuromuscular junction are highly organized. The specialized region of the muscle plasma membrane where innervation occurs is called the **motor end plate**.

The electrical AP is converted into a chemical signal by the presynaptic cell. The AP depolarizes the presynaptic membrane and opens voltage-sensitive Ca^{2+} channels. This influx of Ca^{2+} mobilizes synaptic vesicles, each of which contains approximately 1000–50,000 molecules of the neurotransmitter acetylcholine. Approximately 10% of the vesicles are docked at the plasma membrane in the active zone and immediately exocytosed. The remaining vesicles are located adjacent to the active zone, where they are reversibly tethered to the actin cytoskeleton via synapsins, a family of phosphoproteins. These vesicles move into the active zone and are primed for exocytosis as needed. The maintenance of two synaptic vesicle pools ensures that acetylcholine will be available for release if another AP arrives in quick succession. The exocytosis of neurotransmitters involves many of the same vesicle targeting and fusion events that occur during regulated exocytosis in nonneuronal cells (see Chapter 4). However, secretion of neurotransmitters differs in that it is tightly coupled to the arrival of an AP at the axon terminus, and synaptic vesicles are recycled locally. The entire process of retrieving and refilling synaptic vesicles takes less than 1 min.

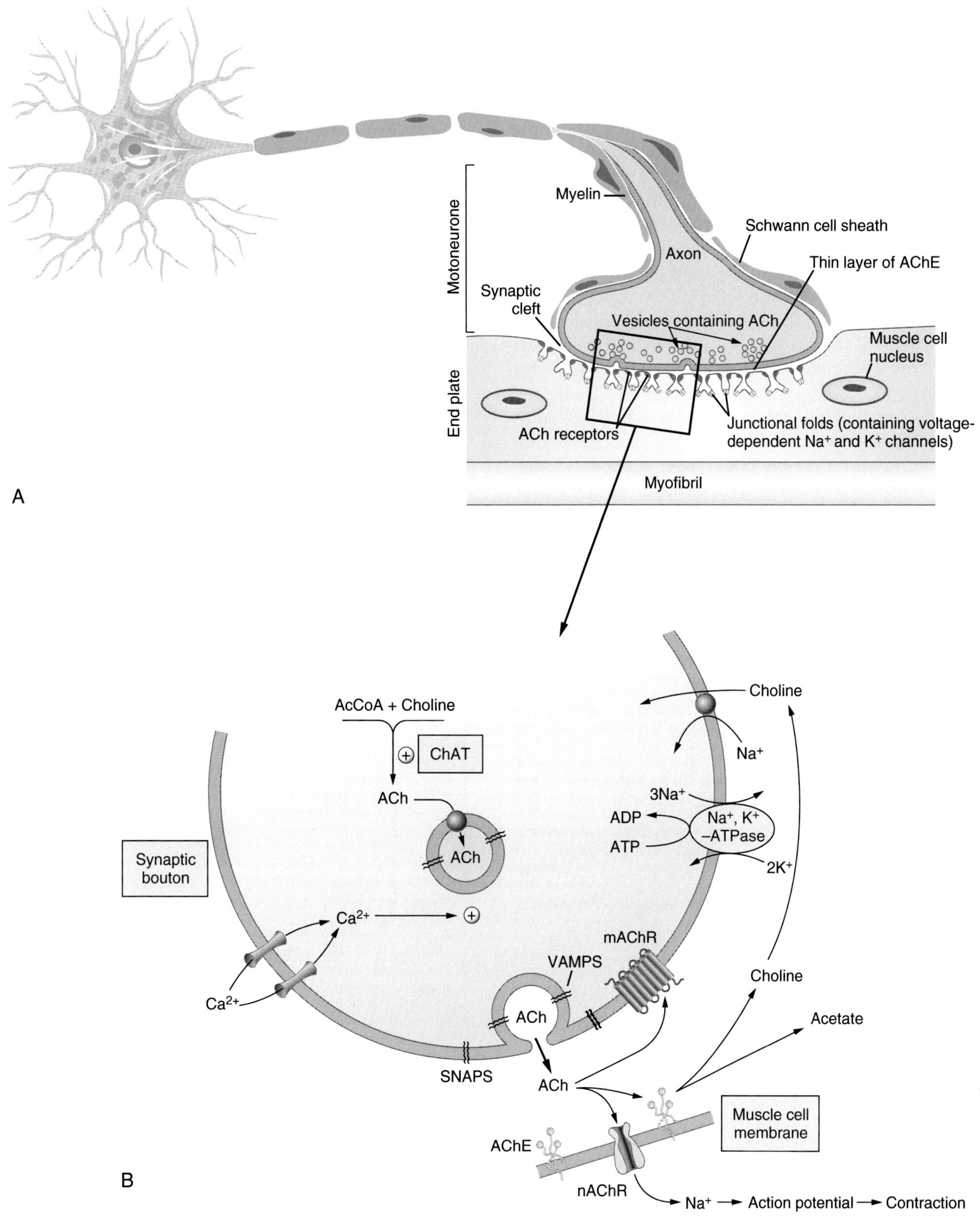

Figure 8–12. **Synaptic transmission at the neuromuscular junction.** (A) The neuromuscular junction is composed of synaptic boutons from the innervating α-motor neuron and the specialized motor end plates on the muscle fiber. Acetylcholine is synthesized in the motor neuron terminal and taken up into synaptic vesicles *(yellow)*. (B) On release the acetylcholine diffuses across the synaptic cleft and binds to a ligand-gated ion channel *(red)*. The number of Na^+ molecules that enter the muscle cell is dependent on the number of acetylcholine molecules released from the neuron. If there is a sufficient depolarization, then action potential will be generated in the muscle cell, and contraction will occur. Acetylcholinesterase, located in the synaptic cleft, hydrolyzes acetylcholine to acetate and choline. The choline is taken back up into the presynaptic cleft and used to refill synaptic vesicles.

The exocytosed neurotransmitter then transmits the signal to the postsynaptic target cell. Acetylcholine and all neurotransmitters diffuse across the synaptic cleft and bind to specific receptors on the postsynaptic cell membrane. In the case of the neuromuscular junction, acetylcholine receptors are concentrated at the beginning of the junctional infolds of the muscle plasma membrane. Binding of acetylcholine to its receptor opens an ion channel for approximately 1 ms, during which time approximately 50,000 Na^+ molecules enter the muscle cell. These Na^+ molecules depolarize the end-plate potential, and the response/depolarization is proportional to the amount of transmitter released. If the depolarization exceeds the threshold, then voltage-dependent Na^+ channels open, and the AP spreads throughout the muscle membrane and initiates the intracellular events needed for muscle contraction (see Chapters 2 and 3). Myasthenia gravis is an autoimmune disorder in which autoantibodies are made against the acetylcholine receptor subtype expressed on muscle cells. These antibodies block the receptor and promote its removal from the motor end plate. In the absence of the receptor, the signal is not transmitted even though the chemical signal is still generated.

It is imperative that the chemical signal/neurotransmitter does not influence neighboring cells and be removed before the next pulse of neurotransmitter is released. In the case of acetylcholine, an enzyme in the synaptic cleft, acetylcholinesterase, hydrolyzes the neurotransmitter into acetate and choline. The time required to hydrolyze acetylcholine is less than a millisecond. A large number of nerve gases and neurotoxins inhibit acetylcholinesterase and prolong the action of acetylcholine. If the concentration of these agents is high enough, they will prevent the relaxation of muscles necessary for respiration, resulting in lethality. A Na^+/choline symporter on the nerve cell membrane transports the choline back into the nerve cell, where it will be used to synthesize more neurotransmitter (Clinical Case 8–2).

Clinical Case 8–2 **Myasthenia Gravis**

Jane Crawford is a 37-year-old married mother who came to her internist reporting fatigue. She reports that she is overwhelmed by even minor physical effort that she could easily perform when she was raising her children. She also reports that she frequently has "heavy eyelids" and falls asleep while watching television in the morning. Her husband says that Jane's voice has changed and has recently become much more nasal.

Mrs. Crawford has been comparatively well physically previously. When she was in her teens, she had sore and swollen fingers bilaterally, which her college physician thought might be early rheumatoid arthritis. However, after a 2-month trial of high-dose aspirin, which bothered her stomach, those symptoms remitted and have not reoccurred.

On physical examination, Mrs. Crawford's blood pressure was 130/85 mmHg, her pulse was normal and regular, her heart normal sized without murmurs, and her thyroid was normal to palpation. She is not obese, and her abdominal examination was normal. She has no obvious signs of anemia, and her gross neurologic examination, including deep tender reflexes, was normal. She had difficulty, however, following the examiner's finger up and down during the eye motion tests. This was exaggerated with repetition. Furthermore, when asked to extend her arms out in front of her with her palms up, she could hold the position for only about 30 s. After receiving her permission for a brief test involving an intravenous injection and preparing a small amount of atropine if needed, her physician gave her an injection of 2-mg edrophonium in normal saline. Five minutes later, he asked her to repeat the arm extension maneuver. To her pleasure, Jane found that she could now hold her arms out for at least 6 min. Furthermore, she reported that her eyelids no longer felt as heavy.

Cell Biology, Diagnosis, and Treatment of Myasthenia Gravis

Mrs. Crawford has early myasthenia gravis. This is an acquired disorder caused by the loss of the normal number of acetylcholine receptors on her muscle membranes adjacent to the end plates of the innervating nerves, usually due to autoantibodies against these receptors. With decreased numbers of receptors, muscles can initiate contraction but cannot maintain it for long periods. As such, muscle fatigue occurs rapidly. Early signs of myasthenia include eyelid drooping (ptosis), which reflects fatigue in a muscle that requires constant contraction. Generalized muscle fatigue, including rapid tiring out, can follow. If untreated, respiratory muscles can be effected leading to respiratory failure and death. An autoimmune origin of this defect in Mrs. Crawford was suggested by her previous history of rheumatoid symptoms and confirmed by the finding of a high-titer IgG anticholinesterase receptor-antibody in her serum.

Before any effort at treatment, her physician ordered some blood tests for lupus erythematosus, rheumatoid arthritis, and thyroid diseases to rule out those disorders that may coexist with myasthenia. He also ordered a mediastinal-computed tomography scan to rule out a possible thymoma, as thymoma is a common cause of myasthenia.

When all of those test results were negative, he started Mrs. Crawford on low-dose oral pyridostigmine, an acetylcholinesterase mimetic drug, three times a day. Mrs. Crawford has had an excellent response, with a marked return of muscular strength and reduction of fatigue. For the present, her physician does not think she will need any antiimmune therapy. He warns her, however, that she will need comparatively frequent follow-up for the rest of her life and that a change in dose or other therapies may eventually be needed.

TABLE 8–4. Major Neurotransmitters and Their Receptors

Transmitter	Receptor Families/Subtypes	Signaling Mechanism
Amino acids		
Glutamate (Glu)	NMDA, AMPA, kainate	Ion channel
	$MGluR_1$, $MGluR_2$, $MGluR_3$, $MGluR_4$, $MGluR_5$, $MGluR_6$, $MGluR_7$	G-protein coupled
Glycine (Gly)	GlyR, NMDA	Ion channel
γ-Aminobutyric acid (GABA)	$GABA_A$	Ion channel
	$GABA_B$	G protein–coupled/Ion channel
	$GABA_C$	Ion channel
Nucleotides		
ATP	$P2X_1$, $P2X_2$, $P2X_3$, $P2X_4$, $P2X_5$, $P2X_6$, $P2X_7$P2Y	Ion channel
		G-protein coupled
Adenosine	A_1, A_{2A}, A_{2B}, A_3	G-protein coupled
Amines		
Acetylcholine	Nicotinic (N_m and N_n)	Ion channels
	M1, M2, M3, M4, M5	G-protein coupled
Norepinephrine	α_{1A}, α_{1B}, α_{1D}	$G_{q/11}$ coupled
	α_{2A}, α_{2B}, α_{2C}	$G_{i/o}$ coupled
	β_1, β_2, β_3	G_s-protein coupled
Dopamine	D_1, D_2, D_3, D_4, D_5	G-protein coupled
Serotonin	$5HT_3$	Ion channel
	$5HT_1$, $5HT_2$, $5HT_4$, $5HT_5$, $5HT_6$, $5HT_7$	G-protein coupled
Histamine	H_1, H_2, H_3, H_4	G-protein coupled
Neuropeptides		
Cholecystokinin (CCK)	CCK_1, CCK_2	G-protein coupled
Neuropeptide Y	Y_1, Y_2, Y_3, Y_4, Y_5	G-protein coupled
Opioid (enkephalins, dynorphin)	δ, κ, μ	G-protein coupled
Lipids		
Endogenous cannabinoids	CB	G-protein coupled

Characteristics, Synthesis, and Metabolism of Neurotransmitters

Neurotransmitters can be divided into three categories: amino acids, amines, and peptides (Table 8–4). The amino acids and amines mediate fast synaptic transmission (submillisecond to millisecond) in the CNS. Acetylcholine mediates fast synaptic transmission at all neuromuscular junctions. The amino acid and amine transmitters are synthesized in the axonal terminal and stored in synaptic vesicles. The amino acids glutamate and glycine are found in proteins and are abundant in the cytosol of all cells. γ-Aminobutyric acid (GABA) and the other amines are found only in neurons that contain the specific enzymes needed to synthesize them from precursor molecules. Choline acetyltransferase uses acetyl coenzyme A from the mitochondria as a donor and catalyzes the acetylation of cytosolic choline. Once synthesized, these neurotransmitters are concentrated in **synaptic vesicles** by transporters embedded in the vesicle membrane. With the exception of acetylcholine, neurotransmitters are removed from the cleft by reuptake into the presynaptic cell via Na^+/neurotransmitter symporters. These transporters are a major site of action for therapeutic and recreational drugs: Cocaine inhibits reuptake of norepinephrine, serotonin, and dopamine; older antidepressants inhibit norepinephrine and serotonin reuptake; and newer antidepressants (fluoxetine hydrochloride [Prozac]) specifically block serotonin reuptake.

Neuropeptide transmitters are found in the gut and throughout the nervous system. These peptides/polypeptides are synthesized by the endoplasmic reticulum in the cell body, packaged into secretory vesicles, and transported to the axonal terminal via fast axonal transport. The opioid neuropeptides are grouped into three distinct families: enkephalins, endorphins, and dynorphins. All three are synthesized as large precursor molecules that are subjected to complex cleavage and posttranslational modifications to generate the biologically active molecule. Individual secretory vesicles can contain multiple neuropeptides. **Secretory vesicles** are larger and more randomly distributed in the axonal terminal than synaptic vesicles (see Fig. 8–11). These vesicles are also referred to as large dense core vesicles because of their dark appearance in electron micrographs. Like synaptic vesicles, secretory vesicle exocytosis is triggered by an increase in Ca^{2+}. However, secretory vesicle fusion with the plasma membrane occurs at random sites on the periphery of the axon, not within the active zone/synaptic cleft. Because secretory vesicles are located some distance from the synaptic cleft, several high-frequency APs may be needed for the Ca^{2+} levels in these areas to reach the threshold required to trigger secretory vesicle release. Thus the time frame for neuropeptide release is much longer, 50 ms or more, than the time frame for amino acid/amine release. Neuropeptides are degraded by extracellular proteases, and this terminates signaling.

Neurotransmitter Receptors

Two classes of neurotransmitter receptors exist: **transmitter-gated ion channels** and **G protein–coupled receptors. Fast chemical synaptic transmission** is mediated by transmitter-gated ion channel receptors. These receptors are multimeric proteins that form a transmembrane pore. When a neurotransmitter binds to the receptor, the pore opens, and ions flow into the cytoplasm. Receptors for excitatory neurotransmitters (acetylcholine and glutamate) are cation channels that permit an influx of Na^+, or Na^+ and K^+, that depolarizes the postsynaptic membrane toward the threshold potential for firing an AP. Receptors for inhibitory neurotransmitters (GABA and glycine) are K^+ or Cl^- channels that hyperpolarize the postsynaptic membrane and suppress APs. Barbiturates and tranquilizers, such as diazepam (Valium), bind to GABA receptors and potentiate the inhibitory actions of GABA by reducing the concentration of GABA needed to activate the ion-gated channel.

Slow synaptic transmission is mediated by G protein–coupled receptors. These receptors are often referred to as metabotropic receptors because of the widespread metabolic effects they trigger. Neurotransmitters from all three categories interact with G protein–coupled receptors and mediate slow synaptic transmission in the CNS. G protein–coupled receptors are a single polypeptide containing seven transmembrane domains. Ligand binding to the receptor activates a small G-protein located on the intracellular side of the postsynaptic membrane. This activated G-protein then directly opens an ion channel, or it can initiate a cascade of intracellular events that indirectly results in the opening of ion channels (see Chapter 9). For example, the three classical opioid receptors μ, Δ, and κ inhibit adenylate cyclase, which reduce cAMP levels, resulting in activation of K^+ currents and inhibition of Ca^{2+} currents. Morphine produces its analgesic effects via interaction with these opioid receptors.

Divergence and Convergence of Neurotransmitter Function

Neurotransmitters exhibit both divergence and convergence. **Divergence** allows a single neurotransmitter to activate multiple receptors. Acetylcholine is an excellent example of the different effects that can be attributed to a single neurotransmitter—depending on the receptor that is activated. Nicotinic acetylcholine receptors are ligand-gated channels and result in excitatory responses that last only milliseconds. An example is the neuromuscular junction detailed in the previous section. In contrast, muscarinic acetylcholine receptors are G protein–coupled receptors and result in a variety of different responses: The M2 subtype found in heart opens a K^+ channel and produces a hyperpolarization that lasts for a few seconds; the M1, M3, and M5 subtypes activate a second messenger phospholipase; and the M4 subtype inhibits another second messenger adenylate cyclase (see Fig. 8–2). **Convergence** allows multiple transmitters acting through their individual receptors to activate the same downstream signaling pathway.

Spatiotemporal Summation

The CNS must integrate information from a variety of sources as it balances the needs of the organism and the demands from the environment. In the CNS a single postsynaptic neuron will receive inputs from thousands of synaptic inputs. The process of integrating all of these inputs over time into a single neuronal output is called **spatiotemporal summation.** At each synapse, neurotransmitter release causes a local change in the membrane potential of the postsynaptic target cell. Some of these synapses are excitatory and produce small depolarizations called **excitatory postsynaptic potentials (EPSPs).** Other synapses are inhibitory and produce small hyperpolarizations referred to as **inhibitory postsynaptic potentials (IPSPs).** Both IPSPs and EPSPs vary in magnitude and duration. These signals travel to the axon hillock where they are spatially and temporally integrated. If excitatory inputs dominate, then the cell body will depolarize, and voltage-sensitive Na^+ channels will open. If a sufficient number of channels are opened, an AP will be generated. If inhibitory signals dominate, then there will be a hyperpolarization and inhibition of an AP (Fig. 8–13).

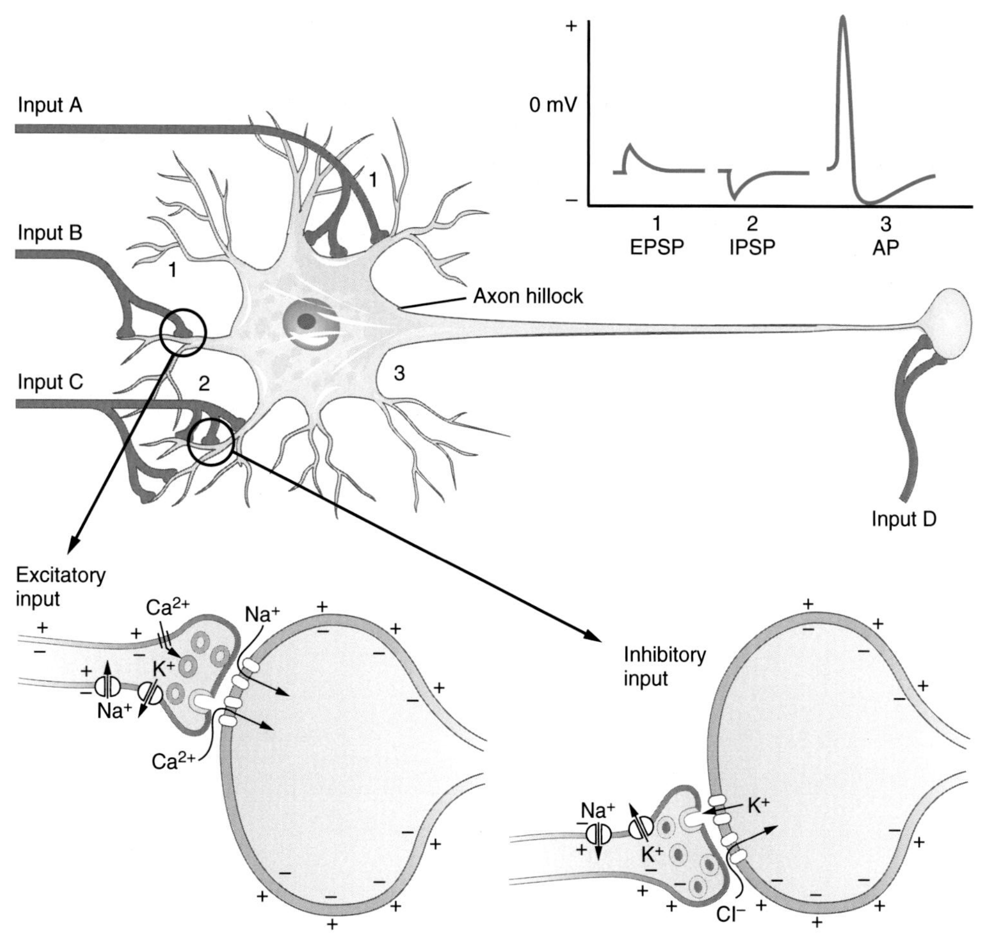

Figure 8–13. **Synaptic transmission in the central nervous system.** Neurons in the CNS have many synaptic contacts distributed on the cell body (Input A), dendritic tree (Inputs B and C), and terminal boutons (Input D). Some of these contacts are excitatory, resulting in EPSPs (Inputs A and B), and others are inhibitory, resulting in IPSPs (Input C). The ultimate decision of whether to fire an AP is determined by the weighted integral of all of these inputs over time.

SUMMARY

The coordination of day-to-day physiology and behavior in multicellular organisms requires that individual cells sense and respond to changes in their environment. To accomplish this task, cells use hundreds of different intercellular signaling molecules to generate an almost unlimited number of spatially and temporally coordinated signals. A basic understanding of the cellular/molecular events involved in intercellular signaling is essential for the diagnosis and treatment of human diseases. Deficiencies in intercellular communication are the precipitating event in many diseases, including endocrine disorders, neurodegenerative disorders, and all types of cancers. In addition, agents that mimic or antagonize the actions of cell-cell signaling molecules are common drug targets. These agents include the widely used over-the-counter analgesics such as aspirin, acetaminophen, and ibuprofen; prescription medications for hormone replacement therapy such as insulin and estrogen; abused medications for improved athletic performance such as GH and erythropoietin; and recreational drugs such as marijuana and cocaine.

Suggested Readings

Emer M, Smyth EM, Grosser T, FitzGerald GA. Chapter 37: Lipid-derived autacoids: eicosanoids and platelet-activating factor. In Brunton LL, Hilal-Dandan R, Knollmann BC, editors. *Goodman & Gilman's: The Pharmacological Basis of Therapeutics*. New York: McGraw-Hill Medical Publishing Division, 2017.

Free BR, Clark J, Amara S, Sibley DR. Chapter 14: Neurotransmission in the central nervous system. In Brunton LL, Hilal-Dandan R, Knollmann BC, editors. *Goodman & Gilman's The Pharmacological Basis of Therapeutics*. New York: McGraw-Hill Medical Publishing Division, 2017.

Kandel ER, Schwartz JH, Jessell TM, Siegelbaum SA, Hudspeth AJ, Mack S. Chapter 8: Overview of synaptic transmission. In *Principles of Neural Science*. 5th ed. The McGraw-Hill Companies, Inc, 2013.

Molitch ME, Schimmer BP. Chapter 42: Introduction to endocrinology: The hypothalamic-pituitary axis. In Brunton LL, Hilal-Dandan R, Knollmann BC, editors. *Goodman & Gilman's: The Pharmacological Basis of Therapeutics*. New York: McGraw-Hill Medical Publishing Division, 2017.

Sibley DR, Hazelwood LA, Amara SG. Chapter 13: 5-Hydroxytryptamine (serotonin) and dopamine. In Brunton LL, Hilal-Dandan R, Knollmann BC, editors. *Goodman & Gilman's The Pharmacological Basis of Therapeutics*. New York: McGraw-Hill Medical Publishing Division, 2017.

Skidgel RA. Chapter 39: Histamine, bradykinin, and their antagonists. In Brunton LL, Hilal-Dandan R, Knollmann BC, editors. *Goodman & Gilman's The Pharmacological Basis of Therapeutics*. New York: McGraw-Hill Medical Publishing Division, 2017.

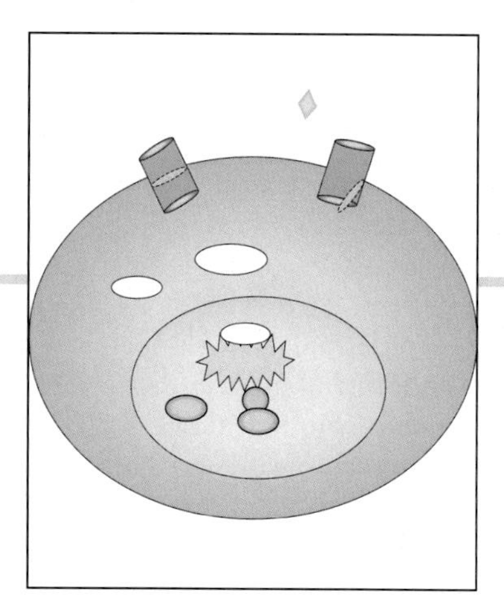

Chapter 9

Cell Signaling Events

INTRODUCTION

The ability of multicellular organisms to evolve into complex organisms would have been drastically hindered if cells lacked the ability to communicate with one another.

All complex organisms, vertebrates or invertebrates, must create and organize cells into higher-order structures like tissues, which further organize together to create organs. For these ultimately complex processes to occur, cells must have distinct signals to send specific messages to other cells around them, as well as ways to sense changes in their environment. Cell communication relies on two steps: first, a signaling cell sends an instructive signal, and second, a responding cell receives the instructive signal and changes its behavior accordingly. Signal transduction pathways allow signals from external stimuli to be moved from the outer cell membrane of the responding cell to its nucleus for a response and/or change to take place. Signal transduction pathways' complexity increases in number and regulation proportionally to the organism's complexity. However, a lot of the basic principles that allow these processes to take place are preserved. Generally, an external stimulus is sensed by a receptor on the plasma membrane or cytoplasm of the responding cell, leading to receptor activation, setting in motion a cascade of events that ultimately leads to activation of transcription factors and alterations in gene expression. This chapter provides examples of key signaling pathways involved in cell differentiation, survival, and proliferation. It depicts the diversity and complexity of cell signaling, along with their roles in normal physiological development, as well as their implications in cancer development and progression.

Mediation of Signaling by Cell-Surface Receptors

Effective cell-to-cell communication relies not only on signaling molecules secreted by the signaling cell, but also on the responding cell's ability to sense the signal. There are a wide variety of cell surface receptors that allow for this to occur in a regulated and precise manner.

KINASE RECEPTOR SIGNALING

Cell-surface receptors with catalytic activity are kinases that utilize phosphorylation as their mode of signal transduction. Activation of a catalytic receptor by its ligand leads to the transfer of a phosphate group from the receptor to an amino acid on the ligand. Phosphorylation typically occurs by covalent linkage of a phosphate group to a serine, threonine, or tyrosine amino acid residue. In a variety of signal transduction pathways, especially those that utilize tyrosine kinases, the first phosphorylation occurs at another molecule of the same receptor type. Most often, two molecules of the same receptor bind the ligand, which allows these two molecules to dimerize and form a *homodimer*. Dimerization allows the receptors to phosphorylate each other, a process referred to

Goodman's Medical Cell Biology. https://doi.org/10.1016/B978-0-12-817927-7.00009-0

as *cross-phosphorylation*, and subsequent receptor activation. Once activated, receptor kinase activity can be directed to proteins of the signaling transduction cascade in close proximity to the plasma membrane, which can be present on their own or have been recruited by other molecules to the cell membrane receptors. The intracellular signal transducers, activated via phosphorylation by cell-surface receptor kinases, are often kinases themselves, which set up a chain of phosphorylation events eventually leading to activation of transcription factors and alterations in gene expression.

In many higher vertebrates, signal transduction through catalytic receptor tyrosine kinases is the most commonly used form of communication between cells. This kind of signal transduction is vital in myriad physiologic and cellular processes in both an adult and developing organism, especially during induction and morphogenesis of tissues. A lot of what is currently understood about catalytic receptor kinases comes from studies done on developing tissues, where extracellular signaling molecules were identified and their signal transduction pathways were elucidated. These studies have helped deepen our knowledge of the molecular basis of tissue morphogenesis, organ function, and cell type induction in both normal and pathologic states.

Receptor Tyrosine Kinases

Receptor tyrosine kinases are catalytic cell-surface receptors that bind signaling molecules and phosphorylate tyrosine residues on their substrates. These receptors are important in many different kinds of cellular interactions. Primarily, the ligands associated with receptor tyrosine kinases are growth factors such as fibroblast growth factor (FGF), epidermal growth factor (EGF), insulin, and vascular endothelial growth factor (VGEF), which have a wide range of biological activities. Signal transduction via these receptors is typically carried through any of the following three pathways: the Grb2/Sos pathway, the phosphatidylonsitol-3-kinase (PI3K) pathway, or the phospholipase C (PLC) pathway (Fig. 9–1). Given that all receptors use the same pathways for signal transduction, a cell's ability to respond to a specific growth factor is dependent on the expression of the correct tyrosine kinase receptor on its cell surface.

Fibroblast Growth Factors

FGFs are crucial in many biological events such as regulating cell migration and proliferation, specification and differentiation of cells, axial patterning of the embryo

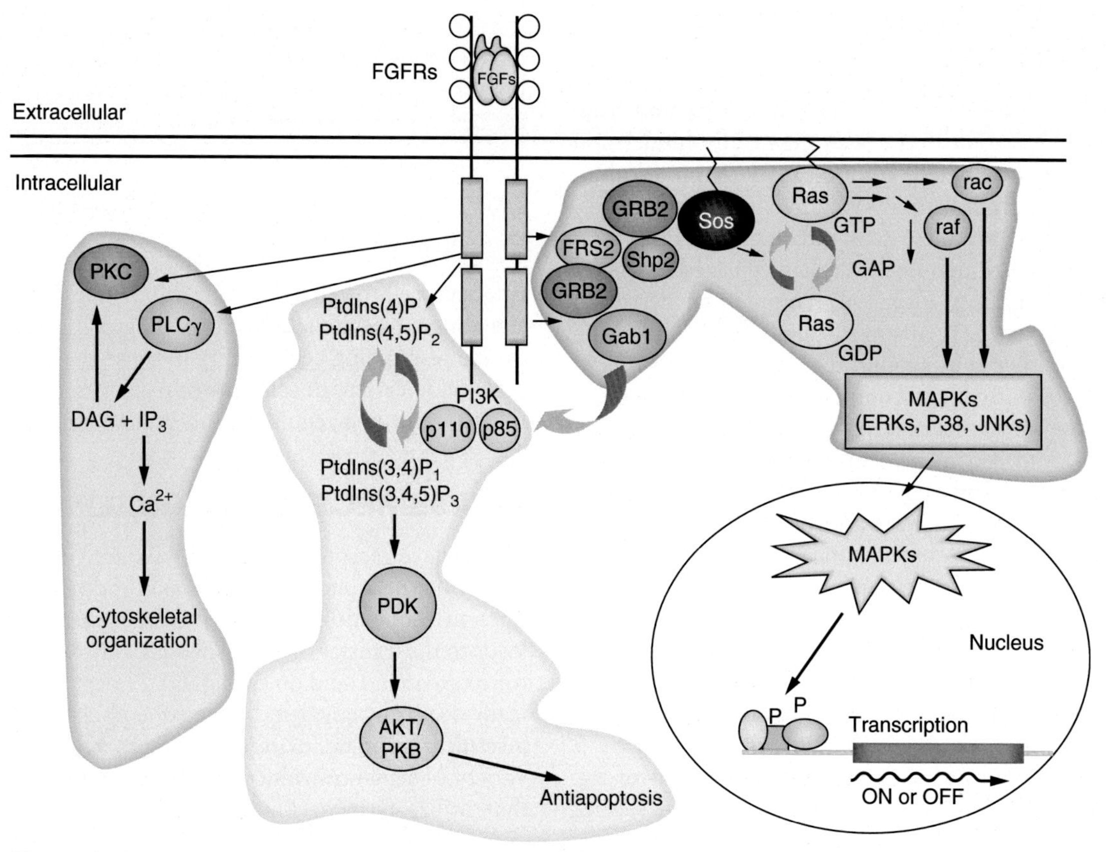

Figure 9–1. **Fibroblast growth factor (FGF) signal transduction pathways.** Activated FGF receptors (FGFRs; *red rectangles*) stimulate the phospholipase Cγ (PLCγ) pathway *(blue highlight)*, the phosphatidylinositol 3-kinase (PI3K)-AKT/-protein kinase B (PKB) pathway *(yellow highlight)*, and the FRS2-RAS-mitogen-activated protein kinase (MAPK) pathway *(green highlight)*. The activated MAPKs (extracellular signal-regulated kinases (ERKs), p38, or c-Jun N-terminal kinases (JNKs)) are translocated to the nucleus where they phosphorylate (P) transcription factors, thereby regulating target genes. *(Modified from Dailey L, et al.* Cytokine Growth Factor Rev *2005;16:233, by permission.)*

and embryonic tissues, and morphogenesis of organs and organ systems. Furthermore, FGF signal transduction pathways are important due to their role in development and cell physiology and thus make a good example to illustrate signaling through receptor tyrosine kinases.

FGF signal transduction pathway activation starts when an FGF ligand binds to the FGF receptors (FGFRs) on a cell surface (Fig. 9–1). This binding leads to FGFR dimerization and *autophosphorylation* of the receptor in its intracellular domain. This autophosphorylation event allows for recruitment and assembly of downstream signaling complexes. FGF signals are transmitted through one of three pathways: the PI3K/AKT pathway, the PLCγ/Ca^2 pathway, and the retrovirus-associated DNA sequences (RAS)/mitogen-activated protein kinase (MAPK) pathway, with the latter being the most commonly used pathway (Fig. 9–1).

RAS/MAPK Signaling Pathway

Activation of the RAS/MAPK pathway occurs when an activated FGFR binds to and phosphorylates tyrosine residues on a membrane-anchored docking protein called FGFR substrate 2α (FRS2α). Phosphorylation of FRS2α facilitates the binding of Grb2 (a small adaptor molecule) complexed with Sos (a guanine nucleotide exchange factor (GEF), essential for RAS activation.

RAS activation occurs when the guanosine diphosphate (GDP) bound to RAS is replaced by guanosine triphosphate (GTP), a reaction catalyzed by GEFs, such as Sos. RAS activation then sets in motion a phosphorylation/activation cascade involving Raf, MAPK/extracellular signal-related kinase (MEK), and the MAPK's extracellular signal-regulated kinase 1 (ERK1) and ERK2. Once the ERKs are inside the nucleus, they can complete the FGF signal transduction through phosphorylation and activation of transcription factors, which will then transcribe FGF-responsive genes.

In vertebrates, the RAS/MAPK pathway is one of the most commonly used signal transduction pathways for cell communication, and has been chosen by numerous signaling molecules and their receptors to affect their signal transmission (Fig. 9–2). Similarly to the catalytic tyrosine kinase receptors and pathways, the RAS/MAPK pathway can be directed to serve different needs by working with a variety of receptor-ligand pairs.

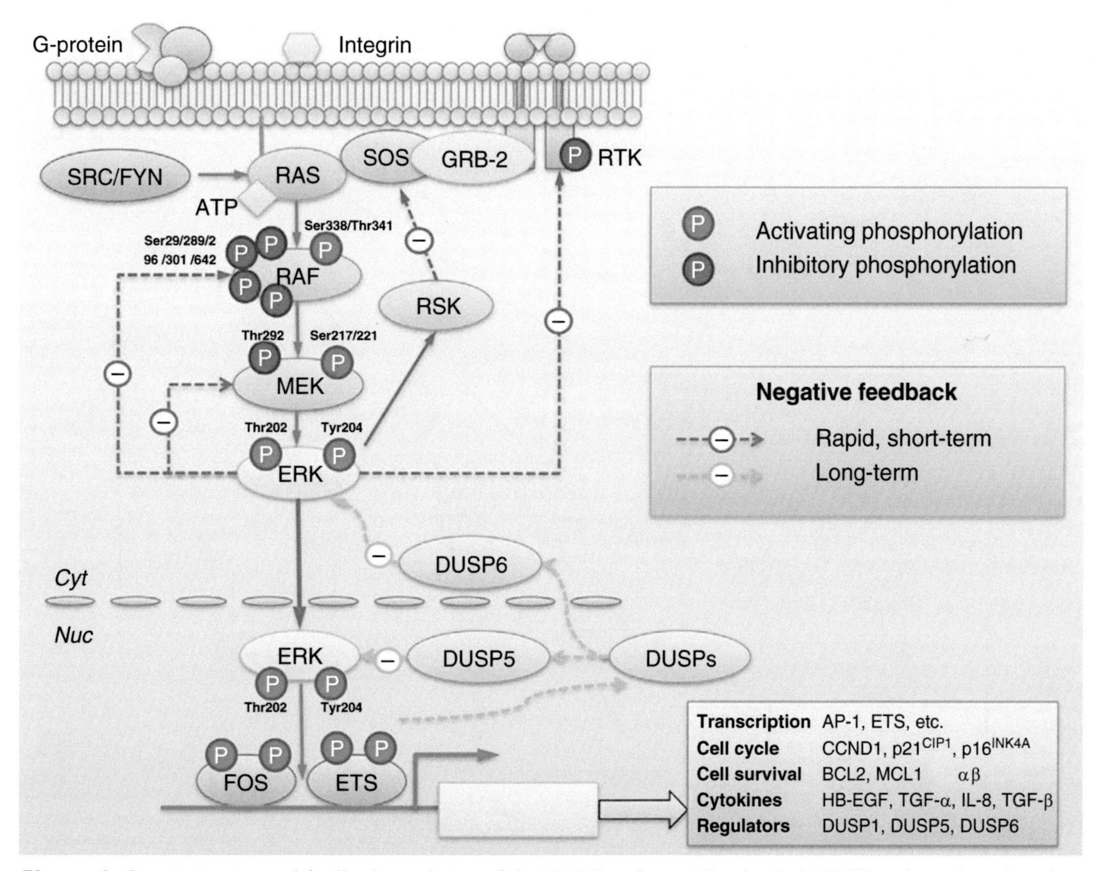

Figure 9–2. **Activation and feedback regulation of the MAPK pathway.** The classical MAPK pathway is activated in human tumors by upstream receptor tyrosine kinases (RTKs) or by mutations in RAS, BRAF, and MEK1. RTKs activate RAS by recruiting adaptor proteins (e.g., GRB-2) and exchange factors (e.g., Sos). RAS activation promotes the formation of RAF dimers, which activate MEK-ERK cascade through phosphorylation. ERK pathway activity is regulated by negative feedback at multiple levels, including the transcriptional activation of DUSP proteins that negatively regulate the pathway. ERK also phosphorylates and thus regulates CRAF and MEK activity directly. ERK, or its immediate substrate RSK, also phosphorylates Sos at several residues, inhibiting its activity and thus negatively regulating RAS activity. *(From Liu, et al.,* Acta Pharm Sin B *2018;8(4):552–562.)*

Alterations in the RAS/MAPK pathway are found in many tumor types. RAS is one of the most frequently mutated genes in human cancer. Oncogenic mutations of RAS are found in the majority of pancreatic, bladder, nonsmall cell lung, colorectal, papillary thyroid, and prostate cancer. BRAF is also frequently mutated in many cancers: BRAF V600E mutation is a classic example of RAS/MAPK pathway activation in more than 50% of melanoma and in papillary thyroid carcinomas. BRAF V600E inhibitors are used in the clinic for the treatment of melanoma with high efficacy. RAS/MAPK pathway activation leads to enhanced proliferation and increased cell motility through increased activity of ERK on several actin polymerization regulators such as the WAVE complex. ERK also connects the RAS/MAPK pathway to activation of several transcription pathways.

Inhibition of the RAS/MAPK pathway is frequently associated with paradoxical downstream pathway activation. For example, inhibition of BRAF leads to activation of MAPK or ERK. This phenomenon is due to multiple negative feedback loops within the pathway. Hence, inhibition of one component of the pathway may lead to activation of a downstream player of the pathway. This is also indicative of the existence of compensatory mechanisms within the RAS/MAPK pathway. Moreover, inhibition of the RAS/MAPK pathway often leads to cross-activation of another signaling pathway, namely the PI3K pathway. These mechanisms underlie tumor resistance mechanisms for single target therapies.

PI3K/AKT Pathway

Activation of PI3K/AKT can happen through several upstream signals such as receptor tyrosine kinases, G-protein coupled receptors (GPCRs), integrins, B- and T-cell receptors, and also through activation of RAS. The PI3K/AKT pathway is the major survival pathway as it controls cell growth, proliferation, and apoptosis (Fig. 9–3). PI3K has a major impact on protein biosynthesis through its role in glucose metabolism regulation.

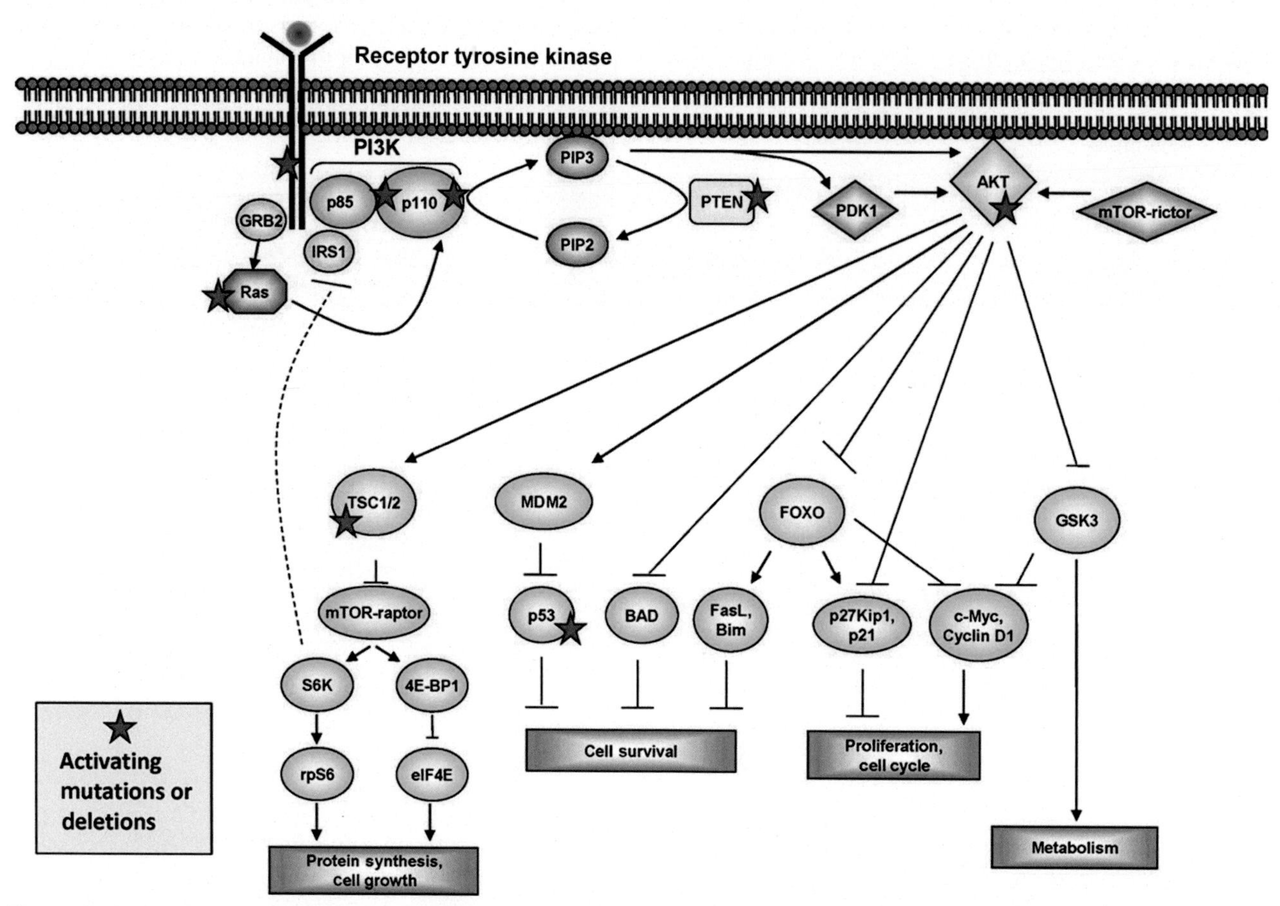

Figure 9–3. **PI3 kinase-AKT pathway mutations in cancer.** Mutations in PI3 kinase (p85 and p110) have widespread effects through activation of AKT to promote cell survival, proliferation, enhanced metabolism, and protein synthesis. The PI3 kinase pathway collaborates with the oncogenic RAS and is negatively regulated by the tumor suppressor PTEN. *Red stars* indicate mutations in the key pathway regulators. Abbreviations for protein in the pathway: *BAD*, Bcl-2-associated death promoter; *Grb2*, growth factor receptor-bound protein 2; *IRS1*, insulin receptor substrate 1; *MDM2*, murine double minute 2; *mTOR*, mammalian target of rapamycin; *PDK1*, 3-phosphoinositide-dependent protein kinase 1; *PI3K*, phosphoinositide-3 kinase; *PIP_2*, phosphatidylinositol bisphosphate; *PIP_3*, phosphatidylinositol triphosphate; *PTEN*, phosphatase and tensin homolog deleted on chromosome ten; *RAPTOR*, regulatory associated protein of TOR; *RICTOR*, rapamycin-insensitive companion of mammalian target of rapamycin; *TSC*, tuberous sclerosis. *(From Baselga J. Targeting the phosphoinositide-3 (PI3) kinase pathway in breast cancer.* The Oncologist *2011;16:12–19.)*

The classical activation of the pathway involves PI3K activity leading to production of a lipid called phosphatidylinositol-3,4,5-trisphosphate (PIP_3). PIP_3 associates with the inner side of the plasma membrane leading to activation of several proteins, including AKT through binding to a pleckstrin homology (PH) domain on their targets. In addition, AKT is phosphorylated and activated on threonine 308 (T308) by PDK1. Complete AKT activity is achieved by mammalian target of rapamycin complex 2 (mTORC2)-dependent phosphorylation on serine 473 (S473).

PI3K activity is negatively regulated by phosphatase and tensin homolog (PTEN) protein. PTEN is a lipid phosphatase, which dephosphorylates PIP_3 and thus antagonizes the action of PI3K and AKT activation. AKT is a major inhibitor of apoptosis through inhibition of BAD and transcription factor FOXO. AKT promotes proliferation by inhibition of cell cycle inhibitors such as p21 and p27.

The role of the PI3K/AKT pathway in human disease and many types of cancer is significant. PI3K is a key pathway implicated in diabetes, cardiovascular and neurological diseases, and cancer. Activating mutations of PI3K or its pathway are frequently found in breast, prostate, glioblastoma, and many other tumor types. Importantly, PTEN is one of the most frequently mutated tumor suppressors in human cancer, making the PI3K/AKT pathway a critical target in antitumor therapy (Clinical Case 9–1).

Clinical Case 9-1 ***Prostate Cancer***

Peter Miller is a 69-year old Caucasian male who has been diagnosed with prostate cancer 10 years ago. He had no symptoms at that time, but his prostate-specific antigen (PSA) test was over 10.3, which was performed following a positive digital rectal examination (DRE). The biopsy test confirmed the suspicion of prostatic adenocarcinoma in 7 out 12 magnetic resonance imaging (MRI)-guided biopsy cores and high prostate tumor volume. The Gleason score was 4 + 3, thus he was considered high risk. Following the visit with the urological oncology surgeon Peter had undergone radical surgical prostatectomy, with postoperative negative tumor margins. His postoperative PSA level was 0.01, indicating complete removal of the primary prostate tumor. Peter recovered quickly from the operation and returned to regular life as a retired accountant. Peter liked traveling; he enjoyed frequent trips to his vacation house in Florida, and an occasional trip to Europe or South Pacific. However, based on the advice from his surgeon, Peter has continued to monitor his PSA levels for tumor recurrence. Five years ago, one of Peter's PSA tests was 1.2 and rose up to 7.8, 6 months later. Peter visited his surgeon, who told him that his tumor was coming back. Peter was referred to a medical oncologist, who placed him on androgen deprivation therapy (ADT) drug Lupron. PSA monitoring indicated that the ADT-based therapy worked as it went down below 0.2 and stayed in this range for about 2 years. Peter did not like the ADT therapy because he lost his energy, sex drive, and felt some pain in his breasts. Unfortunately, after about 2 years of Lupron, two subsequent PSA tests demonstrated continued increases. Peter was put on a novel antiandrogen therapy drug, enzalutamide, and continued to be monitored by a medical oncologist. Unfortunately within 14 months, Peter developed metastatic bone tumors which was confirmed on prostate-specific membrane antigen (PSMA)-PET scan. The sites of metastasis included his spine and one of his pelvic bones. His current treatment options are chemotherapy, radiation or radiopharmaceuticals to curtail metastatic tumor growth, and bisphosphonates or Denosumab for bone loss to prevent clinical fractures. Novel approaches under clinical trials include immunotherapy.

Cell Biology, Diagnosis, and Treatment of Prostate Cancer Metastasis

Peter was diagnosed with prostate cancer, which is the most common nonskin cancer in men worldwide. Risk factors for prostate cancer include older age, race, and family history (familial prostate cancer); less established factors include diet, obesity, smoking, local inflammation, and exposures to chemicals.

Prostate cancer is characterized by genomic instability, which is exemplified by common TMPRSS-ERG gene fusions and frequent somatic gene alterations (mutations, copy number variations, mRNA alterations) of tumor suppressors and oncogenes. Key tumor suppressor genes are PTEN, RB1, p53, and ABI1. Common pathways involved in prostate cancer are PI3 kinase-, MAP kinase-, WNT-, chromatin remodeling, and DNA repair pathways. Frequently mutated genes are IDH1, SPOP, and FOXOA1 with androgen receptor (AR) being the most common.

Localized prostate cancer has high cure rates with radical treatment (prostatectomy or radiation). In contrast, metastatic prostate cancer is incurable. Development of metastatic prostate cancer often takes many years. Patients undergo several treatment regimens before they succumb to the metastatic prostate cancer.

Prostate tumor growth is androgen dependent. Historically, castration to remove the source of androgen, that is, testosterone production was the first treatment of prostate cancer. Subsequent treatment modalities were developed to chemically target the androgen production and androgen receptor pathway. As exemplified by Peter Miller's case, these treatments are applied at different stages of the disease and depend on the observed levels of androgen pathway activity in conjunction with disease progression stage.

Prostate epithelial cells express PSA. The use of PSA as a biomarker for prostate cancer screening has been questioned, but PSA is a great marker for monitoring tumor remission and recurrence status in response to treatment. When Peter's primary tumor, which was a source of PSA secretions, got removed, the PSA level went down. The increases of PSA following prostatectomy and removal of Peter's tumors indicated that tumor cells secreting PSA are coming back. This is clinically termed "biochemical recurrence." At the time of biochemical recurrence, Peter was put on the ADT drug Lupron. Lupron acts as an agonist of luteinizing hormone-releasing hormone (LHRH), by blocking LHRH production, which subsequently leads to lower testosterone levels in the body. This resulted in the reduction of Peter's tumor cells lowering levels of the PSA. During ADT Peter developed side effects: fatigue, loss of sex desire, and growth of breasts, and nipple pain. Other side effects of ADT include mood swings, erectile dysfunction, hot flashes, loss of muscle mass and strength, bone mass loss, and bone fractures. About two years into the treatment Peter's PSA increased again indicating tumor cell reappearance. This phase of the disease is called castration-resistant prostate cancer (CRPC), which is characterized by the lost response to standard ADT treatment. A majority of CRPC tumors are characterized by high levels of AR, including aberrant spliced forms of AR such as ARv7. Peter was put on the second-generation antiandrogen drug enzalutamide, Xandi. Enzalutamide is anti-AR antagonist; it competes with DHT for ligand binding site. Clinical trials indicated high effectiveness of enzalutamide toward tumor shrinkage, and improvement of quality of life such as decreased pain from bone metastasis. During ADT, Denosumab can be prescribed to prevent bone density loss and reduce bone fractures in men receiving ADT. Denosumab is a recombinant humanized antibody that binds to the receptor activator of nuclear factor-KB ligand (RANKL), which is a critical mediator of osteoclast formation, function, and survival. When Peter developed bone metastasis, the disease entered the stage of metastatic CRPC. Unfortunately, although additional treatment options are available (see above) that are likely to extend Peter's life, currently there is no curable treatment for metastatic CRPC prostate cancer.

JAK/STAT Pathway

Unlike growth factor receptors previously discussed, the Janus kinase/signal transducer and activator of transcription (JAK/STAT) pathway provides a slight variation in tyrosine kinase receptor signaling to explore. While FGFR has intrinsic kinase activity, in the JAK/STAT pathway the catalytic activity resides on a separate protein than the ligand-binding receptor (Fig. 9–4).

JAKs help transduce extracellular signals to modify behavior within the targeted cells, usually through phosphorylation-mediated regulation of STAT proteins. The JAK protein family is composed of four members: JAK1, JAK2, JAK3, and TYK2, all of which have catalytic kinase activity. Binding of an extracellular ligand to its cognate receptor results in receptor dimerization, leading to recruitment of the receptor-associated JAKs. JAKs can phosphorylate each other, which increases their catalytic activity and leads to JAK-mediated phosphorylation of the receptors. The phosphorylated site of the receptor serves as a recruitment site for STAT proteins. STAT proteins then bind to the phosphorylated receptors, leading to JAK-mediated phosphorylation of STAT proteins, causing STAT proteins to dissociate from the receptor. Phosphorylated STAT proteins can dimerize to form homodimers or heterodimers in the cytosol and translocate to the nucleus. Once in the nucleus they can initiate transcription of target genes specific to the initial extracellular signal detected by the cell-surface receptor. There are seven members of the STAT family of transcription factors: STAT1, STAT2, STAT3, STAT4, STAT5A, STAT5B, and STAT6, all of which can promote transcription of specific target genes.

JAK/STAT signaling has many roles in normal animal development. For example, JAK/STAT signaling regulates hematopoietic cell division and can even promote cell differentiation (e.g., activating genes to differentiate into one cell type versus another). The JAK/STAT pathway is also necessary for early embryonic development: for example, mutation or loss of either JAK or STAT proteins can cause unstable eye formation or irregular body segmentation in *Drosophila melanogaster*.

The JAK/STAT pathway is one of the most important signaling pathways in immune cells, as it is crucial for cytokine signaling, and thus impacts activation and recruitment of immune cells in response to stimuli. For example, activation of STAT1 can promote transcription of genes involved in inflammation while inhibiting cell division. Other STAT proteins can activate specific cell types within the immune system, thus determining what kind of immune response is mounted depending on the type of extracellular ligand bound to the receptor. For example, STAT5 can activate and mobilize white blood cells, while STAT6 can activate B cells and promote production of immunoglobulin E.

Abnormal JAK/STAT signaling has been implicated in tumor growth and cancer. Since the JAK/STAT pathway can promote transcription of genes involved in cell division, tumor growth can result from dysregulations in this pathway. For example, STAT3 and STAT5 are currently considered markers for aggressive tumors in melanoma and prostate cancers, respectively. Due to the dynamic nature of mammary gland tissue, JAK/STAT signaling is active within mammary tissue to prevent apoptosis during pregnancy and lactation. However, increased levels and/or activity of STAT3 can inhibit apoptosis and sustain cell proliferation in mammary tissue through increased transcription of c-Myc and BCL-2 genes that promote cell division and prevent apoptosis, respectively, resulting in neoplastic changes that ultimately form malignant breast tumors. Hyperactivity in JAK/STAT, as well as other kinase-dependent pathways, is suggestive of increased production of JAK or STAT proteins, increased activity in either proteins, or a combination of both scenarios.

JAK2 mutations are associated with a wide variety of blood disorders such as leukemia, polycythemia vera (PV), myelofibrosis (MF), and essential thrombocythemia (ET). Thus detection of JAK2 mutations is considered a risk factor for these diseases. Additionally, mutations in the STAT genes can increase JAK/STAT signaling in immune cells, promoting their proliferation and activity, which can also result in blood disorders (Fig. 9–4).

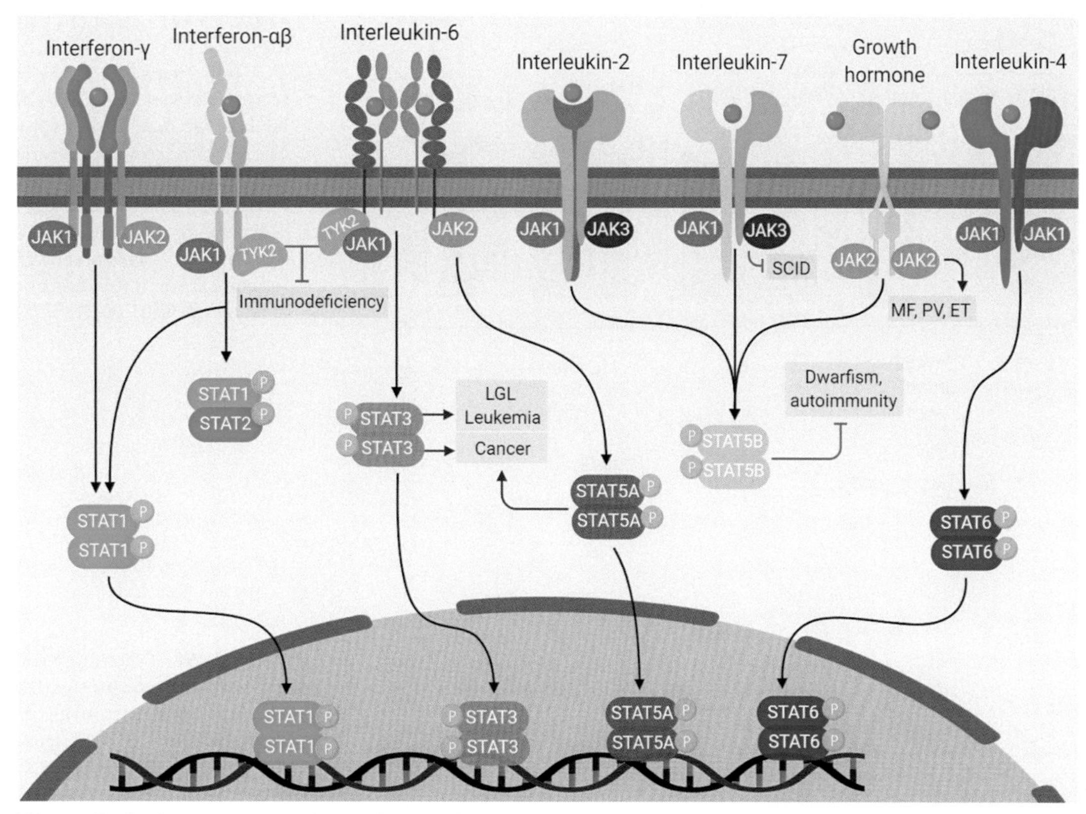

Figure 9–4. **JAK/STAT signaling pathway and associated disorders.** Mutations in JAK and STAT protein dysregulation have been associated with many human diseases with the most prominent being immunodeficiency and cancer.
(By Baylee Porter, Maria Ortiz, and Leszek Kotula, chapter authors' figure.)

Receptor Serine/Threonine Kinases

Similar to receptor tyrosine kinases, serine/threonine kinases are cell-surface receptors that are activated to transmit signals via phosphorylation. However, instead of phosphorylating a tyrosine residue, these kinases phosphorylate their substrates at a serine or threonine amino acid residue. Catalytic receptor serine/threonine kinase signal transduction pathways are used during many different points of an organism's development, as well as in a number of different biological contexts, dependent largely on the type of ligand or signaling molecule inducer, the time and place of the ligand being produced, and control over signaling.

TGF-β Receptor

The importance of serine/threonine receptors becomes apparent in their role in signal transduction of members of the transforming growth factor-β (TGF-β) family. Subfamilies of the TGF-β family are comprised of signaling molecules organized by structural and functional homologies that include TGF-β, bone morphogenetic protein (BMP), and activin (Fig. 9–5).

TGF-β Signaling

TGF-β receptor formation and activation greatly differs from the formation and activation of receptor tyrosine kinases. First, TGF-β receptors are comprised of two receptor subtypes, type 1 (TβR-I) and type II (TβR-II), neither of which is able to bind and transduce the TGF-β signal alone. Instead, ligand binding must first induce a complex of both of these receptors to form so that a functional receptor can form (Fig. 9–6). Generally, it seems as though TβR-I is the signal-transducing receptor and TβR-II is the ligand-binding receptor. First, TGF-β binds to the type II receptor, which causes a conformational change in the TGF-β molecule that allows it to be recognized by the type I receptor. Binding of TGF-β to the type I receptor brings the receptor in close proximity to the type II receptor. This allows the type II receptor to phosphorylate the type I receptor at serine and threonine residues, which activates type I receptor kinase activity. Lastly, type I receptor transduces the TGF-β signal through downstream phosphorylation of signal transducer molecules.

BMP Signaling

An important signaling molecule during organogenesis and cell induction events is BMP. Signaling of BMPs through the TGF-β receptor superfamily is a major signaling pathway in early developmental events, as well as in osteoblast differentiation for bone formation.

Activation of BMP signal transduction via the TGF-β receptor requires BMP binding to both type I and type II

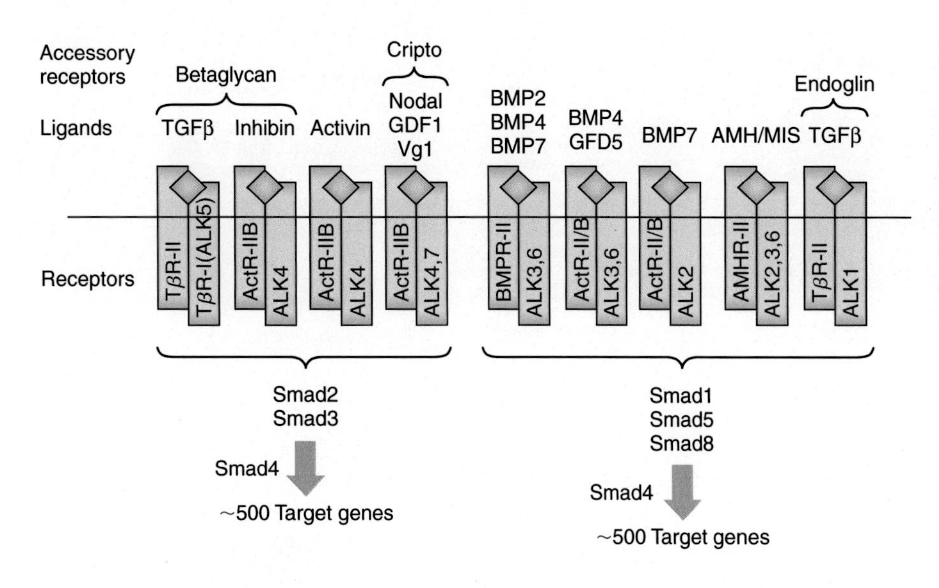

Figure 9–5. **Relations between transforming growth factor-β (TGF-β) and TGF-β-like ligands and their type I and II receptors in vertebrates.** The Nodal ligand binds to the ActR-IIB-Alk4 heterodimer. The activated receptor transmits the Nodal signal via Smad2 and 3, which heterodimerize with Smad4 to activate target genes. The bone morphogenetic protein (BMP) signaling pathways are shown on the right for comparison. *(Modified from Shi Y, Massague J.* Cell *2003;113:685, by permission.)*

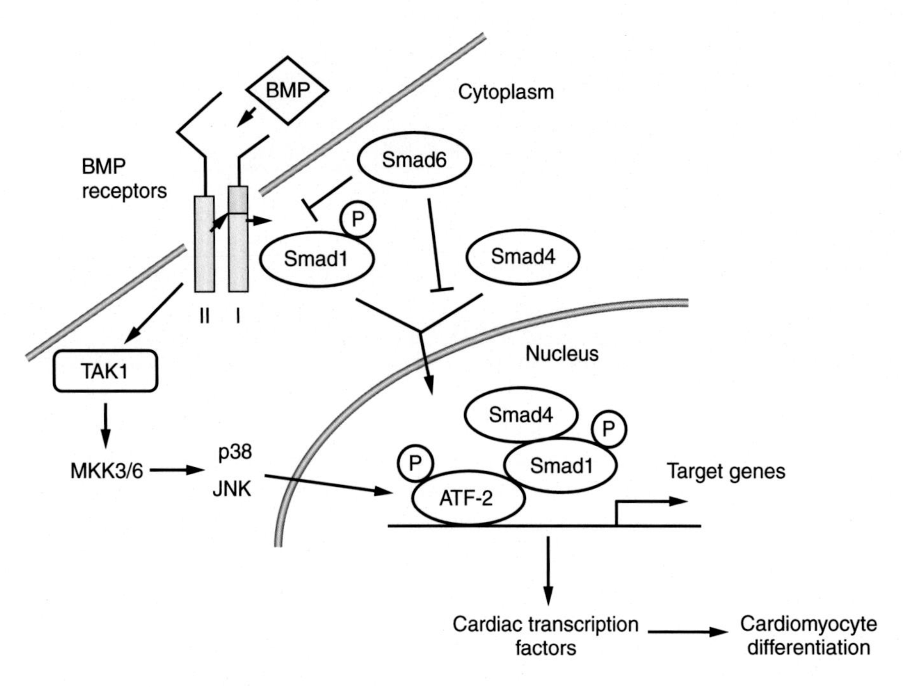

Figure 9–6. **Schematic representation of bone morphogenetic protein (BMP) signal transduction pathways involved in cardiogenic induction.** The BMP signal can be transmitted via the TAK1 signaling pathway or via Smad proteins, in particular Smad1 and 4. The Smad1/4 heterodimer can bind the ATF-2 transcription factor activating it to transcribe BMP-responsive genes. The same can be achieved by the alternate TAK1 pathway via the mitogen-activated protein kinases MKK3/6, which phosphorylate and activate the stress-activated protein kinases p38 and c-Jun N-terminal kinase (JNK) to go on and activate ATF-2. *(Modified from Monzen K, Nagai R, Komuro I.* Trends Cardiovasc Med *2002;12:263, by permission.)*

receptor subtypes. BMP serves as a bridge to bring type I and type II receptors into close proximity for the type II receptor to be able to phosphorylate type I. Once phosphorylated, the type I receptor acts as a kinase to transduce the BMP signal to downstream effectors. Now, activation of either of these two signal transduction pathways can occur: TAK1-MMK3/6-p38/JNK and/or Smad pathways (Fig. 9–5).

TAK1 is a member of the mitogen-activated protein kinase kinase kinase (MAPKKK) superfamily. When the BMP signal is transmitted through the TAK1-MMK3/6-p38/JNK pathway, TAK1 is recruited and phosphorylated by an activated BMP receptor. This leads to a number of phosphorylation reactions taking place, which ultimately lead to the activation of the transcription factor ATF-2 and upregulation of responsive genes.

If the BMP signal is transmitted through the Smad signal pathway instead, Smad1, Smad5, or Smad8 proteins are recruited by the type I receptor (Figs. 9–6 and 9–7). Here, they are phosphorylated and released into the cytosol. The Smad proteins, which are now BMP ligand specific, associate with Smad4, a protein that does not bind with the receptors directly. This new complex, the Smad1/5/8-Smad4 complex, moves from the cytoplasm to the nucleus. The Smad1/5/8-Smad4 heterodimer then binds to and activates ATF-2 in the nucleus. The activated ATF-2 can now transcribe BMP-responsive genes.

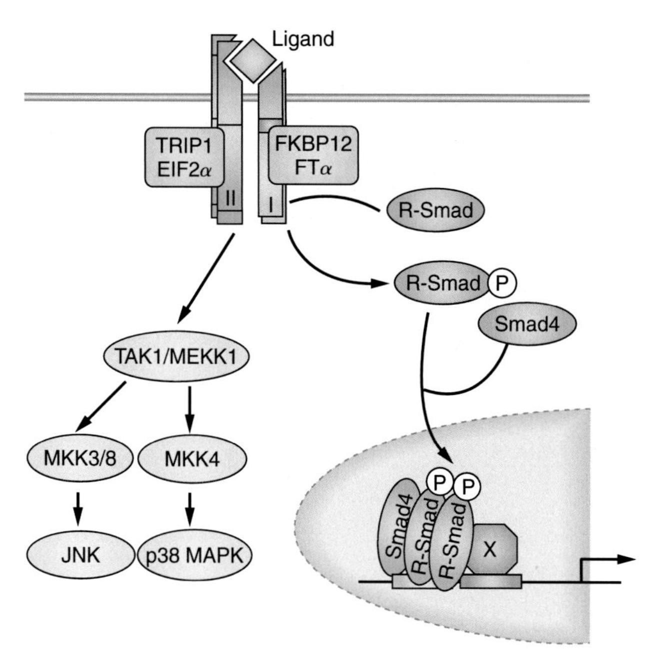

Figure 9–7. **Transforming growth factor-β (TGF-β) receptor signaling through Smad-independent pathways.** The TGF-β signal can be directed to different signaling pathways such as the TAK1/MEKK1 or Smad pathways. This will activate presumably different gene programs through the activation of different transcriptional effectors such as c-Jun N-terminal kinase (JNK), p38, mitogen-activated protein kinase (MAPK), or Smad. *(Modified from Derynck R, Zhang YE.* Nature *2003;425:577, by permission.)*

NONKINASE RECEPTOR SIGNALING

While kinase-dependent cell signaling plays a large role in signal transduction, cells are not limited to these signal transduction pathways. There are indeed cellular pathways that can transduce signal without the use of receptor kinase activity. These pathways include the Wnt/β-catenin, Hedgehog (HH), and Notch signaling pathways. Each of these pathways can be found in most cell types and plays a role in various cell responses, all without direct use of kinase activity to facilitate signal transduction. Instead, they behave to either block degradation of a transcriptional activator (Wnt/β-catenin), prevent transcriptional activators from being converted into transcriptional inhibitors (HH), or function as a transcriptional activator *in cis* upon interacting with the signal-transmitting cell by displacing transcriptional repressors from their target genes (Notch).

Wnt/β-Catenin Signaling

The Wnt/β-catenin signaling pathway is involved in a number of cellular pathways and is essential for embryonic development in vertebrates. Wnt signaling is especially important in cell differentiation/cell fate determination, axis patterning, cell migration, and cell polarity. Wnt proteins are glycoproteins that can associate with the extracellular matrix *in cis* to act on the same cell that produced it and/or a neighboring cell. Alternatively, Wnt proteins can be secreted to act on another cell over a distance, likely by eliciting changes in one cell to transmit signal to a neighboring cell in a manner that resembles a cell-signaling relay to transmit signals to the distant target cell.

Wnt signaling can elicit a number of cellular responses that are primarily dependent on which Wnt protein and receptors are involved. There are 19 Wnt family isoforms in humans with two different families of Wnt receptors: Frizzled and lipoprotein-receptor related proteins (LRPs), which function as coreceptors with Frizzled. Taken together with the fact that Wnt proteins can elicit activating or inhibitory signals, the Wnt/β-catenin signaling pathway requires delicate coordination between the proteins involved to elicit the desired biological response. The wide array of Wnt-induced cellular responses is suggestive of Wnt participation in a number of signaling pathways. Indeed, some Wnt signaling can occur in one of three ways: canonical Wnt signaling, noncanonical planar cell polarity pathway, and noncanonical Wnt/calcium pathway (Fig. 9–8).

In the *canonical Wnt/β-catenin signaling pathway*, Wnt signaling is turned off in the absence of a Wnt ligand. Without a Wnt ligand, the protein β-catenin is targeted for proteosomal degradation by a destruction complex comprised of other proteins: glycogen synthase kinase-beta (GSK3β), Axin, adenomatosis polyposis coli (APC), Dishevelled (Dsh), casein kinase 1α (CK1α), and GBP/Frat. Upon interacting with this complex, β-catenin is phosphorylated by CK1α and GSK3β, targeting β-catenin for ubiquitination by a ubiquitin ligase complex (β-TrCP) and is subsequently degraded by the proteosomal machinery. Degradation of β-catenin prevents its cytosolic accumulation and nuclear translocation, effectively inhibiting transcription of its target genes through this pathway (e.g., c-myc, cyclin D). Wnt signaling is initiated upon binding of a Wnt protein ligand to Frizzled and coreceptors LRP5/6, which disrupts the destruction complex, allowing for cytosolic accumulation of β-catenin (Fig. 9–8, A). β-Catenin then translocates to the nucleus where it can alter cell behavior through changes in gene transcription. Nuclear localization of β-catenin activates latent transcription factors (T-cell factor/lymphoid enhancer-binding factor (TCF/LEF) family) to modify gene transcription in response to the initial signaling stimulus.

The *noncanonical Wnt/calcium pathway* functions to help regulate calcium levels in the cell and acts independently of β-catenin. Similar to the canonical Wnt pathway, Wnt/calcium signaling begins upon binding of Wnt to the Frizzled receptor. Wnt/Frizzled binding activates a trimeric G-protein and PLC enzyme that produces diacylglycerol (DAG) and inositol triphosphate (IP_3) proteins, ultimately resulting in release of calcium from the endoplasmic reticulum. Increase in cytosolic calcium activates calcineurin (CaN) and CamKII (a calcium-dependent phosphatase and kinase, respectively; discussed later), as well as protein kinase C

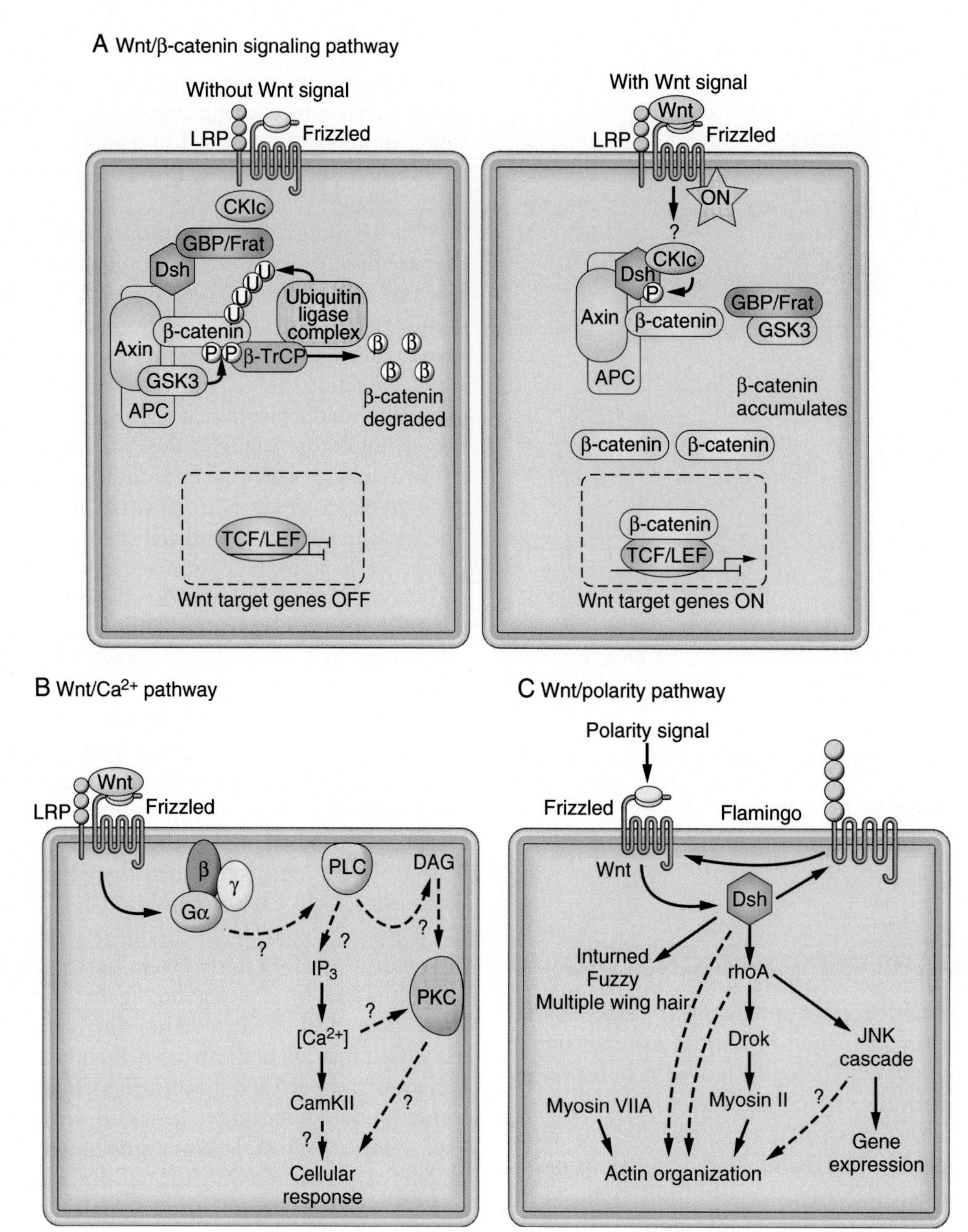

Figure 9–8. **Wnt signaling pathways are diverse.** (A) The canonical Wnt/β-catenin signaling pathway is highly dependent on availability of β-catenin. In the absence of Wnt ligand, β-catenin is marked for proteosomal degradation upon ubiquitination by β-TrCP. Binding of Wnt ligand results in disruption of the destruction complex, preventing ubiquitination of β-catenin. β-Catenin can accumulate in the cytosol for nuclear translocation, activating TCF/LEF transcription factors to promote gene transcription. (B) Noncanonical Wnt/Ca^{2+} signaling relies on GPCR activation to release intracellular Ca^{2+} from the endoplasmic reticulum, which activates Ca^{2+}-dependent enzymes such as calmodulin, calcineurin, and CaM kinases to facilitate Wnt/Ca^{2+} signaling response. (C) Wnt polarity signaling can polarize cells to modulate cell motility especially during embryogenesis. *(Modified from Miller JR. The Wnts.* Genome Biol *2001;3:3001.1, by permission.)*

(PKC). Activation of CamKII modifies cell migration and cell adhesion in response to signals. CaN activation, however, can inhibit TCF/LEF-mediated gene transcription, illustrating that Wnt signaling can induce opposing cellular responses that are dependent on Wnt proteins, their receptors, and the pathway through which signals are transmitted (Fig. 9–8, B).

Noncanonical planar cell polarity Wnt signaling also acts on the cell independently of β-catenin. Wnt ligand binds to Frizzled and another coreceptor, resulting in activation of G-protein Rho, which, in turn, activates Rho-associated kinase (ROCK). ROCK is a major regulator of actin cytoskeleton dynamics within cells and Wnt-mediated activation is a major means of driving cell

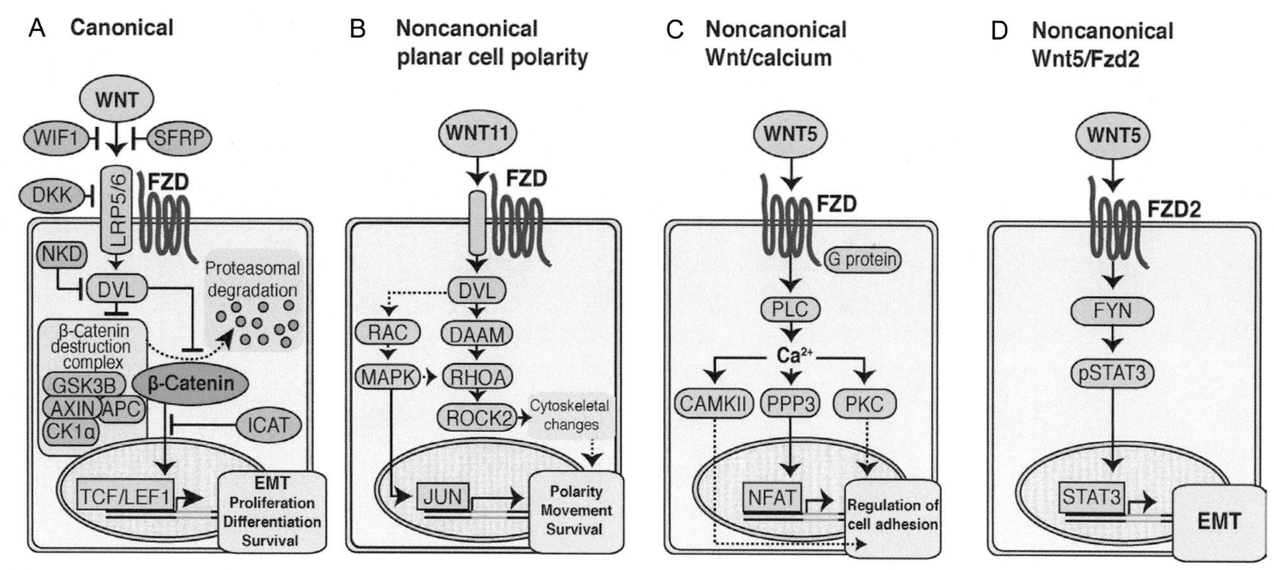

Figure 9–9. **Wnt signaling in cancer.** (A) Canonical Wnt pathway. In the absence of Wnt signaling, the β-catenin destruction complex labels β-catenin for proteasomal degradation. In the presence of Wnt signaling, the destruction complex is inhibited, resulting in stabilization and nuclear translocation of β-catenin, activating transcription of target genes. (B) The noncanonical planar cell polarity (PCP) pathway activates signaling cascades resulting in cytoskeletal changes, as well as alterations in cell polarity, movement, and survival. (C) Noncanonical Wnt/calcium pathway signaling activates intracellular calcium, which, in turn, reduces cell adhesion through further signaling. (D) Noncanonical Wnt5/Fzd2 pathway. Wnt5 signals via the Fzd2 receptor and FYN activates STAT3 transcription leading to epithelial-mesenchymal transition (EMT) in cancer cells. *(From Sandsmark E, et al.* Oncotarget *2017;8:9572–9586.)*

polarity in response to cell signaling events such as those to promote motility and migration or even gastrulation during early embryonic development.

The Wnt protein family has been implicated in cancer since its initial discovery. Canonical Wnt signaling is frequently implicated in breast cancers and increased Wnt pathway activity can be determined by elevated levels of β-catenin in the cell and generally yields a poor prognosis of breast cancer patients. While increased β-catenin can be caused by a number of abnormalities, increased availability of Wnt ligands is a known precursor of β-catenin overexpression and is implicated in breast cancer cell metastasis.

Wnt overexpression can be found in a number of cancer types such as breast, glioblastoma, esophageal, and ovarian cancer, suggesting that increased availability of Wnt ligands may be considered a universally pro-oncogenic event. Furthermore, Wnt signaling dysregulations and its association with epithelial-mesenchymal transition (EMT) make this pathway a key regulator of cancer progression and metastasis (Fig. 9–9).

HH Signaling

HH signaling proteins were first discovered in the late 1970s: mutations in these genes resulted in a "hedgehog"-like appearance in the mutant fruit flies. To date, the HH signaling pathway is one of the most extensively studied signaling cascades in vertebrates. HH signaling is initiated upon translation and posttranslational processing of the Sonic hedgehog (SHH) ligand: SHH undergoes autocatalysis to yield an N-terminal signaling domain (dubbed "SHH-N") and a C-terminal fragment that currently does not have known signaling function and freely dissociates. The SHH-N fragment is further processed to covalently add a cholesterol group to its carboxyl terminus, which facilitates its trafficking, secretion, and binding to its cognate receptor (Fig. 9–10).

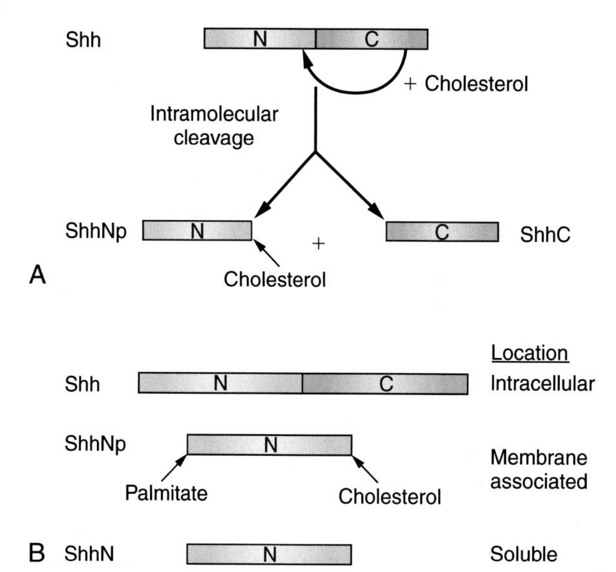

Figure 9–10. **SHH ligand can be cleaved into functionally distinct peptides.** (A) SHH ligand undergoes intramolecular cleavage, yielding two functionally distinct products: SHH-N, to which a cholesterol group is added and can translocate to the nucleus to block gene transcription, and SHH-C, which diffuses freely into the cytosol. (B) Depending on its form, SHH can localize to different parts of the cell. Uncleaved SHH ligand localizes intracellularly while palmitoylated SHH-N remains membrane bound. SHH-N without palmitoylation or cholesterol groups remains soluble and can translocate to the nucleus to regulate gene transcription. *(Modified from Goetz JA, et al.* Bioessays *2002;24:157, by permission.)*

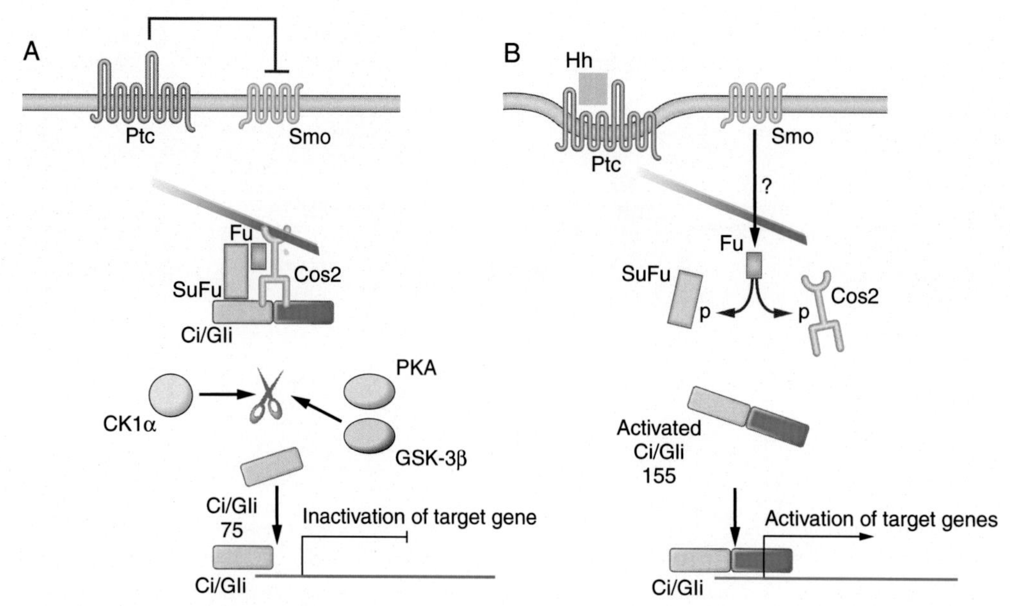

Figure 9–11. **Hedgehog (HH) signaling impacts Gli transcription factor activity.** (A) In the absence of HH ligand, Ptch exerts inhibitory function on Smo, allowing formation of a complex comprised of Fused (Fu), suppressor of Fused (SuFu), and Costal 2 (Cos2). The Fu-SuFu-Cos2 complex cleaves Gli transcription factors, producing a Gli fragment that contains no transcriptional activity. The transcriptionally null Gli fragment translocates to the nucleus and presents a physical hindrance to other transcription factors, thus preventing transcription of Gli target genes. (B) Binding of HH signal nullifies the inhibitory impact of Ptch on Smo, resulting in disruption of the Fu-SuFu-Cos2 complex. Gli transcription factors thus remain uncleaved and can translocate to the nucleus while retaining their transcription-activating features, promoting transcription of Gli target genes. *(Modified from Bijisma MF, et al.* Bioessays *2004;26:387, by permission.)*

The SHH-N ligand can act *in cis* on the producing cell or *in trans* on a nearby cell. Upon addition of the cholesterol group, the SHH-N stays tightly associated to the surface of the cell that produced it, effectively creating an SHH-N gradient that leads away from the SHH-producing cell. The nearby cells respond accordingly to this SHH-N gradient: signal-receiving cells that detect high levels of SHH-N begin production of transcriptional activators to prepare for gene transcription. The opposite effect also happens in other cells: cells that detect low or no levels of SHH-N begin production of transcriptional repressors and rely on non-HH pathways.

Under steady-state conditions, the SHH signaling pathway is inherently inactive, largely in part due to the inhibitory effect of PTCH1 transmembrane protein on SMO, the cognate receptor for SHH-N. Inhibition of SMO results in the formation of a multiprotein complex made of at least three proteins: the kinase Fused (Fu), suppressor of Fused (SuFu), and the kinesin motor protein Costal2 (Cos2). This multimeric protein complex binds to the cell's microtubules to physically sequester the Gli transcription factors, effectively preventing their translocation to the nucleus. Sequestration of Gli transcription factors is closely followed by their cleavage into two parts, one of which is the zinc-finger DNA binding domain (with no transcription-activating domains). This Gli fragment can translocate to the nucleus but, due to lack of transcriptional activity, acts as a physical repressor, and prevents binding of other transcription factors to the Gli target genes, effectively inhibiting their transcription (Fig. 9–11).

Binding of SHH-N to PTCH1 relieves the latter's inhibitory effect on SMO. Activation of SMO results in disruption of the inhibitory impact of SuFu on Gli, thereby preventing cleavage of the GLI family of transcription factors: activators Gli1 and Gli2 along with the repressive Gli3. Gli transcription factors are then activated and retain their transcription-activating domains for translocation into the nucleus to modulate gene transcription (Fig. 9–10, B). The Gli family of transcription factors can modulate a number of target genes, namely those involved with cell proliferation (n-MYC, cyclin D1, and cyclin D2), cell adhesion (Snail), angiogenesis (VEGF), and can even activate transcription of HH signaling proteins to promote a positive feedback loop of the pathway.

HH signaling is of utmost importance in early embryogenesis, especially during vertebrate limb development. Indeed, mutations in HH genes can impact digit formation in mouse embryos, negatively regulate cell fate determination, and even impact adult stem cell proliferation in mammary, neural, and hematopoietic tissues. For example, the pattern and duration of SHH protein expression in specific sites of embryonic tissues will determine whether digits will grow and the length of these new limbs (Fig. 9–12). In addition to mutations, ingestion of exogenous teratogens and/or HH pathway inhibitors can also impact normal embryogenesis. Embryos that

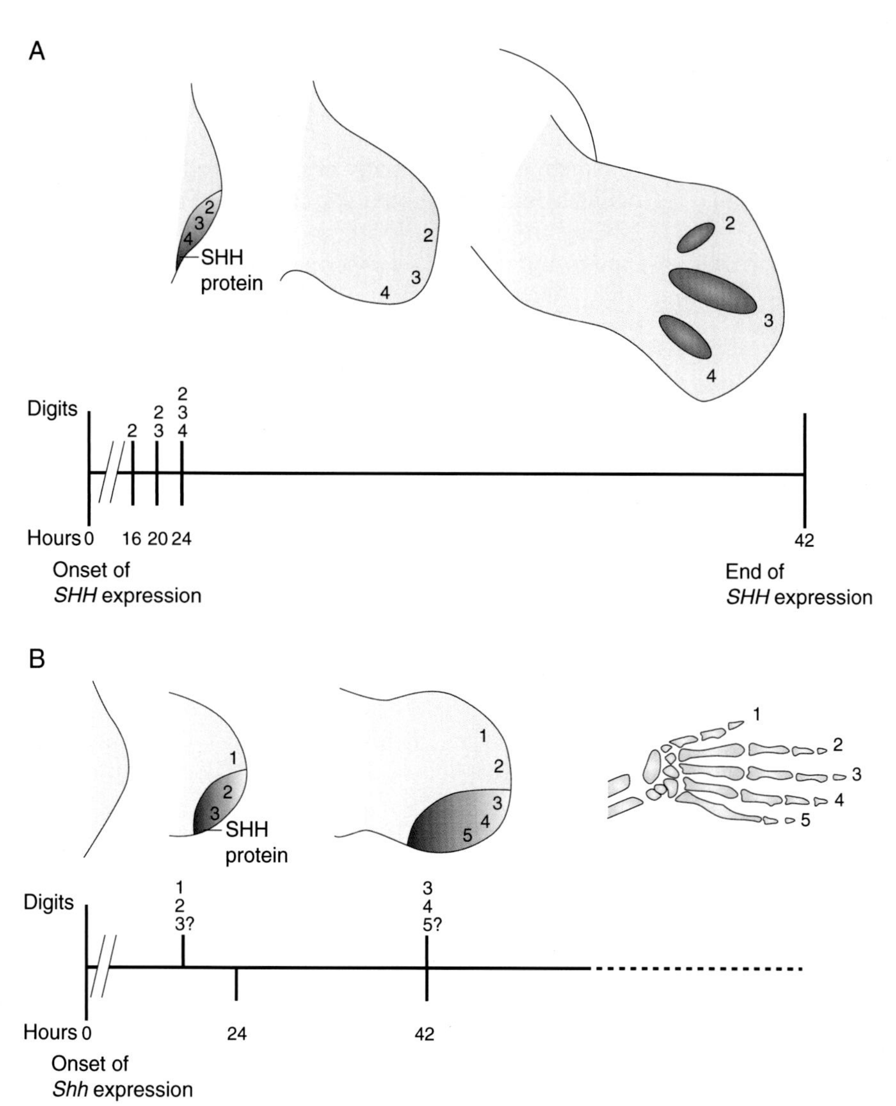

Figure 9–12. **SHH ligand is imperative for proper limb development.** Digit formation is highly dependent on SHH ligand expression. During embryonic development, distal areas with sustained SHH ligand expression will form longer digits, while areas with shorter and/or reduced SHH ligand will form shorter or no digits. *(From Tickle C,* Nat Rev Mol Cell Biol *2006;7: 45–53.)*

have disrupted HH signaling can have a variety of physical defects if they make it to term: as many as 1 in 200 embryos will self-abort if HH signaling is defective. The most severe example of congenital defects arising from abnormal HH signaling is cyclopia: the embryonic prosencephalon fails to divide the orbits of the eye into two distinct cavities, thus the fetus is born with one eye and usually with a missing nose (another consequence of this defect).

In addition to congenital defects, defective HH signaling can also result in cancer. As discussed, active HH signaling can modulate gene transcription through the Gli family of transcription factors. Aberrant activation of HH signaling can occur at many different levels: increased release/availability of the SHH-N ligand, increased availability and/or activity of the SMO receptor, increased availability of Gli transcription factors, or a combination of these possibilities. Indeed, HH signaling has been a topic of cancer therapeutics since its initial discovery. Many cancer types have exhibited increased HH signaling in these tissues, including breast, lung, brain, prostate, and skin.

HH signaling is most frequently cited in basal cell carcinoma, the most commonly diagnosed form of malignant cancer. Patients with basal cell carcinoma often present with loss of the repressive PTCH and/or increased activity in the SMO receptor. Thus increased activity in the HH signaling pathway likely causes tumorigenesis by promoting transformation of adult stem cells into cancer stem cells that ultimately contribute to tumor formation. Additionally, increased HH signaling can also impact expression of proteins involved in metastasis: loss of E-cadherin in epithelial cells results in loss of tight junctions, resulting in increased motility of aggressive cancer cells. While many antagonists against the HH pathway are being designed and tested, the most common treatment modality is to target the SMO receptor for the HH family of ligands.

Notch Signaling

Notch is a highly conserved family of signaling proteins that can be found in most vertebrates and is composed of members Notch1, Notch2, Notch3, and Notch4.

The *Notch* gene was first identified in the fruit fly species *D. melanogaster* as mutations in this gene resulted in holes, or notches, in the fly wings. Notch proteins share similar structure and function: a transmembrane protein that contains an extracellular receptor domain (Notch extracellular domain—NECD) and an intracellular domain (Notch intracellular domain—NICD).

Notch proteins play an important role in embryonic development, especially in processes involved in cell fate determination. The role of Notch signaling can be observed in the developing heart: during embryogenesis, valve formation within the nascent heart requires endocardial cells to delaminate and migrate into the space separating the myocardial and endocardial cell layers, also known as the cardiac jelly (Fig. 9–13, A). Delamination of cells from their original tissue site requires changes in physiological attributes not normally found in these cell types. Thus endocardial cells must undergo a transition from endothelial to mesenchymal, also known as EndMT, that gives these cells motility to migrate to the cardiac jelly. Establishment of endocardial cell colonies in the cardiac jelly promotes their proliferation and formation of heart valves.

Notch signaling is initiated when a closely apposed cell bearing a Notch ligand (Jagged or Delta) comes into proximity with a cell bearing the Notch transmembrane protein. The close proximity allows for the ligand to bind to Notch, which essentially primes the cell for Notch signaling by inducing proteolytic cleavage of the NICD. The NICD can then translocate to the nucleus where it can bind to and regulate a transcription factor complex called CSL, thereby effectively altering gene expression in a manner that promotes EndMT (Fig. 9–13, B). Proteolytic cleavage of Notch upon ligand binding occurs in two steps: binding of Notch to its ligand induces cleavage of the NECD at site S2 by the extracellular protease called ADAM/TACE, releasing the active Notch extracellular truncation (NEXT), which is further cleaved at sites S3 and S4 to release the NICD into the cytosol. In the absence of Notch, CSL acts as a repressive DNA-binding complex by recruiting transcriptional repressors, effectively preventing transcription. However, CSL binding to NICD causes it to become an activating protein, thus promoting transcription of Notch target genes (Fig. 9–14).

Initiation of EndMT as a consequence of Notch signaling gives cells characteristics such as increased

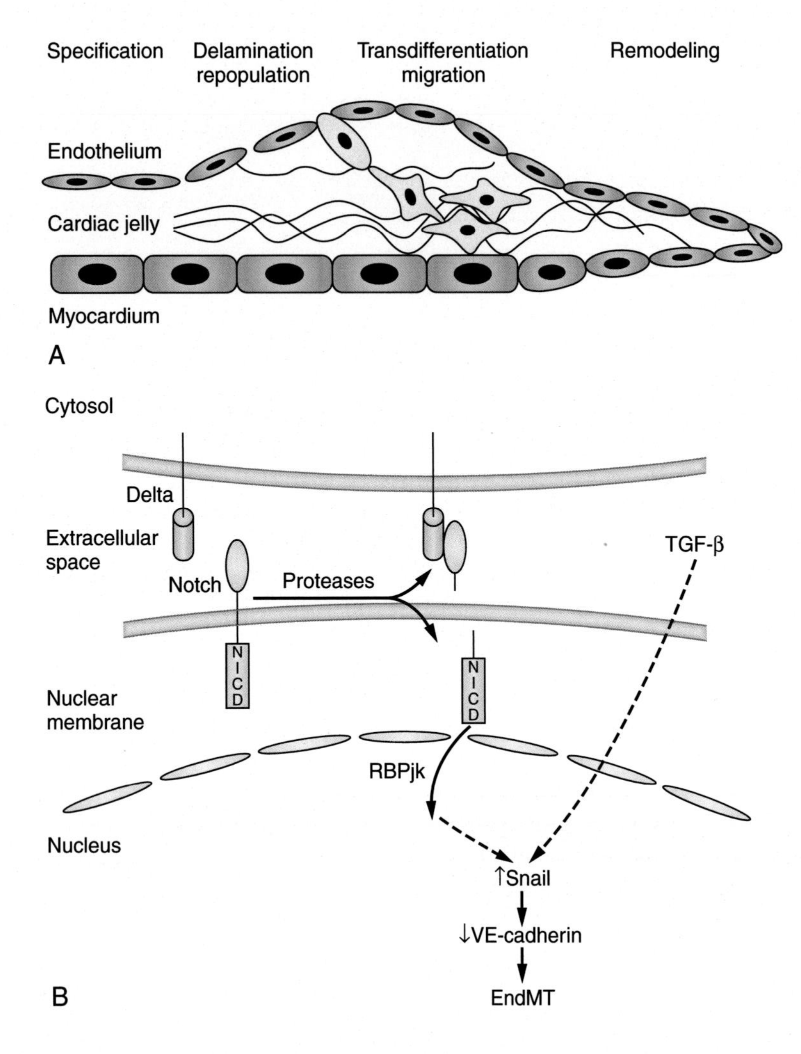

Figure 9–13. EndMT via Notch signaling during endocardial cushion and heart valve formation. (A) Anatomic overview of heart valve development. The developing heart tube contains an outer layer of myocardium and an inner lining of endothelial cells separated by an extracellular matrix referred to as the cardiac jelly. During heart valve formation, a subset of endothelial cells overlying the future valve site are specified to delaminate, differentiate, and migrate into the cardiac jelly, a process referred to as *endothelial-to-mesenchymal transition (EndMT)*. (B) In the developing cardiac cushion, Notch signaling increases the level of transforming growth factor-β2 (TGF-β2), which is known to increase the activity of the transcription factor Snail (or Slug). Snail activity may lead to downregulation of VE-cadherin, an adhesion molecule needed for binding cells together. Downregulation of cell-cell adhesion within the endothelial cell layer may be the first step in the delamination of endothelial cells and their migration into the cardiac jelly. Some evidence exists that Notch signaling may activate Snail independent of TGF-β signaling (*dashed line*). Note that this signaling is between endocardial cells, that is, autocrine signaling, and not between endocardial and myocardial cells. *(From Armstrong EJ, Bischoff J.* Circ Res *2004;95:459, by permission.)*

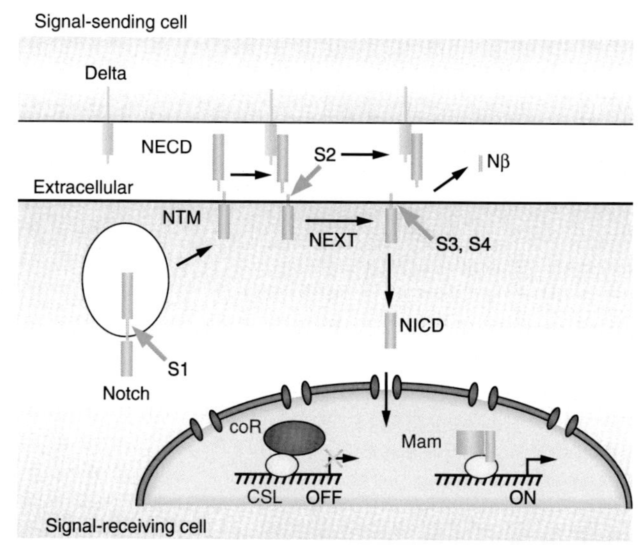

Figure 9–14. **A model for Delta-dependent Notch signaling to the nuclear transcription factor CSL.** Delta at the surface of the signaling cell binds S1-cleaved Notch at the surface of the responding cell. Ligand-dependent S2 cleavage of Notch generates an activated membrane-bound form of Notch called Notch extracellular truncation (NEXT), which is further processed at the S3 and S4 sites. This releases the Notch intracellular domain (NICD), which translocates into the nucleus where it derepresses CSL by displacing the corepressor coR. *(From Schweisguth F.* Curr Biol *2004;14: R129, by permission.)*

proliferation, increased migration, and motility. These acquired physiological characteristics in Notch target cells highlight the impact of Notch-induced gene transcription in driving early heart development. However, it is important to note that hyperactive Notch signaling can also be deleterious in vertebrates and can lead to neoplasias in certain cohorts. For example, while aberrant Notch signaling can be found in many cancer types, it is frequently found in T-cell acute lymphoblastic leukemia (T-ALL). Increased activity in Notch signaling can be due to inherent mutations in the *Notch* gene itself or mutations within the canonical Notch repressors. In cases of T-ALL, Notch signaling can cooperate with c-myc, a known oncogene, to promote de novo protein synthesis to promote leukemia cell proliferation and survival. Additionally, since Notch proteins play an important role in cell fate determination, their role in maintaining cancer stem cell populations has been discussed at length. Indeed, increased Notch signaling can also activate the genes necessary to sustain a stem cell-like phenotype of cells, thus promoting an aggressive cancer type due to increased Notch signaling.

Hormone Receptors

The third type of intercellular communication that is not reliant upon kinases or proteolytic processing is signaling using nonpeptide and peptide hormones. The signaling cell secretes hormones, and they interact with target cells expressing receptors specific for that hormone.

This process starts with one cell secreting hormones, which causes the target cell to go through biochemical changes, independent of the changes in gene expression that also occur in this cell. This is different from hormonal cell-to-cell communication from the signal transduction mechanisms because the main purpose is not to change target genes in the responding cell. However, a class of hormones exists whose main mode of action is also through alteration of gene expression—steroid hormones. Steroid hormones are nonpeptide molecules that are structurally derived from cholesterol and include the hormones cortisol, progesterone, testosterone, aldosterone, 17β-estradiol, and 1,25-dihydroxyvitamin D_3. Despite structural differences from steroid hormones, hormone-like molecules, including 3,5,3-triiodothyronine (thyroid hormone) and retinoic acid (vitamin A), are actually considered to be steroid hormone-like because of structural and functional similarities in their receptors.

Steroid Hormones

Steroid hormones, much like cytokine or growth factor signaling, change gene expression in their target cells; however, they achieve this through a different signaling mode. Instead of having a cell-surface receptor, steroid hormone receptors exist within the cytoplasm or nucleus in a free state. Thus to interact with their receptors, steroid hormones must diffuse through the cell membrane, not in the cascade of intracellular protein kinase signal transducers described earlier. This means that steroid hormone receptors when bound by their hormone ligand can directly affect gene expression, and can be considered ligand-activated transcription factors. Steroid hormones are able to diffuse through the plasma membrane because they are lipophilic (fat loving), which allows them to diffuse through the plasma membrane and reach their intracellular receptors.

The steroid hormone receptors' structure allows this signaling system to function the way it does. There are three main functional domains of steroid hormone receptors: a DNA-binding domain in the center of the molecule, a transcriptional activation domain located at the N terminus of the protein, and a steroid binding domain in the C terminus (Fig. 9–15).

Steroid hormone signaling starts with the synthesis and secretion of the hormones into the extracellular space by the signaling cell. In vertebrates, both fetus and postnatal, the hormone is carried through the bloodstream bound by proteins that prevent its degradation and increase its solubility in blood and is carried by the circulatory system throughout the body. However, only cells that express the hormone receptor will be able to respond. The receptors for androgen, progesterone, estradiol, and vitamin D_3 are in the nucleus, whereas the receptor for glucocorticoids is in the cytoplasm (Fig. 9–16). There is evidence suggesting that nuclear steroid hormone receptors in their inactive state (ligand unbound) are bound to specific DNA

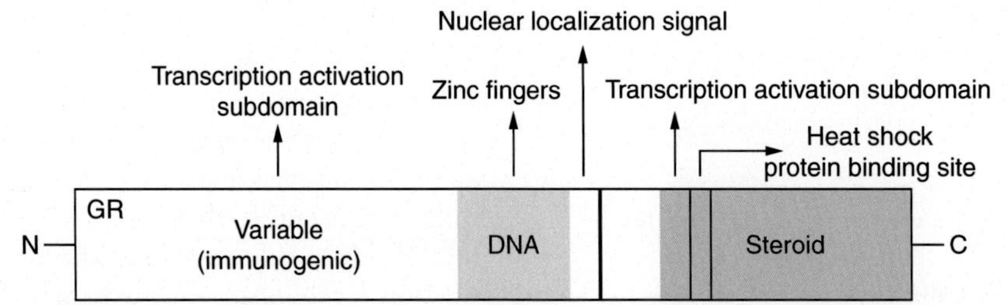

Figure 9–15. **Model of a typical steroid hormone receptor.** The glucocorticoid steroid hormone receptor provides a model for steroid hormone receptor structure. The structural features leading to function are: (1) Steroid: the steroid-binding domain in the C terminus, and (2) DNA: the DNA-binding domain that binds the receptor to specific response elements in the promoters of steroid hormone-responsive genes. Other functional domains include transcription activation subdomain, which recruits molecules of the transcriptional apparatus to the responsive gene's promoter; nuclear localization signal, which is used in translocating hormone-bound receptor to the nucleus; heat shock protein binding site, which binds heat shock protein 90 (Hsp90) in the unbound state to prevent the unoccupied receptor from binding DNA; and zinc fingers, which are protein structural motifs that intercalate into DNA-helical grooves to provide physically tight binding of receptor to DNA. *(From Devlin TM, ed. Textbook of Biochemistry With* Clinical Correlations, *5th ed. New York: Wiley-Liss, 2002.)*

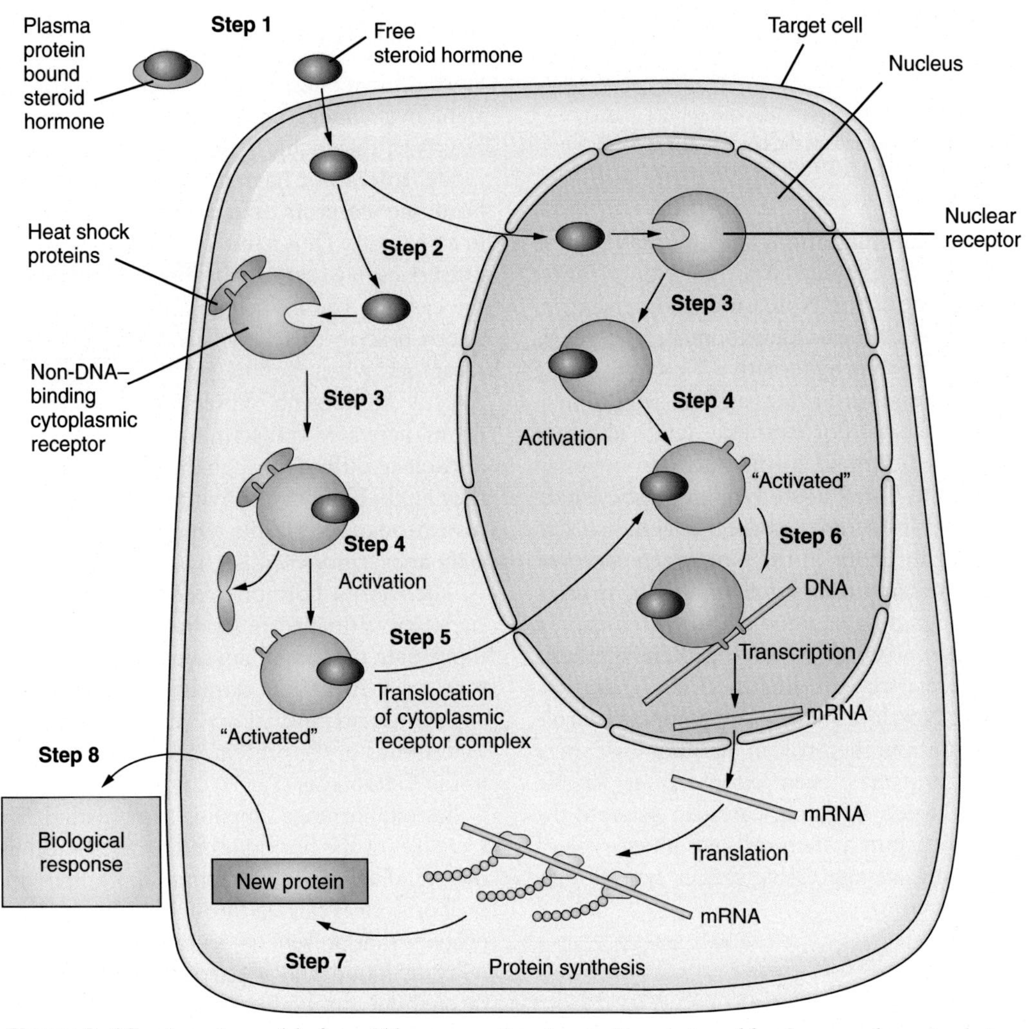

Figure 9–16. **Stepwise model of steroid hormone action.** Step 1: Dissociation of free hormone from circulating transport protein. Step 2: Diffusion of free ligand into cytosol or nucleus. Step 3: Binding of ligand to unactivated cytoplasmic or nuclear receptor. Step 4: Activation of cytosolic or nuclear hormone-receptor complex to activated, DNA-binding form. Step 5: Translocation of activated cytosolic hormone-receptor complex into nucleus. Step 6: Binding of activated hormone-receptor complexes to specific response elements within the DNA. Step 7: Synthesis of new proteins encoded by hormone-responsive genes. Step 8: Alteration in phenotype or metabolic activity of target cell mediated by specifically induced proteins. *(From Devlin TM, ed. Textbook of Biochemistry With* Clinical Correlations, *5th ed. New York: Wiley-Liss, 2002.)*

promoter sequences, potentially acting as inhibitors of transcription.

When a steroid hormone binds to its receptors, cytoplasmic receptors move to the nucleus, and nuclear receptors undergo conformational changes into an active form. In both scenarios, the hormone-bound receptors then bind to their respective hormone response elements in the promoter region of hormone-responsive genes. Once bound to DNA, through its transcriptional activation domain, the hormone receptor is able to recruit and assemble other components of the transcriptional apparatus and transcription of target genes begins. Target genes are transcribed into messenger RNA (mRNA), which is subsequently translated into proteins that change the cell's behavior in response to the steroid hormone.

G-PROTEIN COUPLED RECEPTORS

GPCR signaling is one of the most extensive signaling cascades in vertebrates. GPCRs are normally bound to a heterotrimeric G-protein complex. Two major signaling systems that utilize GPCRs are the autonomic nervous system and the renin-angiotensin-aldosterone system (RAAS).

GPCR Signaling

GPCR signaling is initiated upon binding of a ligand to the GPCR: ligand binding initiates a conformational change in the GPCR and the G-protein dissociates into Gα and Gβγ subunits, and Gα subunit exchanges GDP for GTP, creating an active GTP-bound Gα complex (Fig. 9–17). Both of these subunits are important in regulating the activation of defined second messengers.

There are multiple subtypes of the alpha subunit associated with GPCR, all with different second messengers (Fig. 9–18). For example, the Gαs subunit activates adenylyl cyclase, which, in turn, increases cyclic adenosine monophosphate (cAMP) levels. The increase in cAMP levels activates protein kinase A (PKA), which is a serine/threonine kinase that can phosphorylate several other substrates, including membrane receptors and transcription factors. On the other hand, the Gαq subunit activates PLC-β, which cleaves phosphatidylinositol bisphosphate (PIP_2) into DAG and IP_3, and mediates the activation of PKC. The Gβγ subunit functions as a dimer and activates numerous signaling molecules such as phospholipases, ion channels, and lipid kinases. GPCR signaling is involved in a wide array of cell types and is therefore critical for a variety of bodily functions such as vision, taste, smell, hormone release, immune response, and cell density sensing.

Not surprisingly, overactive GPCR signaling can cause a variety of human diseases, the most prominent of which is cancer. GPCRs are involved in over 30% of all human diseases. Many aberrations within GPCR signaling are likely due to increased availability of the receptors or increased activity, both of which can result in hyperactivation of cellular pathways involved in proliferation and survival. Increased GPCR activity can trigger increased cAMP, which, in turn, can activate ERK to promote expression of genes involved in cancer stemness. Meanwhile, increased GPCR activation can initiate signaling through phosphatidylinositols, which can ultimately promote expression of prosurvival genes.

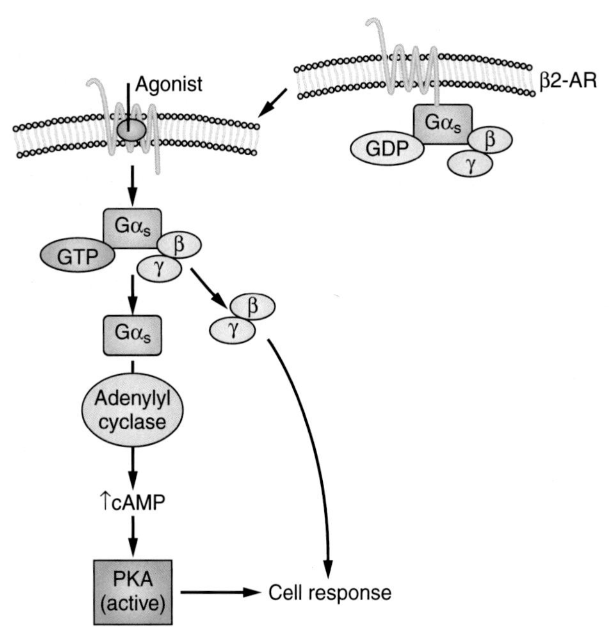

Figure 9–17. **GPCR signaling.** Upon binding of a ligand or agonist, a conformational change in the GPCR protein releases the Gα- and Gβγ-subunits, giving GEF function to the GPCR. GEF activity allows the GPCR protein to stimulate release of GDP molecules, allowing for binding of GTP molecules in their place. Release of GDP allows for the Gα- and Gβγ-subunits to continue GPCR-mediated signaling by activating several effector proteins such as adenylyl cyclase, which, in turn, increases cAMP levels. Increased cAMP activates protein kinase A (PKA), which can phosphorylate several substrates such as other 7-transmembrane receptors and transcription factors. *(Modified from Pierce, et al.* Nat Rev Mol Cell Biol *2002;3:639–650, by permission.)*

This surge in signaling activity can cause abnormal cell proliferation, survival, increased motility, and reduction of apoptosis—all of which are characteristics exhibited by cancer cells. GPCRs have been the target of many clinical trials, many of which aim to antagonize the receptor activity through use of competitive molecules or antibodies (Fig. 9–18).

RAAS Signaling

The octapeptide angiotensin II (Ang II) is the most active component of the RAAS, which is obtained by the cleavage of angiotensinogen by angiotensin converting enzyme (ACE). AT1 and AT2, part of the GPCR superfamily, mediate Ang II signaling. While little is known about the AT2 receptor, it is thought to act as a

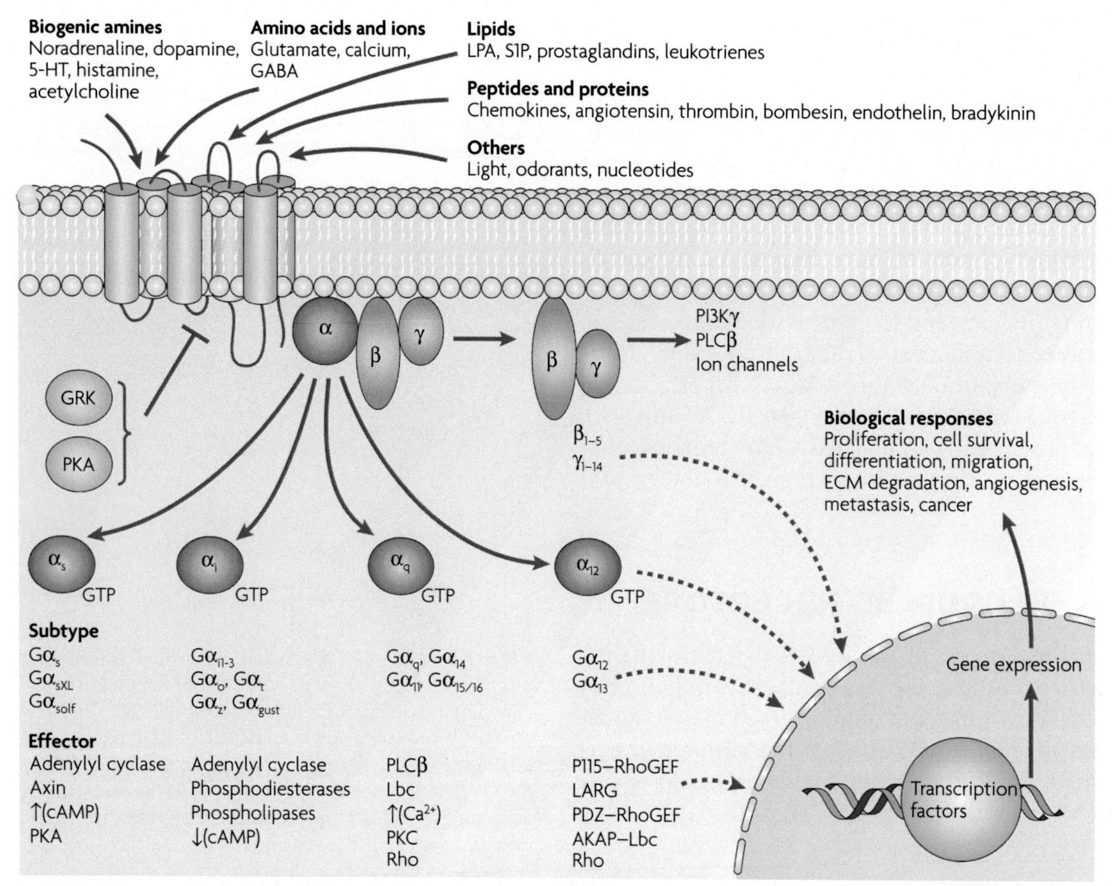

Figure 9–18. **Diversity of G-protein coupled receptor (GPCR ligands and subunits.** A wide variety of ligands use GPCRs to activate different signaling pathways in the cell. The α-subunit of G-proteins is divided into four subfamilies ($G\alpha_s$, $G\alpha_i$, $G\alpha_q$, and $G\alpha_{12}$), and a single GPCR can couple to any of these interchangeably, all activating different downstream effectors. Alterations in GPCR signaling pathways can result in cancer progression and metastasis. *(From Dorsam, et al.* Nat Rev Cancer *2007;7:79–94.)*

counterbalance to AT1, which seems to be the main receptor used as a signaling channel by Ang II (Fig. 9–19). Ang II is responsible for triggering a strong intracellular Ca^{2+} transient. This is important both for excitation-contraction coupling, and for initiating Ca^{2+}-dependent signal transduction pathways. One of the signaling pathways used by AT1 receptors involves Gαq. As previously described, Gαq activates PLC, which cleaves PIP_2 into IP_3 and DAG. Subsequently, IP_3 stimulates the IP_3 receptor (IP_3R), which causes the release of Ca^{2+} from the sarcoplasmic reticulum (SR). Additionally, RAAS activates several transduction pathways such as RAS/MAPK, JAK/STAT, and Ca^{2+}/calmodulin (more later), all of which are associated with cardiac cell hypertrophy and increased cell survival (Fig. 9–19).

ION CHANNEL RECEPTORS

In addition to ligand-receptor interactions, this chapter has emphasized the importance of ion-dependent cell signaling events such as those that occur in response to changes in cytosolic calcium concentrations. In the context of Ca^{2+}, these ions can utilize channels, also known as voltage-gated ion channels, which have selective permeability to calcium ions in response to changes in membrane potential in response to stimuli. Since extracellular Ca^{2+} levels are several orders of magnitude higher than intracellular levels, depolarization of the membrane activates (i.e., opens) the Ca^{2+} channels, allowing for influx of the Ca^{2+} into the cell. Additionally, cells can also have intracellular calcium storage within specialized compartments that can be released into the cytoplasm. Increased cytosolic Ca^{2+} results in myriad cellular changes such as release of neurotransmitters, contraction of muscle cells, and increase in gene transcription, thereby altering expression of specific proteins.

Ligand-gated ion channels are activated upon binding of a specific protein. One of the most prominent examples of ligand-gated ion channels is the IP_3R, which is activated upon binding of IP_3, a by-product of PIP_2 hydrolysis by PLC (discussed earlier in this chapter). Binding of IP_3 to its cognate receptor activates the channel, allowing for release of Ca^{2+} from the sarcoplasmic and endoplasmic reticulum into the cytosol. Release of Ca^{2+} through these ligand-gated ion channels can alter cell proliferation, cell growth, apoptosis, development, as well as learning and memory.

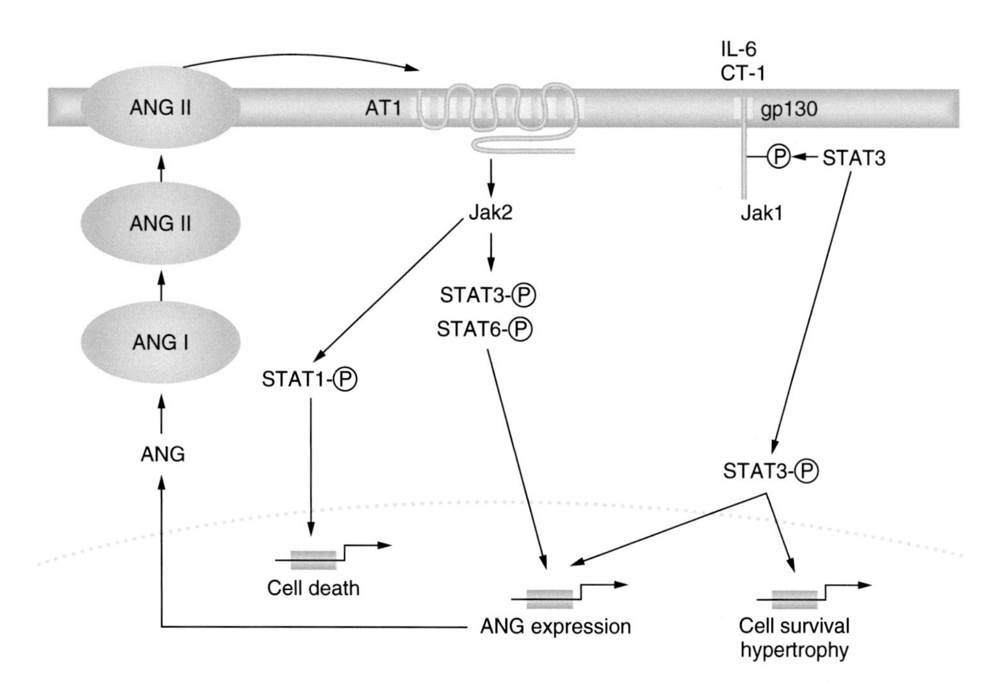

Figure 9–19. **Model for the agonists-induced JAK/STAT signal transduction pathway.** Agonist binding to the respective receptors triggers the tyrosine phosphorylation and activation of tyrosine kinase JAK2, which associates with the AT1 receptor and activates downstream signaling components such as STATs. Activated STATs are heterodimerized and translocated into the nucleus, where they activate transcription of target genes.

Calcium/Calmodulin

As discussed earlier in this chapter, calcium ions are integral for certain cell-signaling cascades as they can modulate enzymatic activity of proteins, such as kinases, phosphatases, proteases, and endonucleases, to impact intracellular activity. Sourcing calcium ions is a feasible task as cells in most organisms contain intracellular calcium concentrations that are orders of magnitude lower than their extracellular environment. Calcium influx into the cytoplasm from the extracellular space is often mediated by a primary transducer (e.g., receptors or hormones). The sudden increase of Ca^{2+} in the cytosolic space, in turn, can modulate activities of calcium-dependent enzymes. However, calcium concentrations within the cell must be tightly regulated, otherwise sustained high Ca^{2+} concentrations in the cytosolic space can result in calcium-dependent enzyme hyperactivity that can lead to deleterious cell signaling events such as neoplastic changes or apoptosis. Thus the cell contains proteins that act as Ca^{2+} transporters, sensors, or buffers to tightly regulate Ca^{2+} levels in the intracellular space. Here, we discuss calmodulin and CaN, two types of secondary messengers within the calcium/calmodulin signaling pathway.

Calmodulin is a Ca^{2+}-binding regulatory protein that is activated upon binding to a calcium ion and can, in turn, act as a Ca^{2+} chaperone to facilitate their interactions with, and thereby promote the function of, calcium-dependent enzymes that have specific roles within different cell types. These calcium/calmodulin-dependent enzymes are: Ca^{2+}/calmodulin-dependent protein kinase (also known as CaM kinase or CaMK), CaN, and myosin light chain kinase (MLCK). There are currently four known CaMK (I–IV), each of which are found in most cell types to modulate Ca^{2+}-mediated signaling. CaMKs can regulate a variety of cell functions, including gene transcription and protein synthesis, or even activate certain cell types. For example, calcium ions and CaMKs can translocate to the nucleus to phosphorylate and activate both CaMKII and CaMKIII, which can promote transcription and negatively regulate protein synthesis, respectively. Similarly to other enzyme families, activity of one enzyme appears to be antagonistic to the other and is likely dependent on the source of the initial stimulus and abundance of each enzyme.

In addition to kinases, increased concentrations of intracellular Ca^{2+} can also activate phosphatases such as CaN (also known as protein phosphatase 2B, PP2B). Unlike CaMKs, which are serine/threonine kinases and add a phosphate group to these residues, CaN is a serine/threonine phosphatase and removes phosphate groups from phosphoserine/threonine residue. For example, CaN dephosphorylates transcription factors of the nuclear factor of activated T cells (NFAT) family (described in the next section) to promote their nuclear translocation, thus promoting gene transcription. Taken together, increased cytosolic concentrations of calcium ions can be regulated by calcium-responsive proteins such as CaN and calmodulin, preventing over- or under-activation of Ca^{2+}-dependent pathways to elicit a specific cellular response (Clinical Case 9–2).

Clinical Case 9–2 Hypertrophic Obstructive Cardiomyopathy

David Miller is a 19-year-old African American male college student who recently decided to try out for the college track team. David has been regularly exercising by leisurely swimming, but had not regularly engaged in vigorous physical activity. During the first tryout for the track team David began to feel some chest pressure, palpitations, and mild shortness of breath, but after resting and drinking water he began to feel better. David made the track team and at the first track meet while running the 440, he again had chest pressure/shortness of breath and felt a sensation of lightheadedness quickly followed by fainting and falling to the track. He regained consciousness quickly, stating that his symptoms had resolved and he felt fine. His teammates related that they witnessed no seizure activity. As a precaution, his coach decided to have him sit out the rest of the track meet and required that he go to the team doctor for an evaluation and medical clearance prior to restarting track practice. David went to the team doctor the next week after having no recurrence of symptoms. Upon evaluation the team doctor found that David had a normal blood pressure of 110/70 mmHg without an orthostatic tilt, a normal pulse of 72 BPM, and a normal respiratory rate. On physical exam, the physician noted that David had a mid-systolic crescendo/decrescendo murmur along the left sternal border with a laterally displaced apical impulse. Upon asking David to perform a Valsalva maneuver he noted that the murmur increased. David had a normal lung, abdomen, extremity, and neurologic exam. David stated that he had no history of a heart murmur and that neither heart disease nor sudden death ran in his family, but his father had died in a car accident when he was 4 so he did not remember what health conditions he had. The team physician obtained an electrocardiogram (ECG) while David was in the office and drew blood for routine blood chemistries, including a complete blood count. The ECG in the office revealed left ventricular hypertrophy (LVH) and the team physician scheduled David for an echocardiogram to further evaluate the murmur and the LVH. The echo revealed increased left ventricular wall size, increased flow velocity along the left ventricular outflow track, and mild mitral valve regurgitation suggesting a hypertrophic obstructive cardiomyopathy (HOCM). The team doctor prescribed David a beta blocker for his condition and strongly advised him to avoid strenuous exercise and heavy lifting. David followed the team doctor's advice and had complete resolution of all symptoms. Unfortunately, he had to resign from the track team.

Cell Biology, Diagnosis, and Treatment of HOCM

HOCM, a prevalent form of Hypertrophic Cardiomyopathy (HCM), is a significant cause of sudden death in younger people, including athletes. It occurs at a rate of 1:500 in the general adult population affecting men and women equally across all races. In the United States, HOCM is responsible for fewer than 100 deaths per year primarily due to a rate of 1:220,000 among athletes. HOCM most often is inherited as an autosomal dominant condition and is the most common genetically transmitted cardiomyopathy. HOCM most commonly results from a mutation in genes coding for sarcomere proteins, which cause structural abnormalities in myofibrils and myocytes that lead to abnormal cardiac force generation and conduction abnormalities.

The cardiac sarcomere is the contractile unit of cardiomyocytes. The proteins involved in sarcomere formation are organized into thick and thin filaments, composed primarily of actin filaments, myosin proteins, and titin. Contraction occurs by sliding and interdigitating of the thick and thin filaments of sarcomeres, which require a highly regulated and precise set of interactions between these two sets of filaments.

Over 800 mutations in nine primary genes encoding sarcomere proteins have been identified as causing Hypertrophic Cardiomyopathy (HCM) in large family cohorts. Two of these causal genes MYH7 (beta myosin heavy chain) and MYBPC3 (myosin-binding protein C) are the two most common being responsible for half of the HCM cases. The other seven major gene defects are in TNNT2 (cardiac troponin T), TNNI3 (cardiac troponin I), TPM1 (alpha tropomyosin), ACTC1 (cardiac alpha actin), MYL2 (Regulatory myosin light chain), MYL3 (Essential myosin light chain), and CSRP3 (cysteine and glycine rich protein 3). There are another 17 causal genes that been observed in smaller families and sporadic cases. While these are the primary defects, they result in proximal defects in response to the changes in sarcomere protein structure and function. An example of a proximal effect is altered calcium sensitivity. In animal models of HCM abnormal intracellular calcium has been demonstrated, with decreased calcium in the sarcoplasmic reticulum and increased calcium in the cytosol. This then leads to important changes in cell signaling events. The increased intracellular calcium associates with calmodulin which in turn increases the calcineurin phosphatase. Calcineurin dephosphorylates the NFAT transcriptional factor that turns on genes that increase cardiac hypertrophy. This has been demonstrated to be true in HCM animal models and a human clinical trial. Diltazem, an L-type calcium channel inhibitor, was first demonstrated to decrease intracellular calcium and inhibit cardiac hypertrophy in a mouse HCM model. More recently it was demonstrated to do the same in a small randomized human trial involving HCM subjects with a mutation in MYBPC3.

Loss of mechanical force due to molecular impairment of sarcomere function to support myofibril contraction is a major cause of muscle thickening. These phenotypic changes result in a mode of compensation by the heart muscle to be able to generate enough mechanical force to maintain adequate tissue perfusion throughout the body.

Clinical presentations associated with these mutations range from mild to severe, depending on the location and type of mutations present, which dictate the effect on the overall function of the sarcomere unit. However, it is difficult to correlate a particular mutation to a clinical outcome due to multifactorial effects on disease onset and clinical presentation. In many cases there are no symptoms and the disease manifests itself only upon heart function tests or upon strenuous physical activity, as seen in this case. Over time the condition leads to cardiac hypertrophy, most often presenting as LVH. Enlarged heart muscle might obstruct heart valve function resulting in heart murmurs and/or abnormal electrical activity, which can cause arrhythmias and subsequent complications such atrial fibrillation and even heart failure.

Treatment options include beta blockers and selective calcium channel blockers along with avoidance of strenuous exercise and heavy lifting. ACE inhibitors and nitrates should be avoided as they decrease afterload and may worsen the left ventricular outflow track obstruction. Septal myomectomy is reserved for persons whose symptoms are not well managed with medications and lifestyle changes.

CaN/NFAT Pathway

The CaN/NFAT pathway behaves downstream of GPCR signaling in a calcium-dependent manner. When the cell is stressed, the GPCR recruits PLC to the membrane, which processes PIP_2 into DAG and IP_3. IP_3 stimulates release of calcium ions (Ca^{2+}) from the SR, increasing cytosolic Ca^{2+} (Fig. 9–20). As a consequence of increased Ca^{2+} in the cell, calmodulin, a calcium-dependent serine/threonine kinase, activates CaN, a serine/threonine phosphatase, which acts upon, and therefore activates, NFAT, a transcription factor normally found in the cytosol. There are five members of the NFAT family of transcription factors: NFATc1, NFATc2, NFATc3, NFATc4, and NFATc5 (all but NFATc5 function in a calcium-dependent manner). Activated NFAT translocates to the nucleus to promote transcription of its target genes such as interleukin-2 (IL-2), which can activate T-cell response, along with increasing production of other cytokines, as a consequence of cellular stress. Since NFAT weakly binds DNA, it must cooperate with other transcription factors, most notably AP-1, to promote transcription of its target genes. In addition to adaptive immune response, CaN/NFAT signaling has also been implicated in development of other systems such as nervous systems as well as skeletal and cardiac muscles. For example, calcium-dependent NFAT activity can regulate axon outgrowth in some populations in cooperation with other transcription factors while also promoting formation of neural networks in the developing brain.

Due to their roles in the transcription of target genes, CaN-mediated activation of NFAT promotes transcription of genes involved in cell proliferation, survival, migration, and differentiation—all of which are functions usually associated with malignant cell types. Use of CaN inhibitors in preclinical studies shows increased apoptotic rate in cancer cells, likely through halting their cell division, suggesting that the CaN/NFAT pathway may be a potential therapeutic candidate in malignant tumors. However, these findings do not apply to all members of the NFAT family: some NFAT proteins appear to play protumorigenic roles while others are tumor suppressive. For example, NFAT1

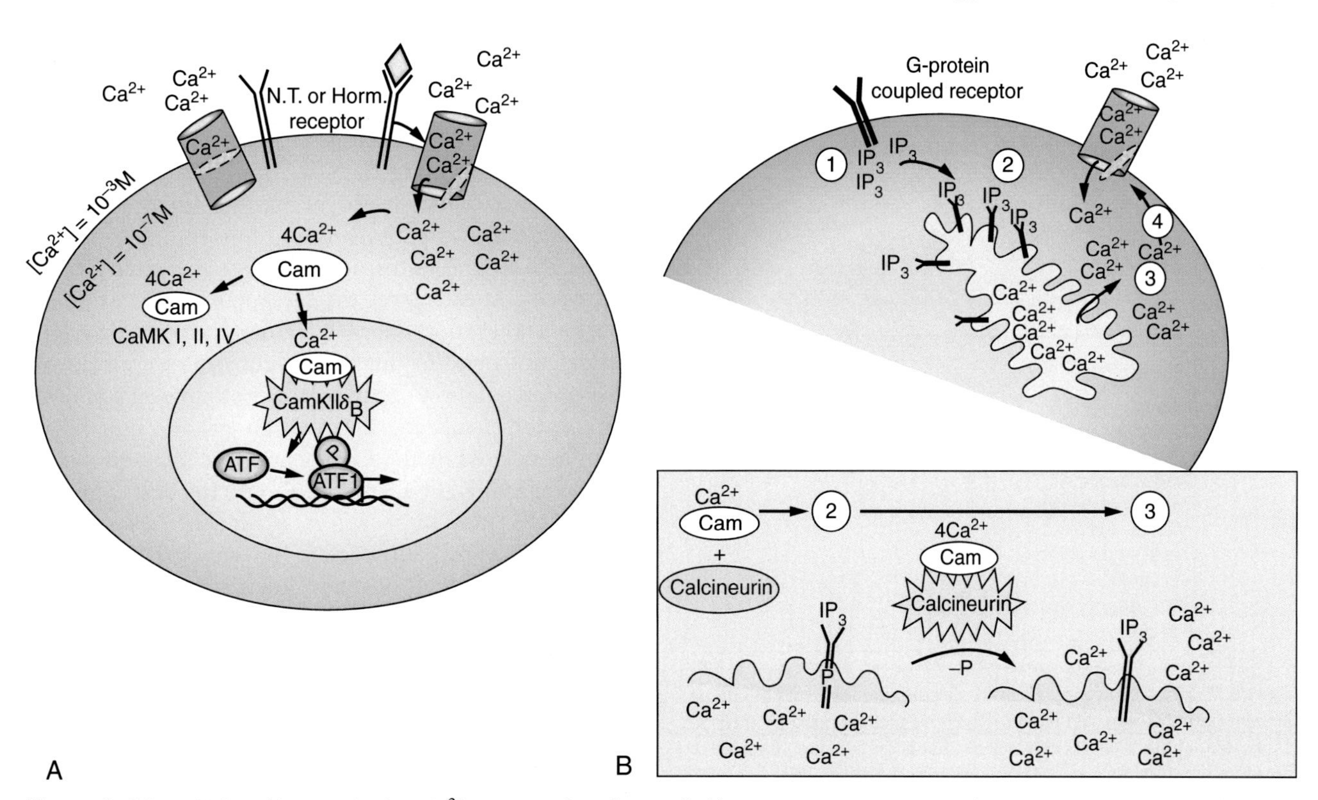

Figure 9–20. **(A) CaM kinase activation.** Ca^{2+} entry into the cell controlled by neurotransmitter (N.T.) or hormone (Horm.) receptor activation increases the intracellular Ca^{2+} concentration. Ca^{2+} ions are bound by calmodulin (CaM), and the Ca^{2+}/calmodulin complex activates kinases such as CamKI, -II, and -IV. Ca^{2+}/calmodulin can also translocate to the nucleus and activate a nuclear isoform of CaMKII, CaMKIIδB, to phosphorylate and activate certain widely used transcription factors (e.g., ATF1). **(B) Calcineurin activation.** Ca^{2+}-dependent activation of calmodulin (as in Panel A) leads to activation of the phosphatase calcineurin. Activation of G-protein coupled receptors causes formation of IP_3 (1) that binds to and activates IP_3 receptors (2) on the endoplasmic reticulum (ER). Calcineurin dephosphorylates the IP_3 receptor, facilitating release of Ca^{2+} stored in the ER into the cytosol (3), and subsequent activation of plasma membrane-bound Ca^{2+} channels (4).

appears to play oncogenic roles in breast cancer by promoting invasion of cancer cells in vitro; however, NFAT5 modulates motility and migration but does not promote invasion (Clinical Case 9–2).

ABERRANT CELL SIGNALING IN CANCER

Tumor growth requires collaboration of signaling pathways to drive specific cellular processes. There are now *10 hallmark cellular processes* recognized as critical for tumorigenic progress: resisting cell death, genetic instability, sustained proliferation, enabling replicative immortality, escaping growth suppressive signals, triggering invasion and metastasis, promoting angiogenesis, altering cellular metabolism, inflammation, and evading immune response (Fig. 9–21). Local tumor growth can be contained or controlled for most tumor types and therefore does not result in patient death. Mortality in cancer most often results from metastatic tumors, which biologically involve four processes: motility and invasion, altered microenvironment, plasticity, and colonization (Fig. 9–22). Through descriptions within this chapter, one can appreciate the delicate balance that must be maintained between and within cell signaling cascades and thus alterations in these regulations can result in the development of cancer.

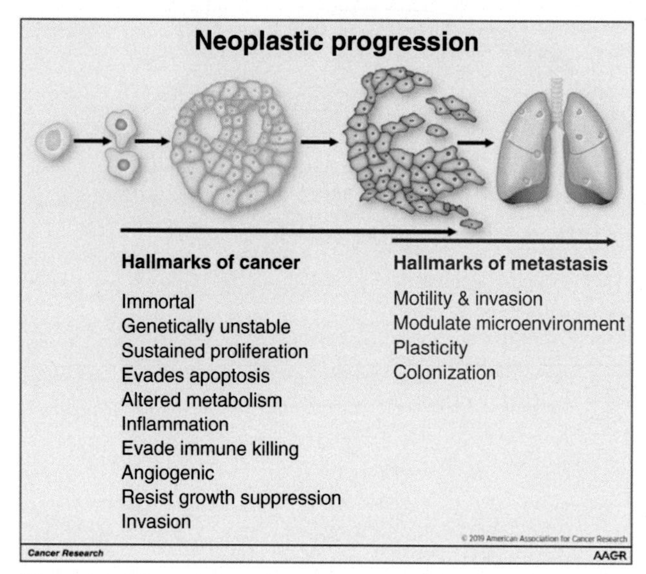

Figure 9–21. Cancer cells require specific characteristics to initiate the metastatic cascade. Malignant tissues are characterized by their ability to adapt to a microenvironment that is designed to eliminate threats to the host. As such, cancer cells gain the ability to thrive in their environment even when they are depleted of nutrients and attacked by the immune system and can even colonize distal tissues to spread. *(From Welch DR, Hurst DR.* Cancer Res *2019, https://doi.org/10.1158/0008-5472.CAN-19-0458, based on Hallmarks of Cancer, Hanahan and Weinberg, 2011* Cell *144 https://doi.org/10.1016/j.cell.2011.02.013.)*

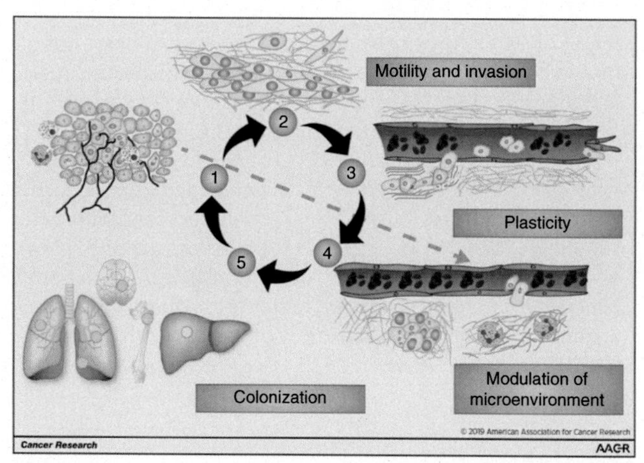

Figure 9–22. Pathogenesis of cancer cell metastasis. Patient mortality results in failure to contain tumor cells to the site of the primary tumor, otherwise known as metastasis. The metastatic cascade involves four processes: cell motility and invasion, alteration of the tumor microenvironment, cell plasticity, and colonization of proximal and distal tissues. *(From Welch DR, Hurst DR.* Cancer Res *2019, https://doi.org/10.1158/0008-5472.CAN-19-0458.)*

GENERAL MECHANISMS OF DYSREGULATED SIGNALING IN CANCER

Aberrations in these pathways can have deleterious effects on the homeostasis of the organism and can lead to various cellular and physical manifestations such as cancer. Here, we briefly discuss how dysregulations in some of the described pathways can lead to neoplastic changes in the cell (Fig. 9–23).

There are several general mechanisms that can lead to neoplastic transformation and/or support tumor growth. Let us start with intrinsic pathways activated within tumor cells. At each level, the deregulation may originate from gene alterations (such as mutations, amplifications, and deletions and/or might be a consequence of downstream upregulation of mRNA or protein-level stabilization). At the membrane, activation of growth signaling is often associated with amplification of growth factor receptors (EGFR, FGFR, or HER2 are the best examples here). Activation of receptors often involves receptor dimerization and subsequent phosphorylation. This results in recruitment of intercellular signaling partners. Typical examples here are Grb2, Sos, and PI3K. Involvement of PI3K in cancer cannot be overstated, as most tumor development involves its activation. PI3K is also a classic example of an oncogene, whose increased activity is typically positively associated with tumor growth. The activation of PI3K, for example, PI3KCA, occurs through two major mechanisms, gene amplification or kinase (catalytic) domain-activating mutations. Increased kinase activity, in turn, leads to an enhanced PIP_3 production and subsequent activation of multiple signaling nodes through AKT.

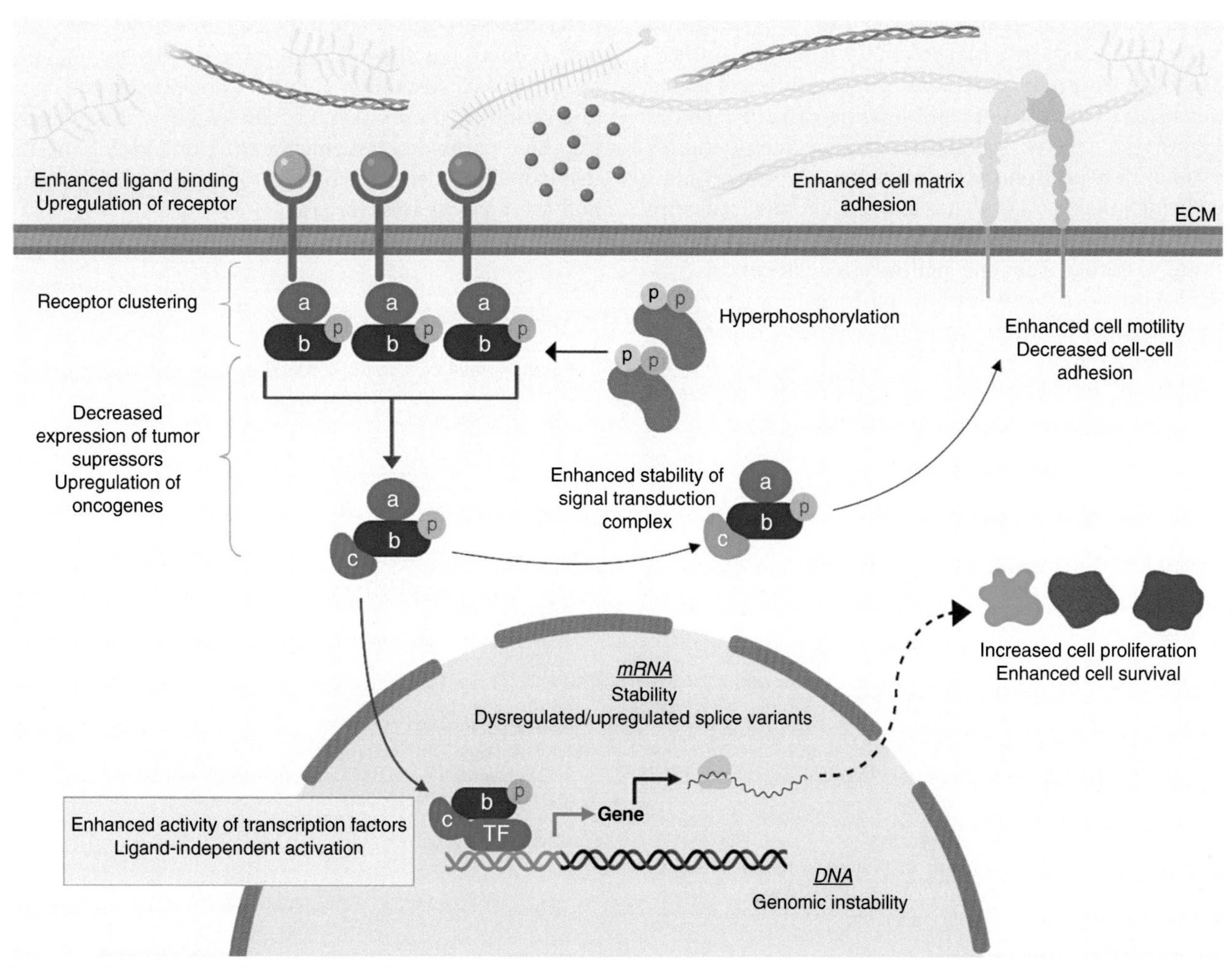

Figure 9–23. **Cancer cell-signaling amplification is a multistep process that drives a positive feedback loop to support tumor growth.** Increased availability of extracellular ligand and cognate receptor causes receptor aggregation, hyperphosphorylation, and activation. In turn, there is a surge in nuclear translocation of transcription factors to transcribe protumor genes, which can regulate their own as well as other gene transcriptions, promoting neoplastic changes within the cells. Alternatively, amplification of certain signaling cascades can also downregulate expression of specific tumor suppressors, either at the transcriptional level or by altering protein stability after the protein has been translated. *(By Baylee Porter and Leszek Kotula, chapter authors' figure.)*

In this generalized schematic (Fig. 9–23), it is important to point out that amplification of intracellular signaling often occurs through upregulation of proteins with oncogenic function (e.g., PI3K or mTORC2), and/or downregulation of tumor suppressors (e.g., PTEN). Gene amplifications and function-activating mutations are frequent abnormalities involving oncogenes; alternatively, tumor suppressor function can be made deficient due to loss-of-function mutations and/or gene deletions. Signaling pathways in cancer cells usually contain several tumor suppressors: oncogene nodes that control tumor progression though cell proliferation, survival, cell motility, and cell-cell and cell-matrix adhesion.

Most signals that originate outside the cell and are amplified and modified in the cytoplasm end up in the nucleus to promote transcription factors to regulate expression of a specific set of genes responsible for cellular functions. Activation of transcription factors, through gene amplification, structural mutations, ligand, or DNA binding site mutation, plays a critical role in tumor progression. The best examples here are myc and androgen receptor in prostate cancer, or STAT3 in leukemia. The signals from transcriptional factors transmitted through multiple gene expression profiles reach out back to cytoplasm and plasma membrane to modulate cellular phenotypes.

SUMMARY

Signal transduction describes how a cell receives a signal, the intracellular changes that occur as a consequence of signal reception, and how the cell changes its behavior in response. Signal transduction is imperative for early development, function, and survival of multicellular organisms. Signal is received from the environment or another cell, and through interactions between several proteins, regulation of nuclear activity and gene transcription, and additional intracellular signaling, the receiving cell can respond. Many cell-signaling cascades are intricately woven with each other to sustain homeostasis of cells through functions of many different signal transducers. One category of signaling cascade includes receptors to receive signals, kinases (serine/threonine/tyrosine) that phosphorylate proteins, adaptor proteins that bring kinases and substrates together, and transcription factors that alter gene expression as a response to stimuli. Another signaling pathway regulates transcriptional repressors through proteolytic cleavage of regulatory precursors. Other signaling cascades depend on hormone signaling and their cognate receptors that function as transcriptional regulators. The pathways described here are a small example of how inter- and intracellular signaling is important for development of an organism, sustaining cellular homeostasis, and how aberrations in these pathways can lead to disease. Studying these pathways can elucidate the underlying mechanisms behind disease and lead to improved diagnostics and therapies against disease.

Acknowledgment

The authors would like to thank Maria A. Ortiz and Baylee Porter for their contributions to the editing and generation of some of the figures.

Suggested Readings

Brechbiel J, Miller-Moslin K, Adjei AA. Crosstalk between hedgehog and other signaling pathways as a basis for combination therapies in cancer. *Cancer Treat Rev* 2014;40:750–759. https://doi.org/10.1016/j.ctrv.2014.02.003 Epub 2014 Feb 24.

Gocek E, Moulas AN, Studzinski GP. Non-receptor protein tyrosine kinases signaling pathways in normal and cancer cells. *Crit Rev Clin Lab Sci* 2014 Jun;51(3):125–137. https://doi.org/10.3109/10408363.2013.874403 Epub 2014 Jan 22.

Nemere I, Pietras RJ, Blackmore PF. Membrane receptors for steroid hormones: signal transduction and physiological significance. *J Cell Biochem* 2003;88:438–445.

Rawlings JS, Rosler KM, Harrison DA. The JAK/STAT signaling pathway. *J Cell Sci* 2004;117:1281–1283.

Roskoski R. The ErbB/HER family of protein-tyrosine kinases and cancer. *Pharmacol Res* 2014;79:34–74. https://doi.org/10.1016/j.phrs.2013.11.002.

Sheppard K, Kinross KM, Solomon B, Pearson RB, Phillips WA. Targeting PI3 kinase/AKT/mTOR signaling in cancer. *Crit Rev Oncog* 2012;17(1):69–95.

Shi Y, Massague J. Mechanisms of TGF-β signaling from cell membrane to the nucleus. *Cell* 2003;113:685–700.

Thisse B, Thisse C. Functions and regulations of fibroblast growth factor signaling during embryonic development. *Dev Biol* 2005;287:390–402.

Welch DR, Hurst DR. Defining the hallmarks of metastasis. *Cancer Res* 2019; https://doi.org/10.1158/0008-5472.CAN-19-0458.

Wodarz A, Nusse R. Mechanisms of Wnt signaling in development. *Annu Rev Cell Dev Biol* 1998;14:59–88.

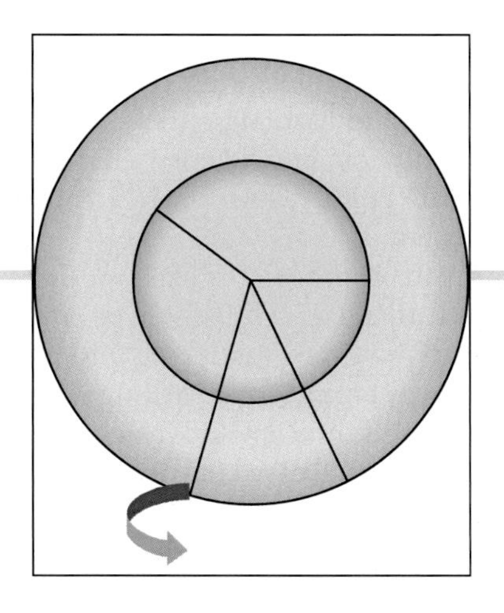

Chapter 10

Cell Cycle and Cancer

CELL CYCLE: HISTORY

The adult human body is composed of approximately 200,000 billion (2×10^{14}) cells, all of which are derived from a single cell, the fertilized egg or zygote. In adults, there are subpopulations of cells that continuously divide or retain the capacity to divide and replace cells that die or are otherwise lost. The process of cell multiplication and division requires numerous intricately regulated steps. The cells must increase in size, duplicate their genetic content contained in their DNA and chromosomes, and upon cell division precisely segregate their chromosomes so that each daughter acquires an exact copy of the parental chromosome complement.

Until the middle of the 20th century, the prevailing view, based on visualization of cells by microscopy, was that most cellular activity occurred during active cell division when microscopists could visualize cells in the various stages of mitosis. The majority of cells were in an "interphase" or nondividing state between mitotic divisions and appeared to the microscopist to be more or less quiescent. That perception changed abruptly in 1953 when Alma Howard, a radiation biologist working in a small laboratory in Hammersmith Hospital in London, and her physicist colleague Stephen Pelc published a seminal article that profoundly changed our fundamental understanding of cell-cycle compartmentalization. By exposing rapidly dividing rootlets of the broad bean, *Vicia faba*, to ^{32}P-orthophosphate and performing autoradiography at subsequent intervals, Howard and Pelc demonstrated that DNA synthesis occurred during a discreet interval of interphase (S phase) that was preceded by a gap of no DNA synthesis and followed by a gap with no DNA synthesis that came before mitosis. These gaps came to be known as G1 and G2, respectively (Fig. 10–1). For many the discovery of cell-cycle compartmentalization made by Alma Howard and Stephen Pelc was just as seminal as the description of the structure of DNA published in the same year by Watson, Crick, Wilkins, and Rosalind Franklin.

The mechanisms that control the progress of cells through the cell cycle are intricate and highly conserved throughout evolution. Hence, lessons regarding how the cell cycle works have been drawn from a wide spectrum of single-cell and multicellular organisms ranging from yeast, to plants, to sea urchins, to clams, to frogs, to mammalian cells. In 2001 the Nobel Prize in Physiology or Medicine was awarded to Leland Hartwell, Paul Nurse, and Tim Hunt, all of whom used nonmammalian model organisms and a combination of genetic and molecular biology approaches to uncover the mechanisms that regulate the cell cycle. With contributions from many other deserving investigators using a variety of model organisms, they found that proteins, designated cyclins, and cyclin-dependent kinases (CDKs) drive cells from one cell-cycle phase to the next in a highly regulated fashion. One might compare the CDKs to an engine and the cyclins to the gearbox that determines whether the engine will idle so that the cells remain stationary or

Goodman's Medical Cell Biology. https://doi.org/10.1016/B978-0-12-817927-7.00010-7

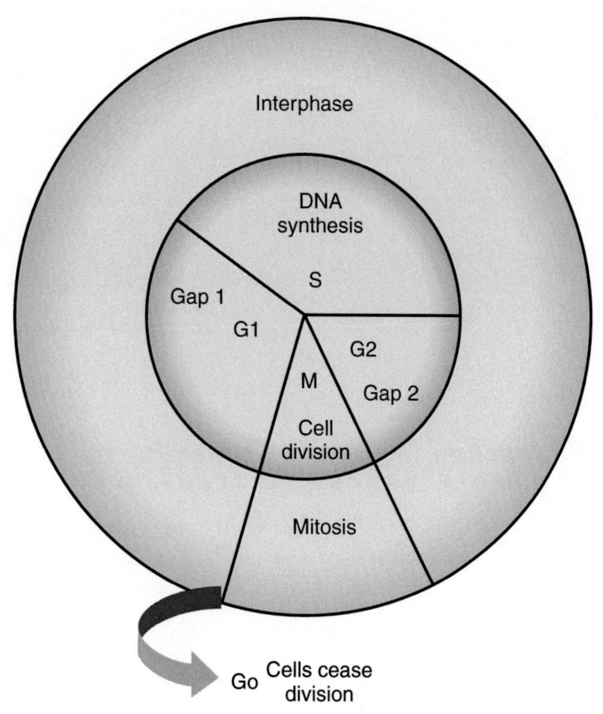

Figure 10–1. Schematic depicting the major phases of the cell cycle.

whether the engine will be in gear and drive the cells forward through the cell cycle. If any component in either the engine or the gearbox becomes compromised, the cell is at increased risk of losing its intrinsic regulatory mechanisms and becoming cancerous.

By the end of the 1960s, Hartwell was utilizing the power of genetics for dissecting the cell cycle. The model system that he used was baker's yeast, *Saccharomyces cerevisiae*, which proved highly amenable for cell-cycle analysis. In an elegant series of experiments, he isolated yeast cells in which genes controlling the cell cycle were conditionally mutant, allowing cell proliferation at permissive temperature, but arrested cells at various stages of the cell cycle at nonpermissive temperature. By this approach, he successfully identified more than 100 genes directly involved in cell-cycle control, which were designated cell division cycle (CDC) genes. One of these genes, cdc28 in *S. cerevisiae*, is a CDK that controls the first step in the progression of cells through the G1 phase of the cell cycle and was therefore also called *start*.

Another important concept in cell-cycle regulation, also introduced by Hartwell, is that of cell-cycle checkpoints, which are activated after stress or environmental challenge. In addition to dissecting the genetics of cell-cycle progression in unperturbed cells, Hartwell studied the sensitivity of yeast to DNA damage by ionizing radiation. Based on his observations that cells are transiently arrested when their DNA is damaged, he coined the term *checkpoint* in 1989 to underscore the notion that cells arrest transiently, repair the DNA damage, and then proceed to the next phase of the cell cycle. It is worth mentioning that, 25 years earlier, Alma Howard, the radiation biologist at Hammersmith Hospital, had also astutely noted that rapidly proliferating cells in the *V. faba* rootlet were delayed in entering mitosis following exposure to ionizing radiation, but did not name the phenomenon. Loss of checkpoint regulation is a common hallmark of most cancers.

Paul Nurse used an approach similar to that of Hartwell's but took advantage of a different type of yeast, *Schizosaccharomyces pombe*, as a model organism. This yeast is distantly related to baker's yeast, having diverged early during evolution. In the mid-1970s Nurse identified the *cdc2* gene in *S. pombe* and showed that it played a key role in the transition from G2 to mitosis (M). He later discovered that it participates in additional cell-cycle transitions, that it is identical to the *cdc28* ("start") gene identified earlier by Hartwell in baker's yeast, and that it controls the transition from G1 to S. Subsequently, Nurse isolated the corresponding gene from humans, which encodes a CDK and was later designated CDK1 (cyclin-dependent kinase 1). Nurse then showed that CDK1 activation is dependent on reversible phosphorylation (i.e., its reversible modification by the addition or removal of phosphate groups). Since this initial finding, several different human CDKs have been described.

The CDKs are activated by forming complexes with one of several cyclins. The cyclins were discovered serendipitously by Tim Hunt as part of the physiology course at the Marine Biological Laboratory (MBL) at Woods Hole in the early 1980s. He noticed that, during synchronous cleavage divisions of the sea urchin *Arbacia punctulata* embryo, a specific protein was destroyed at each cell division and resynthesized in the next cycle. He coined the term *cyclin* to reflect the recurring sequential synthesis and degradation of this protein. His finding was subsequently confirmed by Joan Ruderman in the cleavage embryos of the clam and later in mammalian cells, where there are multiple cyclins that interact with different CDK molecules at different times of the cell cycle. These interactions confer specificity to cell-cycle regulation by defining and selecting the regulatory proteins that are activated or deactivated by phosphorylation. Notably, most of the important principles underlying cell-cycle regulation are derived from studies on plants, two evolutionarily diverged yeast, sea urchin embryos, clam embryos, and frogs, reinforcing the importance of nonmammalian systems for discovering and analyzing fundamental biologic processes.

One of the hallmarks of cancer is uncontrolled cell proliferation. In most, if not all, human tumors, one or more of the cell-cycle checkpoints is compromised such that cells proliferate in an unregulated manner. A detailed but still incomplete understanding of the mechanisms that coordinate events that govern cell proliferation has come from molecular analyses of human tumors, normal tissues, and animal models. In addition to the cyclins and the CDKs, there are numerous proteins and pathways that regulate cell-cycle kinetics, many of which are categorized as oncogenes and tumor suppressors. Oncogenes are mutant versions of normal genes (protooncogenes) whose products are important in facilitating normal cell-cycle progression. When they are mutant the oncogenes promote uncontrolled growth. Tumor suppressors serve to restrain uncontrolled proliferation, so when they are mutant or lost, the cells will proliferate abnormally. Thus oncogenes can be viewed as accelerators of cell proliferation and tumor suppressors as the brakes. Oncogenes are generally activated by activating mutations, such as those found in the *Ras* genes. In contrast, tumor suppressor gene function is lost by mutation, by epigenetic modification (e.g., DNA methylation), and/or by loss of heterozygosity (LOH) as a consequence of mitotic recombination (recombination between chromosome homologs), deletion, or chromosomal loss.

THE CELL CYCLE IS REGULATED BY CYCLINS AND RELATED PROTEINS

Cyclins

In the summer of 1982, Tim Hunt was invited to teach the annual physiology course at the Marine Biological Laboratory at Woods Hole, Massachusetts. His model organism was the sea urchin, *A. punctulata*, whose eggs are fertilized externally. Following fertilization the fertilized eggs coordinately undergo synchronous divisions. Hunt and the physiology students labeled the eggs with ^{35}S-methionine shortly after fertilization and followed the levels of proteins at 10-min intervals by acrylamide gel electrophoresis and autoradiography. While most proteins accumulated over time as expected, the level of one protein oscillated. It accumulated following cell division and then rapidly disappeared as the cells entered mitosis (Fig. 10–2; taken from the original paper by Hunt and colleagues). Based on its oscillating behavior, this protein came to be called "cyclin."

The cyclins are a family of proteins that are centrally involved in cell-cycle regulation and are structurally identified by conserved "cyclin box" regions. The cyclin boxes are composed of about 150 amino acid residues, which are organized into five helical regions and are important in binding partner proteins, including the CDKs. More than 20 cyclins or cyclin-like proteins have been identified, many of whose functions are now becoming known. Those whose functions have been defined are about 60 kDa in size and play critical roles in allowing the progression of cells through all phases of the cell cycle, including mitosis (Fig. 10–3).

Cyclins are the regulatory subunits of holoenzyme CDK complexes that control progression through cell-cycle checkpoints by phosphorylating and inactivating target substrates. The cyclins associate with different CDKs to provide specificity of function at different times during the cell cycle (see Fig. 10–3). The involvement of aberrant cyclin expression with cancer was first realized

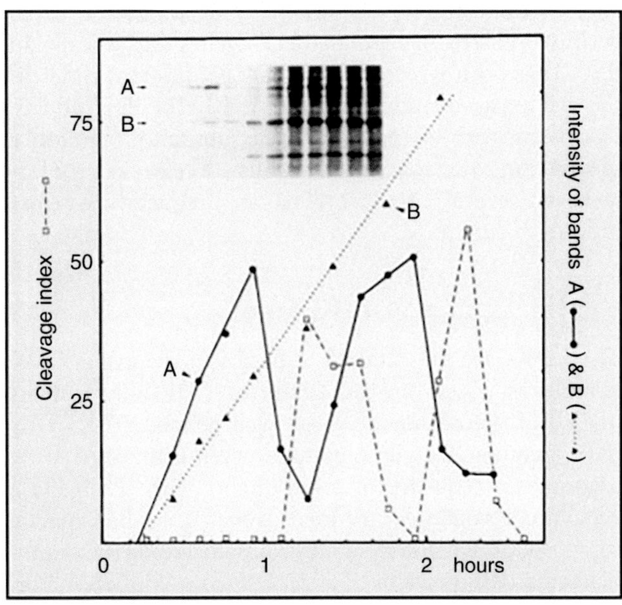

Figure 10–2. **Discovery of a cyclin in fertilized sea urchin eggs.** Fertilized eggs from the sea urchin *Arbacia* were labeled with radioactive methionine, and extracts were resolved on SDS-polyacrylamide gels. The autoradiograph of the protein gel shows the clear result that one protein, called cyclin, accumulates and is precipitously destroyed in mitosis. The *bold oscillating black line (solid circles)* represents the presence of cyclin, the reciprocal *dashed line (open squares)* indicates time of mitosis, and the *linear dashed lines (triangles)* represents the accumulation of protein B. *(Reproduced from Evans, T et al.* Cell *1983;33:389–396.)*

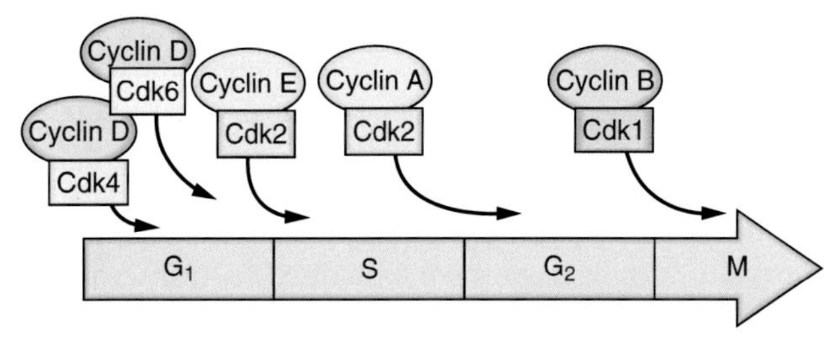

Figure 10–3. **The association of cyclin-dependent kinases (CDKs) with their respective partner cyclins during different phases of the cell cycle.** During the cell cycle the various cyclins pair with their cognate partner CDKs to perform specific functions. The figure depicts the association of specific cyclins with their CDK partners at times of the cell cycle when they are at their peak abundance and performing their cell cycle-specific activities.

when a chromosome breakpoint common in B-cell lymphomas was cloned and shown to encode cyclin D1. At about the same time, cyclin D1 was independently cloned at a chromosome 11 inversion junction, common in parathyroid adenomas that juxtaposed parathyroid hormone gene regulatory elements with the body of the cyclin D1 gene. Subsequently, overexpression of several cyclins, particularly cyclin E, has been associated with cell-cycle deregulation in tumors. Cyclin E overexpression can shorten the time spent in G1 and allow premature entry of cells into the S phase. Cyclin E is also important in facilitating centrosome replication, the precise regulation of which is important in maintaining chromosome stability. Figure 10–3 shows the principal cyclins involved in regulated cell-cycle progression and the CDKs with which they partner at different times of the cell cycle, including cyclins A, B1, D1, D3, and E. Consistent with the notion that disruption of regulatory mechanisms that govern cell proliferation leads to tumorigenesis, several of the cyclins are frequently overexpressed in tumors.

Cyclin-Dependent Kinases

The CDKs are the catalytic subunits of the cyclin/CDK complexes. There are 26 CDKs or CDK-like proteins not all of whose functions are well defined. The CDKs have a common cyclin-binding domain characterized by a consensus sequence of seven amino acids (PSTAIRE). They are serine/threonine kinases that have the capacity to phosphorylate multiple substrates, including the retinoblastoma (Rb) proteins and the E2F transcription factor. Progression through G1 and entry into S phase is regulated by sequential involvement of at least three CDKs: CDK4, CDK6, and CDK2 (see Fig. 10–3). Among other activities, CDK4 in complex with cyclin D appears to mediate phosphorylation of the Rb family of proteins. Further phosphorylation of the retinoblastoma protein (pRb) is mediated by cyclin D/CDK6 and finally by cyclin E/CDK2. These sequential phosphorylations, coupled with other cellular activities, allow the ordered progression of cells through the G1 phase. The activity of cyclin E/CDK2 is essential for the transit of cells from G1 to S phase and for the duplication of centrosomes whose function is required for proper chromosome segregation at mitosis.

Progression through S phase is mediated principally by cyclin A/CDK2. During S phase, cyclin A/CDK2 phosphorylates numerous proteins involved in transcription and DNA replication and repair and proteins thought to be necessary for the completion of S phase and the entry of cells into G2. Among the proteins phosphorylated by this complex are transcription factors such as E2F1 and B-Myb; proteins like RPA, the DNA single-strand binding protein that is involved in DNA replication; and proteins that participate in DNA repair such as BRCA1, BRCA2, and Ku70. As cells complete the S phase and begin to transit into G2, the A-type cyclins associate with CDK1 (also designated Cdc2 for historical reasons). As cells proceed through G2, cyclin A is degraded by a ubiquitin-mediated proteolytic mechanism at the same time that cyclin B is actively synthesized. During mid-to-late G2, cyclin B/CDK1 complexes form and play multiple essential roles during the G2/M transition and during the progression of cells through mitosis (see Fig. 10–3). At least 70 proteins have been identified as substrates for the cyclin B/CDK1 complexes. In the cytoplasm, cyclin B/CDK1 complexes associate with centrosomes during prophase (see later) and facilitate centrosome separation by phosphorylating a centrosome-associated motor protein. These complexes also participate in chromosome condensation by phosphorylating a subset of histones, in the fragmentation of the Golgi apparatus during prophase, and in the breakdown of the nuclear lamina by phosphorylating both lamins A and B. Inactivation of this same complex is also important for the exit of cells from mitosis to G1. This inactivation is achieved by ubiquitination of cyclin B and its proteolytic degradation mediated by the anaphase-promoting complex (APC), an E3 ubiquitin ligase.

Cyclin-Dependent Kinase Inhibitors

If kinases remained constitutively active and kinase substrates remained permanently phosphorylated, the cell-cycle regulatory circuit would be compromised. It is not surprising, therefore, that there are naturally occurring cellular inhibitors of the CDKs. There are at least two distinct families of CDK inhibitors (CKIs). The prototype of the CIP/KIP family is p21, which was identified almost simultaneously by three different groups and consequently had multiple separate designations. It is now most commonly referred to as p21 or p21$^{waf1/cip1}$, which is reflective of its molecular weight. In addition to p21, this family of kinase inhibitors includes related proteins p27 and p57, which are all capable of binding to and forming complexes with CDK1/cyclin B, CDK2/cyclin A, CDK2/cyclin E, CDK4/cyclin D, and CDK6/cyclin D and inhibiting the kinase activity of these complexes, thereby inhibiting cell-cycle progression. The prototypical member of the second CKI class is p16, one of the INK4 class of proteins, which is dependent on Rb or one of the related "pocket" proteins, p107 or p130. Other members of this class of CKIs, based on structure and function, include proteins designated p15, p18, and p19. The designation of this family of kinase inhibitors commonly contains a superscript "INK4" (e.g., p16^{INK4}). Unlike the p21 class of kinase inhibitors, the INK4 proteins bind exclusively to CDK4 and CDK6, prevent the association of these CDKs with their regulatory cyclins, and thereby inhibit their kinase activities (Clinical Case 10–1).

Clinical Case 10–1

Tom Jones is a 68-year-old lawyer who had been in good health until the last 6 months when he noticed he had begun to lose weight even though he was not dieting and had increased satiety. He also had noticed that despite the weight loss his pants seemed tighter in the waist than they had been, and he had to loosen his belt to feel comfortable especially after a meal. Tom also noted that he had been feeling a little more tired lately and no longer wanted to work those 14-h days at the law firm. Tom's wife began to be concerned about his health and urged him to go to his internist for a checkup. When Tom went to his checkup, his doctor noticed on abdominal exam that he had an enlarged spleen (splenomegaly) that extended 5 cm below the left costal margin. Tom's complete blood count (CBC) also showed his white blood cell (WBC) count to be elevated to 70,000 cells/UL and his hematocrit to be low at 30. His doctor sent his blood for a peripheral blood smear that showed an elevated number of myeloblast, myelocytes, and nucleated red blood cells. Tom was referred to a hematologist who performed a bone marrow biopsy, which showed hypercellularity with expansion of the myeloid cell line and prominence of megakaryocytes.

Cell Biology, Diagnosis, and Treatment of Chronic Myelogenous Leukemia

Tom's marrow also was positive for the Philadelphia chromosome, an abnormally small chromosome that is characteristic of chronic myelogenous leukemia (CML). His hematologist diagnosed Tom with CML and started him on a tyrosine kinase inhibitor, imatinib mesylate, in hopes of inducing a hematologic remission. Tom returned to his hematologist in 3 months and had a normal CBC and no longer showed signs of splenomegaly. The hematologist was glad to see this response because Tom was an orphan and did not have a matched donor for an allogeneic bone marrow transplant.

CML is responsible for about 20% of all leukemias affecting adults. The current treatment of choice for CML is the tyrosine kinase inhibitor imatinib mesylate, but allogeneic bone marrow transplant is the only proven cure. Persons treated with imatinib have an overall survival rate of over 83%, but blast crises do occur after initial treatment and carry a poor prognosis. The Philadelphia chromosome, the hallmark of CML, results from the translocation of the *ABL* oncogene from the long arm of chromosome 9 to a specific breakpoint cluster region (*BCR*) on the long arm of chromosome 22. The *BCR/ABL* gene fusion results in production of a fusion protein with tyrosine kinase activity leading to the development of the CML phenotype. The cause of the translocation is unknown. The Sokal scoring system is used as a prognostic index, and low-risk patients with a Sokal score <0.8 have a 91% likelihood of achieving a complete cytogenetic response.

The constitutive tyrosine kinase activity of the BCR/ABL fusion protein that is causal for CML has a profound disruptive impact on the normal progression of the cell cycle. The effect is multifold and causes loss of the G1/S phase checkpoint and leads to genomic instability. One mechanism in CML involves phosphorylation of the CDKi p27, converting it from its nuclear-localized tumor suppressive function to a cytoplasmic-localized oncogenic function. An alternative mechanism is the elevated level of the cyclin D family, which is an inhibitory target of the CDKi p27 and which pushes cells through the G1/S phase boundary.

Cdc25 Phosphatases

A cellular strategy to ensure that the phosphorylation state of proteins is reversible involves the activity of phosphatases with specificity for defined substrates. An example of phosphatases that participate in cell-cycle regulation is the Cdc25 family of phosphatases. Their primary function in cell-cycle control is to dephosphorylate the various phosphorylated CDKs to allow sequential passage through the phases of the cell cycle. For example, CDK2 is phosphorylated on threonine 14 and tyrosine 15, and when so phosphorylated the cells do not transit from G1 to S phase. Cdc25A is a bifunctional phosphatase that is able to dephosphorylate phosphotyrosines and phosphothreonines and phosphoserines. When Cdc25A dephosphorylates CDK2 at phosphothreonine 14 and phosphotyrosine 15, the cells will enter S phase. When DNA damage occurs in the form of double-strand breaks, Cdc25A is degraded by a ubiquitin-mediated process resulting in persistent phosphorylation of CDK threonine 14 and tyrosine 15 accompanied by cell-cycle arrest at the G1/S phase boundary. Similarly, CDK1 is phosphorylated at the same two residues by the specific kinases wee1 and CDK-activating kinase. Unless Cdk1 is dephosphorylated at tyrosine 15, the cells arrest at the G2/M boundary and do not progress through mitosis. The Cdc25C phosphatase is instrumental in dephosphorylating CDK1, thereby allowing the cells to progress through mitosis. Notably, in the absence of Cdc25C, its close relative Cdc25A can assume this activity.

p53

One of the tumor suppressors most commonly mutated or lost in human tumors is p53, which is mutant in more than 50% of all human tumors. The p53 protein is a key regulator of the G1/S and G2/M checkpoints. Its functions are so important that it has been called the "guardian of the genome" and "the cellular gatekeeper for growth and division." In humans, individuals who are born heterozygous for p53 develop multiple types of tumors during early childhood. This syndrome is known as Li-Fraumeni syndrome and is characterized by tumors that have lost function of both *p53* alleles as a consequence of mitotic recombination and loss of heterozygosity (LOH). The amino-terminal sequences of p53 protein serve as a transcriptional activation domain, and the carboxy-terminal sequences appear to be required for p53 to form homodimers and homotetramers. The p53 protein activates the transcription of numerous genes that participate in cell-cycle control, including those encoding *p21*, *GADD45* (a growth arrest, DNA damage-inducible gene), and MDM2 (a protein that is a known negative regulator of p53). One of the essential roles performed by p53 is to participate in the arrest of cells in G1 after genotoxic damage. Presumably, this arrest provides opportunity for DNA repair to occur before DNA replication and cell division. If the damage is sufficiently great, p53 triggers a programed cell death through an apoptotic pathway instead of allowing severely damaged cells to progress through the cell cycle. Tumor cells that lack normal p53 function do not arrest efficiently in G1 and are more likely to progress into S or G2/M. Occasionally, some of these damaged cells will escape cell death and contribute to tumor progression.

The p53 protein, when activated by appropriate phosphorylation, can function as a transcription factor that in turn activates many target genes. In the context of cell-cycle checkpoints, one of its most important targets is the *p21* gene. After damage to DNA, p53 is activated by phosphorylation on specific amino acids, which allows it to induce transcription of the *p21* gene and results in increased levels of the p21 messenger RNA (mRNA) and protein. The p21 protein, in turn, binds to the cyclin E/CDK2 and cyclin D/CDK4 or cyclin D/CDK6 complexes and inhibits the kinase activity of the CDKs. One target of CDK2 is the pRb protein, another prominent tumor suppressor whose phosphorylation by CDK is required for progression from G1 to S phase. Thus, by binding to the CDK/cyclin complex, p21 inhibits phosphorylation of pRb and serves as a second mechanism to produce a G1/S phase cell-cycle arrest.

pRb

As indicated earlier the retinoblastoma (pRb) protein is another major tumor suppressor that is mutant in almost half of all human tumors. Its name is derived from the childhood retinal tumor, retinoblastoma. In 40% of the cases, the disease is hereditary, and most of these patients have bilateral tumors. The remaining cases are predominantly sporadic, and the patients usually have tumors only in one eye. Cytogenetic studies had shown that deletions within chromosome 13, particularly in the long arm, are common in patients with hereditary retinoblastoma who inherit one defective chromosome 13. In the early 1970s this observation led Alfred Knudson to propose the "two-hit" theory of cancer, for which he subsequently received the Lasker Award. This theory argues that, when an individual inherits a mutant tumor suppressor allele, only one additional (or second) "hit" is required to facilitate tumorigenesis and tumor progression. The second hit can be in the form of an independent mutation in the functional allele or as loss of the functional allele because of mitotic recombination (recombination between chromosome homologs) and resultant LOH, similar to what was described for *p53* and Li-Fraumeni syndrome. This hypothesis has been validated for retinoblastoma and for several other hereditary and nonhereditary cancers. In nonhereditary cases the first "hit" occurs in a somatic cell and is followed by a second "hit" in the same cell at a later time.

The pRb protein has multiple and diverse functions in the cell cycle. One prominent role is as a regulator of progression from G1 to S phase. The pRb protein in its hypophosphorylated state (low level of phosphorylation) binds the cyclin D/CDK4 and cyclin E/CDK2 complexes. The pRb protein, in its hypophosphorylated state, also binds and inactivates the transcription factor E2F, which controls the transcription of a subset of genes whose products are important for the transition from the G1 to S phase. Therefore pRb inhibits the progression of cells from the G1 to S phase when in its hypophosphorylated state. The pRb protein is phosphorylated at a subset of sites by cyclin D/CDK4 in early to mid G1 and by cyclin E/CDK2 near the end of G1. Phosphorylation of pRb at its carboxy-terminal region in late G1 by cyclin D/CDK4 releases E2F from its complex, making it available to act as a transcription factor. The consequence of pRb hyperphosphorylation is transcription of E2F responsive genes whose products are required for the transition from G1 to S phase (Fig. 10–4; Clinical Case 10–2).

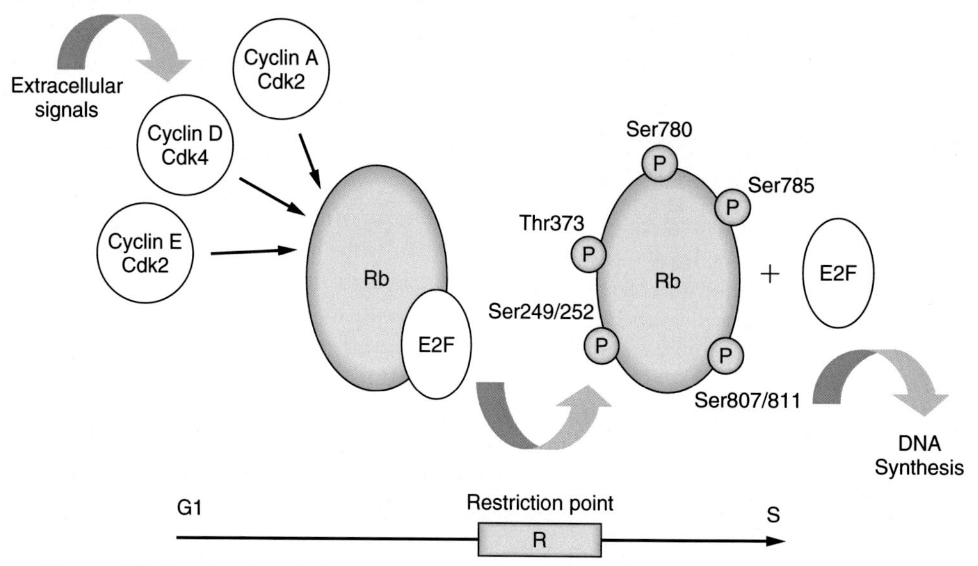

Figure 10–4. **Phosphorylation of pRb by cyclin/cyclin-dependent kinase (CDK) complexes, inactivating pRb and releasing the E2F transcription factor.** The figure illustrates the hyperphosphorylation of pRb by cyclin/CDK complexes, which results in the release of the E2F transcription factor and consequent transcription of a large number of cell-cycle control genes.

Clinical Case 10–2

Mary Moore is 56-year-old postmenopausal Caucasian woman who had been having terrible hot flashes and night sweats to the point of disrupting her sleep. Her gynecologist suggested that she starts estrogen plus progestin for these symptoms, and she had been taking these medications for approximately 4 years. Her doctor insisted that she gets annual mammograms since she was taking a combination postmenopausal hormone replacement therapy (HRT), but she recently skipped this test as she was traveling to Europe on a river cruise. On return to the United States, she forgot to reschedule the mammogram. She noticed that her left breast seemed particularly tender, and she thought she might feel a lump near the nipple. When she noticed her nipple on the left side retracting and dimpling near the area where she thought she felt a lump, she decided to go to her gynecologist for an evaluation.

Cell Biology, Diagnosis, and Treatment of Postmenopausal Breast Cancer

On physical examination, her doctor felt a firm 2-cm lump in her left breast and was concerned that she had lymphadenopathy in her left axilla. The doctor advised Mary to stop taking the HRT and ordered a diagnostic mammogram with ultrasound to evaluate Mary's physical findings. The mammogram revealed a 2–2.5-cm mass in Mary's left breast, and she was referred to a surgeon for an excisional breast biopsy. The pathology report from the biopsy revealed infiltrating ductal carcinoma (IDC) that had receptors for estrogen (ER+) and progesterone (PR+) but was negative for human epidermal growth factor receptor 2 (HER2 negative).

The surgeon recommended a modified radical mastectomy with removal of the lymph nodes in the axilla, and Mary underwent the surgery as soon as possible. The pathology from the mastectomy revealed IDC in two lymph nodes from Mary's axilla, and she was referred to a radiation oncologist for radiation therapy. Mary's oncologist also recommended that she takes ribociclib, a cyclin-dependent kinase inhibitor (CDK 4/6) along with an aromatase inhibitor. Ribociclib is a small molecule inhibitor that selectively binds to the ATP-binding cleft in CDK4 and CDK6, thereby inhibiting the kinase activity of these enzymes that is needed for cell-cycle progression. Mary's initial treatment included ribociclib 600 mg every morning on days 1–21 of 28-day cycle plus the aromatase inhibitor letrozole 2.5 mg every day. Knowing that common toxicity of ribociclib includes neutropenia, thrombocytopenia, and hepatotoxicity, the oncologist regularly checked a complete blood count and liver function tests for Mary. The oncologist also regularly asked Mary to get an ECG to look for a prolonged QT interval—another known toxicity of CDK 4/6 inhibitors. Mary had a good response to treatment but once again began experiencing hot flashes and night sweats. IDC is the most common type of breast cancer, and the incidence of this type of breast cancer increases as women age. Use of combination postmenopausal hormone replacement therapy (estrogen plus progestin) increases the risk of breast cancer, with higher risk associated with longer use.

MITOSIS

What Is Mitosis?

Mitosis is the tightly regulated process of cell division that includes both nuclear division (karyokinesis) and the division of cytoplasm to two daughter cells (cytokinesis). This process can be divided into distinct phases including prophase, prometaphase, metaphase, anaphase, telophase, and finally cytokinesis. The products of mitosis are two daughter cells that have identical DNA content that is also identical to the DNA content of the original parental cell. Depending on the nature of the parental cell, the daughters may be essentially identical in phenotype, or they may differ. In the case of undifferentiated adult stem cells or progenitor cells, one daughter may remain undifferentiated (self-renewing), whereas the other becomes committed to a differentiated lineage.

Interphase and Mitosis

Until the early 1950s histologists and cytologists believed that interphase was essentially a period of cell quiescence and that most of the "action" in cells occurred during mitosis. This concept was based on visual images of cells in histological sections. In most cases, more than 95% of cells were in "interphase," and a minority of cells were in various phases of physical cell division (i.e., mitosis). It was not until the seminal experiments of Alma Howard and Stephen Pelc (see earlier) that scientists appreciated the extent of cellular activity that occurred during interphase. This activity involves general cellular metabolism, including replication, transcription, and translation, much of which is preparatory for the mechanics of mitosis and cell division in proliferating cells (see later and also Fig. 10–5).

Mitotic Stages

Prophase

In prophase the nuclear envelope begins to disaggregate, and chromatin in the nucleus begins to condense and becomes visible by light microscopy as elongated, spindly chromosomes. The nucleolus disappears. Centrosomes begin migrating to opposite poles of the cell, and microtubule fibers extend from centrosome to centrosome and from centrosome to the kinetochore of the centromere of each chromosome to form the mitotic spindle.

Prometaphase

In prometaphase, the nuclear envelope disaggregates, marking the beginning of prometaphase. Multiple proteins associate with the kinetochore within the centromeres, and microtubules attached at the kinetochores begin to move the chromosomes to the center of the cell.

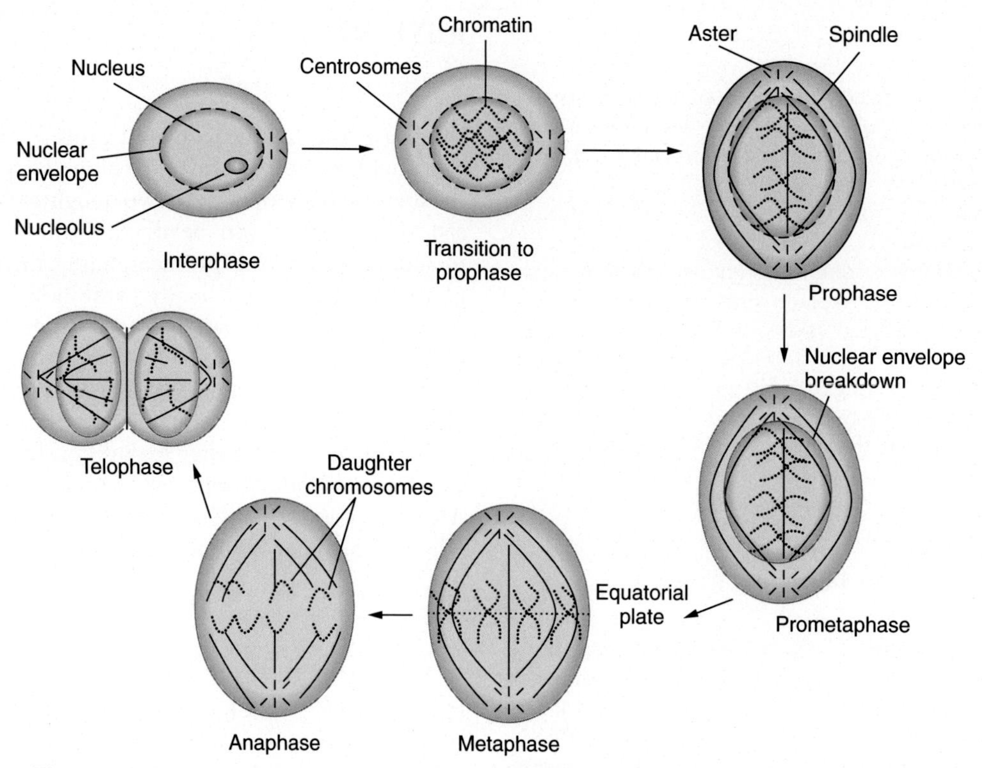

Figure 10–5. **Progression of cells through the various phases of mitosis.** Following the exit from G2, the cells enter the multiple stages of mitosis beginning with a transition to prophase.

Metaphase

As the chromosomes continue to condense and become more compact, they align themselves along a central axis of the cell, which is sometimes referred to as the *metaphase plate.* At this stage, each chromosome is composed of paired chromatids that are the ultimate product of DNA replication. This alignment and organization of chromosomes helps to ensure that, in anaphase, the next phase of mitosis at which chromatids segregate, each daughter nucleus will receive a complete copy of each chromosome.

Anaphase

In anaphase the paired chromatids separate at the kinetochores by which they were previously tethered and randomly segregate and move to opposite poles of the cell. The migration of the chromatids (now designated chromosomes) results from a combination of kinetochore movement along the spindle microtubules away from the center of the cell and toward the cell poles and the physical interaction of polar microtubules.

Telophase

At telophase the chromosomes (formerly chromatids) arrive at opposite poles of the cell, and new membranes form around the daughter nuclei. The chromosomes disperse and are no longer visible as discrete chromosomes by light microscopy. The spindle fibers disappear, and cytokinesis is initiated.

Cytokinesis

In animal cells, cytokinesis results when a ring of actin fibers encircles the cells roughly equidistant from the two poles and participates in the cytokinesis process by contracting and constricting the cell into its two daughter cells, each with one nucleus. The newly formed daughter cells are now ready to resume a new cell cycle or to initiate a program of cell differentiation and exit the cell cycle in a G0 state.

MEIOSIS

Mitosis differs from meiosis in both function and mechanics. The purpose of mitosis is to ensure that both daughter cells receive identical complements of their genomes that are also identical to that of their parent cell. Thus the DNA complement of a dividing somatic cell goes from 2N (where N is the haploid complement of DNA) to 4N after DNA replication and back to 2N after mitosis. In meiosis, spermatogonia and oogonia enter meiotic division as cells with a 4N complement and undergo two successive reduction divisions, with no intervening S phase, to produce four gametes with a 1N (haploid) complement of DNA. The products of spermatogenesis are four functional spermatocytes, whereas those of oogenesis are one functional oocyte and three polar bodies. Besides reducing the DNA complement from 4N to 1N, a primary function of meiotic division is to ensure genetic diversity so that none of the gametes produced are genetically identical to any other gamete produced by that series of meiotic divisions. This is accomplished by recombination between chromosome homologs that appears to be required for subsequent proper chromosome segregation.

Meiotic divisions begin with proliferating germ cells that have completed replication and that have a 4N DNA complement. In males, they are designated spermatogonia, and in females, they are oogonia. In the first meiotic division, the nuclear envelope disaggregates, and the two chromatids of each chromosome remain attached to each other by a common kinetochore (meiotic prophase I). The homologous chromosomes pair and form a "bivalent" composed of two homologous chromosomes, each with two paired chromatids and align themselves at the center of the cell. Thus there are four chromatids in a bivalent. The paired chromosomes, one of paternal and the other of maternal origin, associate with one another along their entire length in what is known as the *synaptonemal complex*. This stage of meiotic prophase I is known as pachytene because the paired chromosomes are condensed and appear fat. During the next stage of the first meiotic prophase (diplotene), the chromosome pairs begin to separate except at a limited number of discreet points called *chiasmata*, which are the sites of reciprocal recombination between the paired chromosomes. These recombination events are required for proper chromosome segregation in meiosis I and form the basis for evolutionary diversity.

The synapsed homologous chromosome pairs, now aligned at the cell center (metaphase I), begin to separate randomly to the opposite poles of the cell during meiotic anaphase I. Telophase I follows, resulting in two cells whose paired chromatids are now chromosomes that are homozygous throughout except in the region distal to the site of crossover. Nuclear membranes may reform, or the cells may rapidly enter meiosis II. The second meiotic division proceeds in the absence of DNA replication and follows the morphologically recognizable steps that one observes for a mitotic division. Because there is no intervening S phase, the resulting gametes are haploid (1N complement of DNA). A generic depiction of the meiotic reduction divisions and the outcome of meiotic recombination is shown in Figure 10–6.

Although the mechanics of meiosis are essentially the same for males and females, significant differences do exist. In both cases, meiotic prophase I is the most protracted of each of the phases. In the human female, meiotic divisions begin during embryogenesis, and by the fifth month of gestation, the oocytes become arrested in the diplotene stage of meiotic prophase I and remain arrested at that stage for decades. The meiotic divisions of individual oocytes are not completed until just before ovulation, at which time the oocyte undergoes two successive meiotic divisions to produce a mature ovum and three polar bodies. The division of cytoplasm is unequal so that only the future egg receives the majority of the cytoplasm, whereas the polar bodies receive sufficiently little cytoplasm that ultimately they do not survive. The sequence of events during oogenesis involves proliferating oogonia, some of which arrest in G2, with a 4N complement of DNA and begin to differentiate. The cells arrest at the diplotene stage of meiotic prophase I and are designated a primary oocyte. Before ovulation, meiotic division is triggered, producing a secondary oocyte (2N) and a 2N polar body that eventually will be discarded. Meiosis II creates a haploid (1N) egg and another polar body. The 2N polar body from the first meiotic division also divides, ultimately yielding an egg and three polar bodies, all of which are 1N. As indicated

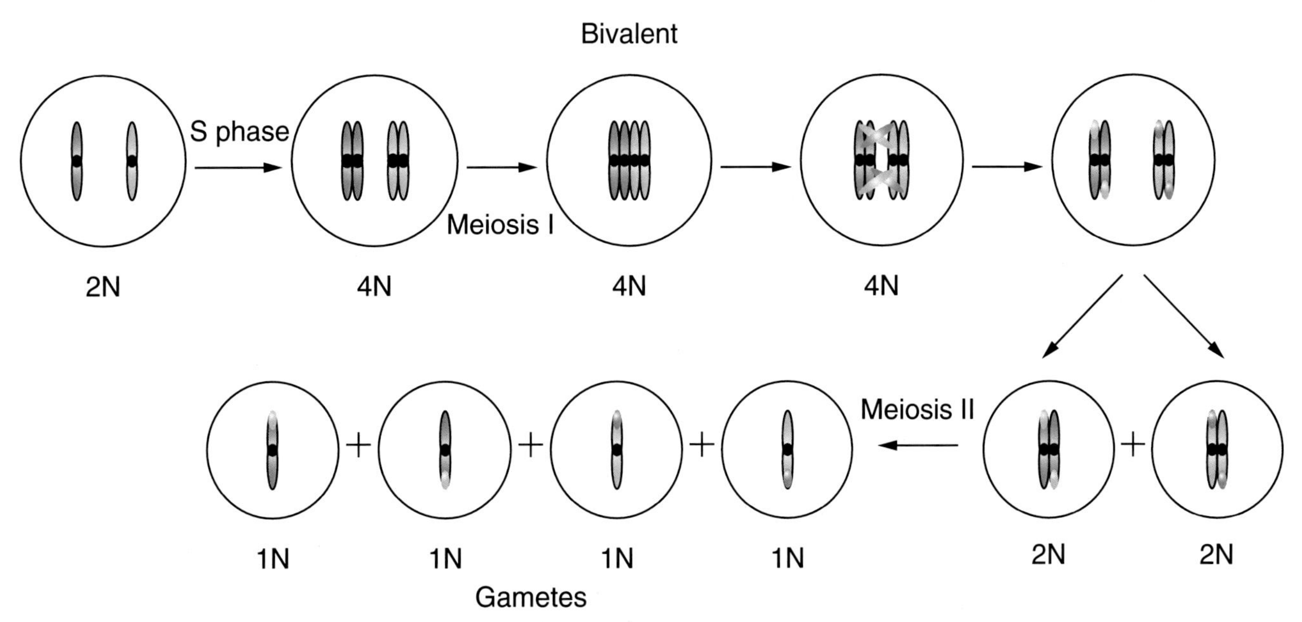

Figure 10–6. **Schematic depiction of reduction divisions during meiosis.** The figure illustrates the pairing of chromosome homologs during meiosis I and recombination between maternally and paternally derived chromosomes to produce genetically different gametes.

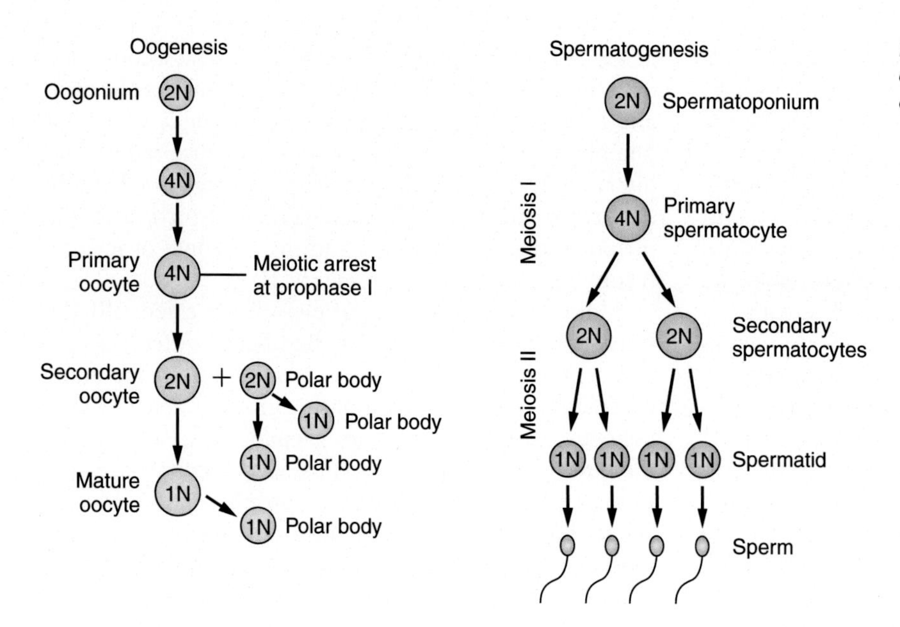

Figure 10–7. **Comparison between oogenesis and spermatogenesis and their meiotic outcomes.**

earlier, meiosis in the male is mechanistically similar, but its timing and outcome are different. Meiosis in male germ cells does not begin until puberty and continues for many decades. Once the meiotic process is initiated, it proceeds with no periods of interruption except for a somewhat protracted first meiotic prophase. The proliferating spermatogonia produce a primary spermatocyte that undergoes the first meiotic division to produce two secondary spermatogonia. Each of the secondary spermatogonia undergoes the second meiotic division to produce a total of four functional haploid spermatocytes. The progression of the meiotic process and the fundamental differences between oogenesis and spermatogenesis are outlined in Figure 10–7.

Some of the players that regulate the mitotic cell cycle also regulate meiosis. In the 1970s Yoshio Masui discovered that a factor that became known as maturation-promoting factor (MPF) was critical for frog oocyte maturation. Later, it was discovered that MPF is a protein complex that contains Cdk1 and cyclin B, the same factors that control entry of somatic cells into mitosis. Notably the MOS protooncogene, a serine/threonine kinase first isolated as an oncogene from the Maloney murine sarcoma virus, is highly expressed in mouse and human germ cells and appears to be required for proper meiotic progression. When its activity is inhibited, meiotic maturation does not occur.

Cell-Cycle Checkpoints

The concept of cell-cycle checkpoints was introduced by Leland Hartwell in the 1980s. The principle of a checkpoint is that, after damage to a cell, the cell will arrest in a nonrandom manner to allow for repair of the damage before proceeding through the cycle. The various cell-cycle checkpoints are the end result of inhibitory signaling pathways that produce a delay or arrest of cell-cycle progression at specific points in the cell cycle in response to DNA damage or other cellular stress. Historically, checkpoints were defined as specific junctures in the cell cycle during which the integrity of DNA is assessed before cells progress to the next phase of the cell cycle. The term *checkpoint* has recently become more ambiguous because it has been applied to the entire ensemble of cellular responses to DNA damage, including the arrest of cell-cycle progression, induction of DNA repair genes, and apoptosis. This expanded definition is not unreasonable because activation of proteins involved in cell-cycle arrest also leads to expression of genes that participate in DNA repair and apoptosis. It is important, however, to appreciate that DNA repair pathways are functional in the absence of damage-induced cell-cycle arrest and that apoptosis can occur independently of the cell-cycle arrest machinery. Accordingly the term *DNA damage checkpoint* should be reserved for events that specifically retard or arrest cell-cycle progression in response to DNA damage.

As discussed earlier the G1/S and G2/M transition points and S phase progression are tightly controlled in unperturbed cells. Many of the same proteins involved in regulating the orderly progression through the cell cycle are also involved in checkpoint responses. Thus the DNA damage checkpoints are not unique pathways activated by DNA damage, but rather are biochemical pathways that operate under normal growth conditions whose effects are amplified after an increase in DNA damage. The principal checkpoints are at the G1/S phase boundary, during S phase of intra-S phase checkpoint, at the G2/M boundary, and during mitosis or the spindle assembly checkpoint.

Molecular Components of the DNA Damage Checkpoints

The DNA damage checkpoints are the end result of complex signaling pathways composed of three principle

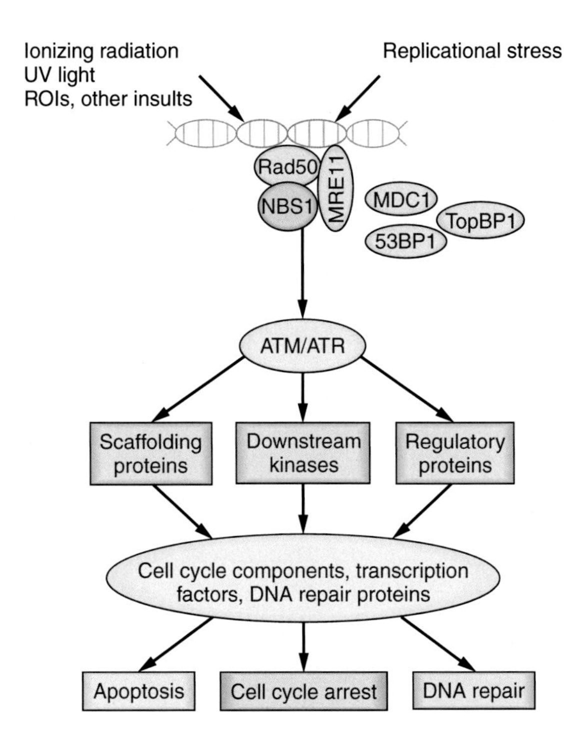

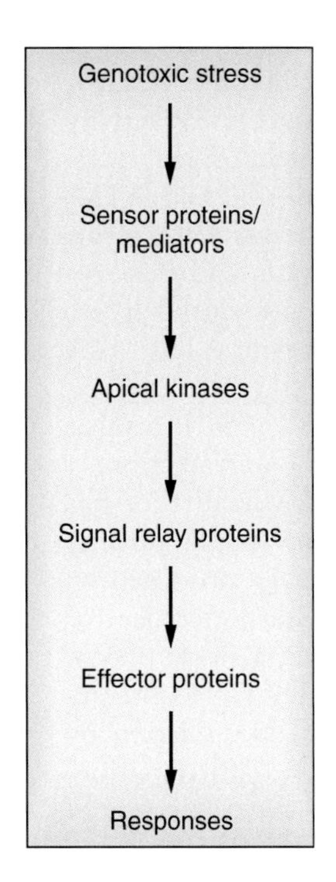

Figure 10–8. **Signaling cascades initiated by damage to DNA.** The class of signaling molecule at each level of the cascade is shown in the panel on the right. *ROI*, region of interest; *UV*, ultraviolet.

components: sensors of damage, signal transducers, and effectors. Genetic and biochemical studies initially performed in yeast and more recently in mammalian cells, including human cells, have identified several proteins involved in sensing DNA damage, in transducing the signals from the sensors, and in carrying out the effector steps of the DNA damage checkpoints (Fig. 10–8). These functional classifications, however, are not absolute because there is considerable overlap in function between each of these components. The DNA damage sensor ATM, for example, also functions as a signal transducer. Furthermore a fourth class of checkpoint proteins, sometimes referred to as *mediators*, has been identified that includes BRCA1, Claspin, 53BP1, and MDC1. Conceptually, this class of proteins has been positioned between the sensors and signal transducers. As in the case of ATM, which functions both as a sensor and transducer, these mediator proteins also appear to participate in more than one step of the checkpoint response.

Although the G1/S, intra-S, G2/M, and spindle assembly checkpoints represent distinct checkpoints within the cell cycle, the DNA damage sensor molecules that activate the various checkpoints appear to either be shared by each of the checkpoints or to play a primary sensor role in one or more cases and a backup role in the others. Similarly the protein kinases and phosphatases that participate as signal-transducing molecules are shared to varying degrees by the different DNA damage response pathways. Although the effector components (proteins that directly inhibit cell-cycle phase transition) of the checkpoints are what define checkpoint specificity, the various sensors, mediators, and signal transducers may play more prominent roles in one checkpoint pathway than in others.

SENSORS RECOGNIZE SITES OF DNA DAMAGE

ATM and ATR

Following damage to DNA, cell-cycle arrest first requires the sensing and recognition of the sites of damaged DNA by sensors such as ataxia-telangiectasia mutated (ATM) to initiate a checkpoint. Mutations in ATM are responsible for a rare genetic syndrome, ataxia-telangiectasia (A-T), which is characterized by cerebellar degeneration, immunodeficiency, genome instability, clinical radiosensitivity, and cancer predisposition. ATM is a 350-kDa oligomeric protein that exhibits significant sequence homology to the phosphatidylinositol 3-kinases (PI3Ks) but lacks lipid kinase activity. It does, however, have protein kinase activity that is stimulated by agents that induce double-strand DNA breaks. ATM preferentially binds free DNA termini at the site of DNA breaks, apparently in monomeric form. After exposure of cells to ionizing radiation, ATM is activated by autophosphorylation and phosphorylates and activates a large number of target proteins, including Chk2, p53, NBS1, and BRCA1, at a serine (S) or a threonine (T) that precedes a glutamine (Q) in the sequence motif of SQ and TQ. Neighboring sequences must also confer some specificity because p53 is phosphorylated by ATM at $S_{15}Q$ but not at $S_{37}Q$.

A related PI3K, ATR, was discovered in the human genome database as a gene with sequence homology to ATM and to SpRad3 (from *S. pombe*), hence the name ATR (*ATM* and *Rad3* related). The gene encodes a large protein of 303 kDa with a C-terminal kinase domain and regions of homology to other PI3K family members. Absence of ATR in mice results in embryonic lethality. In humans, mutations causing partial loss of ATR activity are associated with Seckel syndrome, an autosomal recessive disorder that shares features with A-T. Like ATM, ATR is a protein kinase that has specificity for serine and threonine residues in SQ/TQ motifs and that can phosphorylate essentially all of the proteins that are phosphorylated by ATM. In contrast with ATM, ATR is activated by ultraviolet (UV) light rather than by ionizing radiation, and it is the main PI3K family member that initiates signal transduction after UV irradiation.

The relative specificity of ATR for UV irradiation has raised the possibility that the enzyme might directly recognize specific types of DNA damage produced by UV. ATR binds directly to DNA, with preference for sites of UV-damaged DNA containing a (6–4) photoproduct. Electron microscopy of ATR–DNA interactions has shown that the extent of DNA binding by ATR increases with increasing dose of UV irradiation. Unlike ATM, however, ATR rarely binds termini of linear DNA, indicating that ATR does not effectively recognize DNA double-strand breaks. In vitro phosphorylation of p53 by ATR is stimulated by the addition of undamaged DNA to such a kinase assay and is even further stimulated by the addition of UV-damaged DNA in a dose-dependent manner. Thus ATM is both a sensor and transducer molecule that responds to double-strand breaks, and ATR serves an analogous role for a cellular response to base damage incurred by UV irradiation.

Mediators Associate with Sensors and Signal Transducers

The mediator proteins simultaneously associate with DNA damage sensor proteins and with signal transducers at defined phases of the cell cycle. Consequently, they help provide specificity to the various signal transduction pathways that are activated. The prototype mediator protein is the yeast scRad9 protein, which functions in the signal transduction pathway from yeast scMec1 (ATR in mammals) to yeast scRad53 (CHEK2 or Chk2 in mammals). Another mediator, Mrc1 (*m*ediator of *r*eplication *c*heckpoint) that is found in both types of yeast, *Saccharomyces cerevisiae* and *Schizosaccharomyces pombe*, is expressed only during S phase and is necessary for S phase checkpoint signaling from scMec1/spRad3 (ATR) to scRad53/spCds1 (CHEK2). Humans have at least three proteins that contain the conserved BRCT motif involved in protein-protein interaction that fit into the mediator class of proteins. These include the p53 binding protein, 53BP1; the topoisomerase binding protein, TopBP1; and the *m*ediator of *D*NA damage *c*heckpoint 1, MDC1. These proteins interact with DNA damage sensors such as ATM, with DNA repair proteins such as BRCA1 and the Mre11/Rad50/NBS1 (M/R/N) complex, with signal transducers such as CHEK2, and with effector molecules such as p53. The DNA damage checkpoint response is compromised in cells that have decreased levels of or lack of these proteins.

Signal Transducers CHEK1 and CHEK2 Are Kinases Involved in Cell-Cycle Regulation

Humans have at least two kinases, CHEK1 and CHEK2, whose function is primarily that of a signal transducer in cell-cycle regulation and checkpoint responses. These kinases were identified based on homology with yeast scChk1 and scRad53/spCds1, respectively. Both CHEK1 (Chk1 in mice) and CHEK2 (Chk2 in mice) are serine and threonine kinases that have limited homology to each other and that display moderate substrate specificities. In mammalian cells the double-strand break signal sensed by ATM is transduced predominantly by CHEK2. The UV damage signal sensed by ATR is mainly transduced by CHEK1. There is, however, some overlap between the functions of the two proteins. It is notable that mice lacking Chk1 ($Chk1^{-/-}$) exhibit embryonic lethality, whereas mice deficient in Chk2 ($Chk2^{-/-0}$) are viable and display near-normal checkpoint responses. Mutations in human CHEK2 increase risk for breast cancer, for a rare multitumor Li-Fraumeni (LFS)-like cancer syndrome.

The p53 and Cdc25 Phosphatases Are Important Effector Proteins in Cell-Cycle Regulation

Two of the principal effector proteins are p53 and the Cdc25 phosphatases (see earlier descriptions). The Cdc25 family of proteins dephosphorylates the various CDKs to allow cell-cycle progression. The Cdc25 proteins are themselves targets for phosphorylation, which causes their sequestration by binding to 14-3-3 proteins, causes their exclusion from the nucleus, or causes their ubiquitination and proteasome-mediated degradation. In each case, after DNA damage, the Cdc25 phosphatase is rendered unavailable to dephosphorylate its target CDK protein resulting in a cell-cycle arrest. As examples the hypophosphorylated form of Cdc25A promotes the G1/S transition by dephosphorylating CDK2, and the hypophosphorylated form of Cdc25C promotes the G2/M transition by dephosphorylating CDK1 (Fig. 10–9). Phosphorylation of Cdc25A and Cdc25C at specific residues results in loss of their phosphatase activity due to translocation or proteasome-mediated degradation and arrest at G1/S and G2/M, respectively.

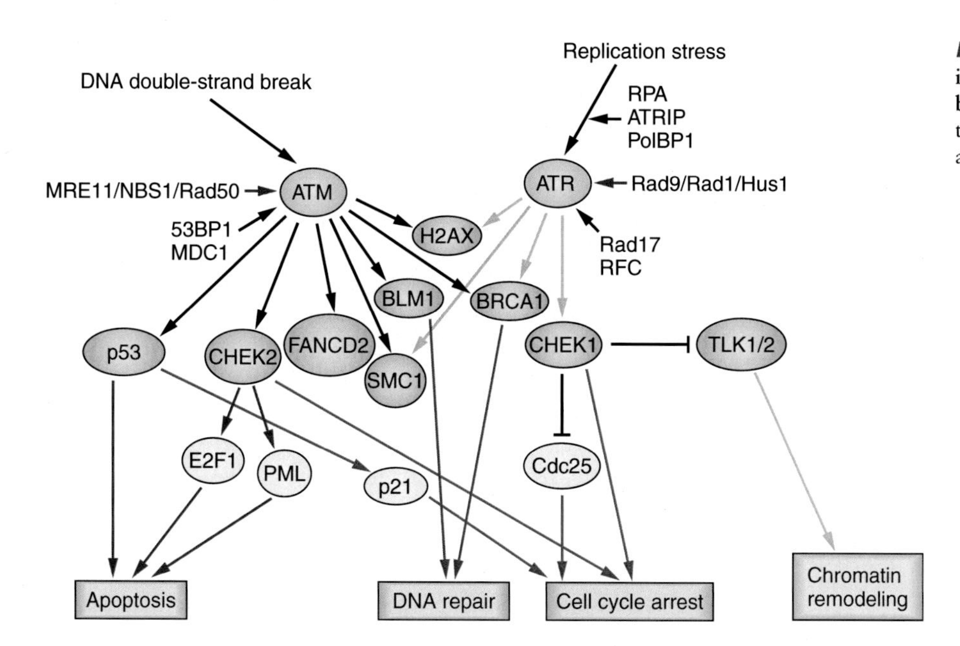

Figure 10–9. **Signaling cascades induced by DNA double-strand breaks or by replication stress.** The figure illustrates the various pathways activated following activation of ATM and ATR, respectively.

G1/S Checkpoint

In an unperturbed population of proliferating cells, the cells that are in G1 become committed to entering the S phase at the "restriction point" (mammalian cells) or "start" (budding yeast). In proliferating human cells the restriction point precedes the actual start of DNA synthesis by about 2 h. When there is DNA damage, however, entry into S phase is prevented regardless of whether or not the cells have passed the restriction point. The cells arrest at the G1/S checkpoint presumably to allow time for the cells to repair the damage.

If the damage to DNA involves DNA double-strand breaks (DSBs) caused by ionizing radiation or by radiomimetic agents such as some chemotherapeutics, ATM becomes activated and phosphorylates and activates numerous target molecules, notably p53 and CHEK2. One consequence of these phosphorylation events is the arrest at the G1/S checkpoint. Following DNA damage in the form of DSBs, ATM becomes phosphorylated and activated. In one case, it activates CHEK2, which in turn phosphorylates Cdc25A phosphatase. This phosphorylation by CHEK2 and other kinases causes the inactivation of Cdc25A by nuclear exclusion and by ubiquitin-mediated proteolytic degradation. Loss of active Cdc25A results in the accumulation of the phosphorylated (inactive) form of CDK2, which is incapable of phosphorylating its associated cyclin and also of Cdc45, a protein whose phosphorylation is required to initiate replication at sites of preformed replication origin complexes. In the second slower but more sustained case, activated ATM phosphorylates p53, as does the newly activated CHEK2. The former phosphorylates p53 at serine 15 and the latter at serine 20. These phosphorylation events inhibit nuclear export and degradation of p53 and allow p53 to serve as a transcriptional activator that induces transcription of p21, the inhibitor of CDK2/cyclin E and of CDK4/cyclin D. Both pathways are functional components of the G1/S checkpoint and allow cells to accumulate at the G1/S phase boundary.

If the DNA damage is caused by UV light or by UV-mimetic agents, rather than by ionizing radiation, the types of DNA damage are different. The cellular response to this type of damage is similar, but not identical, to that induced by double-strand DNA breaks. Damage to DNA that is caused by UV irradiation is not sensed by ATM, but rather by ATR, Rad17-RFC, and the 9-1-1 complex (a complex containing Rad9, Rad1, and Hus1). Once the damage is recognized, ATR phosphorylates the transducing kinase CHEK1, which, in turn, phosphorylates and inactivates Cdc25A, leading to G1 arrest (see Fig. 10–9).

Intra-S-Phase Checkpoint

The intra-S phase checkpoint is activated by damage that occurs while cells are replicating their DNA or by unrepaired DNA damage in cells that escape the G1/S checkpoint. In both cases, cells manifest a block in replication. Because the original definition of a checkpoint implied the involvement of a biochemical pathway that ensured the completion of a reaction before progression to the next cell-cycle phase, the intra-S phase checkpoint does not strictly meet that definition. Nevertheless, as our understanding of molecular events that underlie checkpoints has become clearer, the intra-S checkpoint does meet a definition that is modified to include this checkpoint. A cell-cycle checkpoint is the end product of a molecular regulatory pathway or signaling cascade that ensures an ordered succession of cell-cycle events, which when perturbed results in cell-cycle arrest (Fig. 10–10).

The damage sensors for the intra-S checkpoint encompass a large set of checkpoint and DNA repair proteins. When DNA damage is a frank double-strand break or a double-strand break resulting from replication of a nicked or gapped DNA, the checkpoint is activated by contributions of ATM, the M/R/N complex, and BRCA1. These repair proteins presumably function as sensors because they bind to either double-strand DNA breaks (ATM) or to special branched DNA structures (the M/R/N complex, BRCA1, and BRCA2). They participate in the intra-S checkpoint by initiating a kinase signaling cascade that proceeds through pathways involving CHEK1 and CHEK2 (see Fig. 10–9).

A striking characteristic of the intra-S phase checkpoint is radioresistant DNA synthesis (RDS), which illustrates the intimate relation between this checkpoint and double-strand DNA break repair. In wild-type cells, ionizing radiation causes immediate cessation of active DNA synthesis. In contrast, cells from A-T patients challenged in like manner continue to proceed into and through S phase without delay. This RDS also occurs in cells with mutations in subunits of the M/R/N complex and in FANCD2 (BRCA2) and represents a compromised intra-S checkpoint.

When DNA is damaged by UV radiation that produces pyrimidine dimers or by chemicals that produce bulky DNA adducts, it is the ATR protein or more precisely the ATR-ATRIP heterodimer that serves as the primary sensor of these types of DNA damage. ATR binds to chromatin, preferentially to the UV-induced lesions, and is activated. Activated ATR phosphorylates CHEK1, which in turn phosphorylates and downregulates Cdc25A by causing its degradation and inhibiting the firing of replication origins (see Fig. 10–10).

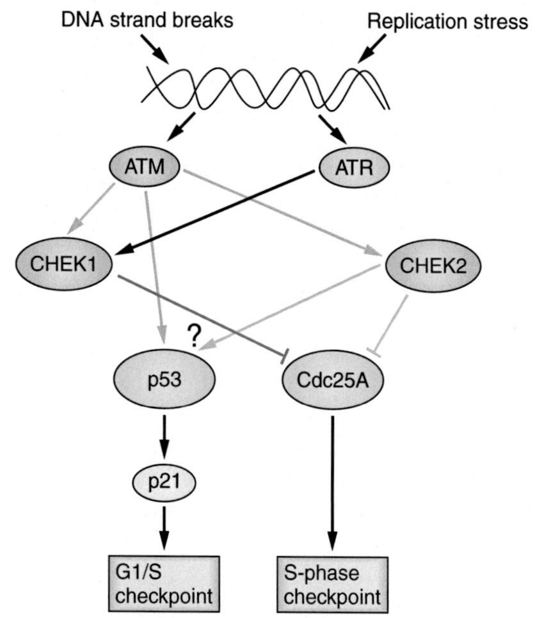

Figure 10–10. Major signaling pathways leading to checkpoint arrest in S phase and G1/S boundary.

G2/M Checkpoints

The G2/M checkpoints actually represent multiple checkpoints that regulate cellular entry into mitosis, transit through mitosis, and exit from mitosis. When DNA is damaged during the late S phase, after cells are committed to divide, the cells will arrest in late G2. Again the arrest is mediated by ATM after introduction of DNA double-strand breaks or by ATR if the lesions are a consequence of chemical challenge or UV exposure. Signaling occurs through CHEK 2 and CHEK 1, respectively, resulting in the phosphorylation and activation of p53 and the induction of the CDK inhibitor p21, mediating a cell-cycle checkpoint arrest. A second pathway, also mediated by ATM or ATR and by CHEK 2 or CHEK1, respectively, inhibits Cdc25A and Cdc25C phosphatase activity. The wEE1 kinase normally phosphorylates CDK1 complexed with cyclin B on tyrosine 15 to prevent the entry of cells into mitosis. This activity is counterbalanced by the phosphatase activity of CDC25. When CDC25 activity is diminished or lost in response to DNA damage, the phosphorylated status of the CDK1/cyclin B complex persists, the cells arrest, and the mitotic checkpoint is enforced. Once CDC25 activity is restored and the inhibitory CDK1 phosphate groups are removed, the cells can progress from G2 to mitosis (Fig. 10–11).

Although the inhibitory pathways mentioned earlier are those that we understand best, other less well understood regulatory proteins, such as the aurora and polo kinases, play significant roles in the transition from late G2 to mitosis. The aurora kinases (A, B, and C) are a family of serine/threonine kinases that localize to the centrosome during mitosis and play an integral role in chromatid segregation and cytokinesis. Loss of aurora A, for example, results in abnormal spindle morphology, while loss of aurora B causes chromosome misalignment and failure of proper cytokinesis. Another key protein that

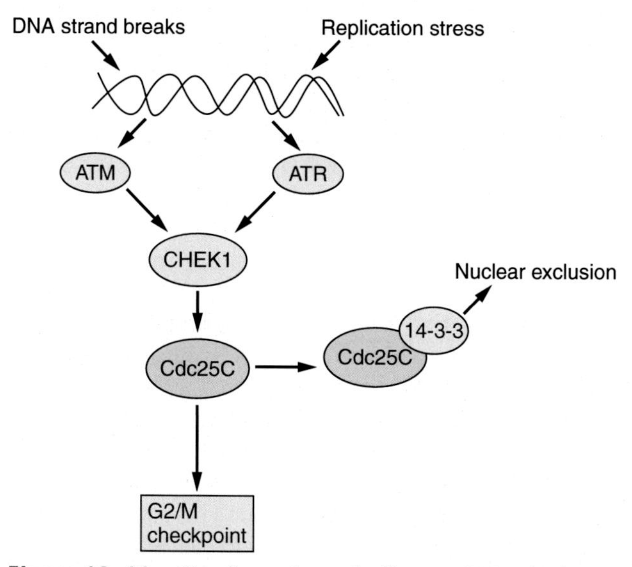

Figure 10–11. Signaling pathways leading to a G2/M checkpoint.

controls the onset of prophase by monitoring microtubule and centrosome integrity is designated CHFR, another ubiquitin E3 ligase. When cells are challenged with drugs that disrupt microtubules or their dynamics, chromosome condensation and centrosome separation are delayed. After such drug treatment or other type of stress, CHFR may function by interfering with the import of CDK1/cyclin B into the nucleus, which is necessary for progression of cells into and through prophase. The polo-like kinases (PLKs) form a family of five kinases that play a variety of roles in cellular metabolism. The PLK1 kinase is overexpressed in a broad range of tumors, and high levels of PLK1 correlate with poor prognosis. PLK1 has major regulatory activity during mitosis. In proliferating cells, its abundance is low during the G1 and S phase of the cell cycle but increases during G2, reaching a peak at mitosis. Its overexpression leads to spindle assembly defects and overcomes the G2/M checkpoint that is normally activated after DNA damage.

Once mitosis is initiated, cell progression is subject to yet another checkpoint, the "spindle assembly checkpoint." This checkpoint is complex and is mediated by inactivating a large macromolecular complex known as the anaphase-promoting complex/cyclosome (APC/C) that has E3 ubiquitin ligase activity that is necessary for the degradation of several key proteins such as cyclin B and securin. Degradation of cyclin B is required for exit from mitosis, and destruction of securin is needed for chromatid separation. Additional proteins that associate with or regulate APC/C activity in selective protein degradation include MAD, BUB1, BUBR1, and Cdc20. Loss of any of these factors compromises the ability of cells to arrest properly when spindle microtubule assembly is perturbed. One consequence of the failure to arrest is "mitotic catastrophe," whereby aberrant mitotic figures are formed, and ultimately cells with aberrant chromosome numbers emerge and undergo cell death.

CELL-CYCLE PERTURBATIONS AND CANCER

The preceding discussion describes many of the signaling molecules and pathways that are required for regulated cell-cycle progression in normal proliferating cells. When mutations or epigenetic modifications perturb any one or combinations of these events or pathways, the risk for developing a tumor is increased. Almost every human tumor is accompanied by one or more mutations in genes that affect control of cell cycle, leading to unregulated cell proliferation. In principle the cell-cycle lesions can occur at any cell-cycle checkpoint or in any phase of the cell cycle, and they represent an important step in the progression to full-blown cancer. Thus human tumors can be characterized in part by the type of cell-cycle control that the tumor cells have lost.

Perturbations in Regulating the G1-to-S Transition in Cancer

When cyclin D1 forms aberrant complexes with CDK4 or CDK6 due to overexpression or other mechanisms that render these complexes hyperactive, the cells behave like those in which pRb has been lost or inactivated and, therefore, are insensitive to mitogenic signals. When such cells are subjected to physiological insults that normally result in pRb-mediated inhibition of progression into S phase, they continue to proliferate. Such aberrant CDK activation or loss of pRb has obvious implications for cancer, and indeed, pRb loss or hyperactivation of CDK4, CDK6, or both occurs in a large fraction of human tumors. Hyperactivation of CDK4 and CDK6 can occur through deregulated expression of the D-type cyclins, loss or epigenetic silencing of $p16^{INK4a}$, or mutations that compromise the inhibitory functions of $p16^{INK4a}$. Hence, every element of the core pRb pathway ($p16^{INK4a}$, D-type cyclins, CDK4, CDK6, and pRb itself) is a potential oncogene or a tumor suppressor and a potential therapeutic target.

Molecular analyses of human cancers fully support this notion. Amplification or rearrangement of the *cyclin D1* gene, for example, and overexpression of the cyclin D1 protein have been described in a wide spectrum of human cancers such as squamous cell carcinomas of the head and neck, carcinomas of the uterine cervix, astrocytomas, nonsmall cell lung cancers, and soft-tissue sarcomas. A well-documented example is the frequent involvement of *cyclin D1* in human breast cancer. The *cyclin D1* gene is amplified in about 15%–20% of human mammary carcinomas and is overexpressed in more than 50% of such tumors. Overexpression of cyclin D1 is common at the earliest stages of breast cancer, such as ductal carcinoma in situ (DCIS), but not in premalignant lesions, such as atypical ductal hyperplasia. Thus overexpression of cyclin D1 can serve as a marker of malignant transformation of mammary epithelial cells. When cyclin D1 is overexpressed in tumor cells, this overexpression is maintained at the same level throughout breast cancer progression, from DCIS to invasive carcinoma, and is preserved even in metastatic lesions.

Other members of the cyclin D gene family, such as those encoding cyclins D2 and D3, are also often amplified, and the encoded proteins overexpressed in many human cancers. Cyclin D2 overexpression is common in B-cell lymphocytic leukemias, lymphoplasmacytic lymphomas, chronic lymphocytic leukemias, and in testicular and ovarian germ cell tumors. Cyclin D3 overexpression is seen in glioblastomas, renal cell carcinomas, pancreatic adenocarcinomas, and several B-cell malignancies such as diffuse large B-cell lymphomas or multiple myelomas.

Overexpression of CDK4 occurs in breast cancers, in gliomas, glioblastoma multiforme, sarcomas, and meningiomas, often as a consequence of gene amplification.

Small molecule inhibitors of CDK4 and CDK6 have now become FDA-approved anticancer therapies. Because of the loss, mutation, or silencing of the gene encoding the CDK4 inhibitor, p16^{INK4a}, a different subset of tumors, including retinoblastoma, osteosarcoma, small-cell lung carcinoma, and bladder carcinoma, is associated with the loss of the pRb protein.

pRb Pathway in Cancer

The retinoblastoma gene (*RB1*) is lost in several types of cancers. The pRB protein is complexed with cyclin/CDK complexes, particularly CDK4/CDK6/cyclin D and CDK2-cyclin E, in a hypophosphorylated form. During G1 the CDK4/6 cyclin D complex monophosphorylates pRB at a single amino acid residue, which maintains the cells in G1. At late G1, pRB becomes hyperphosphorylated that results in the release of the E2F transcription factor from the same complex and the transit of cells from G1 to the S phase. The activation of the E2F transcription factor results in the activation of several hundred E2F target genes, many of which have cell-cycle regulatory function.

Notably the majority of human tumors show loss or mutation in only a single step or member of the pRb pathway in any given tumor. For example, if cyclin D1 is overexpressed in a tumor, pRb and p16^{INK4a} are usually unaffected and expressed normally. This appears to be due to the fact that excess cyclin D1 in complex with CDK4 or CDK6 is sufficient to evade the inhibitory effects of p16^{INK4a} and neutralizes the function of pRb. This observation suggests that the pRb pathway is linear and that CDK4 and CDK6 converge on pRb as their primary critical substrate, at least as far as cancer cell proliferation is concerned. In this view, inactivation of pRb or any component in this pathway, with subsequent loss of regulation of the E2F transcription factor, is critical to cell-cycle control and is sufficient to produce unregulated cell cycle and cell proliferation kinetics in tumors. Loss of this negative control mechanism contrasts with the mechanism by which cyclin E is oncogenic and contributes to the pathobiology of breast tumors. The frequency with which cyclin E is overexpressed in tumors suggests that it may accompany other mutations in the core pRb pathway. This apparent selection for cyclin E overexpression, even in the presence of pRb pathway mutations, is likely due to "downstream" roles of cyclin E1 in replication control and serves to emphasize the more pRb-centric roles of D-type cyclins and CDK4 and CDK6, whose functions are generally dispensable in cells that lack pRb.

ATM in Cancer

As discussed earlier, different types of genotoxic agents evoke distinct and overlapping cellular responses that correspond to the particular type of DNA damage. The different responses are mediated by specific molecules involved in the detection of distinct lesions, such as bulky DNA adducts or DNA strand breaks. Double-strand DNA strand breaks are generated by a variety of genotoxic agents, including ionizing radiation, radiomimetic agents, and inhibitors of topoisomerase, all of which activate the ATM kinase. Cells from patients with ataxia-telangiectasia (A-T) are extremely sensitive to these genotoxic agents and display several impaired signal transduction pathways that mediate cellular responses to this type of genotoxic stress. A-T is characterized by multiple phenotypic aberrations including lymphoreticular malignancies and extreme sensitivity to ionizing radiation. The mutant *ATM* gene that is responsible for this pleiotropic phenotype is inactivated in the majority of A-T patients by mutations that cause frameshifts that truncate its protein product and impair its activity. Disruption of the corresponding gene in the mouse creates a phenotype that reflects most of the features of A-T in humans.

The possible role of the mutant *ATM* gene in cancer predisposition in heterozygous individuals is of concern. The specific issue is whether female A-T carriers are at increased risk for cancer after low-dose radiation and therefore should avoid diagnostic mammography. LOH at *ATM* induced by the low-dose X-irradiation may be sufficient to increase risk for cancer if a woman is otherwise predisposed. The association of mutant ATM with at least one type of malignancy, T-prolymphocytic leukemia (T-PLL), is apparent and has implicated *ATM* as a tumor suppressor gene in somatic cells. Tumor tissue from T-PLL patients shows somatic inactivation of both *ATM* alleles caused by rearrangements or point mutations in greater than 50% of cases, supporting the role of *ATM* as a tumor suppressor and linking the biology of T-PLL to ATM function.

Mice genetically engineered to be ATM deficient ($Atm^{-/-}$) develop aggressive malignant thymic lymphomas and usually die by 4 months of age. In a similar mouse model but in a different genetic strain, the average time of tumor onset was considerably longer, with more than 50% of the animals surviving to 10 months. The difference in cancer susceptibility between the two mouse models is probably due to differences in genetic background between the two mouse strains and reflects the variability that one sees in A-T patients.

p53 in Cancer

The p53 protein, which is mutant in more than half of all cancers, functions as a tetrameric transcription factor that is present at low levels in normal, unperturbed cells. After cells are stressed, p53 undergoes posttranslational modification and is stabilized, resulting in its accumulation. After cells are stressed the p53 protein acts in its capacity as a transcription factor and facilitates the transcription of multiple genes involved in cell-cycle arrest or apoptosis, depending on the cellular context, the extent of damage, or other unknown parameters. In about 90% of cases in which cells or tumors have lost p53 function, the protein is mutant. In the remaining 10% of cases, the protein is completely absent. Most tumors with mutant p53 have lost heterozygosity at that locus, consistent with the behavior of a classic tumor suppressor. The model, which has been substantiated for several tumor types, is that mutation occurs at one allele of p53 and that the remaining wild-type allele is lost as a consequence of LOH caused by mitotic recombination (recombination between chromosome homologs). This finding is consistent with the observation that tumors with LOH at p53 also have lost heterozygosity at multiple linked loci on the short arm of chromosome 17, which houses the *p53* gene.

The p53 protein is an integral component of several signaling pathways. Thus it is not surprising that mutations or changes in expression of other genes within these pathways can affect p53 function and consequently the signaling cascades regulated by p53. Predominant among these is MDM2, a protein that regulates the stability of p53 by physically binding p53 and causing its ubiquitination and transport to the proteasome for degradation. The level of the MDM2 protein, which is elevated in many tumors, especially sarcomas, is the result of amplification of the *MDM2* gene, enhanced transcription of the gene, or enhanced translation of its mRNA. When p53 is phosphorylated by ATM in normal cells, its affinity for MDM2 is reduced, which allows p53 to accumulate and to act in its capacity as a transcription factor. When MDM is overexpressed, it presumably depletes the cell of available p53 and mimics a phenotype consistent with the loss of p53.

Several virally encoded proteins have a mode of action that mimics that of MDM2 and renders the cell effectively p53 deficient. Some human papillomavirus (HPV) subtypes, particularly HPV16 and HPV18, are associated with cervical and laryngeal cancer and produce two oncogenic proteins, E6 and E7. The E6 protein specifically binds p53 and causes its degradation, which explains why p53 mutations in cervical cancers are rare. Notably the second viral protein, HPV E7, binds and inactivates cellular pRb.

THE CHECKPOINT KINASES AND CANCER

CHEK1 and Cancer

Although CHEK1 mutations in tumors are extremely rare, they do occur. They appear to be restricted to carcinomas of the colon, stomach, and endometrium. Some colon and endometrial carcinomas with microsatellite instability have been reported to have frameshift mutations in *CHEK1* caused by insertion or deletion of a single adenine in a polyadenine tract. The resulting truncated CHEK1 proteins are defective because of the lack of the C-terminal end of the catalytic domain and the complete loss of the SQ-rich regulatory domain. Also a subset of small-cell lung carcinomas produces a CHEK1 mRNA that encodes a shorter CHEK1 isoform lacking a conserved sequence in the catalytic domain. The deleted region appears to be involved in substrate selectivity, but the significance of the predominant expression of this alternative CHEK1 isoform in fetal lung and in small-cell carcinomas of the lung, but not in normal adult lung tissue or other types of lung tumors, remains to be established. Notably the complete absence of Chk1 in genetically engineered mice results in early embryonic lethality. Deletion of *CHEK1* in a p53-deficient chicken tumor cell line is tolerated by these cells, as is the earlier mentioned heterozygous truncation mutation of CHEK1 in some human tumors. Given the role of CHEK1 in the DNA damage response, it is likely that either hypomorphic mutations in CHEK1 or loss of function mutations that may occur during the progression of cancer contribute to enhanced genetic instability in some tumors.

CHEK2 and Cancer

The first evidence that genetic alteration in *CHEK2* may predispose to cancer was the detection of rare *CHEK2* germline mutations in families with a form of LFS. LFS is a familial cancer syndrome first associated with germline mutations in p53. The syndrome is characterized by multiple types of tumors, predominantly breast cancer and sarcomas, that arise during childhood. The *CHEK2* polymorphic variant, *CHEK2**1100delC, has a single C deletion that produces a truncated protein and that was first identified in a small subset of families in Finland that manifested an LFS-like syndrome. The fact that *p53* was wild type in all of these cases suggested that germline mutations in *CHEK2* represented an alternative genetic defect predisposing to LFS.

The *CHEK2**1100delC allele is a *CHEK2* variant in which a single nucleotide is deleted, resulting in premature protein chain termination and loss of CHEK function. The observation that the *CHEK2**1100delC variant occurred in LFS stimulated large epidemiologic studies asking whether *CHEK2**1100delC was overrepresented in hereditary breast cancer. The major finding of these studies is that *CHEK2**1100delC is a low-penetrance breast cancer susceptibility allele with an allele frequency of about 1.5% in the general human population. Notably, *CHEK2**1100delC confers no increased cancer risk in breast cancer families that already have mutations in the two previously identified breast cancer susceptibility genes, *BRCA1* and *BRCA2*. This observation is consistent with the concept that *CHEK2*, *BRCA1*, and *BRCpA2* all participate in the same DNA damage response network, whose function can be perturbed by mutations in any one of these genes. At the molecular level the 1100delC truncation eliminates the kinase domain and therefore the CHEK2 kinase activity, and the truncated protein is unstable. The remaining wild-type allele is often lost in the tumors. Thus CHEK2 function is often lost accompanied by a marked reduction in or absence of CHEK2 protein. Interestingly the *CHEK2**1100delC variant is particularly common in those rare families predisposed to combined breast and colon cancer.

Unlike the *CHEK2**1100delC variant whose allele frequency is similar in Western European, North American, and Finnish populations, the missense variant CHEK2 I157T, also originally detected in rare LFS families, is much more common in normal Finnish population (at 5%–6%) than elsewhere. This variant allele is significantly overrepresented in an unselected cohort of breast cancer patients, suggesting that it contributes to the cause of breast cancer in many patients within the Finnish population. Mechanistically the I157T exerts its effect differently than the *CHEK2**1100delC variant. The CHEK2*1100delC variant protein is unstable and when homozygous behaves much like a null mutant. The I157T variant, in contrast, is stable and behaves like a dominant-negative mutant, exerting an effect even when the remaining allele is wild type. The I157T mutant protein interferes with the ability of the wild-type CHEK2 to associate with many of its physiological substrates after exposure to ionizing radiation.

Micro-RNAs in Cell Cycle Control

The cell-cycle regulatory mechanism is governed by a multifaceted and layered set of exquisite controls that include gene expression and translation at different times of the cell cycle, activation and inhibition of cell-cycle regulatory proteins by phosphorylation and dephosphorylation, and removal of activating or inhibitory proteins by degradation. Superimposed on this intricate regulatory machinery is a set of regulatory micro-RNAs (miRNAs) that generally have inhibitory effects on their respective activities. The micro-RNAs, which number is in the thousands, are between 20 and 25 nucleotides in length and are derived from much larger RNA polymerase 2 transcripts that originate from both coding and noncoding regions of the genome. Figure 10–12 presents an overview of some of the micro-RNAs that impinge upon multiple cell-cycle regulatory components at the transcriptional and posttranscriptional levels. While this

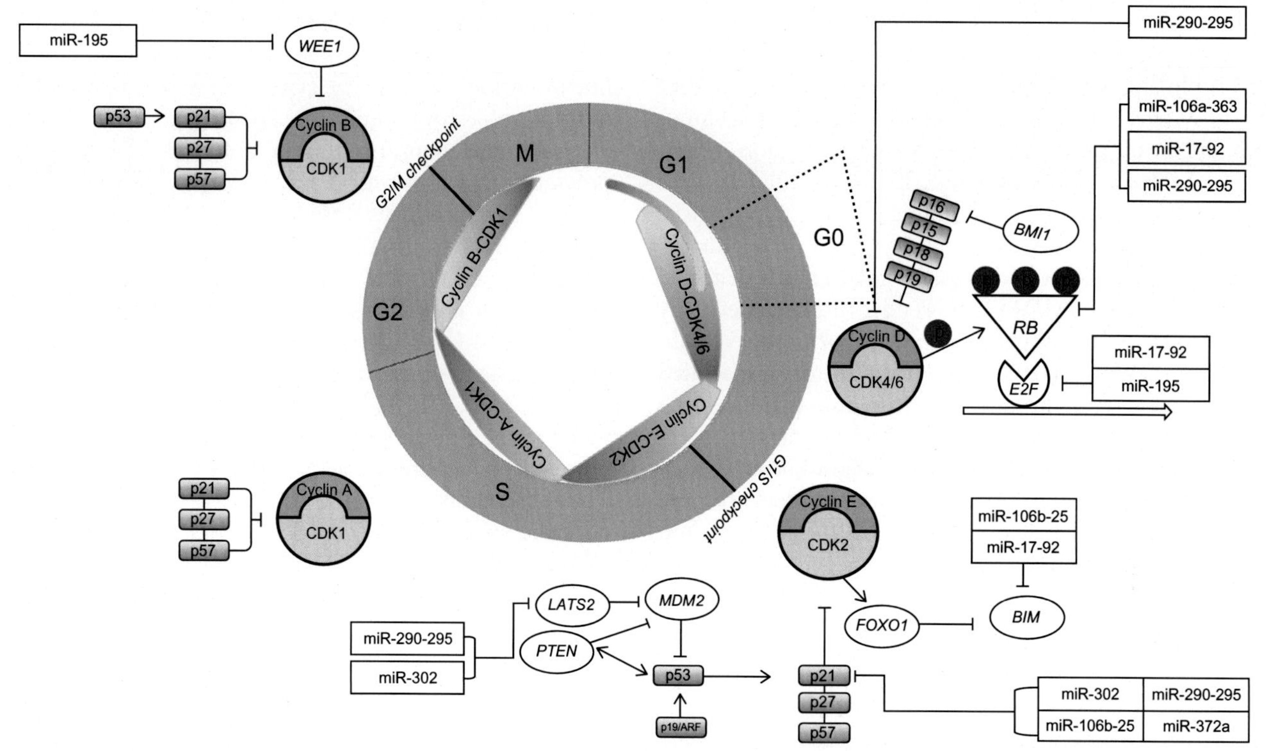

Figure 10–12. **Involvement of micro-RNAs in cell-cycle control in embryonic stem cells.** The figure illustrates the cell-cycle progression in embryonic stem cells (ESCs). Key regulatory elements, including cyclins, CDKs, and CDK inhibitors, form a set of signaling networks that drive cells through the different phases of the cell cycle. Individual miRNAs and clusters of miRNAs participate in regulating cell-cycle progression by directly or indirectly inhibiting or stimulating cell cycle-associated components (e.g., cyclin D, cyclin E, *Rb*, p53, p21, *LATS2*, and *PTEN*). Among them, miR-17-92, miR-290-295, miR-302, miR-106b-25, and miR-106a-363 are abundantly expressed in ESCs. Inhibition of *E2F* by miR-92 and miR-195 decreases transcription of multiple cell cycle-associated transcription factors and proteins (e.g., members of the *E2F* family, *CDK2*, and *CDC25A*), thereby modulating the duration of the G1 phase. Furthermore the expression of p53, the major G1/S and G2/M checkpoint regulator, is decreased due to indirect targeting by miR-290-295 and miR-302, which facilitates the G1/S transition. Similarly, p21 expression is reduced due to inhibition by miR-290-295, miR-372a, miR-302, and miR-106b-25, which inhibits cyclin E-CDK2 activity. *(Mens MMJ, Ghanbar M. Cell cycle regulation of stem cells in ES cells.* Stem Cell Rev *2018;14:309–322.)*

image presents an overview of cell cycle-related micro-RNA activity, the scheme presented is incomplete since many of the micro-RNAs are produced in a cell type-specific manner.

Cell-Cycle Targets for Cancer Therapy

One of the hallmarks of cancer cells is the deregulation of cell-cycle controls. Because of their importance in regulating cell proliferation, one would expect that several cell-cycle regulators would have emerged as effective anticancer targets. In fact, however, the number of cell-cycle regulatory proteins that have been effectively exploited by targeted inhibition is very limited, despite extensive efforts. The general lack of success is probably due to the multiple activities that most of these regulatory components exhibit, thereby generating increased risk for off-target inhibition that gives rise to deleterious side effects. The most promising targets to date are CDK4 and CDK6 for which three inhibitors (palbociclib, ribociclib, and abemaciclib) have received the Food and Drug Administration (FDA) approval for clinical use in the United States. These CDKs complex with cyclin D drive cells through the G1/S phase boundary. Overexpression of the cyclin D by gene amplification or by other mechanisms is common in human cancers, as is the overproduction of CDK4/CDK6. The three CCDK4/CDK6 inhibitors are each approved for clinical use in hormone receptor-positive, HER2-negative breast cancers and are in clinical trials for use in other types of cancers. Small molecule inhibitors of PLK1 are also in clinical use but have not yet been approved by the FDA. Volasertib is a small molecule ATP competitive kinase inhibitor that is a highly selective inhibitor of PLK1. Volasertib has been tested in a phase III trial in combination with low-dose cytarabine for elderly patients with acute myelogenous leukemia (AML). The trials, however, have been unsuccessful due to adverse effects attributable to the PLK1 inhibitor.

SUMMARY

With the emergence of new technologies, our understanding of the cell cycle and its regulation continues to grow exponentially. We now have an appreciation of the pathways and circuits that govern the normal cell-cycle progression and the cellular checkpoint responses to DNA damage and other forms of stress. In addition, the deciphering of the biochemical and genetic pathways that maintain the integrity of cell-cycle progression and cell division provides the foundation for understanding how mutations that impair these pathways can compromise cell-cycle control and lead to genomic instability and cancer. Despite the enormous complexity of these pathways, various components of the cell-cycle circuitry are beginning to prove clinically valuable as diagnostic or prognostic markers, while others are emerging as promising therapeutic anticancer targets. The pathways that lead to programmed cell death, apoptosis, are mediated by many of the same molecular players that participate in cell-cycle control and are also providing therapeutic targets for cancer. The basic strategies are to find compounds that arrest tumor cells so that they will not contribute to the growth of the tumor or to push the tumor cells into apoptosis to rid the lesion of tumor cells. Most of the proteins that have been discussed in this chapter are currently considered to be potential therapeutic targets and are under active investigation in many academic laboratories and pharmaceutical companies. Tumor cells, however, have evolved to escape most antitumor therapies, indicating that most new therapeutic approaches will have to involve combination therapies rather than a single therapeutic strategy. Nevertheless, as the cell cycle and apoptotic pathways are more thoroughly dissected, there is hope that in the near future many cancers for which we currently have no effective treatments will become manageable with concomitant improvement in quality of life.

Suggested Readings

Bertoli C, Skotheim JM, de Bruin RAM. Control of cell cycle transcription during G1 and S phases. *Nat Rev Mol Cell Biol* 2013; 14:518–528.

Harashima H, Dissmeyer N, Schnittger A. Cell cycle control across the eukaryotic kingdom. *Trends Cell Biol* 2013;23:345–356.

Howard A, Pelc S. Synthesis of deoxyribonucleic acid in normal and irradiated cells and its relation to chromosome breakage. *Heredity* 1953;6(Suppl):261–273.

Lukas J, Lukas C, Bartek J. Mammalian cell cycle checkpoints: signaling pathways and their organization in space and time. *DNA Repair* 2004;3:997–1007.

Mens MMJ, Ghanban M. Cell cycle regulation of stem cells by micro-RNAs. *Stem Cell Rev* 2018;14:309–322.

Murray A, Hunt T. *The Cell Cycle: An Introduction.* New York: W. H. Freeman and Company; 1993.

Sánchez I, Dynlacht BD. New insights into cyclins, CDKs, and cell cycle control. *Semin Cell Dev Biol* 2005;16:311–321.

Santamaria D, Ortega S. Cyclins and CDKs in development and cancer: lessons from genetically modified mice. *Front Biosci* 2006; 11:1164–1188.

Teixeira LK, Reed SI. Ubiquitin ligases and cell cycle control. *Annu Rev Biochem* 2013;82:387–414.

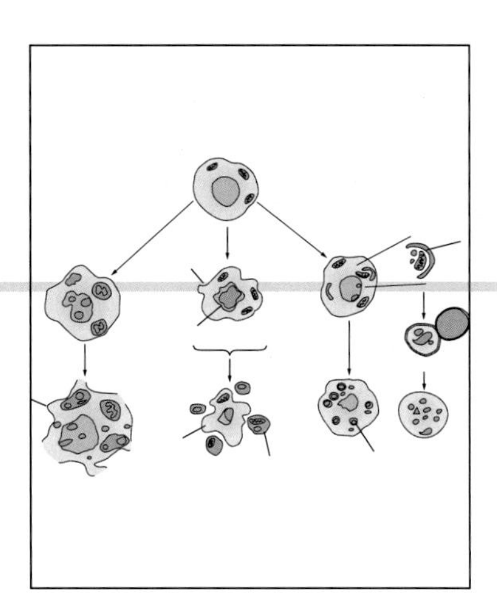

Chapter 11

Regulated Cell Death

CELL DEATH IS CRITICAL IN DIVERSE ASPECTS OF ANIMAL DEVELOPMENT AND THROUGHOUT ADULT LIFE

Normal animal development depends on the coordinated regulation of cellular processes to build and maintain diverse cell populations present in diverse tissue types that together compose complex organisms. An obvious example of such coordinated regulation is the proliferation of precursor cells or progenitor cells and their subsequent specification or differentiation into distinct cell types. Less obvious is the role of programmed cell death, another cellular fate critical to the formation and maintenance of most tissue types. The term, **programmed cell death**, initially was used to emphasize that these cell deaths occur in predictable body locations, at predictable developmental periods, and as part of a predetermined developmental plan of the organism. During development, programmed cell death sculpts body parts and eliminates transitory structures. Examples of this include the elimination of cells within the tail of a tadpole as it changes into a frog, the deletion of interdigital tissue in the formation of fingers, and the hollowing out of tissue to create lumina in vertebrates (Fig. 11–1).

Additionally, programmed cell death rids the developing animal of superfluous or unwanted cells. For example, large numbers of self-reactive lymphocytes or lymphocytes that fail to produce useful antigen-specific receptors are deleted by programmed cell death. In the developing nervous system, about half of the neurons produced undergo programmed cell death soon after they mature in a process that ensures that optimal connectivity between newly generated neurons is established. Programmed cell death continues to play a critical role into adulthood where it maintains a balance with cell division, thus controlling tissue, organ, and body size. Cells that are irreversibly damaged or that are dangerous to the organism, such as transformed cells, or those infected with pathogens such as viruses are also eliminated by the activation of programmed cell death. Lymphocytes that are produced in large numbers in response to a specific pathogen are eliminated by programmed cell death after the pathogen is eliminated. Because regulated cell death is rapid and often occurs without provoking an immunoresponse, the extent to which it occurs was grossly underappreciated until recently. Although initially believed to be unique to multicellular organisms, PCD has been found to occur in unicellular eukaryotes, such as many yeast strains, and even in some prokaryotes.

It is now known that PCD is part of a larger family of morphologically distinct modes of cell death, collectively referred to **as regulated cell death**, all of which are controlled by genetic mechanisms. Whereas programmed cell death is used to define physiological death that occurs during specific stages of development and as part of normal tissue turnover, regulated cell death also includes forms of cell death triggered by intense intracellular or

Goodman's Medical Cell Biology. https://doi.org/10.1016/B978-0-12-817927-7.00011-9

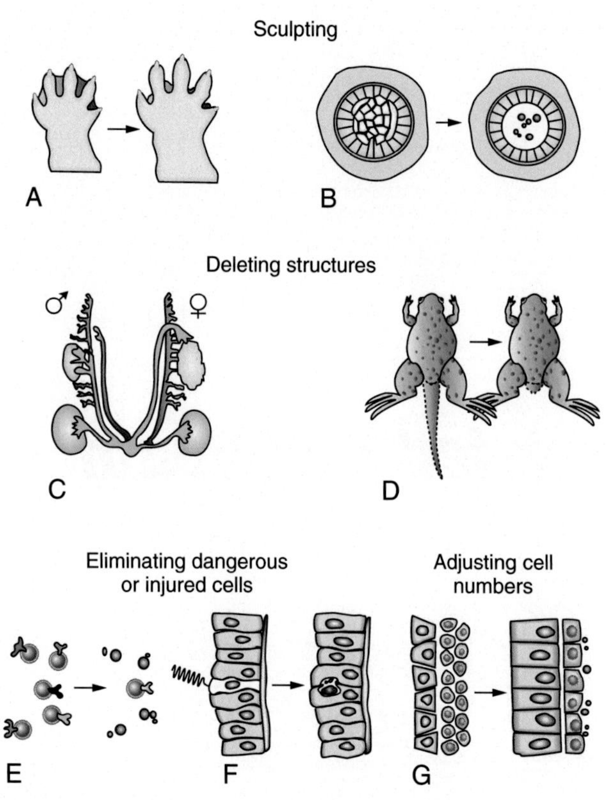

Figure 11–1. **Programmed cell death during animal development.** Programmed cell death serves a number of different functions during development including (A,B) the sculpting of organs and tissues, (C,D) the deletion of unwanted structures during development, (E,F) the elimination of injured or harmful cells, including those that are transformed, damaged, or infected, and (G) the adjustment of proper cell numbers.

extracellular perturbations but that are mediated by specific molecules and a precise sequence of events.

In contrast to regulated cell death in which a dedicated set of genes and signaling pathways are activated, accidental cell death generally results from severe insults of a physical (e.g., high temperature or pressure), chemical (e.g., substantial pH variation), or mechanical (e.g., shear forces) nature that damages cellular constituents. This form of cell death cannot be regulated and is virtually instantaneous. In most cases the cells and the organelles within them swell and rupture spilling their contents in a process called **necrosis**. Because the release of intracellular material includes lysosomal and other degradative enzymes, necrotic cell death damages neighboring cells and provokes an inflammatory reaction in the surrounding tissue. The activities and secretions of macrophages and other cells of the immune system contribute to the damage of tissue in the vicinity.

DISTINCT FORMS OF PROGRAMMED CELL DEATH

The first type of programmed cell death to be described was **apoptosis** (which in Greek means the seasonal falling of leaves from plants or trees). Initial characterization of apoptosis was reported by Kerr, Wylie, and Currie in 1972. Cells undergoing apoptosis shrink and display nuclear and chromatin condensation. Externally the cell appears to boil as the membrane becomes convoluted, an event referred to as membrane "blebbing." The cell may fragment into membrane-bound bodies called apoptotic bodies containing cytoplasmic material, intact organelles, and the occasional chunk of condensed chromatin. Either the whole apoptotic cell or apoptotic bodies are instantaneously engulfed by phagocytosis with no leakage of cellular components and hence no inflammatory response.

The next to be described was **autophagic cell death,** which is characterized by the appearance of large numbers of cytoplasmic vacuoles. In autophagic death, double-membrane sheets, derived largely from the endoplasmic reticulum, form cytoplasmic vacuoles that engulf intracellular organelles and cytoplasmic materials. These vesicles fuse with lysosomes where the sequestered cellular components are digested. Although some amount of autophagy occurs normally and served essential roles, excessive levels result in the cell being eaten from the inside resulting in cell death. In contrast to apoptosis, autophagic cell death is independent of phagocytosis (Fig. 11–2).

Over the past decade, three other forms of regulated cell death have been described and characterized—necroptosis, ferroptosis, and pyroptosis. Although sharing morphological features with necrosis, death by these modes is regulated by genes and signaling pathways. **Necroptosis** is characterized by swelling of intracellular organelles and lysosome-independent formation of "empty spaces" in the cytoplasm and has some similarities to necrosis. No chromatin condensation occurs, although chromatic clustering to loose speckles is sometimes seen. **Pyroptosis** is induced in cells that are infected with microbial pathogens and is characterized by the flattening of the cell ultimately leading to membrane rupture. **Ferroptosis** is caused by iron-dependent lipid perodixation resulting in membrane rupture. A characteristic morphological feature of ferroptosis is the shrinkage of mitochondria along with the disappearance of cristae. Thus and in contrast to apoptotic or autophagic cell death, necroptosis, pyroptosis, and ferroptosis all culminate in the rupturing of the plasma membrane releasing immunogenic molecules, thus provoking an inflammatory response. More recently, other forms of cell death have been described although these remain to be characterized and whether these are actually distinct modes or simply variants of the aforementioned major forms of cell death has yet to be firmly determined.

In the succeeding text, we describe the key molecules and signaling pathways that regulate apoptosis, autophagy, necroptosis, pyroptosis, and ferroptosis. It is estimated that together, 200–300 billion cells, or on average about one million cells/second die through these, and other uncharacterized forms of regulated cell death. Since apoptosis is the most studied and best characterized form of regulated cell death, most emphasis has been placed on it.

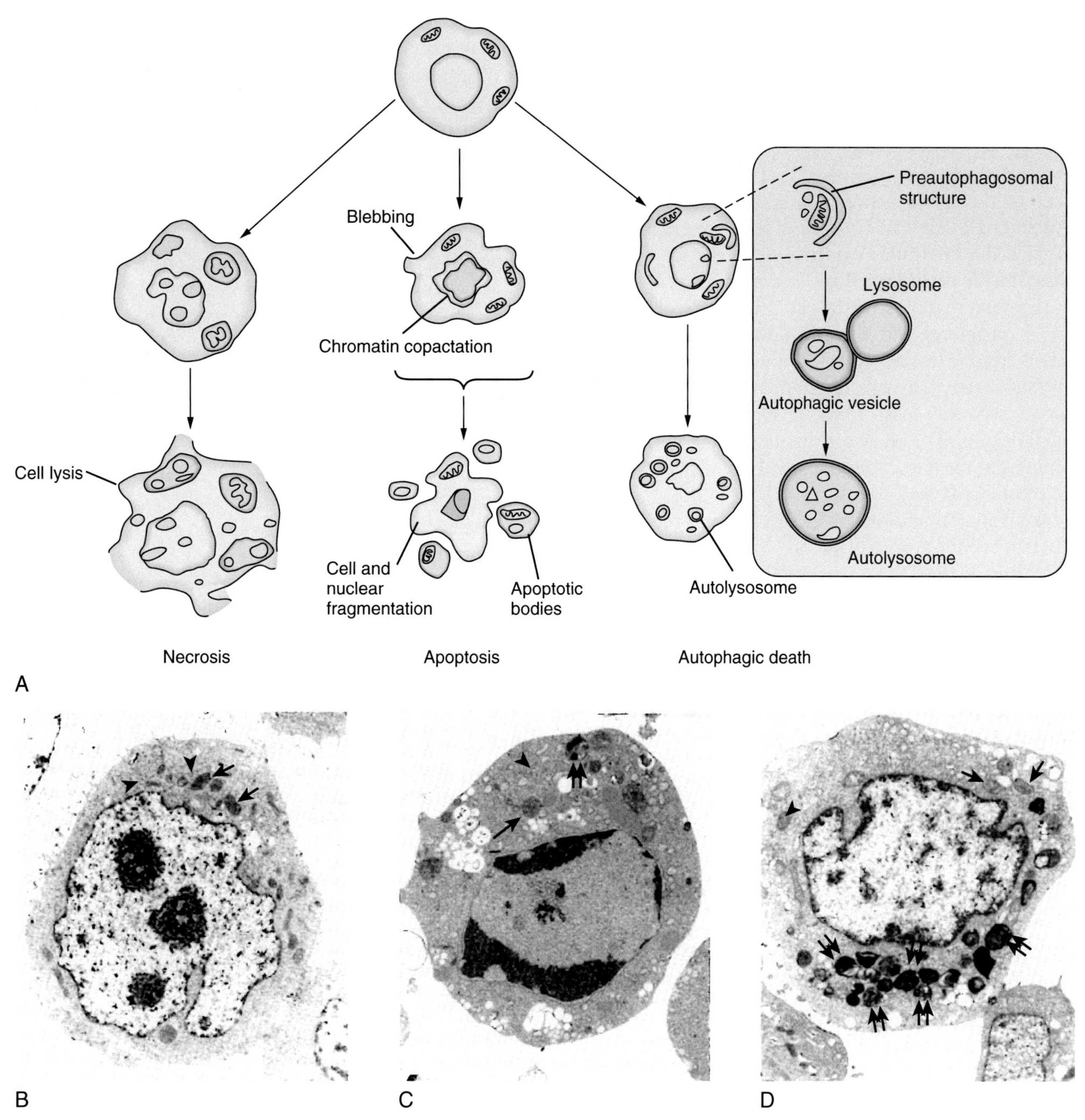

Figure 11–2. **Different forms of cell death.** (A) Schematic illustration of morphological changes that characterize cell death by necrosis, apoptosis, and autophagic cell death. While necrotic cell death culminates in cell lysis and provokes inflammation, apoptotic cells are packaged into apoptotic bodies that are then engulfed by adjacent cells without an inflammatory response. Autophagic cell death is characterized by the appearance of cytoplasmic vesicles engulfing bulk cytoplasm and organelles. The contents of the vesicles are digested by the lysosomal system of the same cell after the fusion of the autophagic vesicles with lysosomes. (B–D) Ultrastructural features of cells undergoing apoptotic and autophagic cell death. (B) A normal cell, (C) an apoptotic cell, and (D) a cell undergoing autophagic cell death are shown. Polyribosomes *(arrowhead)*, mitochondria *(arrow)*, and autophagic vacuoles *(double arrows)* are indicated. While autophagic vacuoles can be seen in healthy cells and apoptotic cells, they are much more abundant during autophagic cell death. *(From Bursch et al.* J Cell Sci *2000;113:1189–1198.)*

APOPTOSIS IS REGULATED BY A CELL-INTRINSIC GENETIC PROGRAM

The first insight into the regulation of apoptosis came from genetic studies in the nematode worm *Caenorhabditis elegans* that were critical in the initial characterization of the core components of the cell death program. During the development of this worm, exactly 1090 cells are generated of which 131 die. Each of the 131 cell deaths occurs at a specific time and location that is essentially invariant from animal to animal. Two genes, CED-3 and CED-4 (CED stands for cell death defective), were identified as genes required for cell death: inactivating mutations in

EGL-1 ⊣ CED-9 ⊣ CED-4 → CED-3

Figure 11–3. **The basic genetic pathway of programmed cell death in *C. elegans*.** Activation of CED-3 and CED-4 lead to cell death in the nematode. CED-9 can inhibit cell death by preventing CED-4 from activating CED-3. EGL-1 can promote cell death by binding to CED-9, thus displacing it from CED-4.

either of these two genes prevents all developmentally occurring cell death in *C. elegans*. This observation also provided the first direct evidence that cells die by an intrinsic suicide program. CED-3 is a protease, while CED-4 is a protein that oligomerizes and through its binding to CED-3 promotes the dimerization and activation of CED-3 leading to cell death. Another protein, CED-9, which is associated with the mitochondrial outer membrane, protects cells from programmed cell death (Fig. 11–3). CED-9 does this by interacting with CED-4 keeping it at the mitochondrial membrane, thus sequestering it from CED-3.

Inactivating mutations of CED-9 leads to the death of cells that would normally survive through development, so worms with CED-9 mutations die early. Genetic studies have shown that the overexpression of another gene, EGL-1 (for egg-laying defective-1), or mutations that hyperactivate EGL-1 (gain-of-function mutations) can induce ectopic cell death in *C. elegans* that can be suppressed by gain-of-function mutation of CED-9 or loss-of-function mutations of CED-3 or CED-4. In contrast a loss-of-function mutation of EGL-1 cannot suppress cell death caused by the loss of CED-9 activity (loss-of-function CED-9). These results are consistent with EGL-1 acting upstream of the CED-3, CED-4, and CED-9 genes in the genetic pathway regulating cell death (Fig. 11–3).

The identification of key components of the genetic program in the nematode led to the search for their homologs in mammals. These studies have resulted in the identification of a family of proteins, collectively called **caspases**, as the mammalian homologs of CED-3. The homolog of CED-4 in mammals is a protein called Apaf-1, while proteins related in sequence to CED-9 comprise the Bcl-2 family. In contrast to the situation in *C. elegans*, however, where CED-9 is antiapoptotic, the Bcl-2 family of proteins is composed of both antiapoptotic and proapoptotic members. Some of these proapoptotic Bcl-2 proteins function as the mammalian counterparts of EGL-1. Consistent with the evolutionary conservation of the mechanisms underlying apoptosis, the overexpression of the nematode proteins CED-3 or CED-4 in mammalian cells results in apoptosis, while overexpression of Bcl-2 can compensate for the lack of CED-9 in the nematode. Given their importance to the regulation of apoptosis in mammals, we will now take a closer look at the caspases and Bcl-2 family of proteins.

THE CASPASES

Caspases are a family of cysteine proteases (have a cysteine in the active site) that represent the mammalian homologs of CED-3. Although members of this family differ in primary sequence and substrate specificity, they are all present normally in the cytosol of healthy cells as zymogens. The procaspases are cleaved to form heterodimers of two small (approximately 10–13 kDa) and two large subunits (approximately 17–21 kDa), which associate to form the active enzyme (Fig. 11–4, A). Active caspases recognize a tetrapeptide sequence within substrates cleaving after an aspartic acid residue within the sequence. For example, the preferred recognition sequence for caspase-1 is Tyr-Val-Ala-Asp (Y-V-A-D), while the recognition sequence preferred by caspase-3 is Asp-Glu-Val-Asp (D-E-V-D).

While most caspases function to regulate apoptosis, some members of the caspase family (e.g., caspases 1, 4, and 5) are involved in cytokine maturation. Caspases regulating apoptosis have been classified into two groups referred to as **initiator caspases** and **effector caspases.** Initiator caspases (e.g., caspase-8, caspase-9, and caspase-10), characterized by long prodomains, have low intrinsic enzymatic activity even in their zymogen form. In this state the initiator caspases are unable to harm the cell. When these procaspases oligomerize, they come into close proximity, and they can now cleave and activate themselves. Oligomerization requires the prodomain because deletion of the prodomain results in the absence of oligomerization and abolishes the apoptotic activity of the initiator caspases. Structural studies have revealed that the prodomain of initiator caspases contain one of two types of domains—**death effector domain (DED)** present in caspase-8 and caspase-10 and **caspases recruitment domain (CARD)** present in caspase-9. Oligomerization of caspases occurs through homotypic DED-DED or CARD-CARD interactions, which leads to their intermolecular cleavage (Fig. 11–5). Once activated, initiator caspases cleave effector caspases (e.g., caspase-3, caspase-6, and caspase-7) leading to their activation. The effector caspases, also referred to as **executioner caspases**, cleave a variety of cellular substrates, thus destroying normal cellular functions and leading to the demise of the cell. More than 300 cellular substrates for caspases have been identified to date. Among these are proteins involved in scaffolding of the cytoplasm and nucleus, signal transduction and transcription regulatory proteins, proteins involved in DNA replication and repair, and cell cycle components. The characteristic morphological features of apoptosis are the direct result of caspase activation. For example, membrane blebbing, a characteristic feature of apoptotic dell death, is caused by caspase-mediated cleavage of the actin-binding protein gelsolin and specific kinases involved in cytoskeletal function.

Activation of the caspases represents a pivotal step in the cell death process. Two major pathways of caspase-mediated cell death have been described in mammals: the **extrinsic pathway** (also called the death receptor-mediated pathway), which plays an important role in the maintenance of tissue homeostasis, especially in the immune system, and the **intrinsic pathway** (or the

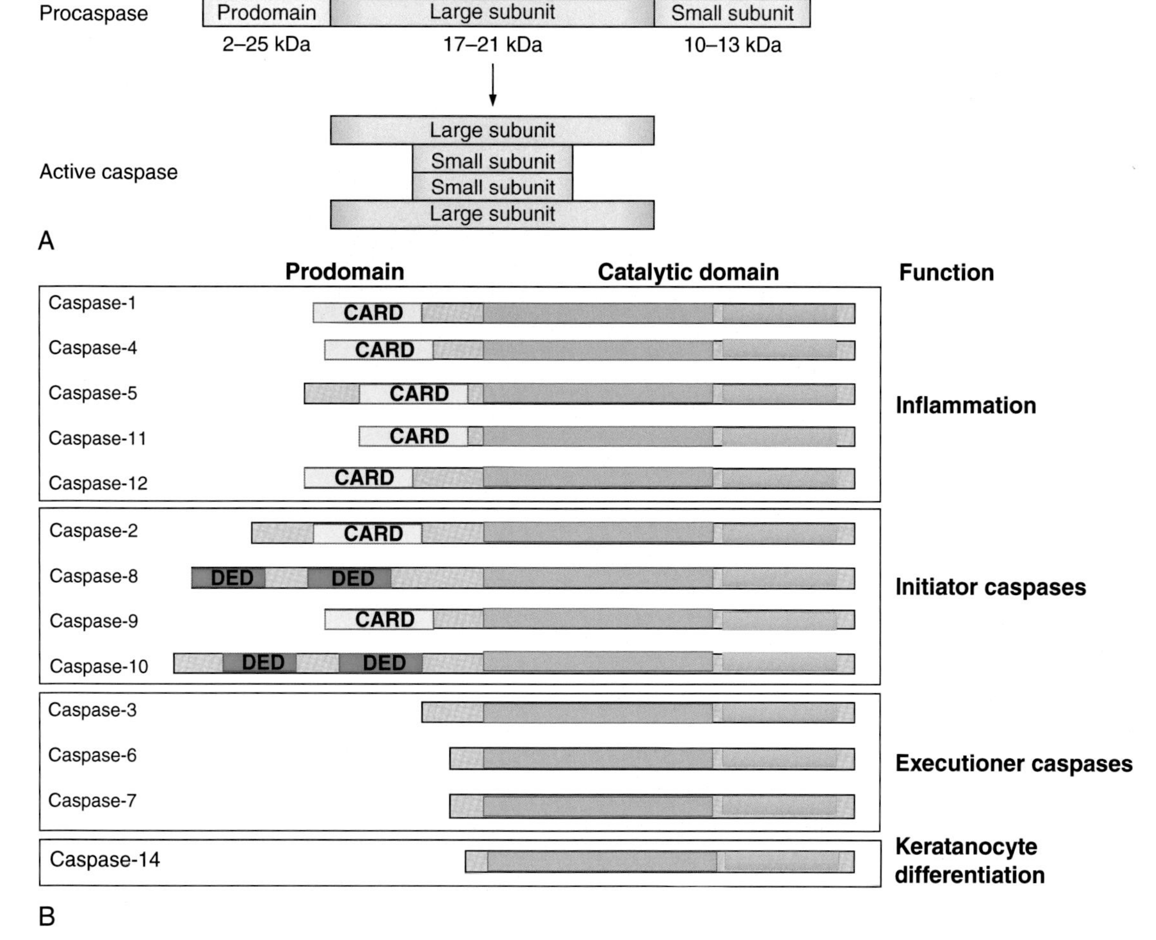

Figure 11–4. **Structure and activation of caspases.** (A) Activation of caspases. Caspases are synthesized in their zymogen form requiring proteolytic processing for activation. Processing of the proenzyme at specific aspartate residues removes the N-terminal prodomain. Cleavage between the large and short subunits and association of two heterodimers to form a tetramer results in the generation of active caspase. (B) Members of the vertebrate caspase family. Caspases include families of initiator caspases and executioner or effector caspases that mediate apoptotic cell death. Other caspases have different functions, including inflammation and differentiation. Many caspases have CARD or DED through which homotypic interactions occur.

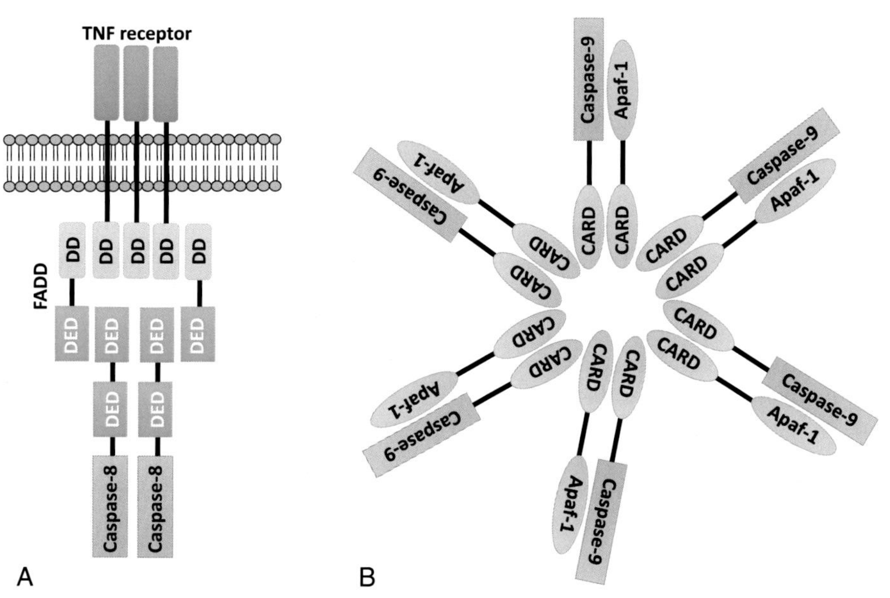

Figure 11–5. **Homotypic DED-DED or CARD-CARD interactions bring initiator caspases in close proximity.** (A) Linking TNF family death receptors to DED family caspases through the bipartite adapter FADD. The DD of FADD binds either directly to the DD in the cytosolic tails of TNF family receptors or indirectly through other DD-containing adapter proteins (not shown). The DED of FADD binds the DEDs of procaspases-8 (and 10), bringing procaspase-8 molecules in close proximity. (B) The CARD of Apaf-1, which oligomerizes into a heptamer, binds to the CARD of procaspase-9 forming a heptameric complex in which procaspase-9 molecules are in close proximity.

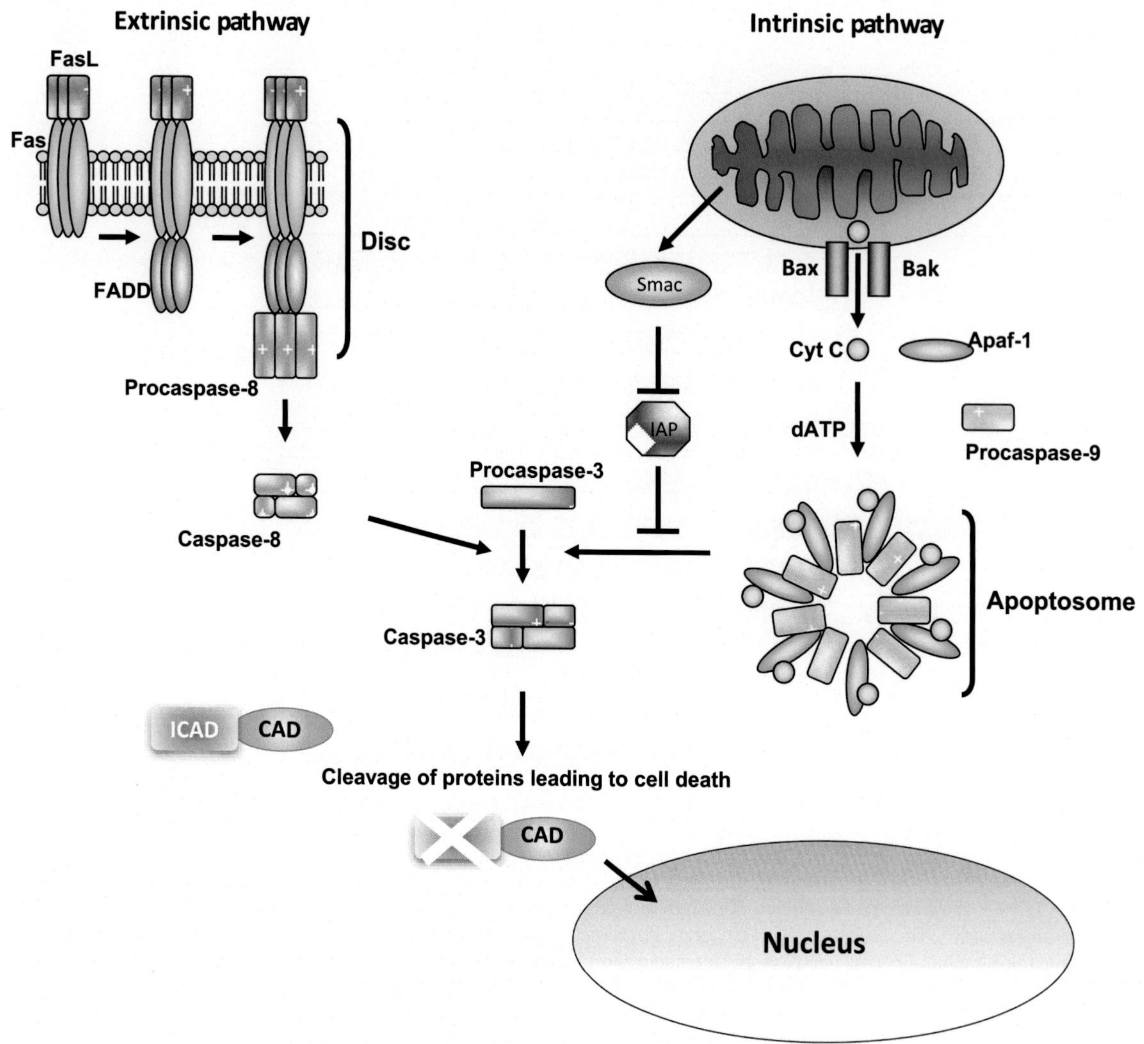

Figure 11–6. **Caspase activation by the extrinsic and intrinsic pathways.** The extrinsic or death receptor pathway (left half of the figure) is activated by the binding of a ligand such as FasL to its receptor Fas followed by the sequential binding of the adaptor protein FADD and procaspase-8, resulting in the formation of a protein complex referred to as a death-inducing signaling complex (DISC). The formation of the DISC permits the activation of procaspase-8, which then proteolyzes and activates caspase-8. The mitochondrial pathway (right half of the figure) also leads to the activation of caspase-3. This pathway involves the release of cytochrome *c* from the mitochondria. Once in the cytoplasm, cytochrome *c* promotes the assembly of Apaf-1 and procaspase-9 into a large complex containing seven molecules each of Apaf-1 and procaspase-9. This complex containing Apaf1, procaspase-9, cytochrome *c*, and ATP is referred to as an apoptosome. Active caspase-9 within the apoptosome can cleave caspase-9. In addition to cytochrome *c*, other proapoptotic proteins such as Smac/Diablo are released. Smac/Diablo acts by sequestering IAPs away from caspases, thus permitting cell death to occur. One of the many targets of caspase-3 is a cytoplasmic protein called ICAD (inhibitor of CAD) which binds to CAD (caspase-activated DNAse) sequestering it. Cleavage of ICAD by caspase-3 liberates CAD to translocate to the nucleus where it cleaves DNA into fragments.

mitochondrial pathway), which is used generally in response to external cues and internal insults such as DNA damage (Fig. 11–6).

The extrinsic pathway is initiated at the cell membrane by the binding of extracellular ligands to members of a small family of about eight receptors called "**death receptors.**" These are all transmembrane receptors on the surface of cells that share a domain of approximately 80 amino acids in their cytoplasmic region commonly referred to as the **death domain.** Among the best studied of the death receptors are the tumor necrosis factor receptor-1 (TNF-R1) and the Fas-ligand receptor (Fas-IC). Binding of ligand to these receptors (TNFα to TNF-R1 or FasL to Fas) leads to their trimerization and conformational change that allows them to interact with the death domain of an adaptor protein (an example of which is TRADD for TNFR1-associated death domain protein or FADD for Fas-associated death domain protein). In addition to a death domain, the adaptor proteins possess a DED that interacts with procaspase-8 recruiting it to the trimerized receptor at the plasma membrane. The receptor-adaptor-procaspase-8 protein complex is called a **death-inducing signaling complex** (DISC). The principal role of the adaptor protein is to bring multiple procaspase-8 molecules into close proximity, promoting their oligomerization, thus allowing autocatalytic cleavage to occur (Fig. 11–6). Although Figure 11–6 shows the core components of the DISC, there are several other proteins associated with the complex that participate in the activation of caspase-8. One of these is a protein called RIP1 (for receptor-interacting serine/threonine protein kinase-1), which, as we will see later in the chapter, also plays a key role in inducing necroptosis, another form of regulated cell death.

The intrinsic pathway of caspase activation on the other hand is initiated by activation of procaspases within the cytosol. Since the mitochondria play an essential role in this mechanism of cell death, this is also called the mitochondrial pathway. In response to many death stimuli, cytochrome *c* is released into the cytoplasm from the intramembrane space of mitochondria, where it is normally involved in electron transport into the cytosol. Once in the cytoplasm, cytochrome *c* binds to Apaf-1 (the mammalian homolog of the *C. elegans* CED-4 protein), which is in an inactive conformation with the CARD inaccessible for interaction. Binding of cytochrome *c* facilitates binding of ATP with Apaf-1 altering its conformation, which exposes the CARD promoting its oligomerization. Seven molecules of procaspase-9 bind to the oligomerized Apaf-1 to form a heptameric complex (Fig. 11–5). Such a high-molecular weight complex of cytochrome *c*, Apaf-1, caspase-9, and ATP is referred to as an **apoptosome**. Proximity-induced cleavage of caspase-9 within the apoptosome leads to its self-activation. Activated caspase-9 cleaves and activates effector caspases such as caspases-3, 6, and 7, which proteolyze a number of cellular proteins leading to cell death (Fig. 11–6).

More recent work has revealed that besides the mitochondria, the endoplasmic reticulum (ER) may be a second compartment that participates in the triggering of apoptosis in mammals. A major function of the ER is to ensure that only properly folded and modified proteins are passed along to the Golgi network for transport to their final destinations. However, prolonged stress to the ER caused by chemical toxicity or by oxidative stress overwhelms its ability to properly fold proteins leads to an accumulation of misfolded proteins within the ER. This stress is relieved through the activation of specific transcription factors, the best studied of which are CHOP and ATF3, that stimulate the production of chaperone proteins. If these chaperone proteins cannot successfully refold the misfolded protein, the cell transports the protein out of the ER for degradation by proteasomes or, as described later, through autophagy. If, however, the misfolded protein cannot be eliminated through degradation, the cell activates the expression of proapoptotic genes, which promote death of the cell. Although the death of the cells suffering irreversible ER stress is beneficial, chronic ER stress could lead to excessive cell loss. Indeed, there is growing consensus that chronic ER stress caused by the build of specific misfolded proteins may contribute to the abnormal loss of neurons in neurodegenerative diseases, such as Alzheimer's disease, amyotrophic lateral sclerosis (ALS), and Parkinson's disease. Chronic ER stress has also been suggested to be involved in cancer and diabetes mellitus. Some viruses have evolved mechanisms not only to escape the consequences of ER stress but also to make use of factors induced by ER stress for their own advantage. For example, the hepatitis B virus (HBV) uses ER stress to promote its own replication.

In addition to the extrinsic and intrinsic pathways, cytotoxic T cells can trigger apoptosis of their target cells through a mechanism called the perforin-dependent/granzyme B-dependent pathway. This pathway utilized mainly to eliminate transformed or virally infected cells involves the injection by the activated cytotoxic T cell of a pore-forming protein, perforin, and a serine protease, granzyme B, into the cytoplasm of the target cell. Granzyme B can directly cleave and activate procaspase-3, thus inducing apoptosis in the target cell independently of mitochondria and death receptors.

CASPASE INHIBITION

The regulation of apoptosis is subjected to a complex system of checks and balances. The activity of caspases can be negatively regulated within cells by endogenously produced proteins. Evidence of the existence of such caspase inhibitory proteins first came from viruses when it was found that CrmA, a protein produced by the cowpox virus, bound caspases inhibiting them. Herpes virus inhibits apoptosis by producing a protein called viral-FLICE-inhibitory protein (v-FLIP) that binds the adaptor protein FADD preventing it from recruiting caspase-8 to the TNF death receptor. Baculoviruses synthesize a potent caspase inhibitor called p35, which prevents the death of insect cells that are infected by the virus. In addition to p35, baculovirus express a second antiapoptotic protein that contain a motif of about 70 amino acids called the baculovirus inhibitory repeat (BIR) domains that is also found in a class of proteins produced in both vertebrate and invertebrate cells and referred to as **inhibitor of apoptosis proteins (IAPs)**. Mammals express eight different IAP proteins containing one to three BIR domains. Although some IAP proteins have other functions unrelated to the regulation of apoptosis, most mammalian IAPs do inhibit effector caspase activity by binding to them through the BIR domain and inhibiting their ability to cleave their substrates. The best studied of these are c-IAP1, c-IAP2, and XIAP. XIAP also binds the proform of caspase-9 blocking its interaction with Apaf-1 and the processing of procaspase-9 into an active enzyme. The antiapoptotic activity of IAPs may itself be negatively regulated by other proteins that are normally localized in the intramembrane space of mitochondria but released along with cytochrome *c* during intrinsic apoptosis signaling. One of these proteins has been called second mitochondrial activator of caspases (Smac), which was also identified by another group and called Diablo. Smac/Diablo binds to the IAPs preventing them from inhibiting caspases. Among other proapoptotic proteins that are released by the mitochondria are apoptosis-inducing factor (AIF) and endonuclease G (EndoG). Upon their release from the mitochondria, these two proteins translocate to the nucleus where they mediate the cleavage of DNA, a hallmark of apoptosis. The action of these proteins does not require caspases and, in fact, occurs after the initiation of apoptosis. Another endonuclease called caspase-activated deoxynuclease (CAD) is responsible for DNA cleavage at earlier stages of the apoptotic process. CAD is normally kept inactive through association with a protein called inhibitor of CAD (ICAD), which is one of the substrates of caspase-3. Cleavage of ICAD by caspase-3 liberates CAD allowing it to translocate to the nucleus and cleave DNA.

As described earlier, each caspase recognizes a tetrapeptide sequence within its substrates. Researchers have coupled these tetrapeptide sequences to chemical groups such as fluorodiazomethyl ketones (FMK) and modified the peptide to make them membrane permeable. Such modified but uncleavable tetrapeptides are recognized by caspases and are bound by the enzyme irreversibly. When administered to cells, these pseudosubstrate peptides function as potent and highly selective inhibitors of caspases and can inhibit cell death in response to a variety of different apoptotic stimuli (Fig. 11–7).

Several caspase-deficient mice have been generated that exhibit tissue and cell-specific or stimulus-dependent

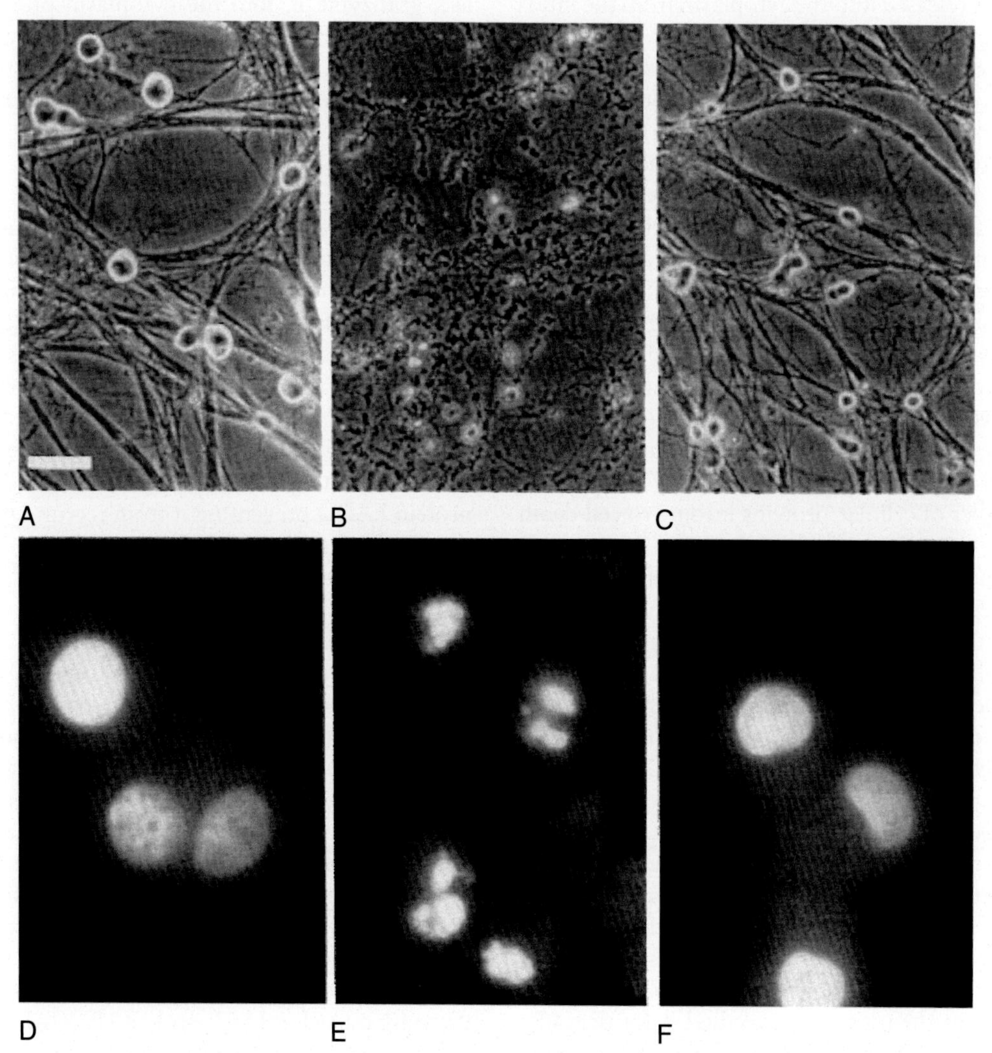

Figure 11–7. **Inhibition of caspases protects neurons from death.** Sympathetic neurons cultured from the superior cervical ganglion of rats or mice will survive in tissue culture medium supplemented with NGF (Nerve Growth Factor). Removal of NGF from the medium leads to apoptosis. Panels (A)–(C) show images from phase contrast microscopy. When maintained in the presence of NGF (Panel A), the neurons appear bright, and their processes are smooth and abundant. When deprived of NGF (Panel B), the cell bodies and the processes of the neurons degenerate. Neuronal cell death induced by NGF deprivation is prevented if a pan-caspase peptide inhibitor, z-VAD-fmk, is added to the culture medium (Panel C). Panels (D)–(F) show images on nuclei stained with a fluorescent DNA-binding dye. Whereas chromatin staining is diffuse within the nuclei of neurons receiving NGF (Panel D), chromatin within the nuclei of cultures deprived of NGF is condensed and fragmented (Panel E), hallmarks of apoptotic cell death. Inhibition of caspases by the addition of z-VAD-fmk (Panel F) prevents the condensation and fragmentation of chromatic even in the absence of NGF. *(From Deshmukh et al.* J Cell Biol *1996;135:1341–1354.)*

defects in apoptosis. Mice lacking caspase-9 have a highly enlarged brain and die in utero. Caspase-3 is not active in these mice. The brain deformity indicates that caspase-9 is required for the developmentally-regulated death of neurons that occurs during normal brain development and which eliminates about 50% of all neurons produced. Not unexpectedly, given that it acts as an essential partner of caspase-9, mice lacking Apaf-1 also have a larger brain and die prenatally. Mice lacking caspase-3 are smaller than normal, display ectopic masses containing excess neurons in several brain regions, and only survive until 3 weeks of age. The milder phenotype of the caspase-3 knockout mice as compared with the caspase-9 knockout mice is likely to be due to the functioning of other caspase-9-activated effector caspases, such as caspase-7. Caspase-8 and FADD-deficient mice have profound cardiac defects and die in utero indicating that the extrinsic pathway of caspase activation is critical for embryonic development. Several other caspase knockout mice, including mice lacking caspases 1, 2, 6, 11,12, and 14, develop normally, although adult mice can show minor phenotypes. For example, mice lacking caspase-1 are more susceptible to virus infections but display reduced apoptosis following heart failure. The relatively minor phenotype in mice lacking these caspases may reflect functional redundancy among some caspase proteins.

The Bcl-2 Proteins

The Bcl-2 family of proteins is named after Bcl-2, which was the first member of the family to be discovered. Members of the Bcl-2 protein family are critical apoptosis regulators that act by controlling release of cytochrome *c* and other apoptotic factors such as Smac/Diablo from the mitochondria. Humans contain over 20 Bcl-2 family proteins all of which possess at least one of four conserved Bcl-2-homology (BH) domains, designated BH1–BH4. Whereas some Bcl-2 proteins are antiapoptotic, other members are proapoptotic. Based on the BH domains they possess and whether they are pro- or antiapoptotic, Bcl-2 proteins are subdivided into three groups (Fig. 11–8). Group 1 Bcl-2 proteins are antiapoptotic and possess all four BH domains. This subfamily includes Bcl-2 itself and Bcl-4 or Bcl-5 other structurally related proteins, such as Bcl-XL and Mcl-1. Antiapoptotic Bcl2 proteins are often regarded as functional homologs of the *C. elegans* CED-9 protein. Group 2, also referred to as multidomain Bcl-2 proteins, also possess four BH domains but are proapoptotic and referred to as "executioner proteins." Two major multidomain Bcl-2 proteins, called BAX and BAK, are particularly crucial for cell death as cells lacking these two proteins fail to undergo apoptosis in response to a wide range of apoptotic stimuli. Group 3 is composed of proapoptotic Bcl2 proteins with only the BH3 domain and is generally referred to as BH3-only proteins. BH3-only proteins are considered to be functional homologs of C. *elegans* EGL-1 and serve to sense and transmit signals resulting from intense stress or serious defects in the functioning of the cell. These proteins interact directly with pro- and antiapoptotic proteins through their BH3 domains. Interestingly, while interaction with BH3-only proteins inhibits the protective activity of antiapoptotic Bcl2 proteins, such interaction with multidomain Bcl2 proteins stimulates their apoptotic activity. At least 10 different BH3-only proteins have been identified in vertebrates permitting signal- and cell-specific activation of specific BH3-only proteins. The activity of some BH3-only proteins increases in response to apoptotic stimuli as a result of increased expression, whereas others are activated by posttranslational modifications, including phosphorylation and proteolytic cleavage. The best studied of the

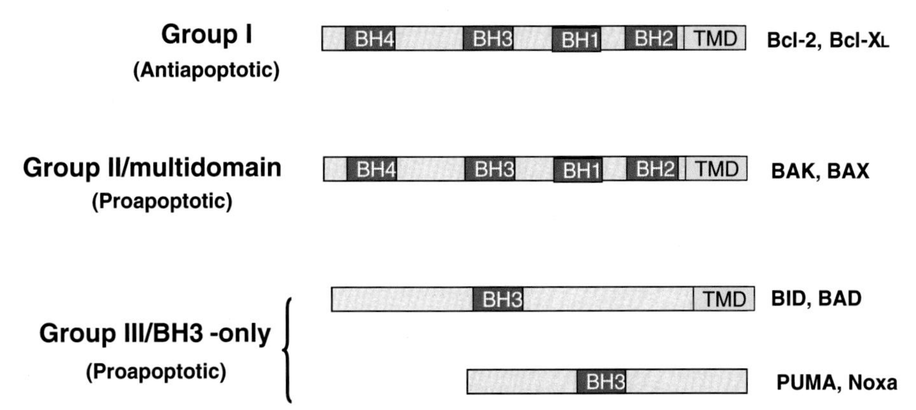

Figure 11–8. **The Bcl-2 family of proteins.** On the basis of functional and structural criteria, the Bcl-2 family has been divided into three groups. Members of the Group 1 family are antiapoptotic and possess all four BH domains (BH1–BH4). Group II Bcl-2 proteins, also referred to as multidomain Bcl2 proteins, also contain four BH domains. In contrast, Group III consists of a large group of proapoptotic proteins that contain the BH3 domain only and are therefore referred to as BH3-only Bcl2 proteins. Many Bcl-2 proteins also have a C-terminal transmembrane domain (TMD) with which they attach to various intracellular membranes such as the mitochondrial membrane. The figure also lists representative members of each subfamily.

BH3-only subfamily are BAD, BID, Puma, and Noxa. BAD activity is inhibited by its phosphorylation at specific sites by AKT, a kinase that potently promotes cell survival. In contrast, BID is activated through its cleavage by caspase-8 generating tBID, the active form of BID. Once produced, tBID stimulates the proapoptotic activity of BAX and BAK. Puma and Noxa are not expressed in healthy cells, but their expression can increase dramatically under specific apoptotic conditions, such as exposure to genotoxic agents and DNA damage.

In response to an apoptotic stimulus, the expression of the antiapoptotic Bcl-2 proteins generally decreases. This, along with the increased expression and activity of BH3-only proteins, results in the efficient sequestration of antiapoptotic Bcl-2 proteins by BH3-only proteins freeing up BAX and BAK. BAX and BAK then oligomerize and insert themselves into the outer mitochondrial membrane to form pores resulting in the release of cytochrome *c* and other apoptotic stimulators from the intermembrane space. Despite much effort, exactly how this happens remains unclear. It is believed that in most cells, BAX is predominantly cytosolic and translocates to the outer mitochondrial membrane after an apoptotic stimulus, whereas BAK is localized constitutively at the mitochondrial membrane where it is kept inactive by antiapoptotic Bcl-2 proteins. The presence of specific proteins and lipids, one of which is cardiolipin, helps the targeting of Bcl-2 proteins to the outer mitochondrial membrane. Furthermore, conformational changes of both BAX and BAK are necessary for their oligomerization and pore formation.

Many viruses encode functional homologs of Bcl-2. Among the best studied of these is the E1b 19-kDa protein encoded by adenovirus. Host cells infected with viruses are recognized by the immune system leading to the release of many inflammatory cytokines including TNF. Through activation of its receptor, TNF induces apoptosis in the infected cell by activating BAX and BAK. The ability of TNF to kill the host cell is blocked by the E1b 19-kDa protein, which acts like an antiapoptotic Bcl-2 protein. By preventing apoptosis, E1b allows the adenovirus to replicate in the host cell.

The apoptotic pathway is generally similar across the animal phyla (Fig. 11–9). In *Drosophila melanogaster* a protein called ARK acts similarly to *C. elegans* CED-4 and vertebrate Apaf-1 in that oligomerizes and activates a CARD-containing protein called DRONC, which is considered as the functional homolog of vertebrate

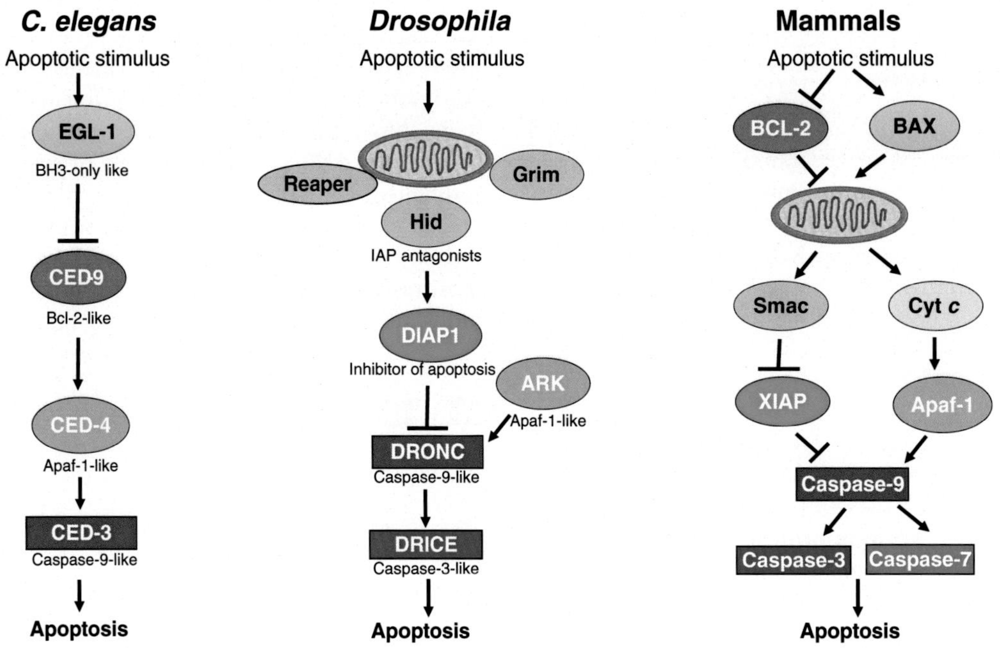

Fig. 11–9. **The apoptosis pathway in *C. elegans*, *Drosophila*, and vertebrates.** In vertebrates, antiapoptotic proteins, such as Bcl-2, preserve the integrity of the mitochondria and prevent release of cytochrome *c* (Cyt *c*). Activation of proapoptotic Bcl-2 proteins, such as Bax, promotes Cyt *c* release as well as the release of other apoptotic proteins, such as Smac. While Cyt *c* promotes Apaf-1 oligomerization and the recruitment of the initiator caspase, caspase-9, Smac promotes apoptosis by inhibiting IAPs. Following its activation, caspase-9 cleaves and activates effector caspases, such as caspase-3 and caspase-7, which induce cell death through the cleavage of multiple cellular proteins. In *C. elegans*, apoptosis is initiated by the inhibition of the Bcl-2-like protein, CED-9, by the BH3-only-like protein, EGL-1, through direct interaction. This liberates the Apaf-1-like protein, CED-4, which oligomerizes and activates the caspase-9-like protein, CED-3, leading to cell death. There is effector caspase or Cyt *c* release in *C. elegans*. In *Drosophila*, also Cyt *c* is not released, but apoptosis is triggered by the increased expression of the RHG proteins, Reaper, Hid, and Grim, which inactivate the IAP, DIAP1. This permits the Apaf-1-like protein, ARK, to activate the caspase-9-like protein, DRONC. Activated DRONC activates effector caspases like DRICE, to kill the cell.

caspase-9. Besides DRONC, *Drosophila* express seven caspases, including DREDD, which contains a DED resembling vertebrate caspase-8. However, there are some important differences in apoptosis signaling between these species too, particularly with regard to how the initiator caspases are activated. For example, in vertebrates, cytochrome *c* released from the mitochondria is necessary for Apaf-1 activation. In contrast, CED-4 is constitutively active but requires disassociation from CED-9 for caspase activation. Unlike vertebrates, therefore, there is no need for initiator caspases. In *Drosophila* also, activation of caspases does not require release of cytochrome *c* from the mitochondria. Instead, DRONC is normally kept inactive through interaction by an IAP protein called DIAP1 (Fig. 11–9). Apoptosis requires the expression of key proapoptotic genes, Reaper, Hid, and Grim (referred to as the RHG family), which are associated with the mitochondrial membrane and which disable DIAP1, permitting an Apaf-1-like protein called ARK to activate DRONC. DRONC kills the cell through activation of an effector caspase called DRICE (Fig. 11–9). Thus, in contrast to the central role played by mitochondria in vertebrates, there is no known role for the mitochondria in regulating apoptosis in worms and flies).

Inhibition of Apoptosis in Healthy Mammalian Cells by Extracellular Factors

The machinery for executing the cell death program through the intrinsic and extrinsic programs exists in the cell normally. However, in healthy and functional cells the apoptotic program is inhibited. This is mostly because of the actions of a variety of extracellular factors, most commonly growth factors, such as insulin-like growth factor (IGF-1); vascular endothelial growth factor (VEGF); and, in the case of neurons, members of the neurotrophin family of protein, including NGF and brain-derived neurotrophic factor (BDNF), as well as cytokines such as certain interleukins and interferons. These growth factors and cytokines bind cell surface receptors to activate antiapoptotic signaling pathways. Thus the survival fate of a cell is often dictated by other cells through soluble signals, the absence of which enables the apoptotic program.

Two well-characterized and powerful antiapoptotic signaling pathways are the **PI-3 kinase-Akt** and/or the **Raf-MEK-ERK pathways** (see Chapter 9). In the PI-3 kinase-Akt pathway, receptor activation leads to activation of Akt by PI-3kinase, which then phosphorylates proapoptotic BH3-only proteins BAD and BIM. This causes cytosolic 14-3-3 proteins to interact with them preventing their localization to the mitochondrial membrane and the stimulation of cytochrome *c* release. Another well-established target of Akt is FoxO, a proapoptotic member of forkhead family of transcription factors, which induces transcription of other proapoptotic genes including BAX, BIM, and FasL. Again, phosphorylation of FoxO by Akt promotes its sequestration by 14-3-3 in the cytoplasm, preventing its translocation to the nucleus. At the same time, Akt can translocate to the nucleus to activate survival-promoting transcription factors such as CREB and the NF-κB subunit, p65/RelA, which induce the expression of genes encoding antiapoptotic Bcl-2 proteins and other antiapoptotic genes. Similarly, in the Raf-MEK-ERK pathway, activated ERK can phosphorylate BAD and BIM inhibiting their proapoptotic activity. While Akt-mediated phosphorylation of BIM reduces is apoptotic activity, phosphorylation of BIM by ERK promotes its proteasomal degradation. ERK also phosphorylates caspase-9, which contributes to its inactivation. Like Akt, activated ERK can also translocate to the nucleus where it phosphorylates transcription factors such as CREB and c-Fos, resulting in increased expression of antiapoptotic genes.

CLEARANCE OF THE APOPTOTIC CELL

Perhaps the most important aspect of apoptosis is the rapid clearance of the dying cell, a phagocytic process sometimes referred to as "efferocytosis." Broadly, two types of phagocytes are involved: professional phagocytes, such as macrophage and immature dendritic cells, and nonprofessional phagocytes, including endothelial cells of blood vessels and tissue-resident cells close to the apoptotic cell, such as epithelial cells and fibroblasts, and in the brain, microglia and astrocytes. In comparison with professional phagocytes, which have a large capacity to engulf dying cells, nonprofessional phagocytes are less capable. But although displaying a lower capacity to engulf, because of their proximity to the apoptotic cell, nonprofessional phagocytes contribute substantially to their clearance. The removal is so swift and effective that cells undergoing apoptosis are difficult to detect in vivo. Impairment of apoptotic cell clearance during development leads to defects in morphogenesis and tissue function and in the adult to disruption of tissue homeostasis. Additionally, because these uncleared apoptotic cells ultimately lose membrane integrity releasing their components, inflammation that damages surrounding tissue as well as autoimmune disorders generally ensue. For example, in systemic lupus erythematosus (SLE), an autoimmune disorder in which patients experience chronic systemic inflammation, lymph node germinal centers display an increase in uncleared apoptotic cells. Similarly, alveolar macrophages isolated from patients with chronic obstructive pulmonary disease (COPD) show reduced engulfment capacity consistent with a highly elevated level of free apoptotic cells in the lungs of these patients. As described earlier a key function of apoptosis is to rid the body of cancerous cells. In this context, it is interesting that the expression of one of the best characterized engulfment receptors, BAI1, is severely reduced or entirely lost in certain glioblastomas. It is likely that this results in the survival of cancerous cells that would normally have been eliminated by apoptosis.

Interestingly the clearance process is initiated by the apoptotic cells itself. This involves "find me" and "eat me" signals. "Find me" signals are released at very early stages of apoptosis and sensed by professional phagocytic cells resulting in their migration toward the dying cell. The best characterized of these "feed me" signals are two lipid-based signaling molecules: the phospholipid, lysophosphatidylcholine (LPC), and the sphingolipid, sphingosine-1-phosphate (S1P). LPC is produced through a type of phospholipase A2 that is activated by caspase cleavage, whereas excessive SIP is produced following caspase activation through the increased expression of an enzyme called sphingosine kinase. A more recently identified "feed me" signal is the chemokine protein fractalkine/CX3CL1, which when cleaved by caspases produces a soluble fragment that is released by apoptotic cells attracting phagocytes. In addition to these caspase-dependent signaling molecules, small particles of about 5 μm called apoptotic blebs that are often shed from the blebbing apoptotic cell attract phagocytes to the apoptotic cell. Interestingly, such circulating apoptotic blebs originating from vascular dysfunction and increased cell death are found abundantly in patients with cardiovascular disease and correlate with disease severity.

The "eat me" signals are either molecules that are normally intracellular but become exposed to the outside upon activation of apoptosis, or preexisting molecules on the cell's surface that are chemically modified (Fig. 11–10). These lipid or protein molecules specific for apoptotic cells are bound by **engulfment receptors** (also referred to as efferocytic receptors) located on the plasma membrane of professional and nonprofessional phagocytic cells. In some cases the binding involves molecules that bridge the receptor and the cell surface ligand on the apoptotic cell. Recent research has found that distinct repertoires of engulfment receptors are expressed by phagocytes in different tissues or microenvironments. Binding of the receptors either directly or through bridging molecules to the "eat me" signals activates signaling pathways within the phagocytic cell that leads to the reorganization of the actin cytoskeleton, which permits the engulfment of the dying cell (Fig. 11–10). Once engulfed the dead cell is inserted into membrane organelles called phagosomes through the coordinated effort of a large number of signaling proteins. In addition, V-ATPases and acidic proteins, such as cathepsins, are brought to the phagosome leading its acidification and lysis of the dead cell or particles of it. Through fusion, the contents of the phagosome are transferred to lysosomes where the dead cell is completely digested to generate nucleotides, amino acids, and lipids that can be reused by the phagocyte.

Different types of transmembrane engulfment receptors on phagocytic cells and a variety of secreted bridging molecules that promote tethering of phagocytic cells to the dying cell have been identified in vertebrates. Genetic deletion of a single receptor or bridging molecule in mice does not affect apoptotic clearance demonstrating redundancy in the recognition machinery, consistent with the physiological importance of swift and efficient clearance of apoptotic cells. Besides initiating the process of engulfment, activation of the engulfment receptor (perhaps along with other signals from the apoptotic cell) stimulates the engulfing cell to release antiinflammatory proteins, such as transforming growth factor-β (TGFβ), prostaglandin E2 (PGE2), and IL-10, that suppress the

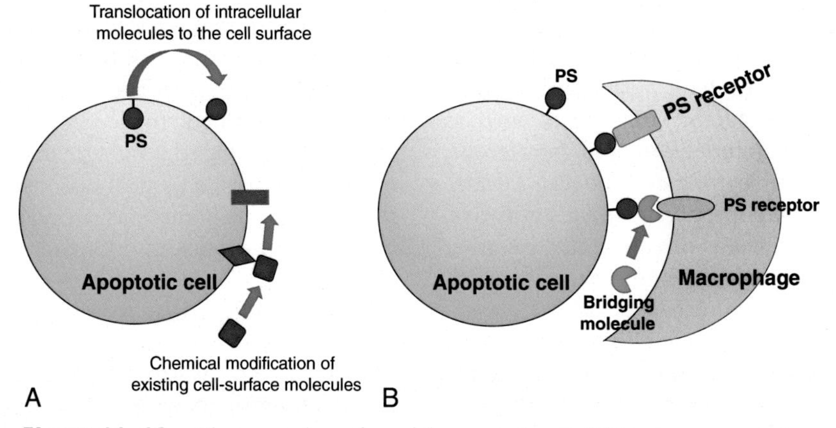

Figure 11–10. **Changes at the surface of the apoptotic cell.** (A) During apoptosis the structure of the cell surface changes because of the translocation of some molecules that are normally intracellular to the outside surface of the plasma membrane and the modification of existing molecules. (B) Engulfment receptors on the phagocytic cell bind to "eat me" signals either directly or indirectly. The figure shows that phosphatidylserine (PS) is translocated from the inner layer of the plasma membrane lipid bilayer to the outer layer through the action of a flippase, a lipid translocating enzyme. PS is recognized by phagocytic receptors either directly or indirectly through the binding of bridging molecules to PS.

production of proinflammatory cytokines. Hence, while inducing the production of proinflammatory cytokines when encountering pathogens or bacterial toxins, the same phagocytes instead actively suppress the release of such cytokines following interaction with apoptotic cells ensuring a complete absence of an inflammatory response during the clearance process.

One of the best characterized "eat me" signal is phosphatidylserine (PS), a lipid that is recognized by a variety of phagocytic receptors. In healthy cells, PS is localized in the inner membrane of plasma membrane lipid bilayer (Fig. 11–11). This is accomplished by a member of a family of ATP-dependent enzymes called flippases that translocate any PS localized on the outer leaflet of the plasma membrane to the inner leaflet (see Chapter 2). When apoptosis is initiated, however, caspases cleave the flippase resulting in their inactivation. At the same time caspases cleave another set of enzymes called scramblases, which in contrast to flippase, activates them. This promotes the scrambling of PS such that much of it is now on the outer leaflet of the plasma membrane (Fig. 11–11). Thus exposure of PS on the surface of apoptotic cells, which promotes its engulfment by phagocytes, involves flippase inactivation and scramblase activation, both of which are mediated by activated caspases.

Modifications, such as glycosylation and oxidation, and changes in charge of preexisting proteins on the surface of the apoptotic cell also serve as "eat me" signals. Although a variety of other "eat me" signals have been identified, many lines of evidence indicate the exposure of PS is necessary and sufficient for engulfment of apoptotic cells. Indeed, PS masking inhibits apoptotic cell clearance while inactivating flippase activity in otherwise healthy cells is sufficient for their engulfment by phagocytes. The existence of several other "eat me" signals may indicate that the apoptotic cell has many different ways to promote its own clearance pointing to the importance of efficient engulfment to the health of the organism.

How does activation of the engulfment receptors lead to phagocytosis? In *C. elegans*, two parallel and redundant signaling pathways are activated following binding of engulfment receptors to PS: one that is initiated by CED-1 and the other that is activated by CED-2. Engulfment of apoptotic cells is largely unaffected in *C. elegans* mutants in which any one of these pathways is inactivated, but when both CED-1 and CED-2 initiated pathways are inactivated, engulfment is blocked. Both signaling pathways converge at CED-10, a GTPase, activation of which leads to the polymerization of actins to form a phagocytic cup around the dead cell. The pathway is evolutionary conserved although multiple functional homologs have been identified in mammals for CED1 (such as MEGF10) and CED-2 (such as CRKII) as well as other components of these engulfment signaling pathways. In mammals, Rac1 is the major GTPase that mediates rearrangement of the cytoskeleton. Following internalization of the cellular corpse, Rac1 is downregulated causing the closure of the phagocytic cup to form the phagosome, which then undergoes maturation.

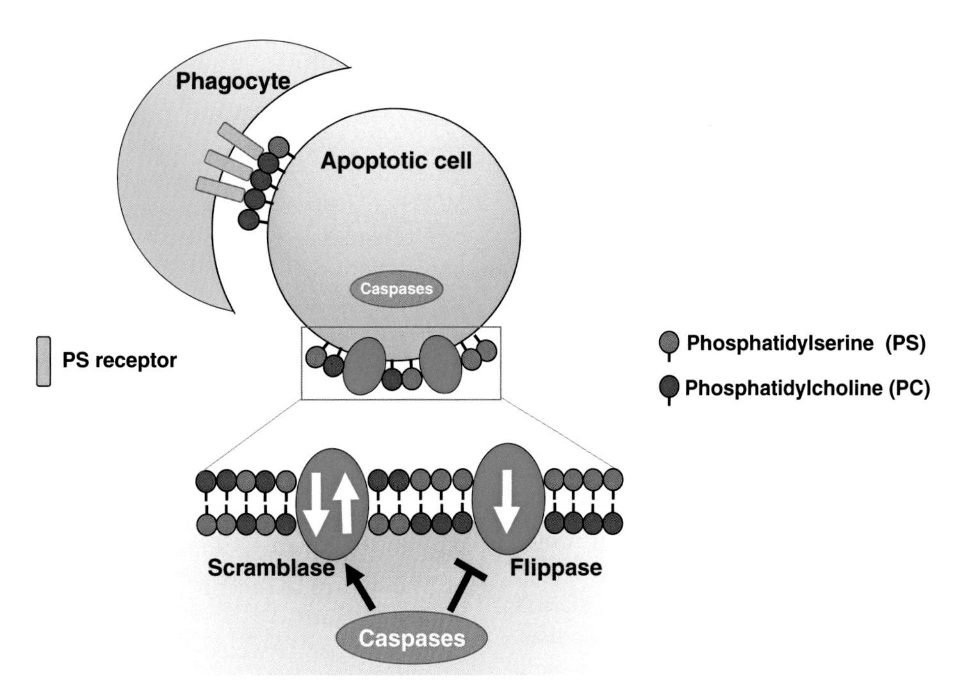

Figure 11–11. **Exposure of phosphatidylserine (PS) is a key "eat me" signal in apoptotic cells.** In healthy cells, flippase is active, and scramblase is inactive causing PS to localize to the inner leaflet of the plasma membrane. In apoptotic cells, effector caspases cleave and inactivate flippase while activating scramblase causing the translocation of PS to the outer leaflet of the plasma membrane. PS is recognized and bound by PS receptors on phagocytic cells that leads to the engulfment of the apoptotic cell.

APOPTOSIS AND HUMAN DISEASE

While critical for normal tissue homeostasis, the failure of cells to regulate apoptosis appropriately can have severe consequences to the health of the organism. Diseases involving deregulated apoptosis can be divided into two groups—those in which apoptosis fails to occur resulting in abnormally prolonged cell survival, and those in which apoptosis occurs prematurely leading to increased and undesirable cell death. Among the diseases resulting from insufficient apoptosis are a variety of cancers. Although initially believed to be a consequence of uncontrolled cell proliferation, it is now well established that many types of cancers result from the inability of cells to undergo apoptosis. Bcl-2 was originally identified as proto-oncogene whose chromosomal translocation is responsible for lymphomas. Increased Bcl-2 expression has been observed in many types of cancers and is currently considered to be a predictive factor for worse prognosis in prostrate and colon cancer. Mice engineered to overexpress Bcl-2 specifically in the B-cell lineage develop follicular lymphoma at 12 months of age providing proof that cancers can arise solely from the inhibition of apoptosis. Akt was also initially identified as a proto-oncogene. Gene amplifications and increased activity of Akt are found in several human cancer diseases. PI-3 kinase-Akt signaling is negatively regulated by phosphatase and tensin homolog deleted on chromosome 10 (PTEN), a lipid phosphatase that dephosphorylates the substrates of PI-3K, thus preventing Akt activation. PTEN is a potent tumor suppressor, inactivating mutations of which promotes tumorigenesis. Several studies have shown that the blockade of Akt signaling using molecular biological approaches can reverse the transformed cellular phenotype in tissue culture and animal models of oncogenesis. Mutations in Ras and Raf proteins that cause these proteins to become overactive are also a common cause of cancers, including colorectal cancer and melanoma skin cancer, as a result of reduced apoptosis. B-Raf inhibitors are being used in the treatment of melanoma, and several other Ras and Raf inhibitors are being tested clinically. Several studies have shown that proapoptotic Bcl2 proteins are downregulated in cancers. For example, Puma, which is normally induced following DNA damage and genotoxic stress, is mutated, or its expression is lost in several types of cancers. A majority of chemotherapeutic drugs used to treat human cancers work by activating the apoptotic machinery in tumor cells. Disturbingly the resistance of some cancers to conventional chemotherapeutic drugs and irradiation is also due to their enhanced resistance to apoptosis. For example, most chemotherapeutic drugs stimulate the expression of Puma although this does not happen in cancer cells when its expression is completely lost. In malignant melanoma, a deadly cancer that fails to respond to conventional chemotherapy, reduced expression of Apaf-1 has been observed. Restoration of Apaf-1 expression by treatment with the methylation inhibitor 5-aza-2′-deoxycytidine enhances the sensitivity of cultured malignant melanoma cells to standard chemotherapeutic drugs and rescues the apoptotic defects in them. The IAPs are also an attractive target for cancer therapy. Preclinical studies have shown that reducing the levels of specific IAP proteins by the expression of antisense oligonucleotides against their mRNAs reduces tumor growth.

Decreased apoptosis also contributes to certain autoimmune diseases. During the normal development of the immune system, self-reactive lymphocytes are eliminated by apoptosis. Failure of these self-reactive lymphocytes to die causes autoimmunity. Some autoimmune diseases such as lupus erythematosus are associated with mutations in proapoptotic molecules rendering the autoreactive lymphocytes resistant to apoptosis. Developmental defects such as cleft lip are also caused by the failure of cells to undergo apoptosis during embryonic development.

Premature or excessive apoptosis, on the other hand, occurs in neurodegenerative diseases such as Alzheimer's disease, Parkinson's disease, ALS, and retinitis pigmentosa. Human clinical trials are underway to examine whether infusion of neurotrophic factors into the brains of patients with these diseases can slow down neurodegeneration by activating the neuron's survival-promoting signaling pathways. Similarly, chemical drugs that block apoptosis in tissue culture and animal models of neurodegeneration are being tested in clinical trials involving patients with Parkinson's disease and other neurological pathologies. Familial ALS, a progressive disorder associated with the death of motor neurons in the spinal cord and brain, is often caused by mutations in the gene encoding Cu/Zn superoxide dismutase (SOD1). Mice that are genetically engineered to express the mutated human SOD1 gene develop a motor neuron degenerative pathology resembling ALS. The mutant mice suffer paralysis like humans with ALS and have therefore served as a useful model to understand the mechanisms underlying disease pathogenesis and to develop potential treatment strategies. If Bcl-2 is overexpressed in these "ALS mice," disease symptoms are delayed. Likewise the administration of a pharmacological caspase inhibitor or the overexpression of IAP proteins in these mice attenuates disease progression and increase life span. These encouraging results raise the possibility that drugs and strategies that block apoptosis may also be useful in the treatment of ALS as well as other neurodegenerative pathologies in human patients.

Outside the nervous system, increased apoptosis has also been implicated in the pathological depletion of $CD4^+$ cells in HIV-infected patients and in diseases involving tissue damage such as myocardial infarction, congestive heart failure, renal damage, and cirrhosis of the liver (Clinical Case 11–1).

Clinical Case 11–1

Mr. Moore is a 65-year-old Caucasian farmer who has grown cotton and soybeans for the past 40 years. He has been the primary person responsible for planting, harvesting, and applying pesticides to his crops as he runs a small family farm. The farm has been doing well and making a profit. To celebrate he had been eating second helping at meals and always eating dessert, and he had gained so much weight that he had gotten up to a weight of 270. In the last 6 months, however, he had begun to notice that he had unintentionally lost over 30 pounds and was often tired. His wife is 55 years old and has been going through the menopausal change and has been suffering with hot flashes and night sweats. During one of her nightly hot flashes, she noticed that Mr. Moore was also sweating even though he did not describe feeling hot. While giving him a neck rub, Mrs. Moore notice a lump on the right side that had not been there 1 month ago. She urged him to go to the doctor to see about this new lump and the fatigue that he had been complaining about. Finally, Mr. Moore decided to go to his doctor after the cat bit him, and he thought the lump in his neck was getting bigger.

Cell Biology, Diagnosis, and Treatment of B-Cell Lymphoma

Initially when Mr. Moore went to the doctor, he was diagnosed with an infection from a cat bite he had recently received and treated with oral antibiotics, but his symptoms did not resolve over the next month and continued to get worse. During the second visit to the doctor, Mr. Moore was scheduled for a biopsy of the enlarged lymph node, and the pathology report revealed that Mr. Moore had B-cell lymphoma. He was referred to an oncologist who also ordered a CT scan of the chest and abdomen to assist with staging and an LDH level. After staging, Mr. Moore was found to have tumor in only one lymph node area that classified him as Stage I localized disease. A SEER stage with localized disease has a 72% 5-year survival rate. B-cell lymphomas make up 85% of non-Hodgkin lymphomas in the United States. About 74,200 people each year are diagnosed with this disease, and over 19,900 will die from the cancer. It is thought that in B-cell lymphoma a translocation between chromosomes 14 and 18 turns on the Bcl2 gene that is an antiapoptotic gene that favors tumor cell survivorship and resistance to therapy.

Autophagy

Like apoptosis, autophagy (in Greek, "to eat oneself") is an evolutionarily conserved process that serves critical functions. In healthy cells, autophagy clears protein aggregates and eliminates proteins that have served their function (Fig. 11–12). It is also the major mechanism to rid the cell of entire organelles that are damaged or dysfunctional. Therefore a basal level of autophagy is crucial for maintaining healthy cells. Autophagy is upregulated under conditions of severe nutrient deprivation where it helps cells survive by degrading macromolecules to their building blocks so that these can then be used to synthesize cellular components essential for its survival.

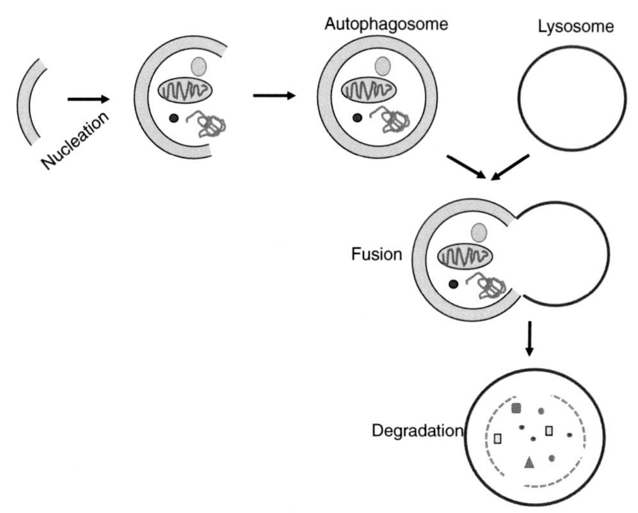

Figure 11–12. **Schematic of major steps in autophagy.** Expansion of a membrane derived from the ER membrane, or possibly other sources, leads to the formation of a cuplike structure that engulfs cytoplasmic material, including organelles such as mitochondria. This extends further to form a double-membrane vesicle called the autophagosome, which fuses with the lysosome. The cargo of the autophagosome is digested by lysosomal enzymes.

Autophagy during conditions of cellular starvation is triggered by the activation of AMP kinase (AMPK) that inhibits the mammalian target of rapamycin complex I (mTORC1), a complex that is activated under nutrient-rich conditions (please refer to Chapter 4 for a detailed description of the autophagic pathway). The inhibition of mTORC1 results in its disassociation from a complex called the ULK-complex, a component of which is the ULK, a kinase that phosphorylates key components of another complex called the class III PI-3 kinase complex. Two important components of the class III PI-3 kinase complex are a protein called Beclin 1 and a protein called Vps34, which, through its ability to produce phosphatidylinositol-3-phosphate (PI3P), promotes the formation of the isolation membrane. The isolation membrane elongates to form the phagophore membrane, a cuplike structure that starts engulfing cytoplasmic material, including entire organelles. The cuplike structure matures to form a vesicle, the autophagosome, which fuses with the lysosome where the contents of the autophagosome are broken down by lysosomal degradative enzymes to yield the building blocks that are translocated out of the lysosome and used for the synthesis of molecules essential for the maintenance of the cell's survival through the period of nutrient deprivation.

Although autophagy occurs normally and can be upregulated during conditions of cellular starvation, an extreme upregulation of autophagy leads to cell death by self-consumption. The mechanism by which the extreme upregulation occurs, converting nonlethal autophagy to lethal autophagy, is poorly understood. Furthermore and although recognized as a form of programmed cell death, autophagic cell death is considerably less prevalent than apoptosis. While extreme upregulation of autophagy leads to cell death, its impairment causes the accumulation of misfolded proteins that can also have fatal consequences to the cell. There is broad consensus that failure to efficiently clear toxic protein aggregates through autophagy contributes to the abnormal loss of neurons in many brain degenerative diseases. Indeed, stimulating autophagy through genetic or pharmacological approaches has a strong beneficial effect in a variety of experimental models of neurodegenerative disease.

Recently, several different subclasses of autophagy have been described in which specific organelles or molecular structures are enclosed in the autophagosome for clearance by lysosomal degradation. In contrast to nonselective or bulk autophagy, these forms of autophagy are referred to as selective autophagy based on the selective cargo that is enclosed (Table 11–1). Growing evidence suggests that distinct autophagic receptors are involved for recognition and encapsulation of different cargo. Among these, **mitophagy** has been well characterized. Mitophagy serves to remove dysfunctional or superfluous mitochondria, thus preserving mitochondrial number and function. Two proteins that play a critical role promoting mitophagy are a kinase called PINK1 and a ubiquitin ligase called Parkin (Fig. 11–13). In normally functioning mitochondria, PINK1 is transported into the mitochondria where it is degraded by proteases. However, when the mitochondrial membrane potential dissipates, mitochondrial import is prevented that leads to the accumulation of PINK1 on the outer mitochondrial membrane where it phosphorylates ubiquitin on membrane proteins and Parkin. This causes the recruitment of Parkin to the mitochondrial membrane and stimulation of its E3 ligase activity. As a result, several proteins on the surface of the dysfunctional mitochondria are polyubiquitinated, which is recognized by adaptor proteins. The adaptor proteins connect the dysfunctional mitochondria to autophagosome resulting in their encapsulation. This is followed by the fusion of the autophagosome with lysosomes in which the mitochondria are degraded. Impaired mitophagy has been implicated as a contributing factor in several diseases, including cardiovascular diseases, cancer, and neurodegenerative diseases. Indeed, mutations in the genes encoding PINK1 and Parkin that reduce their functional activity cause familial Parkinson's disease.

TABLE 11–1. Selective Autophagy Types and Cargo

Type	Cargo
Aggrephagy	Protein aggregates
Ferritinophagy	Ferritin
Mitophagy	Mitochondria
Glycophagy	Glycogen
Lipophagy	Lipid droplets
Lysophagy	Lysosomes
Nucleophagy	Nucleus, nuclear proteins
Peroxyphagy	Peroxisomes
Reticulophagy	Endoplasmic reticulum
Xenophagy	Bacteria, viruses

Necroptosis

Necroptosis is a mode of regulated cell death that has been characterized more recently, which, while sharing morphological features with necrosis, is regulated by signaling molecules. It is often triggered by pathogenic infection and serves to stimulate the response of the immune system such that the spread of the pathogen is prevented. Cells undergoing necroptosis display cytoplasmic swelling and swelling of cytoplasmic organelles, including the mitochondria.

In necroptosis, perturbations of the extracellular environment caused by pathogenic infection are detected by cell surface receptors, including the death receptors, T-cell receptor, and Toll-like receptors (TLRs). The best characterized of these receptors is the receptor for tumor necrosis factor (TNFR1), a death receptor. As described previously in this chapter, TNFR1 and other death receptors also trigger apoptosis through the extrinsic pathway, a process requiring the activation caspase-8.

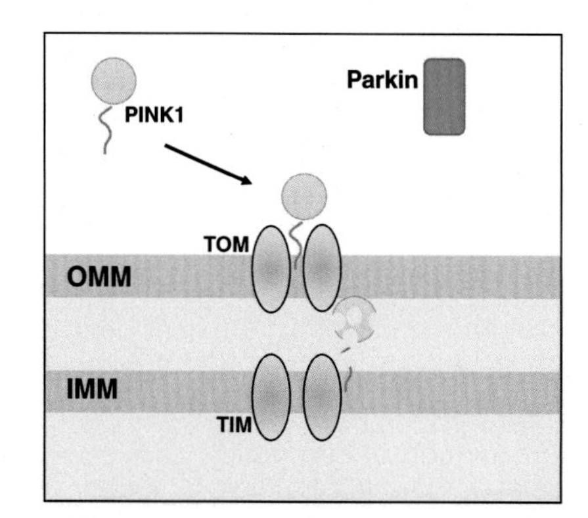

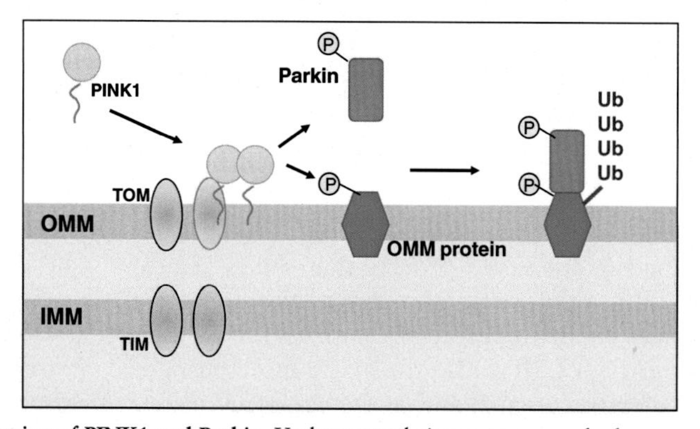

Figure 11–13. **Mitophagy involves the function of PINK1 and Parkin.** Under normal circumstances and when mitochondria are healthy, PINK1 is transported into the mitochondria through the TOM/TIM complex and degraded. However, when the mitochondria are dysfunctional, PINK1 accumulates at the surface of the outer mitochondrial membrane (OMM). PINK1 phosphorylates the E3 ligase, Parkin, and ubiquitin in OMM proteins. This causes the recruitment of Parkin to the OMM where is polyubiquitinated membrane proteins, which is recognized by autophagy receptors. The dysfunctional mitochondria are enclosed in autophagosomes and transported to lysosomes where they are degraded.

Three proteins play a key role in promoting necroptosis—the serine/threonine kinases receptor-interacting protein (RIPK1 and RIPK3) and the mixed-lineage kinase domain-like (MLKL) protein (Fig. 11–14). In TNF-induced necroptosis the binding of TNF to TNFR1 causes the recruitment of the adapter protein TNF receptor-associated death domain (TRADD), which recruits (RIP1), a protein kinase that is also a component of the DISC in apoptotic death. RIP1 interacts with the TNFR1 through a death domain that it possesses, which the TNFR1 also possesses. Following binding, RIP1 undergoes two modifications, polyubiquitination and phosphorylation, which result in its activation. Activated RIP1 associates with RIP3, another kinase, via RHIM (interaction of RIP homotypic interaction motif) domains in both proteins to form a protein complex called the **necrosome.** Interaction of RIP1 with RIP3 results in the phosphorylation of RIP3, which then recruits MLKL to the necrosome within which it is phosphorylated by RIP3. MLKL phosphorylation promotes its oligomerization, and these oligomers translocate to associate with phosphoinositide species on the cytoplasmic side of the plasma membrane (Fig. 11–14). By a process that is not well understood, MLKL oligomers cause disruptions in the plasma membrane leading to the eventual lysis of the cell. Exactly how lysis occurs is not fully understood but is believed to be due to the influx of sodium through Na^+ and Ca^{2+} channels resulting in increased osmotic pressure rupturing the plasma membrane. Substantial amounts of inflammatory cytokines are released from the ruptured necroptotic cell, which stimulates the immune system.

It is known that many proteins regulate the RIP1-RIP3-MLKL core pathway and their identities and extent of involvement are beginning to be understood. One protein that plays a pivotal role in suppressing necroptosis is caspase-8. Active caspase-8, which promotes apoptotic death through the extrinsic pathway, suppresses necroptosis by cleaving RIP1, thus destroying its catalytic activity and its ability to recruit RIP3. Hence, inhibition of caspase-8 is generally regarded as a requirement for the activation of necroptosis. The activation of the necroptosis by inactivation of caspase-8 has led to the idea that necroptosis may have evolved as a backup mechanism to trigger programmed cell death if the apoptotic machinery is somehow disabled. Indeed, several viruses encode different types of proteins that inhibit caspase-8 activation in an attempt to suppress apoptosis permitting their replication. In these situations, necroptosis may act as a backup mechanism to limit viral replication.

More recently, other cell surface receptors have been identified that trigger necroptosis in response to

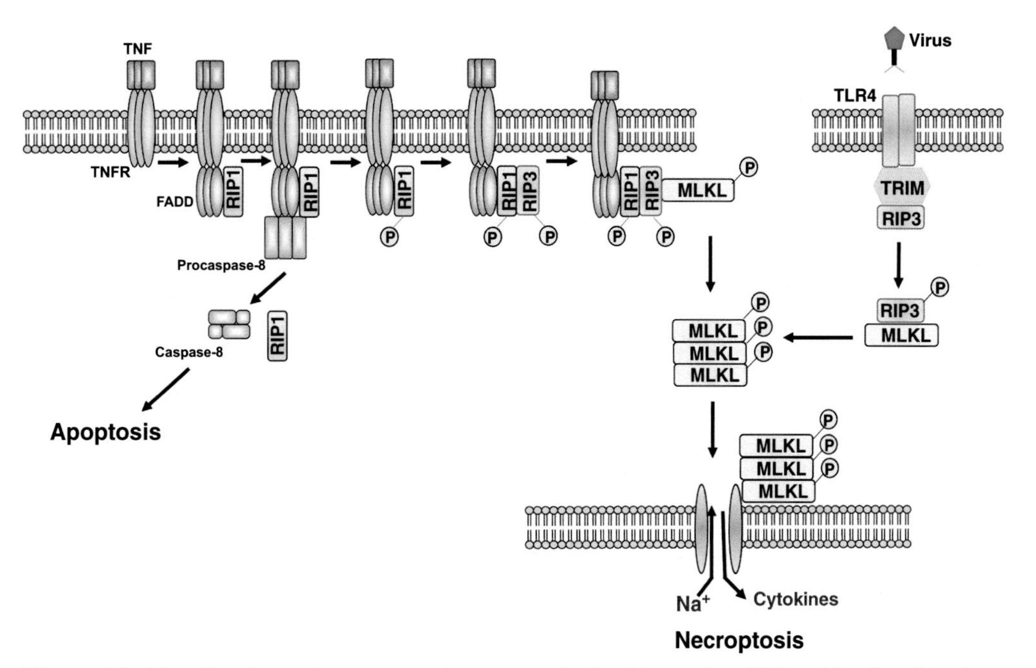

Figure 11–14. **Signaling pathways promoting necroptosis.** Cytokines such as TNF and Fas ligand bind to their receptors causing their trimerization and recruitment of adaptor proteins, procaspase-8, RIP1, and other proteins. This can result in the activation of caspase-8, which proteolyzes RIP1 and other substrates to promote apoptosis. However, if caspase-8 is not activated, RIP1 is activated by phosphorylation. Through its RHEM domain, activated RIP1 recruits RIP3, which also has a RHIM domain. RIP1 phosphorylates RIP3, which leads to its activation, causing the recruitment of MKLK, which is then phosphorylated by RIP3. This leads to the oligomerization of MKLK, which binds to phosphoinositides in the plasma membrane to form pore through which Na^+ comes into the cell followed by water. This causes swelling of the cell and membrane rupture. The leakage of cellular contents stimulates inflammation. Necroptosis can also result from the activation of Toll-like receptors (TLR), like TLR3 and TLR4. In this case, RIP3 is activated through the binding of an adapter protein to the TLR. In this RIP1-independent mechanism, RIP3 activates MLKL, which then oligomerizes and forms pores leading to membrane rupture.

extracellular stimuli, including members of the **pathogen recognition receptor (PRR)** family, such as TLR4 (Toll-like receptor-4). PPRs are expressed mainly by cells of the innate system, including macrophage, monocytes, and dendritic cells, and function to detect molecules typical for pathogens, such as liposaccharides (LPS), a major outer surface membrane component present in almost all gram-negative bacteria, and a group of proteins that are typically released from cells during pathogenic infection collectively termed as **damage-associated molecular patterns (DAMPs)**. In the case of necroptosis induced by TLR4, the activation of RIPK3 does not require RIP1. Instead, it involves an adaptor protein called TIR-domain-containing adapter-inducing interferon-β (TRIF), which also possesses a RHIM domain (Fig. 11–14). TRIF can directly interact with RIPK3 to activate MLKL. Interestingly, some viruses, like CMV and HSV, encode RHIM domain-containing proteins that can subvert necroptotic cell death by interacting with RIPK1 or RIPK3 preventing the activation of MLKL. In addition to extracellular cues, necroptosis can be activated within the cell by double-stranded RNA (often viral nuclei acids) through the activation of other PRRs that are localized within the cell. One such receptor is TLR3.

Although mostly described in the context of unplanned cellular perturbations, such as those induced by microbial infection, necroptosis plays an important role during normal development, particularly in the nervous system and immune system. Demonstrating the importance of necroptosis in development is the finding that mice lacking RIPK (Ripk −/−) die at 1–3 days of age. However, unwanted necroptosis is involved in many pathological conditions including ischemic injury in the heart and brain and in inflammatory diseases caused by injury. Accumulating evidence points to a role for necroptosis in the pathological loss of neurons in neurodegenerative diseases.

Pyroptosis

Pyroptosis is a form of cell death that functions to protect against various bacterial and viral infections by eliminating the infected cell, thus destroying the intracellular replication niche of the pathogen. In addition, it also stimulates the host's defensive responses. Pyroptosis was first described in 2001 when it was observed that macrophages infected with bacteria die through a process involving membrane rupture that was dependent on the activity of the inflammatory caspase, caspase-1, a caspase that is not involved in promoting apoptosis. Following infection of the cell by a microbial pathogen, caspase-1 is activated and cleaves members of a family of six diverse proteins called **gasdermins (GSDMs)**. Of the GSDMs, GSDMD plays a key role in pyroptosis (Fig. 11–15).

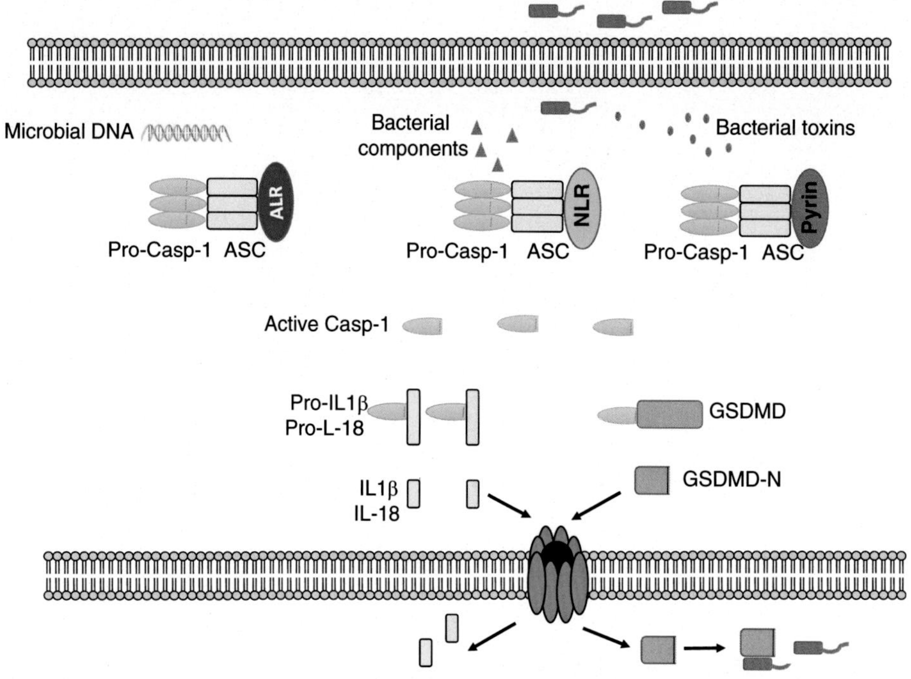

Figure 11–15. **Pyroptosis eliminates microbial pathogens within and outside cells.** Viral DNA, components of microbial pathogens, or toxins released by microbial pathogens are recognized by intracellular pattern recognition receptors (PRRs), which form protein complexes called inflammasomes. While viral DNA is recognized by PRRs called absent in melanoma 2-like receptors (ALRs), microbial components and toxins are recognized by NOD-like receptors (NLRs) and pyrin, respectively. Within the inflammasome, caspase-1 is brought in close proximity and cleaves itself to become activated. Activated caspase-1 processes pro-ILβ and pro-IL-8 to generate their functional forms. Activated caspase-1 also cleaves GSDMD to generate an N-terminal fragment, GSDMD-N, which oligomerizes to form pores in the plasma membrane. ILβ and pro-IL-8 are released through these pores and attract phagocytes, which engulf the infected cell. GSDMD-N is also released from the cell and binds to the membrane of microbial pathogens damaging or killing the pathogen outside the cell.

In the infected cell, activated caspase-1 cleaves GSDMD to generate an N-terminus fragment, GSDMD-N, which can oligomerize. Oligomerized GSDMD-N binds phosphoinositides on the inner leaflet of the plasma membrane to form pores. Structural studies indicate that each oligomeric pore is 10–15 nM in diameter and composed of 16 GSDMD-N fragments. Other intracellular proteins that caspase-1 cleaves during pyroptosis are IL-1β and IL-18, which produces the mature form of these cytokines, both of which lack a normal secretion signal. Although too small for bacteria to exit, the oligomeric GSDMD-N pores allow the release of mature IL-1β and IL-18. Neutrophils and macrophage are attracted to these cytokines and clear the infected cell in which the bacteria or other microbial pathogens are now trapped. In addition to cytokines, ATP and small-molecular-weight proteins are also released from GSDMD-N pores. Because proteins are released, osmolarity is not lost, and the cell does not burst, as in necroptosis, but instead flattens while its contents are released. Some membrane rupture occurs in late stages releasing other host immunogenic molecules, which further enhance the stimulation of the immune system assisting in the removal of the pathogen. GSDMD-N is itself released and can bind cardiolipin, a lipid present in bacterial membranes resulting in the destroyal of bacteria outside of the cell. Thus, in addition to promoting engulfment of infected cells by phagocytic cells, GSDMD-N kills microbial pathogens outside the cell after being released from pores in the membranes of the pyroptotic cells.

How does the cell connect entry of invasive microbial pathogens and toxins produced by them to caspase-1 activation? Pyroptosis occurs mostly in cells of the innate immune response that express PRRs, such as macrophage, monocytes, and dendritic cells, or epithelial cells in the skin and mucosa where infection is first encountered. During pyroptosis, cytosolic PRRs recognize viral DNA, microbes, microbial components, and microbial toxins that are within the cells. The PRRs recruit other proteins, including procaspase-1, to form a multiprotein signaling complex called the **inflammasome**, a molecular hallmark of pyroptosis. The composition of inflammasome can vary depending on the inflammatory pathogen or pathogenic molecule, but well-established components are PRRs, such as absent in melanoma 2-like receptors (ALRs), a protein called pyrin, and members of the **NOD-like receptor** (**NLR**) family of proteins. Whereas some NLRs recognize bacterial toxins, others recognize microbial components. Pyrins recognize microbial toxins and modifications of host proteins by bacteria, while ALRs recognize double-stranded DNA. Inflammasomes also contain an adaptor protein called apoptosis-associated speck-like protein containing CARD (**ASC**), which recruits procaspase-1 stimulating its autoprocessing to produce the active cleaved enzyme (Fig. 11–15).

Although the binding to phosphoinositides at the plasma membrane is best described, recent observation indicate that GSDMD-N fragment may also bind to phosphoinositides on membranes of intracellular organelles, including endosomes, lysosomes, and mitochondria. Whether the membranes of these organelles are also permeabilized and the extent to which this may contribute to cell death remains to be clarified.

It is now known that human caspase-4 and caspase-5 (and caspase-11 in mice), which are expressed more widely that caspase-1, also trigger pyroptosis. These caspases form a complex, referred to a noncanonical inflammasome, that detects and binds cytosolic bacterial endotoxins. Rather than GSDMD, activated caspase-11 cleaves the GSDMDE protein to trigger pyroptosis. It is not known if GSDMs, other than GSDMD and GSDME, which share similar N-terminus domains, also promote pyroptosis and, if so, what signals and upstream mechanisms regulate their pyroptotic activities.

While pyroptosis is a protective mechanism for the host, abnormal activation could have detrimental effects. Indeed, gain-of-function mutations in some of the inflammasome sensor proteins, such as pyrin and certain NLR proteins, are responsible for the inflammatory pathology in autoinflammatory genetic diseases, including familial Mediterranean fever (FMF), a condition in which patients suffer recurrent episodes of painful inflammation in the abdomen, chest, or joints (Clinical Case 11–2). Such autoinflammatory diseases are generally treated by inhibiting IL-1β activity. Overactivation or chronic inflammasome activation can result in multiorgan failure sepsis syndrome, a common cause of childhood death, and a contributing factor in a significant number of adult hospital death is autoinflammatory disease, which is also treated by inhibiting IL-1β activity. The recently discovered noncanonical inflammasome is thought to be especially important in sepsis. Indeed, mice in which the caspase-11 gene is knocked out are resistant to death from LPS-induced sepsis, a resistance not offered by deletion of caspase-1 (presumably because caspases 4 and 5 are still active). There is some evidence suggesting that increased activity of the NLR inflammasome and accelerated destruction of pancreatic beta cells contributes to type 2 diabetes.

Despite the differences in the molecular machinery regulating pyroptosis and necroptosis, there are similarities between these two modes of death. Both pyroptosis and necroptosis protect cells against microbial infections through their sensing and activation by immunogenic microbial molecules. Both mechanisms are initiated by the formation of large protein complexes—the inflammasome for pyroptosis and the necrosome for necroptosis. Both modes of cell death provoke an inflammatory response caused by membrane damage, which in the case of pyroptosis is mediated by GSDMD, while for necroptotic cell death, it is mediated by MLKL. In both pyroptosis and necroptosis, the inflammatory response can be enhanced by the release of DAMPs and inflammatory cytokines such as IL-1β. And finally the dysregulation of both pyroptosis and necroptosis have been suggested to play a key role in the development of autoimmune and other antiinflammatory disorders.

It is likely that pyroptosis evolved as a backup mechanism to necroptosis to protect against pathogenic infections.

Clinical Case 11–2

Rebecca Abrams is a 7-year-old Caucasian female who has been living with her adopted family in Orlando, FL, but she was originally from Turkey. She loves to go to the theme park and ride the roller coaster. She also commonly eats dessert at every meal to the point of being overfull and has occasionally gotten sick while riding the roller coaster. On the latest trip to the theme park, Rebecca's mother noticed she felt hot and began complaining of abdominal pain, knee pain on the right side, and a headache. When they returned home her mother took her temperature and noted that Rebecca had a fever of 101°F. This episode of illness lasted about 24h and resolved with over-the-counter ibuprofen. Unfortunately, similar symptoms recurred in about 2 months and then again at 4 months, at which point her mother took her to the pediatrician.

Cell Biology, Diagnosis, and Treatment of Familial Mediterranean Fever

At the pediatrician, Rebecca was found to have a high white blood cell (WBC) count and a high sed rate but normal renal function. On physical exam, she was noted to have a very swollen right knee and a red rash over her legs. Upon further questioning, it was discovered that a similar set of symptoms had recurred in Rebecca's older brother before they were separated by adoption and he had been diagnosed with familial Mediterranean fever (FMF). At that point the pediatrician ordered a genetic test for Mediterranean fever, which was positive, and Rebecca was started on colchicine. FMF is mostly inherited in an autosomal recessive pattern. FMF primarily affects persons from the Mediterranean region, especially persons from Armenia, Turkey, or who have Jewish ancestry. The prevalence of FMF in these population groups can range from 1 in 200 to 1 in 1000 but is much less common in other groups. Mutations in the MEFV gene are thought to cause FMF. The MEFV gene codes for the protein called pyrin, which is found in WBCs and plays a role in the regulation of inflammation. Mutations in the MEFV gene can reduce the activity of the pyrin protein resulting in disruption of the inflammatory process. This inappropriate or prolonged inflammatory response leads to the symptoms of FMF, fever, and pain in the abdomen, chest, and joints. Triggers for FMF include infection, trauma, strenuous exercise, or psychological stress. Amyloidosis is one of the most common complications of FMF and is associated with progression to end-stage renal disease and proteinuria. Colchicine prevents amyloid deposition and may cause it to regress.

Ferroptosis

Ferroptosis refers to cell death caused by excessive iron-dependent lipid peroxidation. The only distinctive morphological feature in cells undergoing ferroptosis is the condensation of mitochondria, disappearance of cristae, followed by the rupturing of the outer mitochondrial membrane. Ferroptosis as a distinct form of programmed cell death was first described in 2012 through a screen for anticancer drugs that identified erastin, a chemical that inhibits the X_c^- system. X_c^- system is the glutamate/cysteine antiporter in the plasma membrane through which cystine is imported into the cell. Within the cell, cystine is converted to cysteine, which is used to synthesize glutathione. Glutathione serves as a substrate for glutathione peroxidase (GPX4), which protects cells against oxidative stress by converting hydrogen peroxide to water. Additionally, GPX4 coverts toxic lipid peroxides to nontoxic lipid alcohols (Fig. 11–16). Thus erastin induces lipid peroxidation and consequently ferroptosis through its inhibition of the X_c^- system, which indirectly reduces GPX4 activity. Since then, it has been shown that direct pharmacological inhibition of GPX4 also triggers lipid peroxidation and ferroptosis.

Iron is necessary for the toxic lipid perodixation that causes ferroptosis. Free iron exists in cells either in the ferrous ion (Fe^{2+}) state or the ferrous (Fe^{3+}) state. A disruption in iron homeostasis can result in Fe^{2+} elevation. Changes to lipid metabolism that increases the pool of polyunsaturated lipid also sensitize cells to ferroptosis. Through Fenton chemistry, Fe^{2+} combines with peroxides to generate highly reactive hydroxyl radical and peroxyl radicals promoting nonenzymatic oxidation of lipids. In addition to nonenzymatic peroxidation, lipid peroxidation is caused by an elevation in the activity of lipoxygenase (LOX), an iron-containing enzyme that catalyzes the deoxygenation of lipids. In all cases the buildup of toxic lipid peroxides kills the cell through ferroptosis.

The subcellular location of lethal lipid peroxidation has yet to be firmly established. Because peroxidation of lipids of the plasma membrane would compromise integrity of the cell, it is considered a likely source. The impaired mitochondrial morphology during ferroptosis has led to the suggestion that it may also be a key source of the toxic lipid peroxides. Indeed an isoform of GPX4 is localized in mitochondria. However, recent studies have found that nonmitochondrial lipid peroxidation precedes peroxidation in the mitochondria and that the alterations in mitochondrial morphology leading to the outer membrane rupture are a late event in the process of ferroptosis. Additionally, mitochondria-deficient cells have been found to retain the ability to undergo ferroptosis, arguing against a necessary role for the mitochondria in this form of cell death. Some indications are that the ER may also produce toxic lipid peroxides when stressed, linking chronic or intense ER stress to ferroptosis.

As described earlier, toxic lipid peroxides in ferroptosis could be generated either through a reduced ability of GPX4 to detoxify them (which may be caused by depletion of its substrate, glutathione), through the excessive production of these lipid peroxides by the elevation in

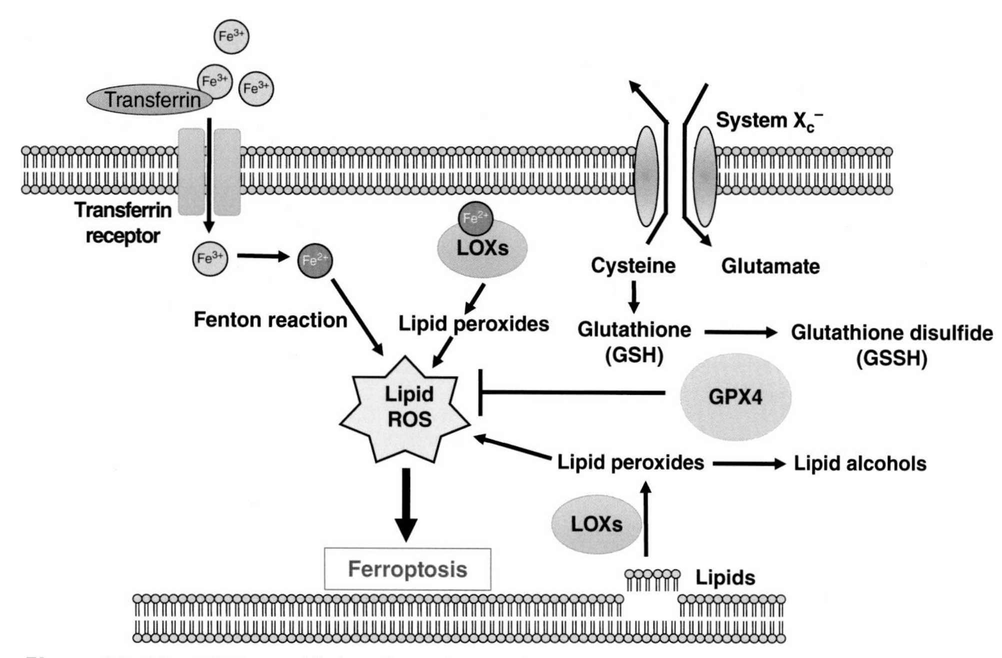

Figure 11–16. **Lipid peroxidation triggers ferroptosis.** System X_c^- imports cystine, which is reduced to cysteine within the cell, and used in biosynthesis of glutathione (GSH), a necessary substrate of glutathione peroxidase (GPX4). Toxic lipid hydroperoxides, derived from membrane lipids through the action of lipoxygenases (LOXs), care converted by GPX4 to their corresponding nontoxic alcohols, thus protecting the cell. Uncleared lipid hydroperoxides triggers the formation of lipid reactive oxygen species (ROS) that induces ferroptosis. Ferroptosis can therefore result from decreased levels of GSH or from reduced activity or the inhibition of system X_c^- or GPX4. Also necessary for the induction of ferroptosis is disruption in iron homeostasis. Ferric iron (Fe^{3+}), imported through the transferrin receptor following binding with transferrin, is reduced to ferrous iron (Fe^{2+}). Accumulation of Fe^{2+} not only produces lipid ROS through the Fenton reaction but also can catalyze lipid peroxidation by combining with cytosolic LOXs, which then leads to the production of lipid ROS and ferroptosis.

the level of Fe^{3+}, an increase in polyunsaturated lipid, or by an increase in LOX activity. A number of modulators of ferroptotic death have been identified that block these different mechanisms, and these agents protect against ferroptosis in experimental systems. For example, chemical iron chelators and radical-trapping antioxidants, such as ferrostatin-1 and vitamin E, inhibit ferroptotic death by preventing autooxidation of polyunsaturated lipids. Iron chelators such as deferoxamine and deferiprone protect against ferroptosis by inhibiting Fenton chemistry or indirectly inhibiting LOX. Direct LOX inhibitors also protect against ferroptosis. Cells lacking LOX are resistant to ferroptosis caused by chemical inhibition of GSH demonstrating that LOX plays an important role in promoting ferroptosis.

Ferroptosis inhibitors could be useful in the treatment of several human pathologies, including degenerative diseases of the kidney, liver, and brain. Strong evidence from animal models suggest that ferroptosis contributes to neurodegeneration in Parkinson's disease, Alzheimer's disease, Huntington's disease, and traumatic brain injury. Several ferroptosis activators have also been identified and used in experimental systems, including inhibitors of X_c^- system, direct inhibitors of GPX4, and inducers of lipid peroxidation. These activators of ferroptosis could be useful in the treatment of cancer. Indeed, it is believed that one physiological function of ferroptosis is to suppress the development of tumors in mammals. Emerging evidence suggests that the tumor suppressor p53 may use ferroptosis in its function as a tumor suppressor.

SUMMARY

Death of cells can be triggered accidentally via necrosis or in a regulated manner involving evolutionarily conserved signaling molecules and pathways. Apoptosis, the first and best characterized form of regulated cell death, plays an essential role during embryological development, following birth, and during adulthood. Apoptosis is a regulated by complex mechanisms involving a large number of molecules located in the nucleus, cytosol, ER, mitochondria, and the plasma membrane. Over the past decade, four other forms of regulated cell death have been characterized—autophagic cell death, ferroptosis, necroptosis, and pyroptosis. Autophagic cell death involves the formation of autophagosomes that transport cellular material for degradation by the lysosome resulting in the cell consuming itself from the inside. Necroptosis and pyroptosis are distinct forms of cell death that play a crucial role in protecting cells from infection by microbial pathogens. Ferroptosis results from the iron-dependent buildup of toxic lipid peroxides. In necroptosis, pyroptosis, and ferroptosis, the cell membrane ruptures resulting in the release of inflammatory proteins. Unraveling the mechanisms regulating these various modes of cell death will lead to the identification of drugs that can modulate them, which would have great value in the treatment of a variety of serious human diseases.

Suggested Readings

Anding AL, Baehrecke EH. Autophagy in cell life and death. *Curr Top Dev Biol* 2015;114:67–91.

Cory S, Adams JM. The Bcl2 family: regulators of the cellular life-or-death switch. *Nat Rev Cancer* 2002;2:647–656.

Dhuriya YK, Sharma D. Necroptosis: a regulated inflammatory mode of cell death. *J Neuroinflammation* 2018;15:199.

Jacobson MD, McCarthy N, eds. *Apoptosis*. USA: Oxford University Press, 2002.

Kaufmann SH, Hengartner MO. Programmed cell death: alive and well in the new millennium. *Trends Cell Biol* 2001;11:526–534.

Levine B, Kroemer G. Biological functions of autophagy genes: a disease perspective. *Cell* 2019;176:11–42.

Liu X, Lieberman J. A mechanistic understanding of pyroptosis: the fiery death triggered by invasive infection. *Adv Immunol* 2017;135:81–117.

Scorrano L, Korsmeyer SJ. Mechanisms of cytochrome c release by proapoptotic BCL-2 family members. *Biochem Biophys Res Commun* 2003;304:437–444.

Shan B, Pan H, Najafov A, Yuan J. Necroptosis in development and diseases. *Genes Dev* 2018;32:327–340.

Stockwell BR, Friedman Angeli JP, Bayir H, et al. Ferroptosis: a regulated cell death nexus linking metabolism, redox biology and disease. *Cell* 2017;171:273–285.

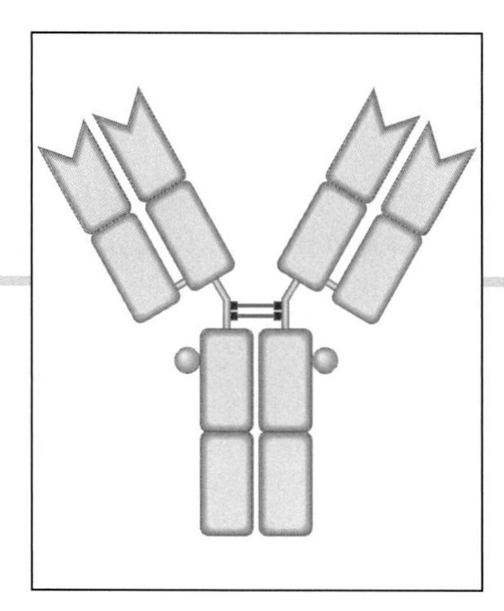

Chapter 12

Cell Biology of the Immune System

INTRODUCTION

Immunity is the ability of an organism to resist infection. The immune system specializes in defense against pathogens. It is composed of a collection of cells and tissues dispersed throughout the body. Once an immune cell or a molecular receptor detects evidence of infection, a signaling process ensues, either within the same cell or relayed to other cells or molecules. This signaling results in the so-called **immune response** that ultimately leads to the formation of **effector immune mechanisms** that destroy the pathogen that triggered the immune response in the first place.

It is customary to divide the immune system into two components, **innate** and **adaptive**, for descriptive purposes. While convenient, this is an artificial divide because the two components work in concert and in an interdependent fashion to provide immunity. Cells of the immune system and their products generate the immune response. Most immune cells are formed in the bone marrow from stem cells that give rise to all blood cells, hence the name pluripotent hematopoietic stem cells (HSCs). Innate immune cells are derived from the myeloid lineage of these stem cells. In contrast, adaptive immune cells are derived from the lymphoid lineage.

Innate Immunity

The first line of defense is **physical barriers** made of specialized tissues covering the whole organism, for example, the skin or lining cavities exposed to the external environment, for example, the respiratory and gastrointestinal tracts. These nonimmune parenchymal cells can also respond to infections by secreting antimicrobial peptides and intercellular signaling soluble molecules known as cytokines, which transmit signals to immune and other cells.

Cells of the innate immune system comprise neutrophils, macrophages, dendritic cells, eosinophils, basophils, and mast cells. **Neutrophils** engulf microorganisms by phagocytosis and mediate microbial killing within intracellular organelles. **Macrophages** are long-lived tissue-resident cells that act as scavengers removing dead cells, debris, and microbes by phagocytosis. Also, they play a critical role in mediating inflammation by releasing inflammatory mediators and chemokines. **Dendritic cells (DCs)** are so-called because their membranes have dendrites similar to those of neurons. They are a class of phagocytic cells that specialize in the initiation of adaptive immunity through their ability to activate **T lymphocytes (a.k.a. T cells)**, one of the types of lymphocytes that play a critical role in the

Goodman's Medical Cell Biology. https://doi.org/10.1016/B978-0-12-817927-7.00012-0

adaptive immune response. **Basophils** and **eosinophils** defend against parasites by discharging the toxic content of their cytoplasmic granules onto them. **Mast cells** are tissue resident, mainly around potential portals of entry of microbes, for example, under the skin and mucosal surfaces. When triggered, they discharge inflammatory mediators, which also contribute to defense against parasites.

The innate immune response is triggered when pathogens, or associated tissue injury, are detected by specialized receptors called **pattern recognition receptors (PRRs)**, which recognize common patterns of microbial structures or metabolism, the so-called **pathogen-associated molecular patterns (PAMPs)**. This recognition event activates innate immune cells to produce effector mechanisms that can directly eliminate the pathogen or signal to other immune cells to amplify the response. Local **inflammation** is one important outcome of innate immune activation. It results from the secretion of **cytokines** and **chemokines**. Cytokines are secreted protein molecules that serve intercellular communication. Chemokines are a subset of cytokines that attract cells to the source of their secretion, a phenomenon known as **chemotaxis**, a process that is important for aggregating immune cells at the site of infection or injury. Cytokines, chemokines, and many other different types of mediators secreted by innate immune cells cooperate to cause inflammation. This comprises vasodilatation enhancing blood flow, increased capillary permeability allowing protein-rich fluid to shift to the tissue, and the extravasation of circulating leukocytes through capillary walls to accumulate at the inflammation site.

Adaptive Immunity

Lymphocytes, the main cellular components of the adaptive immune response, possess two unique features: antigen specificity and immunological memory. An **antigen** is a molecule that can be a target for an adaptive immune response. **Immunological memory** refers to the ability to make faster and stronger responses against antigens that have been encountered before.

There are two types of lymphocytes: **B lymphocytes (a.k.a. B cells)** and **T lymphocytes**. Both have similar but distinct antigen-specific receptors. The B cell receptor (BCR) is a membrane-bound glycoprotein produced by the same gene that encodes for soluble **antibodies** or **immunoglobulins (Igs)**. The T-cell receptor (TCR) differs from Igs in structure and antigen binding. During lymphocyte development the genes for their receptors are generated by a random and irreversible combination of genomic DNA segments. This process allows for an infinitely broad repertoire of antigen specificity. However, each mature lymphocyte carries a unique receptor that is passed on to its progeny, a **clone** of cells sharing the same receptor and descending from a single lymphocyte. Upon recognition of antigen via their specific receptors, lymphocytes become activated, proliferate, and differentiate into different functional subtypes of cells, which bring about the "effects" of the immune response, hence the name **effector cells**.

Both types of lymphocytes originate in the **bone marrow**, where B lymphocytes complete their development. Progenitors of T lymphocytes migrate to the **thymus** to undergo additional maturation. The bone marrow and thymus are the **primary lymphoid organs**. The spleen and lymph nodes dispersed throughout the body are the **secondary lymphoid organs** where lymphocytes aggregate and perform their immune response function. Under physiological conditions, mature lymphocytes circulate through the blood, peripheral tissues of the body, and lymph, constantly scanning for evidence of infection. An adaptive immune response starts when lymphocytes recognize an antigen via their antigen-specific receptor. The BCR can recognize the antigen directly. The T-cell receptor recognizes a fragment of an antigen on the surface of so-called **antigen-presenting cells**, the prototype of which is the **dendritic cell** responsible for initiating all T-cell responses. The antigen-specific receptors of lymphocytes are responsible for the exquisite antigen specificity, first of the two distinguishing features of adaptive immune responses. Following antigen stimulation a lymphocyte clone carrying the same antigen-specific receptor expands by proliferation. Some of the cells also differentiate into so-called **memory cells**, endowed with a lower threshold for activation. The two phenomena combined mediate the second distinguishing feature of adaptive immune responses, namely, **immunological memory**. A distinct lineage of cells that is lymphocyte-like but lacks an antigen-specific receptor comprises **natural killer (NK)** cells and **innate lymphoid cells (ILCs)**. These cells fall between the innate and adaptive types. They serve to amplify certain types of immune responses by secreting specific cytokines. NK cells are also capable of directly killing infected cells.

Lymph is the extracellular fluid that is being constantly drained from parenchymal tissues via afferent lymphatic vessels to lymph nodes and then via efferent lymphatics to the blood. Foreign antigens introduced into the tissues are transferred in soluble form and within dendritic cells to the lymph nodes via afferent lymph. Lymph nodes are encapsulated lymphoid tissues present all over the body along lymphatic vessels, which allow lymphocytes to encounter antigen brought from the peripheral tissues. The spleen is an encapsulated abdominal organ that removes senescent cells and microbial particles from the blood. The spleen is the site of immune responses to blood-borne infections. Lymph nodes and the white pulp of the spleen contain B-cell zones (follicles) and T-cell zones.

INNATE IMMUNE CELL BIOLOGY

Detection of pathogens by **pattern recognition receptors (PRRs)** initiates the innate immune response. PRRs are germline-encoded receptors expressed on innate immune cells or present in soluble forms in the extracellular fluid. PRRs recognize a range of molecular structures characteristic of microbial patterns known as **pathogen-associated molecular patterns (PAMPs)**. Another set of molecular structures detected by innate receptors are self-derived markers of infection, stress, or transformation that are known as **damage-associated molecular patterns (DAMPs)**. Triggering receptors on innate immune cells activates their defense functions: (i) ingestion of microbes (**phagocytosis**); (ii) intracellular **microbial killing**; (iii) release of proinflammatory cytokines and chemokines, thus mediating **inflammation**; (iv) secretion of **antiviral** cytokines, for example, type I interferons; and (v) expression of **costimulatory** molecules that help initiate the adaptive immune response.

Phagocytosis

The main phagocytic cells are macrophages and monocytes, neutrophils, and dendritic cells. Under baseline conditions, macrophages are resident in most tissues as a self-renewing population. During inflammation, circulating monocytes may also give rise to tissue macrophages. Neutrophils possess a greater phagocytic capacity than macrophages but only circulate in blood under the steady state. Neutrophils are the first innate cells to be recruited to sites of infection. There are two main classes of dendritic cells: conventional dendritic cells (cDCs) and plasmacytoid dendritic cells (pDCs). While phagocytic, the main function of cDCs is antigen processing and presentation to T cells. pDCs are the major producers of type I interferons, the antiviral cytokines.

Phagocytosis starts with the recognition of a microbe by certain PRRs on the surface of a phagocyte leading to internalization of the microbe in a large membrane-bound vesicle called the phagosome. Lysosomes fuse with the phagosome to form the phagolysosome, which becomes acidic and acquires antimicrobial peptides and enzymes. Several types of receptors initiate phagocytosis. The first class is the C-type lectins, for example, dectin-1, which binds to fungal beta-glucans and the mannose receptor, which recognizes mannan ligands on bacteria and fungi. The second class of phagocytic receptors is the scavenger receptors, which bind bacterial anionic polymers and lipoproteins. The third type of receptors is Fc and complement receptors, which bind to antibody- or complement-coated microbes, thus facilitating phagocytosis, a process known as **opsonization**.

Microbial Killing

Neutrophils are specialized microbial killers endowed with additional cellular mechanisms to achieve this. They possess **primary granules**, which are membrane-bound intracellular organelles that contain **defensins, elastase**, and other antimicrobial molecules. In their **secondary granules**, neutrophils store an inactive form of **cathelicidin**, another antimicrobial peptide. Upon microbial detection, both primary and secondary granules fuse with the phagosomes where neutrophil elastase, prestored in primary granules, activates cathelicidin by proteolytic cleavage. Active cathelicidin is a cationic amphipathic peptide that disrupts microbial cell membranes. The granules contain other acid hydrolases, for example, cathepsins, which contribute to antimicrobial killing.

In addition to phagocytic receptors, two **G protein-coupled receptors (GPCRs)** (Fig. 9–17, p. 287) directly stimulate microbial killing in neutrophils and other phagocytes. The **N-formylmethionine-leucine-phenylalanine (fMLF) receptor** detects a unique product of bacterial protein synthesis, the *N*-formylmethionine amino acid residue present at the beginning of polypeptide chains in prokaryotes, not in eukaryotes. The **C5a receptor** binds a byproduct of the activation of the complement system. Signaling through these receptors leads to the production of **reactive oxygen species (ROS)** in the phagolysosome and also the attraction of the cells to the source of the receptor-bound ligand, that is, **chemotaxis**.

NADPH oxidase, also known as phagocyte oxidase, is a membrane-associated multicomponent enzyme that generates ROS in phagocytes. In the resting state the enzyme is not assembled and hence inactive. Two components, p22 and gp91, are located in the plasma membranes of neutrophils and macrophages. The other subunits, p40, p47, and p67, reside in the cytosol. When phagocytes are activated, p22 and gp91 move to the membrane of the phagolysosome, where they are joined by the cytosolic subunits to form a fully assembled functional NADPH oxidase. The activity of NADPH oxidase causes a rapid and transient increase in oxygen consumption by phagocytes leading to the generation of superoxide anion (O_2^-) in the lumen of the phagolysosome (the **respiratory burst**). Superoxide dismutase (SOD) converts O_2^- into H_2O_2. Further reactions lead to the formation of other ROS, for example, hydroxyl radical and hypochlorite (Fig. 12–1). ROS are toxic products to both microbes and host cells. Recent evidence suggests that their microbicidal effect is because they lead to potassium cation (K^+) influx into the lumen of the phagolysosomes, which lowers the pH, making it optimum for the antimicrobial enzymes, for example, elastase and cathepsin G. Other microbicidal mechanisms include the formation of reactive nitrogen species, for example, nitric oxide (NO), catalyzed by the inducible nitric oxide synthase. Failure of the neutrophil to perform the respiratory burst due to a genetic deficiency in NADPH oxidase causes chronic granulomatous disease (Clinical Case 12–1).

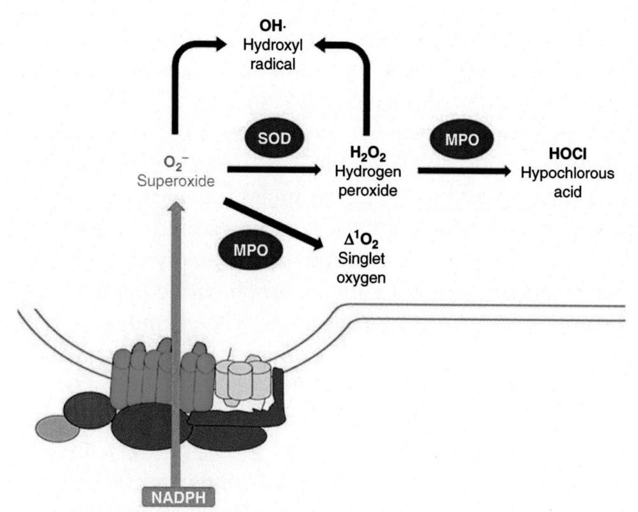

Figure 12–1. Generation of reactive oxygen species (ROS) in the respiratory burst. Activated NADPH oxidase catalyzes the transfer of electrons from NADPH to molecular oxygen, generating superoxide anions (O_2^-). Superoxide dismutase (SOD) converts O_2^- to hydrogen peroxide (H_2O_2). O_2^- is converted to other reactive oxygen species that can damage nucleic acids, proteins, and cell membranes. Myeloperoxidase (MPO) converts H_2O_2 to hypochlorous acid (HOCl), which can enhance the clearance of invading pathogens. MPO can also directly convert O_2^- into singlet oxygen. Components of the NADPH oxidase are gp91phox *(green)*, gp22phox *(light green)*, and regulatory factors *(purple)*. *(Modified from* Front Cell Infect Microbiol *2017;7:373.* https://doi.org/10.3389/fcimb.2017.00373, *Fig. 2.)*

Clinical Case 12–1

Roy Smith is a 4-year-old who has been in the 10th percentile on growth charts over his entire lifetime. He and his father are very close and recently had been creating a new flower bed planting rose bushes and mulching the surrounding area. While handling the rose bush, Roy got a thorn injury of his right forearm. His mother washed the area and put antibiotic ointment on the spot but noticed the area become increasingly red with swelling and warmth. One day later, Roy began to run a fever of 101°F and appeared to be breathing faster and coughing. His mother and father became increasingly concerned and decided to take him to the emergency room. A chest x-ray revealed a necrotizing pneumonia, and a complete blood count (CBC) showed an elevated white blood cell count. Roy began to have increased difficulty with shortness of breath. His pulse oximetry was low at 75 mmHg leading to a decision to intubate him. Roy was started on broad spectrum antibiotics and underwent a bronchoscopy that revealed *Aspergillus fumigatus* infection. Amphotericin B was added to the medical regimen, and surgery was considered to remove the fungal mass in Roy's lung. Blood cultures were also positive for *Staphylococcus aureus*. With a history of recent yard work with mulch that led to an *Aspergillus* lung infection, the physicians became concerned that Roy may have chronic granulomatous disease (CGD) and decided to order a dihydrorhodamine 123 (DHR) test to see how well his neutrophils were functioning, and they also ordered genetic testing for CGD.

Cell Biology, Diagnosis, and Treatment of Chronic Granulomatous Disease

CGD is an inherited primary immunodeficiency disease that results in increased susceptibility to infections caused by certain bacteria and fungi. CGD is caused by mutations in *CYBA*, *CYBB*, *NCF1*, *NCF2*, or *NCF4* genes that code for the five protein components of the enzyme complex NADPH oxidase in neutrophils and other phagocytic cells. Under resting conditions, NAPDH oxidase is not functional. Upon activation of neutrophils the five proteins assemble onto the membrane of the phagolysosome to form a fully functional NADPH oxidase that catalyzes the generation of superoxide anions. This important neutrophil function is known as the respiratory burst because it results in the rapid generation of reactive oxygen species (ROS). Originally, ROS were thought to kill microbes directly. We now know that ROS work indirectly by raising the pH inside the phagolysosome to be optimum for antimicrobial enzymes such as elastase and cathepsin G. CGD is diagnosed by demonstrating a defect in the neutrophil respiratory burst, for example, with the DHR test, and confirmation of a mutation in one of the five genes that encode the NADPH oxidase. The most common form of CGD occurs in 1 in 200,000 persons due to mutations in the *CYBB* gene and is inherited in an X-linked manner. Treatment for CGD is primarily aimed at preventing infections and often includes prophylactic antibiotics such as trimethoprim and sulfamethoxazole to prevent bacterial infections and itraconazole to prevent fungal infections. Interferon gamma may also be used to prevent infections, and bone marrow transplant is used in some cases to cure CGD.

Neutrophils can also contribute to extracellular microbial killing. After infection, some neutrophils undergo a particular form of cell death whereby the nuclear chromatin and DNA come out of the cell intact, instead of being digested in the way that would occur during apoptosis. The extruded material then forms a fibrillar web known as **neutrophil extracellular traps (NETs)**, which entrap microorganisms rendering them more easily phagocytosed by other cells (Fig. 12–2). The generation of NETs by neutrophils requires an intact respiratory burst function.

Recognition of microbes at the site of infection causes **inflammation.** Resident tissue macrophages secrete cytokines that dilate local blood capillaries and induce their endothelial lining to upregulate **cell adhesion molecules.** Consequently, leukocytes move out of circulation into the infected tissue, a process known as **extravasation.** Chemokines released by parenchymal cells and macrophages guide leukocytes to the infection site. Increased capillary permeability shifts plasma fluid and proteins from blood into the tissue. Collectively, all of the aforementioned changes mediate the cardinal signs of inflammation: pain, swelling, heat, and redness.

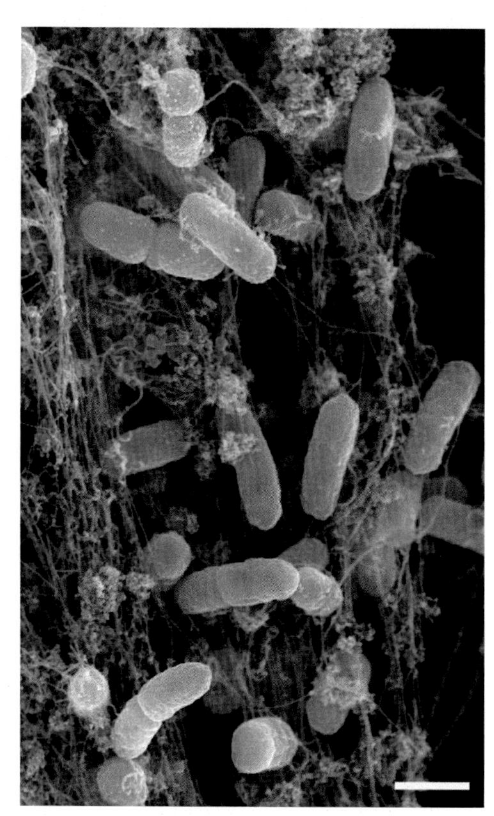

Figure 12–2. Scanning electron micrograph of human neutrophil extracellular traps (NETs) with entrapped *Salmonella*. Bar = 1 μm. *(Modified from Brinkmann V, Zychlinsky A. Neutrophil extracellular traps: is immunity the second function of chromatin?* J Cell Biol *2012;198(5):773–83.* https://doi.org/10.1083/jcb.201203170.*)*

Pattern Recognition Receptors (PRRs)

Toll-like receptors (TLRs) are the prototype PRRs. TLRs are transmembrane proteins. The extracellular domain comprises multiple **leucine-rich repeats (LRRs)** that bind ligands. The cytoplasmic tail contains the signaling **Toll-IL1 receptor (TIR)** domain. TLRs sense microbes in the extracellular space. TLR1, 2, 4, 5, and 6 are located in the outer cell membrane, whereas TLR3, 7, 8, and 9 are in the membrane of the endosomal compartment facing toward the lumen. Each TLR recognizes a particular class of PAMPs (Table 12–1). **Triacyl lipoproteins** of Gram-negative bacteria activate TLR1, 2, and 6. TLR4 recognizes the outer bacterial wall structures **lipoteichoic** acid and **lipopolysaccharide (LPS)** of Gram-positive and Gram-negative bacteria, respectively. Recognition of LPS by TLR4 requires three other helper molecules, two are the soluble proteins MD2 and LPS-binding protein, and the third is cell surface CD14 (CD stands for cluster of differentiation, a nomenclature of leukocyte surface molecules). TLR5 binds to **flagellin**, a repeating protein component of bacterial flagella. The **unmethylated CpG** dinucleotides (abundant in bacterial DNA) trigger TLR9. In mammals, methyltransferases (which bacteria do not have) methylate genomic CpG DNA heavily at cytosine, thus making unmethylated CpG a marker of bacterial DNA. TLR3 binds to double-stranded RNA,

TABLE 12–1. TLRs: Localization, Typical Ligands, and Recognized Pathogens

TLR	Localization	Natural Ligands	Recognized Pathogens
TLR1	Extracellular	Triacyl lipopeptides	Bacteria
TLR2	Extracellular	Lipoproteins, peptidoglycan, LTA, zymosan, and mannan	Bacteria
TLR3	Endolysosomal compartment	dsRNA	dsRNA
TLR4	Extracellular and endolysosomal compartment	LPS, RSV, and MMTV fusion protein, mannans, and glycol-inositol phosphate from *Trypanosoma* spp.	Gram-negative bacteria and viruses
TLR5	Extracellular	Flagellin	Bacteria
TLR6	Extracellular	Diacylipopetides, LTA, and zymosan	Bacteria
TLR7	Endolysosomal compartment	GU-rich ssRNA and short dsRNA	Viruses and bacteria
TLR8	Endolysosomal compartment	GU-rich ssRNA, short dsRNA, and bacterial RNA	Viruses and bacteria
TLR9	Endolysosomal compartment	CpG DNA and hemozoin from *Plasmodium* spp.	Bacteria, viruses, and protozoan parasites

DC, dendritic cell; *dsRNA*, double-stranded RNA; *LPS*, lipopolysaccharide; *LTA*, lipoteichoic acid; *MMTV*, mouse mammary tumor virus; *RSV*, respiratory syncytial virus; *ssRNA*, single-stranded RNA; *TLR*, Toll-like receptor.

Modified with permission from Broz P, Monack DM. Newly described pattern recognition receptors team up against intracellular pathogens. *Nat Rev Immunol* 2013;13:551–65. https://doi.org/10.1038/nri3479.

a byproduct of viral replication. Mutations in the TLR3 gene predispose to recurrent **herpes simplex encephalitis** in children. TLR7 and 8 recognize single-stranded RNA of some viral genomes. Mammalian single-stranded RNA is only present in the nucleus and cytosol, but not in endosomes. The pioneering experiments that led to the discovery of TLRs were carried out by **Hoffmann** and **Beutler** (Nobel Prize, 2011, shared with Steinman for the discovery of dendritic cells).

Ligand binding to TLRs causes their cross-linking and dimerization, which results in bringing the cytoplasmic TIR domains together. This allows their binding to TIR domains of adaptor molecules, for example, **MyD88**. Most TLR signaling activates the **NFκB, interferon regulatory factor (IRF), activator protein 1 (AP-1)**, and **mitogen-activated protein kinase (MAPK)** protein signaling families. The adaptor MyD88 contains two domains, a TIR and a **death domain**. The latter interacts with the IRAK1/IRAK4 complex, which in turn recruits the ubiquitin ligases TRAF6 and UBC13. The polyubiquitin chain thus formed recruits **TAK1**, which phosphorylates the IκB kinase (IKK) complex comprising **IKKα**, **IKKβ**, and **IKKγ** (**NEMO**, for NFκB essential modulator). The IKK complex phosphorylates **inhibitor of kappa B (IκB)**. The NFκB transcription factors p65 and p50 are sequestered in the cytoplasm by IκB under baseline conditions. Phosphorylation of IκB causes its degradation, which permits the heterodimer p65/p50 to translocate to the nucleus where they drive gene transcription of proinflammatory cytokines. **IRAK4 deficiency** is characterized by recurrent bacterial infections. **X-linked anhidrotic ectodermal dysplasia with immunodeficiency** is caused by NEMO mutations. These two inborn errors of immunity highlight the importance of the aforementioned pathway for defense against bacteria.

TLRs that specialize in sensing nucleic acids in endosomes (TLR3, 7, 8, and 9) signal through the IRF family members. TLR3 triggering activates **IRF3**, whereas TLR7 activates **IRF7**. After their phosphorylation, IRF3 and IRF7 move to the nucleus and activate the gene transcription of antiviral cytokines **type I interferons** (**interferon-α** and **interferon-β**). This pathway is particularly important in **plasmacytoid dendritic cells (pDCs)**, the main source of type I interferon during viral infections.

Another protein family of PRRs contains a **nucleotide-binding oligomerization domain (NOD)**; hence, they are termed **NOD-like receptors (NLRs)**. This family of proteins specializes in monitoring the cytosol for bacterial products. In addition to NOD, NLR proteins contain LRR (for bacterial recognition) and CARD (for signaling) domains. Like TLRs, NLRs signal through NFκB. **NOD2** recognizes **muramyl dipeptide (MDP)**, a component of peptidoglycans present on most bacteria. Mutations in the LRR domain of NOD2 compromise the natural defense barrier function of intestinal epithelia resulting in an inflammatory bowel disease known as **Crohn's disease.**

A subfamily of NLRs contains a **pyrin** domain instead of CARD and is called the **NLRP** family, for example, **NLRP3**. NLRP3 is present in the cytosol bound to the chaperone protein HSP90, which keeps it inactive. NLRP3 is triggered by K^+ efflux that may be caused by pore-forming bacterial toxins or nearby cellular death, releasing ATP molecules that stimulate the K^+ channel P2X7. NLRP3 signaling leads to the formation of a multicomponent protein complex known as the **inflammasome**, which consists of multimers of NLRP3, **ASC**, and **procaspase 1**. Activation of the inflammasome results in proteolytic cleavage of procaspase 1, releasing the active enzyme caspase 1, which in turn cleaves the proforms of cytokines **IL-1β** and **IL-18**, thereby releasing the active forms of these inflammatory cytokines. **Gout** is a disease characterized by joint inflammation due to the deposition of monosodium urate crystals, which stimulate the NLRP3 inflammasome causing the release of inflammatory cytokines responsible for the symptoms and signs of the disease. Mutations in NLRP3 are associated with inherited periodic fever syndromes, for example, **Muckle-Wells syndrome**, which is a rare episodic autoinflammatory condition due to the inappropriate activation of the NLRP3 inflammasomes.

RIG-I-like receptors (RLRs), another family of PRRs, sense intracellular viral RNA. They bind to viral RNA with an RNA helicase-like domain. The first member discovered is **retinoic acid-inducible gene I (RIG-I)**, which senses the difference between viral and self RNA by recognizing the uncapped 5′ end of single-stranded RNA transcripts characteristic of virus metabolism. Another member of the RLR family is **MDA-5 (melanoma differentiation-associated 5)**, which senses double-stranded viral RNA transcripts in the cytosol. Sensing of viral RNA activates RIG-I and MDA-5 to stimulate the production of type I interferon through interaction with a downstream adaptor molecule known as **mitochondrial antiviral signaling protein (MAVS)**, which transmits the signal through a chain of multiple signaling proteins resulting in activation of IRF3 and type I interferon production.

A key component of the cytosolic DNA-sensing cellular machinery is the protein **STING, stimulator of interferon genes.** It recognizes cyclic dinucleotides formed by enzymes present in most bacteria leading to the activation of IRF3 and production of type I interferon. STING also recognizes other cyclic dinucleotides, for example, **cyclic guanosine monophosphate-adenosine monophosphate (cGAMP)** made from any cytosolic DNA by the action of a mammalian enzyme **cyclic GAMP synthase (cGAS)**. There are several other cytosolic DNA sensors; their mechanisms of action are currently under investigation.

Innate Cell Activation Induces Adaptive Immunity

Activation of the innate dendritic cells and macrophages via microbial recognition by PRRs triggers the induction of **costimulatory molecules** on these cells. **B7.1 (CD80) and B7.2 (CD86)** are two important costimulatory molecules that bind costimulatory receptors on T cells and activate adaptive immune responses. For years, microbial products have been used in vaccination alongside protein antigens to enhance the adaptive immune response. These products, so-called vaccine adjuvants, for example, LPS, work by inducing costimulatory molecules on resident tissue macrophages and dendritic cells and stimulating the production of cytokines that favor the development of specific types of T-cell immunity. **Steinman** discovered dendritic cells and pioneered experiments to understand their function (Nobel Prize, 2011, shared with Hoffmann and Beutler).

Cytokines are small messenger proteins released by cells that act by binding to specific receptors. Cytokine families are the IL-1 family, the interferons, the TNF family, and the hematopoietin superfamily. The **IL-1 family** members, for example, IL-1β, are produced as inactive proproteins that have to be cleaved to be activated. The IL-1 receptor signals through the NFκB and inflammasome pathways. **Interferons** are either type I, which includes IFN-α and IFN-β, or type II, which is IFN-γ. Their receptors signal through the JAK-STAT pathway. The **TNF family** members are usually trimeric transmembrane proteins, for example, TNF-α. TNF receptors are of two types, TNFR-I, which is expressed widely, and TNFR-II, which is restricted to lymphocytes. The **hematopoietin superfamily** is the largest. It includes GM-CSF, IL-3, and IL-6. Their dimeric receptors often share a common polypeptide chain, for example, the common gamma chain (γ_c) shared by the receptors for IL-2, IL-4, IL-7, IL-9, IL-15, and IL-21. Mutations in the γ_c gene account for 75% of cases of **X-linked severe combined immunodeficiency (SCID)**. The hematopoietin receptors signal through the JAK-STAT pathway.

Cytokine receptors, which signal via the JAK-STAT pathway, are formed of heterodimers. Each chain is noncovalently bound to a different member of the **Janus kinase (JAK)** family. Cytokine receptor binding brings the JAKs closer allowing mutual activation. JAKs phosphorylate the cytoplasmic signaling tails of both cytokine receptor chains allowing for binding by the SH2 domain of a dimer of the **signal transducers and activators of transcription (STATs)**. JAKs further phosphorylate the STAT dimer resulting in its activation and translocation to the nucleus where it induces gene transcription. Each JAK or STAT seems to be associated with a particular cytokine receptor and is involved in a specific function in immunity.

A subset of cytokines, known as **chemokines**, specializes in mediating **chemotaxis** of cells. All chemokines signal through G protein-coupled receptors resulting in changes to the cellular cytoskeleton of responsive cells and a change in the pattern of cell adhesion molecules nearby. These changes combined bring about directed cellular migration. Chemotaxis is important in regulating the function of immune and nonimmune cells. In immunity, it plays an important role in inflammation, adaptive immunity, and immune cell development. Chemokines are classified into two groups based on their structure. As the names imply, **CC chemokines** contain two adjacent cysteine residues, whereas **CXC chemokines** have a single amino acid separating the two cysteines. Many cell types produce chemokines in response to microbes or agents that cause damage, for example, urate crystals. Many substances may act as chemoattractants, for example, the complement activation products C3a and C5a.

During inflammation, leukocytes extravasate from blood vessels to the inflammatory site. **Cell adhesion molecules** play a crucial role in this process as they regulate the interaction between leukocytes and endothelial cells lining the blood vessels. Three families of adhesion molecules are important for leukocyte recruitment and activation, selectins, integrins, and the immunoglobulin (Ig) superfamily. **Selectins** are membrane lectins expressed on activated endothelium. They bind specific carbohydrate moieties, for example, sialyl Lewis^x, on the surface of circulating leukocytes. **Integrins** are expressed on leukocytes and have two protein chains, α and β. Different members have different types of α- and β-chains. LFA-1 is an α_1,β_2-integrin that binds tightly to the Ig-like proteins **intercellular adhesion molecules ICAM-1** and **ICAM-2** on the vascular endothelium. Integrins, for example, LFA-1, become activated (capable of strong binding to its ligand) when cells are activated. Defects in integrins cause leukocyte adhesion deficiencies manifesting as delayed wound healing and recurrent bacterial infections, which highlights the importance of integrins in the inflammatory response.

Extravasation of leukocytes from blood vessels occurs due to signals from the infection site. Neutrophils constitute the first wave of defense cells that go through this process. There are four stages to neutrophil extravasation. First, selectins are induced on the endothelium of blood vessels by inflammatory mediators from the site of infection, for example, histamine, TNF-α, and C5a. Selectins bind to sulfated sialyl Lewis^x on the neutrophil surface, which slows them down, a phenomenon known as **rolling**. Second, LFA-1 on neutrophils is activated by CXCL8 (IL-18) and other chemokines, allowing tight binding between the now activated integrin LFA-1 and its ligand ICAM-1 on the vascular endothelium. This interaction stops rolling and firmly attaches the neutrophil to the endothelium, hence the term **tethering**. The third step involves phagocytes squeezing through the endothelial cells and traversing the endothelial basement membrane. This process is called **diapedesis** and involves LFA-1 and other adhesion molecules, for example, PECAM. In the fourth step, **leukocyte migration** to the site of infection occurs along the extracellular matrix-associated chemokine gradient (Fig. 12–3).

Cytokines secreted by resident tissue macrophages and dendritic cells in response to infection exert their action mainly on the nearby surrounding cells. However, IL-1β, IL-6, and TNF-α act on distant target cells like hormones to initiate the **acute phase response**. They stimulate hepatocytes to increase the production of the acute phase reactants, for example, C-reactive protein (CRP) and mannan-binding lectin (MBL). Both proteins activate the complement cascade upon recognition of PAMPs. Cytokines of the acute phase response also increase the temperature of the body and stimulate the bone marrow to increase the production of leukocytes.

Type I interferons play a crucial role in cellular defense against viral infections. They are mainly produced by pDCs in response to viral nucleic acid recognition by PRRs. Interferons create a condition of resistance to viral infections in cells by activating endonucleases, suppressing the translation of viral RNA, and restricting virus entry into the cytosol. They also enhance chemokine secretion and other immune cell functions.

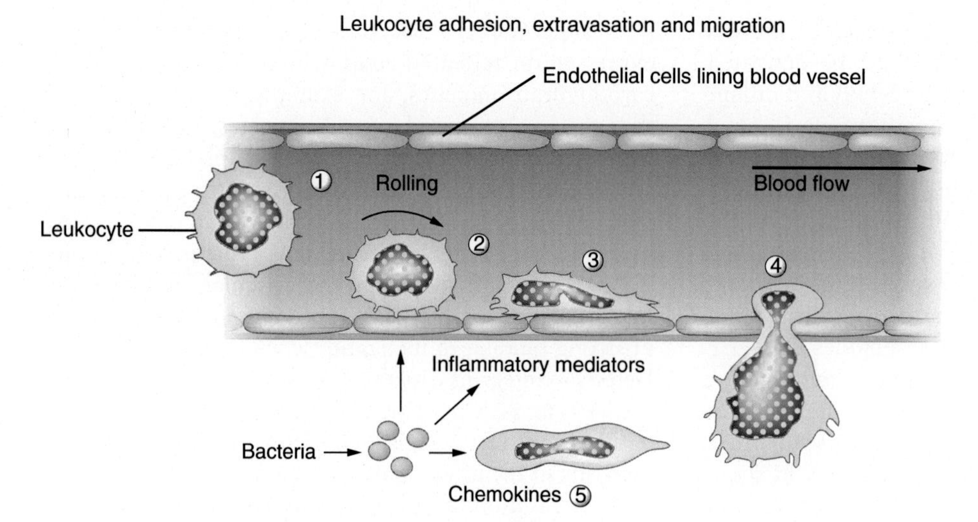

Figure 12–3. Leukocyte adhesion and migration. *(Modified from Garrod DR. Cell to cell and cell to matrix adhesion. In: Latchman D, ed. Basic Molecular & Cell Biology, 3rd ed. Oxford, United Kingdom: Blackwell BMJ Books, 1997:80–91, by permission.)*

Innate Lymphoid and Natural Killer Cells

Innate lymphoid cells (ILCs), including **natural killer (NK)** cells, are cells possessing lymphoid features without a clonal antigen-specific receptor. These cells develop in the bone marrow from a common precursor with B and T cells. ILCs amplify innate signals transmitted by cytokines secreted by macrophages and dendritic cells. ILCs are all resident in tissues and can be classified into three groups. **Group I ILCs (ILC1s)** make INF-γ in response to IL-12 and IL-18 secreted by macrophages and dendritic cells. Thus they help with protection against viruses and intracellular pathogens. **NK** cells are related to ILC1s with unique additional features. NK cells can be found both in tissues and blood circulation, have cytotoxic properties, and require IL-15 for differentiation. ILC2s secrete IL-4, IL-5, and IL-13 in response to thymic stromal lymphopoietin (TSLP) and IL-33. They are important in mucosal and parasite immunity. ILC3s secrete IL-17 and IL-22 in response to IL-1β and IL-23. They contribute to immunity against fungi and extracellular bacteria.

NK cells kill tumors and infected cells. Macrophages and dendritic cells secrete type I interferons and IL-12, which enhance NK cytotoxicity. NK cells induce cell death by apoptosis via three mechanisms. They release onto target cells the content of their cytotoxic granules containing pore-forming proteins, for example, perforin. They use surface tumor necrosis factor-related apoptosis-inducing ligand (TRAIL). They express Fc receptors, which allow antibody-mediated cellular cytotoxicity (ADCC). In response to IL-12 and IL-18, NK cells secrete large quantities of INF-γ. They are particularly important in early defense against virus infections.

NK cells use sets of activating and inhibitory receptors to differentiate between infected and healthy cells. The activating receptors recognize endogenous molecules associated with cellular stress (stressed self). The inhibitory receptors recognize the ubiquitously expressed class I HLA molecules, which are downregulated during infection (missing self), thus releasing the inhibition of NK cell function. The balance of signaling between these two sets of receptors determines whether or not an NK cell will be activated to kill a cell. Structurally different families of receptors regulate NK cell function: killer cell immunoglobulin-like receptors (KIRs), killer cell lectin-like receptors (KCRs), and natural cytotoxicity receptors (NCRs). Each structural family contains activating and inhibitory receptors. NK cell receptors bind to class I HLA, nonconventional class I HLA, and stress-induced class I HLA-like molecules. Inhibitory receptors signal through an **immunoreceptor tyrosine-based inhibition motif (ITIM)** in their cytoplasmic tail. Activating receptors signal through associating with cell surface proteins, which contain an **immunoreceptor tyrosine-based activation motif (ITAM)** in their cytoplasmic tail. NCRs also recognize cell stress-induced molecules, for example, MIC-A and MIC-B.

ANTIGEN RECOGNITION

Antibody Structure

Antibodies (immunoglobulins) are B cell-derived circulating plasma proteins that bind with great strength to and discriminate between different antigens. There are two forms of antibodies: secreted forms in the plasma and extracellular fluid and transmembrane forms, which act as cell surface receptors. An antibody has two **light (L)** and two **heavy (H)** polypeptide chains. The structural unit is the **Ig domain**, about 110 amino acids long. An Ig domain is a sandwich of two β-pleated sheets stabilized together by a disulfide bond. Each sheet has about four antiparallel polypeptide chains connected by loops. Both the antibody heavy and light chains have an N-terminal **variable** Ig domain (V). The light chain has a C-terminal **constant** Ig domain (C). The heavy chain contains at least three C-terminal C domains. Adjacent variable domains of the heavy and light chains (V_H and V_L) form an antigen-binding site. An antibody molecule has two antigen-binding sites. The C region of the heavy chains mediates the functions of the soluble form of antibodies. The C regions of IgA, IgD, and IgG contain three Ig domains, whereas the C regions of IgE and IgM contain four Ig domains.

The antigen-binding part of an antibody molecule is called **fragment antigen binding (Fab)**. The part made up of the C-terminal heavy chain C regions is known as **fragment crystallizable (Fc)**. Three polypeptide loops protrude from the N-terminal part of each V region and form the **complementarity-determining regions (CDRs)**, CDR1, CDR2, and CDR3. CDRs are the most variable stretches of the V regions, hence the name hypervariable regions (Fig. 1-6, p. 5). Antibodies are formed of five classes depending on the type of the heavy chain: IgA, IgD, IgE, IgG, and IgM. Their subclasses are IgA1, IgA2, IgG1, IgG2, IgG3, and IgG4. The heavy chain types are α1, α2, δ, ε, γ1, γ2, γ3, γ4, and μ. Different antibody types undertake different functions based on the interaction between their Fc portion and other cells and molecules of the immune system. There are two classes of light chains, κ and λ, that differ at the C-terminal part. Approximately 60% of antibodies carry κ, and 40% carry λ light chains. The ratio between the two light chains is disturbed in conditions with lymphocyte clonal expansion, for example, lymphoma, a clinically useful observation. All surface antibodies and secreted IgE and IgG are monomers, whereas secreted IgA is a dimer, and secreted IgM is usually a pentamer. The heavy chain of multimeric IgA and IgM contains an additional disulfide-linked polypeptide, the joining (J) chain, which stabilizes the complex and facilitates transport across the epithelial barrier in the case of IgA.

Antigen Recognition by Antibody

Antigens are defined as molecules that can be bound by the receptors of adaptive immunity: B- and T-cell receptors. **Immunogens** are molecules that could stimulate an adaptive immune response. The part of an antigen (usually, macromolecules such as proteins, polysaccharides, lipids, and nucleic acids), which is bound by antibodies, is known as the **antigenic determinant** or **epitope.** Some macromolecules (usually not globular proteins) have repeating identical epitopes (multivalency) that can cross-link the BCR and thus initiate B-cell activation. For proteins, sometimes only the primary structure (amino acid sequence) defines the epitope, but the conformation (tertiary structure) may be important for other epitopes.

The six hypervariable CDRs of the Ig V regions form an antibody's antigen-binding site. The interaction with an antigen occurs through multiple **reversible noncovalent** interactions: electrostatic, hydrogen bonds, van der Waals forces, and hydrophobic interactions. These relatively weak forces mediate tight binding because of the complementarity of shapes (like a lock and key) between the antibody and antigen at the epitope. The strength of binding between the two molecules is called **affinity** and is often measured by the **dissociation constant (K_d)**. When an antibody binds a multivalent antigen, the overall strength of attachment is called the **avidity.** A low-affinity antibody may have a high-avidity interaction with a multivalent antigen, for example, IgM. When antibodies bind antigens, they form **immune complexes,** which can be quite large, depending on the relative concentration of antibodies and antigens. Some immune-mediated inflammatory diseases are caused by large immune complexes coming out of solution and depositing in tissues.

Antigen Recognition by the T-Cell Receptor

Unlike antibodies, T cells cannot recognize antigen directly. T cells recognize fragments of foreign antigen presented on the surface of cells in conjunction with the major histocompatibility complex (MHC) molecules, which are referred to as the human leukocyte antigen (HLA) molecules in humans. In most T cells the T-cell receptor (TCR) is a heterodimer of two polypeptide chains, TCRα and TCRβ. In a small fraction of T cells, two other polypeptide chains form the heterodimer, TCRγ and TCRδ. A disulfide bridge connects the two chains. In contrast to Igs a TCR has only one antigen-binding site and is never secreted. The extracellular portion of the TCR has two V and C domains, analogous to those of Igs (Fig. 12–4).

Two classes of **HLA molecules,** class I and II, share a similar overall structure; two domains are close to the cell membrane, and two are more distal. In both classes the two distal domains fold together into a groove, which binds a peptide fragment of antigen. Class I HLA molecules consist of two polypeptide chains that are noncovalently linked, the **α-chain** (a member of the Ig superfamily comprising three Ig domains: α_1, α_2, and α_3) and **β_2-microglobulin.** The α_1- and α_2-domains form the peptide-binding groove (Fig. 12–5). Class II HLA molecules consist of a noncovalent complex of two membrane polypeptides that also belong to the Ig superfamily,

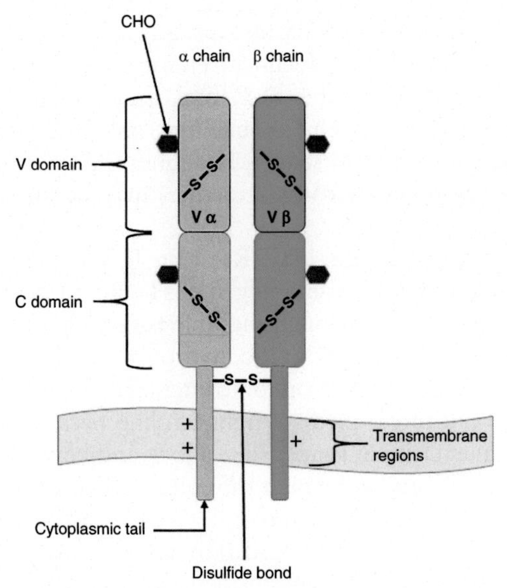

Figure 12–4. A diagram depicting the T-cell receptor structure. Each of the TCR α- and β-chain comprises a V and a C domain. The α- and β-chains are joined by disulfide bonds. Carbohydrate side chains are indicated (CHO). Positive amino acid residues in the transmembrane region are depicted by +. *(Based on Al-Lazikani B, Lesk AM, Chothia C. Canonical structures for the hypervariable regions of T cell alpha-beta receptors.* J Mol Biol *2000;295(4):979–95. Fig. 1.)*

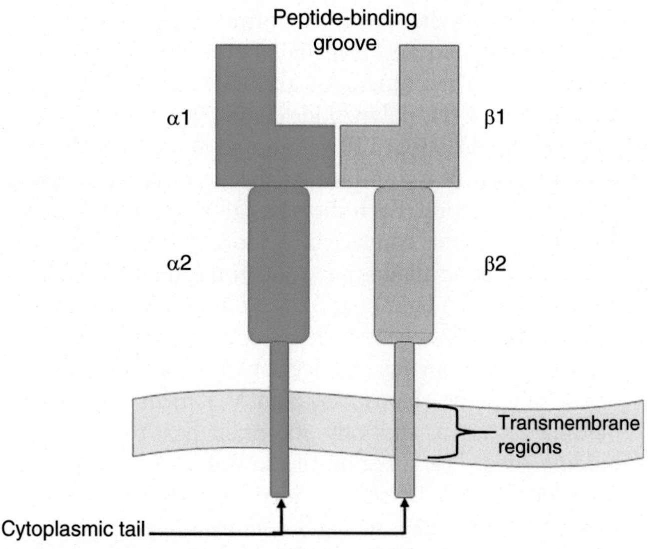

Figure 12–6. A diagram depicting the structure of a class II HLA molecule. Class II HLA molecules are heterodimers composed of two polypeptide chains that span the cell membrane, the α- and β-chains. Each chain comprises two domains, as shown. The α1- and β1-domains form the peptide-binding groove. *(Based on Dessen A, Lawrence CM, Cupo S, Zaller DM, Wiley DC. X-ray crystal structure of HLA-DR4 (DRA*0101, DRB1*0401) complexed with a peptide from human collagen II.* Immunity *1997;7(4):473–81.* https://doi.org/10.1016/s1074-7613(00)80369-6, *Fig. 1.)*

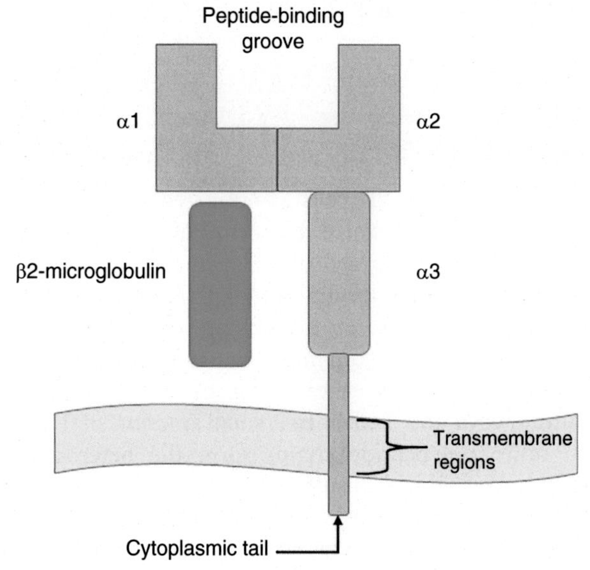

Figure 12–5. A diagram depicting the structure of a class I HLA molecule. Class I HLA molecules are heterodimers of an α-chain and β2-microglobulin. The α-chain comprises three domains, α1, α2, and α3. The α1- and α2-domains form the peptide-binding groove. β2-microglobulin is noncovalently attached to the α-chain, which anchors the structure to the cell membrane. *(Based on* Cell *1992;70(6):1035–48.* https://doi.org/10.1016/0092-8674(92)90252-8, *Fig. 1.)*

α- (different from the class I α chain) and β-chains. The α_1- and β_1-domains fold together to form the peptide-binding groove (Fig. 12–6), as with class I. The TCR interacts with the complex of peptide plus HLA molecules, making contact with ^residues in both. Most polymorphisms in the HLA are within the peptide-binding region affecting the walls of the groove, where the TCR interacts, and the groove, where antigen peptides bind (Fig. 12–7).

HLA molecules bind tightly to a wide array of pathogen and self-derived intracellular peptides. The stability of HLA molecules is dependent on peptide binding. Peptides do not come off HLA molecules except when denatured in acidic pH. This property ensures that when HLA molecules reach the cell surface, they do not loose or exchange peptides, ensuring that the TCR interacts with HLA molecules loaded with an accurately representative sample of intracellular peptides.

T cells have two major subsets differentiated by the surface molecules CD4 and CD8. CD4 is expressed by helper T cells, whereas cytotoxic T cells express CD8. Both molecules bind to nonpolymorphic parts of the HLA molecules away from the peptide-binding groove but simultaneous with the TCR interaction with peptide-HLA complex; hence, CD4 and CD8 are named T-cell **coreceptors**. The interaction of CD4 or CD8 with HLA molecules enhances signaling through the TCR.

Class I HLA molecules are expressed on all nucleated cells, thus enabling CD8 cytotoxic T cells to recognize peptides from intracellular pathogens on any such cell. CD8 T cells can, therefore, specifically kill infected cells. CD4 T cells specialize in helping immune effector cells. Class II HLA molecules are present only on the surface of dendritic cells, B cells, and macrophages, but not on parenchymal cells. When CD4 T cells recognize antigen peptides on class II HLA of B cells, they release cytokines

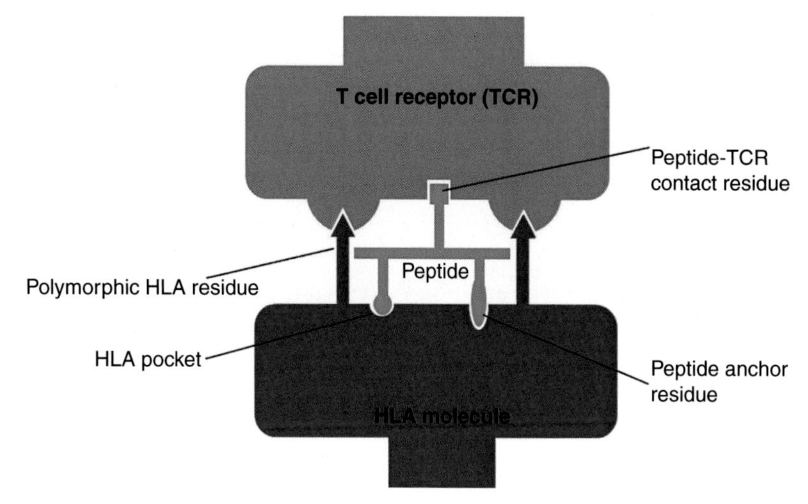

Figure 12–7. A diagram showing the T-cell receptor interacting with peptide and HLA. The peptide lies in the HLA groove anchored by residues that fit snugly into HLA pockets. The TCR makes contact with both peptide and polymorphic regions of the HLA molecule. *(Based on Sundberg EJ, Deng L, Mariuzza RA. TCR recognition of peptide/MHC class II complexes and superantigens.* Semin Immunol *2007;19(4):262–71.* https://doi.org/10.1016/j.smim.2007.04.006, *Fig. 1.)*

that help B cells secrete antibodies. When CD4 T cells recognize peptides on class II HLA of macrophages, they release cytokines that enhance macrophage killing of pathogens.

T cells that bear the γδ TCR recognize antigen directly and not as a peptide bound to HLA molecules, in contrast to αβ T cells. γ/δ T cells recognize self molecules induced by cellular stress as a consequence of infection, rather than pathogen-derived antigens, including nonclassical HLA molecules and some lipids.

ANTIGEN RECEPTOR GENE REARRANGEMENT

Lymphocytes express highly diverse antigen receptors that can recognize a large variety of foreign substances. Lymphocytes generate such receptor diversity during their development before they encounter antigen. Each lymphocyte clone produces a unique antigen receptor. A relatively small number of genes generate a huge set of distinct receptors through a process of random rearrangement of gene segments leading to the formation of a large number of variable region exons.

A unique antigen receptor gene is formed by the fusion of a specific variable (V) gene segment with downstream diversity (D) and joining (J) gene segments. This highly specialized process is known as **V(D)J recombination.** It occurs only in developing B and T cells and not in any other type of cell.

Germline Structure of Immunoglobulin and T-Cell Receptor Genes

Three different loci of genomic DNA encode genes of the Ig heavy chain and κ light and λ light chains. From 5′ to 3′, each Ig gene locus contains a group of V segments, D segments, J segments, and C segments (Fig. 12–8). For example, there are 45 V gene segments in the heavy chain locus, 35 in κ, and 30 in λ. D segments are only present in Ig heavy chain locus. The number of J and C segments varies according to the Ig gene locus. The κ light chain has one Cκ segment. The λ light chain has four Cλ segments. The heavy chain has nine C_H segments for the nine isotypes and subtypes. One exon encodes the C domain of Ig light chains. Five or six exons encode each C_H gene, in addition to two small exons that encode the transmembrane region and cytoplasmic tail of membrane forms of Igs.

The structure of each TCR locus (α, β, γ, or δ) is similar to the Ig loci. D segments are present in the β and δ loci only. The α locus has 45 V and 50 J segments, whereas the β locus has 50 V, 2 D, and 12 J segments. The C region of the TCR has four exons encoding for the extracellular Ig domain, a hinge, a transmembrane portion, and a cytoplasmic tail.

V(D)J Recombination

At a given BCR or TCR gene locus, V(D)J recombination in developing lymphocytes brings together one V genomic DNA segment, one D segment (if present), and one J segment. The specific segment used from each group is selected randomly. This rearrangement results in the formation of a single V(D)J exon that encodes the variable region of an antigen receptor. The rearranged gene is then transcribed into a primary RNA transcript. RNA splicing to form mRNA combines the V(D)J and C region exons. Antigen receptor gene rearrangement occurs differently in each developing lymphocyte, which allows each lymphocyte clone to make a distinct antigen receptor.

Recombination signal sequences (RSSs) are present at the 3′ end of each V gene segment, the 5′ end of each J gene segment, and at both ends of every D gene segment. Each RSS has a very conserved heptamer (CACAGTG, right next to the coding sequence), a nonconserved spacer

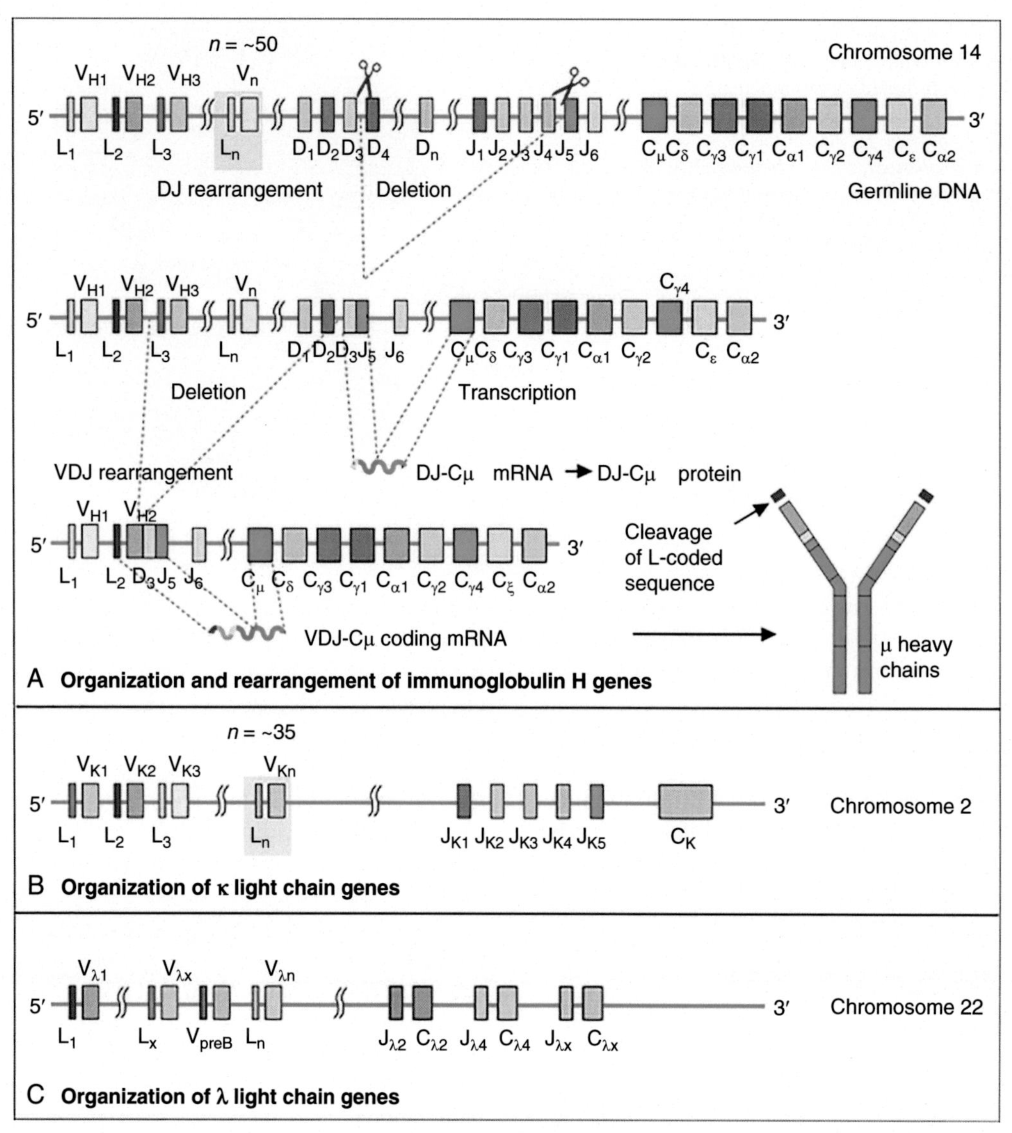

Figure 12–8. Genomic organization of immunoglobulin genes. The structure of germline Ig genes is depicted for the heavy (A) and light chain (B and C) genes. The boxes represent V, D, J, and C DNA segments. Boxes labeled L represent leader sequences. An example is shown of a V(D)J recombination event leading to the formation of the VH_2-D_3-J_5-Cμ coding sequence, with the deletion of intervening DNA, which codes for an IgM heavy chain (μ). *(Data from Burmester G-R.* Color Atlas of Immunology. *New York: Thieme, 2003.)*

that is formed of either 12 or 23 nucleotides, and a conserved nonamer. During V(D)J recombination, double-stranded DNA breaks take place between the RSS and the immediately neighboring coding sequence of gene segments. This DNA break allows a V segment to join either a D or J segment, while the intervening DNA gets removed as a circle (Fig. 12–8). Recombination between two segments occurs only if one of the segments is adjacent to a 12-nucleotide spacer and the other is adjacent to a 23-nucleotide spacer (the 12/23 rule). V(D)J recombination can only take place in Ig and TCR genes because these are the only genes where RSSs are present.

Rearrangement of the antigen receptor in developing lymphocytes is a special case of nonhomologous DNA repair present in all cells conducted by enzymes. The complex of **recombination-activation gene-1 (RAG-1)** and **recombination-activation gene-2 (RAG-2)** is specific to developing lymphocytes and essential for starting the DNA break. ARTEMIS is another enzyme involved in this process. RAG-1/RAG-2 or **ARTEMIS** mutations are responsible for some cases of SCID, highlighting the importance of these enzymes for successful antigen receptor rearrangement.

Several mechanisms contribute to the diversity of antigen receptors in lymphocytes. Different combinations of different gene segments generate different receptors. Removal or addition of nucleotides at V-D, D-J, or V-J junctions during the joining process, so-called junctional diversity, accounts for the majority of diversity in receptors.

ANTIGEN PRESENTATION TO T CELLS AND THE HUMAN LEUKOCYTE ANTIGENS (HLA)

Antigen-presenting cells (APCs) capture foreign antigen at the site of its entry and migrate to the draining lymph nodes. Naïve T cells (never encountered antigens before) recognize antigens presented by APCs and trigger an adaptive T-cell response. HLA molecules on the surface of APCs are essential for antigen presentation to T cells, which cannot recognize antigen directly.

Qualities of T-Cell Antigens and the Role of Antigen-Presenting Cells

Most T cells recognize short peptides derived from foreign protein antigens displayed on the surface of APCs in complex with HLA molecules. This mechanism is known as MHC (or HLA) restriction. A high degree of variation of HLA molecules between individuals (polymorphism) affects peptide binding and TCR interaction. CD4 T cells recognize peptides in the context of class II HLA, whereas CD8 T cells recognize peptides in the context of class I HLA.

Different cell types may function as APCs. DCs are the most efficient APCs at inducing T-cell responses and are the only APCs that can initiate naïve T-cell responses. Macrophages and B cells can also act as APCs, but mostly to previously activated, but not naïve T cells. Microbial products enhance the function of antigen presentation by APCs through recognition of PAMPs by PRRs. This mechanism is how adjuvants of vaccines work. Adjuvants are usually either microbial products or substances that mimic the effect of microbial products on APCs. In the course of antigen presentation, T cells activate APCs by cell contact-dependent mechanisms and secretion of cytokines. For example, activated T cells express CD40L, which cross-links CD40 on APCs. Activated T cells also secrete IFNγ. Both mechanisms combine to increase the efficiency of APCs further following T-cell activation, thereby acting as a positive feedback loop.

Several qualities of DCs enable them to be the most efficient APCs for the induction of primary T-cell responses. DCs are present in almost every tissue and are interspersed with the tissues' parenchymal cells. DCs are present at high density in areas that are likely to be portals of entry for microbes, for example, epithelial cells of the skin and mucosal surfaces. In the tissues, DCs exist in an immature state specialized in antigen capture via receptor-mediated endocytosis and phagocytosis, using receptors such as C-type lectins, and receptor-independent engulfment of surrounding fluids, a process known as pinocytosis (cell drinking). When they encounter microbes, DCs mature by losing their adhesion to surrounding parenchymal cells and upregulating the chemokine receptor CCR7. This maturation allows DCs to migrate inside the lymphatic vessels toward lymph nodes where CCR7 ligands CCL19 and CCL21 are generated in the T-cell zones, the same area where naïve T cells circulate. This common chemokine response maximizes the chances of interaction in the lymph nodes between circulating naïve T cells and antigen-loaded maturing DCs. Mature DCs possess high levels of peptide-loaded HLA and costimulatory molecules, as well as secreted cytokines, which are all required for initiating naïve T-cell responses.

The Human Leukocyte Antigen (HLA) Locus

Early animal experiments have shown that the MHC is not only involved in transplantation rejection but also played a fundamental role in the generation of adaptive immune responses. MHC genes became known as the immune response (Ir) genes because certain class II MHC alleles determined whether or not an immune response occurred. **Zinkernagel** and **Doherty** provided experimental evidence that T cells recognized antigen peptides in conjunction with self MHC molecules (Nobel Prize, 1996), a phenomenon known as **MHC restriction.** They showed that virus-specific T cells responded only to antigen presented by cells from animals that shared the same MHC as the T cells.

MHC genes are the most polymorphic in the genome. This polymorphism allows different individuals in a population to present a different set of peptides from the same protein antigen, ensuring that at least some individuals of the population will be able to respond to any pathogen.

HLA (the human MHC) genes are present on the short arm of chromosome 6. There are three class I HLA molecules, HLA-A, HLA-B, and HLA-C. There are also three class II HLA molecules, HLA-DP, HLA-DQ, and HLA-DR. An individual expresses HLA alleles inherited from both parents, that is, HLA genes exhibit codominant expression. Each chromosome carries a set of HLA alleles that are inherited together as the HLA haplotype of which each human has two.

The pattern of expression of HLA molecules relates directly to their function. Class I HLA molecules are expressed on all nucleated cells. This pattern allows CD8 cytotoxic T cells, which recognize antigen in the context of class I HLA molecules, to kill any cell that might harbor an intracellular pathogen, for example, a virus or a malignant transformation. Class II HLA molecules are expressed on DCs, macrophages, and B cells. This pattern allows CD4 helper T cells, which recognize antigen in the context of class II HLA molecules, to be primed (sensitized) by DCs and initiate a T cell response. When so activated, CD4 T cells can then help B cells to differentiate into antibody-secreting cells and activate macrophages to kill intracellular organisms.

Cytokines can increase the expression of class I and II HLA molecules, which provides a mechanism for both innate and adaptive immune responses to enhance the expression of these important antigen-presenting molecules. Type I interferons produced by many cells as an innate response to viral infections upregulate class I HLA molecules, one of the mechanisms by which innate responses promote adaptive immunity. Similarly, IFNγ induces the upregulation of class II HLA molecules on APCs, thus providing an amplification feedback loop.

Transcriptional regulation is the main factor influencing the levels of HLA molecules on the cells' surface. **Class II transcription activator (CTTA)** acts as a master transcription factor for class II HLA genes as it stabilizes a complex of all other necessary transcription factors at the class II gene promoter. Mutations in CTTA cause the **bare leukocyte syndrome**, a rare form of immunodeficiency in humans in which cells do not express class II HLA molecules, even when stimulated with IFNγ.

The ability of constituent peptides of a protein to bind to HLA molecules determines the immunogenicity of the protein. T cells cannot respond to a protein unless HLA molecules present some of the peptides of said protein. Peptides that bind class I HLA molecules are 8–11 amino acid residues in length, whereas peptides that bind class II HLA molecules are slightly longer reaching 12–16 amino acid residues in length. This difference in bound peptide length between class I and class II HLA molecules is because the antigen peptide-binding groove that each possesses is more open at the end in class II molecules, thus allowing the peptide to dangle out of the groove beyond its floor. The specificity of HLA molecules to peptides is much broader than that of the TCR. For a given HLA-bound peptide, the amino acid residues that bind to HLA are different from those that bind to the TCR. Peptides load onto HLA molecules during their biosynthesis. Class I molecules are loaded with peptides derived from the digestion of cytosolic proteins, whereas class II molecules capture peptides derived from the digestion of antigens from the extracellular milieu in the endosomes. Although noncovalent, peptide-HLA binding is very stable due to a very slow off-rate. The HLA groove wraps around the peptide, and the latter is essential for the stability of the HLA protein. Most of the peptides bound to HLA are derived from self-proteins, yet this does not elicit a T-cell response because T cells are usually tolerant of self-antigens through mechanisms of immunological tolerance. Only a very small proportion of surface HLA molecules will have the same loaded foreign peptide, but this is sufficient to initiate a T-cell response.

Antigen Processing of Proteins

Antigen processing is a cellular mechanism in which protein antigens are digested into short peptides to be presented by HLA molecules on the surface of APCs for recognition by T cells. Antigen processing occurs differently between antigen presented by class I and antigen presented by class II HLA molecules. These differences have a functional significance to the way T cells respond to antigen. Cytosolic proteins are processed for presentation by class I HLA molecules, whereas internalized extracellular proteins are processed for presentation by class II HLA molecules.

For class I presentation, **proteasomes** partially degrade cytosolic microbial and damaged self-proteins (p. 129). Peptides generated in the cytosol by proteasomal degradation of proteins translocate to the endoplasmic reticulum (ER) with the help of **transporter associated with antigen processing (TAP)**, a member of the ABC transporter family of proteins. In the lumen of the ER, newly synthesized class I α- and β_2-microglobulin form a complex with the TAP dimer aided by **tapasin**. When a suitable peptide binds to the cleft of class I HLA molecules, they lose affinity to tapasin, exit the ER, and reach the cell surface. Class I HLA molecules complexed with peptides are very stable. Class I molecules that do not bind peptide are unstable and do not get to the cell surface. On the cell surface, peptide class I HLA molecules present the peptide to CD8 T cells.

For class II presentation, APCs, for example, DCs and macrophages, internalize extracellular protein antigens using antigen capture receptors, such as C-type lectins, Fc receptors for antibodies, C3b receptors, and surface Ig (on B cells). Internalized proteins localize to **endosomes** and microbes are ingested into **phagosomes**. Both endosomes and phagosomes fuse with **lysosomes**, which are highly acidic and contain many proteolytic enzymes. In lysosomes and late endosomes, enzymes, mainly **cathepsins**, digest protein antigens into peptides. During the synthesis of class II HLA molecules in the ER, the **invariant chain (I_i)** occupies the peptide-binding groove so that it does not bind peptides from the ER. The I_i leads newly synthesized class II molecules to late endosomes and lysosomes, where they encounter peptides derived from internalized extracellular proteins. In the endosomes, cathepsins cleave I_i, leaving a short **class II-associated Ii peptide (CLIP)** to occupy the antigen-binding groove. Only high-affinity peptides in the endosomal/lysosomal compartment can then dislodge CLIP from the peptide-binding groove. **HLA-DM**, a nonpolymorphic class II HLA molecule that does not get to the cell surface, facilitates this peptide exchange. The complex of peptide-class II HLA molecules is stable and translocates to the cell surface to present the peptide to CD4 T cells (Fig. 12–9).

As an exception to the aforementioned, DCs are capable of capturing infected and tumor cells and presenting microbial and tumor antigens to CD8 T cells on class I HLA molecules. This pathway is known as **cross-presentation**. Some phagosomes allow antigens to the cytosol where proteasomal degradation and transport of peptides to the ER occur as usual for class I presentation. The phenomenon of cross-presentation is

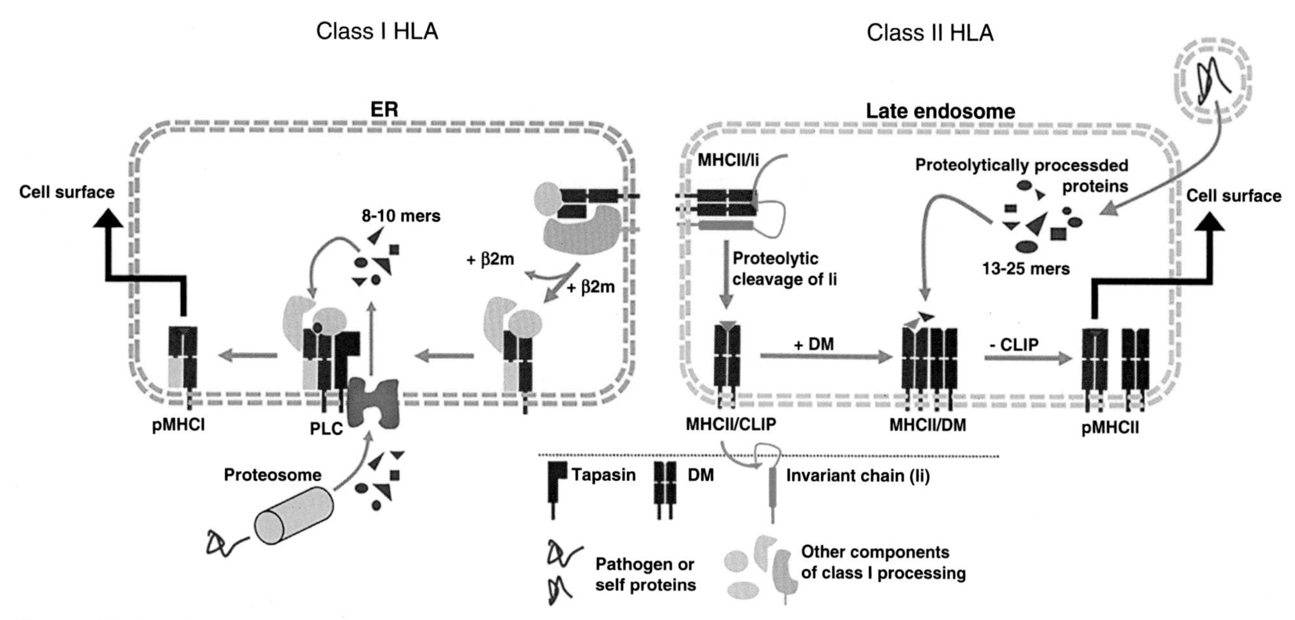

Figure 12–9. Illustration of class I HLA *(left)* and class II HLA *(right)* antigen processing pathways. *CLIP*, class II-associated invariant chain peptide; *ER*, endoplasmic reticulum; and *PLC*, peptide loading complex. *(Data from Wieczorek M, Abualrous ET, Sticht J, Álvaro-Benito M, Stolzenberg S, Noé F, Freund C. Major histocompatibility complex (MHC) Class I and MHC Class II proteins: conformational plasticity in antigen presentation.* Front Immunol *2017;8:292.* https://doi.org/10.3389/fimmu.2017.00292.)

fundamental to allow DCs, the most efficient APCs, to prime CD8 T cells against pathogens that do not infect DCs and against tumor antigens.

The recognition of cytosolic antigens, for example, virus and tumor antigens, by class I-restricted CD8 cytotoxic T cells allows the killing of infected or tumor cells. In contrast the recognition of exogenous antigens by class II-restricted CD4 helper T cells provides help to B cells to make antibodies and activates phagocytosis by macrophages. MHC-restricted CD4 and CD8 T cells may recognize small chemicals, for example, drugs or metals, which modify self-peptides or MHC. The combined chemical/protein is then seen as foreign by T cells. γδ T cells are not MHC-restricted and recognize lipids or modified self-proteins associated with cellular stress. NKT cells recognize endocytosed lipids presented by CD1, a nonclassical class I-like MHC molecule.

SIGNALING THROUGH ANTIGEN-SPECIFIC RECEPTORS

The binding of T- and B-cell receptors to their specific antigen is fundamental to adaptive immunity. However, antigen binding to a receptor is not sufficient for a response. Nonpolymorphic proteins associated with the antigen receptors help to transduce the signal to the inside of the cell.

T-Cell Receptor Signaling

The TCRαβ heterodimer is associated with the CD3 complex formed of CD3γ, CD3δ, CD3ε, and CD3ζ. CD3 subunits are membrane proteins with a short ectodomain and a cytoplasmic tail containing **immunoreceptor tyrosine-based activation motifs (ITAMs)**. Each ITAM has two tyrosine residues. The Src family kinase, **Lck**, is associated with the T-cell coreceptors CD4 or CD8. Lck phosphorylates the tyrosines on the CD3 ITAMs. When both tyrosines of an ITAM are phosphorylated, they recruit **ζ-chain-associated protein-70 (ZAP-70)**, which contains a pair of SH2 domains (one SH2 binds one tyrosine). Lck phosphorylates ZAP-70 when it is bound to an ITAM, resulting in the activation of the kinase domain of ZAP-70. Activated ZAP-70 forms a complex with other proteins and activates **phosphatidylinositol 3-kinase (PI3K)**. Following these critical early interactions after receptor engagement, signaling proceeds along several parallel downstream pathways that lead to transcriptional activation, cytoskeleton reorganization, increased cellular adhesion, and enhanced metabolism.

An important signaling pathway of T-cell activation involves the enzyme **phospholipase Cγ (PLCγ)**, which is activated by phosphatidylinositol triphosphate (PIP_3) generated by the phosphorylation of phosphatidylinositol diphosphate (PIP_2) with the enzyme PI3K. Activated PLCγ cleaves PIP_3 into **diacylglycerol (DAG)** and **inositol triphosphate (IP_3)**, which causes the influx of extracellular Ca^{2+}, **ras**, and **protein kinase Cθ (PKCθ)** activation. Ca^{2+} activates the transcription factors **nuclear factor of activated T cells (NFAT)**. This pathway is a calcineurin dependent and leads to the transcriptional activation of many T-cell genes, especially **interleukin-2 (IL-2)**. DAG activates ras, which in turn activates a series of mitogen-activated protein kinases (MAPKs) ending with **Erk**. The result is the generation of the transcription factor AP-1, which stimulates the transcription of many genes in T cells, especially IL-2. Active PKCθ stimulates TRAF6 resulting in the degradation of IκB and the translocation of NFκB to the nucleus to activate transcription.

Another effect of TCR engagement is increased LFA-1 adhesiveness due to a change in the conformation of this integrin, leading to increased affinity for its ligand ICAM-1. This effect on integrins is mediated by the recruitment of the adaptor molecule **ADAP** and is also dependent on the formation of the **immune synapse**, a region of the T-cell membrane in stable interaction with molecules on the APC. A reorganization of the cytoskeleton, which follows TCR engagement, also contributes to the formation of the immune synapse. PI3K generates PIP_3, which activates **Vav** that in turn changes the conformation of the **Wiskott-Aldrich syndrome protein (WASP)**. Activation of WASP leads to actin polymerization. Mutations in WASP lead to a severe form of immunodeficiency with functional defects in T-cell activation and leukocyte mobility.

The activation of PI3K also results in the phosphorylation and activation of the kinase **Akt,** which in turn phosphorylates several downstream proteins. Akt activation induces cellular metabolism through the activation of several glycolytic enzymes, promotes the antiapoptotic function of Bcl-2 that enhances cell survival, and stimulates the **mammalian target of rapamycin (mTOR)** that enhances anabolism.

B-Cell Receptor Signaling

The antigen-specific Ig heavy chain on the surface of B cells is associated with nonpolymorphic membrane Igα and Igβ proteins. They form a disulfide-linked heterodimer and have ITAMs on their cytoplasmic regions. The Src family kinases Fyn and Lyn phosphorylate ITAMs on the IgαIgβ heterodimer when the BCR is engaged. The tyrosine kinase **Syk** is recruited to the phosphorylated ITAMs through its tandem SH2 domains. Three surface proteins CD19, CD21, and CD81 form the B-cell coreceptor. CD21 binds to the complement fragment C3dg on the surface of microbes and phosphorylates the cytoplasmic tail of CD19, augmenting the signal through the BCR. PI3K contributes to signaling downstream of the BCR like in T cells. **Bruton's tyrosine kinase (Btk)** is a B cell-specific tyrosine kinase that helps PLCγ to catalyze the formation of DAG and IP_3. The deficiency of Btk causes **X-linked agammaglobulinemia** characterized by failure of BCR signaling during B-cell development. BCR signaling shares further downstream pathways with TCR signaling involving Vav, WASP, Akt, and mTOR.

Regulation of Antigen Receptor Signaling

To induce naïve T-cell activation, signaling through the TCR alone is not sufficient. Further signaling through the **costimulatory receptor CD28** is required. The ligands for CD28 are the costimulatory molecules **B7.1 (CD80) and B7.2 (CD86)**, which are highly expressed on mature DCs and other specialized APCs. CD28 is expressed on naïve T cells. When it binds to B7.1 or B7.2 during antigen presentation, it is phosphorylated on its cytoplasmic domain by Lck, leading to the activation of PI3K and the generation of PIP_3. CD28 signaling acts in concert with TCR signaling to maximize PCLγ activation with the recruitment of Akt and Vav activation. One of the major functions of CD28 costimulation is to enhance the transcription of the **IL-2 gene**, essential for T-cell activation and proliferation.

Similar to naïve T cells, naïve B cells also require signals additional to TCR engagement for full activation. **CD40**, a member of the **TNF receptor superfamily**, is expressed by B cells and engages CD40L on activated CD4 T cells. CD40 activates both the NFκB and PI3K pathways. CD40-CD40L interaction maximizes Akt activation mediated by BCR signaling. CD40 also signals through pathways different from the BCR. CD40 recruits the adaptor proteins **TNF receptor-associated factors (TRAFs)**, which serve as E3 ubiquitin ligases. CD40 signaling is uniquely capable of activating the **noncanonical pathway of NFκB**, which is distinct from NFκB activation by the BCR.

Some inhibitory receptors on lymphocytes dampen immune responses by interrupting costimulation. **CTLA-4 and PD-1** are CD28-related molecules expressed by activated T cells. CTLA-4 binds the same B7.1 and B7.2 costimulatory ligands of CD28 but with a higher avidity. Therefore it blocks the interaction between CD28 and B7 ligands and reduces costimulation. Surface expression controls the function of CTLA-4. When not phosphorylated, CTLA-4 binds to clathrin adaptor molecules that remove it from the cell surface. When phosphorylated, CTLA-4 remains on the cell membrane. Regulatory T cells express high levels of CTLA-4, which is crucial for their immunosuppressive function. Targeting the inhibitory receptors CTLA-4 and PD-1 to improve T-cell responses using engineered proteins is a strategy for cancer immunotherapy called **checkpoint blockade**, which was pioneered by **Allison and Honjo** (Nobel Prize, 2018).

Other inhibitory receptors on lymphocytes regulate their function by interfering with receptor phosphorylation. These inhibitory receptors contain in their cytoplasmic tails **immunoreceptor tyrosine-based inhibitory motifs (ITIMs)**. Tyrosine phosphorylation of an ITIM motif recruits phosphatases. **SH2-containing phosphatase (SHP)** is a tyrosine phosphatase, so it reverses the effect of protein tyrosine kinases. SH2-containing inositol phosphatase (SHIP) is an inositol phosphatase, so it reverses the action of PI3K by removing phosphate from PIP_3 to form PIP_2. The inhibitory receptor PD-1 is expressed on activated T cells and interacts with the B7-family members PD-L1 and PD-L2. PD-1 contains an ITIM motif in its cytoplasmic tail. Control of PD-1 expression plays a critical role in the regulation of T-cell responses. The inhibitory receptor, **FcγRIIB** on B cells, binds IgG antibodies resulting in the recruitment of SHIP, which interrupts the action of PI3K and thus blocks signaling downstream of the BCR. **CD22** is another inhibitory receptor on B cells that binds nonmicrobial glycoproteins on the cell surface. An ITIM motif on the cytoplasmic domain of CD22 recruits SHIP, which reverses the action of tyrosine kinases, thereby blocking signaling by the BCR.

LYMPHOCYTE DEVELOPMENT

The maturation of B and T cells occurs in primary lymphoid organs. Progenitor cells derived from the multipotent HSCs commit to the lymphoid lineage. These early lymphocytes proliferate and rearrange the antigen-specific receptor. Following successful receptor rearrangement the selection of clones with a useful receptor takes place. Clones with a rearranged receptor that has the potential to bind self-antigens are eliminated. The lymphocytes then differentiate into subtypes. B cells become B-1 or follicular B cells. T cells differentiate into CD4 and CD8 αβ T cells, γδ cells, NKT, and mucosal-associated invariant T cells (MAIT cells). HSCs give rise to the common lymphoid progenitor cells from which B and T lymphocytes derive. The bone marrow is the primary lymphoid organ where lymphocytes develop in adults, with further maturation of T cells in the thymus.

B-Cell Development

The earliest committed cells to the B-cell lineage in the bone marrow are the **pro-B cells** that express CD19 and CD10, as well as RAG-1 and RAG-2 leading to V(D)J recombination of the Ig heavy chain. Only cells that make productive rearrangements survive and differentiate into **pre-B cells,** which make Ig μ-chains. Pre-B cells express a **pre-BCR** formed by the association of Ig μ-chains with surrogate light chains (similar to κ and λ chains but nonpolymorphic). The pre-BCR associates with the signaling polypeptides Igα and Igβ and drives the proliferation of pre-B cells in a ligand-independent manner. Pre-BCR signaling requires Btk. Successful rearrangement of an Ig heavy chain gene on one chromosome inhibits gene rearrangement on the other chromosome (allelic exclusion), which prevents B-cell clones from expressing more than one BCR. Pre-BCR signaling helps with the rearrangement of Ig light chain genes.

The next stage of B-cell development is the **immature B cell,** which carries a fully assembled IgM receptor. Immature B cells that recognize self-antigen in the bone marrow with high avidity undergo **receptor editing** or cell death by **apoptosis**. Receptor editing involves reactivation of RAG genes and rearrangement of another Ig light chain gene to let the immature B cell express another BCR. High-avidity recognition of self-antigens despite receptor editing leads to apoptosis, also known as **negative selection**. B-cell tolerance of self-antigens is critically dependent on both receptor editing and negative selection. Immature B cells that do not have strong avidity for self-antigens leave the bone marrow as **mature B cells** expressing both IgM and IgD and able to proliferate in response to antigen encounter in peripheral lymphoid organs. Most mature B cells are follicular B cells (B-2 cells). Naïve follicular B cells recirculate through secondary lymphoid organs, for example, lymph nodes, and would die in a few months if they do not encounter an antigen. They enter lymphoid tissues through interactions with high endothelial venules. The B-1 subset of B cells differs from follicular B cells in that their BCR is less diverse. B-1 cells secrete IgM antibodies specific to microbial nonprotein antigens that are called natural antibodies. B-1 cells make the IgM antibodies specific to the ABO blood group antigens.

T-Cell Development

Early bone marrow-derived progenitors enter the thymus via the bloodstream. In the thymus, developing T cells are known as thymocytes, which migrate through thymic stromal cells of the cortex toward the medulla. Thymic stromal cells secrete IL-7, an important growth factor for developing thymocytes. The entry of progenitors into the thymus and migration toward the medulla are controlled by regulating chemokine receptor expression and chemokine gradients. In the first stage of development, thymocytes lack both CD4 and CD8 on their surface; hence, they are called **double-negative** cells. These cells express RAG-1 and RAG-2 and start with the rearrangement of TCRβ locus. A productive rearrangement in a double-negative thymocyte leads to the surface expression of the TCRβ protein chain associated with the nonpolymorphic pre-TCRα chain, thereby forming a **pre-TCR** complex capable of signaling. Pre-TCR signaling drives the expansion of double-negative thymocytes in a ligand-independent fashion, inhibits any further TCRβ rearrangement (β-chain allelic exclusion), and initiates the rearrangement of the TCRα locus. Both the CD4 and CD8 coreceptors reach the surface at the next stage of development of thymocytes, the so-called **double-positive** cells, in which they express the TCRαβ heterodimer. Rearrangement of the TCRα gene deletes the δ-chain locus, thus confirming the commitment of developing thymocytes to the αβ T-cell lineage.

Double-positive thymocytes with a rearranged receptor that can interact with self-peptides complexed with self-HLA will differentiate further into **single-positive** cells (either CD4 or CD8 positive). This process is known as **positive selection**, in which self-peptides play an important role and the affinity of interaction is low. Thymocytes that recognize self-peptide-class I HLA develop into CD8 single positive, whereas those that recognize self-peptide-class II HLA develop into CD4 single positive. Thymocytes with a rearranged receptor that recognizes peptide-HLA with high avidity undergo cell death by apoptosis (**negative selection**) or develop into natural regulatory T cells. Negative selection is the main mechanism of **central tolerance** because it occurs in primary lymphoid organs. In contrast, **peripheral tolerance** occurs in secondary lymphoid tissues. Death of immature self-reactive thymocytes occurs both at the double-positive and single-positive stages. APCs at the thymic cortex are thymic cortical epithelial cells, whereas APCs in the medulla are medullary thymic epithelial cells and bone marrow-derived DCs and macrophages. In medullary thymic epithelial cells, the **autoimmune regulator** (**AIRE**) gene drives a low-level expression of tissue-restricted antigens, for example, antigens specific to endocrine glands. Expression of such antigens in thymic medulla ensures the clonal deletion (negative selection) of any immature T-cell clones with rearranged receptors specific to these antigens. Recognition of self-antigen in the thymus also leads to the generation of natural regulatory T (**nTreg**) cells, which help to maintain immunological tolerance, thus preventing autoimmunity. If a developing thymocyte rearranges the γ and δ chains before TCRβ rearrangement, the cell becomes committed to the γδ lineage. γδ T cells share a common precursor with their αβ counterparts. Finally a small population of NKT cells also develop in the thymus.

T-CELL ACTIVATION AND DIFFERENTIATION

In the cortex of a lymph node, naïve T cells encounter activated APCs loaded with antigens from the local tissue. The two cells bind transiently with the help of adhesion molecules, for example, LFA-1 on T cells and ICAM-1 and ICAM-2 on APCs. This intercellular adhesion ceases if the TCR does not recognize any of the peptide-HLA complexes on the surface of the APC. However, if the TCR does recognize the peptide-HLA determinant for which it is specific, the affinity of LFA-1 increases considerably, stabilizing the adhesion between T cells and APCs.

Activated APCs, for example, mature DCs, provide three different signals to naïve T cells in the course of their activation. The **first signal** is the interaction between the **TCR** and a specific peptide-HLA complex stabilized by the coreceptors CD4 or CD8. The **second signal** is the so-called **costimulatory** signal delivered by members of the B7 family of molecules on APCs through the engagement of the CD28 on T cells. The **third signal** is secretion by APCs of **cytokines** that determine the differentiation of subsets of effector T cells. These three signals together stimulate the expansion and proliferation of T cells and their differentiation into effector T-cell subsets. Activated T cells secrete large amounts of IL-2, which is important for their proliferation.

In addition to CD28, other costimulatory molecules are involved in the activation of T cells. **ICOS** on T cells binds to ICOSL on activated DCs, B cells, and macrophages. ICOS is important for the interaction between T cells and B cells in the germinal center of lymph nodes and for Ig isotype switching. CD70 on activated DCs engaging **CD27** on naïve T cells is a powerful costimulation signal early in the interaction between the two cell types. **CD40L** on T cells interacting with CD40 on DCs or B cells provides important activation signals for T cells. CD40L-CD40 engagement is essential for B cells switching Ig isotypes. The importance of this pathway has been shown by the **X-linked hyper IgM syndrome** in which mutations in CD40L cause the failure of B cells to switch from IgM to IgG.

Once T cells differentiate into effector cells, their activation requirements change such that they no longer depend on costimulatory signals to respond. CD8 cytotoxic T cells are thus able to kill infected cells even if they cannot express costimulatory molecules. Similarly, CD4 T cells can now activate B cells and macrophages even without costimulatory molecules. Another change to effector T cells is that they downregulate L-selectin, so they stop circulating through lymph nodes. Instead, effector T cells increase the expression of ligands of P-selectin and E-selectin and LFA-1, which enables them to roll on the vascular endothelium and extravasate at inflammatory sites.

Cytotoxic CD8 T cells require additional help for the optimum killing of target cells (infected cells or those with malignant transformation). Activated effector CD4 T cells bind to DCs in the course of antigen presentation. CD40L cross-links CD40 on DCs, which in turn stimulates the DCs to upregulate costimulatory molecules, further providing the necessary signals for naïve CD8 T-cell activation that is also enhanced by the IL-2 secreted by activated CD4 T cells.

There are several subsets of CD4 T helper cells. These subsets are T helper 1 (Th1), Th2, Th17, T follicular helper (Tfh), and T regulatory (Treg) cells. Different types of ILCs act in concert with the different Th subsets to ensure that the dominant immune response type is appropriate for the type of pathogen. **Th1** cells are distinguished by the secretion of **IFNγ**, whereas **Th2** cells secrete **IL-4**, **IL-5**, and **IL-13**. **Th17** cells secrete **IL-17A**, **IL17F**, and **IL-22**. **Tfh** cells help B cells switch their Ig isotype class. **Treg** cells regulate the immune response to avoid undue tissue damage from an overexuberant immune response and help maintain tolerance to self-antigens.

Different Th subsets coordinate different types of immune responses to different categories of pathogens. Th1 cells combat **intracellular pathogens**, namely, viruses, protozoans, and intracellular bacteria, for example, *Mycobacterium tuberculosis* and *M. leprae*. INFγ, the main product of activation of Th1 cells, accentuates the ability of the macrophage to kill intracellular pathogens (Clinical Case 12–2). Th1 responses also favor B-cell IgG class switching to opsonizing IgG antibodies. Th2 responses facilitate defense against **extracellular parasites**, especially helminths, by activating eosinophils and mast cells and inducing IgE responses. IgE is a dominant immune mediator of **allergic diseases.** Th17 cells are helpful to clear extracellular **bacteria and fungi** and enhance neutrophil-mediated responses. Th17 cells also help B-cell class switching to opsonize IgG antibodies and improve epithelial cell barrier function. Tfh cells help with clearing **all classes of pathogens**, in collaboration with the Th1, Th2, and Th17, by providing critical help to B cells to differentiate into high-affinity antibody-producing cells in the germinal centers of lymphoid tissue. Tfh cells express CXCR5 and PD-1 on the cell surface. The function of Treg cells is to downregulate the function of other Th cells. Natural Treg (**nTreg**) cells are derived from the thymus, whereas inducible Treg (**iTreg**) cells are derived from peripheral naïve mature CD4 T cells in the periphery.

Clinical Case 12–2

Mary Williams is a 3-year-old Caucasian girl born to unrelated parents. She presented with gradual onset of low-grade fever, night sweating, weight loss, red rash on her arms and trunk, axillary and inguinal lymphadenopathy, and hepatosplenomegaly. She was admitted to hospital for investigation. A complete blood count (CBC) revealed an elevated white blood cell count. Blood cultures on three separate days showed acid-fast bacilli other than tuberculosis, most likely *M. avium* complex (MAC). Mary tested negative for the human immunodeficiency virus (HIV), and she had never received any immunosuppressive drug. Due to her presentation with a disseminated infection with nontuberculous mycobacteria (NTM), in the absence of immunosuppression, Mary's case was discussed with a tertiary referral center to consider a primary immunodeficiency disorder. Sequencing of a panel of genes involved in these disorders revealed that Mary's genome harbored a deleterious homozygous mutation in the IL-12Rbeta1 (*IL12RB1*) gene. Mary's MAC isolates were susceptible to clarithromycin, which her physician prescribed for a year with repeated blood cultures until negative. Her treatment was supplemented with subcutaneous injections of interferon (IFN)-gamma because of her rare genetic condition.

Cell Biology, Diagnosis, and Treatment of Mendelian Susceptibility to Mycobacterial Disease (MSMD)

Mary Williams has a rare primary immunodeficiency disorder called Mendelian susceptibility to mycobacterial diseases (MSMD), an inherited disease caused by genetic defects in the IFN-gamma pathway. Defense against *M. tuberculosis* and nontuberculous mycobacteria (NTM) is critically dependent on a successful interaction between macrophages and T cells. In response to mycobacteria, macrophages produce interleukin-12 (IL-12) that stimulates T cells and natural killer cells via the IL-12 receptor to secrete IFN-gamma. IFN-gamma binds to its receptor on the macrophages leading to their further activation and IL-12 secretion. Activation of the macrophage by INF-gamma results in mycobacterial killing through an unknown mechanism. MSMD patients have disease-causing mutations in one of the following genes: *IFNGR1*, *IFNGR2*, *IL12RB1*, and *IL12B*, mostly inherited in an autosomal recessive fashion. The diagnosis of MSMD should be suspected when a child suffers from recurrent or disseminated NTM infection or disseminated infection with the Bacille Calmette-Guérin (BCG) strain of *M. bovis* in countries where BCG vaccination is used routinely. Failure to detect one of the proteins involved in the IFN-gamma pathway and demonstration of a mutation in a corresponding gene by sequencing confirm the diagnosis. Treatment of MSMD hinges on the use of prompt and prolonged antibiotic therapy of the mycobacterial infection. NTM are widespread in the environment. The specific source of infection is usually unknown. MAC is the most common species of NTM that causes disease. IFN-gamma is an additional treatment option for all MSMD except complete defects of the IFN-gamma receptors where the drug, understandably, does not work.

The differentiation of naïve CD4 T cells into distinct Th cell subsets is cytokine mediated at the time of T-cell priming. Th1 differentiation occurs when the cytokines **IFNγ** and **IL-12** dominate. Activation of STAT1 by interferons and STAT4 by IL-12 is crucial for Th1 differentiation. STAT1 activation induces the expression of the transcription factor **T-bet**, the master regulator of Th1 differentiation, which in turn induces the genes of IFNγ and the inducible IL-12 receptor protein IL12Rβ2. Group I ILCs, for example, NK cells, and Th1 cells themselves secrete IFNγ, which serves to amplify Th1 responses as a positive feedback loop. Th2 differentiation is dependent on **IL-4**. IL-4 receptor signaling activates STAT6, which switches on the transcription factor **GATA3**, the master regulator of Th2 development. GATA3 induces the expression of the IL-4 and IL-13 genes. Group II ILCs, as well as Th2 cells themselves, can also produce IL-4, thus acting again as a positive feedback loop to enhance Th2 responses. Th17 cells develop when the cytokines **IL-6** and transforming growth factor (**TGF-β**) dominate. IL-6 receptor signaling activates **STAT3**. Th17 differentiation also requires **IL-23**. Group III ILCs amplify Th17 responses by secreting IL-17 and IL-22. The master regulator of Th17 development is the transcription factor **RORγt**. Tfh development requires **IL-6**. An important transcription factor for Tfh is **Bcl-6**, which induces the expression of CXCR5, the receptor for the chemokine CXCL13. Stromal cells of lymphoid follicles secrete **CXCL13**, which brings together Tfh and B cells. The interaction between **ICOS** on Tfh cells and **ICOSL** on B cells helps B cells to make antibodies. Tfh cells also secrete **IL-21**, which helps B cells to differentiate into plasma cells.

Regulatory CD4 T (**Treg**) cells are crucial for the regulation of adaptive immune responses. nTreg cells develop in the thymus from thymocytes that rearrange an αβTCR that can recognize self-antigens. The development of iTreg cells depends on the stimulation of naïve CD4 T cells with TGF-β without proinflammatory cytokines, for example, IL-6. In the intestine, APCs secrete **retinoic acid,** which acts with TGF-β to support iTreg development. Both nTreg and iTreg cells express the transcription factor **FoxP3** (master regulator of Treg cell development and function), which blocks the transcription of the IL-2 gene. Treg cells possess high levels of surface **CD25** (the α-chain of the high-affinity IL-2 receptor), which removes IL-2 from the developing naïve T-cell environment. Both nTreg and iTreg cells secrete **TFG-β** and **IL-10**, which are immunosuppressive cytokines. TGF-β inhibits T-cell activation, and IL-10 downregulates HLA and B7 molecules on APCs and inhibits their production of the cytokines IL-12 and IL-23. Treg cells express high levels of CTLA-4, which competes with CD28 for B7 molecules on APCs and could also remove B7 molecules from the surface of APCs.

CELLULAR EFFECTOR MECHANISMS

Effector T cells perform their function through the interaction with a target cell that displays specific antigen on the surface. The TCR, coreceptors, and adhesion molecules form a **supramolecular activation complex (SMAC)** or **immunological synapse** at the site of contact with target cells. The outer region of the SMAC contains cell adhesion and cytoskeletal molecules, for example, LFA-1 and talin. The central region includes the TCR, CD4, CD8, and CD28. The effector molecules secreted in response to signaling by the TCR are concentrated into the intercellular space confined by the SMAC. CD4 T effector cells secrete mainly cytokines (soluble and membrane bound) as their effector molecules, and they act mainly on cells displaying class II HLA molecules and the receptors for such cytokines. CD8 T effector cells secrete both cytokines and cytotoxic substances as their effector molecules.

CD8 cytotoxic T cells kill their target cell upon activation via their TCR by releasing preformed proteins from their granules into the immunological synapse, for example, granzyme and perforin. Granzyme induces target cell apoptosis by activating caspases and damaging mitochondria. Both effects lead to the triggering of the intrinsic apoptotic pathway. Perforin helps deliver granule content into the cytoplasm of target cells by punching holes in the cell membrane as it polymerizes. CD8 cytotoxic T cells kill their target cells quickly and accurately by apoptosis. Activated CD8 T cells also secrete IFNγγ which inhibits viral replication and increases the expression of class I HLA molecules necessary for the presentation of antigen to CD8 T cells.

B-CELL ACTIVATION AND DIFFERENTIATION

Antigens travel from tissues to draining lymph nodes by afferent lymphatic vessels. In the lymph node follicles, naïve B cells differentiate into antibody-secreting long-lived plasma cells and memory B cells when activated by antigen through the BCR (surface IgM and IgD). B-cell activation through the BCR is enhanced by stimulation through the coreceptor CR2 (CD21) and TLRs. There are two types of antibody responses. Protein antigens usually stimulate T-dependent B-cell responses, whereas multivalent nonprotein antigens usually induce a T-independent B-cell response.

In lymphoid tissues, naïve CD4 T cells recognize antigen processed and presented by DCs. Naïve B cells respond to the same antigen in native conformation. The activated T and B cells migrate toward each other because they both upregulate **CXCR5**, so they respond to the gradient of its ligand **CXCL13** released in the lymphoid tissue by **follicular dendritic cells (FDCs)**. Activated B cells process and present the protein antigen to activated Th cells, which in turn provide help to B cells. This phenomenon is known as **T-B collaboration.** During this process, B cells recognize a conformational determinant (epitope) that is linked to a peptide determinant that is presented to T cells by the class II HLA molecules of B cells. A special case of T-B collaboration results in the so-called **hapten-carrier effect.** A hapten is a small chemical that is linked covalently to a protein antigen (carrier). B cells recognize the hapten and present a peptide from the linked protein to Th cells. This linked recognition forms the basis of **conjugated vaccines** in clinical practice where a nonprotein microbial determinant is covalently linked to a foreign protein carrier to enhance vaccine effectiveness.

The **germinal center** is a specialized area of lymphoid follicles in which T-dependent antibody responses occur. In the germinal center, activated B cells proliferate rapidly undergoing affinity maturation and Ig isotype switching and differentiate into long-lived antibody-producing plasma cells and memory B cells. A germinal center has only one or a few B-cell clones. The rapidly proliferating B cells mutate the rearranged V region of their antigen-specific receptor at a very high rate (**somatic hypermutation**), particularly at the antigen-binding CDRs. The enzyme **activation-induced deaminase (AID)** plays a critical role by removing an amino group from cytosine, changing it to uracil at specific mutation hotspots in the V region. Then, by coming into close contact with antigen-bearing FDCs and Tfh cells, B cells with BCRs possessing the highest affinity for antigen are selected. This process is repeated several times and is, therefore, known as **affinity maturation.** The interaction of ICOS and CD40L on Tfh cells with ICOSL and CD40 on activated B cells, respectively, is critical for the formation of germinal centers. FDCs play an important role in the formation of germinal centers. Unlike bone marrow-derived DCs, FDCs are lymph node stromal cells that express high levels of complement and Fc receptors on their surface. Captured by these receptors the unprocessed whole antigen is persistently presented on the surface of FDCs to be recognized by B cells in the course of the germinal center development.

Tfh cells are also essential for germinal center formation. After the early stages of activation, some T cells express CXCR5, which directs them to germinal centers under the influence of the chemokine CXCL13 released by FDCs. In germinal centers, recently activated T cells interact with activated B cells and differentiate into Tfh cells. Tfh cells secrete IL-21, which helps the differentiation of activated B cells into plasma cells.

During T-dependent activation, B cells switch Ig isotype from IgM and IgD to IgA, IgG, IgE, and IgM, which allows antibodies to undertake different effector functions depending on the infective organism. Ig isotype switching requires **CD40-CD40L** interaction, **AID**, and **cytokines**. Tfh-secreted cytokines determine the Ig isotype that B cells switch to, in response to different infections. IgG dominates responses to many bacterial and viral infections, whereas IgE is the main Ig isotype secreted in response to helminthic infections. The anatomical site of the B-cell response also influences Ig isotype switching, for example, IgA is the dominant isotype in mucosal antibody responses. The genetic mechanism of Ig isotype switching is **switch recombination.** It involves cutting the Ig heavy chain DNA at a unique **switch region** at the 5′ end of the Cμ and a similar switch region at the 5′ end of one of the downstream C_H loci (α, ε, or γ). Two enzymes play a central role in catalyzing this double-stranded DNA breaks, AID, and **uracil N-glycosylase (UNG)**. The cellular machinery for DNA break repair by nonhomologous end-joining ligates the two double-stranded DNA ends. This recombination excludes the intervening DNA and brings a previously rearranged V(D)J segment close to the 5′ end of a downstream C_H gene. Cytokines control which C_H region is favored, for example, IL-4 promotes V(D)J rejoining with the Cε gene, which results in the production of IgE with the same V region as that of the IgM previously produced by the B cell before Ig isotype switching.

The result of T-dependent antigen-specific B-cell activation in the germinal center is the formation of long-lived **plasma cells** and **memory B cells.** Plasma cells are terminally differentiated B cells that secrete large quantities of antibodies. BCR and IL-21R signaling help to induce **plasmablasts**, which migrate from secondary lymphoid tissues to the bone marrow, where they differentiate into long-lasting plasma cells. This differentiation involves the development of an elaborate endoplasmic reticulum to help with Ig secretion. Plasma cells also increase the production of secreted antibodies, relative to membrane-bound antibodies, through alternative splicing of Ig heavy chain mRNA. Some activated B cells develop a capacity to survive for a long time and become memory B cells by increasing the expression of the antiapoptotic protein **Bcl-2**. The main feature of memory B cells is the rapid production of a large quantity of high-affinity antibodies upon secondary antigen activation. During the germinal center reaction, B cells express high levels of the transcription factor **Bcl-6,** which helps with rapid cell cycle entry. Bcl-6 also represses p53, which promotes DNA damage-induced apoptosis, thus protecting germinal B cells during Ig isotype switching and somatic hypermutation. **Blimp-1** promotes plasma cell differentiation by repressing Bcl-6, thus switching off the germinal center reaction.

HUMORAL EFFECTOR MECHANISMS

In contrast to cell-mediated effector mechanisms of adaptive immunity, humoral immunity is mediated by secreted antibodies that are specific to foreign antigens. Antibodies deliver their effector function through the Fc portion. Different isotypes of the Ig heavy chain perform different effector functions. IgG **opsonizes** microbes for phagocytosis by neutrophils and macrophages through binding Fcγ receptors. Furthermore, IgG **neutralizes** microbes and their toxins by blocking binding to their cellular receptors. NK and other cells recognize cells coated with IgG with Fcγ receptors and kill them via **antibody-dependent cellular cytotoxicity** (ADCC). IgG and IgM aggregate in immune complexes with antigen to activate the **classical pathway of complement.** IgA contributes to mucosal immunity by neutralizing microbes and their toxins in the lumen of the gut. IgE induces mast cell degranulation and helps with eosinophil-mediated defense against helminths. Effector functions of antibodies are induced only upon antigen binding to ensure this solely occurs when required.

Fc receptors on leukocytes bind the constant part of Ig heavy chain and are classified into three types depending on the Ig heavy chain they bind, Fcγ receptors (**FcγRs**), the Fcα receptor (**FcαR**), and Fcε receptors (**FcεRs**). The FcγRs are FcγRI (high affinity), FcγRII (medium affinity), and FcγRIII (low affinity). Most FcRs are activating, except for **FcγRIIB**, which is **inhibitory** and provides a negative feedback loop to regulate myeloid and B-cell function. It is through this mechanism that high-dose intravenous Ig infusion helps to treat some autoimmune diseases. Parasitic worms are too large to be engulfed by phagocytosis. However, mast cells and eosinophils use the high-affinity FcεRI to recognize IgE-coated helminths and kill them via the discharge of their toxic granule content.

THE COMPLEMENT SYSTEM

The complement system is a set of plasma proteins that plays an important role in immunity. When triggered, complement components initiate a sequential proteolytic amplification reaction that rapidly leads to an inflammatory response, opsonization of microbes, and lysis of cell membranes of pathogens. Immune cells detect complement activation by recognition of cleaved complement fragments with complement receptors. There are three pathways of complement activation, the **classical, lectin,** and **alternative** cascades. Antigen binding by **IgM or IgG** (immune complex) triggers the classical pathway. **Mannan-binding lectin (MBL)** or ficolins activate the lectin cascade. **Surface structures** of all pathogens and **properdin (P)**, which binds pathogens, can activate the alternative pathway. The central reaction in complement activation that is shared by all three cascades is the cleavage of C3 to two fragments: **C3a** (small) and **C3b** (large). This enzymatic cleavage exposes a highly reactive thioester group on C3b and other sites for interaction with receptors and complement regulatory proteins.

Complement Pathways and the Membrane Attack Complex

The **classical complement** pathway is initiated by IgM or IgG binding antigen (immune complex), which results in a conformational change in the Fc portion revealing the complement binding site. C1q binds the immune complex, which triggers the sequential activation of C1r and C1s. Activated C1s further cascades the cleavage of the downstream components C2 and C4, forming **C4b2a**, the **classical pathway C3 convertase**, which catalyzes the central reaction of the complement pathway (Fig. 12–10). The next step involves the formation of C4b2a3b, the **C5 convertase**. The resulting C5b triggers the terminal lytic complement complex known as the **membrane attack complex (MAC)**. The **lectin pathway** is initiated by pathogen recognition with MBL or ficolins. This recognition activates **MBL-associated serine proteases (MASPs)**, which cleave C2 and C4, forming C4b2a, the C3 convertase, as in the classical pathway. In the **alternative pathway**, spontaneous hydrolysis of C3 occurs at a low level. Hydrolyzed C3 binds to factor B, which is cleaved by factor D and forms the **alternative pathway C3 convertase, C3bBb**. This low-grade activation process is controlled at cell surfaces and tissues by complement regulatory proteins, especially **factor H**. Absence of such regulation on microbial surfaces allows the alternative pathway C3 convertase to amplify the cascade. The **alternative pathway C5 convertase** is formed by the addition of a second C3b to the alternative pathway C3 convertase, **$(C3b)_2Bb$**. Both of the C3 and C5 convertases of the alternative pathway are stabilized by **properdin (P)**, which also functions as a pattern recognition molecule. All three complement pathways converge with the cleavage of C5 into C5a and C5b. C5b triggers the assembly of the MAC, which is a complex of C5b, C6, C7, C8, and many C9 molecules. The MAC pierces the surface membrane lipid bilayer disrupting the osmotic balance and causing lysis of cells. This complement-dependent cell lysis is one of the antimicrobial mechanisms of complement. Complement activation contributes to antimicrobial defense by multiple other mechanisms. **C3a** and **C5a** are called **anaphylatoxins** because they are potent proinflammatory mediators. C3b and C4b are also potent **opsonins**.

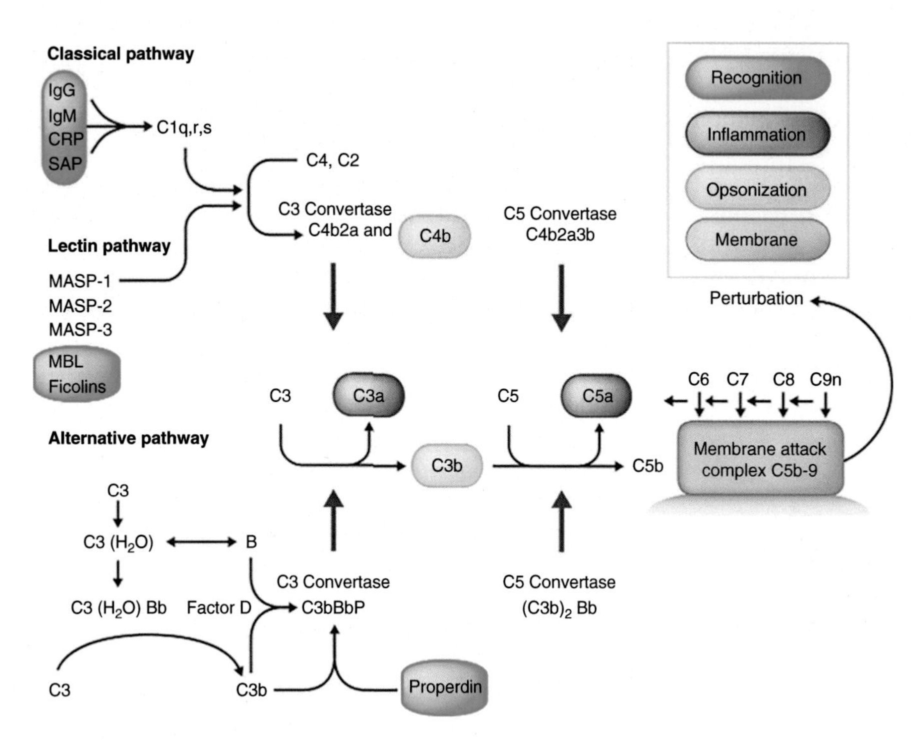

Figure 12–10. A representation of the complement pathways. An overview diagram shows the classical, alternative, and lectin complement pathways, highlighting the stages of recognition, inflammation, opsonization, and membrane lysis. *(Data from Rich RR, Fleisher TA, Shearer WT, Schroeder Jr, HW, Frew AJ, Weyand CM.* Clinical Immunology: Principles and Practice, *5th ed. Philadelphia: Elsevier, 2018. ISBN: 978-0-7020-6896-6, Fig. 21.1.)*

Complement Regulation and Receptors

C1 esterase inhibitor (C1-INH) is a plasma serine protease inhibitor that irreversibly inactivates C1r and C1s, thereby controlling the initiation of the classical pathway. C1-INH also inhibits MASPs and the kallikrein system. Inherited deficiency of C1-INH causes **hereditary angioedema,** a condition that manifests as recurrent episodes of soft tissue edema. **Factor I** inhibits C3 and C5 convertases by cleaving C3b and C4b. **Factor H** is a fluid-phase inhibitor that accelerates the decay of the convertases. The membrane proteins **CD55** and **CD46** protect cells from complement-mediated lysis by helping to inhibit C3 and C5 convertases. **CD59** is the primary inhibitor of the MAC. CD59 binds C9 preventing its assembly into the MAC. **Complement receptor 1 (CR1 or CD35)** is a large protein that binds to C3b, C4b, and C1q. Erythrocytes express CD35, which provides the main pathway of transporting soluble immune complexes in the blood to the liver and spleen where they are removed and destroyed by macrophages. On monocytes and neutrophils, CD35 facilitates phagocytosis of organisms (**opsonization**) coated with C3b and C4b. B cells, FDCs, and some epithelial cells express **CR2 or CD21,** which binds to small complement fragments. CD21 is also the Epstein-Bar virus receptor and acts as a B-cell costimulator. **CR3** is **CD11b/CD18,** and **CR4** is **CD11C/CD18.** Both are β_2-integrins expressed on neutrophils, monocytes, and NK cells. CR3 and CR4 bind small complement degradation fragments and provide an essential function of microbial removal after complement activation. C3a and C5a are complement fragments generated during activation of the complement cascade. They have potent proinflammatory and chemotactic effects. C3a and C5a release vasoactive amines from mast cells and basophils, which leads to increased vascular permeability. C5a is a chemoattractant to neutrophils, monocytes, and macrophages. C3a is a specific chemokine for mast cells and eosinophils. The **C3aR** and **C5aR (CD88)** are seven-transmembrane G protein-coupled receptors. C3aR is expressed on mast cells and basophils. In contrast, C5aR is expressed on neutrophils, monocytes, and macrophages.

IMMUNOLOGICAL TOLERANCE

Immunological tolerance is the state of lack of immunological responsiveness to an antigen. Tolerance to self-antigens is a key physiological feature of a normal immune system. Failure of tolerance to self results in autoimmunity. **Central tolerance** develops during the development of immature lymphocytes in primary lymphoid organs (thymus and bone marrow). Immature developing T cells in the thymus that have high-affinity receptors to self-antigens they encounter in the thymus die by apoptosis (central tolerance by negative selection). Similarly, immature B cells developing in the bone marrow that have a high affinity to self-antigens also die by apoptosis. Unlike with T cells, immature B cells with a high-affinity receptor to self-antigens can also edit the receptor and so change the specificity. **Peripheral tolerance** is the immunological unresponsiveness of mature lymphocytes to self-antigens in the peripheral tissues. Mature CD4 T cells become unresponsive when they receive antigen stimulation in the absence of costimulatory signals or when the inhibitory receptors are engaged, for example, CTLA-4 and PD-1. Both nTreg and iTreg cells contribute to peripheral tolerance by inhibiting immune responses through several mechanisms. Mature B cells become unresponsive when they recognize antigen in the peripheral tissues in the absence of help from T cells. Failure of immunological tolerance to self-antigen is the fundamental cause of autoimmune diseases. Administration of antigen can induce tolerance rather than immunity under certain conditions. This observation may form the basis of antigen-specific treatment of some autoimmune diseases.

LYMPHOCYTE RECIRCULATION

Lymphocyte recirculation is a physiological immune mechanism that ensures a maximum chance for naïve B and T cells to encounter antigen and initiate an immune response. Naïve lymphocytes move from the blood to secondary lymphoid tissues, through lymphatics back into the blood, and then move to secondary lymphoid tissues again and so on continually. Naïve lymphocytes express high levels of L-selectin, which binds to addressins on the high endothelial venules of lymphoid tissues, thus allowing migration from blood to the lymphoid tissues. The expression of the chemokine receptor CCR7 on naïve lymphocytes helps with their migration from the blood into lymphoid tissues in response to the chemokines CCL19 and CCL21 that are released there. After the activation of naïve into effector lymphocytes, the latter leave the lymphoid tissues via the sphingosine 1-phosphate (S1P)/S1P receptor pathway and return to the blood. Effector lymphocytes downregulate CCR7 and L-selectin but upregulate the ligands of P-selectin and E-selectin, chemokine receptors, and integrins, molecules that mediate extravasation through the endothelium at sites of infection or inflammation.

IMMUNOLOGICAL HYPERSENSITIVITY REACTIONS

The aberrant activation of immune cells is the fundamental mechanism of pathology in all immune-mediated inflammatory diseases and results in the inappropriate discharge of effector immune mechanisms (or immune responses). These mechanisms are collectively known as immunological hypersensitivity reactions and can be classified into four types. IgE is the main mediator of **type I (immediate) hypersensitivity,** the dominant mechanism of allergic disease. In type I hypersensitivity, Th2 cells and their cytokines (IL-4, IL-5, and IL-13) play a dominant role. In **type II (antibody-mediated) hypersensitivity**, IgG and IgM specific to antigens in the solid phase (e.g., cell surface or extracellular matrix) cause tissue damage by activating complement and inducing inflammation. In **type III (immune complex-mediated) hypersensitivity**, IgG and IgM specific to soluble antigens in the fluid phase form large immune complexes that deposit in tissues causing inflammation and injury. In **type IV (cell-mediated) hypersensitivity**, CD4 T-cell activation induces inflammation, and CD8 T cells directly kill target cells. Th1 cells and type I cytokine (e.g., IFN-γ) responses play the dominant role. This purposefully simplified classification helps the understanding of the immunological mechanisms of disease.

SUMMARY

Cells of the immune system have adapted their biology for their specialist function of defense against infection. Innate immune cells, such as macrophages and dendritic cells, act as sentinels for the early detection of pathogens with germline-encoded pattern recognition receptors. Macrophages provide fast defense through phagocytosis and microbicidal activity against ingested organisms. Innate immune cells also mediate inflammation through the secretion of cytokines and chemokines that influence other cells through binding to specific receptors. Activation of the complement system results in the direct killing of microbes and the amplification of inflammation. The initial encounter with a pathogen triggers dendritic cell maturation, which initiates naïve T-cell responses, thereby linking innate and adaptive immunity. Adaptive immune cells recognize antigen with the hugely diverse receptors formed during lymphocyte development by a unique process of V(D)J recombination. CD4 T cells activate macrophages and help B cells to make pathogen-specific antibodies. In the extracellular space, antibodies neutralize microbes and their toxins, enhance phagocytosis, and potentiate antimicrobial defense by activating complement. CD8 T cells and natural killer cells eliminate intracellular pathogens by killing the infected cells that harbor them. B cells and antibodies recognize antigen in its native conformation. In contrast, T cells recognize an antigen fragment combined with human leukocyte antigen molecules, only after processing and presentation by antigen-presenting cells. Different effector mechanisms dominate immune responses depending on the invading microbes and their portal of entry. Unlike innate immunity, adaptive immune responses possess exquisite antigen specificity and immunological memory, thus providing long-term protection against microbes after the initial infection. Tolerance mechanisms tightly regulate the immune system to avoid reactivity against self-antigens. Failure of immunological tolerance may cause autoimmune disease. Major defects in immune cell function may lead to severe forms of immunodeficiency, which could be fatal. Immunological mechanisms underpin the pathology of many human diseases.

Suggested Readings

Abbas AK, Lichtman AH, Pillai S. *Cellular and Molecular Immunology*, 9th ed. Philadelphia: Elsevier. 2018. ISBN 978-0-323-47978-3.

Murphy K, Weaver C. *Janeway's Immunobiology*, 9th ed. New York: Garland Science, Taylor & Francis. 2017. ISBN 978-0-8153-4505-3.

Punt J, Stranford S, Jones P, Owen JA. *Kuby Immunology*, 8th ed. Macmillan Learning. 2018. ISBN 978-1-319-11470-1.

Rich RR, Fleisher TA, Shearer WT, Schroeder Jr. HW, Frew AJ, Weyand CM. *Clinical Immunology*. Principles and Practice. 5th ed. Philadelphia: Elsevier. 2018. ISBN 978-0-7020-6896-6.

Rudolph MG, Stanfield RL, Wilson IA. How TCRs bind MHCs, peptides, and coreceptors. *Annu Rev Immunol* 2006;24:419–466.

Samelson L, Shaw A. (eds). Immunoreceptor Signaling. Cold Spring Harbor, NY: Cold Spring Harbor Laboratory Press, 2010.

Schatz DG, Swanson PC. V(D)J recombination: mechanisms of initiation. *Annu Rev Genet* 2011;45:167–202.

Starr TK, Jameson SC, Hogquist KA. Positive and negative selection of T cells. *Annu Rev Immunol* 2003;21:139–176.

Trombetta ES, Mellman I. Cell biology of antigen processing in vitro and in vivo. *Annu Rev Immunol* 2005;23:975–1028.

Zhu J, Yamane H, Paul WE. Differentiation of effector CD4 T cell populations. *Annu Rev Immunol* 2010;28:445–489.

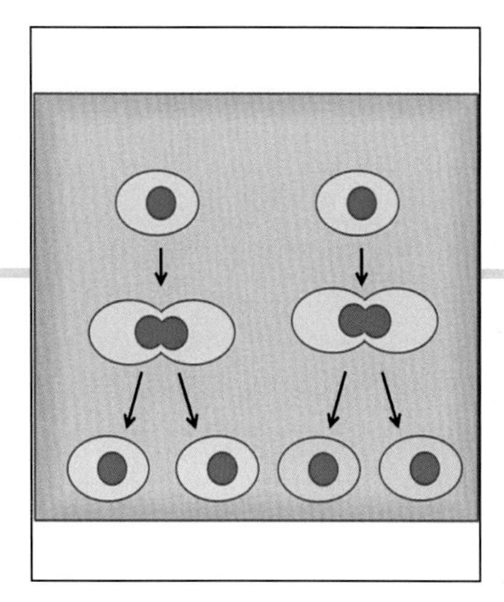

Chapter 13

Stem Cells and Regenerative Medicine

Stem cells, generally defined as clonogenic cells that are capable of both self-renewal and multilineage differentiation, are units of biological organization responsible for the development and regeneration of tissue and organ systems. In terms of their difference in self-renewal ability, stem cells can be divided into two subsets: one is a long-term self-renewal subset with the capacity of indefinite self-renewal, and the other is a short-term self-renewal subset self-renewing for a defined interval. The earliest stem cells in a developing embryo are totipotent cells, giving rise to oligolineage progenitors, which generate progeny that are more restricted in their differentiating potential, and finally become functionally mature cells. In adult organisms, various stem cells, such as mesenchymal stem cells (MSCs) and hematopoietic stem cells (HSCs), act as the body's natural reservoir, replenishing stocks of specialized cells that have been used up or damaged. Besides these naturally existing stem cells in our bodies, differentiated cells can be reprogrammed to an embryonic-like state by transfer of nuclear contents into oocytes or by fusion with embryonic stem cells (ESCs). This chapter provides examples of different types of stem cells, as well as their application in regenerative medicine and disease therapeutics.

TYPES OF STEM CELLS

Embryonic Stem Cells

ESCs are pluripotent stem cells derived from the inner cell mass of the blastocyst, an early-stage embryo. ESCs are distinguished by their ability to differentiate into any cell type and to self-renew indefinitely. These traits make them valuable in the scientific/medical fields.

Human embryos reach the blastocyst stage 4–5 days postfertilization, consisting of 50–150 cells. ESCs of the inner cell mass are capable of differentiating to generate primitive ectoderm, which ultimately differentiates during gastrulation into all derivatives of the three primary germ layers: ectoderm, endoderm, and mesoderm (Fig. 13–1). These include each of more than 220 cell types in the adult human body. Under defined conditions, ESCs are capable of propagating indefinitely in an undifferentiated state. Conditions must either prevent the cells from clumping or maintain an environment that supports an unspecialized state. ESCs also have the capacity, when provided with appropriate signals, to differentiate into nearly all mature cell phenotypes. In this way, ESCs distinguish themselves from adult stem cells, which are multipotent and can only produce a limited number of cell types (Fig. 13–2).

Goodman's Medical Cell Biology. https://doi.org/10.1016/B978-0-12-817927-7.00013-2

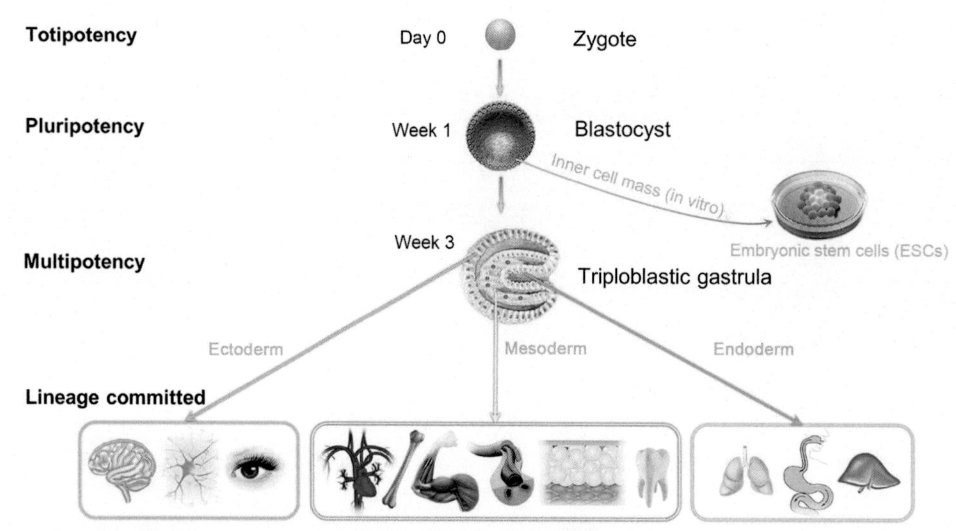

Figure 13–1. The development process from zygote to three germ layers. Through mitosis, the zygote undergoes multiple stages (morula, blastocyst, and triploblastic gastrula) to form an individual. ESCs are derived from the inner cell mass of blastocysts, which will develop into embryos. Trophoblasts are derived from the outer cell mass, which will develop into placenta. Three embryonic germ layers, including ectoderm, mesoderm, and endoderm, are formed in the gastrula period. The formation of these three germ layers lays the foundation of organs and system development. As the embryo develops, cell totipotency decreases. A small number of mesenchymal stem cells maintain their multipotency in the differentiated tissues.

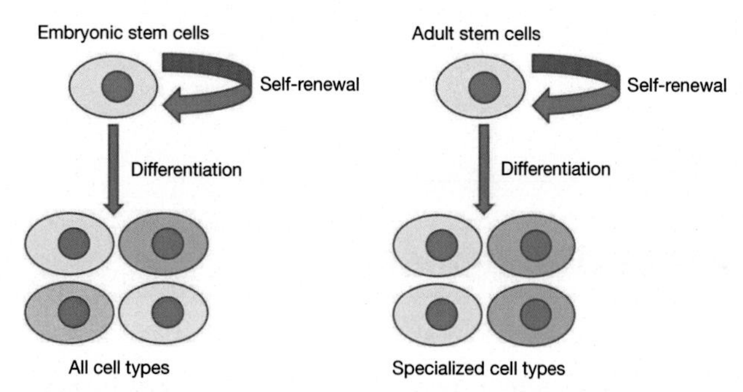

Figure 13–2. Difference in the differentiation potential between ESCs and adult stem cells. ESCs are able to differentiate into nearly all types of mature cells, while adult stem cells can only produce a limited number of cell types that are specific for their organ of origin.

Because of their plasticity and potentially unlimited capacity for self-renewal, ESC therapies have been proposed for regenerative medicine and tissue replacement after injury or disease, which may serve as a possible solution to the donor shortage dilemma. In addition, tissue/organs derived from ESCs are immunocompatible with the recipient. Aside from these uses, ESCs can also serve as tools for the investigation of early human development, the study of genetic disease, and as in vitro systems for toxicology testing.

Adult Stem Cells

Adult stem cells, also known as somatic stem cells, are undifferentiated cells, found throughout the body after development. They multiply by cell division to replenish dying cells and regenerate damaged tissues. Unlike ESCs, they can be found in almost every organ in children as well as in adults. Adult stem cells are not pluripotent like ESCs, but they are multipotent. They can only generate a limited number of specialized cell types that are specific for their organ of origin. Neural stem cells, for example, can only differentiate into brain cells, whereas blood stem cells can only form specialized cells of the blood system.

Scientific interest in adult stem cells is centered on their ability to divide or self-renew indefinitely, and generate all the cell types of the organ from which they originate, potentially regenerating the entire organ from a few cells. Unlike ESCs, the use of human adult stem cells in research and therapy is not considered to be controversial, as they are derived from adult tissue samples rather than human embryos designated for scientific research. Currently, they are mainly studied in humans and model organisms such as mice and rats. Here, we will provide a few examples of the well-studied adult stem cells.

Mesenchymal Stem Cells

MSCs are multipotent stromal cells that can differentiate into a variety of cell types, including osteoblasts, chondrocytes, hepatocytes, cardiomyocytes, pancreatic cells, and adipocytes (Fig. 13–3). They are found in bone marrow, adipose tissue, umbilical cord, placenta, lung, and other organs. With regard to the location in which MSCs are found, these cells are called, for example, bone marrow-derived mesenchymal stem cells (BMSCs) and adipose-derived mesenchymal stem cells (ADSCs). The detailed comparisons of cellular characteristics between BMSCs and ADSCs are summarized in Table 13–1.

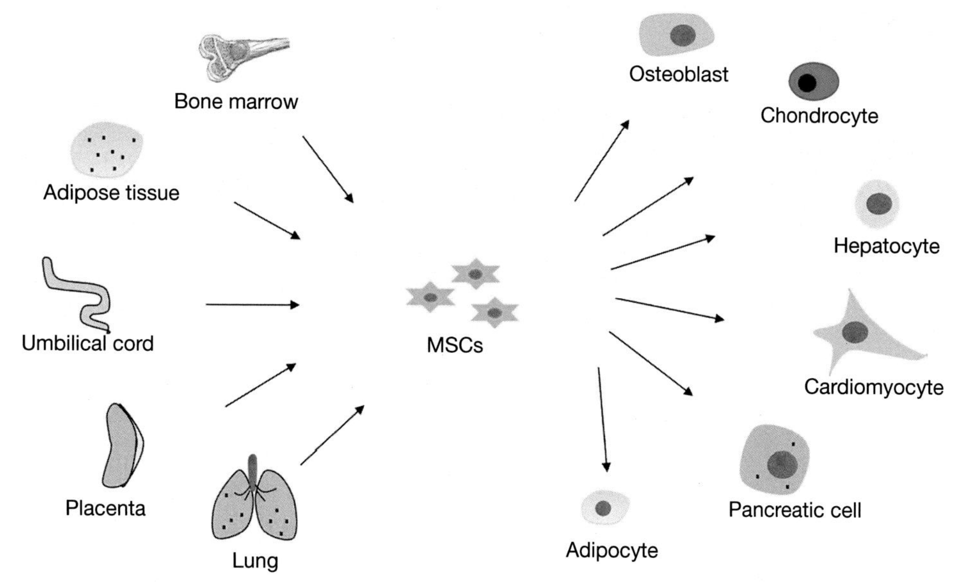

Figure 13–3. The sources of mesenchymal stem cells (MSCs) and their lineage cells. MSCs are derived from a variety of organs, including bone marrow, adipose tissue, umbilical cord, placenta, lung, and other organs. As multipotent stromal cells, they can differentiate into osteoblasts, chondrocytes, hepatocytes, cardiomyocytes, pancreatic cells, adipocytes, etc.

TABLE 13–1. Comparison of Cellular Characteristics between BMSCs and ADSCs

	BMSCs	ADSCs
Source	Bone marrow	Adipose tissue
Abundance	Rare	Abundant
Surface markers	CD73, CD90, CD105, STRO-1	CD13, CD29, CD44, CD71, CD73, CD90, CD105, CD166
Proliferation rate	++	+++
Maximal passage	7–9	8–9
Colony frequency	++	+++
Differentiation potential	Osteoblasts, chondrocytes, myocytes, adipocytes, neural cells hepatocytes, pancreatic cells	Osteoblasts, chondrocytes, myocytes, adipocytes, neural cells hepatocytes, pancreatic cells

STRO-1, stromal cell surface marker-1.

Hematopoietic Stem Cells

HSCs are mainly found in the bone marrow, especially in the pelvis, femur, and sternum. They are also found in umbilical cord blood and, in small numbers, in peripheral blood. HSCs are able to give rise to both the myeloid and the lymphoid lineages of blood cells (Fig. 13–4). The process, hematopoiesis, balances enormous production needs (more than 500 billion blood cells are produced every day) with the need to precisely regulate the number of each blood cell type in the circulation. In vertebrates, the vast majority of hematopoiesis occurs in the bone marrow and is derived from a limited number of HSCs that are multipotent and capable of extensive self-renewal.

Induced Pluripotent Stem Cells

Induced pluripotent stem cells (also known as iPS cells or iPSCs) are a type of pluripotent stem cell that can be generated directly by reprogramming adult cells with defined factors. These iPSCs hold great promise for regenerative medicine as a renewable source of autologous cells. The iPSC technology was first established by Shinya Yamanaka's lab in Kyoto, Japan, which showed in 2006 that the introduction of four specific genes encoding transcription factors (Oct4, Sox2, Kruppel-like factor 4 [Klf4], myelocytomatosis oncogene [c-Myc]) could convert adult skin cells into pluripotent stem cells (Fig. 13–5). This groundbreaking discovery has since led to a variety of methods for reprogramming adult cells into pluripotent stem cells that can become virtually any cell type. This allows researchers to bypass the need for embryos and create patient-specific cell lines that can be studied and used in drug therapies and regenerative medicine. However, the safety of patient-specific iPSCs needs to be further evaluated before any clinical application for patients.

THE PROPERTIES OF STEM CELLS

Self-renewal and potency are the two most important properties of stem cells. Self-renewal is the process by which stem cells undergo mitosis to generate one or two daughter stem cells, which possess the same capacity of self-renewal and differentiation. With this ability, stem cells are able to go through numerous cycles of cell division while maintaining the undifferentiated state. In stem cell self-renewal, two cell division mechanisms exist

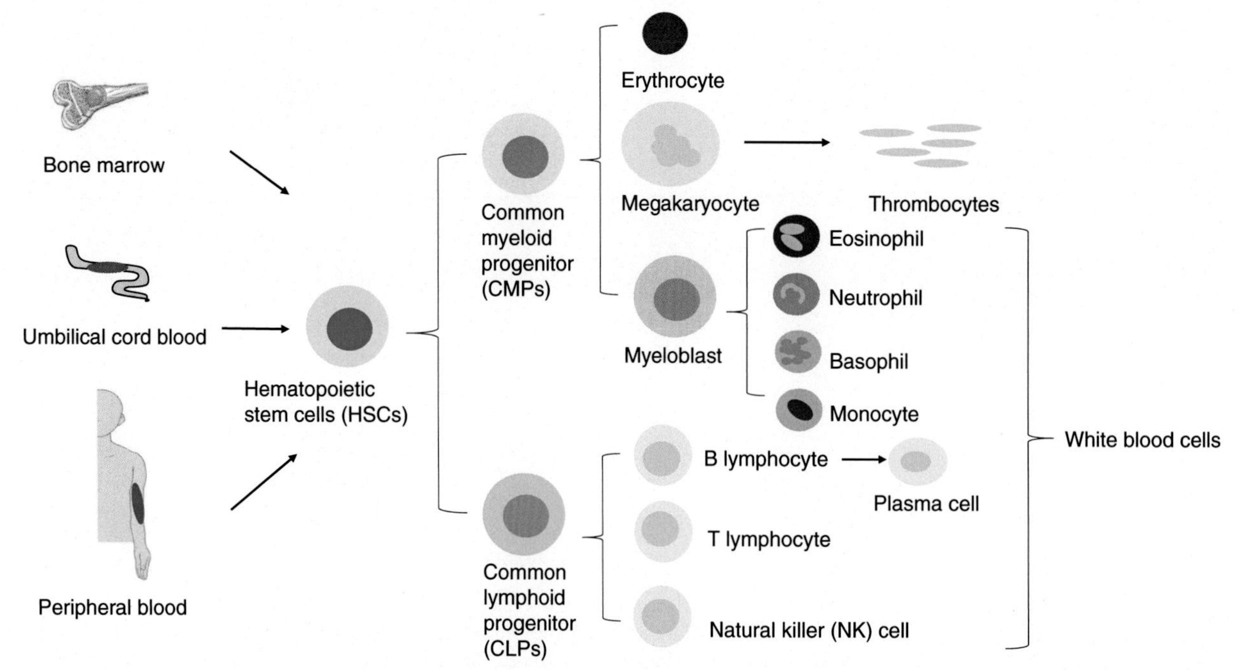

Figure 13–4. The sources of hematopoietic stem cells (HSCs) and their lineage cells. HSCs are mainly derived from bone marrow. They are also found in umbilical cord blood and, in small numbers, in peripheral blood. HSCs give rise to both the myeloid and the lymphoid lineages of blood cells, and then differentiate into erythrocyte, thrombocyte, eosinophil, neutrophil, basophil, monocyte, plasma cell, T lymphocyte, and NK cells.

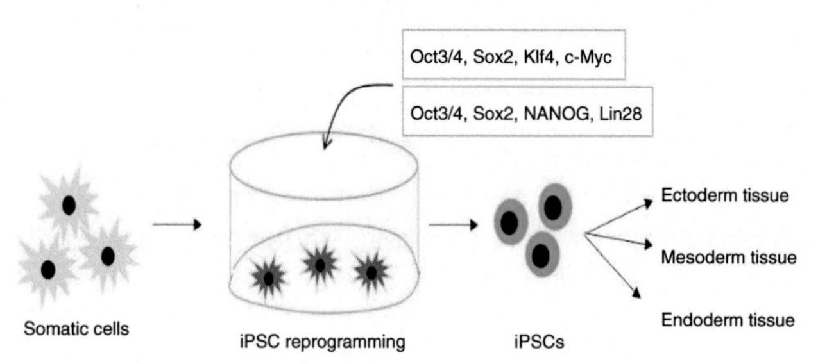

Figure 13–5. Human iPSCs generation and identification. iPSCs are generated from various somatic cells by the introduction of four genes: Oct3/4, Sox2, Klf4, c-Myc or Oct3/4, Sox2, NANOG, Lin28, a process called iPSC reprograming. The utmost characteristic of iPSCs is the same differential totipotency as ESCs, recovering the capacity of differentiating into ectoderm, mesoderm, and endoderm.

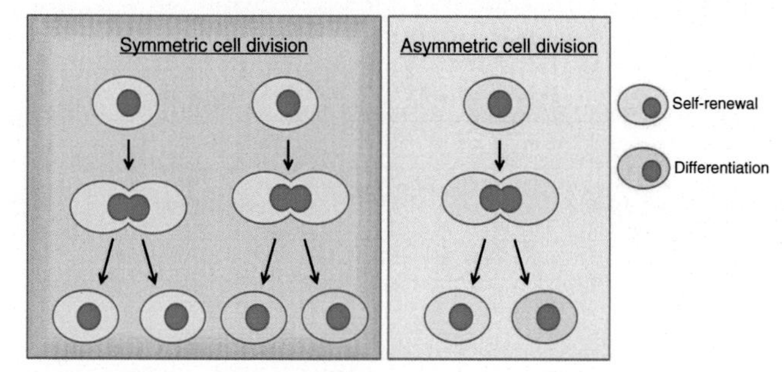

Figure 13–6. Stem cell division. Symmetric cell division indicates that one stem cell develops into two differentiated daughter cells or produces two stem cells identical to the original. The stem cell can also divide into one daughter cell that is identical to the original stem cell, and another cell that is differentiated through asymmetric cell division.

to ensure that a stem cell population is maintained (Fig. 13–6): (1) symmetric division: when one stem cell develops into two differentiated daughter cells, another stem cell undergoes mitosis and produces two stem cells identical to the original, and (2) asymmetric division: a stem cell divides into one daughter cell that is identical to the original stem cell and another daughter cell that undergoes differentiation. Self-renewal by symmetric cell division is often observed in transient stem cells appearing in early embryonic development to increase body size. In contrast, self-renewal by asymmetric cell division can be found in permanent stem cells in embryos at later developmental stages, and also in adults to maintain the homeostasis of the body. Continuous self-renewal is regulated via transcription in the nucleus, as well as triggered by extracellular signals.

Potency specifies the differentiation potential of the stem cells (Fig. 13–7). With regard to their differentiation ability, different terms are used. (1) Totipotent stem cells are stem cells that can differentiate into embryonic and extraembryonic cell types. These cells are produced from the fusion of an egg and sperm cell and can construct a complete, viable organism. Cells produced by the first few divisions of the fertilized egg are also totipotent. (2) Pluripotent stem cells are stem cells that can differentiate into any of the three germ layers. Such cells are the descendants of totipotent cells and can differentiate into nearly all cells (e.g., ESCs). (3) Multipotent stem cells can differentiate into a limited number of cell types, especially those within a closely related family (MSCs, HSCs, etc.). (4) Oligopotent stem cells are stem cells that can differentiate into only a few cell types, such as lymphoid or myeloid stem cells. (5) Unipotent stem cells can produce only one cell type, their own, but have the property of self-renewal, which distinguishes them from nonstem cells (e.g., progenitor cells).

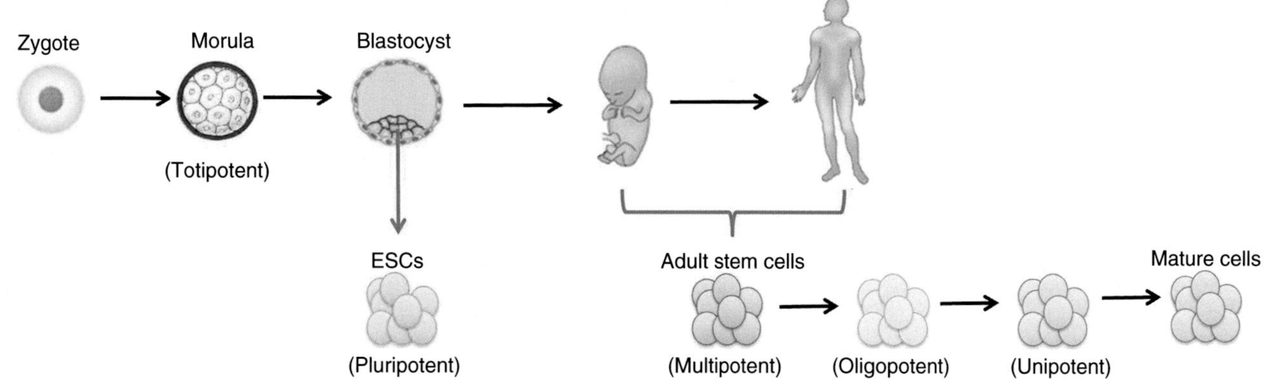

Figure 13–7. Different potency of stem cells during development. After fertilization, the zygote starts to replicate and divide to form a morula, with around 30 totipotent cells that can potentially differentiate into all cell types. ESCs within the blastocyst are the descendants of totipotent cells and can differentiate into nearly all cells. The adult stem cells are multipotent stem cells, being able to differentiate into a limited number of cell types. As the adult stem cells differentiate, oligopotent cells and unipotent cells are sequentially generated with further limited differentiation potential, which finally gives rise to the mature cells.

FUNCTIONS OF ADULT STEM CELLS

In the mid-19th century, Cohnheim first proposed the hypothesis that all the cells involved in tissue repair come from peripheral blood. For now, numerous studies have demonstrated that a small number of adult stem cells can enter the peripheral blood from bone marrow, and the role of these adult stem cells is to rapidly repair the slight injury, as well as protect the structural and functional integrity of tissues and organs. Hence, it is easy to see that in a normal state, the main function of adult stem cells is to maintain the physiologic turnover and homeostasis of different organs. While in pathophysiological status, adult stem cells play a critical role in facilitating the repair of injury, which, however, is uncertain to achieve the result of full functional restoration. Although the function of adult stem cells in mammalian health and regenerative medicine has yet to be fully elucidated, the whole physiological process of adult stem cells participating in tissue repair can be summarized as follows: mobilization from niche, homing to injured tissues, differentiation into functional cells, or secretion of trophic factors to promote tissue repair (Fig. 13–8).

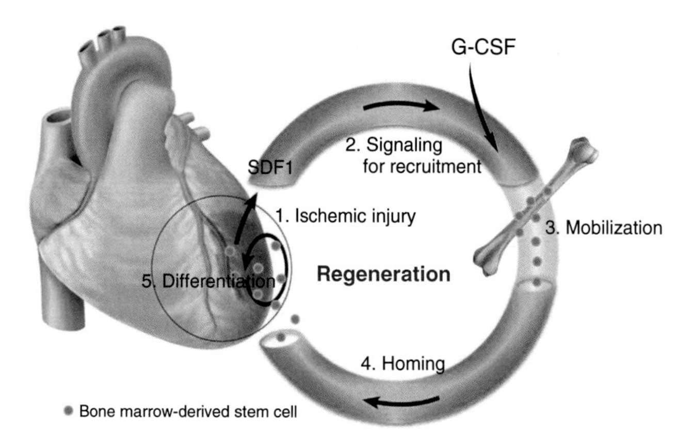

Figure 13–8. Process of tissue repair after injury. The signals released from injury tissue (such as SDF-1 released from ischemic myocardial tissue) enter circulation and induce mobilization of adult stem cells that reside in niches. The mobilized stem cells then home to the injury site and differentiate into functional cells or secrete trophic factors to promote tissue repair.

Stem Cell Niche

Stem cell niche is the specific environment in which adult stem cells reside in the organism. It consists of the extracellular matrix (ECM), various types of cells, solubilized and matrix-bound growth factors, as well as other cytokines. Adult stem cells have limited functions without the niche, and adult stem cells in their specific niche often exhibit an apical-basal axis of polarity. During mitosis, polarity establishes an asymmetrical localization of self-renewal regulators. Within the daughter cells thus formed, one inherits most of the polar and self-renewal determinants to reestablish polarity and reverts to the "stem cell" state, while the other loses these regulators and generates a critical mass of progenies necessary for differentiation facilitating tissue maintenance and repair.

Niches are usually located in the areas with good blood perfusion, where stem cells and/or their secretome can easily transport to the circulation, and signals in the circulation can have easy access to stem cells. The dynamic balance between stem cells and the microenvironment in niche determines the phenotype, status, and functions of stem cells (Fig. 13–9).

Mobilization and Homing of Adult Stem Cells

In the normal physiological state, a small number of adult stem cells in the niche can enter the blood and return to the niche through the peripheral blood circulation. In the pathophysiological state, dormant stem cells are mobilized by injury signals and enter the circulation, which is called mobilization. Many studies have demonstrated

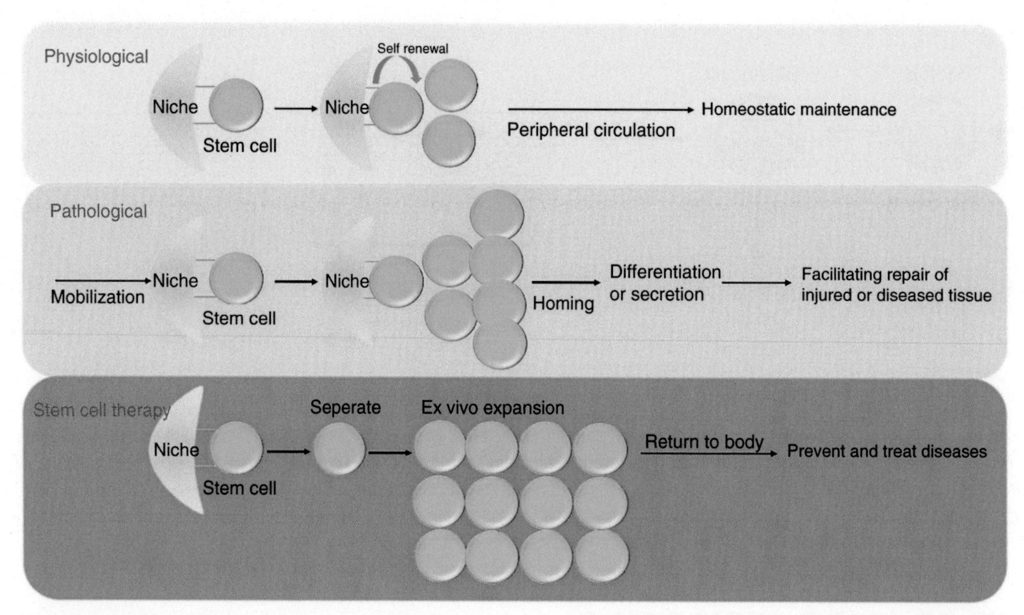

Figure 13–9. Niche and adult stem cells. Signals from the niche determine the status of adult stem cells. In the physiological state, stem cells reside in the niche and keep self-renewal. Under pathological conditions, stem cells are mobilized and home to the injury site, facilitating tissue repair. Stem cell therapy mimics the process of stem cell mobilization and homing.

that endogenous or exogenous stem cells have the character of distributing to damaged regions upon different injuries such as under ischemic or hypoxic conditions. The complex process of stem cell mobilization and homing includes the mobilization of stem cells into the peripheral blood circulation by various inducing factors, adhesion to endothelial cells through adhesion molecules expressed by injury sites, migration through the endothelial cell layer mediated by chemokines, localization in damaged tissues, and differentiation, proliferation, or secretion of related trophic factors to repair injuries together with immune cells.

TECHNIQUES OF STEM CELL PROCESSING

Possessing the self-renewal capacity and multipotential abilities that are important to individual development and tissue repair, stem cells have become the most promising cells used for disease treatment and tissue regeneration. For example, intravascular injection of BMSCs has been proved to be a promising strategy for autoimmune diseases, vascular diseases, graft versus host disease, and diabetes. Hundreds of clinical trials of BMSC cell therapy are now being carried out worldwide. Since stem cells quantity in vivo is very limited, stem cell culture in vitro provides an accessible way for stem cell expansion and amplification.

The methods for culturing adult stem cells in vitro are similar to the way for culturing most of the somatic cells (cardiomyocytes, fibroblasts, etc.). Generally, cells are isolated from the source tissue by mechanical separation or enzymatic digestion, then culture medium containing multiple nutrients, such as glucose, amino acids, vitamins, salts, trace elements, and serum, is used to maintain cell growth and expansion. However, the special properties of stem cells make the process much more challenging. First, the isolation of stem cells is more difficult since the number of some specific stem cells is extremely limited and the specific markers of stem cells are deficient. Second, the culture condition of stem cells in vitro is demanding. Though it is known that the quality of stem cells is closely regulated by its microenvironment in vivo, the exact factors that affect stem cell properties remain largely unknown. Third, stem cells are unstable during culture. In particular, stem cells can expand rapidly in culture, but the stemness of stem cells is inversely related to the incubation time of each passage. The cells are cultured under standard conditions, but they gradually lose their proliferative capacity and differentiate into their lineages. A large number of variables must be considered in the process of stem cell expansion before any experimental and clinical application. Below, ESCs, adult stem cells, and iPSCs are used as examples to introduce the methods of stem cell processing.

Embryonic Stem Cell Culture and Identification

Most human ESCs (hESCs) are usually derived from discarded embryos or embryos donated from a couple who have completed in vitro fertilization treatments and have no desire to utilize the remaining embryos for transplantation. To generate hESCs, the inner cell mass (ICM) must be isolated first as hESCs can only be obtained from the ICM of human embryos. Immunosurgery has been most commonly used for the selective lysis and removal of trophoblast (the outer cell layer of blastocysts) cells through complement-dependent antibody cytotoxicity. However, immunosurgery utilizes animal-derived products such as antihuman antiserum and guinea pig complement, which make this technique unsuitable for the generation of clinical-grade hESCs. To circumvent the use of immunosurgery, mechanical dissection or chemical dissolution of the trophoblast layer was used to isolate the ICMs. Mechanical dissection is a crude method, which is dependent on the operator's technical skills. Chemical dissolution of the trophoblast layer by acid tyroid solution may damage the cells of the ICM due to the acidification of the medium. Hence, laser dissection, a method for creating an opening in the zona pellucida of cleavage-stage embryos for blastomere biopsy, may serve as an alternative approach to chemical or mechanical removal of the trophoblast cells.

After ICM isolation, the stems cells are grown to generate the ESCs using feeder layers, ECMs, proteins, peptides, and synthetic polymers. In particular, cells are collected when they originate from the outgrowth of the ICM. Then, they are dissociated to small cell clumps and transferred to fresh feeder layers after washing. When the cells show ESC-specific morphology (i.e., small size and a high nuclear/cytoplasmic ratio), the area containing the pluripotent stem cells is isolated using the sterile glass capillary, dissociated and passaged. These procedures are repeated approximately six times until homogeneous undifferentiated colonies are obtained. When the ESCs reach a stable proliferation rate with homogeneously undifferentiated cell populations, they are maintained under standard conditions and passaged (Fig. 13–10).

One of the chief characteristics of hESCs is differentiation into all three lineages, including ectoderm, mesoderm, and endoderm. For example, hESCs form embryoid bodies that are basically structured with three germ layers in three-dimensional (3D) culture and express genetic markers for all three germ layers. As hESCs are pluripotent stem cells coming from a single cell-derived clone, each cell has the same karyotype and unique capabilities to differentiate into all kinds of cell types such as adrenal cells, keratinocytes, insulin-producing cells, neuronal cells, cardiac cells, and liver cells. Moreover, hESCs have a continuous self-renewal activity, which could form clones in vitro. Therefore the properties just described provide us with accessible ways to identify the culture cells isolated from the ICM.

Adult Stem Cell Culture and Identification

Adult stem cells are found in most organs, but they only reside in their specific niches. Thus different adult stem cells are found in different organs, such as MSCs and HSCs from bone marrow, adipose stem cells from fat tissue, neural stem cells within the anterolateral ventricle and hippocampus of the brain, and epidermal stem cells in the skin.

Human BMSCs (hBMSCs) are usually aspirated from the sternum or iliac crest of healthy donors or patients. The samples are then generally separated by density gradient centrifugation for mononuclear cell (MNC) isolation. An alternative method for isolating the mononuclear fraction of the whole marrow is to use a red blood cell lysis buffer. This approach uses ammonium chloride to lyse the erythrocyte population without damaging the MNCs. After harvesting the MNCs, the

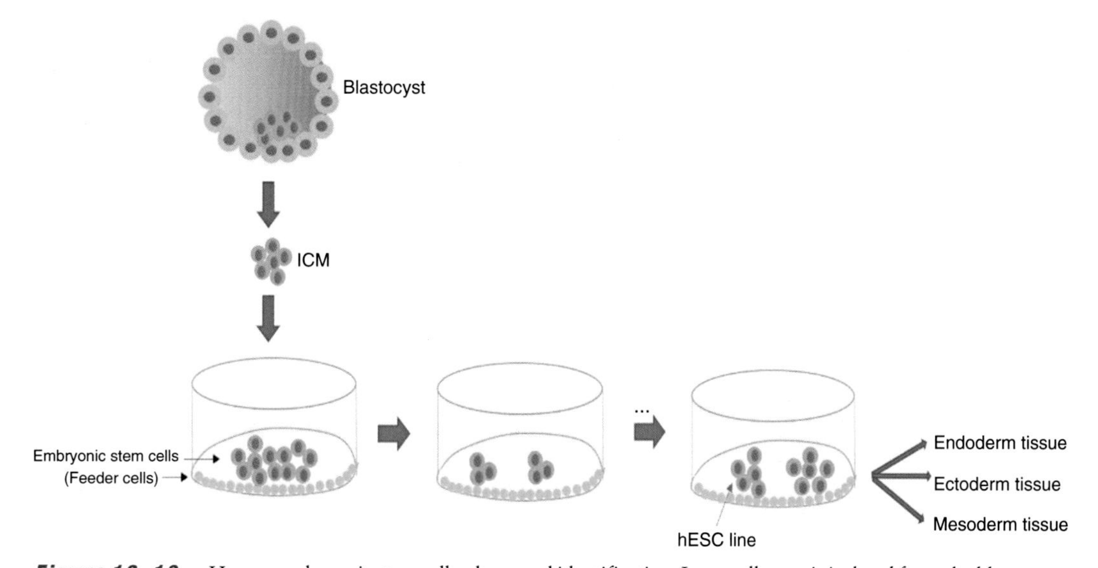

Figure 13–10. Human embryonic stem cell culture and identification. Inner cell mass is isolated from the blastocyst, then the cells are dissociated into small cell clumps and transferred to fresh feeder layers. ESC lines are established after the splitting of the homogeneous undifferentiated colonies. ESCs have potential to differentiate into all three lineages (ectoderm, mesoderm, and endoderm).

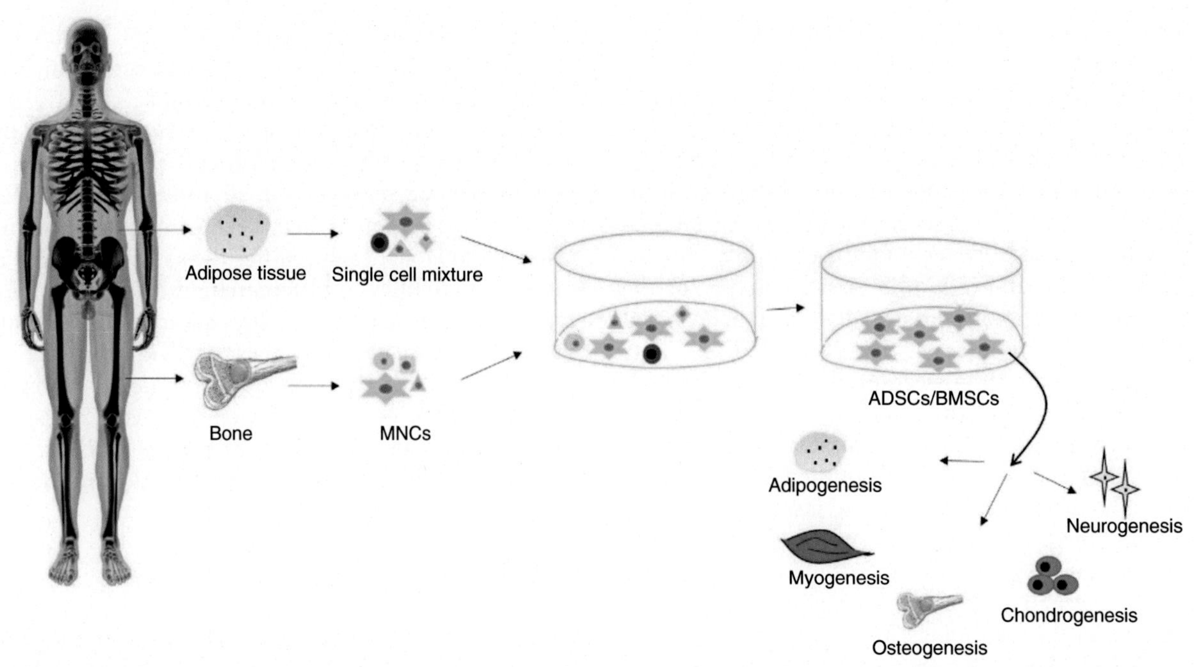

Figure 13–11. Human adult stem cell culture and identification. Adult stem cells can be obtained from a diversity of tissues. For example, adipose-derived stem cells are from fat tissue, and bone marrow mesenchymal stem cells are from bone marrow. After being isolated from the source tissue (single cell mixture for ADSCs, MNCs for BMSCs), cells are seeded in cultural flasks. Following removal of nonadherent cells by changing culture medium and undergoing culture passaging, ADSCs/BMSCs are enriched and purified. ADSCs/BMSCs have multidirectional differentiation potential, such as adipogenesis, myogenesis, osteogenesis, chondrogenesis, and neurogenesis.

hBMSC populations are isolated by the classical adhesion method based on their adhesion ability. Subsequently, the monolayer cells are expanded after being placed into culture flasks, allowing removal of nonadherent cells. In this way, hBMSCs are purified to proliferate for several passages. hBMSCs could also be isolated directly by cell sorting since a variety of markers of hBMSCs are identified (Fig. 13–11).

Human ADSCs (hADSCs) can be extracted from adipose tissue obtained through liposuction or surgeries. The sample is immediately digested by a mixture of enzymes, such as collagenase, giving rise to a single cell mixture. After digestion, the stromal fraction (on the bottom) that consists of a heterogeneous mixture of cells, including ADSCs, is further purified by the change of growth medium and cell passage. In approximate 2 weeks of growth, only ADSCs, the most abundant stem cells, remain. ADSCs can also be isolated by cell sorting with a combined use of positive and negative surface markers (Fig. 13–11).

MSCs isolated from bone marrow or adipose tissue express a similar set of surface antigens, which are positive in CD29, CD73, CD133, CD105, CD54, CD44, CD90/Thy 1, HLA-1, and SH2, and negative in LA-DR, CD31, CD34, CD117, and hematopoietic cell markers. ADSCs are generally similar but not identical to BMSCs. For example, ADSCs express CD49d, but BMSCs do not. BMSCs express CD106, which is absent in ADSCs.

In addition to the surface epitopes listed previously, hBMSCs and hADSCs also express stemness-related transcription factors, such as the sex-determining region Y-box2 (Sox-2), nanog, POU domain class 5 transcription factor 1 (Oct3/4), etc. The expression levels of these transcription factors are usually used to reflect the stemness of cultured cells. Colony formation assay and induced differentiation assay toward their lineage cells (such as osteoblasts, adipocytes, and chondroblasts) are widely used to detect the proliferation rate and differentiation potential of the stem cells. In addition, MSCs isolated from bone marrow or adipose tissue inhibit mixed lymphocyte reactions and produce the cytokines interleukin (IL)-6, IL-10, CXC12, basic fibroblast growth factor, hepatocyte growth factor, platelet-derived growth factor BB, vascular endothelial growth factor, transforming growth factor beta 1 (TGFβ1), and TGFβ2. All of these factors are applicable to the characterization of hBMSCs and hADSCs.

Human iPSC Generation and Identification

hESCs have great potential for self-renewal and differentiation into all tissues of the body, consisting of an important source of material for regenerative medicine and cell therapy. However, the use of ESCs is limited by ethical and religious conflicts, as well as immunological incompatibility. Human iPSCs (hiPSCs) were proposed to be a substitution for ESCs applicable in regenerative medicine.

hiPSCs can be generated from various types of somatic cells (fibroblasts, cardiomyocytes, hematopoietic cells, etc.) by a process called reprogramming. In brief, hiPSCs are generated either by the introduction of four genes, constitutive expression of Oct3/4, Sox2, Klf4, and c-Myc (also referred as OSKM-"Yamanaka's cocktail"), or by an independently defined combination of lentivirally transduced genes Oct3/4, Sox2, NANOG, and Lin28 (Fig. 13–5).

hiPSCs are found to be similar to hESCs in their morphology, proliferation rate, surface antigens, gene expression, epigenetic status of their pluripotent cell-specific genes, and telomerase activity. Furthermore, these cells can differentiate into cell types of all three germ layers in vitro as well as into teratomas in vivo.

CHALLENGES OF STEM CELL PROCESSING AND APPLICATION

Routine medical practice using stem cells has great potential, thus preclinical studies related to stem cell therapy have been undertaken extensively. Besides the rapid evolution in differentiating stem cells into many different cell types, it remains a big challenge regarding how to obtain specialized cells in sufficient amount and high purity for a specific therapeutic application. A thorough understanding of the mechanisms from stem cell culture to transplantation into a patient is of great importance to realize stem cell applications in the clinic.

The difficulties in identifying stem cells, especially adult stem cells, are compounded by the fact that there is no consensus regarding the characteristic surface epitopes that can be used to identify the cells. How to maintain the stemness of stem cells in vitro is another essential question for stem cell culture, as stem cells tend to differentiate into lineage cells spontaneously. Besides, few topics in regenerative medicine have inspired such impassioned debate as the immunogenicity of cell types and tissues differentiated from pluripotent stem cells. While early predictions suggested that tissues derived from allogeneic sources may evade immune surveillance altogether, the pendulum has since swung to the opposite extreme, with reports that the ectopic expression of a few developmental antigens may prompt rejection, even of tissues differentiated from autologous cell lines. Another difficulty lies in how to achieve the directed differentiation of stem cells. Since tissue microenvironmental cues are essential for directing stem cell fate in vivo in both physiological and pathological conditions, many researchers have devoted their efforts to discovering key factors that direct stem cell differentiation. The integration of some soluble factors or ECM components into a biophysical cue-based induction regime can enhance the efficiency, efficacy, and sustainability of the inducing effect, but a complete understanding of this process is yet to be reached.

THE ETHICAL ISSUES ASSOCIATED WITH HUMAN-DERIVED STEM CELLS

It seems morally compelling to explore all the sources that might help to obtain clinically applicable stem cells because of their promise for regenerative medicine. However, some of the manipulations of stem cells, such as embryo destruction and cell reprogramming, have raised deep ethical concerns.

Currently, hESCs cannot be obtained without causing the destruction of a human embryo. This fact makes the application of hESCs extremely problematic. The most promising hope to conclude this debate may be to develop a scientific technique capable of producing ESCs that are suitable for research and therapeutic purposes without destroying a human embryo. For this purpose, hiPSCs were generated by reprogramming somatic cells to mimic the properties of hESCs. However, this also results in some ethical concerns. In particular, it is likely that abnormal reprogramming occurs during the induction of hiPSCs, and that the stem cells may generate tumors in the process of stem cell therapy. Moreover, hiPSCs should not be used to clone human beings, to produce human germ cells, or to make human embryos. Adult stem cells, such as BMSCs and ADSCs, are attracting increased attention from researchers because they can be obtained without causing these ethical issues.

APPLICATION OF STEM CELLS IN REGENERATIVE MEDICINE

The evolution of stem cell biology, developmental biology, and the techniques of processing stem cells (including isolation, expansion, passage, cryopreservation, and transplantation of stem cells) give rise to a medicine that is based on natural or induced regenerative capacity—regenerative medicine. As briefly defined by Mason and Dunnill, "regenerative medicine replaces or regenerates human cells, tissue or organs, to restore or reestablish normal functions."

Medical science has been successful in treating acute injuries and infectious diseases. However, since chronic diseases have been a burden for the global population because of the growth of an increasingly elderly population and the prevalence of age-precipitated degenerative diseases and disabilities, it is urgent to develop next-generation solutions to promote longitudinal wellness and reduce the socioeconomic burden associated with chronic diseases. In this context, regenerative medicine, which addresses functional restoration of specific tissues and/or organs of patients suffering from severe injuries or chronic diseases in a condition where the organism's own regenerative responses do not suffice, provides a great solution.

Multimodal regenerative methods incorporate the transplant of healthy tissues, prompt the body to enact a regenerative self-healing response in damaged tissues, and use tissue engineering to manufacture new tissue. Stem cells and derived products are the distinctively active ingredients used in regenerative regimens that leverage their capacity to form de novo tissue and/or promote innate repair (Fig. 13–12, Table 13–2).

Bone and Teeth

Bone is a rigid organ that provides support to the body, permits locomotion, and protects the internal organs. Bone also serves as the main reservoir for calcium and phosphorus, and along with bone marrow, produces blood cells. Bone has the capacity to regenerate and self-repair. However, this regenerative capacity can be impaired or lost due to bone injury beyond a critical size or under the presence of certain disease states such as complex trauma, inflammation, and tumor removal after surgery. Traditional repair techniques for bone injury are performed using autologous bone, allogeneic bone, or xenograft bone grafts, but they are limited by several factors, including the amount, accessibility of sources, and various postprocedural complications (bleeding, pain, hypoesthesia, hyperesthesia, infections). Adult stem cells, which can be obtained from wide sources with rapid

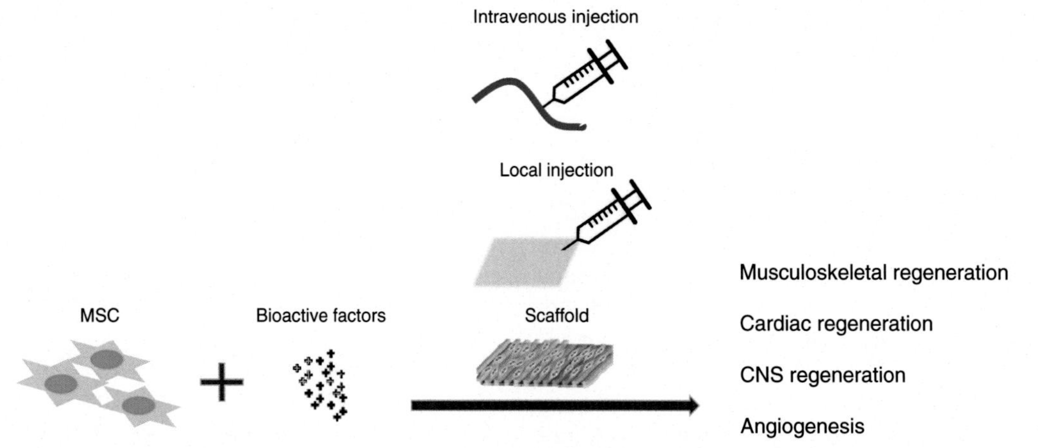

Figure 13–12. Strategies of MSC delivery and regenerative medicine. There are three major strategies of MSC delivery in various settings of regeneration medicine: intravenous injection, local injection, and scaffold-aided stem cell patch.

TABLE 13–2. Clinical Trials of Stem Cell Application in Regenerative Medicine

	Disease	Treatment Strategy
Bone and teeth	Knee osteoarthritis, cartilage degeneration	Bone marrow-derived mesenchymal stem cells, adipose-derived mesenchymal stem cells, placenta-derived mesenchymal stem cells
	Multiple myeloma	Peripheral blood stem cells, umbilical cord blood stem cells
	Chronic paraplegia	Bone marrow stromal cells
	Osteosarcoma	Peripheral blood stem cells
	High-risk Ewing's tumors	Hematopoietic stem cells, peripheral blood stem cells
	Rheumatoid arthritis	Umbilical cord blood mesenchymal stem cells
	Spinal cord injury	Spinal cord-derived neural stem cells
	Osteochondral fracture of talus	Mesenchymal stem cells derived from the umbilical cord
	Primary myelofibrosis, secondary myelofibrosis	Hematopoietic stem cells
	Chronic periodontitis	Gingiva mesenchymal stem cells
Heart	Ischemic cardiomyopathy, myocardial infarction, heart failure	Bone marrow-derived stem cells, umbilical cord blood stem cells
	Dilated cardiomyopathy, left ventricular dysfunction	Bone marrow-derived stem cells, adipose-derived stromal cells
	Refractory angina	Bone marrow-derived mononuclear cells
	Chest pain	Peripheral blood stem cells
Lung	Chronic obstructive pulmonary disease, pulmonary fibrosis, interstitial lung disease	Stem cells from venous, adipose, or bone marrow
	Idiopathic pulmonary fibrosis	Placental mesenchymal stem cells
	Acute respiratory distress syndrome	Bone marrow-derived mesenchymal stem cells
	Bronchiolitis obliterans	Hematopoietic stem cells
	Bronchopulmonary dysplasia	Umbilical cord mesenchymal stem cells
	Pulmonary hypertension	Adipose-derived mesenchymal stem cells

TABLE 13–2. Clinical Trials of Stem Cell Application in Regenerative Medicine—cont'd

	Disease	Treatment Strategy
Central nervous system	Supratentorial primitive neuroectodermal tumors, high-risk medulloblastoma, neuroblastoma, cerebral adrenoleukodystrophy	Hematopoietic stem cells, peripheral blood stem cells
	Neurological disorders, stroke, traumatic brain injury, CADASIL, chronic traumatic encephalopathy, cerebral infarction, cerebral ischemia, cerebral stroke, cerebral hemorrhage, cerebral palsy, Parkinson's disease, multisystem degeneration, multiple system atrophy, progressive supranuclear palsy, amyotrophic lateral sclerosis, neuropathy, diabetic neuropathies	Bone marrow-derived stem cells
	Alzheimer's autism, cognitive impairment	Bone marrow-derived stem cells
Blood	Thrombocytopenia	Umbilical cord-derived mesenchymal stem cells
	Fanconi anemia	Bone marrow hematopoietic stem cells
	Hematologic cancer, including chronic myeloproliferative disorders, leukemia, lymphoma, multiple myeloma, plasma cell neoplasm, myelodysplastic syndromes, myelodysplastic/ myeloproliferative neoplasms, refractory multiple myeloma	Peripheral blood stem cells, umbilical cord blood stem cells, hematopoietic stem cells
	Sickle cell disease	Hematopoietic stem cells
	Alpha-thalassemia	Utero hematopoietic stem cells
	Beta-thalassemia	Hematopoietic stem cells
	Severe aplastic anemia	Peripheral and cord blood stem cells
Liver	Hepatic veno-occlusive disease	Hematopoietic stem cells
	Liver cirrhosis	Umbilical cord-derived mesenchymal stem cells, bone marrow-derived mesenchymal stem cells
	Hepatitis B	Umbilical cord mesenchymal stem cells
	Acute-on-chronic liver failure	Bone marrow-derived mesenchymal stem cells
Metabolism	Type 1 diabetes	Stem cells from human exfoliated teeth, umbilical cord blood-derived multipotent stem cells
	Type 2 diabetes	Intrapancreatic autologous stem cells, stem cells from human exfoliated teeth
	Inborn errors of metabolism, including adrenoleukodystrophy, metachromatic leukodystrophy, globoid cell leukodystrophy, Gaucher's disease, fucosidosis, Wolman disease, Niemann-Pick disease, Batten disease, GM1 gangliosidosis, Tay-Sachs disease, Sandhoff disease	Hematopoietic stem cells
	Systemic amyloidosis	Hematopoietic stem cells
Muscle	Amyotrophic lateral sclerosis	Bone marrow-derived stem cells
	Crohn's anal fistula	Adipose tissue-derived stem cells
	Intractable common extensor tendon injury	Adipose-derived mesenchymal stem cells
	Duchenne muscular dystrophy	Bone marrow-derived stem cells
Others	Kidney cancer, renal cell carcinoma	Hematopoietic stem cells
	Chronic renal failure	Adipose tissue-derived mesenchymal stem cells
	Melanoma	Autologous CD34$^+$ stem cells
	Extragonadal germ cell tumor, ovarian cancer, testicular germ cell tumor	Peripheral blood stem cells
	Oral mucositis	Hematopoietic stem cells
	Systemic lupus erythematosus	Hematopoietic stem cells
	Breast cancer	Hematopoietic stem cells
	Breast cancer-related lymphedema	Adipose tissue-derived mesenchymal stem cells
	Vocal fold scarring	Bone marrow-derived stem cells
	Dry eye, ocular inflammation, ocular surface disease, ocular discomfort, blepharitis	Topical cadaveric-derived corneal epithelial stem cells
	Extraocular retinoblastoma, optic neuropathies, age-related macular degeneration, retinitis pigmentosa, Stargardt disease, optic neuropathy, nonarteritic ischemic optic neuropathy, optic atrophy, optic nerve disease, glaucoma, Leber hereditary optic neuropathy, blindness, night vision loss, partial vision loss, low vision, retinopathy, maculopathy, macular degeneration, retina atrophy, central retinal vein occlusion	Bone marrow-derived stem cells
	Corneal scars and opacities	Limbus-derived stem cells
	Crohn's disease	Mesenchymal stem cells
	Skin scars, cutis laxa	Adipose-derived mesenchymal stem cells

Continued

TABLE 13–2. Clinical Trials of Stem Cell Application in Regenerative Medicine—cont'd

Disease	Treatment Strategy
Multiple sclerosis	Hematopoietic stem cells, neural stem cells
Thin endometrium, intrauterine adhesion	Umbilical cord mesenchymal stem cells
Uterine scar	Umbilical cord mesenchymal stem cells
Diabetic foot ulcer	Adipose-derived mesenchymal stem cells, bone marrow-derived stem cells
Nonobstructive azoospermia male infertility	Bone marrow-derived mesenchymal stem cells
Devic's disease	Hematopoietic stem cells
Ulcerative colitis	Adipose mesenchymal stem cells
Aneurysmal bone cysts	Bone marrow-derived mesenchymal stem cells
Myelodysplastic syndrome	Peripheral and cord blood stem cells
Lower extremity ischemia, leg ulcer, gangrene	Bone marrow-derived stem cells
Premature ovarian failure	Bone marrow-derived stem cells
Hormone deficiency	Adipose-derived mesenchymal stem cells
Aging frailty	Longeveron mesenchymal stem cells

proliferation and multidirectional differentiation potential, have been proven to be capable of overcoming the various drawbacks of traditional bone transplantation and becoming the most promising material in bone regeneration.

There are three key elements of stem cell-based application in bone repair: seed cells that can differentiate into various bone cells to make up the complicated bone tissue, growth factors that are essential for osteogenesis and angiogenesis, and scaffold materials that serve as a structural and logistic support for the developing tissue and sustain cell behavior under certain bone disorders. BMSCs are the most commonly used seed cells for the treatment of bone diseases, because these cells can be guided along multiple mesenchymal lineages, including bone, cartilage, muscle, ligament, tendon, and stroma. ADSCs have also attracted much attention for bone tissue repair at present due to their wide sources, convenience in extraction, rapid proliferation, and multidirectional differentiation potential. Growth factors play an important regulatory role in seed cell proliferation, tissue lineage differentiation, and protein synthesis that is essential during the whole regeneration process. The combined effects of various growth factors are often better than the effect of a single growth factor in stem cell differentiation. Scaffolds are necessary for supporting seed cells due to the large variability in bone defects, for example, under large size bone defects or fractures. The scaffold material is usually a porous 3D structure that provides sufficient space and mechanical support for the newly regenerated bone tissue and also serves as an ECM to mediate cell-to-cell interaction and signal transduction.

The strategies of stem cell application in bone regeneration are various due to the huge complexity of bone architecture and the high metabolic activity of the bone. Local injection of BMSCs to the bone injury site that usually results from tumor removal has been observed since 1990 as the initial stem cell therapy for bone injury. Since then studies have been expanded to other bone diseases such as osteonecrosis, fractures, and osteoarthritis. Studies have shown that stem cell transplantation combined with core decompression for treating femoral head necrosis is more effective than core decompression alone. In addition, bone marrow transplantation for the treatment of severe lower extremity comminuted fractures has shown a successful healing of clinical fracture in patients. In the clinical trial of treatment of ankle joint fracture, the patient group treated with a combination of bone marrow stem cells with external fixator had the shorter healing time compared with the control group with just fixator alone. Osteoarthritis is a chronic impairment of joint cartilage and its surrounding tissues. Direct injection of stem cells into the joint and implantation of stem cell-loaded scaffold are two main ways for stem cell-based therapy for osteoarthritis. Studies have shown that the injection of BMSCs into a patient's knee joint effectively relieves the pain of knee osteoarthritis and improves knee joint function. ADSCs were also widely used in cartilage repair. For example, the injection of autogenous ADSCs from patients who suffered from knee osteoarthritis increases the thickness of the meniscus after a 12-week treatment.

Modification of BMSCs that facilitate bone regeneration has also been reported. For example, the addition of recombinant human bone morphogenetic protein 2 to a silica-coated calcium hydroxyapatite (HASi)-rabbit BMSC construct has been shown to promote bone healing and regeneration in large critical-size-defect rabbits.

Teeth are considered special bone tissues. The application of stem cell-based therapy in the dental field is similar to that of bone regeneration, providing injured areas with seed cells as well as growth factors. At present, the most widely applied stem cell in the oral field is dental stem cells (DSCs). DSCs are MSC-like populations with self-renewal capacity and multidifferentiation potential. Dental pulp stem cells (DPSCs) are the first isolated and characterized DSCs. Later, other types of DSCs were discovered: stem cells from human exfoliated deciduous teeth, periodontal ligament stem cells, dental follicle precursor cells (DFPCs), stem cells from apical papilla (SCAP), and human periapical cyst-mesenchymal stem cells. These DSCs retain proliferation ability and multipotency of differentiation, which can be exploited for providing sufficient seed cells to injured areas, and promote regeneration of injured teeth tissues and local immunomodulation.

Stem cell-based therapy for tooth regeneration includes bone reconstruction, tooth root formation, dentin-pulp regeneration, periodontal regeneration, and neural tissue regeneration. For example, SCAPs can be used as a stem cell source for osteo/odontogenic differentiation and bone formation when they are seeded into synthetic scaffolds and transplanted into immune-deficient mice. DFPCs can develop into root-like tissues with dentin-pulp-like tissues and cementum-periodontal complexes assisting with treated dentin matrix, which serves as a natural biological scaffold.

Heart

The mammalian heart behaves as a terminally differentiated organ with limited regenerative capacity. Most of the mature cardiomyocytes lose the abilities to proliferate and undergo mitotic division. Specifically, the adult human heart has a minimal regenerative capacity with a cardiomyocyte renewal rate of less than 1% per year. Loss of cardiomyocytes caused by heart injury, such as myocardial infarction, is usually replaced by scar tissue, leading to impaired heart function, which is irreversible and eventually develops into heart failure. There is currently no efficient therapy for heart regeneration. Stem cell therapy as a potential solution was applied to heart diseases several decades ago, which was proven to be useful in myocardial repair and regeneration in cardiac diseases in animal models.

The strategies of stem cell delivery to injured heart tissue are various. The first widely used delivery strategy is intravenous injection, also called systematic injection, which is easy to apply and less invasive. However, the drawback is also obvious: injected stem cells are widely distributed throughout the whole body with few cells homing to the heart tissue. Evidence has shown that more than 90% of cells are washed out within the first hour posttransplantation. Instead of staying in the heart tissue, stem cells accumulate in the liver, lungs, and spleen, causing toxicity in the whole body. To overcome this drawback, local injection appeared as an alternative strategy. Stem cells are directly injected into impaired tissue to improve local concentration. This injection is usually at multiple points in or near the damaged site; however, a consensus has not been reached regarding improvement in the therapeutic efficacy of the local injection in comparison with the systemic injection.

To further increase the local concentration of stem cells, several approaches are being tested in preclinical research. For example, stem cells can be encapsulated by a magnetic material, which is responsive to a magnetic field to guide cells to the area of interest. The ultrasound-mediated delivery system is also used as a way to improve local stem cell retention. Stem cells are attached to gas-filled microbubbles, which thereby become highly susceptible to acoustic radiation forces, and can be released in the area of interest by ultrasound treatment. However, to what extent these methods work will still need more testing.

Biomaterials are used with stem cells for the purpose of protecting cells from damage during and after transplantation. Materials used now contain biomaterials and synthetic biomaterials. Biomaterials include biodegradable polypeptide (silk fibroin from worms and insects), polysaccharide-based materials (chitosan from crustacean shells, alginate from brown algae, agarose from red algae, hyaluronic acid, collagen), and fibrin derived from blood plasma. Synthetic biomaterials include peptide amphiphiles incorporating cell adhesive ligands (injectable nanomatrix gel) and other polymer-based materials. Tissue patch, made of scaffold and stem cells, is also applied in animal studies to avoid engrafted cell death in the harsh tissue environment. Current tissue patch approaches include hydrogel-based engineered heart tissue, scaffold-free cell sheets, and 3D-printed cardiac tissue with a complex ECM structure. The commonly used scaffold materials are ECMs such as collagen I, Matrigel, fibrin, and other natural biomaterials such as hyaluronic acid-based hydrogel (Clinical Case 13–1).

Clinical Case 13–1

Mr. Tyler is a 72-year-old Caucasian man who has a history of hypertension, hyperlipidemia, and obesity. Mr. Tyler really likes to eat fast food and has been regularly having French fries for lunch along with his hamburger and strawberry milk shake. He loves to add extra salt to his French fries as he likes to make them look like they have a dusting of snow on them. He routinely walks half a block each day for exercise, but had recently noticed increased ankle swelling and mild shortness of breath. Mr. Tyler had a prescription for a "water pill" but he stopped taking it because it made him have to get out of bed at night to urinate. Since stopping his "water pill" he had noticed increased difficulty lying flat in the bed.

Cell Biology, Diagnosis, and Treatment of Congestive Heart Failure

At his doctor's appointment, his blood pressure (BP) was 150/95 mmHg and his body mass index (BMI) was 32 kg/m^2. On physical exam his doctor heard rales bilaterally in his lung bases and observed 1+ pitting edema in his lower extremities but did not hear a heart murmur. Given Mr. Tyler's medical history and symptoms, the doctor ordered a chest X-ray, which revealed mild pulmonary

Continued

edema and a slightly enlarged heart. After obtaining these results the doctor subsequently ordered an echocardiogram that revealed a left ventricular ejection fraction of 35%, global hypokinesis of the left ventricle, and pulmonary artery systolic pressure of 38 mmHg, but no valvular abnormalities. The doctor told Mr. Tyler that he had congestive heart failure with reduced ejection fraction (HFrEF) and advised him to stop salting his food and to lose weight. The doctor also prescribed Lasix 40 mg each day and losartan 50 mg. Mr. Tyler took these medications and stopped salting his food and noticed a marked improvement in his shortness of breath and ability to walk further for exercise than before. He also stopped eating at fast food restaurants every other day and lost 15 lb. On his next visit to his doctor, Mr. Tyler's BP had decreased to 126/82 mmHg, his lung exam was now normal, and his BMI was down to 28 kg/m^2. His doctor was very happy with these changes and encouraged Mr. Tyler to continue his current practices.

Determination of volume status and appropriate use of diuretics (Lasix) when a patient is volume overloaded is an important part of symptom management of HFrEF. Inhibition of the renin-angiotensin-aldosterone system with medications such as angiotensin receptor blockers (losartan) can improve survival and decrease hospitalizations for patients with HFrEF. Unfortunately, over the next 2 years, Mr. Tyler had multiple recurrences of his HFrEF symptoms with hospitalization and his ejection fraction declined to 15% on echocardiogram. During these episodes he became increasingly refractory to treatment with diuretics and angiotensin receptor blockers.

With this advanced heart failure, therapeutic options for Mr. Tyler remain limited to heart transplantation. Stem cells, particularly MSCs, have a broad range of activities, making them particularly interesting candidates for a new heart failure therapeutic. MSCs may exert paracrine effects by secreting cardioprotective factors. These secreted factors may stimulate angiogenesis and vascular remodeling, attenuate fibrosis, modulate inflammation, regulate cell differentiation and survival, and recruit resident stem or progenitor cells.

Central Nervous System

There is substantial evidence that neural progenitors in the subventricular zone proliferate, differentiate, and become new functional neurons throughout life. However, the capacity of neurogenesis diminishes sharply in adults, becoming very low by middle age. Injuries such as ischemia can stimulate proliferation, migration, and differentiation of neural stem cells and neural progenitors in the brain. Unfortunately, endogenous neurogenesis does not supply enough cells to repair neurological damage from major events, like stroke and age-dependent neurodegenerative diseases. Stem cell transplantation thus holds great promise for these diseases.

Stroke causes damage to different areas of the brain depending upon the area affected. There are two main types of stroke: ischemic and hemorrhagic stroke. Ischemic stroke accounts for over 80% of the total number of strokes, while hemorrhagic stroke accounts for 15% of all strokes, which is responsible for a disproportionate 40% of stroke-related deaths. The drastic damage to brain tissues following strokes includes the destruction of a heterogeneous population of brain cell types, major disruption of neuronal connections and vascular systems, and inflammatory cascade. Thrombolysis and/or thrombectomy is the only validated therapeutic strategy for ischemic stroke, but it works only in a narrow time window of 4.5 h after the initial injury. Surgical treatment or aneurysm treatment could be applied to prevent or stop bleeding at the acute stage of hemorrhagic stroke, and then antiinflammation strategies could be used. Currently, there are no effective therapies for the subacute and chronic stage of strokes.

In 1984, the first stem cell transplantation into a brain was conducted in a rat model of ischemic stroke. To date, neurorestorative stem cell-based therapy has become a top priority for stroke research. Studies using animal models of stroke have shown that stem cells transplanted into the brain not only make the animals survive but also lead to their brain function improving. Possible mechanisms include cell replacement, trophic influences, immunomodulation, and enhancement of endogenous repair processes.

Neurodegenerative diseases are a group of chronic, progressive disorders characterized by the gradual loss of neurons in discrete areas of the central nervous system (CNS). These neurodegenerative diseases include Alzheimer's disease (AD), Parkinson's disease (PD), amyotrophic lateral sclerosis (ALS), multiple sclerosis, Huntington's disease, multiple system atrophy, and age-related macular degeneration. Neurodegenerative diseases usually extend over a decade, and the actual onset of neurodegeneration may precede clinical manifestations by many years. The mechanism that drives chronic progression of neurodegenerative diseases remains elusive. The current treatments are more for relieving symptoms and cannot effectively slow down the relentless neurodegenerative process because of the lack of regeneration of lost cells.

The requirement in treating neurodegenerative disease makes stem cell transplantation a promising therapy and a mainstream technology in AD, PD, and other neurodegenerative disorders because neurons and glia have been generated successfully from stem cells in vitro to develop stem cell-based transplantation therapies for human patients. Sufficient evidence from animal models of stem cell therapy has shown that stem cells can be incorporated into existing neural networks. They also secrete a variety of neurotrophic factors to modulate neuroplasticity and neurogenesis, which appear to increase brain acetylcholine levels, ultimately leading to improved memory and cognitive function in animal models. Intravenous infusion is the most preferred delivery method of stem cells, and MSCs are the most frequently used stem cell source to date (Clinical Case 13–2).

Clinical Case 13–2

Mrs. Smith is a 76-year-old widow who has been living alone for the last 3 years since her husband of 52 years died from lung cancer. Mrs. Smith had been managing her own affairs, including paying her bills, shopping for groceries, and preparing meals for herself and her neighbor. Her son regularly checks on her on the weekends and while talking to the neighbor became aware that his mother seemed to be forgetting things more frequently. While watching a special on the Korean War with her son, Mrs. Smith stated that she was surprised that the United States sent troops to South Korea. This somewhat alarmed the son because he knew that his mother had served as a nurse in the medical units in South Korea and had spent 1 year in the country. Mrs. Smith called her son after becoming confused while driving to her regular grocery store and becoming lost and unable to find her way home. At that point the son took the keys to the car away from his mother and offered to take her shopping and transport her to appointments. The son began to notice that he was having a harder time reaching his mother on the phone as it seemed to always be busy. One Saturday when the phone remained busy for 2 h, the son went to his mother's house to find her talking to a stranger on the phone who hung up immediately when the son got on the phone. Mrs. Smith stated that this stranger had informed her that she owed back income tax and that she would be arrested if she did not immediately mail $4000 to a PO box in Nebraska. She had just put the check in an envelope and was in the process of mailing it. The son had her phone number changed to an unlisted number so that the scammers could not reach his mother and took the check book away from her to avoid scammers taking all of her savings. At this point the son became concerned that his mother was having increased difficulty remembering things like have already paid income tax and with her judgment so he made an appointment at the local memory care center to have his mother evaluated.

Cell Biology, Diagnosis, and Treatment of Alzheimer's Disease

During the evaluation by the memory care center it was found that Mrs. Smith had the e4 allele of the apoliprotein E, which increases the risk of developing AD. A computed tomography scan of Mrs. Smith's head revealed atrophy in the temporal lobe and the frontal cortex. Neuropsychological testing using the mini-mental status examination suggested a clinical diagnosis of probable dementia and blood tests revealed Mrs. Smith to have normal thyroid function tests, normal B_{12} levels, normal renal function tests, normal hematocrit, and a negative test for syphilis. Mrs. Smith was diagnosed with Alzheimer's dementia and since she had been prescribed postmenopausal hormone replacement therapy for hot flashes her doctor stopped this medication as it may increase her risk of developing dementia. She was also prescribed donepezil, an acetylcholinesterase inhibitor, in hopes that this medication would slow the progression of her disease. Unfortunately, Mrs. Smith continued to have increasing difficulty managing her affairs and her son was forced to put her in an assisted living facility.

Most of the treatments for AD are drugs that are designed to temporarily increase acetylcholine. These drugs can improve the symptoms but are not able to inhibit the disease progression, which is an important feature of AD. Therefore repopulation and regeneration of depleted neuronal circuitry by exogenous stem cells is a rational therapeutic strategy. There are five clinical trials using MSCs to treat AD in the United States, two of them (ClinicalTrials.gov identifier: NCT02600130 and NCT02833792) are at present recruiting participants.

Skin and Wound Healing

Skin has the ability of self-healing and renewal upon injury, which is known as wound healing. Wound healing is a dynamic and complex process involving the release of cytokines, growth factors, and chemokines. The main steps of wound healing include hemostasis, inflammation, proliferation, and maturation. Hemostasis occurs immediately after wound injury, followed by inflammation, which contains damage to a certain extent by clearing pathogens and foreign material from the wound. Proliferation starts about 3 days postinjury. It mainly involves angiogenesis, granulation tissue formation, collagen deposition, and epithelial formation, which take about 2–3 weeks. The last step is maturation, during which the collagen type is restored to usual and the wound tissue matures, resulting in full cross-linking and restoration of a somewhat normal structure. The vascular network rapidly regresses as well. Any failure in these steps may lead to a chronic, nonhealing wound or abnormal scar formation.

Acute injury usually heals well even without any treatments; however, severe skin damage, such as a large area of a burn injury, not only requires debridement and anti-inflammation but also requires recovery treatments. Whether prognosis is good mainly depends on injury depth and areas. Other diseases, such as diabetes, can also affect the process of wound healing. Treatment of these severe wounds, no matter acute or chronic, can be long lasting and with little effect. Stem cell therapy was then introduced to wound healing treatment, which has shown a positive influence on animal models with severe wounds.

The simplest approach to applying stem cells is to spray cells onto the skin, or directly inject cells into the dermis. A bioscaffold-based stem cell delivery showed a better treatment effect. This technique provides stem cells with a microenvironment similar to that of skin, which can enhance cell-cell communication and accelerate wound repair by increasing proliferation, differentiation, and paracrine capacity of the stem cells. Materials of scaffold and precondition methods of stem cells are various. For example, studies with stem cell sheet combined with artificial skin showed a better survival rate of stem cells. Pretreatment of stem cells with special media containing growth factors or platelet-rich plasma, or with genetic modification, could enhance proliferation and differentiation of stem cells, eventually favoring the process of wound healing. Mechanisms of stem cell therapy involve residing cell stimulation, biomolecule release, inflammation control, and ECM remodeling.

Liver

Liver cells have a strong ability to regenerate, which leads to liver self-repair after injury. However, when severe damage is beyond the capacity of repair or chronic liver disease causes persistent injury, liver failure happens, leading to high morbidity and mortality. The most effective treatment option for both chronic and acute liver failure is liver transplantation; however, the shortage of available donor organs and high costs of surgery remain a huge drawback. Stem cell therapy as an alternative treatment of liver diseases has shown promising effects.

Nearly all types of stem cells were used for liver disease treatment in studies with animal models. Liver progenitor cells, MSCs, and HSCs are widely used, among which MSC transplantation is already applied to human patients with liver diseases. Multiple strategies have been used to increase stem cell survival and retention, and to lower toxicity. For example, clinical research showed that patients receiving repeated stem cell infusion gained better improvement in liver functions throughout a 12-month follow-up compared with patients who received a single infusion.

There are over 100 clinical trials on stem cell-based therapy for liver disease, which are mostly in the stages of phase I and phase II. MSCs isolated from adipose tissue or bone marrow are the most common cell sources. The major administration route is peripheral blood infusion.

Kidney

The kidney, unlike the liver that has a strong regeneration capacity throughout life, is more analogous to the heart, which has limited regenerative ability after injury. Although most acute kidney diseases are often reversible if found and treated quickly, the progress of chronic kidney diseases eventually leads to severe end-stage renal disease and kidney failure. Current treatments available for patients with kidney failure are dialysis and kidney transplantation. As with other organ transplantations, kidney transplantation is hampered by a shortage of donors and the use of immunosuppressants to prevent rejection. Stem cell transplantation as a therapeutic alternative was then initiated in the late 1950s.

Different types of stem cells, ranging from MSCs, ADSCs, renal progenitors, and so forth, can stimulate renal repair in animal models of kidney failure. Compared to other types, MSCs, especially from nonrenal origin, are the most administrated type for transplantation. Autologous MSCs are preferable because of the low risk of allosensitization, although both autologous and allogeneic MSCs have potential benefits as a therapeutic approach to reduce inflammation and promote endogenous repair. When using allogeneic stem cells, MSC identity and quality require careful evaluation for every derivation to ensure safety.

Compared to the number of preclinical studies, few clinical trials have been conducted so far. There are still a number of uncertain affecting factors that need to be resolved. For example, the identification of specific tissues, organs, or individual-derived MSCs that are most suitable for the treatment of kidney disease; the number of MSC transplantations to achieve renal function reversal; the stage of kidney disease progression at which MSCs can be used; long-term effects; and safety of MSCs after transplantation.

Diabetes

Diabetes is a chronic metabolic disease characterized by high levels of glucose in the blood, which can be divided into two main etiopathogenetic categories: Type 1 diabetes mellitus (T1DM) and Type 2 diabetes mellitus (T2DM). In T1DM the body's immune cells attack and damage insulin-producing β cells in the islets of the pancreas, resulting in the deficiency of insulin secretion. T2DM is a more prevalent category caused by a combination of insulin resistance and inadequate insulin secretory responses and functions. Treatment for patients with T1DM is immunosuppressant agents, which do not repair the function of β cells. Current treatment for hyperglycemia caused by T2DM is to supplement the patients with insulin, which only gives limited control over the patients' blood sugar levels. All of the current treatments for patients with either T1DM or T2DM do not reproduce/repair damaged β cells, indicating that diabetes is irreversible and eventually progresses to severe diabetic conditions. At the end stage of the disease, the only effective treatment is transplantation of the pancreas, but this is limited by the shortage of donors. The use of stem cells in treating diabetes is a new hope for patients.

Numerous studies have been performed on animals using various types of stem cells, especially MSCs. Animal models have shown that stem cell transplantation is safe and effective. In 2007, the first clinical trial assessing stem cell transplantation treatment for T1DM was reported. Stem cells can be pretreated before transplantation, for example, stem cells are cocultured with the patients' blood lymphocytes. Compared with the numbers of clinical trials for T1DM, fewer clinical trials have been conducted in stem cell treatment for T2DM. Most of the clinical trials are not randomized and had a relatively small sample size. Further investigation, especially randomized, controlled, double-blind clinical trials on a large sample population of patients, should be conducted to draw a firm conclusion for stem cell therapy for diabetes.

CELL-FREE THERAPEUTIC STRATEGY IN REGENERATIVE MEDICINE

The key mechanisms of action of stem cells in regenerative medicine include not only the cells "homing" to the injury sites and differentiation into multiple cell types to replace or restore damaged tissues, but also the secretion of bioactive factors to regulate local or systemic physiological processes. Recent studies have paid more attention to the effect of bioactive factors produced by MSCs on the regulation of tissue repair and regeneration. These factors/molecules secreted from cells to the extracellular space are called secretomes, which consist of soluble proteins, free nucleic acids, lipids, and extracellular vesicles. The latter can be subdivided into apoptotic bodies, microparticles, and exosomes. The secretome from MSCs, including BMSC, ADSC, and DPSC, is emerging as a cell-free therapeutic strategy for multiple diseases. There are two major forms of secretome that are used in the studies, including conditioned media (CMs) and exosomes/microvesicles.

CM represents the complete regenerative milieu of cell-sourced secretomes and vesicular elements. The soluble components of the secretome can be separated from the microvesicle fraction by centrifugation, filtration, polymer precipitation, ion-exchange chromatography, and size-exclusion chromatography. Both of these components are capable of independently triggering regeneration and repair as well as of mediating the de novo organogenesis of tissue-engineered organs ex vivo. It has been demonstrated that MSC-derived CM is as effective as the transplantation of the corresponding MSCs in the treatment of diseases in numerous animal models.

An exosome is defined as a specific class of lipid membrane-bound extracellular vesicle characterized by a diameter of 40–150nm and a density of 1.09–1.18g/mL. There is increasing evidence that has shown that MSCs produce massive amounts of exosomes in comparison with other cells and that many of the regenerative properties of MSCs are mediated through secreted exosomes. Exosomes can be internalized by other cells principally by phagocytosis, fusion with the cell membrane, and receptor–ligand interaction, allowing the release of their contents into the cytoplasm. It has been reported that treatment with MSC-derived exosomes and microvesicles improves clinically relevant organ functionality.

The secretome of ADSCs or BMSCs has been widely studied. As they display similar secretory profiles, the secretome of ADSCs is discussed here as an example. There are numerous studies showing the beneficial effects that the ADSC's secretome exerts in the CNS, immune system, heart, muscle, and even in general cell survival. The profile of the secretory products of ADSCs has been characterized by mass spectrometry, which identified more than 100 individual proteins in the CM. Among these proteins, 41% are associated with cytoplasmic components, 15% are known as secretory factors, 11% are derived from nuclei, 4% are from endoplasmic reticulum, 3% are attributed to the ECM, 4% originate from mitochondria, and 22% are not associated with a specific subcellular location. Their functions involve regulation of cell metabolism, cytoskeletal resembling, cell signaling transduction, protein inhibition, protein degradation, protein processing, chaperone activity, DNA repair, ECM formation, iron storage, guanosine diphosphate-binding protein activity, and mitosis.

ADSCs' secretome profile can be modulated by exposure to different agents. For instance, ADSCs exposed to basic fibroblast growth factor or epidermal growth factor gained a significant increase in their release of hepatocyte growth factor (HGF), a cytokine with a role in both hematopoiesis and vasculogenesis. ADSCs also respond to an inflammatory stimulus, such as lipopolysaccharide. Under these conditions, the ADSCs increased their secretion of both hematopoietic (granulocyte-macrophage colony-stimulating factor and IL-7) and proinflammatory (IL-6, -8, -11, and tumor necrosis factor-α) cytokines. These characteristics make the secretome of stem cells a potentially powerful tool in future approaches in developing cell-based therapeutics for regenerative medicine (Fig. 13–13).

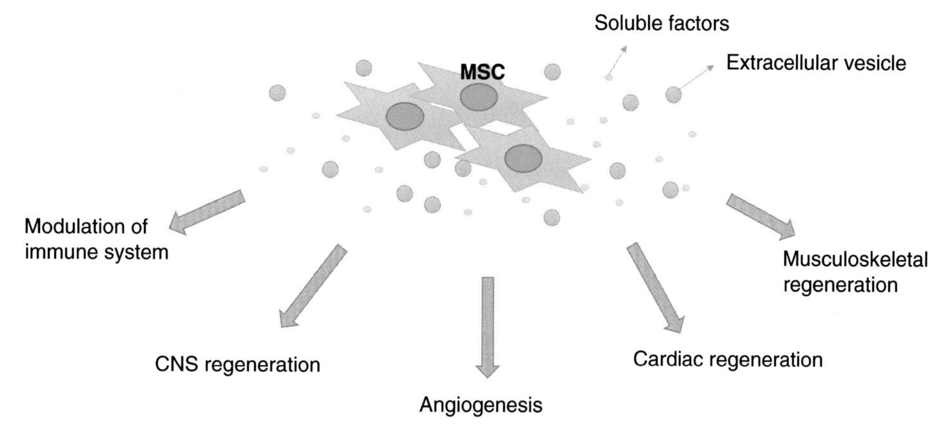

Figure 13–13. MSC secretome and regenerative medicine. The broad repertoire of MSC secretome makes it a possible strategy for regenerative medicine, including immunomodulation, CNS regeneration, angiogenesis, cardiac regeneration, musculoskeletal regeneration, and future enlisted applications.

Modulation of the Immune System

MSCs regulate the proliferation, activation, and function of immune cells. Both preclinical studies in animal models and interventional clinical studies involving MSCs (http://clinicaltrials.gov) have shown their suppressive effects on both innate and adaptive immunity. MSCs typically express major histocompatibility complex-I (MHC-I) but lack the expression of MHC-II, CD40, CD80, and CD86 on the cell surface, and thus they escape T-cell recognition and often fail to induce an immune response by the transplant host. Therefore the most promising potential applications involve the treatment of graft versus host disease, as well as of autoimmune and inflammatory diseases, such as systemic lupus erythematosus, T1DM, multiple sclerosis, and Crohn's disease.

Central Nervous System

The possible validity of MSCs for CNS regeneration has been demonstrated in different models of injury and degeneration. These beneficial actions were initially attributed to the possible differentiation of MSCs toward the neuronal lineage. However, despite all the efforts made on this topic in recent years, there is no conclusive data yet on the in vivo differentiation of functional neurons from MSCs. Therefore the strongest speculation related to the role of MSCs in neuronal regeneration is mainly focused on their possible trophic support through their secretome on the host cells.

It has been shown that ADSCs' secretome has a neuroprotective effect in a rat model of brain hypoxic-ischemic injury. In this study, treatment by ADSC-CM infusion significantly protected against hippocampal and cortical volume loss, and promoted behavioral and learning ability by using a Morris water maze functional test. Particularly interesting is the confirmation of penetration of the blood-brain barrier by protein components of the injected CM, as well as their predominant binding to structures within affected regions of the brain.

The beneficial effects of secretome on CNS include modulation of the inflammatory environment on the site, enhanced vascularization of the regenerating site, increased thickness of the myelin sheaths, modulation of the Wallerian degeneration stages, and reduction of fibrotic scarring. However, studies on more in-depth characterization of the neurotrophic factors present in the medium are mandatory for future application.

Vascularization

Angiogenesis, the process that leads to the formation of new blood vessels, is a key event in any process of regenerative medicine. Various studies have demonstrated the effect of MSC-derived secretome on key steps of angiogenesis. For example, MSCs of different populations (e.g., adipose, amniotic, bone marrow, and Wharton jelly umbilical vein) induce proliferation and migration of endothelial cells to promote tube formation, as well as prevent endothelial cells from apoptosis in vitro. A number of angiogenic stimulators and inhibitors have been identified in MSC secretome, including HGF, vascular epithelial growth factor (VEGF), TGF-β, FGF-2, placental growth factor, tissue inhibitor of metalloproteinase-1, and angiopoietins 1 and 2. These factors in MSC-CM may represent a balanced cocktail that acts in concert to promote angiogenesis.

Moreover, several studies demonstrate that the secretion of these angiogenic factors can be modified depending on chemokines and hypoxic conditions. It was shown that when cultured in hypoxic conditions, ADSCs were able to increase their VEGF secretion fivefold. CM obtained from ADSCs grown under hypoxic conditions were able to increase the growth of endothelial cells and simultaneously reduce the apoptosis rate of this population of cells.

Wound healing is another process for which the formation of new blood vessels is also extremely important, and MSC-CM has potential utility in such conditions. Studies have shown that the secretome of ADSCs could control fibroblast proliferation/migration and protect fibroblasts from oxidative stress, which obviously affect the time required for wounds to heal. Studies using human uterine cervical stem cell (hUCESC)-CM on epithelial healing in animal models of dry eyes after induction of corneal epithelial ulcer have shown promising results. After the injury, dry eyes treated with hUCESC-CM showed improved epithelial regeneration.

Musculoskeletal Tissue Regeneration

The therapeutic effect of MSC secretome in musculoskeletal diseases is a new frontier for regenerative medicine. In the past 10 years, more than 40 studies (including 12 in vivo studies and 1 clinical study) have tested CM or extracellular vesicles from multiple cell types in musculoskeletal tissue regeneration. In vivo studies were performed with CM in bone and periodontal defects, arthritis, and muscle dystrophy pathologies. CM from MSCs induces new bone formation, osteogenic differentiation of the resident MSCs, angiogenesis, suppression of inflammation, reduction in arthritis index, reduction of osteoclastogenesis, and cartilaginous tissue formation. There was a clinical study using CM from human MSCs in patients needing alveolar bone regeneration. Gradual bone formation was observed within a 6-month follow-up, with no systemic or local complications. The possible factors involved include platelet-derived growth factor receptor β, C3a, VEGF, monocyte chemoattractant protein-1 and -3, IL-3 and -6, and insulin-like growth factor (IGF)-1, which are responsible for cell migration, proliferation, osteogenic differentiation, and angiogenesis.

Cardiac Regeneration

Cardiac regeneration has been another area where MSCs have been applied and the effects of the secretome of MSCs have been observed. In fact, there have been a number of studies reporting positive effects of MSCs after they were transplanted/injected into different animal models of myocardial infarction/failure. Similar to the application of MSCs in the models of other tissue ischemic injuries, the improvement in cardiac regeneration in animal models was attributed to an increase in vascularization and/or a cell-protective effect, namely by reducing apoptosis. VEGF and IGF-1 were identified as the contributing factors to the phenomenon. In addition to the growth factors mentioned, it is likely that other MSC-derived proteins also contribute to this favorable outcome; however, additional studies will be necessary to validate this speculation.

Advantages and Questions

The use of cell-free therapies such as MSC-sourced secretome in regenerative medicine provides a number of advantages over stem cell-based applications. (1) Application of the secretome resolves several safety considerations potentially associated with the transplantation of living and proliferative cell populations, including immune compatibility, tumorigenicity, emboli formation, and the transmission of infections. (2) MSC-sourced secretome may be evaluated for safety, dosage, and potency in a manner analogous to conventional pharmaceutical agents. Storage can be done without the application of potentially toxic cryopreservative agents for a long period of time without loss of product potency. The biological product could be modified to desired cell-specific effects. (3) The time and cost of expansion and maintenance of cultured stem cells could be greatly reduced, and off-the-shelf secretome therapies could be immediately available for the treatment of acute conditions such as cerebral ischemia, myocardial infarction, or military trauma.

Up to now, there are still questions to be answered before the application of MSCs' secretome. A better understanding of the secretome should be attained with the possibility of discovering new potential factors and by application of unbiased, global discovery techniques such as liquid chromatography/mass spectrometry. In addition to this information, it is advisable that serial studies be made on how the secretome changes along with the passage of cells. Additionally, it remains unclear if it is preferable to transplant MSCs or, alternatively, their CM; both approaches may have a therapeutic potential best suited to a particular pathology dependent on the target disease/injury. If a prolonged release of factors is needed, then MSC transplant may be beneficial. The use of genetically or culture condition-modified MSCs capable of releasing induced levels of a particular cytokine or growth factor merits consideration. Alternatively, if an initial burst of factors is desired, then the injection of concentrated CM from MSCs would be the most logical approach.

Another relevant topic is how the secretome of MSCs can be modulated. A possible approach could be through the use of bioreactors, as MSCs are sensitive to dynamic environments. Another possibility would be the use of defined media that specifically trigger the expression of the desired trophic factors. Alternatively, the growth of these cells in suspension-based bioreactor systems and in contact with engineered biomaterials surfaces could also be possible routes to follow. With these approaches, it will probably be possible to expand the applications of MSCs, and their secretome, to a wider range of regenerative medicine protocols in a rational, evidence-based manner.

SUMMARY AND PERSPECTIVE

ESCs constitute the entire organ systems of our body through proliferation and differentiation starting from a single fertilized egg at the beginning of the embryonic stage. Adult stem cells maintain the structural and functional integrity of the organ systems throughout the lifespan through mobilization from the niche, self-renewal for propagation, homing to the destination, and target tissue lineage differentiation when an injury occurs in an organ system. It could be speculated that ESCs determine the quality of our organ systems at birth and adult stem cells maintain our health status after birth. Therefore the quality and quantity of endogenous adult stem cells would determine our healthy lifespan. For this reason, the focus in the future of stem cell research and technology development would be shifted to the promotion of healthy longevity although our efforts currently focus more on the treatment by stem cells of chronic diseases. In this context, the most challenging technology of regenerative medicine will be how to ensure the availability, either endogenously or exogenously, of stem cells to the needed organ systems.

Suggested Readings

Golchin A, Farahany TZ. Biological products: cellular therapy and FDA approved products. *Stem Cell Rev Rep* 2019;15(2):166–175.

Harasymiak-Krzyżanowska I, Niedojadło A, Karwat J, Kotuła L, Gil-Kulik P, Sawiuk M, Kocki J. Adipose tissue-derived stem cells show considerable promise for regenerative medicine applications. *Cell Mol Biol Lett* 2013;18(4):479–493.

Kang YJ, Zheng L. Rejuvenation: an integrated approach to regenerative medicine. *Regen Med Res* 2013;1(1):1–8.

Khan FA, Almohazey D, Alomari M, Almofty SA. Isolation, culture, and functional characterization of human embryonic stem cells: current trends and challenges. *Stem Cells Int* 2018;2018:1429351. https://doi.org/10.1155/2018/1429351.

Robert L, Anthony A. Essentials of Stem Cell Biology. 3rd ed. Elsevier; 2005.

Salgado AJ, Gimble JM. Special section: the mesenchymal stem cell secretome in regenerative medicine. *Biochimie* 2013;95(12):2195–2464.

Scadden DT. The stem-cell niche as an entity of action. *Nature* 2006;441(7097):1075–1079.

Trávníčková M, Bačáková L. Application of adult mesenchymal stem cells in bone and vascular tissue engineering. *Physiol Res* 2018;67(6):831–850.

Walmsley GG, Ransom RC, Zielins ER, Leavitt T, Flacco JS, Hu MS, Lee AS, Longaker MT, Wan DC. Stem cells in bone regeneration. *Stem Cell Rev Rep* 2016;12(5):524–529.

Wright DE, Wagers AJ, Gulati AP, Johnson FL, Weissman IL. Physiological migration of hematopoietic stem and progenitor cells. *Science* 2001;294(5548):1933–1936.

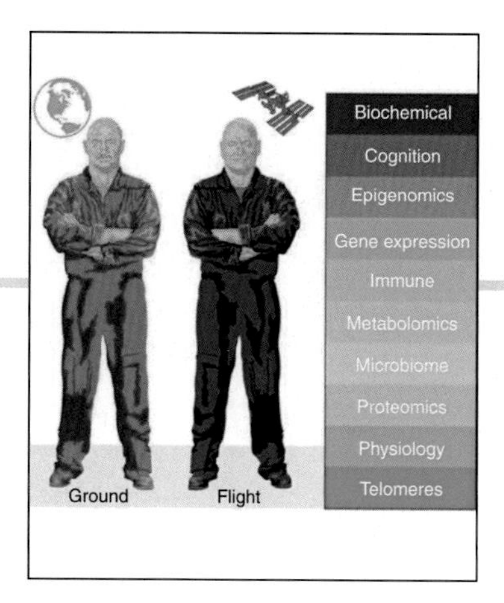

Chapter 14

Omics, Informatics, and Precision Medicine

INTRODUCTION

The pace of precision and genomic health care continues to accelerate. Clinical applications based on differences among individuals are critical in cancer therapeutics, anticoagulation therapy, and organ transplantation. A maxim of health care has always been to *treat the patient, not the disease*, but this integrative and personalized approach is demanding, particularly in fast-paced and hard-pressed healthcare systems. The drive toward more individualized care, known as precision health care, has many roots, but the Human Genome Project has been the major modern catalyst. Due to relentless progress in genetics, genomics, molecular and cellular biology, and imaging we have far better ways to define and quantify individual differences at many levels—from DNA to biomarkers, to body and brain composition, to environmental exposures. Statistical methods, database systems, and computational power have also increased at a matched pace. Heterogeneous clinical data are now routinely integrated into sophisticated electronic health record systems that are becoming the computational partner in diagnostic and therapeutic decision making. The convergence of classical diagnostic methods with these new types of dynamic and multiscalar data is beginning to deliver improved care and treatment in a way that accounts to a greater extent for each individual's unique genetics, biochemistry, demographic characteristics, exposure history, and lifestyle.

The term *omics* in this chapter refers to the suite of diverse and often large datasets collected at different levels or scales of biological organization—DNA sequence (genomics), DNA and histone modifications (epigenomics), RNA expression (transcriptomics), protein and peptide levels and distribution (proteomics), metabolite concentrations (metabolomics), microbial communities throughout the body (metagenomics, see Chapter 15), and the vast constellation of classic healthcare data types such as weight, height, blood pressure (phenomics), and environmental exposure (the exposome). Compared to single-level snapshots, this dynamic stack of data types has the potential to provide a far more holistic understanding of an individual, their unique risks, and better ways to prevent and treat diseases and conditions. The NASA Twin Study is a preview of this new wave of complex multiomics data (Fig. 14–1). In this case, monozygotic twins were studied over 2 years with the goal of determining the impact of nearly 1 year in orbit (microgravity) on health. The same general type of multiomic longitudinal data are now being used to fine tune care for common diseases such as diabetes. We are moving beyond the generic one-size-fits-all recommendations to specific predictions. However,

Goodman's Medical Cell Biology. https://doi.org/10.1016/B978-0-12-817927-7.00014-4

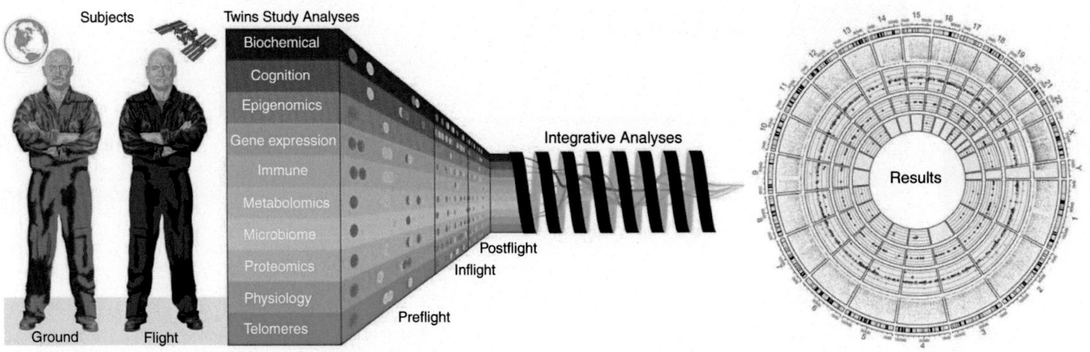

Multidimensional, longitudinal assays of the NASA Twins Study

Figure 14–1. **Longitudinal multiomics data integration.** This process is at the frontier of biomedical research and clinical care. To the *left*, the Kelly monozygotic twins—Mark to the far *left in green* served as experimental control for his brother and astronaut Scott Kelly *(blue)*. The experimental design and many data types that were acquired are listed in the center of the figure. The twins were studied at many of the same timepoints using the latest omics methods before, during, and after Scott's 340 days on the International Space Station. To the *far right* is a circular plot of data types that have been aligned by chromosome number (chromosome 1 at the 3 o'clock position). For example, the wider outer band in *red* summarizes gene expression levels in blood. *(From Garrett-Bakelman FE, et al. The NASA Twins Study: a multidimensional analysis of a year-long human spaceflight.* Science *2019;364(6436). pii: eaau8650. PMID: 30975860.)*

establishing the accuracy of specific recommendations is still a major challenge.

There have been notable successes, particularly in choosing appropriate doses for chemotherapeutic drugs based on the patient's genotypes. For example, the dosage for thiopurines often used to treat childhood leukemias in now set based on each child's *TPMT* gene variants. The effective dose varies more than 10-fold. A series of decisions based on specific leukemia subtypes and well-defined genotype-by-drug interactions is the single largest factor in the striking improvement in pediatric leukemia survival rates—from 65% up to 90% over the last three decades. But in general, we do not have an adequate understanding of complex relations among many layers of genetic and omic data that lead to better or worse outcomes. The more complex the disease or condition, and the more the disease is modulated by environmental cofactors, the harder it becomes to predict the best treatment options. While the human genome is made up of about 22,000 pairs of genes and millions of DNA variants, the great majority of these variants have no known function that we can yet link to diagnosis, treatment, and prognosis. We are at the beginning of this process of precision, and more individualized health care is needed. Overpromising is a serious risk given the probabilistic nature of even the best of predictions. Humans are far more complex than the weather. This does not necessarily mean that clinically useful predictions will be harder than predicting the weather in a week, but it is a good reminder to be humble.

GENOMES, GENES, AND DNA VARIANTS

There is substantial evidence that genetic variation of many different types and frequencies affect disease risk, progression, and treatment. We often estimate the cumulative role of DNA difference on traits such as eye color, stature, and disease risk by a deceptively simple and misunderstood measure called *heritability*. Rare gene variants and mutations that almost always lead to clear-cut diseases or disorders have heritabilities close to 1 (or 100%)—a strong concordance of DNA variants (alleles) to the occurrence of trait or disease in a particular environment. In contrast, conditions that that are modulated by many developmental, stochastic (random) events, and that are more sensitive to environmental factors—blood pressure, body weight, cardiovascular and pulmonary diseases, metabolic disorders, substance abuse, and personality traits—typically have heritabilities of 0.5 or less. But the key point to make in this context is that high heritability does not limit treatment options. In fact, the right treatment can potentially lead to a cure even of a monogenic Mendelian mutation. Phenylketonuria is a classic example of a Mendelian metabolic mutation that inactivates the *PAH* gene and leads to the accumulation of dangerously high levels of the amino acid phenylalanine. While this disorder has a nominal heritability close to 1, it can nonetheless be treated effectively by carefully eliminating phenylalanine from the diet. This highlights the crucial role of gene-by-environmental or gene-treatment effects even for traits with high heritability—effective treatment trumps heritability. And we now are entering an era in which it is becoming practical to repair DNA sequences for severe genetic diseases such as sickle-cell anemia using precisely targeted DNA editing methods—in particular variants of CRISPR-*Cas9* gene editing. While there are serious ethical challenges, there are also many patients for whom fixing otherwise lethal mutations, such as those in *CFTR* (cystic fibrosis) and *HTT* (Huntington's disease), is warranted.

The human genome consists of about 6.5 billion base pairs (bp) of DNA—3.25 billion bp inherited from each parent—two sets of 22 autosomes, chromosomes X and Y, and the maternally inherited mitochondrial genome. Roughly 5% of the nuclear genome encodes well-defined exons and introns of genes that produce mRNAs, rRNAs, tRNAs, and microRNAs, but only 1.3% corresponds to a protein-coding sequence that is translated into a core set of approximately 22,000 proteins, but actually hundreds of thousands of major and minor protein variants—so-called isoforms—per cell type. Most of the remaining DNA sequence is made of noncoding and repetitive nucleic acid sequence—often called junk DNA. However, an important fraction of the nonprotein coding genome is made up of crucial stretches of DNA that modulate gene expression, mainly enhancer regions and insulators. These regions are often encrusted with proteins, and collectively these DNA-protein aggregates make up a complex genomic–epigenomic switch box that selects a subset of genes that should be most actively transcribed or most strictly suppressed in each cell type under different conditions.

Maternal and paternal chromosomes typically differ at 2–4 million DNA sites in any one human—roughly one difference per 1000 bp. The most common type of DNA sequence variant is a *single nucleotide polymorphism* or SNP—a simple substitution of one nucleotide for another. Roughly 90 million SNPs have already been defined in human populations, but only 10% are common and detected in 5% or more individuals. Each human genome also includes hundreds of thousands of small DNA insertions and deletions (indels), a few thousand larger copy number variants that may include one of more whole duplicate genes, and other even more massive rearrangements, inversions, losses, and duplications of entire chromosomes (Fig. 14–2). Finally, we all harbor a small number of spontaneous de novo germline variants—the ultimate source of a subset of genetic problems, but also one of the key sources for evolutional change. There are probably fewer than 100–200 de novo variants in any one human. Most will be neutral and have no detectable effects, but 10 or fewer may be the types of mutations that affect RNA expression or protein sequence. Finally, as cells divide they acquire rare somatic mutations. These mutations accumulate as a function of mitotic frequency, environmental exposures, and the fidelity of DNA repair. They are of particular importance in cancer biology.

Most DNA variants—those that have a frequency above 5%, do not contribute to disease risk in isolation. However, a combination of many genetic differences that individually have modest effects can contribute to striking differences in disease risk. Most common diseases, particularly those jointly affected by many environmental factors, are modulated by 100–1000 common DNA variants. Type 2 diabetes is an example of a so-called polygenic disorder, the severity of which is influenced by hundreds of common DNA variants, and of course diet and lifestyle. The same is true for many cardiovascular, neurological, and renal diseases. While many cancers are triggered by somatic mutations and chromosomal rearrangements, rates of tumor growth and metastasis are strongly modulated by the inheritance of common germ-line variants. It is possible to sum up the effects of all of these common DNA variants to derive a cumulative polygenic risk score—a rough prediction of the

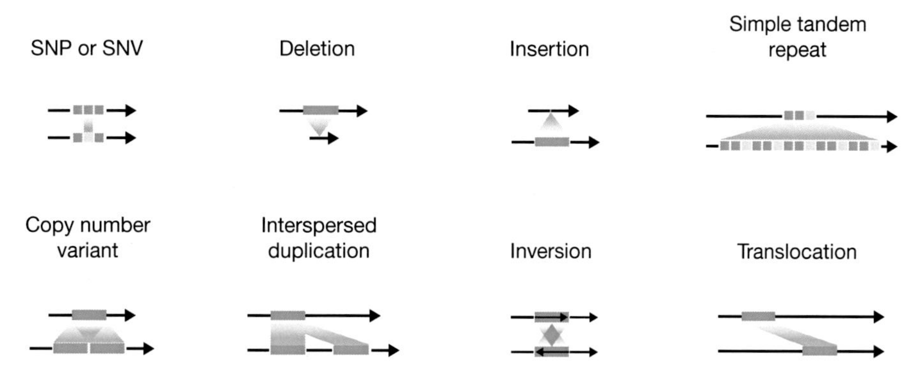

Figure 14–2. **Types of DNA sequence variants.** The DNA variants highlighted in the first row are from single nucleotide polymorphisms (or single nucleotide variants) to difference of up to 200 nucleotides, for example, simple tandem repeats (also known as microsatellites) such as that which causes Huntington's disease. The lower row of variant types generally involves much longer segments of DNA (>200 nt)—from repetitive DNA fragments and mobile DNA elements (sometimes called jumping genes) to inversions, duplications, translocations, or deletions. These latter types of mutations are common in cancer cells.

likelihood of disease given comprehensive data on genotype across all chromosomes.

There are two major approaches to discovering the causal relation between DNA differences and differences in disease risk—family-based linkage studies and genome-wide association studies (GWAS). The first method has been used for nearly a century to find chromosome regions containing mutations that cause Mendelian diseases such as phenylketonuria, sickle-cell anemia, albinism, and Huntington's disease. These rare mutant alleles have large effects, and knowing the genotype is a nearly perfect predictor of risk. While it may be possible to predict disease, this does not mean that mechanisms are known. Huntington's disease, for example, is caused by a simple tandem repeat (Fig. 14–2, top right) of 36 or more CAG codons that are translated into a long stretch of glutamine amino acid residues in the first exon of the HTT protein. Yet, it is still unclear how this dominant mutation specifically disrupts survival of neurons in specific parts of the forebrain, and there is not yet an effective treatment.

GWAS methods (Fig. 14–3) now dominate human genetics, and the same GWAS techniques are now common in some areas of clinical practice as an effective first-pass way to improve patient care—particularly the initial dosing of drugs known to interact with key enzymes and transporters in gut and liver. The techniques involve extracting DNA from saliva or blood and hybridizing fluorescent DNA fragments to millions of complementary oligonucleotides that have been attached to the surface of a small coated piece of glass (about 1 cm on edge). The intensity of fluorescence at each of millions of coordinates on this glass microarray is a measure of the binding affinity of a DNA fragment. These images are then converted into the probable SNP genotype of the DNA fragments. Many clinics, hospitals, companies, and direct-to-consumer genotyping services currently use SNP microarray technologies that generate from 0.5 to 5 million SNP readouts per sample. That is enough to derive a very good first approximate of a patient's genome without sequencing all 6.5 billion bp. Although it is somewhat more expensive, as sequencing becomes less expensive the trend now is to sequence entire exomes (all protein-coding sequences in the genome) or even the entire genome—so-called *full genome sequence*. Sequencing can provide a more complete compendium of DNA variants in each individual.

GWAS studies, whether based on SNPs or genome sequence, typically rely on large populations, from a thousand to a million subjects. Subjects are categorized either as *cases* with disease or as matched *controls* (Fig. 14–3, middle). The relative frequencies of each of the DNA variants, or allele variants, is compared between cases and controls. If one or more alleles are much more common in cases than the controls, then these variants and the surrounding region of the chromosomes are causally linked to disease risk. Perhaps the single most important result of GWAS has been the proof that most common diseases and traits are modulated by hundreds of distinct DNA variants, many of which do not even overlap gene or protein-coding sequences. Most of these variants are in the large intergenic stretches of DNAs between genes, and they are likely to influence expression of genes by affecting DNA conformation and by affecting interactions of DNA with protein complexes, polymerases, and cofactors (Fig. 14–4).

Despite the extraordinary throughput of sequencing technologies, there are limits to the ability of current technologies to resolve some extremely important and relevant DNA variants: i.e., genome rearrangements, insertions or deletions of mobile elements (often derived from viral genomes), gene amplifications, and inversions and translocations (Fig. 14–2, bottom row). These variants can have important effects, and new long-read technologies that can sequence up to 200 kb of DNA in a single test will soon give clinicians a much more complete understanding of the first layer of the omics stack.

Clinical Case 14–1 Targetable Gene Fusion in Cancer Treatment

Gary Gardner is a 55-year-old male who first came to his physician complaining of lower back pain. While he initially was diagnosed with a slipped disk, the pain became referred to the left flank area. After an emergency department visit, a chest X-ray was performed and he was referred to a hematologist/oncologist for concern for a pulmonary embolus but came in with incomplete records. He continues to complain of severe back pain and recent loss of appetite. He tells the hematologist/oncologist that he has never smoked and he denied prior alcohol use. He has no family history of cancer. On exam he was not short of breath, but he had point tenderness over the T9–T12 vertebrae. Laboratory tests revealed positive fecal occult blood but otherwise no significant findings. Computed tomography (CT) of the chest showed diffuse esophageal wall thickening together with a trace left pleural effusion. There were multiple lesions in the liver as well as a 3.5-cm right adrenal lesion. There were also bony lesions at the T10 vertebra. A magnetic resonance image of the brain was performed that showed a 4-mm lesion in the left parietal lobe. Liver biopsy was performed that showed metastatic poorly differentiated carcinoma. Despite extensive pathology testing, the site or origin of the cancer was unable to be determined though it was considered consistent with an esophageal, lung, or biliary tract cancer. The remaining tissue from the biopsy was sent out for comprehensive molecular testing. Based on liver RNA-sequencing (RNA-seq), an NTRK1-IRF2BP2 gene fusion was noted. While awaiting the results, the patient's status declined precipitously, and he was ultimately hospitalized on IV fluids with wasting and severe pain. The patient was about to be hospitalized when the NTRK1-IRF2BP2 gene fusion was identified. Larotrectinib was rush ordered for the patient while in hospital. After 2 days of oral treatment,

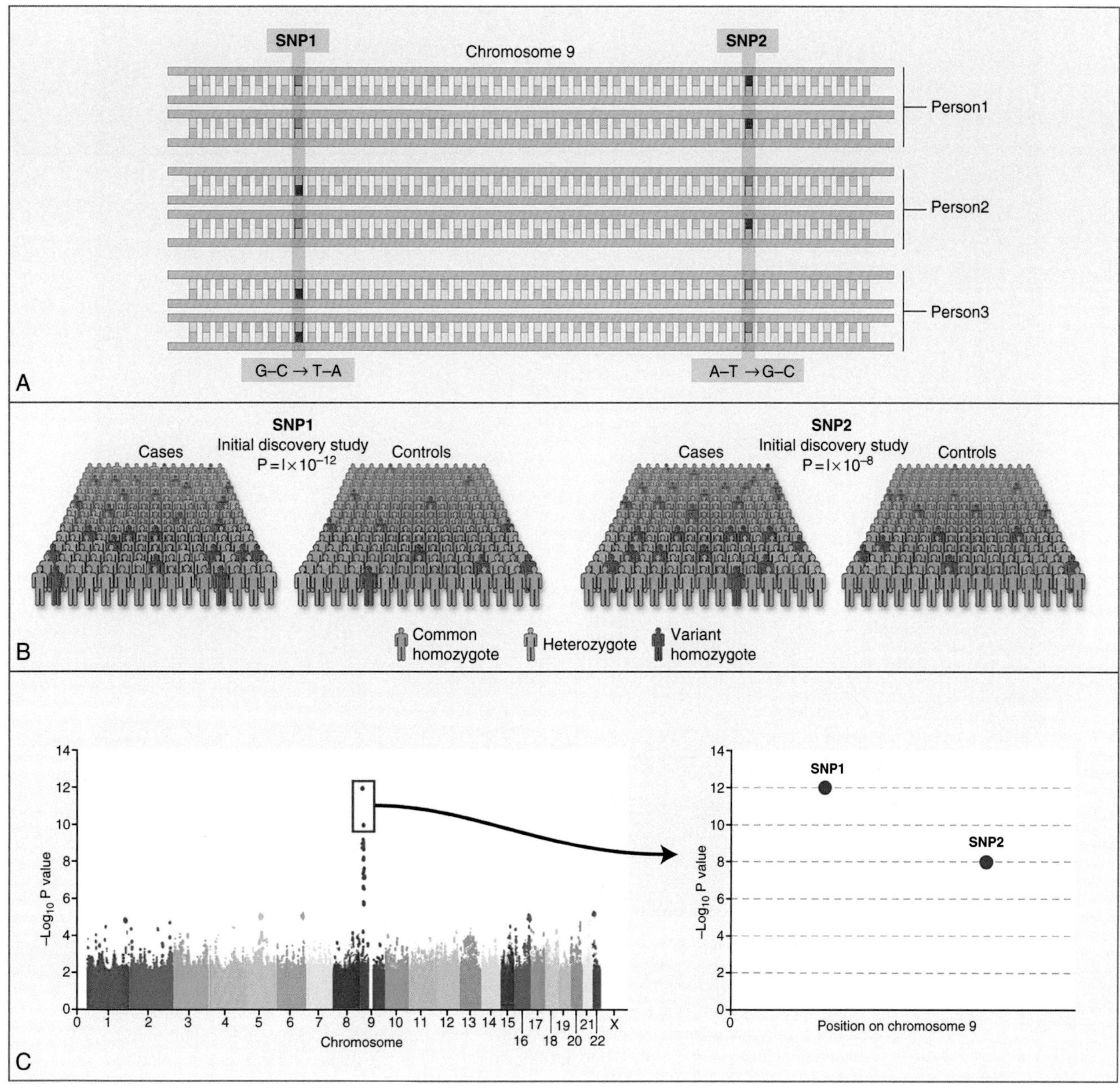

Figure 14–3. Genome-wide association studies (GWAS). GWAS are typically based on a case-control design in which a million or more SNPs are genotyped for each of 1000s of subjects. (A) depicts DNA sequence and genotypes for three subjects on chromosome 9. Two highlighted SNPs are used to genotype all subjects. In (B) the strength of linkage between SNPs and cases versus controls is calculated as a probability. In this example, SNP1 and SNP2 are associated with disease, and with the best *P* values of 10^{-12} and 10^{-8}. The *P* values are often converted to –log *P*—in this case values of 12 and 8, respectively. (C) shows the *P* values for all genotyped SNPs with each chromosome shown in a different color. The results define a small part of chromosome 9 that causally contributes to disease risk. *(From www.nejm.org/doi/full/10.1056/NEJMra0905980.)*

Mr. Gardner felt significantly better and was able to leave the hospital, though he still required pain control. A repeat CT scan 6 weeks later showed near complete remission of liver, esophageal, and adrenal lesions, partial resolution of bony lesions, and complete resolution of brain lesion.

Cell Biology, Diagnosis, and Treatment of NTRK Fusions in Cancer

Most genes that ultimately cause cancer are not abnormal in all cells. Instead, these genes become mutated only in cells that ultimately go on to become cancerous. These genes, known as oncogenes, can become activated in a number of different ways. The DNA within the gene may become mutated at a single point in the DNA, resulting in a gene that stimulates cell division and growth without effective negative feedback control. Alternately, genes can become oncogenic when part of the gene becomes spliced onto part of another gene during cell division in a process known as chromosomal rearrangement. This can result in the loss of the portion of the gene that prevents indiscriminate transcription and its replacement with a portion of another gene that may activate transcription, turning what was formerly a benign gene into an active cancer-promoting gene.

Continued

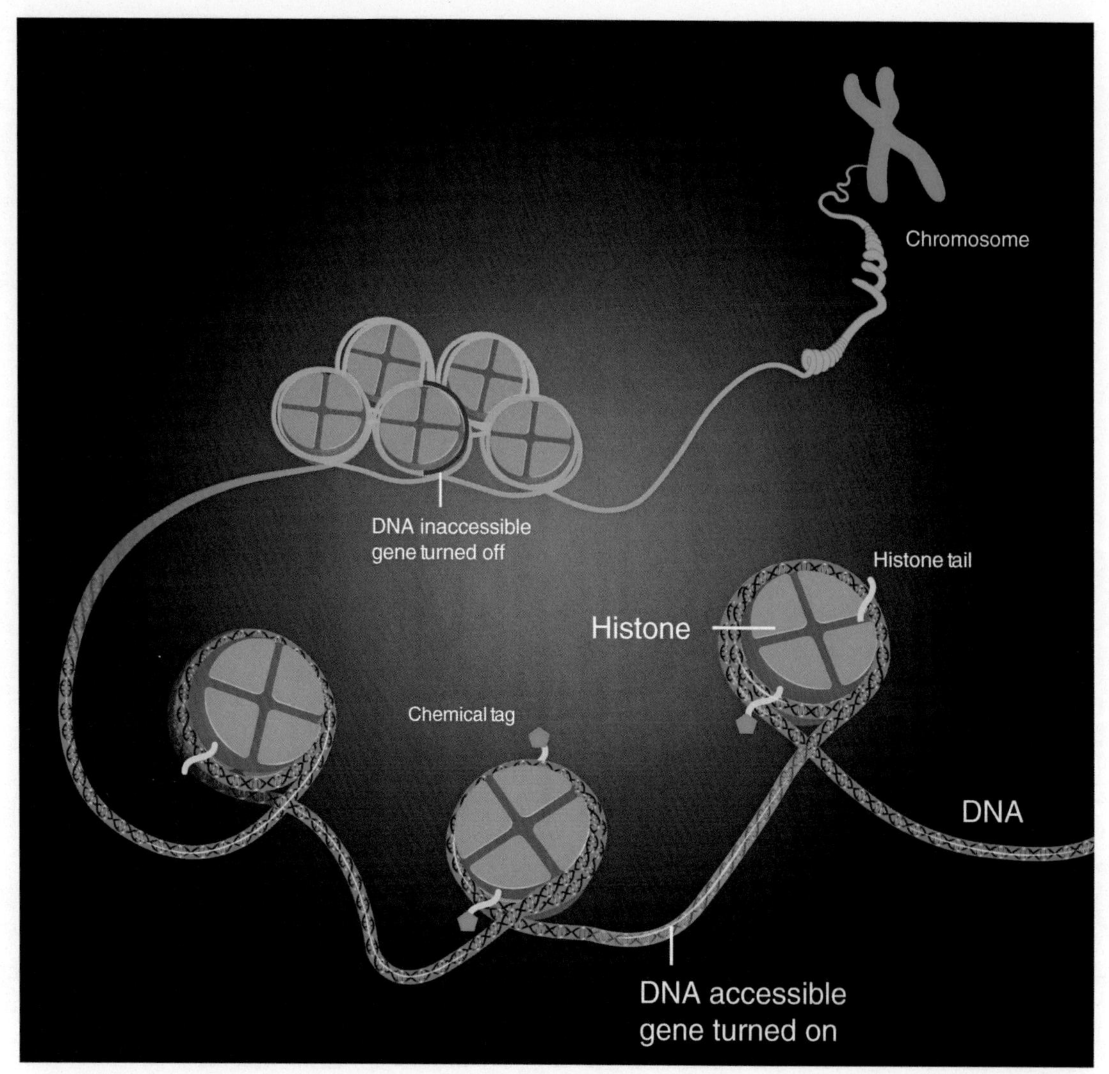

Figure 14–4. Epigenetics. Chromosomes have a complex and dynamic DNA-protein architecture that controls both DNA replication and the production of specific subsets of RNAs. The noncoding DNA is highlighted in *green*, whereas two protein-coding DNA regions are labeled *red*. One of these genes toward the bottom is accessible by transcription factors and RNA polymerase (not shown) and is in the on-state. DNA is also wrapped around sets of histone proteins, and in this state the DNA cannot be transcribed. The short *yellow* peptide tails of the histone H3 proteins can be modified in several ways—by phosphorylation, acetylation, methylation, glycosylation, or ubiquitination—and this, in turn, modifies chromatin structure and dynamics. *(Taken from the NIH Genome.gov web site).*

Neurotrophic receptor tyrosine kinase genes *NTRK1*, *NTRK2*, and *NTRK3* (encoding the proteins TRKA, TRKB, and TRKC) when appropriately expressed result in cell differentiation and may help specify certain types of neuronal tissue. However, when the *NTRK* genes are aberrantly fused with other chromosome regions, this can lead to the expression of constitutively activated tyrosine kinase receptors. Such *NTRK* fusion genes can occur in almost any cancer but are found in only a very small proportion (<1%) of most common cancers.

The expression and activation of the tyrosine kinase receptors by the NTRK fusion protein also results in a phenomenon known as "oncogene addiction," meaning the cancer is largely dependent on signaling through the abnormal pathway created by fusion gene products. As such, tumors with these fusions are exquisitely sensitive to targeted inhibitors of these pathways. Because of the rarity of these alterations, research teams developing targeted agents against these cancers needed to enroll patients onto clinical trials regardless of cancer type, as long as they had one of these fusions. This resulted in the Food and Drug Administration's first "tumor agnostic" approval for cancer treatment in 2019, when larotrectinib, a small molecule tyrosine kinase inhibitor, was approved in any type of cancer that harbored an *NTRK* fusion. In the trial, significant tumor responses were seen with this oral drug in up to 75% of patients, and remission could occur rapidly, as was the case with Mr. Gardner.

EPIGENETICS AND THE MODULATION OF TRANSCRIPTION

Epigenetics and epigenomics refer to the study of long-lasting changes in DNA and histone structure that influence gene expression (Fig. 14–4). Major modifications include DNA methylation of cytosine nucleotides or the acetylation of histone proteins. The prefix *epi* (*upon* or *over*) indicates these epigenetic effects do not change DNA sequence. Epigenetic modifications are crucial during development and cellular differentiation, and while each cell inherits almost precisely the same DNA complement, each cell type will ultimately have a unique epigenome and unique gene expression patterns. Even after maturity, environmental factors, toxicants, drug, stressors, and age produce long-lasting epigenetic changes.

DNA methylation is a chemical process that adds a methyl group to DNA. It can alter proteins from binding to specific DNA motifs by strengthening interactions, weakening them, or blocking them all together. This, in turn, can affect cellular metabolism and gene expression. On a broader scale, it can influence an organism's development, aging, and disease processes. DNA methylation has a well-established role in modulating transcription. Certain transcription factors are sensitive to CpG methylation, exhibited by decreased binding to their preferred DNA sequence motifs when methylated compared to unmethylated. The addition of methyl groups can change the structure of DNA and block transcription factors from binding. Another type of epigenetic alteration is histone modification. Histones can be modified by the chemical addition of an acetyl or methyl group. The post-translational modification made to histones can impact gene expression by altering chromatin structure or recruiting histone modifiers.

Genome-wide maps of cytosine methylation and various histone modifications can be identified with current sequencing technologies coupled with bisulfite conversion. While epigenetic modifications are required for normal development, they can also contribute to disease states. Cytosine methylation is a conserved epigenetic mark that is strongly associated in the modulation of gene expression and with the silencing of mobile elements and other "junk" repeat sequences. The role of epigenetic modifications such as DNA methylation associated with fetal growth restriction, placental dysfunction, and development are increasingly being investigated. The first human disease linked to epigenetic change was cancer. Diseased tissues from patients with colorectal cancer had less intense DNA methylation than neighboring healthy tissue from the same patient. Because methylation of genes typically blocks the initiation of transcription, the loss of methylation can dysregulate expression and increase the proliferation of cells.

Although different sequence-based assays are available, bisulfite conversion of 5-methylcytosine is the most informative strategy to analyze DNA methylation patterns. Several large consortia, including Roadmap Epigenomics and International Human Epigenome Consortium, have been created to establish comprehensive epigenomic maps of human tissues and cell types. The information generated from these consortia serve as an important resource in advancing the field of epigenetics.

TRANSCRIPTOMES AND THE RNA WORLD

The transcriptome represents the genetic code that is transcribed into RNA molecules and is an essential link between genotype, cell function, normal higher-order phenotypes, and, of course, disease risk. Understanding how the transcriptome is controlled is a crucial first step in understanding disease risk and progression. Advances in RNA-seq technologies are providing a more accurate view of the impressive diversity and dynamics of many types of RNA molecules. Overall, RNA-seq has the ability to identify differentially expressed genes and detect differences in expression and in messenger RNA (mRNA) isoforms and the many mRNA splice variants that can be produced from a single gene. It is also possible to accurately quantify changes in expression levels under development and different conditions, even in single cells. We can begin to assemble a fundamental, but at the same time a highly dynamic, atlas of RNA expression of all human cell types.

The basic RNA-seq workflow begins with RNA isolation, mRNA enrichment, or ribosomal RNA depletion, converting the RNA to the complementary cDNA, preparing for sequencing by creating a sequencing "library," and then actual cDNA sequencing. The sample is typically sequenced to a depth of 20–100 million cDNA sequence reads, each of which is about 100 nucleotides long. These short reads are good enough to computationally align to a human "reference" genome.

Analysis of differentially expressed genes can be broken down into several steps. The first steps involve estimating the number of original RNA molecules that correspond to known genes, using a standard reference transcriptome for all tissues—essentially, the 5% of the genome that corresponds to exons and introns. Once transcripts are counted, they are filtered and normalized to account for the length of a gene and the total number of cDNA sequences that were produced—the read depth. The last step involves modeling the experiment to identify significant differentially expressed genes. Differences in the computational approaches used at each stage can have significant effects on the conclusions drawn from the data, and computation procedures are often tweaked depending on technical problems and the key questions.

Recent advances in RNA-seq now make it possible to isolate and sequence single cells. The first step is single cell isolation, which can be performed by various mechanical and enzymatic methods. Following lysis, a microfluidic device combines the single cell suspension with oligonucleotides hybridized to beads into single droplets. Each bead contains unique identifiers to identify the single mRNA after amplification. Once each cell is labeled in the droplet environment, the cells are released and sequenced. During data analysis, mRNA transcripts are separated based on their unique barcodes. Single cell sequencing promises higher resolution of cellular differences and a better understanding of the function of an individual cell. Another emerging technology is long-read direct RNA-seq. To date, RNA-seq methods rely on converting mRNA to cDNA before sequencing. Oxford Nanopore recently demonstrated that their technology can be used to sequence RNA directly, without modification. Although still in its infancy, this approach removes the biases introduced in sample preparation steps and has the huge potential to enable new insights.

Since its inception, RNA-seq has had challenges moving into the clinic. This is in part due to the lack of consensus and overabundance of bioinformatics tools for reproducible analysis. Recent studies suggest the value of combining two or more tools to reduce false-positive results. Ongoing efforts toward establishing reference standards, assay optimization, and reproducibility for clinical conditions are required to facilitate adoption of RNA-seq tests into the clinical setting.

PROTEOMICS—THE BUSINESS END OF OMICS

Understanding the proteome of a cell, tissue, or organism requires knowing the identity, sequence, copy number, posttranslational modifications, and interactions of every protein component. The study of the proteome, called proteomics, requires sophisticated mass spectrometers with sensitivity at the attomole level and accuracy to 1 part per million. The proteome is far more dynamic and complex than the genome, in part because protein isoforms can result from alternative splicing of mRNA, and proteins harbor single amino acid polymorphisms arising from nonsynonymous single nucleotide polymorphisms and undergo numerous posttranslational modifications. Furthermore, protein-protein interactions and posttranslational modifications can serve as regulators affecting the properties, functions, and interactions of proteins, thereby affecting cellular function.

To unify the understanding of the human proteome, the Human Proteome Project was formed in 2010 to map the entire human proteome using current and emerging technologies. Significant advances in mass spectrometry (MS) instrumentation and protein network prediction have allowed the identification of approximately 16,000 proteins and their binding partners within the human protein interactome (Fig. 14–5). The combination of proteomics and interactomics has been a powerful set of tools in identifying candidate biomarkers that are predictive of disease or progression of a diseased state. In cancer diagnostics, proteomics studies have used body fluid samples to identify protein biomarkers for bladder cancer, pancreatic cancer, ovarian cancer, and hepatocellular carcinoma.

METABOLOMICS AND LIPIDOMICS

Metabolomics quantifies multiple small molecule types, such as amino acids, fatty acids, carbohydrates, or other products of cellular metabolic functions. In the last decade, metabolomics has become a well-established "omics" approach for biomarker discovery and in combination with genomics has provided insight to the biochemical responses of an organism to intrinsic and extrinsic factors. Some examples include discovery of biomarkers for different diseases such as risk to cardiovascular disease and diabetes, and identification of environmental contaminants. This is accomplished through untargeted- and targeted-based approaches. MS can determine the type and abundance of chemicals through the accurate measurement of their mass-to-charge ratios (m/z). Gas chromatography-mass spectrometry and liquid chromatography-mass spectrometry (GC-MS or LC-MS) are the most commonly used analytical platforms for metabolic profiling. Although other techniques are available, GS-MS offers several advantages, including robustness, reproducibility, selectivity, and sensitivity, and importantly it offers a large variety of well-established metabolic libraries. For targeted metabolomics that only focus on smaller, known groups of metabolites, one instrumental platform is typically enough. However, for untargeted metabolomics, where the goal is to collect a comprehensive dataset, sometimes multiple preparation protocols as well as multiple analytical instruments are required to capture the diverse repertoire of metabolites in a given sample.

The metabolome is very complex and contains thousands of metabolites that are highly variable in terms of chemical diversity. Metabolites and metabolism are in constant flux where metabolite concentrations are known to vary significantly in terms of their stability. In addition, many secondary metabolites are unstable in the presence of oxygen when stored for long periods of time or when removed from the cell, which can cause significant challenges for analysis. Collectively, this emphasizes the importance of experimental design when setting up a metabolomics research study.

In the clinical setting, metabolomics is closely linked to many diseases and has emerged as a useful tool for monitoring drug response. Drug metabolism varies with many physiological variables, including ethnicity, age, gender, weight, and diet, making it challenging to predict an individual's response. However, because metabolomics measures the cumulation of intrinsic and extrinsic factors, it can be used to directly monitor drug responses and adjust drug dosing to an individual. In cancer, many

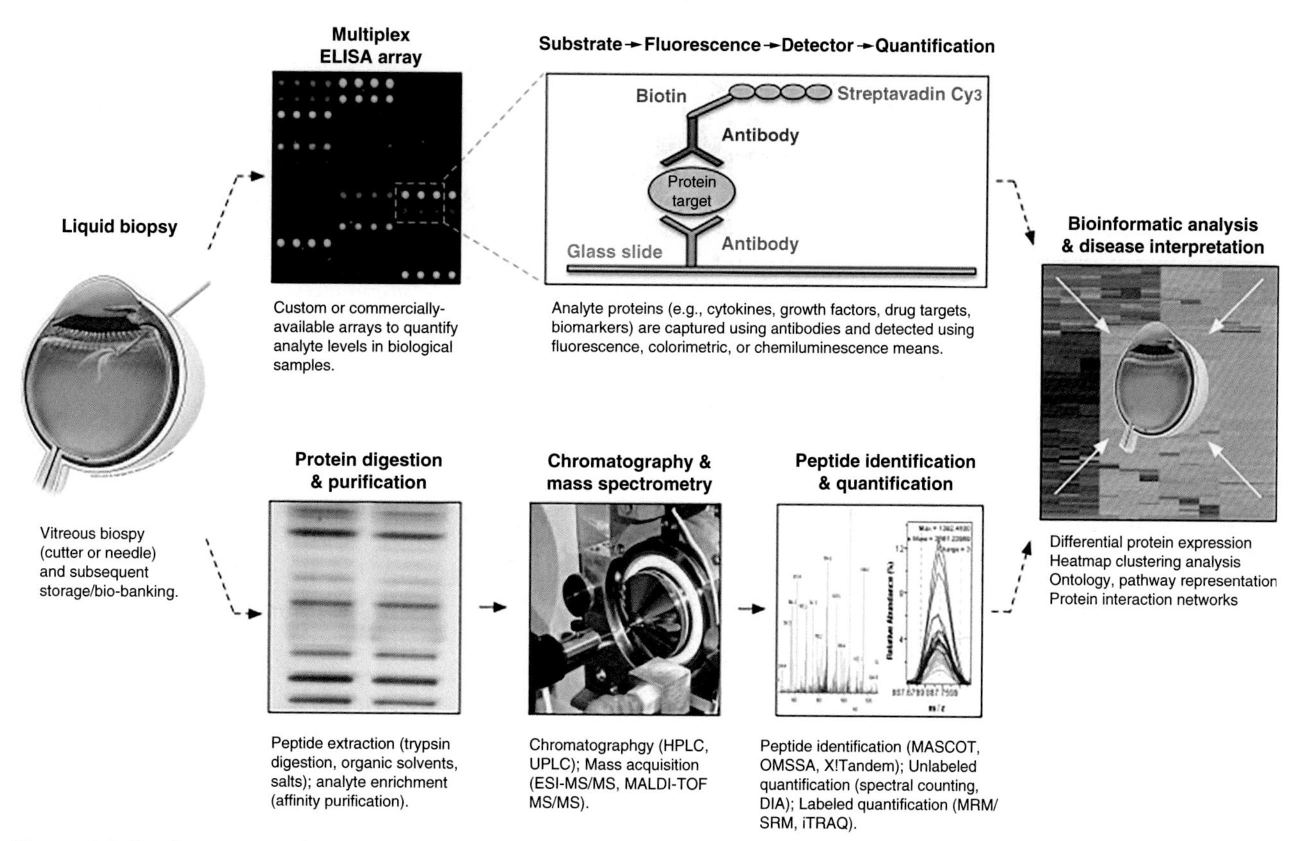

Figure 14–5. Proteomics and personalized health care. Proteomics offers powerful platforms for observing changes in sequence, amount, and posttranslational modifications of the protein content of fluids from individual human subjects. One can start with various fluids, including, but not limited to, plasma, cerebral spinal fluid, saliva, or any tissue. In the case of this figure the starting material is a liquid vitreous biopsy. The key to successful analysis is that all samples be prepared and stored in identical fashion. The samples can be analyzed using ELISA arrays (above platform) where specific antibodies against proteins of interest (growth factors, cytokines, biomarkers, drug targets, etc.) are immobilized on the array. These antibodies can associate with the proteins of interest and quantify them in the vitreous liquid biopsy by fluorescence, colorimetric, or chemiluminescence detection. In the platform below, the complex set of proteins in the vitreous fluid are initially separated and enriched by any protein purification step. In this figure the approach is SDS PAGE, but often various HPLC approaches are used. Individual protein bands are excised from either one-dimensional SDS-PAGE (shown here) or two-dimensional isoelectric focusing SDS-PAGE. Then, in-gel digestion is performed, often using trypsin. The tryptic peptides can be further enriched or injected directly through an HPLC column into a tandem mass spectrometer. Collision-induced dissociation, caused by the collision of the peptide with gas particles, results in the formation of smaller charged fragments, which allows determination of the sequence of the tryptic peptides. Unlabeled (spectral counting or data-independent acquisition) or labeled (isotope-coded affinity tags, isobaric tags for relative and absolute quantification, and multiple reaction monitoring) approaches allows quantification of the peptides. For the upper or lower platforms, the final step is bioinformatic analyses that allow determination of differences from person to person in individual proteins by heat map analysis, protein networks by centrality measurements, and pathways by ontology platforms. *(From PMID: 30271679.* https://www.ncbi.nlm.nih.gov/pmc/articles/PMC6159735/.*)*

tumors exhibit distinct metabolic phenotypes. The fundamental aspects of cellular metabolism are reprogrammed in cancer cells to support their aberrant growth and proliferation. By utilizing various forms of statistical modeling, researchers have used metabolomics to identify metabolic signatures that differentiate tumor from noncancerous tissues and distinguish tumor stages. For example, choline and related metabolites are elevated in prostate cancer, while lipids are elevated in kidney tumors. Another area where metabolomics has made an impact is the field of organ transplantation. Rapid MS-based monitoring of immunosuppressant drugs and their metabolites is being used to individualize immunosuppressant drug dosage following transplantation. As technologies and methods continue to become faster, more sensitive, and more reliable, metabolomics has the potential to make a major impact on patient care. Likewise, the development of software tools for data processing, interpretation, and quality control are becoming more standardized and widespread, which is important for reproducibility and patient care.

BIOINFORMATICS AND SYSTEMS BIOLOGY

Comprehensive and more holistic approaches to understanding disease processes and an individual's best care options require data of many types (Fig. 14–6). Genetic and even genomic data can be generated efficiently, and multiomic and imaging datasets from blood, biopsies, urine, and stools are both practical and common. The DNA data are relatively stable and except in the case

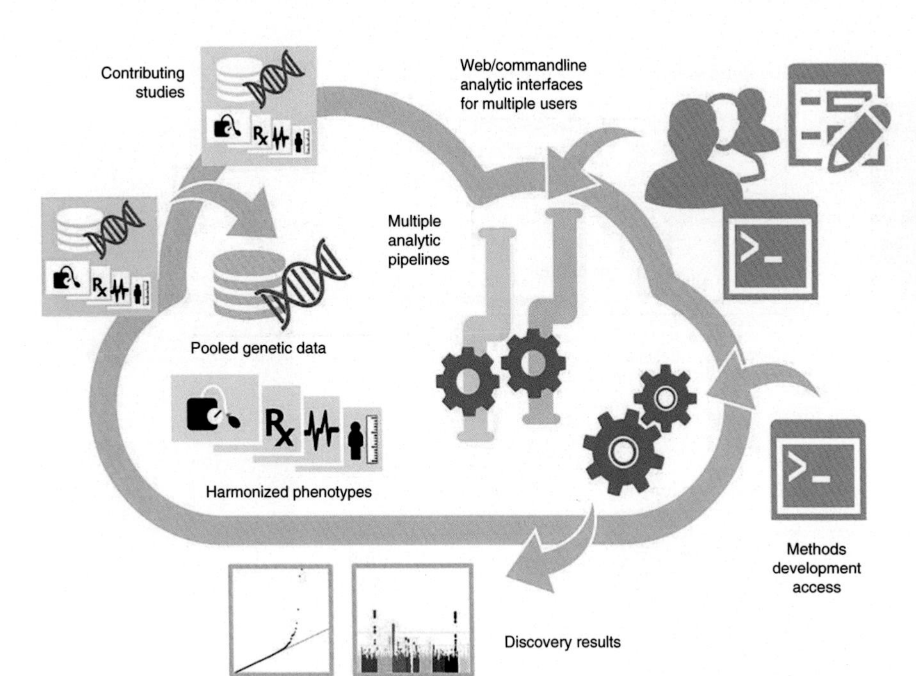

Figure 14–6. Electronic web and cloud-based healthcare data analysis and integration. Developing extensible, powerful, and secure systems for healthcare delivery is critical. Some of the key components are illustrated in and around the *light blue* data and analysis cloud. In the *upper left*, some of the key data are entered into the system, both individual healthcare data and key results from studies and trials. *Upper right* highlights diverse users of these data, from patients and clinicians to healthcare system administrators. Finally, the *center* and *lower* parts of the figure summarize several of the technical and computational approaches used to convert datasets into better prevention and treatment. *(From PMID 29074945 and analysiscommons.com.)*

of cancers, do not need to be followed over time. However, all of the other tiers of data are unique to particular tissue types, cell types, and timepoints. These levels are dynamics, and changes, of course, provide critical insights into many molecular, cellular, drug, and even social interactions that modulate health—the whole is referred to as an interactome.

The synthesis of interactome data in a way that produces reliable insights and predictions is in its infancy, and this is even true for model organisms such as yeast, fruit flies, and mice. This is one reason why we generally do not yet have effective interventions for most chronic diseases and even many monogenic Mendelian diseases such as Huntington's. A single snapshot of one data type offers limited insights into the whole biological interactome. For example, GWAS have identified hundreds of variants linked to type 2 diabetes, but given the key role of environment and diet, this is not enough to provide strong recommendations. However, a full interactome improves both diagnosis and treatment. This is particularly true if data can be acquired at multiple timepoints for each individual—starting with baseline interactomes when healthy. Acquiring the rich data types is both hard and expensive, but a major part of cutting-edge biomedical research. The work of Schüssler-Fiorenza, Snyder and colleagues on type 2 diabetes risk and early intervention (see Suggested Readings) is one superb example of the interface between interactome research and effective clinical application of individualized health care.

Systems biology involves the use of statistical analysis to identify discrete, individual features—sometimes called biomarkers—that highlight disease status or progression. However, due to inherent data differences across omics techniques, integrating multiple platforms remains an ongoing challenge. This is because the different analytical tools and experimental designs used for individual omics do not necessarily permit comparisons or integration, an area that becomes increasingly difficult with greater variability and complexity of measurements. For example, RNA-seq expression studies can only be compared if the same genome version, transcript annotation, and quantification tools are used for all datasets.

Clinical Case 14–2 Recurrence and Typing of Cancer in Liver Transplantation

R.J. Smith, a 46-year-old male, underwent deceased-donor liver transplantation in 2007. He had been diagnosed with Child-Pugh B cirrhosis related to a hepatitis C virus (HCV) infection and early-stage hepatocellular carcinoma (HCC). His alpha-fetoprotein (AFP) serum level was 21 U/mL. While waiting 7 months for transplantation, he underwent two sessions of transarterial chemoembolization, and this successfully stabilized his condition. There was no preoperative biopsy of the liver nodule, which was instead diagnosed as HCC by radiologic criteria. His posttransplant course was uneventful, and cyclosporine-based immunosuppression was initiated without steroids. A 4-cm diameter encapsulated and moderately differentiated (stage G2) HCC, with microvascular invasion, was found on the explanted liver. There was no necrosis in the treated nodule, and malignant cells were not found in the thrombus, or detected intraoperatively in the main portal vein.

HCV graft hepatitis occurred 6 months later and as a result combined interferon and ribavirin treatment was initiated and continued at several dosages, formulations, and intervals for 5 years. Periodic elevation in transaminases and HCV-RNA serum levels were managed. At multiple biopsies, Ishak scores of 11–13 with mild fibrosis (F2) and mild to severe steatosis were noted from the second posttransplant year. The patient showed adequate liver function and no signs of portal hypertension throughout the entire follow-up.

Nine years after transplant, a 2-cm solid nodule in the subcutaneous tissue of the lower abdomen was detected on the right side of the umbilicus. This was surgically removed and was a G2 HCC. Less than 2 months later, multiple and bilateral liver lesions of <1 cm in size were detect by CT scan, but the rest of the metastatic workup was negative, and AFP remained at 6.2 U/mL. The presence of two sites of peritoneal tumor spread at the subsequent laparotomy meant that initial plans of combined resection and ablation of the intrahepatic HCC recurrence had to be abandoned. In February 2017, considering the presence of extrahepatic disease not amenable to intraarterial treatments, the patient was placed on sorafenib (Nexavar-Bayer at a 400-mg BID dosage) and with a grade one hand-foot toxicity (NCI-CTCAE v3.0), and 3 months later a progression in tumor burden was assessed, due to diffuse multinodular intrahepatic HCC growth. Meanwhile, the expression of surrogate markers of activation of the principal pathways deregulated in HCC were investigated in the resected specimen.

Cell Biology, Diagnosis, and Treatment of Recurrent HCC

DNA was sequenced from both tumor tissue from the explanted liver (the recipient liver) and from the new tumor derived from the donor liver. The *CTNNB1* gene that encodes β-catenin had the same point mutation in exon 3 (GGA → AGA G34R) in both primary and recurrent HCC samples collected 9 years apart. To confirm the recurrent nature of HCC versus a de novo mutation, DNA variants that differed between recipient and donor were genotyped from cirrhotic tissue surrounding the primary tumor and from the recurrent HCC. All genotypes confirmed that the evaluated specimens belonged to the same individual. Conversely, there was a mismatch between nontumoral tissue in the recipient's "native" liver and the donor liver sample. The mutation in exon 3 of the *CTNNB1* gene and the β-catenin positive nuclear immunostaining caused aberrant activation of the Wnt pathway: a rather frequent condition indicating the utility of genome sequence in both primary and recurrent tumors to confirm the same point mutation and therefore the diagnosis of a true recurrent HCC. Seeding of cancer cells was excluded by the absence of any percutaneous tumor biopsy preceding transplantation. Recurrence of HCC more than 10 years after transplantation is very rare and the pathogenesis of such events may be linked to liver stem cells that can have long-lasting latency. Their immunophenotypical characterization as well as their role in HCC development warrant further studies.

GENOMIC MEDICINE: PROSPECTS AND ETHICS

Faced with the rapid expansion of our knowledge of the impact of DNA sequence difference on human health, and despite significant controversy, advocates of personalized health care are pushing the envelope of clinical care, producing both discoveries of genotype/phenotype relations, and also applications of omics data and biomarkers in more precise treatment of individual patients. These advances have led to concerns over the impact of these new types of information on individuals, families, and societies. Of significant concern are ethical questions surrounding the return of information that is often not well understood or that is not actionable yet. For example, do teenagers need to know they have trinucleotide expansions associated with high risk of Huntington's disease when they reach 50 years of age? The American College of Medical Genetics and Genomics has developed guidelines for the return of secondary findings to participants in genomic studies. The most recent list includes ~59 gene variants for which data should be returned to patients or families.

In 2008, GINA—the Genetic Information Nondiscrimination Act—was enacted in the United States to protect against genetic discrimination in the workplace and through one's health insurance. GINA prevents health insurers from making decisions regarding eligibility, cost, coverage, or benefits on a health insurance policy based on an individual's genetic information. In addition, GINA supports research by protecting the privacy of participants and guarding against the misuse of genetic information obtained through research. Genetic data can sometimes be combined with openly available online personal data—perhaps soon even with just a good photograph—to relink genotype files to particular individuals. As genomics and the sophistication of the understanding of genome-to-phenome relations improve, we will need to contend wisely with complex repercussions on individuals, families, and the broader social and ethical impact of omics.

SUMMARY

An enormous amount of omics data is being generated due to the rapid development of new quantitative methods. More powerful systems for analyzing data are currently being developed, but in the context of health care it is still a major challenge to generate and integrate data, and to translate information into effective prevention and optimal treatment. The effort to combine quantitative time series data of many types with technical and clinical expertise is far from trivial. There will always be the expediency of applying generic one-size-fits-all solutions, rather than genuinely individualized care. Further improvement in the synthesis of many data types, and in the development of quantitative metrics on the accuracy of prediction, will lead to both a deeper understanding of health and disease and to better individualized prevention and treatment.

Suggested Readings

Committee on a Framework for Developing a New Taxonomy of Disease. *Toward Precision Medicine: Building a Knowledge Network for Biomedical Research and a New Taxonomy of Disease*. National Academics Press; 2011. www.nap.edu/read/13284/.

Deisseroth CA, Birgmeier J, Bodle EE, Kohler JN, Matalon DR, Nazarenko Y, Genetti CA, Brownstein CA, Schmitz-Abe K, Schoch K, Cope H, Signer R, Undiagnosed Diseases Network, Martinez-Agosto JA, Shashi V, Beggs AH, Wheeler MT, Bernstein JA, Bejerano G. ClinPhen extracts and prioritizes patient phenotypes directly from medical records to expedite genetic disease diagnosis. *Genet Med* 2019;21(7):1585–1593. https://doi.org/10.1038/s41436-018-0381-1. PubMed PMID:30514889. PubMed Central PMCID:PMC6551315.

Denny JC, Bastarache L, Roden DM. Phenome-Wide Association Studies as a tool to advance precision medicine. *Annu Rev Genomics Hum Genet* 2016;17:353–373. https://doi.org/10.1146/annurev-genom-090314-024956. Review. PMID:27147087. PMCID: PMC5480096.

Katada S, Imhof A, Sassone-Corsi P. Connecting threads: epigenetics and metabolism. *Cell* 2012;148(1–2):24–28. https://doi.org/10.1016/j.cell.2012.01.001. 22265398.

MacArthur DG, Manolio TA, Dimmock DP, Rehm HL, Shendure J, Abecasis GR, Adams DR, Altman RB, Antonarakis SE, Ashley EA, Barrett JC, Biesecker LG, Conrad DF, Cooper GM, Cox NJ, Daly MJ, Gerstein MB, Goldstein DB, Hirschhorn JN, Leal SM, Pennacchio LA, Stamatoyannopoulos JA, Sunyaev SR, Valle D, Voight BF, Winckler W, Gunter C. Guidelines for investigating causality of sequence variants in human disease. *Nature* 2014;508(7497):469–476. https://doi.org/10.1038/nature13127. PMID:24759409. PMCID:PMC4180223.

Mukerjee S. *The Gene: An Intimate History*. New York: Scribner; 2016.

Schüssler-Fiorenza Rose SM, Contrepois K, Moneghetti KJ, Zhou W, Mishra T, Mataraso S, Dagan-Rosenfeld O, Ganz AB, Dunn J, Hornburg D, Rego S, Perelman D, Ahadi S, Sailani MR, Zhou Y, Leopold SR, Chen J, Ashland M, Christle JW, Avina M, Limcaoco P, Ruiz C, Tan M, Butte AJ, Weinstock GM, Slavich GM, Sodergren E, McLaughlin TL, Haddad F, Snyder MP. A longitudinal big data approach for precision health. *Nat Med* 2019;25(5):792–804. https://doi.org/10.1038/s41591-019-0414-6. PMID:31068711. PMCID:PMC6713274.

Willsey AJ, Morris MT, Wang S, Willsey HR, Sun N, Teerikorpi N, Baum TB, Cagney G, Bender KJ, Desai TA, Srivastava D, Davis GW, Doudna J, Chang E, Sohal V, Lowenstein DH, Li H, Agard D, Keiser MJ, Shoichet B, von Zastrow M, Mucke L, Finkbeiner S, Gan L, Sestan N, Ward ME, Huttenhain R, Nowakowski TJ, Bellen HJ, Frank LM, Khokha MK, Lifton RP, Kampmann M, Ideker T, State MW, Krogan NJ. The Psychiatric Cell Map Initiative: A convergent systems biological approach to illuminating key molecular pathways in neuropsychiatric disorders. *Cell* 2018;174(3):505–520. https://doi.org/10.1016/j.cell.2018.06.016 Review. PMID:30053424. PMCID:PMC6247911.

Zeevi D, Korem T, Zmora N, Israeli D, Rothschild D, Weinberger A, Ben-Yacov O, Lador D, Avnit-Sagi T, Lotan-Pompan M, Suez J, Mahdi JA, Matot E, Malka G, Kosower N, Rein M, Zilberman-Schapira G, Dohnalová L, Pevsner-Fischer M, Bikovsky R, Halpern Z, Elinav E, Segal E. Personalized nutrition by prediction of glycemic responses. *Cell* 2015;163(5):1079–1094. https://doi.org/10.1016/j.cell.2015.11.001. 26590418.

Chapter 15

The Microbiome

INTRODUCTION TO THE MICROBIOME

Microorganisms can be found almost everywhere, including in and on the human body (Fig. 15–1). The largest population of microorganisms on the human body resides in the gastrointestinal tract. Known as the gut microbiota, this complex ecosystem is comprised of a range of microorganisms, including bacteria, fungi, archaea, helminths, and viruses. The total genetic material of a specific microbiota is known as the microbiome, though the terms microbiota and microbiome are often used interchangeably. The bacterial fraction of the microbiome, or bacteriome, has been studied the most and has therefore been characterized the best, though our knowledge of other components within the microbiome, especially the virome and the mycobiome, has been increasing steadily over the last few years.

In recent years, largely due to advances in sequencing technology, several large-scale projects such as the Human Microbiome Project have made great progress developing protocols for investigating the microbiome (see Box) and mapping the microbiome using these methods. One of the major strategies by which these projects acquire knowledge about the microbiome is by sequencing the microbiota of several body sites of hundreds to upward of a thousand volunteers. Then, the differences in the microbiomes of the participants can be compared and cross-referenced with other data from these participants, such as age, BMI, or diet. From the mid-2000s, the field has mapped out an estimated 87% of the metagenomes found in the microbiome.

Recent estimates put the total bacterial count on an average human at around 3.0×10^{13}, which is slightly more than the estimates of human cells in the body. According to estimates, the gut microbiota of a 70-kg individual would weigh in at approximately 0.2 kg. Because of the wide taxonomic variety found in the microbiome, more than 99% of genes found in a human are microbial in origin.

The microbiome is in constant communication with the host and is essential for host health. There is a growing realization that the microbiome plays a key role in maintaining homeostasis throughout the lifespan. Moreover, several large patient cohorts, including inflammatory bowel disease and diabetes, have been characterized, greatly improving our available toolset by mapping the microbiome not only in health, but also in disease. However, many questions remain unanswered, including some of the precise mechanisms by which the microbiome influences the host and vice versa. Moreover, a growing emphasis is now placed on the metabolic capacity of the microbiota as it is capable of producing a variety of metabolites that can communicate with molecular targets on host cells (Fig. 15–2).

Goodman's Medical Cell Biology. https://doi.org/10.1016/B978-0-12-817927-7.00015-6

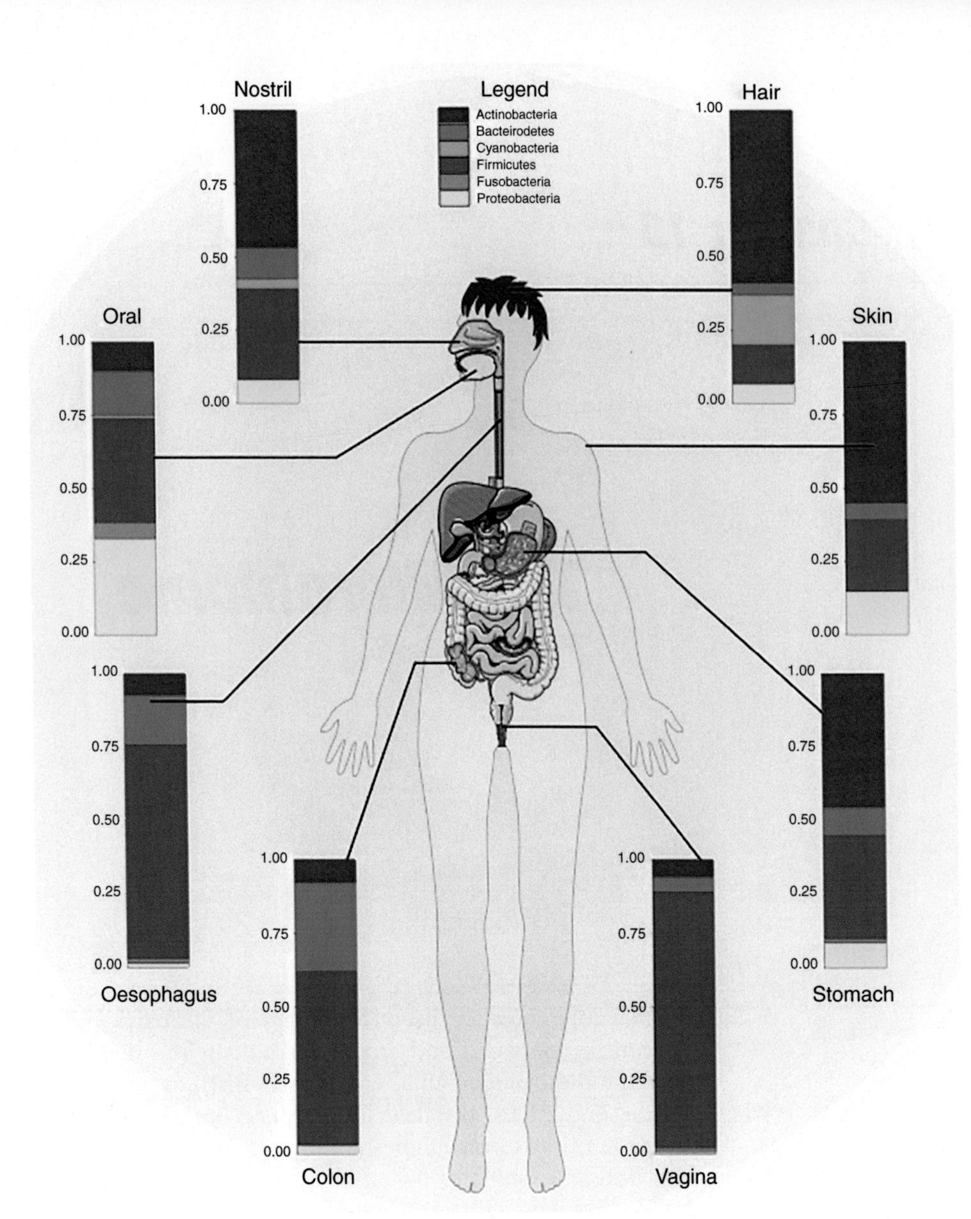

Figure 15–1. **The human microbiome biogeography.** Microbiota compositions vary per body site. In general, the microbial compositions from the same sites over different individuals are more alike than the microbial compositions of different sites of the same individual.

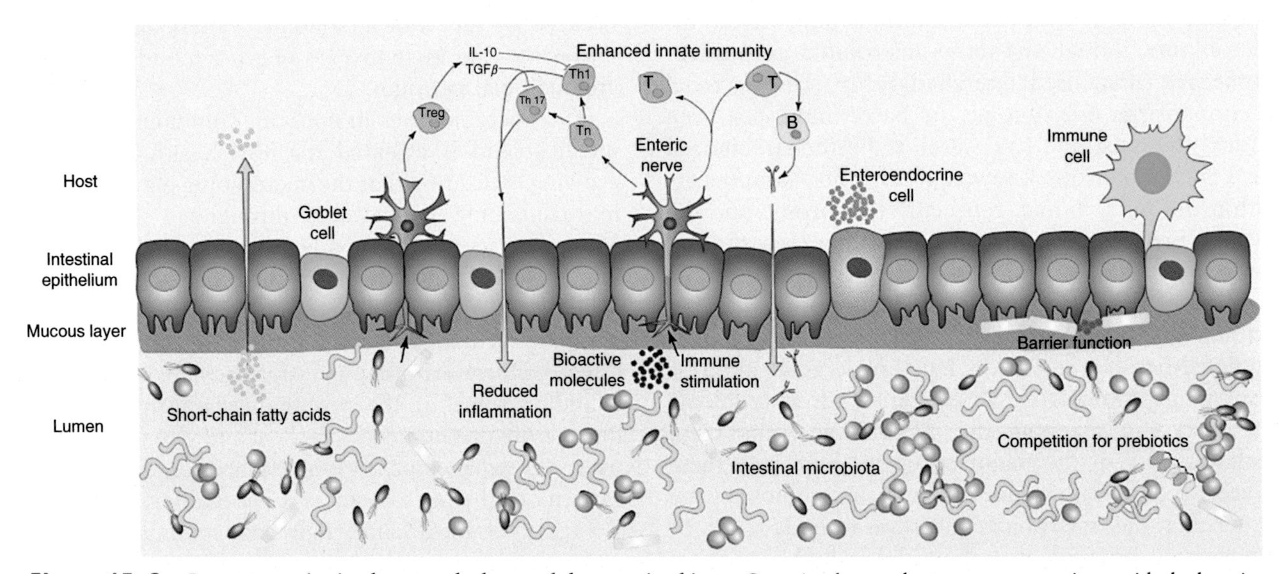

Figure 15–2. **Intercommunication between the host and the gut microbiome.** Gut microbes are known to communicate with the host in numerous ways including through the production of metabolites, the modulation of the immune system, and altering gut permeability. In turn, the host influences the gut microbiome in a variety of ways including altering the gut lumen environment through diet, and immune modulation.

HOW TO ASSESS THE MICROBIOME

Next-Generation Sequencing

The composition of the gut microbiota can be assessed from stool samples or biopsies (Fig. 15–3). Nowadays, there are two main methods to quantify the microbiome: 16S ribosomal RNA (or simply 16S) sequencing and whole genome shotgun (WGS) sequencing. The former involves amplifying and sequencing the highly conserved 16S ribosomal RNA subunit found only in all bacteria. Because the 16S ribosomal subunits are so well conserved, they can serve as a fingerprint sequence, enabling us to compare them to a database of 16S sequences and accurately assign a taxonomic classification. In contrast, WGS typically involves no amplification step; all DNA in the sample is sequenced and computationally assembled, often with the aid of presequenced microbial reference genomes. The major advantage of WGS is that one can elucidate differences down to a strain level, which allows for ascertaining the metabolic capacity of the bacteria. For a comparison of the two methods, see the following table.

	16S Ribosomal RNA Sequencing	WGS Sequencing
Amplification step	Yes	No
Access to metabolic profile	No; only through computational inference	Yes
Resolution	Generally genus level	Strain level
Cost	Relatively cheaper	Relatively more expensive
Target material	Only bacteria	Any microbe, even host material

Bioinformatic Tools

Two main features of the microbiome that are often assessed in scientific literature are the relative abundance of specific microbes and the diversity of the microbiome. Ecology distinguishes roughly two types of metrics of diversity: *alpha-diversity* and *beta-diversity*.

Alpha-diversity metrics refer to the degree of diverseness in a single microbiota. For instance, the microbiota of a patient with *Clostridium difficile* infection, a condition where the microbiota can be almost entirely overgrown with a single

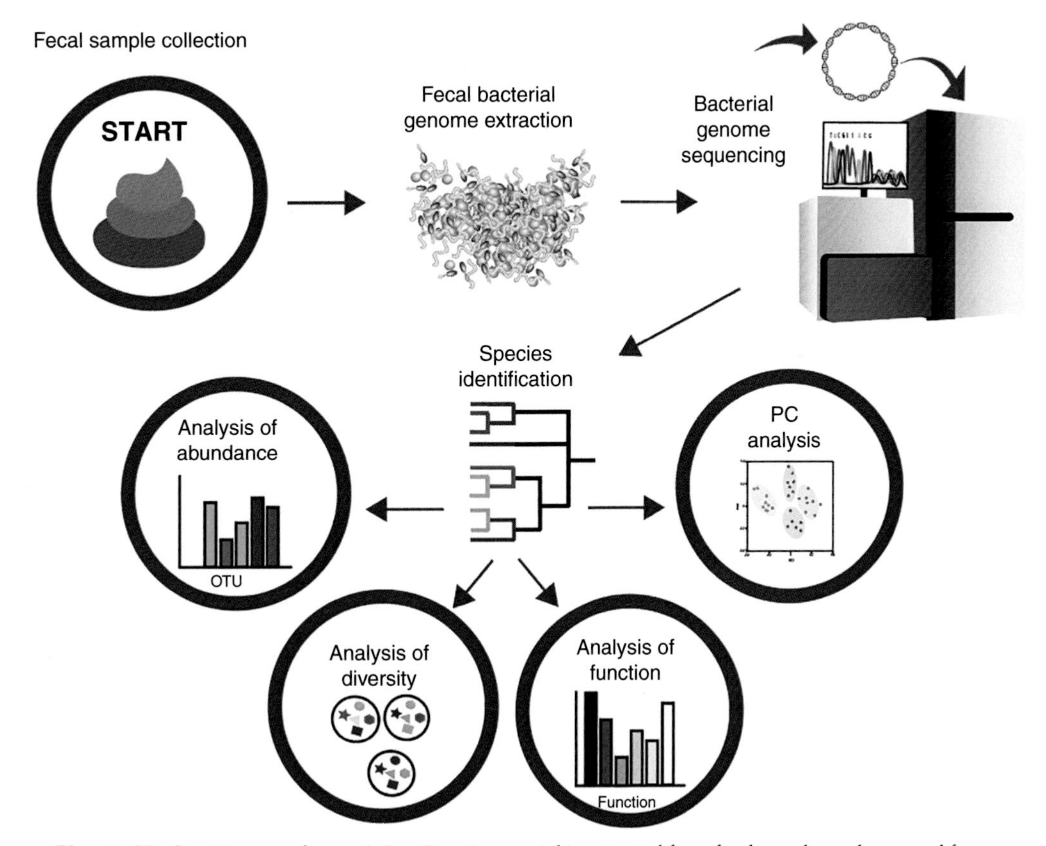

Figure 15–3. **From stool to statistics.** Genetic material is extracted from fecal samples and prepared for sequencing. Digitized genetic sequences are filtered for quality and compared to a database to identify which microbes the sequences belong to. From there, many different multivariate statistical analyses can be performed, including principal component analysis, to characterize a sample and compare it to other samples.

pathogenic bacterium, will have a comparatively low alpha-diversity. In adults it is thought that a lower alpha-diversity is associated with inflammation and illness, though certain diets can lower alpha-diversity in a harmless way.

Beta-diversity metrics refer to the degree of difference between two microbiota groups. There is a large amount of variance in the microbiome in the general population. A common way to assess the state of a microbiota is to compare it to that of microbiota that have the lowest beta-diversity compared to it, i.e., are more similar to each other. In general, the beta-diversity between healthy humans is higher than the beta-diversity of a single human microbiome over time.

Certain microbes or groups of microbes are thought to fulfill important metabolic functions in the host organism. This concept is known as an *ecological guild*: taxonomically unrelated but functionally similar organisms that fulfill a certain function within their ecosystem. The specific bacteria fulfilling the role of the guild can vary depending on geographical location of the host, but the effects they have on the microbiome and host will be similar. An overview of common approaches to investigate the microbiome and its relation to health is shown in Figure 15–4.

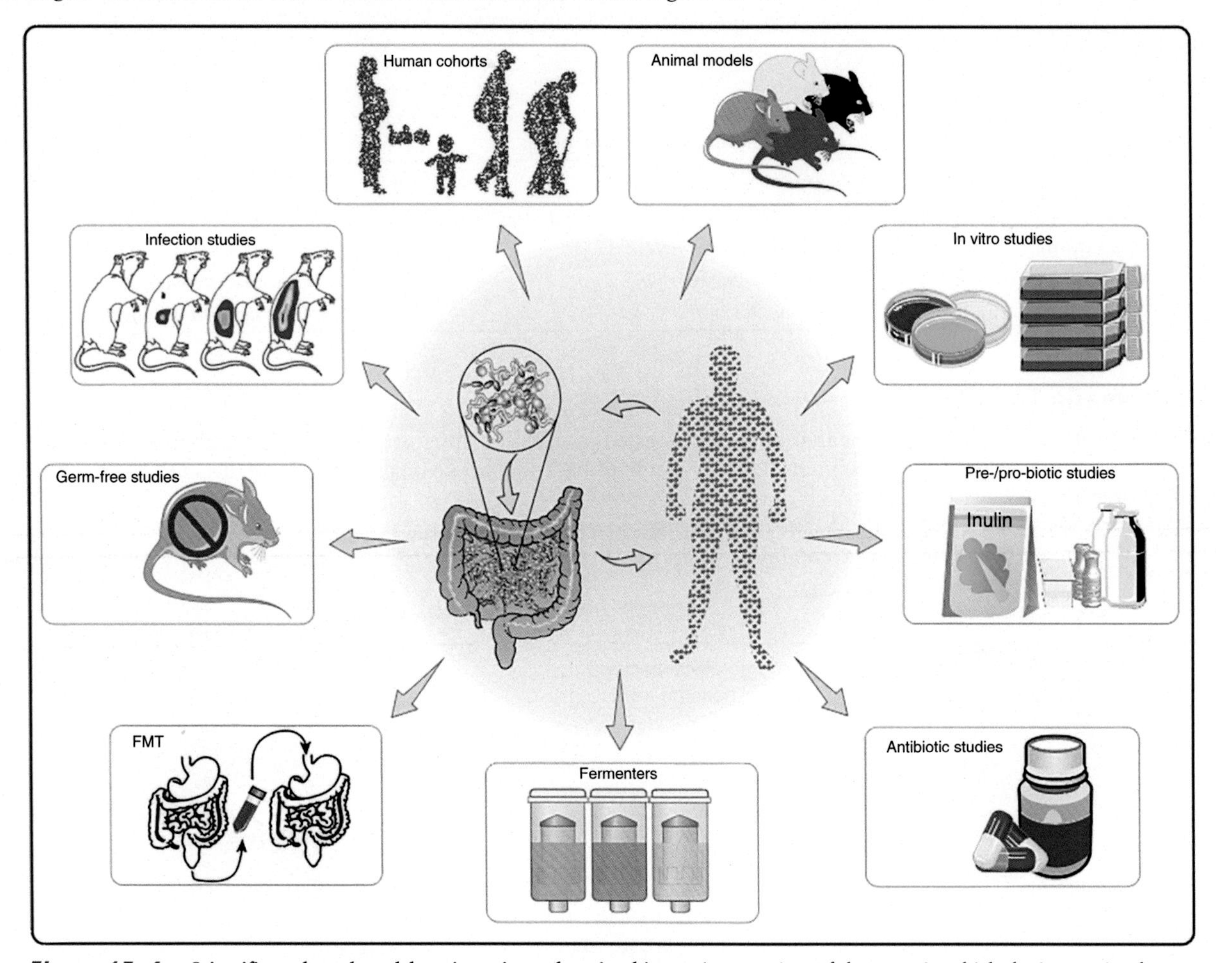

Figure 15–4. **Scientific tools and models to investigate the microbiome.** An overview of the ways in which the interaction between the microbiome and the host is investigated. Examples of techniques currently available include human clinical trials, conventional or germ-free animal studies, intervention studies where humans or animals are given an FMT, anti-, pre-, pro-, or synbitoics and their effects studied, followed by laboratory experiments designed to supplement the information garnered from the interventional treatment. Typically, a combination of these techniques is used in modern research laboratories to reach a conclusion.

Where it Begins

Typically, the human body is colonized by microbes during birth where the infant comes into contact with microbes from the maternal vaginal and gut microbiome. During normal development, the infant microbiome then develops in tandem with the developing infant. Human milk oligosaccharides (HMOs) enriched in breastmilk and later solid foods selectively shape the developing microbiome.

It is worth noting that prokaryotes and archaea have existed for much longer than eukaryotes and thus eukaryotes have never existed in a world without microbes. This has led to the conceptualization of humans as "holobionts"—a dynamic ecosystem comprised of a host and its associated microorganisms that can vary with time, localization, and function. Collectively, the host and microbial genomes of a holobiont are termed a hologenome, and variation in the hologenome caused by changes in the host and/or microbes may impact phenotypes that may be subject to natural selection.

Sometimes in the context of microbiome and host health, the term *dysbiosis* is used. Dysbiosis refers to an unhealthy state of the microbiome and while widely used, the term can be problematic in a scientific context. An unhealthy state of the microbiome suggests that we know what a healthy microbiome looks like or that it even exists. In reality, there is quite some variance between microbiomes and there is no consensus on the existence of a core normal microbiome. Furthermore, a microbiome can be in a perfectly stable state but still cause health problems to the host or be quite unstable but cause no health issues for the host.

MAJOR FACTORS INFLUENCING THE MICROBIOME

The composition of the microbiome is influenced by many factors, including its own composition and environment (Fig. 15–5). The diversity of the human gut microbiome varies greatly by lifestyle, age, and biogeographical status. Generally, the highest microbial diversity can be found in indigenous tribes with little contact with Western society. Conversely, the lowest diversity can generally be found in individuals with a poor-quality diet high in salt, processed sugars, and saturated fats.

Diet

The most important factor shaping the healthy gut microbiome is diet. Diet can change both the composition and *alpha-diversity* of the microbiome rapidly. For example, a person switching from an animal-based diet to a plant-based diet will exhibit a strong change in their composition within 24 h. After reverting to the original diet, the microbiome tends to move toward its original composition. In terms of *beta-diversity*, healthy adults with the same diet tend to be comparatively close to each other, indicating that they are compositionally more similar. Two types of diet are often discussed in the context of how they impact microbiome composition: the Western diet and the Mediterranean diet. The Western diet refers to a diet rich in sugar, salt, and fat and is generally considered to be a major contributor to a generally unhealthy lifestyle. In terms of microbiota, the Western diet is associated with a reduction in *Bacteroidetes*, an increase in *Proteobacteria* and *Firmicutes*, and a general increase in systemic inflammation of the host. In contrast, the Mediterranean diet refers to a diet rich in whole grains, legumes, nuts, vegetables, and a moderate amount of fish. The diet has been shown to have health benefits that reduce the risk of cardiovascular disease and depression. In terms of microbiota, it is associated with the exact opposite of the Western diet: a reduction in *Proteobacteria* and *Firmicutes*, an increase in *Bacteroidetes*, and a general antiinflammatory effect. Dietary components like polyphenols and dietary fibers, like inulin, fructooligosaccharides, and galactooligosaccharides, have a profound effect on the composition of the microbiome, generally shifting it in favor of microbes with the metabolic capacity to catalyze it. Fermented foods are another dietary strategy to boost the composition of the microbiota. On the other hand, many of the components of processed food, including emulsifiers and sweeteners, have been shown to have negative effects on the microbiota composition in animal models (Fig. 15–6).

Therapeutic Drugs

Among the various classes of drugs, antibiotics represent the clearest and most effective way to directly target the microbiome. Antibiotics, especially in cocktails, can be

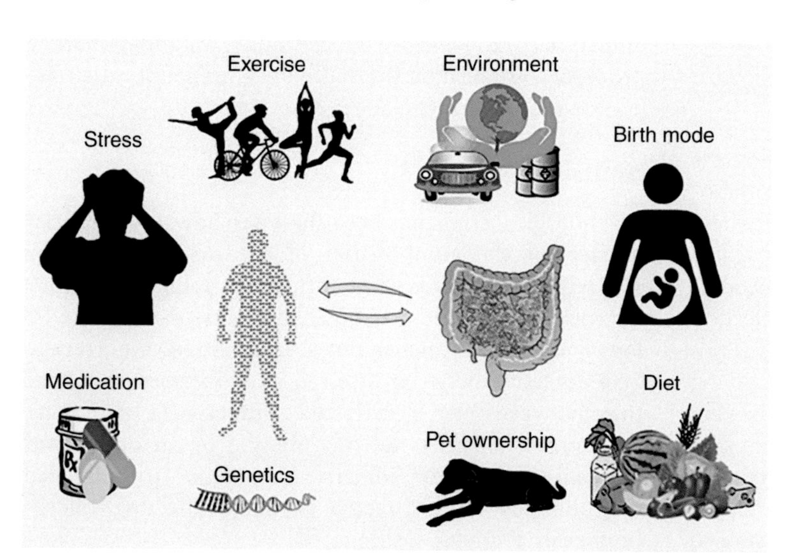

Figure 15–5. **Factors influencing the composition of the microbiome.** A wide array of factors are known to influence the microbiome, ranging from environmental factors to genetics. The most relevant are shown here.

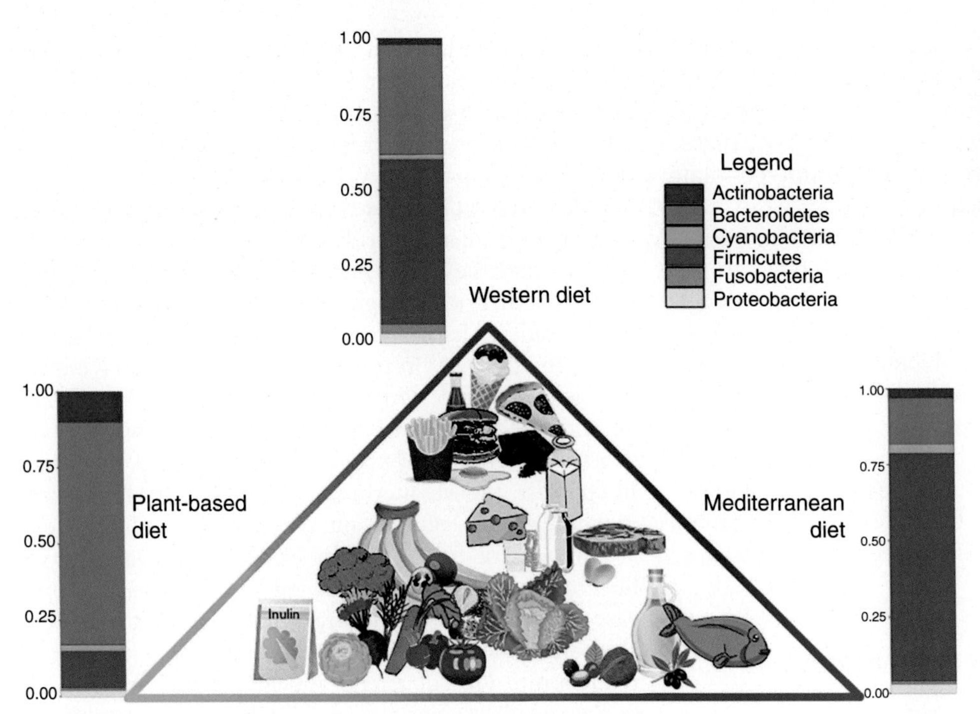

Figure 15–6. The effect of different diets on the composition of the gut microbiome. It has been shown that the relative abundances of the main phyla of the gut microbiota differ greatly depending on the main types of foods consumed in the diet. Fish and oil are associated with a higher relative abundance of Firmicutes, whereas plant-based food stuffs drive a greater relative abundance of Bacteroidetes.

and are used to almost entirely wipe out a microbiome if necessary, for example, during a *C. difficile* infection. Apart from antibiotics, a vast array of drugs has an impact on the microbiome. Notably, proton pump inhibitors (PPIs) have a profound effect on the microbiome. This is in part because PPIs by design reduce the acidity of the stomach, which has the added effect of changing the selectivity of the stomach and allowing more microbes to enter the gut. Many other drugs have an effect on the microbiome. Often, this is because these drugs either display antimicrobial activity or they can be metabolized by certain microbes, giving them an advantage and thus increasing their abundance. As an important consequence of this, the microbiome has the potential to alter the pharmacology of drugs, potentially changing their concentration and half-life.

Stage of Life

The composition and alpha-diversity of the microbiome is dependent on the age of the host. The healthy infant microbiome has a low alpha-diversity, containing only a few very specific microbes, most of which are *Lactobacillus* and *Bifidobacterium*. HMOs are preferentially metabolized by these microbes, resulting in a strong selection process in favor of the infant-associated microbes. Peak alpha-diversity is attained during adolescence and young adulthood. During the aging process, alpha-diversity drops again. Unhealthy aging, for instance, in tandem with a poor diet, stress, and illness, is coupled with a sharper decline in alpha-diversity and an increase in the abundance of proinflammatory-associated microbes and a sustained low-level inflammation. In tandem, this process is known as *inflammaging*.

Exercise

Exercise has been shown to affect the microbiome, with moderate levels of exercise increasing both alpha-diversity and the abundance of antiinflammatory microbes like those in the *Akkermansia* genus. Because of the often-differing diet and lifestyle between athletes and their non-athlete counterparts, it is often problematic to tease apart the precise driver of the differences we observe in the microbiome. For instance, in a study of the Irish rugby team, their microbiome exhibited a higher alpha-diversity, correlating with a higher protein and creatine kinase consumption, which can be explained by a diet high in animal-derived protein. Furthermore, antiinflammatory properties are also associated with a reduction in stress, a known effect of regular exercise.

Stress

Psychological stress has been shown to have a substantial impact on the composition of the microbiome. The stressed microbiota is often enriched in proinflammatory microbes and pathogens. Alpha-diversity tends to be lower in stressed microbiomes. In addition, the average beta-diversity between stressed microbiomes is higher than between their nonstressed counterparts, indicating that stress perturbs the microbiome in an undirected manner. In contrast, dietary fiber tends to influence microbiomes in a directed manner, affecting microbiomes in a similar fashion.

GENETICS

While there are several large human cohorts with microbiome data, some of which also have human genome data, it is very difficult to tease out general relationships between the microbiome and the host genome. Nonetheless, the human host genome has been shown to affect the composition of the microbiome in very specific manners. For instance, *Bifidobacterium*, a genus associated with lactose metabolism, is more prevalent in individuals with the human lactase nonpersister genotype, a genotype that usually leads to lactose intolerance. Another type of instance of genetic impact on the microbiome can be seen in the case of certain metabolic disorders, where the metabolic gut lumen environment is altered as a result, which in turn alters the microbiome. In mice, where genetics factors are more easily controllable than in human studies, gut microbiome composition is related to the strain of mouse investigated.

THERAPIES TARGETING THE MICROBIOME

Because the microbiome is altered in many diseases (Fig. 15–7 and Table 15–1) and disorders and because of its malleability, it is a promising therapeutic target. As a consequence, several approaches have been developed to target the microbiome.

Probiotics and Prebiotics

Prebiotics and probiotics are two of the most studied microbiome-based therapeutic strategies. Prebiotics are defined as foods not digestible by humans (such as fibers) that have a beneficial effect on the microbiome for the host. Probiotics, on the other hand, are live microbes that, when taken in adequate amounts, confer a health benefit on the host. When both pre- and probiotics are coadministered, which happens increasingly, the term "synbiotic" is used. Some synbiotics have been shown to be more effective than that of the probiotic and prebiotic component individually. Microbiome interventions specifically designed to improve mental health have been coined as psychobiotics. It should be noted that psychobiotics do not necessarily have to target a clinical population but may also be intended for general use. Indeed, an increasing number of studies are emerging reporting the beneficial effects of specific probiotics on behavior.

Fecal Microbiota Transplantation

The concept of fecal microbiota transplantation (FMT) as a therapeutic intervention is disrupting Western medicine completely (Fig. 15–8). The procedure involves introducing fecal microbiota from a selected donor to the gastrointestinal tract of the recipient, with the aim of making the recipient microbiome more similar to

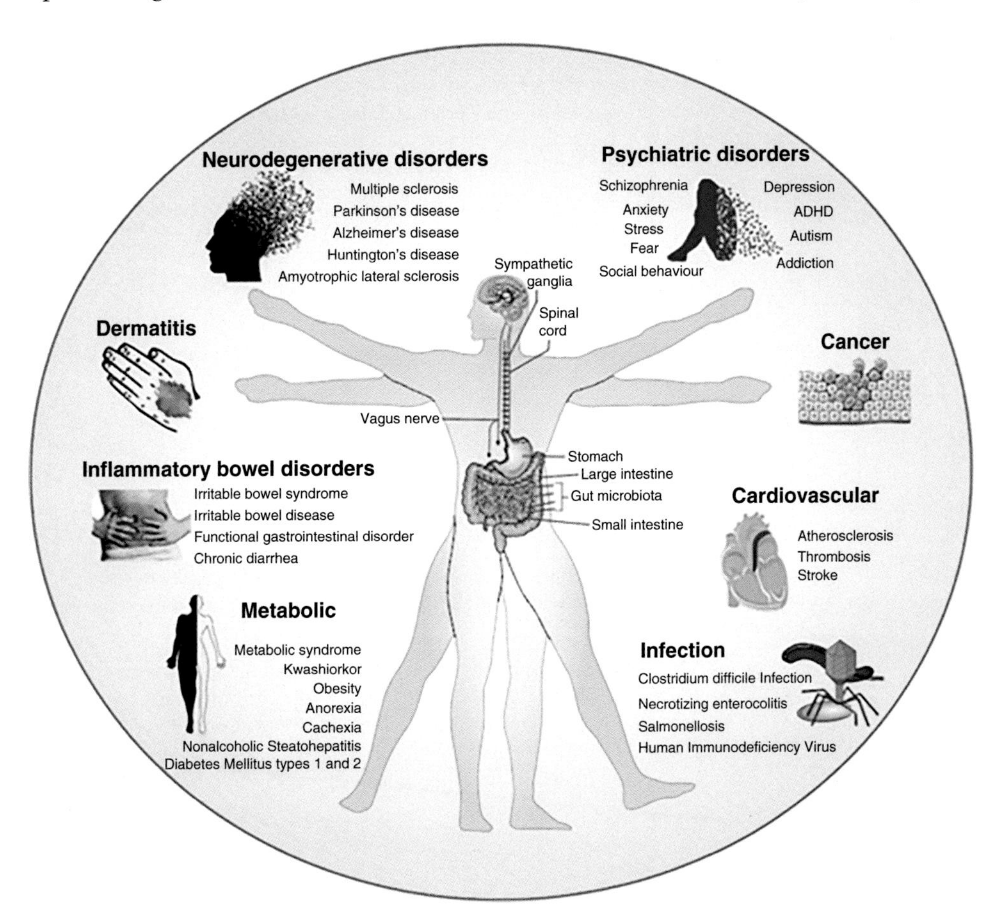

Figure 15–7. The microbiome is implicated in a variety of disorders. Much recent research has found evidence for a role for the microbiome in many diseases including skin, metabolic, and cardiovascular disease, as well as cancer, infection, and neurodegenerative and psychiatric disorders.

TABLE 15–1. Microbiome and Disorders

Diseases		Evidence
Gastrointestinal disorders	Inflammatory bowel disease (IBD)	The microbiome signature in IBD is well characterized in humans and involves a lower alpha-diversity and a deficiency in short-chain fatty acid producers in the gut mucosa. Degree of compositional change over time in the microbiome, or volatility, is higher in IBD than in healthy controls.
	Irritable bowel syndrome (IBS)	IBS is linked to a wide array of microbiome alterations. Alpha-diversity tends to be lower and volatility tends to be higher, reflecting the unstable state of the IBS gut.
Immune and inflammatory system	Allergies	Early life microbiome alterations, for example, as a result of antibiotics exposure or a cesarean section, have been associated with the development of allergies.
	Multiple sclerosis (MS)	The MS microbiome differs from that of healthy controls. This difference is associated with an altered gut immune system. Mice that underwent fecal microbiota transplantation (FMT) from MS patients started exhibiting autoimmune encephalomyelitis, a main symptom of MS.
	Rheumatoid arthritis (RA)	Certain rare microbes were found to be associated with the onset of RA.
	Psoriasis	Psoriasis patients have a lower alpha-diversity than controls.
Cancer		Both proinflammatory carcinogenic as well as antiinflammatory microbes with tumor suppressor properties are known. While most strongly associated with colon cancer, most types of cancer are associated with the microbiome. Some tumors host their own microbiome. Certain microbiota are known to affect immunotherapeutic efficacy (see Clinical Case 15–2).
Infection	*Clostridium difficile* infection (CDI)	CDI is characterized by a vast overgrowth of *C. difficile* in the gut microbiome. This condition can be treated by FMT to reinstate a stable microbial ecosystem in the gut.
	Necrotizing enterocolitis (NEC)	NEC is a disease of preterm infants and has a high morbidity and mortality rate. There have been reports of proteobacteria overgrowth in patients.
	Salmonellosis	Certain commensal microbiota are known to be protective against salmonellosis, stopping its colonization in the gut. In patents with salmonellosis, the gut microbiome composition and motility are drastically altered as a result.
	Human immunodeficiency virus (HIV)	Following HIV infection, the gut microbiome loses alpha-diversity. A microbiome signature, previously associated with a proinflammatory phenotype, has been identified in patients. Rare elite controllers, who spontaneously control HIV infection, display no differences in their microbiome compared to healthy controls.
Metabolic disorders	Type 1 diabetes mellitus (T1DM)	Antiinflammatory bacteria like *Bifidobacterium* and *Lactobacillus* are negatively associated with β-cell autoimmunity and are found in lower abundance in T1DM patients. Conversely, proinflammatory microbes, like *Bacteroides* and *Clostridium*, are associated with an accelerated onset of disease.
	Type 2 diabetes mellitus (T2DM)	The T2DM microbiome is compositionally distinct from healthy controls. Several microbial species have been found to have an effect of insulin sensitivity by way of metabolite production.
	Obesity	In humans, the obese microbiome has a distinct signature. In rodent studies, high fat diet-induced obesity is always linked to a clear shift in microbial composition compared to controls.
	Metabolic syndrome	Alterations in the microbiome in functions typically associated with the modulation of metabolism are often found in patients with metabolic syndrome.
	Kwashiorkor	In a twin study, as a consequence of malnutrition and the resulting host physiology, kwashiorkor patients have been found to have a widely altered microbiome compared to their twin.
	Nonalcoholic steatohepatitis (NASH)	The gut microbiome plays an important role in metabolite degradation and is in constant communication with the liver. Diseases of the liver are commonly associated with a corresponding signature in the microbiome. In NASH, bacteria known to produce alcohol are present in higher abundances.

TABLE 15–1. Microbiome and Disorders—cont'd

Diseases		Evidence
Cardiovascular disorders	Atherosclerosis, thrombosis	In patients with atherosclerosis a depletion of short-chain fatty acids has been found. The gut microbiome modulates platelet activation, which in turn impacts arterial thrombosis development.
	Stroke	Ischemic stroke can trigger an altered microbiome composition. Higher abundance of short-chain fatty acid producing bacteria has been associated with an improved recovery rate after the stroke.
Neurodegenerative, neurodevelopmental, and psychiatric disorders	Major depressive disorder (MDD)	MDD and quality of life have been linked to specific functional modules and compositional states of the gut microbiome. Furthermore, depressive-like mood has been transferred to rodents by way of FMT from depressed patients.
	Alzheimer's disease (AD)	The AD microbiome sports more pathogenic microbes and has a generally lower alpha-diversity. In germ-free and antibiotic-treated mouse models of AD, the hallmark amyloid plaques fail to accumulate.
	Parkinson's disease (PD)	The PD microbiome composition is altered compared to healthy controls. In addition, vagotomy, the removal of the nervus vagus, which represents one of the routes of communication between the gut and brain, has been shown to be protective against PD.
	Autism spectrum disorder (ASD)	The microbiome is altered in ASD, having a lower alpha-diversity and a higher proportion of proinflammatory microbes. In addition, gut motility is often altered in ASD.
	Schizophrenia	The abundance of *Lactobacillus* and *Bifidobacterium*, two microbial genera that are generally associated with health, have shown to be altered in patients with schizophrenia and to be predictive of the severity of symptoms after the first episode.
	Amyotrophic lateral sclerosis (ALS)	In a mouse model that is genetically susceptible to ALS, germ-free and antibiotic-treated mice succumbed to the disease much faster than their counterparts with an unaltered microbiome.
	Epilepsy	The ketogenic (very low carbohydrate) diet is commonly used as a treatment to control epilepsy. In a mouse model of epilepsy, the beneficial effects of the ketogenic diet were lost in germ-free and antibiotics-treated mice.

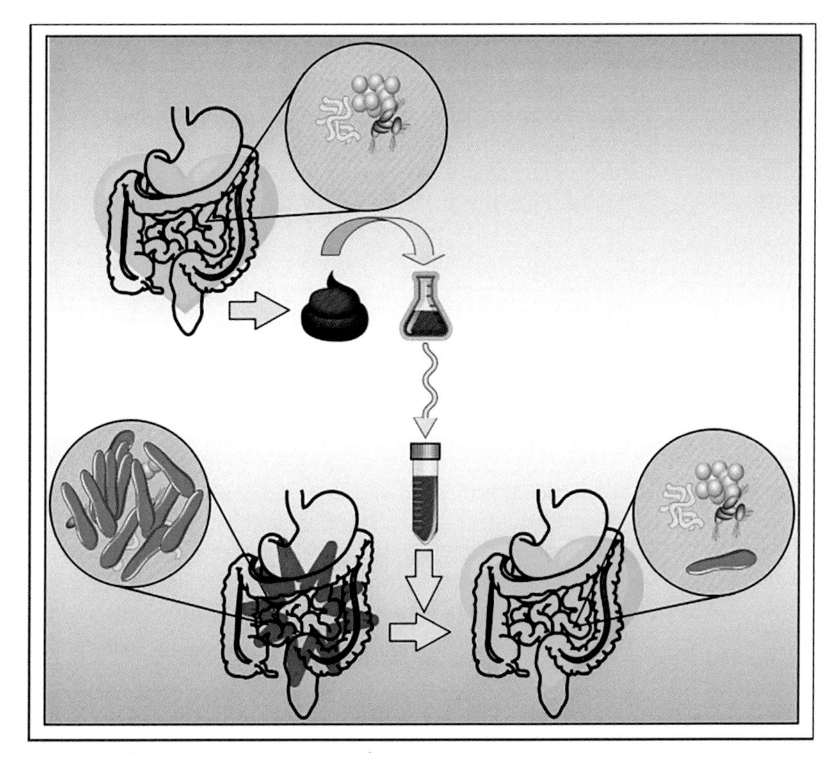

Figure 15–8. **Fecal microbiota transplantation (FMT) as a therapy.** Stool from a healthy donor is collected and screened for safety. Then, the FMT is given to the patient, sometimes after antibiotic-treated to wash out the old microbiota. The FMT restores host health by instating a microbiome similar to that of the healthy donor.

that of the donor. The sample is usually delivered by a nasogastric tube or capsules, but enemas are not uncommon. FMT has been shown to be highly effective in restoring the microbiome to a stable and healthy state, with a beta-diversity closer to the donor microbiome. In some cases, an autologous FMT can be used, where the donor sample is collected from the donor, perhaps before an intensive antibiotics course. A crucial part of FMTs is ensuring the quality of the donor sample. For instance, the sample should be pathogen free (see Clinical Case 15–1).

Clinical Case 15–1

Margie Smith is a 56-year-old Asian woman, who had recently been diagnosed with an upper respiratory tract infection and insisted that her doctor give her amoxicillin because she was planning on traveling and did not want her symptoms to worsen. Despite little evidence that her symptoms were bacterial in origin, the doctor relented and gave Margie a 7-day course of the antibiotic. She began feeling better and went on her trip to her high school reunion for the next week in San Francisco, where she toured the Ghirardelli Chocolate Factory and bought a 2 lb bag of assorted dark chocolates. Two weeks after returning home, Ms. Smith began having mild watery diarrhea 4–7 times each day that continued for 3 weeks, and a mild fever. Initially, she felt that the excessive chocolate she had been eating had caused the diarrhea but stopping the chocolate did not make the symptoms resolve so she made an appointment to see her doctor.

Cell Biology, Diagnosis, and Treatment of *C. difficile* Colitis

Her physical examination revealed a low-grade fever of 101°F and mild tenderness of her abdomen with normal bowel sounds. Her complete blood count showed a slightly elevated white blood cell count and normal renal function. She had a negative test for occult blood in her stool. With the recent use of antibiotics, the doctor suspected that Ms. Smith may have an antibiotic-induced diarrheal condition such as *C. difficile* colitis and obtained samples of her diarrheal stool and sent them for an enzyme immunoassay for *C. difficile* toxin A and B and for stool culture. The mild nature of her disease did not lead the doctor into thinking that Ms. Smith had pseudomembranous colitis due to *C. difficile* so he did not order a computed tomography (CT) scan of her abdomen. He empirically prescribed her a 10-day course of metronidazole 500 mg three times a day and advised her not to drink alcoholic beverages with this medication. After 4 days of the metronidazole, Ms. Smith's symptoms had markedly improved and she no longer had either fever or diarrhea. The doctor called to let her know that her stool tests were positive for *C. difficile* and had advised her that the colitis may recur and an alternate therapy for multiple recurrence was fecal transplantation. Ms. Smith was somewhat appalled by the thought of fecal transplantation, so after her course of metronidazole she started a diet high in probiotics, including an over-the-counter supplement containing probiotics. She vowed never to ask for antibiotics for a viral upper respiratory tract infection again.

C. difficile is a major nosocomial pathogen and it is recommended that universal precautions be implemented to reduce risk of infection. *C. difficile* colitis is often managed by stopping the causative antibiotic as soon as possible and this may affect recurrence of the disease. Antidiarrheal agents such as diphenoxylate with atropine should not be used in *C. difficile* colitis as they may increase the duration and severity of symptoms. In mild cases of *C. difficile* colitis the first-line therapy is oral metronidazole 500 mg TID for 10–14 days. In severe cases, oral vancomycin 125 mg QID for 10–14 days may be used. Relapse occurs in over 20% of cases and typically happens 3 days to 3 weeks after the treatment has stopped. Management of the first recurrence is the same as that for the initial bout of the disease. There are limited data about whether probiotics are beneficial, but they may improve the diarrheal symptoms of some patients. Fecal transplantation is a newer therapy involving transfer of stool (and thus microbiota) from a healthy donor to the person with *C. difficile* with efficacy rates of over 90%. Samples for fecal transplantation need to be very well screened in order not to allow transmission of other known human pathogens such as HIV and hepatitis. Although it seems like a new therapy and is challenging regulators and practitioners alike as to what one considers a medicine to be, it actually dates back to ancient China. It is by far the best example of how the microbiome can be harnessed to save lives.

Clinical Case 15–2

Earl Haggard is a 63-year-old Caucasian farmer from Arkansas with a history of chronic urinary tract infections after bladder surgery 10 years ago. He went to see a dermatologist for the first time with an enlarging blue-black lesion on his left hand. He stated that he always had a mole in that location, but that over the last 2 years it has been getting bigger, with the borders becoming more undefined, the color becoming less brown and more reddish-orange, and the appearance of both a bump and some skin sloughing and weeping. He also stated that he has become increasingly short of breath recently, with occasional coughs and a couple of episodes of blood-tinged sputum.

On physical exam, there was a 5 × 4-cm ulcerated irregular reddish brown lesion on the dorsum of his left wrist. He has a number of small satellite lesions surrounding this that appear to track up his wrist and onto his forearm. He has marked fullness in the left axilla with palpable enlarged lymph nodes, and he has a slight wheeze over the right middle lung field.

The dermatologist performed a shave biopsy of the hand lesion that revealed malignant melanoma with ulceration and a thickness of at least 4 mm, with tumor present at the deep margin. Mr. Haggard was sent to a surgical oncologist who again palpated lymph nodes in the axilla and performed a core needle biopsy on the largest, which also revealed melanoma. In preparation for lymph node dissection, the surgical oncologist ordered a CT scan of the chest, abdomen, and pelvis. The CT revealed numerous enlarged "matted" lymph nodes in the left axilla as well as a large 4-cm lesion in the right middle lobe of the lung slightly impinging on the bronchus. Interventional radiology biopsied this lesion, which was also found to be melanoma. Genetic testing on the tumor revealed a mutation in the BRAF gene at V600. Laboratory tests were notable for a lactate dehydrogenase level two times the upper limit of normal.

Mr. Haggard was sent to a medical oncologist who prepared to start him on immunotherapy to treat his metastatic melanoma. A history was taken and revealed that Mr. Haggard had been on regular courses of antibiotics for his urinary tract infections for many years, and took chronic antibiotics that were broadened at the time of active infections. He recently had a week-long episode of urinary pain that resulted in a 2-week course of ciprofloxacin and metronidazole. His medical oncologist considered whether to start him on immunotherapy or therapy targeted against the BRAF mutation.

Cell Biology, Diagnosis, and Treatment of Melanoma and Impact of Dysbiosis

Melanoma is a cancer of the pigment-producing cells of the skin. The most dangerous form of skin cancer, it occurs in approximately 90,000 Americans yearly, and almost 10,000 people in the United States die yearly. Lifetime sun exposure is a contributing factor. Staging of melanoma is dependent on depth of invasion of the primary lesion, presence of ulceration, lymph node involvement, presence of in-transit or satellite lesions, and spread to distant organs. Stage IV (metastatic) melanoma, such as Mr. Haggard has, is considered incurable, though since 2010 a number of treatments have been approved that can significantly extend life even beyond 5 years.

About 40%–50% of melanoma has a mutation in the BRAF gene at position 600. Studies have shown that targeted therapy with a combination of a BRAF inhibitor and a MEK inhibitor can produce response rates as high as 75% in these patients. However, the responses tend to be less durable than responses seen with immunotherapy. However, immune therapy is an alternative treatment for patients with melanoma. There is no consensus yet on whether a patient with a BRAF mutant melanoma should be started on a targeted therapy or immunotherapy as an initial treatment.

Melanoma, unlike many other cancers, is highly dependent on evasion of the normal host immune system. It can do so by expression of various factors on tumor cells (such as PD-L1) that inhibit the activation of the humoral immune system, mainly CD-4 and CD-8 T cells. Despite displaying antigens that the immune system would consider foreign, these "checkpoint" antigens keep the immune system from activating. As such, melanoma is also considered immunoresponsive—if the immune system can be activated despite the tumor evasion, it can often clear the tumor without the use of cytotoxic drugs. The most common treatments for melanoma currently include immune checkpoint inhibitors, including monoclonal antibodies against CTLA-4 or PD-1 that inhibit the inhibitors—resulting in normal activation of the immune system by tumor antigens.

Reliance on the host immune system to kill cancer comes with a great deal of uncertainty. Each host has a different immune system with different sensitivities to foreign antigens, slightly different abilities to distinguish self from nonself, and a different state of baseline activation. It is becoming increasingly clear that one of the main mechanisms by which the activity of the baseline host immune system is determined, is through interaction with the host microbiome.

The microbiome refers to the totality of all the commensurate lifeforms that colonize the human body. Humans generally live with at least as many microbes in or on their body as they have human cells. The overwhelming majority of these microbes are symbiotic with the human host. Occasionally, overgrowth or colonization by an infectious agent can lead to a parasitic relationship leading to disease. The resident microbes in humans may engage in "cross-talk" with the immune system by exchanging chemical signals that ultimately influence immune reactivity and immune targeting. While these relationships are still being elucidated, it seems clear that the microbiome plays a role in the activation of Toll-like receptors, which may allow the immune system to better recognize foreign antigens, such as cancer.

That a dysregulated microbiome may have some influence on the outcomes of immunotherapy has been suggested by studies looking at the relationship between antibiotics, which generally result in depletion of the microbiome and outcomes in patients who receive immunotherapy. Some studies have demonstrated that patients who have received antibiotics in the 30 days prior to starting immunotherapy have worse outcomes and are less likely to respond to immunotherapy. Though the causative relationship with microbiome depletion is far from clear, the association is compelling, and a great deal of research into the interplay between the microbiome and efficacy of immunotherapeutic drugs is ongoing. Moreover, growing research is demonstrating that the microbiome may affect the efficacy of a variety of drugs and also that various medications can impact microbiota composition.

Given that Mr. Haggard has a BRAF mutation, and given the association with worse outcomes for immunotherapy seen in patients receiving antibiotics immediately beforehand, the medical oncologist decides to start treatment with dabrafenib and trametinib (BRAF and MEK inhibitors). Mr. Haggard achieved a partial response and remains without progression 10 months later.

SUMMARY

It is always worth remembering that we are living in a microbial world and that all aspects of human health and disease may be affected by shifts in microbiome especially at vulnerable periods across the lifespan.

While the microbiome field has developed enormously in the last few decades, the field is still in its infancy in some ways. Often, it is difficult to interpret specific changes in microbial composition in the context of impact on the host. Similarly, it can be hard to distinguish cause and effect in the microbiome field. For instance, does obesity change the microbiome, or does a specific state of the microbiome induce obesity? In at least some cases, we suspect it actually goes both ways; phenotype induces microbiome state and microbiome state in turn induces phenotype. There is a growing realization that the microbiome can be harnessed to manage and treat a variety of disease states (see Clinical Case 15–1) and improve the efficacy of drugs (see Clinical Case 15–2).

Suggested Readings

Cryan JF, et al. The microbiota-gut-brain axis. *Physiol Rev* 2019;99(4): 1877–2013.

Gilbert JA, Blaser MJ, Caporaso JG, Jansson JK, Lynch SV, Knight R. Current understanding of the human microbiome. *Nat Med* 2018;25(4):392–400.

Helmink BA, Khan MAW, Hermann A, Gopalakrishnan V, Wargo JA. The microbiome, cancer, and cancer therapy. *Nat Med* 2019;25(3): 377–388.

Huttenhower C, et al. Structure, function and diversity of the healthy human microbiome. *Nature* 2012;486(7402):207.

Lynch SV, Pedersen O. The human intestinal microbiome in health and disease. *N Engl J Med* 2016;375(24):2369–2379.

Pasolli E, et al. Extensive unexplored human microbiome diversity revealed by over 150,000 genomes from metagenomes spanning age, geography, and lifestyle. *Cell* 2019;176(3):649–662.

Zmora N, Suez J, Elinav E. You are what you eat: diet, health and the gut microbiota. *Nat Rev Gastroenterol Hepatol* 2019;16 (1):35–56.

Postlude

While this textbook is in its final stage of production, our world is facing the COVID-19 pandemic, which is unlike anything that most of us have experienced in our lifetime. As of today (April 21, 2020), over 2.5 million people have been confirmed to be infected globally by the novel severe acute respiratory syndrome coronavirus 2 (SARS-COV-2), and over 171,000 people have died from coronavirus disease 2019 (COVID-19). The number of confirmed cases is probably less than 2% of the actual number because of the multitude of asymptomatic people and low-to-moderate severity cases that have not been tested for the presence of the virus or antibodies produced by the immune response to the virus.

SARS-COV-2, like other members of the coronavirus family, is an enveloped single-stranded RNA virus. The 30 kb genome encodes RNA-dependent RNA polymerase (RdRp), proteases, and four major structural proteins. The virion is composed of a helical nucleocapsid (N) protein wrapped around the RNA genome, while the envelope contains an abundant membrane (M) glycoprotein and envelope (E) protein. The trimeric spike (S) protein associates with a receptor on the surface of type 2 pneumocytes. The viral S protein interacts with the angiotensin-converting enzyme 2 (ACE2) receptor on the cell surface, followed by selective cleavage of S protein by the serine protease TMPRSS2. This is referred to as a priming step that allows fusion of the viral and host cell membranes during receptor-mediated endocytosis. The virus enters the host cell via an endosome pathway, and in subsequent steps (translation, selective proteolysis, replication by RdRp, and assembly of the virion in the Golgi complex), SARS-COV-2 is replicated and then released from the host cell. Our bodies attempt to neutralize the virus by the innate and adaptive immune response described in Chapter 12. In severe cases, there is an increased production of inflammatory cytokines and an enhanced movement of monocytes and neutrophils into the lung. This leads to what is called cytokine storm syndrome.

All of the aspects of viral structure, mechanisms of host cell entry through receptor-mediated endocytosis, the ACE2 receptor and its signaling pathways, the transcriptional and translational aspects of viral replication, and the immune response to the virus should be understood by students who have read *Goodman's Medical Cell Biology*.

The COVID-19 pandemic has led to a great demonstration of human empathy and cooperation: heroes who have taken care of COVID-19 patients, those who have watched out for our safety and supplied us with food and other necessities of life; and the superheroes who have worked without stop in academic, industry, and government laboratories to find the treatments that will end this pandemic. I would like to further dedicate this book to these heroes and superheroes but most importantly to those of you who have lost loved ones before the cures could be found. It is my earnest hope that the lessons learned during this pandemic will not be forgotten once it is over. If so, we will emerge living in a kinder and safer world.

Steven R. Goodman, Editor

Index

Note: Page numbers followed by *f* indicate figures, *t* indicate tables, and *b* indicate boxes.

B

D

E

F

G

H

I

J

K

L

M

N

R

Printed in the United States
By Bookmasters